Terry Sessford,
      Plessey Marine,
         Somerset.
October 1982.

page 6: line 34

    C. J. Rice should read C. G. Rice

page 34: Fig. 1.3

$$(\alpha x + \beta y + \gamma z) \text{ should read } \frac{1}{c_0}(\alpha x + \beta y + \gamma z)$$

page 188: Equation 8.2

$$\int_{p_0}^{p} \quad \text{should read} \quad \int_{p_0}^{p_0 + p}$$

page 188: Equation 8.3

$$u = \pm c_0 \pm \frac{\gamma + 1}{2} v \; ; \text{ should read } u = \pm c_0 + \frac{\gamma + 1}{2} v \; ;$$

page 280: Equation (12.35)

    should read $\quad k^2 = (\omega/c)^2 = k_x^2 + k_\phi^2 + k_r^2$

page 341: Equation 15.11

$$\int_{-1}^{+1} \quad \text{should read} \quad \int_{-1}^{+1}$$

page 347: line 6

    tengential should read tangential

page 363: Equation 16.35

    dS should read $dS(\xi)$

page 566: Fig. 22.51a

    Scheffield should read Sheffield

page 762: Fig. 28.4

    countours should read contours

    Contour marked '8' should read '6'

page 763: 3rd line from bottom

    $-(L_{10} - L_{90})$ should read $+(L_{10} - L_{90})$

page 766: line 27

    170 00 should read 170 000

page 799:

    top three lines should be inserted after line 11

NOISE AND VIBRATION

NOISE AND VIBRATION

Edited by

R. G. WHITE, PhD, CEng, FIERE, F InstP
Director of the Institute of Sound and Vibration Research
University of Southampton, England

and

J. G. WALKER, BSc, PhD, M InstP, FIOA
Research Fellow and Short Course Organiser
Institute of Sound and Vibration Research
University of Southampton, England

ELLIS HORWOOD LIMITED
Publishers · Chichester

Halsted Press: a division of
JOHN WILEY & SONS
New York · Brisbane · Chichester · Toronto

First published in 1982 by
ELLIS HORWOOD LIMITED
Market Cross House, Cooper Street, Chichester, West Sussex, PO19 1EB, England

The publisher's colophon is reproduced from James Gillison's drawing of the ancient Market Cross, Chichester.

Distributors:

Australia, New Zealand, South-east Asia:
Jacaranda-Wiley Ltd., Jacaranda Press,
JOHN WILEY & SONS INC.,
G.P.O. Box 859, Brisbane, Queensland 40001, Australia

Canada:
JOHN WILEY & SONS CANADA LIMITED
22 Worcester Road, Rexdale, Ontario, Canada.

Europe, Africa:
JOHN WILEY & SONS LIMITED
Baffins Lane, Chichester, West Sussex, England.

North and South America and the rest of the world:
Halsted Press: a division of
JOHN WILEY & SONS
605 Third Avenue, New York, N.Y. 10016, U.S.A.

© 1982 R. G. White and J. G. Walker/Ellis Horwood Ltd.

British Library Cataloguing in Publication Data
Noise and vibration
1. Noise
I. White, R. G.   II. Walker, J. G.
620.2'3   TD892

Library of Congress Card No. 82-12110

ISBN 0-85312-502-3 (Ellis Horwood Ltd.)
ISBN 0-470-27553-7 (Halsted Press)

Typeset in Press Roman by Ellis Horwood Ltd.
Printed in Great Britain by Butler & Tanner, Frome, Somerset.

Table of Contents

Editors' Preface

We would like to take the opportunity to acknowledge the various contributions from individuals and organisations which have led to the production of this book.

It is common practice to refer to a particular book through the names of those appearing on its covers. However, the real credit for the contents of this book goes to the many individual contributors. The contents represent the cumulative experience of its authors and their students, research workers, engineers and technicians, who formed and work in a unique organisation, the I.S.V.R. As Professor Large says in his introduction, the research carried out at the I.S.V.R. is dependent upon finance from external sources and without such sponsorship the knowledge presented in this book would have never been gained.

The transformation from authors' manuscript to the final production has been achieved through the dedication and cooperation of the secretarial and drawing office staff of the Institute.

We offer our sincere thanks to all who have been involved in so many ways.

R. G. White
J. G. Walker

Introduction

Professor J. B. Large
Dean of the Faculty of Engineering and Applied Science, University of Southampton, formerly Director of the Institute of Sound & Vibration Research.

The Institute of Sound and Vibration Research annually produces a variety of specialist short courses and it has been our objective that the information imparted to the participants should be produced in a more permanent form and made available to a wider readership.

The contents of this book represent the majority of the subjects included in our annual Advanced Course in Noise and Vibration. This course, presented in Southampton during September, has become an important feature in the activities of the Institute of Sound and Vibration Research. The origins of the course are older than the Institute itself, which is now entering its 20th year as an academic and research department within the University of Southampton. The original course was designed for workers interested in acoustic fatigue and noise control problems in the aerospace industry. It was initially given in the University and in 1961 was presented in the USA sponsored by the USAF, Dayton, Ohio. The inital success of the course led to publication of the book 'Noise and Acoustic Fatigue in Aeronautics' edited by E. J. Richards and D. J. Mead. The book, first published in 1968 by John Wiley & Sons, is no longer in print, but it carefully records the state of our knowledge some 15 years ago in aerospace noise control and the vibration response of structures to noise. Professor Richards and Dr. Mead are still both active in the field of noise and vibration control and make important contributions to the course which are evidenced in this book. Dr. Mead, of the Department of Aeronautics and Astronautics, is well known for his work on structural analysis and vibration damping. Professor Richards left Southampton University in 1968 to become Vice-Chancellor at Loughborough University but has since returned as a research professor in the Institute and actively leads a research team in the field of machinery noise and vibration control.

Since 1968 the course has expanded and evolved, reflecting the wider interest in the subject and although problems related to aerospace design are included, they no longer dominate the contents. Some of the newer topics include mechanical noise and vibration control, road vehicle noise control and machinery health monitoring and signal processing. The Advanced Course also includes lectures on the subjective effects of noise and vibration on man and occupational noise effects. We believe it is important that engineers have an understanding of these subjects and be made aware of the relevant noise and vibration criteria and regulations.

The book reflects the structure of our course and is organised so that about half the chapters are directed towards the basic principles of acoustics, vibration and structural response, with emphasis on modern methods of analysis. The remaining chapters deal with specific problems of control and the application of the fundamental knowledge discussed in the earlier sections of the book. Our lecture notes are under continual revision as new research and development advances our knowledge. All subjects included are rapidly changing and therefore it is important to maintain a contemporary outlook. In order to achieve this, we anticipate that the book will receive regular revision, although the basic content of the book will serve as a permanent source of information.

Although the authors have described their topics in great detail, they have been constrained by the physical size of the book; hence each chapter contains an up-to-date list of references which enables the reader to expand his knowledge of a particular topic and gain a deeper insight into the subject background. The size of the book has also caused us to limit our horizons slightly in so far as we have not included the contents of a seminar on environmental planning which always forms part of the advanced course and is presented in the same spirit as the topics covering human effects.

The Institute, although a University department, is highly reliant on its research and development work to support its activities and receives over three-quarters of its income from government research councils, government departments and public and private industries. This places the Institute in an unusual position for it is able to advance the frontiers of knowledge through its research and at the same time it is highly active in the world of applied technology.

The editors have accomplished a difficult task in a short time period and hopefully the book will be a useful reference work for teachers, researchers and engineers.

Symbols

Included in this list are some of the symbols commonly used thoughout the book with an indication in parentheses of the chapter in which they appear.

A: amplitude (1)
area of a plate (25)
cross-sectional area (7), (15), (21), (20)
total absorption (area) (1), (24)

ADC: analogue-to-digital converter (6)

a: half of crack length (20)
radius of a disc (26)
radius of contact area (27)
radius of a piston (11)
radius of a sphere (22)

a_f: flexural wave velocity (9)
a_l: longitudinal wave velocity (9)
a_n: phase velocity (13)
a_o: speed of sound (10)

B: cylinder diameter (18)
effective frequency bandwidth (4)
flexural rigidity (26)
number of blades on a rotor (19)
B_p: bending stiffness of a plate (26)

b: blade chord (19)

C: material constant (20)
C_a: damping constant of a vibration neutraliser (25)
C_f: level of gas force (18)

C_r: generalised damping coefficient in mode r (3), (5)
C_s: structural characteristic (18)
CV: convection velocity (13), (25)
$[C]$:   damping matrix (15)

c: damping constant (3), (11), (18), (25)
c_{c_0}:   speed of sound (1), (7), (14), (19), (21), (22), (24)
c_f: phase velocity (7)
c_g: group velocity (7)
c_0:   wave propagation velocity (1), (8)
c_s: speed of sound in a moving fluid (14)
c_p: flexural wave speed in a plate (25)
c_0:   critical dampling coefficient (3)
c_t: convection speed (2)

D: diameter of a tube (2)
        directivity factor (2)
        dissipation function (15)
        drag force (19)
        fan diameter (18)
        flexural rigidity of a plate (9), (11), (25)
        Nozzle diameter (14)
D_m: modified Doppler factor (14)
D_s: Doppler factor

d: blade span (19)
        depth of cavity (21)
        distance (2)

E: energy density (1)
        isolator effectiveness (26)
        noise energy (22)
        Young's modulus, extensional modulus, modulus of elasticity (1), (9)
        (15), (25), (26), (27)
$E[\ ]$:   expected value (4)
EI: flexural stiffness (3)
$\bar{E}$:   time averaged total energy (7)
E_c: Young's modulus of a composite material (25)

F: force (3), (5), (11), (18), (26)
        interference factor (2)
$F(i\omega)$:   complex spectrum of F(t) (5)

$F_\triangle^2$:   mean square value of force over a bandwidth Δw (25)

$\langle F^2 \rangle$:   mean square value of force (25)

f: frequency, Hz

f_m: vibrational relaxation frequency (1)

f_r: mode shape (3) (13)

$G(f)$:   spectral density function (14)

         Fourier transform of a transient time function (22)

$G(\omega)$:   spectral density (single sided) (4)

G_{FV}:   cross spectral density between force and velocity (26)

$G_T(\omega)$:   source spectrum (19)

GQ: torsional stiffness (26)

g: shear parameter (25)

$g(t)$:   a transient (22)

h: enthalpy (12)

         imaginary part of complex stiffness (3), (4)

         radius of a duct (21)

         thickness (1), (9), (11), (26)

$h(t)$:   impulse response function (3), (4), (5)

I: intensity (acoustic power per unit area) (1)

         polar moment of inertia (7)

         second moment of area of cross-section (15), (25), (26)

i: complex operator

J: moment of inertia (18)

J_m: Bessel function (19)

j: complex operator

j_r: joint acceptance of mode r (3)

K: isentropic bulk modulus of elasticity (1)

         stiffness (3), (5), (7), (18)

         stress intensity factor (20)

         wave number (25)

$[K]$:   stiffness matrix (15)

K_a: spring stiffness of a vibration neutralizer (25)

K_r: generalised stiffness in mode r (3)

k: wave number (1), (9), (10), (11), (16), (19), (22), (25)

         spring stiffness (3), (5), (25), (26)

L: length (21)
       length of a line junction (7)
       lift force (19)
       maximum source dimensions (2)
L_r: generalised force in mode r (3)
L_s: mean square acceleration over surface (24)
L_w: sound power level from a machine (24)
       sound power level measured in an ideal acoustical environment (24)
L_{w_i}:   sound power level measured in-situ (24)
L_p: mean sound power level measured over a surface (24)
$L_{A_{eq}}$: A-weighted L_{eq} (22), (28), (29)
L_A: A-weighted sound pressure level (28), (30)
L_{eq}:   Equivalent continuous sound pressure level (22), (28), (29), (30)
L_{dn}:   Day-night average level (28)
L_N: loudness level (28)
L_N: noise level exceeded for N% of the time period (28)
L_{NP}:   noise pollution level (28)
L_P: sound pressure level (18), (21)
L_{PN}:   Perceived noise level (28)

l: length of a beam (26)
       spanwise correlation length (19)
       stiffener spacing (25)
l_y: airway width (21)

M: harmonic moment (13)
       Mach number (10), (12), (14), (19), (21)
       mass (5), (7), (22), (18)
       mobility (26)
       moment (11)
       point mass (25)
       total mass of a plate (25)
M_a: mass of a vibration neutraliser (25)
M_c: inertial coupling (7)
       Mach number (14)
M_r: generalised mass in mode r (3)
$[M]$:   inertia matrix (15)

m: integer (10), (12), (14), (19)
       mass (3), (22), (25), (26)
       mass per unit area (7), (11)
       mass per unit length (7), (25)

material constant (20)
slope of a regression line (18)
m':   virtual mass of air (22)

N: engine speed (18)
number of cycles (20)
number of samples (4)

$$\frac{|P_{MAX}|}{|P_{MIN}|} \quad (2)$$

n: combustion index (18)
integer (19)
modal density (7)
number of data samples (6)
number of statistical degrees of freedom (4)
shaping factor (20)

P: airway perimeter (21)
force (18), (25)
plate perimeter (25)
power (26)
$P(x)$:   probability function (4)
P_{be}:   blocked pressure (11)
P_g: gas force (18)
P_I: inertia force (18)
P_n: approximate sound power level (24)
P_r: Prandtl number (8), (12)
P_{RAD}:   radiated pressure (11)
P_o: pressure (1)
peak pressure (22)

p: pressure (1), frequency (6)
$p(x)$:   probability density (4), (23)
p_s: back pressure (8)
p_x: distributed load per unit length (15)
p_o: equilibrium pressure (1)

Q: magnification factor (3)
rate of volume displacement (2)
Q_f: power transmission spectrum (26)

Q_r: complex generalised force of mode r (13)

Q: volume velocity (11)

q: generalized coordinate (displacement) (3), (15)

q_r: Fourier coefficient (3)

R: bearing reaction force (18)

      pressure reflection factor (2)

      radius (11)

      radius of sphere (16)

      resistance (7)

      specific acoustic resistance (21)

      stress ratio (20)

R_{rad}:   real part of radiation impedance (11)

$R(\tau)$:   autocorrelation function (4)

$R_{x_1 x_2}(\tau)$: cross-correlation function (4)

R':   airway radius (21)

      distance from source (2)

$\hat{R}$:   acoustic Reynolds number (8)

r: distance (14), (19)

      distance from centre in spherical wave propagation (1)

      distance from crack tip (20)

      number of records averaged (27)

r_p: extent of plastic zone (20)

S: acoustically hard surface area (16)

      area of enveloping surface (22), (24)

      platform area of a blade (19)

$S(\omega)$:   spectral density (4), (5)

$S_F(\omega)$:force spectral density (25)

S_p: power spectral density (8)

S':   flexible surface area (16)

S'':   area of sound absorbing material (16)

T: Kinetic energy (15)

      reverberation time (1), (24)

      temperature (1), (14)

      transmissibility (26)

T_0:   temperature (1)

T':   tension (1)

$\overline{T}$:   time averaged kinetic energy (7)

t: time (1)
t_0:   duration of acceleration (22)

U: convection speed (19)
            strain energy (15)
            velocity (10), (11), (12)
U_A: aircraft forward speed (14)
U_J: jet efflux velocity (10), (14)
U_s: fluid velocity (14)
$\overline{U}$:   time averaged potential energy (7)

u: displacement (15)
            velocity (2), (8), (10)

V: vehicle speed (18)
            velocity (26)
            volume (1), (16), (22), (24)
            volume velocity (21)
V_c: eddy convection velocity (14)

v: velocity (1), (7), (8), (10), (22), (24)
v_r: fluid velocity (14)
v_0:   maximum velocity (22)

W: power (1)
            tyre width (18)
            work done by non-conservative forces (15)
W_{rad}: radiated sound power (18), (22)

w: displacement (3)

X: specific acoustic reactance (21)
$X(i\omega)$: complex spectrum of x(t) (5)

x: spatial coordinate (1)

Y: geometric parameter (25)

Z: point impedance (26)
            specific acoustic impedance (1), (21)
Z_{rad}: radiation impedance (11)
Z_s: acoustic impedance (2), (16)

z: interference parameter (2)
point impedance (7)

α: attenuation coefficient (18)
sound absorption coefficient (1), (21)
receptance (3), (5), (13), (25)
viscothermal dissipation (12)
α_{cl}: classical attenuation constant (1)
α_f: absorption coefficient at frequency f (21)
α_r: reflection coefficient (9)
α_T: heat conduction attenuation constant (1)
α_t: transmission coefficient (9)
α_v: viscous attenuation constant (1)
α_{vib}: molecular vibration attenuation constant (1)
α_1, α_2: volume fractions of components in a composite material (25)

β: complex wave number (12)
mobility of an equivalent infinite structure (26)
nonlinearity coefficient (8)
β_d: shear loss factor (25)

Γ: coupling factor (11)
γ: normalized real component of approximate mobility (26)
ratio of specific heats (1), (8), (12), (14)
shear strain (25)
$\dot{\gamma}$: time rate of change of direction (22)
$\gamma_{x,y}^2(\omega)$: coherence function (5)

Δ: sampling interval (6)
transducer spacing (26)
ΔdB: attenuation (21)
Δ_N: number of modes having a resonance in the bandwidth Δ_ω (25)
Δ_{op}: peak to peak pressure amplitude (8)
Δ_ω: half power point bandwidth (3), (7), (27)
$\Delta N/\Delta_\omega$: modal density (22)
width of a narrow frequency band (25)
Δ_1: near field error (24)
Δ_2: finity error (24)
Δ_3: measurement error (24)

$\delta:$ clearance (18)

         complex propagation constant (13)

         impact parameter (22)

         logarithmic decrement (3)

         normalized imaginary component of approximate mobility (26)

$\delta^*:$ boundary layer thickness (19)

$\delta(r):$ Dirac delta function (3)

$\delta(f):$ Delta function in the frequency domain (6)

$\delta(\phi):$ Dirac delta function (19)

$\epsilon:$ direct strain (25)

         phase constant (13)

$\zeta:$ specific impedance ratio (2)

         viscous damping ratio (3), (5), (27)

$\zeta_a:$ viscous damping ratio of a vibration neutralizer (25)

$\zeta_n:$ hysteretic damping factor (3)

$\eta:$ loss factor (3), (5), (7), (22), (25), (26), (27)

$\eta_c:$ loss factor of a composite material (25)

$\eta_d:$ extensional loss factor (25)

$\eta_m:$ modal loss factor (25)

$\eta_{RAD}:$ radiation loss factor (22), (25)

$\eta_r:$ loss factor of a structural wave (13)

$\dot{\eta}_s:$ structural loss factor (22)

$\eta_\alpha, \eta_\beta:$ modal average internal loss factor (7)

$\eta_{\alpha\beta}:$ modal average coupling loss factor (7)

$\eta_{12}, \eta_{21}:$ coupling loss factor (7)

$\Theta:$ absolute temperature (12)

         angle of incidence (2)

         harmonic angular rotation (13), (26)

$K:$ wave number (11)

$\kappa_r:$ torsional stiffness (25)

$\lambda:$ wavelength (1), (8), (9), (10), (21), (22)

$\lambda_m:$ modal wavelength (3)

$\lambda_t:$ trace wavelength (3)

$\mu:$ mass per unit area (3)

         non-dimensional, complex wavenumber (16)

real part of complex propagation constant (13)
shear viscosity (8)

μ_{acc}: acceleration noise parameter (22)
μ_v: bulk viscosity (8)
μ': mass ratio (25)
$\bar{\mu}$: equivalent longitudinal viscosity (8)

ν: kinematic viscosity (12), (19)
Poisson's ratio (1), (9), (26), (27)

ξ: displacement (1), (11), (26)
random variable (4)

ρ: density of acoustic medium (16)
mass density (1), (8), (9), (14), (18), (22), (25)
volume density (26)
ρ_s: material density of a plate (11)
ρ_0: density (1), (7), (8), (10), (12), (22), (24)
$\bar{\rho}$: mean density (21)
time averaged value of density (10)

σ: direct stress (20), (25)
non-dimensional excess wave-speed (8)
spherical surface (16)
standard deviation (8), (10), (23)
unit-depth flow resistance (2)
σ_{rad}: radiation efficiency (22), (24)
radiation ratio (18)
σ_y: yield stress (20)
σ^2: variance (4), (6)

τ: shear stress (25)
time delay (4)

ϕ: phase angle (3)

Ω: angular velocity (19)
imaginary part of cross spectral density (8)
ω/ω_a (25)
ω/ω_0 (26)
ω: angular frequency, rads/sec
ω_a: natural frequency of an auxiliary system (25)

ω_c: critical frequency
ω_d: damped natural frequency (3), (27)
ω_i: natural frequency of mode i (22)
ω_n: undamped natural frequency (3), (27)
ω_r: undamped natural frequency (25)
ω_0:   undamped natural frequency (7), (26)
$\dot{\omega}$:   rate of change of frequency with time (27)

CHAPTER 1

Theory of acoustics (I)

C. L. Morfey and G. H. Koopmann†

Institute of Sound and Vibration Research, University of Southampton

1.1 INTRODUCTION

Chapters 1 and 2 on the theory of acoustics are intended to provide the scientist whose discipline is not primarily acoustics and vibration with a brief review of the acoustical terminology and methodology assumed as background in the remaining chapters. We begin by reviewing the methods available for defining wave motion in different kinds of media and then go on to consider the influence of boundaries as reflectors and sources of sound.

1.2 INTRODUCTION TO WAVE MOTION

The phenomenon of wave motion occurs in different media in many different forms, e.g. waves on the ocean surface, vibrations on a stretched string, electromagnetic radiation, sound propagation in the air, and so on. When we ask what these various waves have in common, at first we may be tempted to say that they all exhibit oscillation or periodicity. Although this is frequently the case, it is not universal; the so-called tidal wave, for example, being just one exception. A more characteristic property of wave motion emerges when observed from a dynamic rather than a kinematic point of view. The arrival of a wave at a certain point in the medium disturbs the matter in the vicinity of that point, and thus the wave must be regarded as transmitting energy from one point to another. The transfer of energy by this means does not involve the transfer of matter, however, as no element of matter suffers a permanent displacement from its original location. The identification and quantification of wave motion, along with its associated transfer of energy and momentum is the main occupation (and headache!) of the acoustician. Theorists construct elaborate mathematical models to predict this motion for highly idealized conditions, while experimentalists and practitioners construct elaborate devices to measure it physically without interfering with it too much.

---

† Now at Department of Mechanical Engineering, University of Houston, Texas.

Observed kinematically, different kinds of wave motion can be identified in terms of the particular paths which an element of perturbed matter describes when displaced. Expressing this displacement relative to the direction of the wavefront provides a means of classification. Table 1.1 gives a few of the more common wave motions encountered in acoustics together with their propagation velocities.

Table 1.1

Wave motion	Motion relative to wavefront	Medium	Wavefront propagation velocity
(1) Compressional (longitudinal)		Gases	$\left[\dfrac{\gamma p_0}{\rho_0}\right]^{1/2}$
		Liquids	$[K/\rho]^{1/2}$
		Solids	$\left[\dfrac{E(1-\nu)}{\rho_0(1+\nu)(1-2\nu)}\right]^{1/2}$
(2) Shear (transverse)		Solids	$\left[\dfrac{E}{2\rho_0(1+\nu)}\right]^{1/2}$
		Strings (cross-sectional area A)	$\left[\dfrac{T'}{A\rho_0}\right]^{1/2}$
(3) Flexural		Rect. rods (thickness h)	$\left[\dfrac{Eh^2}{12\rho_0}\right]^{1/4}\omega^{1/2}$
		Plates (thickness h)	$\left[\dfrac{Eh^2}{12(1-\nu^2)\rho_0}\right]^{1/4}\omega^{1/2}$
(4) Rayleigh		Surfaces of solids	$0.385\left[\dfrac{E(2.6+\nu)}{\rho_0(1+\nu)}\right]^{1/2}$

p_0 = equilibrium pressure
γ = ratio of specific heats
K = isentropic bulk modulus of elasticity
ρ = mass density
E = extensional modulus (Young's modulus)
ν = Poisson's ratio
T' = tension
ω = angular frequency

These first two chapters are principally concerned with compressional waves in fluids. Such waves involve changes in the state of the fluid (pressure, temperature, etc.), as well as the kinematic variables (particle displacement, velocity and acceleration). Throughout this book the following notation will be used for the two sets of variables.

Rest variables		*Perturbation variables*	
pressure	p_0	displacement	$(\xi_1, \xi_2, \xi_3) = \boldsymbol{\xi}$
density	ρ_0	velocity	$(v_1, v_2, v_3) = \mathbf{v}$
temperature	T_0	pressure	p
		density	ρ
		temperature	T

The pressure, density, and temperature of the fluid locally at any instant are thus given by $p_0 + p$, $\rho_0 + \rho$, and $T_0 + T$ respectively.

To describe a particular wave field, it is often adequate to specify the position, and time dependence of the pressure, p, and the velocity, $\mathbf{v}$. The ratio and product of these two variables define two important dynamic properties of the medium and the wave motion perturbing it, namely the impedance and intensity (or mean energy flux).

Symbolically, the impedance (referred to as the specific acoustic impedance) in a specified direction (i) is:

$$Z = \frac{p}{v_i} \tag{1.1}‡$$

while the intensity, which is the time average of the acoustic power transferred per unit area, is the vector quantity

$$\mathbf{I} = \overline{p\mathbf{v}}. \tag{1.2}$$

With the above definitions in mind, let us next develop briefly three methods which are available for approaching problems in acoustics, namely, wave acoustics, ray acoustics, and energy acoustics.

1.3 WAVE ACOUSTICS

To develop this approach, we begin by assuming that the fluid supporting the acoustic perturbation has the following properties: (1) continuity, (2) homogeneity, (3) no internal friction, and (4) no thermal conductivity. Further, the perturbations themselves will consist of small amplitude displacements.

---

‡In equation (1.1), p and v_i refer to single-frequency complex quantities.

If these idealizations are met to within practical limits, the corresponding wave motion is governed by wave equations of the form

$$\left(\frac{\partial^2}{\partial x_i^2} - \frac{1}{c_0^2}\frac{\partial^2}{\partial t^2}\right)\left\{\begin{array}{c} p \\ \xi_j \\ v_j \end{array}\right\} = 0 \qquad (1.3)$$

where $\partial^2/\partial t^2$ and $\partial^2/\partial x_i^2$ are second-order partial derivatives with respect to time, t, and space coordinates, x_i, respectively. Note that the wave equation can be expressed in terms of any of the perturbation variables, p, ξ_j, v_j, etc. since these are all uniquely related to one another. Solutions to the wave equation include the general form $p(x_i \pm c_0 t)$ where $p(x_i)$ describes a pressure wave profile at the time $t = 0$. For the one-dimensional case $(x_i = x)$, $p(x - c_0 t)$ describes the progression of the pressure wave profile in the positive x direction (a forward propagating wave), while $p(x + c_0 t)$ describes that in the negative x direction (a backward propagating wave).

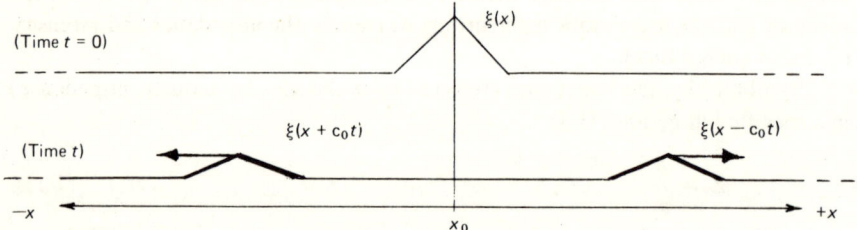

Fig. 1.1 – A stretched string struck at $t = 0$, illustrating forward and backward propagating waves.

A special kind of wave notion is that described by the harmonic, sinusoidal (or exponential) wave given as $\sin k(x \pm c_0 t)$ for a fixed value of c_0. The wave number $k = \omega/c_0 = 2\pi/\lambda$ is proportional to the angular frequency, ω, at which the wave repeats itself in time. It is inversely related to the wavelength, λ, the distance over which the wave repeats itself in space. Sinusoidal wave motion is of fundamental importance in acoustics since every function $p(x_i \pm c_0 t)$ that occurs in practice can be represented by Fourier's theorem as a sum or an integral of sinusoidal functions.

1.3.1 Plane wave propagation in air

In one-dimensional space $(x_i = x)$ a forward-moving sinusoidal pressure wave of amplitude A is given as

$$p = A \sin \omega(t - x/c_0). \qquad (1.4)$$

With changing time, the pressure distribution moves parallel to the x direction. At a particular time, t, the sound pressure is constant in planes perpendicular to the x direction, and thus the waves are referred to as **plane waves**. The impedance and intensity of plane waves can be determined by using Newton's second law (momentum equation) which relates p and v according to

$$\rho_0 \frac{\partial v}{\partial t} = -\frac{\partial p}{\partial x}. \tag{1.5}$$

Substituting a space differentiated form of (1.4) into (1.5) and integrating with respect to time gives

$$v = \frac{A}{\rho_0 c_0} \sin \omega(t - x/c). \tag{1.6}$$

The impedance of a plane progressive wave is then given as

$$Z = \rho_0 c_0 \ (\text{kg m}^{-2} \ \text{sec}^{-1}). \tag{1.7}$$

The product $\rho_0 c_0$ is known as the **characteristic acoustic impedance**, the value of which in air at room temperature (20°C) and standard pressure (1 atm) is 414 kg m^{-2} s^{-1}.

Being purely real (radiative) with no imaginary (reactive) component, this impedance represents physically the acoustic loading that would oppose the motion of a large plate vibrating as an interface between air and a vacuum.

The intensity of a plane sinusoidal wave is given as

$$I = \frac{1}{T} \int_0^T \frac{A^2}{\rho_0 c_0} \sin^2 \omega(t - x/c) \mathrm{d}t \tag{1.8}$$

where T = one period. Hence,

$$I = \frac{1}{2} \frac{A^2}{\rho_0 c_0} = \frac{\overline{p^2}}{\rho_0 c_0} \ (\text{W m}^{-2}) \tag{1.9}$$

where $\overline{p^2}$ is the mean square pressure. In air, the mean square pressure corresponding to an intensity of 1 W m^{-2} is

$$\overline{p^2} = 414 \ (\text{Pa})^2 \tag{1.10}$$

which gives a sound pressure level of

$$L_p = 10 \log_{10} \frac{\overline{p^2}}{p_{\text{ref}}^2} = 20 \log_{10} \frac{\sqrt{\overline{p^2}}}{p_{\text{ref}}} \doteq 20 \log \frac{20.3 \ \text{Pa}}{20 \ \mu\text{Pa}} \doteq 120 \ \text{dB}. \tag{1.11}$$

Two sinusoidal plane waves of equal amplitude travelling in opposite directions combine to give a pressure field of the form

$$p = A[\sin(\omega t - kx) + \sin(\omega t + kx)] = 2A \sin \omega t \cos kx. \tag{1.12}$$

The resultant wave is called a **standing wave** wherein the pressure oscillates at a particular position with constant amplitude, but the amplitude itself is a function of position. The corresponding velocity is

$$v = -\frac{2A}{\rho_0 c_0} \cos \omega t \sin kx \qquad (1.13)$$

and thus the intensity of a standing wave is zero since the pressure and velocity are 90° out of phase.

1.3.2 Spherical wave propagation in air

In considering three-dimensional wave propagation, the most important type of wave is the **spherical wave** for which the sound field quantities are functions only of time, t, and distance, r, from the centre. An outgoing spherical pressure wave, in which the magnitude of the pressure disturbance p is small compared with $\rho_0 c_0^2$, satisfies the spherical wave equation

$$\frac{\partial^2 (rp)}{\partial r^2} - \frac{1}{c_0^2} \frac{\partial^2 (rp)}{\partial t^2} = 0, \qquad (1.14)$$

with a solution at angular frequency ω of the form (in complex notation)

$$p = \frac{A}{r} e^{i(\omega t - kr)}. \qquad (1.15)$$

The corresponding perturbation velocity is obtained from Newton's law as

$$v = \frac{A}{i\omega\rho_0} \left[\frac{ik}{r} + \frac{1}{r^2} \right] e^{i(\omega t - kr)}. \qquad (1.16)$$

As in the plane wave, the velocity has only one component in the direction of propagation. Both pressure and velocity have constant amplitude and phase on spherical surfaces centred on the origin.† This makes the impedance of a spherical wave complex and dependent upon the ratio of the radius, r, to the acoustic wavelength, λ, i.e.,

$$Z = \rho_0 c_0 \frac{ikr}{1 + ikr} = \rho_0 c_0 \frac{2\pi i(r/\lambda)}{1 + 2\pi i(r/\lambda)}. \qquad (1.17)$$

Only when $r \gg \lambda$ does $Z \rightarrow \rho_0 c_0$ in which case the associated pressure and velocity oscillate in phase and decrease inversely with r. When considering propagation from spherical sources, this rate of decrease is commonly referred to as the loss due to spherical spreading, and in decibels represents a decrease in intensity of 6 dB doubling of distance.

To compute the intensity of a spherical wave, the velocity can be written in terms of the real part of equation (1.16) in which both the radiative and reactive parts contribute as

$$v = \frac{Ak}{\omega\rho_0 r} \cos(\omega t - kr) + \frac{A}{\omega\rho_0 r^2} \sin(\omega t - kr). \qquad (1.18)$$

† Note, however, that the velocity field is made up of a radiative (real) part, and a reactive (imaginary) part which is 90° out of phase with the pressure field.

Writing the pressure in a similar manner yields an expression for intensity as

$$I = \frac{1}{T} \int_0^T \frac{A^2}{r^2\rho_0 c} \cos^2(\omega t - kr)\mathrm{d}t \; + \; \frac{1}{T} \int_0^T \frac{A^2}{r^3\rho_0 c} \sin(\omega t - kr)\cos(\omega t - kr)\mathrm{d}t$$
$$(1.19)$$

The second integral is identically zero when T is equal to one period, and thus the intensity becomes

$$I = \left(\frac{A}{r}\right)^2 \frac{1}{2\rho_0 c_0} = (\overline{p^2}) \frac{1}{\rho_0 c_0}. \tag{1.20}$$

Note that the intensity decreases as the inverse square of the distance for a spherical wave. The rate of energy flow, W, from a source is given by

$$W = I.4\pi r^2$$
$$= \left(\frac{A}{r}\right)^2 \frac{4\pi r^2}{2\rho_0 c_0} = \frac{2\pi A^2}{\rho_0 c_0}. \tag{1.21}$$

As expected from energy considerations, the rate of energy flow through any surface surrounding the source is the same.

1.3.3 Propagation losses

In the previous sections, the acoustic wave motion was assumed to propagate in an idealized medium which was free of losses due to viscosity, heat conduction, and molecular absorption. For a wide variety of problems, these assumptions are realized practically, and the solutions obtained from the corresponding wave equation adequately describe the particular wave field under consideration. When wave propagation over long distances is considered, however, losses in the medium become significant and must be accounted for in the overall problem. Basically, the losses inherent in a gas can be classified under three main mechanisms which irreversibly convert acoustical energy into heat energy as follows:

(1) Viscous losses – due to the viscous shearing forces of a gas which oppose the particle motion of a sound wave. The attenuation constant due to viscosity, α_v, at a frequency of f Hz in air is given approximately by (1 atmosphere, 20°C)

$$\alpha_v = 8.5 \times 10^{-8} f^2 \text{ dB/km.} \tag{1.22}$$

(2) Heat conduction losses – due to heat transfer between adjacent compression and rarefaction regions within the gas. The attenuation constant due to heat transfer, α_T, at a frequency of f Hz in air is given approximately by (1 atmosphere, 20°C)

$$\alpha_T = 3.6 \times 10^{-8} f^2 \text{ dB/km.} \tag{1.23}$$

(3) Molecular absorption — energy losses due to a relaxation phenomenon of the molecules of a gas which are excited into resonance by the passage of a sound wave. In air, the principal effect involves an interaction of water vapour with the vibrational resonance of oxygen and nitrogen molecules so that the molecular absorption is highly dependent on the humidity content of air. The corresponding attenuation constant, α_{vib}, is thus a function of the frequency of the sound wave, f, and the molecular vibrational relaxation frequency, f_m, and for oxygen relaxation at 20°C is given by

$$\alpha_{vib}(O_2) = \frac{0.056\, f^2/f_m}{[1 + (f/f_m)^2]} \text{ dB/km.} \tag{1.24}$$

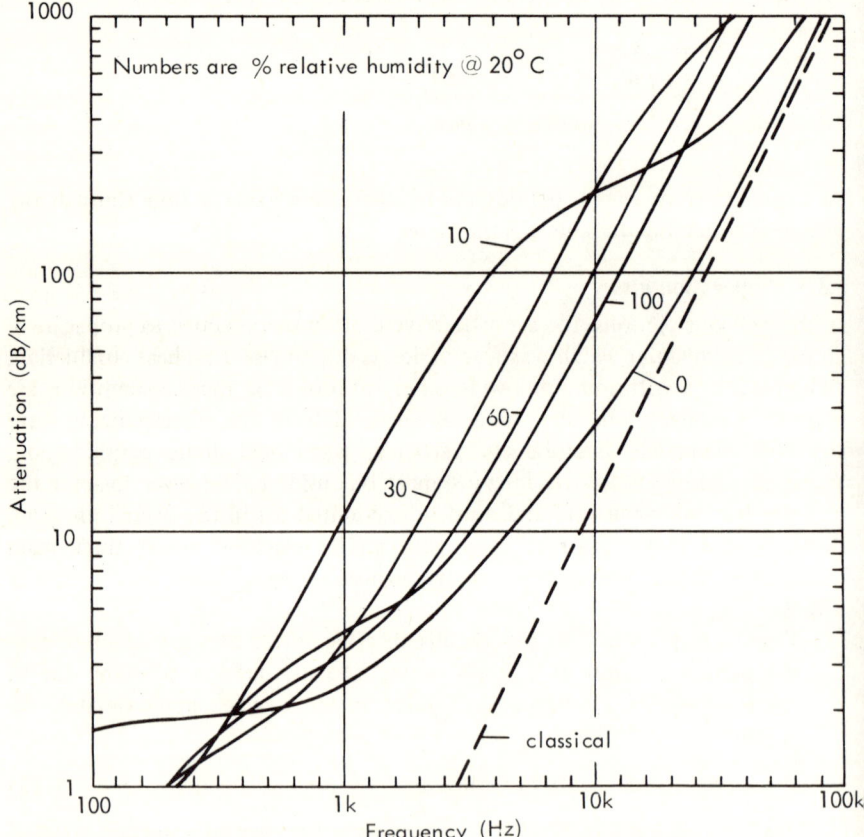

Fig. 1.2 — Air absorption losses at 20°C, 1 atmosphere.†

---

† To allow for pressure variations, the scales may be relabelled as α/p^*, f/p^* where p^* is the pressure in atm. This applies to both α_{cl} and α_{total}.

In combination the losses due to viscosity and heat conduction give the classical attenuation constant, α_{cl}, i.e.

$$\alpha_{cl} = \alpha_v + \alpha_T. \tag{1.25}$$

The functional dependence of α_{cl} and α_{total} on frequency and humidity for air (1 atmosphere, 20°C) is shown in Fig. 1.2. The gap between α_{cl} and α_{total} at 0% humidity fails to close at high frequencies because of *rotational relaxation*, whose contribution is roughly equal to α_T.

1.4 RAY ACOUSTICS

In the previous section, the wave field was defined completely in functions involving time and space. This approach is feasible when used to describe wave propagation in unbounded (and certain bounded) media which are homogeneous. When considering wave propagation over large distances, as in the atmosphere, inhomogeneities due to wind, temperature gradients, etc., make the wave approach too cumbersome, and a simplified approach must be adopted. The ray acoustics approach is one such simplification wherein families of rays are obtained as solutions of a simpler differential equation called the **eikonal equation**. In special cases, these solutions also satisfy the wave equation and under rather broad conditions provide good approximations to the more exact solutions.

In a stationary homogeneous medium (constant c_0 and ρ_0), a solution to the three-dimensional wave equation written in Cartesian coordinates x, y, and z can be expressed in the form of a sinusoidal function

$$p = \bar{p} \exp i\omega \left[t - \frac{(\alpha x + \beta y + \gamma z)}{c_0} \right] \tag{1.26}$$

where α, β, and γ are the directional cosines of a straight line (ray†) perpendicular to the **wavefront** (surface S of constant phase) described by the plane

$$\frac{1}{c_0}(\alpha x + \beta y + \gamma z) = S = \text{constant.} \quad \text{(See Fig. 1.3)} \tag{1.27}$$

When the speed of sound, c, is variable, equation (1.26) is no longer a solution of the wave equation. However, if the variations of c are small over distances comparable to a wavelength of sound, we can construct an approximate solution in which surfaces of constant phase propagate at the local speed of sound $c(x, y, z)$, so that some portions of the wave move faster than others. This is illustrated in Fig. 1.4, where $c\Delta t$ is the perpendicular distance between two surfaces of constant phase. Note that a general function $S(x, y, z)$ has been

---

† However, it should be noted that in moving media, rays are no longer parallel to the local wave normal.

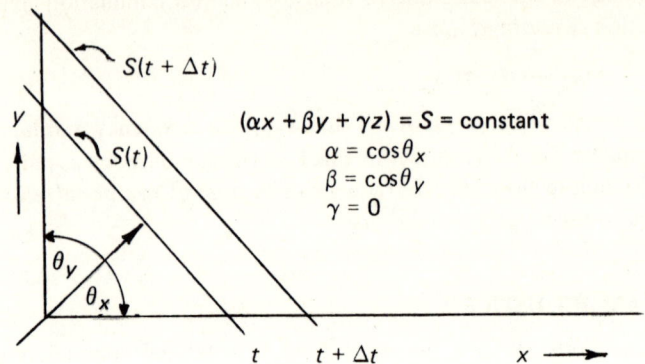

Fig. 1.3 – Sketch showing wavefront surfaces (S = constant) in a uniform medium.

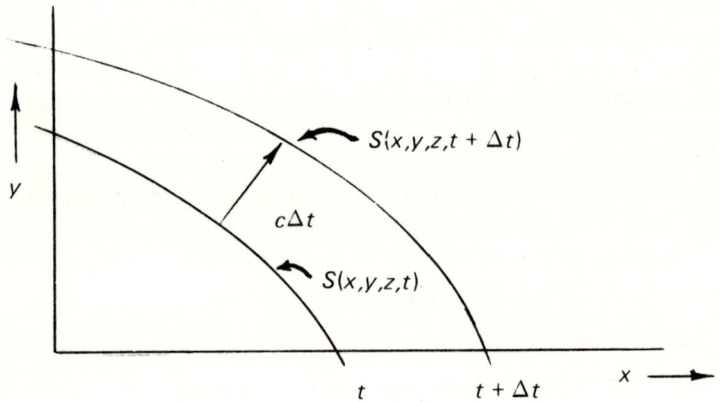

Fig. 1.4 – Surfaces of constant phase in a medium of variable sound speed.

adopted to describe the overall spatial contribution to the phase angle. Thus equation (1.26) is rewritten with the exponential factor replaced by

$$\exp i\omega[t - S(x, y, z)]. \tag{1.28}$$

To write a trial solution to the wave equation which describes propagation in an inhomogeneous medium, the amplitude of the wave must be allowed to vary spatially in addition to the function $S(x, y, z)$, such that

$$p = \bar{p}(x, y, z) \exp i\omega[t - S(x, y, z)]. \tag{1.29}$$

Taking the necessary derivatives of this solution and substituting them into the wave equation produces a complex expression which can be separated into real and imaginary parts. The real part is of most interest and is given as

$$\left(\frac{\partial S}{\partial x}\right)^2 + \left(\frac{\partial S}{\partial y}\right)^2 + \left(\frac{\partial S}{\partial z}\right)^2 - \frac{1}{c^2} = \frac{1}{\omega^2 \bar{p}}\left(\frac{\partial^2 \bar{p}}{\partial x^2} + \frac{\partial^2 \bar{p}}{\partial y^2} + \frac{\partial^2 \bar{p}}{\partial z^2}\right). \tag{1.30}$$

Equating the left-hand side of equation (1.30) to zero gives the eikonal equation. It can be seen that strictly, the eikonal equation is only consistent with the wave equation if the term on the right side of equation (1.30) is zero. This will be true, in general, only in the limit of very high frequencies where $\omega^2 \to \infty$. The approximation will be good, however, whenever the change in the sound speed is small over a local wavelength of sound.

The function $S(x, y, z)$ can be eliminated from the eikonal equation to give three ordinary differential equations which together constitute the equations of ray motion in a stationary medium, namely

$$\frac{\partial(\alpha c^{-1})}{\partial s} = \frac{\partial c^{-1}}{\partial x}, \quad \frac{\partial(\beta c^{-1})}{\partial s} = \frac{\partial c^{-1}}{\partial y}, \quad \frac{\partial(\gamma c^{-1})}{\partial s} = \frac{\partial c^{-1}}{\partial z}, \tag{1.31}$$

or equivalently one vector equation,

$$\frac{\partial(n c^{-1})}{\partial s} = \nabla c^{-1}. \tag{1.32}$$

Here n is the unit normal to the wavefront, with components (α, β, γ), and s is the distance measured along the ray (Fig. 1.5).

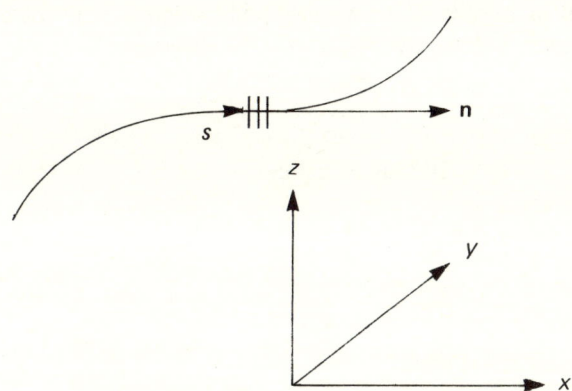

Fig. 1.5 – Sketch of ray geometry, showing wavenormal vector $n = (\alpha, \beta, \gamma)$.

To illustrate the usefulness of these equations, we consider the case where c varies only in the z direction so that

$$\frac{\partial c^{-1}}{\partial x} = \frac{\partial c^{-1}}{\partial y} = 0. \tag{1.33}$$

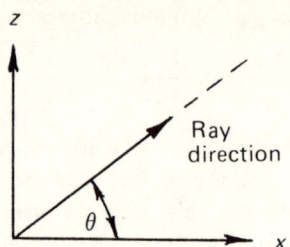

Fig. 1.6 – Ray geometry for $c = c(z)$.

If the ray propagates at angle θ to the horizontal $(x\text{-}y)$ plane, as shown in Fig. 1.6, equation (1.31) gives

$$\frac{\partial(c^{-1}\cos\theta)}{\partial s} = 0 \tag{1.34}$$

and

$$\frac{\partial(c^{-1}\sin\theta)}{\partial s} = \frac{dc^{-1}}{dz}. \tag{1.35}$$

From equation (1.34) it follows that $(1/c)\cos\theta$ has a constant value along a particular ray. If P and P' are two points on the ray, then

$$\frac{1}{c}\cos\theta = \frac{1}{c'}\cos\theta'. \tag{1.36}$$

Moreover, if P' is located at a reference point where $c'(z) = c_0$ and θ_0 is the direction of the ray at this point, equation (1.36) becomes

$$\frac{\cos\theta}{\cos\theta_0} = \frac{c}{c_0} \tag{1.37}$$

which is identical with Snell's law in optics.

The dependence of ray curvature on the change in sound speed can be obtained by writing equation (1.37) as

$$c = K\cos\theta, \left(K = \frac{c_0}{\cos\theta_0}\right) \tag{1.38}$$

and differentiating with respect to z to give

$$\frac{dc}{dz} = -K\sin\theta\,\frac{d\theta}{dz}$$

$$= -K\frac{dz}{ds}\frac{d\theta}{dz} = -K\frac{d\theta}{ds}. \tag{1.39}$$

The ray curvature thus has the opposite sign of the sound speed gradient, i.e., the ray always bends in the direction of a lower sound speed region as illustrated in Fig. 1.7.

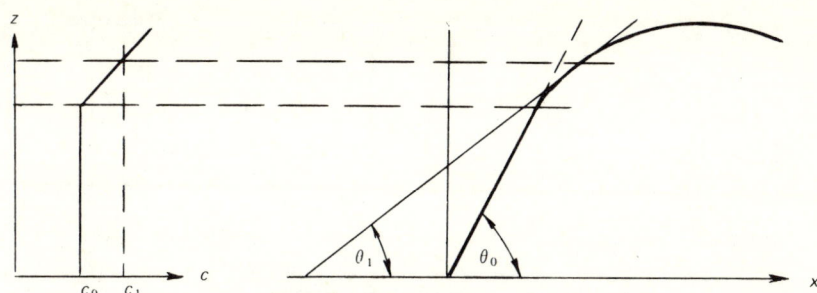

Fig. 1.7 – Sketch showing downward refraction of rays by a layer of positive dc/dz (that is, temperature inversion).

Examples of ray curvature for various sound speed gradients are shown in Fig. 1.8.

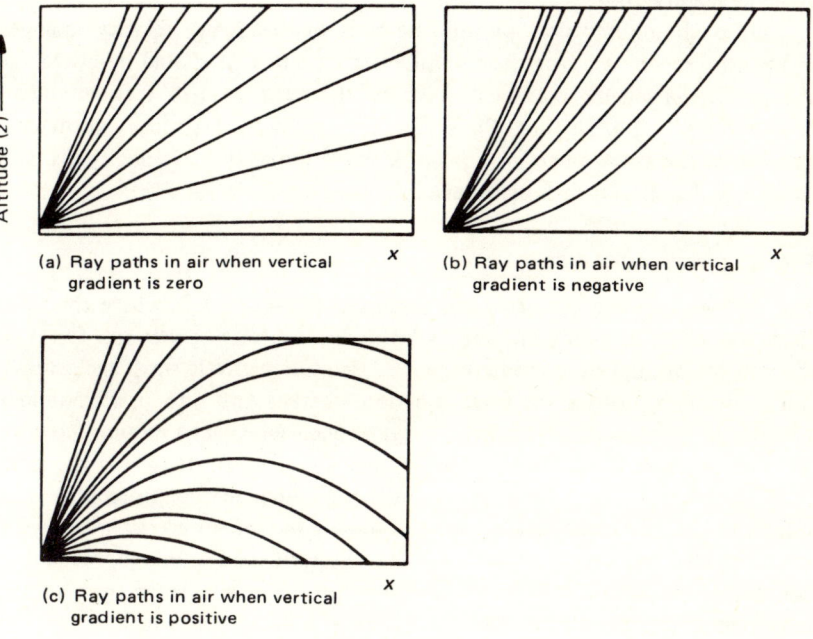

(a) Ray paths in air when vertical gradient is zero

(b) Ray paths in air when vertical gradient is negative

(c) Ray paths in air when vertical gradient is positive

Fig. 1.8 – Sketch showing ray paths in a still atmosphere with different values of the gradient dc/dz.

Sound speed gradients generated by wind are of special interest since the ray curvature is dependent upon the direction of the wind relative to that of the sound, as shown in Fig. 1.9. Under the condition shown, the sound radiating from a source to position A would be little affected by the wind velocity gradient,

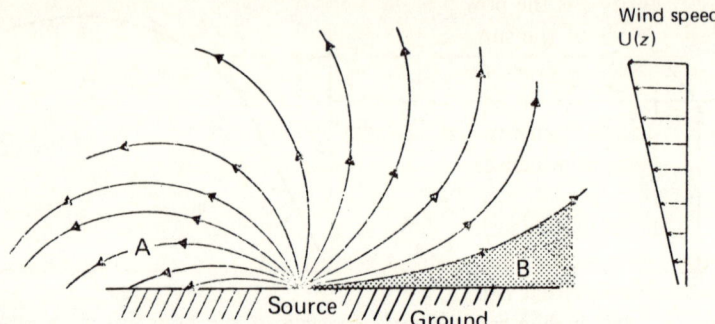

Fig. 1.9 – Shadow zone produced by wind shear.

while that radiating to position B would be greatly reduced owing to the rays bending away from the ground surface. The region around position B is referred to as the **shadow zone**.

Although a discussion on the methods of measuring velocity gradients in the atmosphere is beyond the scope of this chapter, it should be noted that any prediction scheme necessarily involves the acquisition of vast amounts of meteorological data. In turn, these data are processed statistically to describe a region of the atmosphere in terms of a collection of typical velocity gradients and the probability of their occurrence.

1.5 ENERGY ACOUSTICS

The method of wave acoustics was discussed in section 1.3, where the sound field was given in terms of exact solutions to the wave equation. This same method can be applied to predicting sound fields in relatively small enclosures of simple shape, by solving the wave equation together with appropriate boundary conditions. When dealing with large, irregular enclosures such as rooms, however, an exact analysis is no longer possible, and a method of describing the sound field statistically must be adopted. Although such methods preclude exact solutions, they do give fairly accurate average values of sound quantities which are often all an acoustical engineer requires. Fundamental to this approach is the concept of the diffuse field, in which the sound energy emanating from a source in a room over a band of frequencies produces a sound field which is composed of a great many superimposed rays propagating in all directions throughout the room with equal probability. Moreover, the sound pressure level and thus the energy density, E (joules/m^3), is constant throughout the volume of the room. Under these conditions, the intensity, I, incident on any imaginary surface in the room, is related to the local energy density by

$$I = \frac{c_0 E}{4}.$$

$$(1.40)$$

The quantity I is the power incident per unit area of surface from all directions on one side of the surface. The energy approach (from which the Sabine equation derives) is developed by setting up an energy balance in the room between (1) the rate at which energy is absorbed at the boundary surfaces; (2) the rate at which energy contained in the air throughout the room changes; and (3) the rate at which energy is produced by the source. Consider each of these in turn.

First, each boundary surface in the room can be assigned an acoustical energy absorbing property called the **sound absorption coefficient**, α, which is a quotient of the sound energy that is not reflected and the incident sound energy. For an acoustically 'hard' wall, $\alpha = 0$; for an open window, $\alpha = 1$. In general, the sound absorption coefficient is a function of angle of incidence and frequency. Since the boundaries of an enclosure seldom consist of just one kind of material, it is necessary to represent α as an average of all the sound absorption coefficients, α_n, for all of the boundary surfaces, S_n. We therefore write

$$\alpha S = \sum_n \alpha_n S_n \tag{1.41}$$

where

$$S = \sum_n S_n. \tag{1.42}$$

The quantity αS is usually written as A, which gives the total absorption present and has the dimensions of area. (Note that the averaging process described above is only valid where the absorbing surfaces are uniformly distributed along the enclosure boundaries.) Recalling that the acoustic intensity has the units of energy per unit time per unit area, the product of A and the diffuse field intensity, I, gives the total energy absorbed per unit time by the combination of all the room's surfaces.

Next, an expression for the rate at which energy changes in the air throughout the room can be written in terms of a time derivative of the product of the room volume, V, and the energy density, E, as

$$\frac{d}{dt}[VE(t)] \tag{1.43}$$

which in terms of intensity is (from equation (1.40))

$$\frac{4V}{c_0}\frac{d}{dt}[I(t)]. \tag{1.44}$$

Finally, using the above results, a diffuse-field energy balance for an enclosure containing a source of power, W, can be set up as

$$W(t) = AI(t) + \frac{4V}{c_0}\frac{dI(t)}{dt}. \tag{1.45}$$

If the time history of the source is such that

$$W(t) = W_0 \quad t < 0$$
$$W(t) = 0 \quad t > 0 \tag{1.46}$$

the resulting solution to equation (1.45) gives an expression for the decay of the steady state sound field as

$$I(t) = \frac{W_0}{A} \exp\left(-\frac{Ac_0 t}{4V}\right). \tag{1.47}$$

In this equation the quantity W_0/A, which has the units of watts per unit area, gives a measure of the steady state intensity (denoted by I_0) in the room. Note that this quantity is dependent upon room absorption as would be expected, but not on the volume or shape of the room.

For a diffuse field, the steady-state mean square pressure is related to the energy density by

$$\overline{p^2} = \rho_0 c_0^2 E = 4\rho_0 c_0 I, \tag{1.48}$$

and thus the mean square pressure at any point in the room generated by a steady-state source of power, W, is

$$\overline{p^2} = \frac{4W\rho_0 c_0}{A}. \tag{1.49}$$

Conversely, this expression can be used to determine the amount of absorption in a room, provided that the power output of the source is known beforehand.

A more convenient method of determining absorption, however, is to use a transient decay measurement. This method makes use of equation (1.47) which, written in terms of intensity level, has the form

$$10 \log_{10} \frac{I(t)}{I_0(t)} = -\frac{4.34 Ac_0 t}{4V}. \tag{1.50}$$

A measure of the decay rate is taken as the time required for the intensity, $I(t)$, to decrease by 60 dB from the initial steady state value, I_0. This time is called the **reverberation time** of the room and is given by

$$T = 0.163 \text{ sec m}^{-1} \frac{V}{A}. \quad (c_0 = 340 \text{ m/s}) \tag{1.51}$$

Consider an application of the above expression. Assume that a room measuring 10 m × 8 m × 5 m has surfaces with the following absorption coefficients:

	100 Hz		1000 Hz	
	α	αS (m^2)	α	αS (m^2)
Concrete floor	0.01	0.8	0.02	1.6
Brick walls	0.02	3.6	0.04	7.2
Tiled ceiling	0.50	40.0	0.80	64.0
		———		———
		44.4		72.8
Reverberation time, T		1.5 sec.		0.9 sec.

In the above example, it can be seen that the reverberation time of the room is controlled mainly by the choice of absorptive material used for covering the ceiling.

BIBLIOGRAPHY

Acoustical Society of America (1978) *American National Standard* ANSI S1.26. Method for the calculation of the absorption of sound by the atmosphere.

Advisory Group for Aerospace Research and Development (1979) *AGARD Report No. 686*. Special course on acoustic wave propagation.

Bazley, E. N. (1976) *National Physical Laboratory Acoustics Report Ac. 74* Sound absorption in air at frequencies up to 100 kHz.

Clay, C. S. & Medwin, H. (1977) *Acoustical oceanography: principles and applications*. New York: J. Wiley.

Evans, L. B., Bass, H. E. & Sutherland, L. C. (1971) *Journal of the Acoustical Society of America* 51, 1565-1575. Atmospheric absorption of sound: theoretical predictions.

Kinsler, L. E. & Frey, A. R. (1962) *Fundamentals of acoustics* (2nd edn.). New York: J. Wiley.

Kinsler, L. E., Frey, A. R., Coppens, A. B., Sanders, J. V. (1982) *Fundamentals of acoustics* (3rd edn.).

Kuttruff, H. (1973) *Room Acoustics* (2nd edn.). London: Applied Science Publishers Ltd.

Maekawa, Z. (1974) in *Proceedings of 8th International Congress on Acoustics* (London, July 1974). Environmental sound propagation (invited lecture).

Meyer, E. & Neumann, E. G. (1972). *Physical and applied acoustics*. New York: Academic Press.

Officer, C. B. (1958) *Introduction to the theory of sound transmission*. New York: McGraw-Hill.

Piercy, J. E., Embleton, T. F. W. & Sutherland, L. C. (1977) *Journal of the Acoustical Society of America* 61, 1403-1418. Review of noise propagation in the atmosphere.

Wiener, F. M. (1958) *Noise Control* 4, 224-228. Sound propagation outdoors.

CHAPTER 2

Theory of acoustics (II)

C. L. Morfey

Institute of Sound and Vibration Research, University of Southampton

The propagation of plane and spherical sound waves was discussed in Chapter 1 without reference to boundaries. In this chapter we consider boundaries first as reflectors and then as sources of sound.

2.1 REFLECTION OF SOUND FROM PASSIVE BOUNDARIES

The sections which follow are concerned with reflection from rigid and non-rigid plane surfaces, and with the resulting interference between incident and reflected waves. The measurement of surface impedance in a standing-wave tube is discussed, and the first part of the chapter concludes with some practical considerations on the construction of sound-absorbent surfaces.

2.1.1 The rigid plane boundary: plane-wave reflection and interference

Fig. 2.1 illustrates the process of interference between incident and reflected plane waves at a rigid plane boundary of infinite extent. Reflection occurs at the boundary with no change of amplitude or phase; thus the velocity components normal to the boundary associated with the incident and reflected waves cancel at the surface, as required by the boundary condition. The resulting interference

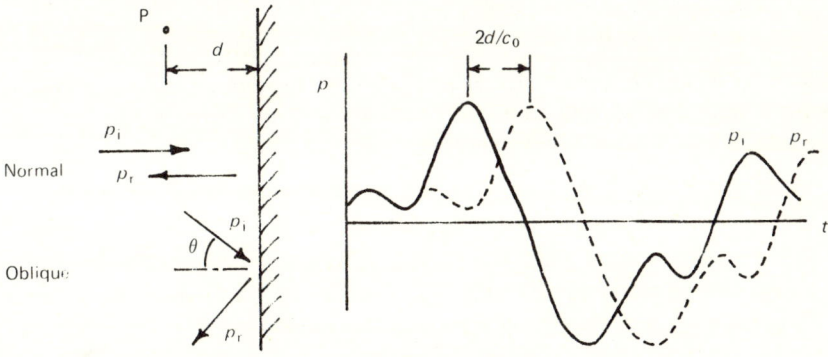

Fig. 2.1 – Plane-wave reflection and interference (rigid boundary).

at any point (e.g. P in Fig. 2.1) may be obtained by adding to p_i (the incident-wave pressure) its value delayed by a time $2d/c_0$, which represents the round-trip time from P to the boundary and back.

For oblique incidence at an angle θ (see Fig. 2.1) the same applies, except that the delay is $2d \cos\theta/c_0$. This can be seen by considering the plane-wave phase speed normal to the surface: for normal incidence the phase speed is c_0, for oblique incidence it becomes $c_0/\cos\theta$.

The generalization to surfaces of finite impedance is greatly simplified by considering waves of a single frequency, and is taken up in section 2.1.4. For the moment, however, we retain the rigid-surface boundary condition in order to examine the interference which results from incident spherical waves.

2.1.2 The rigid plane boundary: spherical-wave reflection and interference

Fig. 2.2 shows a source at Q radiating sound to an observer at P, with a rigid infinite 'ground plane' at distance h below the source.

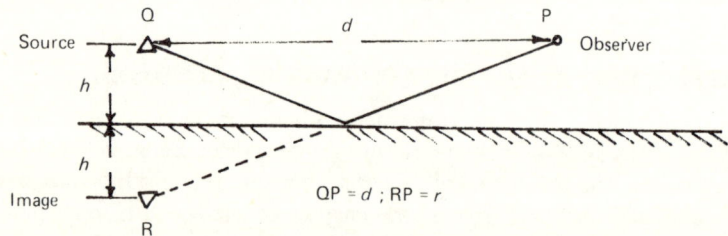

Fig. 2.2 – Spherical-wave reflection and interference (rigid boundary).

The boundary condition imposed by the ground plane, namely zero normal velocity over the entire surface, may be reproduced in the absence of the ground plane by adding an identical image source at R, i.e. in the mirror-image position with respect to Q. The effect of the reflecting surface is completely modelled, in the region above the plane, by the image source.

The pressure received at P may accordingly be written as

$$p = p_d + p_r, \tag{2.1}$$

where p_d is the pressure which reaches P directly from Q (i.e. as if the reflecting surface were absent), and p_r is the contribution from the image source at R. To calculate p_r, it is sufficient to note that

(i) the amplitude† is less by a factor d/r than that of p_d;

(ii) the reflected signal suffers a delay of $(r-d)/c_0$ relative to the direct signal;

(iii) the radiation angle with respect to the source is different for the image and the actual source (see Fig. 2.2); this is important where the source is directional. Note that the image source is inverted.

---

† Or r.m.s. value, in the case of time-stationary noise rather than a single-frequency signal.

The nature of the resulting interference field is illustrated in Fig. 2.3, for the special case of an omnidirectional point source with both Q and P at the same height h. The source is assumed to radiate an octave band-limited white-noise spectrum, with geometric centre frequency f, and the ground-plane interference effect is plotted as a change in level (ΔL) versus the non-dimensional distance $d\lambda/h^2$.

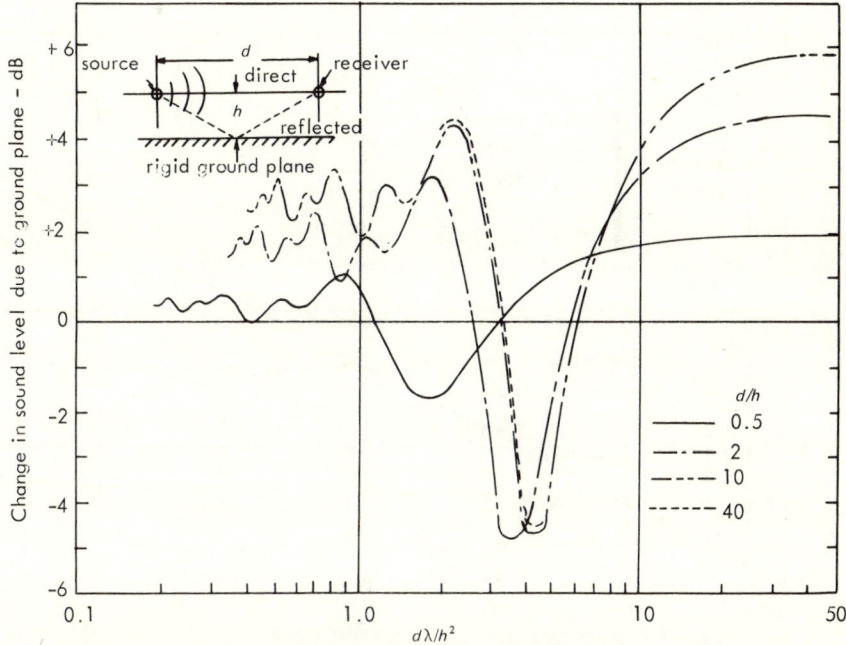

Fig. 2.3 – Ground interference effect: hard surface, omnidirectional point source.

Although in practice ground surfaces are never perfectly plane or acoustically hard (especially at low frequencies), the above model does demonstrate some of the problems inherent in measuring sound propagation near the ground. When the ground impedance is finite, the reflection of spherical waves cannot in general be accounted for by a simple image source, and the analysis is correspondingly complicated. The reader is referred to Delany & Bazley [2.1] for a discussion of the resulting interference phenomena, the most important being a substantial increase in attenuation near the ground at large distances.

2.1.3 Screening of spherical waves by a barrier

The diffraction of spherical waves by a semi-infinite rigid plane is another idealized problem whose solution is of practical interest, this time in relation to barriers. As a method of noise control, barriers are widely used in situations where there is no means of influencing the noise source, one example being road traffic noise.

Fig. 2.4 shows the source/observer/barrier configuration and presents barrier attenuation predictions based on the theoretical and experimental work of Maekawa [2.2]. Note that the key parameter, in this idealized situation, is the ratio of the excess path length $(a + b - d)$ to the acoustic wavelength. When this ratio exceeds about 1, the predicted barrier attenuation increases with frequency at 10 dB per decade.

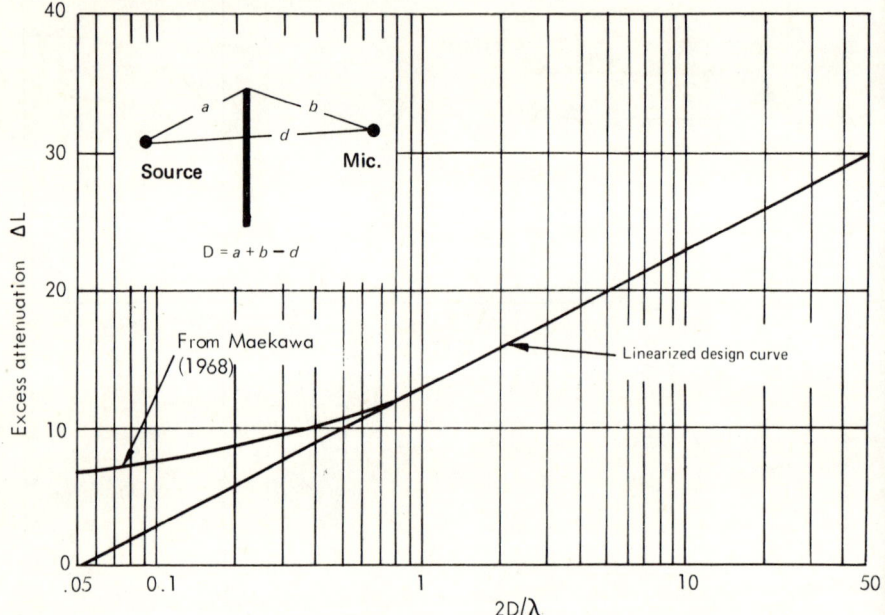

Fig. 2.4 – Excess attenuation due to a semi-infinite barrier.

Maekawa has generalized his findings to provide practical methods for predicting the attenuation due to barriers of finite dimensions, and for including ground reflection effects.

2.1.4 Plane boundary of finite impedance: plane-wave reflection and interference

Fig. 2.5 illustrates the reflection of normally-incident plane waves from a uniform surface of specific acoustic impedance Z_s. The value of Z_s is defined as the ratio of the complex pressure and normal velocity at the surface:

$$Z_s = (p/v)_{surface.} \tag{2.2}$$

Note that Z_s is in general a complex quantity, and refers to sound of a single frequency; its value for any given boundary will be different at different frequencies. On the other hand, within the linear response range Z_s is independent of the surface pressure amplitude.

A further variable which can affect the specific impedance of a given boundary is the angle of the incident sound waves. The reason is that, in principle, the normal velocity at one point on the surface is not determined just by the pressure at that point, but depends on the values of the prsure at all other points. A good example is the interface between two fluids. Provided — as here — we consider only one angle of incidence at a time, this does not matter; but the need to use a different Z_s value for each angle† should be borne in mind.

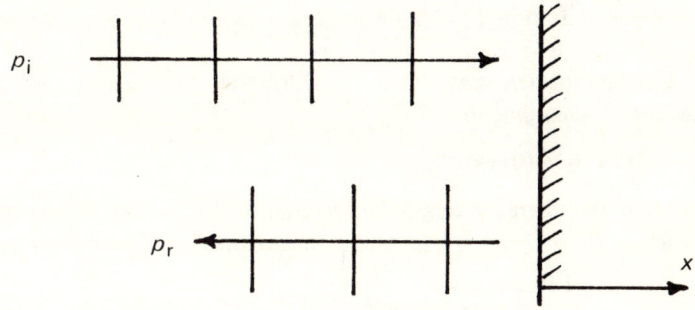

Fig. 2.5 — Reflection of normally-incident plane waves from a plane boundary of finite impedance.

For the normal-incidence situation sketched in Fig. 2.5, the pressure and velocity at any position x (measured normal to the boundary) may be written as

$$p = p_i + p_r; \quad v = \frac{p_i}{\rho_0 c_0} - \frac{p_r}{\rho_0 c_0}, \tag{2.3}$$

where v is taken as positive in the incident-wave direction. The ratio p/v at any value of x is therefore

$$\frac{p}{v} = Z(x) = \rho_0 c_0 \frac{p_i + p_r}{p_i - p_r}. \tag{2.4}$$

If we denote the complex ratio of the reflected and incident pressures, p_r/p_i, by $R(x)$ then equation (2.4) becomes

$$\frac{Z}{\rho_0 c_0} = \frac{1 + R}{1 - R}. \tag{2.5}$$

The complex quantity R is the pressure reflection factor; its modulus r is the **reflection coefficient**. The ratio $Z/\rho_0 c_0$, often denoted by ξ, is the **specific**

† A boundary for which Z_s is independent of the surface pressure distribution is sometimes referred to as a **locally-reacting** or **point-impedance** boundary.

impedance ratio in the x direction. Equation (2.5) is the basis of the standing-wave tube method of impedance measurement (see section 2.1.5 below).

Problems

(1) Show that for obliquely incident waves at angle of incidence θ, the impedance and reflection factor at any point in the standing-wave field are related by

$$\zeta \cos\theta = \frac{1+R}{1-R}, \quad R = \frac{\zeta \cos\theta - 1}{\zeta \cos\theta + 1}; \tag{2.6}$$

i.e. the generalization to oblique incidence simply involves replacing ζ by $\zeta \cos\theta$.

(2) For normally incident waves as in Fig. 2.5, show that the reflection factor varies with x according to

$$R(x) = R(0) \exp 2ikx. \tag{2.7}$$

(3) Show that the *energy transmission coefficient* in the direction normal to the surface, $\alpha_t = 1 - r^2$, is given under normal-incidence conditions by

$$\alpha_t = \frac{4\,\mathrm{Re}\,\zeta}{|1 + \zeta|^2} \text{ (independent of } x\text{).} \tag{2.8}$$

The energy transmission coefficient defined above, which can refer to plane waves at any angle of incidence, corresponds at a boundary to the surface absorption coefficient defined in Chapter 1. Here it is viewed as a property of an axial standing-wave field, not limited to boundary surfaces.

2.1.5 The impedance tube

If a rigid-walled uniform tube is terminated with a sample of unknown acoustic impedance, the sample impedance value can be determined by exciting the tube with a single-frequency source and measuring the axial standing-wave pattern set up in the tube. Fig. 2.6 illustrates the method in principle.

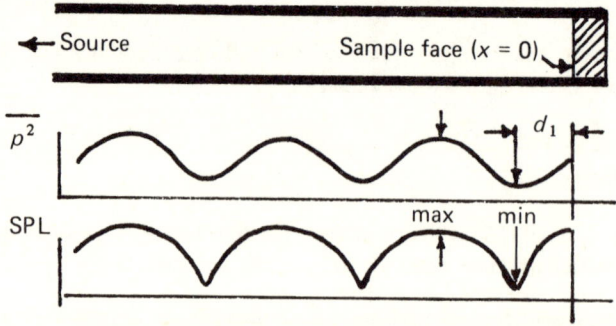

Fig. 2.6 – Standing-wave pattern in an impedance tube.

Provided the higher-order modes are excluded†, the standing-wave field consists of incident and reflected plane waves travelling axially along the tube. Measurement of the corresponding reflection factor R_s at the sample face gives the normal-incidence impedance ratio through equation (2.5). For this purpose it is convenient to work with the amplitude and phase of R_s, as follows.

(i) At the point labelled 'min' in Fig. 2.6, the pressures in the incident and reflected waves are 180° out of phase; at 'max' they are in phase. The ratio of $|p_{max}|$ to $|p_{min}|$, denoted by N, is therefore related to the reflection coefficient r $(= |R_s|)$ by

$$N = \frac{1+r}{1-r}; \quad r = \frac{N-1}{N+1}. \tag{2.9}$$

(ii) At 'min' $(x = -d_1)$, the phase of p_r relative to p_i is π radians; thus at $x = 0$, the relative phase is $\pi + \delta$ (where $\delta = 4\pi d_1/\lambda$), from equation (2.7). It follows that the reflection factor at the sample face is

$$R_s = \frac{N-1}{N+1} \exp i(\pi + \delta). \tag{2.10}$$

Substituting this result into equation (2.5) gives finally

$$\frac{Z_s}{\rho_0 c_0} = \frac{2N - i(N^2 - 1)\sin\delta}{N^2 + 1 + (N^2 - 1)\cos\delta}, \tag{2.11}$$

which allows the real and imaginary parts of Z_s to be calculated from measured quantities.

Note that equation (2.11) refers to a loss-free standing-wave tube; that is one in which the incident and reflected waves suffer negligible attenuation along the tube. Corrections for tube losses are discussed by Beranek [2.3].

2.1.6 Physical realization of acoustically absorbing boundaries

Acoustical design requirements

Typical applications of sound-absorbing surfaces are

(i) in a large enclosure, or room, to reduce the reverberant noise level;

(ii) in a duct, to absorb sound as it propagates along the duct.

In case (i), the sound field can often be treated as diffuse (see Chapter 1), and the most important acoustical characteristic of the surface is its random-incidence absorption coefficient $\bar{\alpha}$.

---

† This implies an upper frequency limit of $f_{max} = 0.586\, c_0/D$ (in a circular tube of internal diameter D), unless special precautions are taken.

If the surface impedance Z_s is known for all angles of incidence θ, the value of $\bar{\alpha}$ follows as

$$\bar{\alpha} = \int_0^{\pi/2} \alpha(\theta) \sin 2\theta \, d\theta,$$ (2.12)

where $\alpha(\theta)$ is the surface absorption coefficient at angle θ given by

$$\alpha(\theta) = \frac{4 \operatorname{Re}(\zeta) \cos \theta}{|1 + \zeta \cos \theta|^2} \quad \text{(compare with equation (2.8))}.$$ (2.13)

For example, if the surface is locally-reacting its ideal impedance (for maximum $\bar{\alpha}$) is purely resistive, with $Z_s \doteq 1.5 \, \rho_0 c_0$; $\bar{\alpha}$ is then 0.95.

In case (ii), however, the attenuation produced by a lined duct is not uniquely determined by $\bar{\alpha}$, since we are not usually dealing with a diffuse sound field — particularly if the dimensions of the duct cross-section are not large compared with the wavelength. The optimum wall impedance for maximum attenuation in a given length depends on the duct dimensions, the frequency, and the spatial characteristics of the sound to be attenuated (see Chapter 12).

Porous materials for sound absorption

Fig. 2.7 shows a type of construction commonly used for sound-absorbent surfaces. The porous material may consist of rockwool or glassfibre blanket, porous plastic foam, or a rigid porous material such as woodwool slabs or acoustic plaster.

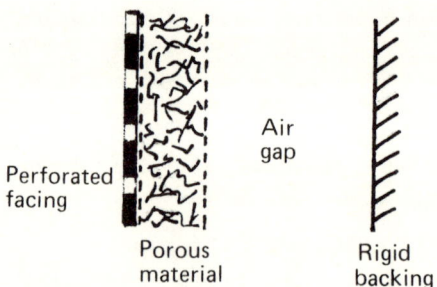

Fig. 2.7 – Components of a sound-absorbing boundary.

The function of the porous material is to allow the acoustic velocity field to penetrate with relatively little obstruction until the energy is dissipated by viscous action in the fine pores or interstices. As a general rule, a thickness in excess of $\lambda/8$ (for material + air gap) is required for optimum absorption. The aim is to achieve the largest possible velocity amplitudes within the porous layer.

Additional energy dissipation is in principle provided by the thermal inertia of the fibrous or solid matrix, but this contribution is small compared with the viscous dissipation mentioned above, unless the total thickness is less than the $\lambda/8$ minimum.

In selecting porous materials for sound absorption, it may also be important to consider such features as flammability and mechanical integrity.

Significance of flow resistance for porous materials

For purposes of maximizing $\bar{\alpha}$, a useful guide is that the porous layer in Fig. 2.7 should have a flow resistance† of around $3\rho_0 c_0$. If the layer is too thin, significant reflection will occur from the rigid backing; if the flow resistance is too high, the layer itself will reflect significantly (unless a very 'open' material is being used).

It should not be inferred that the resistance measured for steady flow applies directly to acoustic velocity fluctuations. The acoustical properties of a porous layer are quite different from the steady-flow properties. There is nevertheless a close connection, for many porous materials, between the acoustical properties at a given frequency and the *unit-depth flow resistance* σ. Delany & Bazley [2.4] found that their measurements of characteristic impedance and propagation constant, in a range of materials, collapsed well on the non-dimensional parameter $f\rho_0/\sigma$ for parameter values between 0.01 and 1.

Physically, the parameter $f\rho_0/\sigma$ − or rather its square root − is a measure of the ratio of the typical interstitial dimensions in a porous material to the viscous acoustic boundary layer thickness.

The effect of varying the air gap

Increasing the air-gap, in constructions of the type shown in Fig. 2.7, generally improves the $\bar{\alpha}$ value up to a limiting depth of $\lambda/4$. An air gap of $\lambda/(2 \cos \theta)$ (or integer multiple thereof) is equivalent to no gap at all. Thus an air gap chosen for low-frequency absorption can degrade the high-frequency absorption. In practice this is seldom a serious problem, since most porous layers absorb more effectively at high frequencies in any case.

Partitioning of the air gap into cells may offer advantages (including a closer approach to point-reacting behaviour); the partitions may have to extend into the porous layer as well to obtain the full effect.

The effect of facing layers

Little effect on the surface impedance results from *perforated covers*, provided the open area exceeds about 15% of the total. For situations where it is necessary to guard against fibre leakage or ingress of contaminants, finely-woven fabric

---

† The **flow resistance** of a layer of porous material is defined as the ratio of the pressure drop across the layer to the volume flow rate per unit area through the layer, under conditions of slow steady flow.

facings are available with low flow resistance. The use of paint, or any other impervious coating applied directly to the surface of the porous layer, destroys the high-frequency absorption properties; lightweight membrane facings (e.g. plastic film) are permissible only if a gap is left between the facing and the porous layer, so that the facing can vibrate independently.

2.2 VIBRATING SURFACES AS SOURCES OF SOUND

A simplified discussion of sound radiation from vibrating surfaces is given below, with emphasis on low-frequency and high-frequency limiting behaviour. The special case of radiation from a plane surface is treated in detail.

2.2.1 Compact acoustic sources

We consider first the radiation of sound in the low-frequency limit, defined by the wavelength of the radiated sound being large compared with the radiating object. The radiating surface is then described as **acoustically compact**, and can be modelled simply as a point source of sound.

The simplest point source of sound occurs at the origin of an outgoing spherical wave of the type already considered in Chapter 1, i.e.

$$p = \frac{1}{r} f(t - r/c_0). \tag{2.14}$$

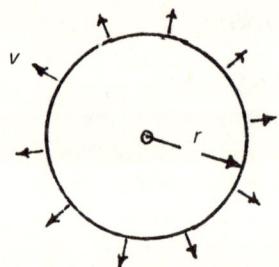

Fig. 2.8 – Outgoing spherical wave.

This form of pressure wave implies that at the origin $r = 0$ (Fig. 2.8), there is a finite rate of volume displacement or volume injection Q. The physical origin of the volume displacement might be a small pulsating solid object, or an explosion, or a pulsating gas bubble in a liquid. Any such disturbance will in reality occupy a finite region and prevent the application of equation (2.14) down to zero radius ($r \rightarrow 0$). For the time being, however, we consider the source region to be vanishingly small and evaluate Q as the limit, as $r \rightarrow 0$, of the volume flow rate $4\pi r^2 v$ across a sphere of radius r centred at the origin. In fact it is simpler to evaluate dQ/dt, since

$$\frac{dQ}{dt} = \lim_{r \to 0} 4\pi r^2 \frac{\partial v}{\partial t}$$

$$= \lim_{r \to 0} 4\pi r^2 \left(-\frac{1}{\rho_0} \frac{\partial p}{\partial r} \right); \tag{2.15}$$

thus $\quad \dfrac{dQ}{dt} = \lim\limits_{r \to 0} 4\pi r^2 \cdot \dfrac{1}{\rho_0} \left\{ \dfrac{1}{r^2} f(t - r/c_0) + \dfrac{1}{c_0 r} \dfrac{\partial}{\partial t} f(t - r/c_0) \right\} \tag{2.16}$

where the last line follows from equation (2.14). As r tends to zero, the second (far-field) term vanishes. The term which is left corresponds to the hydrodynamic near field, where the motion is unaffected by the fluid compressibility (note that c_0 does not appear).

Equation (2.16) gives

$$\frac{dQ}{dt} = \frac{4\pi}{\rho_0} f(t), \quad \text{i.e. } f(t) = \frac{\rho_0}{4\pi} \frac{dQ}{dt}. \tag{2.17}$$

The outgoing spherical wave field is therefore related to the *volume acceleration* dQ/dt at the origin as follows:

$$p(r, t) = \frac{\rho_0}{4\pi r} \frac{dQ}{dt} (t - r/c_0). \tag{2.18}$$

Equation (2.18) shows that the pressure at distance r from the origin, in a small-amplitude outgoing wave with spherical symmetry, depends on the value of dQ/dt at a time r/c_0 earlier. The time $t - r/c_0$ is known as the **emission time** or **retarded time**.

Problems

(1) Show that the sound power radiated from a compact source of volume displacement $V(t)$ is

$$W = (\rho_0/4\pi c_0) \overline{\dot{V}^2}. \tag{2.19}$$

(2) A sphere of 50 mm diameter vibrates with 1 mm surface amplitude at 315 Hz. The vibration is in phase over the whole surface. What power is radiated into the following fluids at 15°C, 1 atm: (i) air, (ii) water, (iii) mercury? The values of density (kg/m^3) and sound speed (m/s) are respectively (i) 1.225, 340; (ii) 999, 1465; (iii) 13 600, 1460.
[0.14 watts, 26 watts, 360 watts].

2.2.2 Interference between multiple sources

The point source model may be used to build up arrays of sources and also continuous source distributions. The resulting sound field is then generally directional, in contrast to the field of each constituent source element. The reason for the directionality is interference.

Consider an array of discrete point sources of strengths $A_1, A_2, \ldots A_n$,[†] situated at distances $r_1, r_2, \ldots r_n$ from an observation point P (Fig. 2.9). Super-

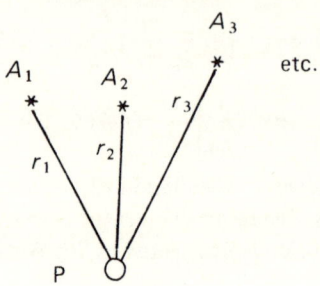

Fig. 2.9 – Multiple sources.

position of the individual pressure fields from each source gives the pressure at P as

$$p = p_1 + p_2 + \ldots + p_n, \tag{2.20}$$

where p_1 etc. are the pressures due to the separate sources. Thus the *mean square* pressure is

$$\overline{p^2} = (\overline{p_1^2} + \overline{p_2^2} + \ldots) + 2(\overline{p_1 p_2} + \overline{p_1 p_3} + \ldots). \tag{2.21}$$

It is the cross terms $(\overline{p_1 p_2}$ etc.) which confer directionality on the sound field, since far from the source region the squared terms are omnidirectional.

Note that if the source strengths A_1, A_2, etc. are *uncorrelated*, the mean square pressure $\overline{p^2}$ is simply the sum of the mean square pressures from the separate sources.

2.2.3 Near and far fields of a source distribution

At sufficiently large distances from any finite source region radiating into a free field, the fluctuating pressure and velocity are related as in a plane wave travelling away from the source. As the source is approached from any given direction, departures from the plane-wave relationship are generally observed. The region in which this occurs is called the **near field** of the source.

For a source region of maximum dimension L, radiating sound of wavelength λ, an observation point at distance R' will be in the far field (i.e. outside the near field) provided the following criteria are met:

(i) $R'/\lambda \gg 1$ (outside hydrodynamic near field);   (2.22)

(ii) $R'/\lambda \gg (L/\lambda)^2$ (outside geometric near field).   (2.23)[‡]

[†] A spherically symmetric outgoing wave of the form (2.14) is said to have a **source strength** $A = 4\pi f(t)$ at the origin. Thus $A = \rho_0 \mathrm{d}Q/\mathrm{d}t$.

[‡] For spatially random (incoherent) source distributions, (ii) is replaced by $R' \gg L_c^2/\lambda$ where L_c is the coherence length scale, and we need to specify (iii) $R' \gg L$.

The distance L^2/λ associated with criterion (ii) is known as the **Rayleigh distance**. Note that this is the controlling criterion when L/λ is large (i.e. the source region extends over several wavelengths).

2.2.4 Source images in a plane boundary

So far in the present chapter we have considered only sources in a free field. The effect of a rigid plane boundary is easily incorporated, however, since it can be represented by image sources (as has already been mentioned in section 2.1.2 of this chapter).

The situation for a single point source is illustrated in Fig. 2.10. Case (a) is the physical situation. Case (b) is a free-field model, which produces an identical field to (a) in the region $x > 0$. The image is identical to the real source; the resulting symmetry ensures that the normal velocity across $x = 0$ is zero, which is the physical boundary condition imposed in case (a).

A simple point source placed *on* a rigid flat surface radiates twice the free-field pressure, as can be seen directly from the argument above. Note that although the surface is strictly infinite for the image argument to hold, a good approximation is obtained for surface dimensions large compared with the acoustic wavelength of interest.

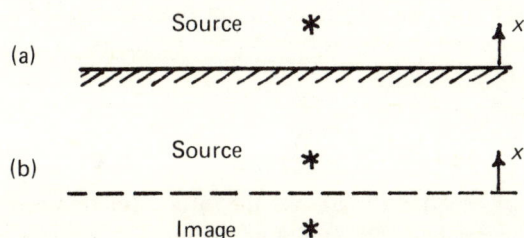

Fig. 2.10 – Source near a plane rigid boundary (a), and the equivalent free-field situation (b).

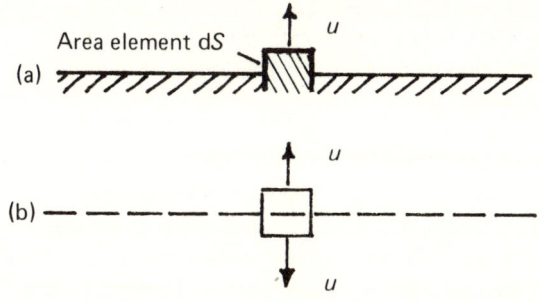

Fig. 2.11 – Area element of a vibrating surface as an elementary source.

Problem

Show that the *sound power* radiated by a point simple source is doubled (relative to the free-field power) by placing the source next to a large rigid surface.

2.2.5 Vibrating plane surfaces

If a large flat surface vibrates with normal velocity u (Fig. 2.11), the resulting sound field can be calculated by regarding each area element dS of the surface as a point acoustic source, which displaces volume at a rate $u \, dS$.

The equivalent free-field situation for each element is shown in the lower part of Fig. 2.11; an identical image source layer is added to the original source layer. Thus the element dS contributes a pressure disturbance

$$dp = \frac{dS}{2\pi r} \rho_0 \dot{u}(t - r/c_0) \tag{2.24}$$

at distance r in any direction (above the surface).

For vibration at a single frequency, it is convenient to substitute

$$p = \hat{p} \, e^{i\omega t}, \quad u = \hat{u} \, e^{i\omega t} \tag{2.25}$$

where $\hat{p}, \hat{u}$ are the complex amplitudes of the pressure and the surface velocity. (The actual pressure and velocity are understood to be the real parts of p and u.) Equation (2.24) then becomes

$$d\hat{p} = \frac{e^{-ikr}}{2\pi r} i\rho_0 c_0 k \hat{u} \, dS. \tag{2.26}$$

Integration of equation (2.26) over the surface (or that part of it which is moving), gives the resultant pressure at any point. Either near-field or far-field pressures may be calculated in this way.

Working in the frequency domain (equation (2.26)) rather than the time domain (equation (2.24)) greatly simplifies the calculation of interference fields from non-compact source distributions. Note that in the present case, interference can arise either through the $\exp(-ikr)$ factor — which varies from point to point on the surface — or from phase variations in $\hat{u}$.

2.2.6 Directivity of extended source distributions

Radiation from a plane vibrating surface provides a simple example of interference leading to a directional sound field. Since the vibrating-surface problem has been reduced to one of calculating the radiation from a plane source layer in a free field, the directionality of plane source distributions will be considered next with the aid of examples. The present section, and the remainder of the chapter, deal entirely with single-frequency excitation.

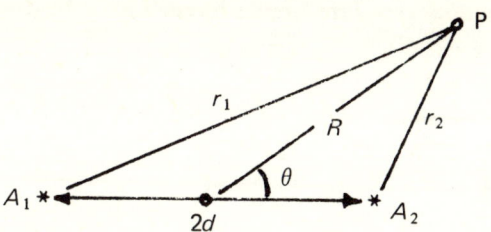

Fig. 2.12 – Definition sketch for radiation from two point sources.

The simplest example of a plane source distribution of finite extent is a pair of point sources (Fig. 2.12). If the source strengths are specified by the complex quantities A_1, A_2, the respective pressures contributed at P are given by

$$p_1 = (\exp(-ikr_1)/4\pi r_1)A_1$$
$$p_2 = (\exp(-ikr_2)/4\pi r_2)A_2. \tag{2.27}$$

In the *far field*, the $\exp(-ikr)$ phase factors may be approximated by writing

$$r_1 \doteq R + d\cos\theta, \quad r_2 \doteq R - d\cos\theta. \tag{2.28}$$

Furthermore, the $(1/r)$ factors may be approximated by $1/R$, giving finally

$$p \doteq (e^{-ikR}/4\pi R)(A_1 e^{-ikd\cos\theta} + A_2 e^{ikd\cos\theta}). \tag{2.29}$$

Problem

Suppose $A_2 = A$, $A_1 = -A$ (equal and opposite sources). Show that when kd is small,

$$p \doteq (e^{-ikR}/4\pi R)(ikd\cos\theta)2A \quad (kd \ll 1). \tag{2.30}$$

The product $2Ad$ is called the **dipole moment** of the source pair defined above.

For *continuous* source distributions, the summation process above is replaced by integration. The next page gives some simple examples chosen to illustrate interference from extended source distributions which are inherently in phase (so that the interference arises solely from path length differences).

At any point in the far-field, the interference factor (F) shown indicates the reduction of the radiated pressure relative to the corresponding value for a compact source ($kb \ll 1$) of the same total strength. Thus the actual far-field pressure in each case is

$$p = (e^{-ikR}/4\pi R)AF, \tag{2.31}$$

where A is the total source strength.

Source distribution	*Interference parameter*	*Interference factor*
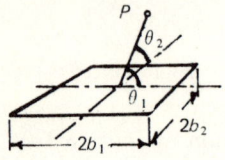 Uniform line source	$z = kb\cos\theta$	$F = \dfrac{\sin z}{z}$
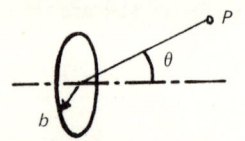 Uniform rectangular source	$z_1 = kb_1\cos\theta_1$ $z_2 = kb_2\cos\theta_2$	$F = \dfrac{\sin z_1}{z_1}\cdot\dfrac{\sin z_2}{z_2}$
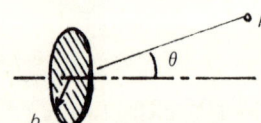 Uniform ring source	$z = kb\sin\theta$	$F = J_0(z)$
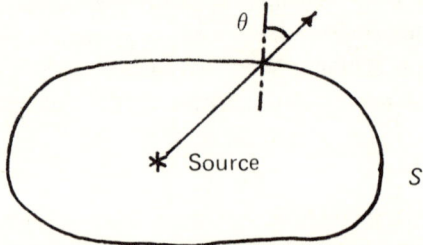 Uniform disc source	$z = kb\sin\theta$	$F = 2J_1(z)/z$

2.2.7 Sound power radiated from directional sources

The total sound power radiated into a free field by a directional source of sound, such as a vibrating surface of finite extent, is related to the far-field pressure by

$$W = \frac{1}{\rho_0 c_0}\int_S \overline{p^2}\,dS\,\cos\theta, \tag{2.32}$$

where θ is the angle which the radiation direction makes with the normal to S (Fig. 2.13). The surface of integration S may be of any shape, as long as it surrounds the source without encroaching on the near field.

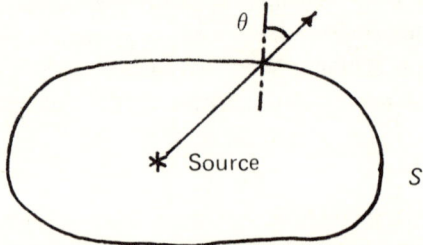

Fig. 2.13 – Sound power integration.

If the mean square pressure $\overline{p^2}$ refers to a particular discrete frequency or band of frequencies, the resulting value of W is the **discrete-frequency** or **band-limited** sound power. Alternatively, use of overall mean square pressures in (2.32) gives the **overall** sound power.

2.2.8 Directivity factor and directivity index

A measure of the degree to which a far-field radiation pattern beams sound in a particular direction is the ratio

$$D = 4\pi R^2 \overline{p^2}(R)/\rho_0 c_0 W, \tag{2.33}$$

known as the **directivity factor**. This implies that the mean square pressure in a given direction from the source region is D times the value for omnidirectional radiation at the same distance.

The **directivity index** is defined as $10 \log_{10} D$, i.e. the value of D in decibels. Note that the directivity factor D, like the sound power W, can be defined for any given band of frequencies.

2.2.9 Radiation impedance of vibrating surfaces

When a surface bounding an expanse of fluid is made to vibrate, a fluctuating pressure field is set up on the surface itself. A method of estimating the surface pressure field, which can be applied to flat surfaces several wavelengths or more in extent, has already been mentioned in section (2.2.5): it consists of integrating either equation (2.24) or equation (2.26) over the vibrating area. However, these results do not apply to curved surfaces. An alternative method of calculation is described next which is valid for spherical or cylindrical surfaces, with plane surfaces as a special case. This is the wave-matching technique: it provides useful insight into the importance of the surface phase speed, as will be illustrated for the plane case.

The fluctuating surface pressure is of interest for two reasons. First, it represents a fluid loading effect which may react back on the surface motion; this is important when coupled structural–acoustic response is being considered (see Chapter 11). Secondly, it offers a means of calculating the radiated sound power without first finding the far-field pressure: the power is the integral of $\overline{pu}$ over the surface,

$$W = \int_S \overline{pu} \, dS, \tag{2.34}$$

or in terms of complex quantities as used for single-frequency excitation,

$$W = \tfrac{1}{2} \operatorname{Re} \int_S p^*u \, dS \quad (*\text{denotes complex conjugate}). \tag{2.35}$$

If the surface pressure p is calculated by the source-element approach described in section 2.2.5, the use of (2.34) or (2.35) for sound power calculations offers no advantage over the far-field method. On the other hand the wave-matching approach, described in the next two sections, gives a rapid estimate of the specific radiation impedance Z_r presented by the adjacent fluid to the vibrating surface.

The definition of Z_r is simply

$$Z_r = (p/u)_{\text{surface}} = (\hat{p}/\hat{u})_{\text{surface}}; \tag{2.36}$$

the difference between Z_r and the specific surface impedance Z_s (defined in section 2.1.4) is that Z_r is the impedance of the *fluid* (looking outward from the surface), whereas Z_s is the impedance of the *surface* (looking inward from the fluid). Just as with Z_s (in general), the value of Z_r at any point on the surface is not a local property but depends on the motion of the entire surface.

2.2.10 Radiation from a pulsating sphere

As a first illustration of wave matching, we examine the radiation impedance of a uniformly pulsating spherical surface (Fig. 2.14). The pressure field set up by

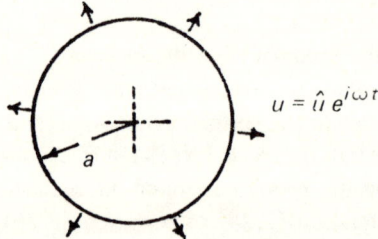

Fig. 2.14 – Definition sketch for pulsating spherical surface.

the surface motion will possess the same spherical symmetry as the surface motion. Thus for single-frequency excitation, we expect a pressure field of the form

$$p = \frac{A}{r} \exp i(\omega t - kr) \quad (r > a), \tag{2.37}$$

which corresponds to an outgoing spherical wave (Chapter 1).

The velocity field implied by equation (2.37) must match the normal velocity of the surface at $r = a$. From the radial momentum equation

$$\frac{\partial v}{\partial t} = -\frac{1}{\rho_0} \frac{\partial p}{\partial r} \quad (v = \text{fluid radial velocity}), \tag{2.38}$$

it follows that p and v at any point in the fluid are related by

$$i\omega\rho_0 v = (1/r + ik)p. \tag{2.39}$$

Matching v at $r = a$ to the surface velocity u gives the surface radiation impedance as

$$Z_r = \frac{i\omega\rho_0}{1/a + ik}\,;\tag{2.40}$$

i.e.

$$\mathrm{Re}(Z_r) = \rho_0 c_0 \frac{k^2 a^2}{1 + k^2 a^2}\,, \quad \mathrm{Im}(Z_r) = \frac{\rho_0\omega a}{1 + k^2 a^2}.\tag{2.41}$$

These results show that at sufficiently low frequencies ($k^2 a^2 \ll 1$, or wavelengths appreciably greater than the circumference of the sphere) the radiation impedance of a pulsating sphere is mainly reactive, and arises from the inertia of the fluid moving to and fro with the surface. At high frequencies, on the other hand, the impedance becomes mainly radiative, and Z_r tends towards $\rho_0 c_0$ as $k^2 a^2$ becomes large (wavelengths much less than the circumference of the sphere).

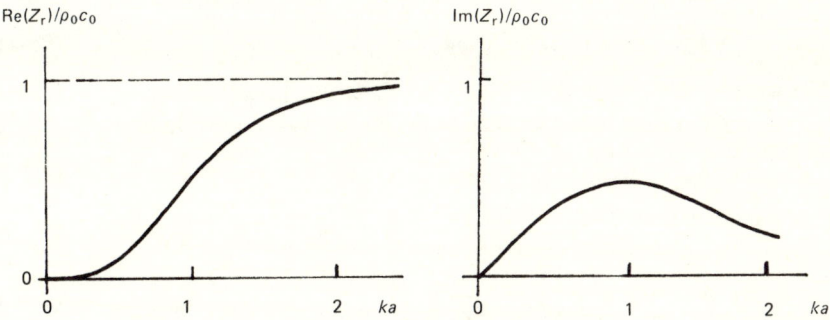

Fig. 2.15 – Radiation impedance of a uniformly pulsating sphere.

The above conclusions apply generally to radiation from finite vibrating surfaces, in that for wavelengths large compared with the surface dimensions the fluid loading is mainly reactive and little sound is radiated for a given surface velocity amplitude; while for wavelengths small compared with the minimum dimension normal to the radiation direction, values of Z_r around $\rho_0 c_0$ are expected and the reactive loading component is comparatively small.

2.2.11 Radiation from travelling waves on a plane boundary

As a second example of the wave-matching technique, we consider the wave field set up by an infinite plane boundary whose surface motion takes the form of sinusoidal travelling waves, or ripples, running in the x direction (Fig. 2.16).

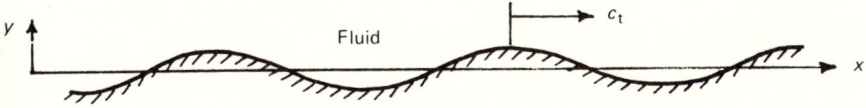

Fig. 2.16 – Travelling-wave vibration pattern on a large plane surface.

The y direction is normal to the undisturbed surface, while the z direction is parallel to the wave crests.

The solution to this problem demonstrates the importance of the travelling-wave speed c_t in relation to the sound speed c_0 in the adjacent fluid.

The normal velocity of the surface at any point (x, z) is represented by

$$u = U \exp i\omega(t - x/c_t);$$

(2.42)

the surface motion is at a single frequency (time factor $\exp i\omega t$), as follows from the assumption of a spatially-sinusoidal pattern convected at constant speed c_t. The expected form of the pressure field in the fluid is

$$p = A \exp(i\omega t - ik_x x - ik_y y - ik_z z) \quad (y > 0),$$

(2.43)

which will satisfy the wave equation $(\partial^2 p/\partial t^2 = c_0^2 \nabla^2 p)$ provided

$$k_x^2 + k_y^2 + k_z^2 = (\omega/c_0)^2.$$

(2.44)

The wavenumber components (k_x, k_z) parallel to the surface are controlled by the surface vibration pattern, which in the present case corresponds to

$$k_x = \omega/c_t, \quad k_z = 0$$

(2.45)

(compare equation (2.42)). It follows from (2.44) that the y wavenumber is

$$k_y = \pm \sqrt{\left(\frac{\omega}{c_0}\right)^2 - \left(\frac{\omega}{c_t}\right)^2}.$$

(2.46)

Evidently k_y may be either real or imaginary. The implications of this result will be discussed once an expression has been found for Z_r.

In order to calculate the surface radiation impedance Z_r, we use the momentum equation in the y direction to give

$$i\omega v = (ik_y/\rho_0)p \quad (v = y \text{ component of fluid velocity}).$$

(2.47)

It follows that, provided the pressure field is described by equation (2.43) with a single value of k_y,

$$Z_r = \frac{\omega \rho_0}{k_y} = \pm \rho_0 \left\{ \left(\frac{1}{c_0}\right)^2 - \left(\frac{1}{c_t}\right)^2 \right\}^{-\frac{1}{2}}.$$

(2.48)

The choice of k_y values in equation (2.46) is determined by the remaining boundary conditions of the problem. On the assumption that no other reflecting boundaries are present, the pressure field for $c_t > c_0$ (i.e. real k_y) must represent an outgoing wave, so k_y is taken as positive. For $c_t < c_0$ (i.e. imaginary k_y), under the same conditions we must choose

$$k_y = -i \sqrt{\left(\frac{\omega}{c_t}\right)^2 - \left(\frac{\omega}{c_0}\right)^2}$$

(2.49)

to ensure exponential decay of the pressure disturbance away from the surface.

The results of the wave-matching calculation may therefore be summarized as follows.

(i) $c_t > c_0$ (fast surface waves)

Pressure disturbances propagate as waves away from the surface. The surface radiation impedance is given by

$$Z_r = \rho_0 c_0 / \sqrt{1 - (c_0/c_t)^2};$$ (2.50)

it is purely real (radiative), and tends towards $\rho_0 c_0$ as c_t/c_0 increases (Fig. 2.17).

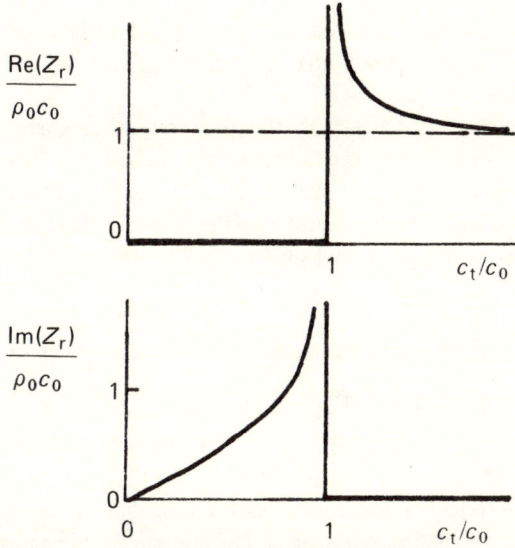

Fig. 2.17 – Radiative and reactive components of radiation impedance, as a function of phase speed.

(ii) $c_t < c_0$ (slow surface waves)

Pressure disturbances decay exponentially away from the surface. The surface radiation impedance is given by

$$Z_r = i\rho_0 / \sqrt{\left(\frac{1}{c_t}\right)^2 - \left(\frac{1}{c_0}\right)^2};$$ (2.51)

it is purely imaginary (reactive), and tends towards $i\rho_0 c_t$ as c_t/c_0 decreases (Fig. 2.17).

Note that the *wavelength* λ_t of the surface vibration pattern is related to the phase speed c_t and the frequency f by

$$\lambda_t = c_t/f.$$ (2.52)

Problem

Show that when $c_t \ll c_0$, the pressure amplitude in the fluid decays by a factor $\exp(-2\pi)$ in a distance from the surface equal to the surface wavelength λ_t.

2.2.12 Radiation efficiency

A useful measure of the effectiveness with which a vibrating surface radiates energy into a fluid is the **radiation efficiency**, or **radiation ratio**, defined as

$$\sigma = \frac{W}{\rho_0 c_0 \int \overline{u^2}\, dS}. \tag{2.53}$$

Expressed in words, the radiation efficiency is the ratio of the sound power actually radiated, to the power that the same surface motion would radiate if the radiation impedance were $\rho_0 c_0$.

For the situation discussed in the previous section, since the radiation impedance is uniform over the surface, the radiation efficiency is simply $\mathrm{Re}(Z_r)/\rho_0 c_0$. Thus σ for an infinite plane surface is zero if the phase speed of the surface motion is subsonic relative to the adjacent fluid, and tends to unity once the phase speed becomes well supersonic.

Problems

(1) Show that a uniformly pulsating sphere has a radiation efficiency

$$\sigma = k^2 a^2/(1 + k^2 a^2). \tag{2.54}$$

(2) In a uniform flat plate of solid material, free bending waves travel at a speed $c_t = (\omega b c_p)^{\frac{1}{2}}$, where $b = (\text{plate thickness})/(12)^{\frac{1}{2}}$ and c_p is a property of the plate material (assumed isotropic). Given that c_p for steel is 5420 m/s, find the minimum thickness of steel plate which will radiate 1 kHz sound at high efficiency (a) into air at $15°C$, (b) into water at $15°C$.

[12 mm, 220 mm (neglecting fluid loading effects).]†

REFERENCES

[2.1] Delaney, M. E. & Bazley, E. N. (1971) *Journal of Sound and Vibration* **16**, 315–322. A note on the effect of ground absorption in the measurement of aircraft noise.

[2.2] Maekawa, Z. (1968) *Applied Acoustics* **1**, 157–173. Noise reduction by screens.

[2.3] Beranek, L. L. (1949) *Acoustic Measurements* New York: Wiley.

[2.4] Delany, M. E. & Bazley, E. N. (1969) *NPL Aero Report* Ac 37 Acoustical characteristics of fibrous absorbent materials.

---

† In the steel/water case, not only is fluid loading important but thin-plate theory is no longer accurate – see Junger & Feit [2.5].

[2.5] Junger, M. C. & Feit, D. (1972) *Sound, Structures and Their Interaction* Cambridge, Mass: M.I.T. Press.

BIBLIOGRAPHY

Brekhovskikh, L. M. (1980) *Waves in layered media* (2nd edn.). New York: Academic Press.

Pierce, A. D. (1981) *Acoustics: an introduction to its physical principles and applications*. New York: McGraw-Hill.

CHAPTER 3

Fundamentals of vibration

B. L. Clarkson

Institute of Sound and Vibration Research, University of Southampton

3.1 INTRODUCTION

Periodic motion is a familiar aspect of our everyday life. Our hearts beat, our lungs oscillate, a periodic rocker motion or a simple pendulum provide our measure of time. In the factory we can hear the steady hum of industrial machinery, or at our leisure watch the daily ebb and flow of tides. In this chapter we shall limit ourselves to a consideration of mechanical vibration, although many of the concepts which are discussed have direct parallels in such fields as electricity or thermodynamics.

Often real-life vibration problems are very complicated. The motion of a simple pendulum is easy to understand, but the flutter of an aircraft or the vibration of a motor car travelling over a rough road are problems which require very sophisticated analysis procedures. From the engineering design point of view it is always necessary to make a simpler representation of the real system. An analysis of the behaviour of this simple model can then be used to give a guide to the main features of the real structure. Thus considerable emphasis in vibration theory is placed on the analysis of the behaviour of simple systems. This is not because simple systems occur in real life but because they can be used to represent real problems. This 'modelling' of an engineering system is a very important part of vibration studies.

Vibration can be classified in several ways. A **free vibration** occurs without externally applied forces. This type of vibration occurs after a system has been given an initial displacement or velocity such as might result from an impact. A **forced vibration** occurs with the application of externally applied forces. Such vibration can either be periodic, aperiodic, or random. **Periodic vibration** repeats itself in regular intervals of time, but in **aperiodic** or **random vibration** there is no such regular repetition. During any motion of practical systems energy is dissipated and damping of the motion occurs. When the vibration involves relative motion of a rivetted joint or sound radiation in higher frequency vibration, there is significant energy dissipation, and so the damping can be considerable. Most mechanical systems are lightly damped unless specific measures have been taken to provide high damping.

Vibrations can also be classified by the number of degrees of freedom of the motion. The number of degrees of freedom corresponds to the number of independent coordinates which are needed to specify the motion completely. An idealized single degree of freedom system (shown diagrammatically in Fig. 3.1) consists of a mass m, a spring of stiffness k, and a viscous damper of rate c. This might be used to represent the predominant low-frequency

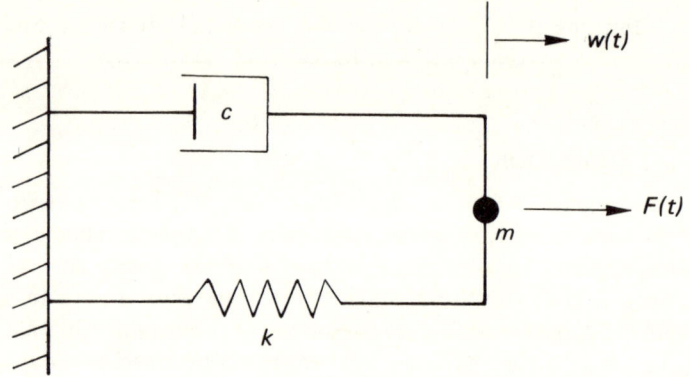

Fig. 3.1 – Single degree of freedom system with viscous damper.

rotational motion of an automobile engine on its rubber mounts. In such a case the mass would represent the rotational inertia of the engine, and the stiffness would represent the rotational restraint provided by the mounts when the engine rocks. The energy dissipated in the rubber mounts provides the damping. The single coordinate description of this type of motion would be the rotation of the engine about its fore and aft axis.

3.2 FREE VIBRATION OF THE SINGLE DEGREE OF FREEDOM SYSTEM

We can now describe the behaviour of the simple system (represented by Fig. 3.1) by considering the equation of motion of the mass m. The displacement w of the free end of the spring relative to the unstrained position, when the mass is subjected to the time-dependent force $F(t)$, is governed by the equation:

$$m \frac{d^2 w}{dt^2} + c \frac{dw}{dt} + kw = F(t). \tag{3.1}$$

It is often convenient to rewrite this equation as follows:

$$\frac{d^2 w}{dt^2} + \frac{c}{m} \frac{dw}{dt} + \frac{k}{m} w = \frac{F(t)}{m}. \tag{3.2}$$

The equation can be simplified by using the following notation:

$$\omega_n = \sqrt{\frac{k}{m}} \quad \text{natural frequency}$$

$$\zeta = \frac{c}{c_0} \quad \begin{array}{l}\text{where } c_0 \text{ is the critical damping coefficient for which}\\ \text{the system would just not be capable of vibrating}\\ \text{freely when disturbed.}\end{array}$$

$$c_0 = 2\sqrt{mk}$$

ζ is called the **viscous damping ratio** or proportion of critical damping.

The solution to (3.2) is:

$$w = A_0 \exp(-\zeta\omega_n t) \cos(\omega_n \sqrt{(1 - \zeta^2)}t - \phi). \tag{3.3}$$

A_0 and ϕ are constants which depend on the initial conditions. If the damping is zero the solution shows that the motion of the system will be harmonic with frequency ω_n. Thus ω_n is the undamped natural frequency of the system. Equation (3.3) shows that when damping is present the motion has a frequency $\omega_n\sqrt{(1 - \zeta^2)}$; this is sometimes called ω_d, the damped natural frequency.

If the system is disturbed from rest and allowed to vibrate freely the resultant vibrations will decay. The envelope of the decaying vibration as shown in Fig. 3.2, is seen to be given by the term $\exp(-\zeta\omega_n t)$. This suggests a simple way in which the damping of a real system can be measured.

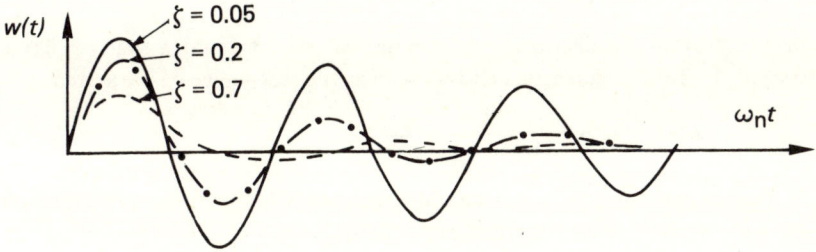

Fig. 3.2 – Response to a delta function impulse.

If we measure the amplitude of successive cycles we have

$$w_1 = A_0 \exp(-\zeta\omega_n t)$$

$$w_2 = A_0 \exp\left(-\zeta\omega_n t + \frac{2\pi}{\omega_n}\right).$$

The ratio of these amplitudes $w_1/w_2 = e^{-2\pi\zeta}$.

This rate of decay is often defined by the **logarithmic decrement** δ which is the logarithm to the base e of the ratio of successive amplitudes:

$$\delta = \ln(w_1/w_2).$$

It can be seen that δ and ζ are connected by the relationship

$$\delta = 2\pi\zeta. \tag{3.4}$$

3.3 HARMONICALLY FORCED VIBRATION

Solutions for different forcing functions can now be obtained by using the appropriate form for $F(t)$. When $F(t)$ represents a harmonically varying force ($F_0 \cos \omega t$) it is convenient to represent it in the complex exponential form

$$F(t) = F_0 \cos \omega t = \mathrm{Re} F_0 e^{i\omega t}. \tag{3.5}$$

In subsequent work it will be assumed that we are only considering the real part of forces and displacements, and so we shall drop the prefix Re.

The solution to equation (3.1) can now be written in the form

$$w(t) = w_0 e^{i\omega t} \tag{3.6}$$

in which w_0 is a complex number. Substituting equations (3.5) and (3.6) into equation (3.1) we find that

$$w(t) = w_0 e^{i\omega t} = \frac{F_0 e^{i\omega t}}{(-\omega^2 m + i\omega c + k)} \tag{3.7}$$

which can be written as

$$w = \alpha(i\omega) F_0 e^{i\omega t}$$

where $\alpha(i\omega)$ is the displacement response of the mass to a unit amplitude harmonic force of frequency ω. It is known as the **receptance** of the system.

$$\alpha(i\omega) = \frac{1}{(-\omega^2 m + i\omega c + k)}. \tag{3.8}$$

The receptance $\alpha(i\omega)$ is a complex number which can be written in the form $|\alpha(i\omega)| \, e^{-i\phi}$ where $|\alpha(i\omega)|$ is the modulus and ϕ is the argument of the number. Thus

$$|\alpha(i\omega)| = \frac{1}{\sqrt{[(k - \omega^2 m)^2 + \omega^2 c^2]}}; \quad \tan\phi = \frac{+c}{(k - \omega^2 m)}. \tag{3.9}$$

Equation (3.8) now becomes

$$w(t) = |\alpha(i\omega)| F_0 e^{i(\omega t - \phi)}. \tag{3.10}$$

Restoring this equation to the real form (by extracting the real part from the right-hand side), we have

$$w(t) = |\alpha(i\omega)| F_0 \cos(\omega t - \phi). \tag{3.11}$$

This states that the displacement w oscillates at the frequency ω, with an amplitude $|\alpha(i\omega)|\,F_0$, and lags by a phase angle ϕ behind the exciting force.

3.3.1 More general form of the receptance

When using this approach to model a more complicated system it is often more convenient to have an expression for $\alpha(i\omega)$ in terms of frequencies and damping ratio. We can do this by writing the right-hand side of equation (3.2)

$$\frac{F(t)}{m} = \frac{k\,f(t)}{m} = \omega_n^2\,f(t) \quad \text{where } f(t) = \frac{F(t)}{k}.$$

Thus writing $f_0 = F_0/k$, equation (3.7) becomes

$$w(t) = \frac{k f_0\,e^{i\omega t}}{-\omega^2 m + i\omega c + k} = \frac{f_0\,e^{i\omega t}}{\left\{1 - \dfrac{\omega^2}{\omega_n} + 2i\zeta\,\dfrac{\omega}{\omega_n}\right\}}. \tag{3.12}$$

The advantage of this form is that the natural frequency ω_n and the damping ratio ζ of the real system can be measured directly, whereas the equivalent mass m may be difficult to determine.

Consider now the variation of the displacement amplitude per unit amplitude of exciting force ($|w|/F_0$) as the forcing frequency changes. From equation (3.11) we have

$$\frac{|w|}{F_0} = |\alpha(i\omega)| = \frac{1}{\sqrt{[(k - \omega^2 m)^2 + \omega^2 c^2]}}. \tag{3.13}$$

This may be written in the form

$$\frac{|w|}{F_0} = \frac{1}{k}\,\frac{1}{\{(1 - (\omega/\omega_n)^2)^2 + 4(\omega/\omega_n)^2\zeta^2\}^{1/2}}. \tag{3.14}$$

3.3.2 The magnification factor Q

When $\omega = 0$ (i.e. the force is steady), $|w|/F_0 = 1/k$. We call this the **static displacement** w_s. As ω approaches the natural frequency, the displacement first increases sharply, and then passes through a maximum — **resonance** occurs. At this maximum, the displacement is Q times the static displacement. We call Q the **magnification factor**. It may easily be shown that $Q = 1/2\,\zeta$. Plotting $|w|/F_0$ against ω, we have the familiar frequency response curve, shown in Fig. 3.3. The 'width' of the frequency response curve is sometimes defined by the frequency $\Delta\omega$, which is the difference between the two frequencies at which the amplitude is $1/\sqrt{2}$ times the maximum amplitude. Provided the damping factor is no greater than, say, 0.3, we find

$$\frac{\Delta\omega}{\omega_n} = 2\zeta = 1/Q. \tag{3.15}$$

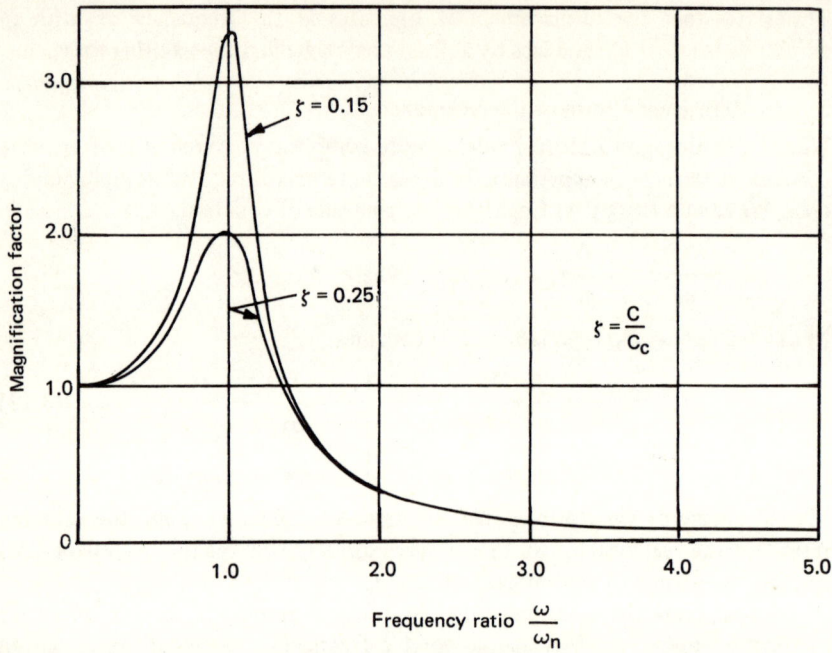

Fig. 3.3 – Amplitude of response.

3.3.3   The frequency response curves

The phase angle ϕ of the response relative to the force varies with frequency as shown in Fig. 3.4. The Figs 3.3 and 3.4 show that the effect of damping is to decrease the width of the resonance curve and to increase the rate of change of phase angle in the region of the natural frequency. The information presented in these two figures can be combined, as in Fig. 3.6, to make a vector diagram

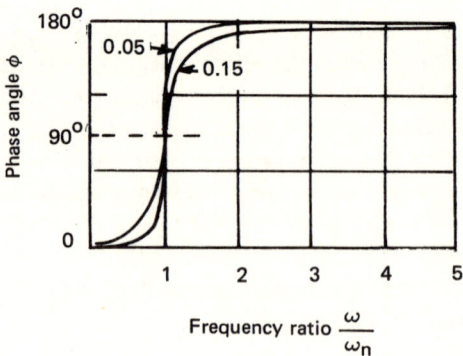

Fig. 3.4 – Phase of response.

of the response. The main characteristics of this diagram are that the curve is approximately circular for the case of light damping. The frequency at which the curve crosses the real axis is the undamped natural frequency of the system, and the rate of change of arc length is greatest at this point.

3.3.4 Frequency response curves for velocity and acceleration

We are often as interested in the amplitudes of the fluctuating velocity, $|\dot{w}|$, or acceleration $|\ddot{w}|$, as in the displacement. It is easily shown that $|\dot{w}| = \omega|w|$ and $|\ddot{w}| = \omega^2|w|$. Using these expressions and equation (3.14), we may show that

$$\frac{|\dot{w}|}{F_0} = \frac{1}{c}\frac{2\zeta(\omega/\omega_n)}{\{(1-(\omega/\omega_n)^2)^2 + 4(\omega/\omega_n)^2\zeta^2\}^{1/2}} = k|\alpha(i\omega)|\frac{2\zeta(\omega/\omega_n)}{c}$$

and (3.16)

$$\frac{|\ddot{w}|}{F_0} = k|\alpha(i\omega)|\frac{1}{m}(\omega/\omega_n)^2. \tag{3.17}$$

As the frequency, ω, varies, each of these expressions varies in a way generally similar to that of $|w|/F_0$, but with certain significant differences, as shown in Figs 3.5a and 3.5b.

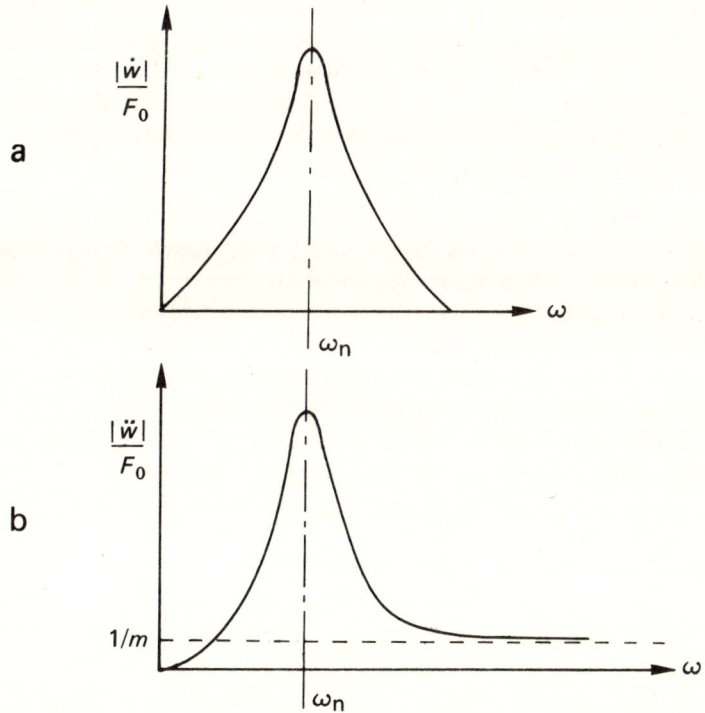

Figs 3.5a and 3.5b – Velocity and acceleration amplitudes of a single degree of freedom system.

Each of the expressions is zero at zero frequency. The velocity function tends to zero as $\omega \to \infty$, but the acceleration function tends to $1/m$.

At the natural frequency, ω_n, the amplitude of the displacement $|w|/F_0$, is $1/\omega_n c$, the amplitude of the velocity $|\dot{w}|/F_0$ is $1/c$, and of the acceleration is ω_n/c, i.e. it is the *damping, c,* which controls the amplitudes at resonance.

At very low frequencies (below ω_n),

$$\frac{|w|}{F_0} \doteqdot \frac{1}{k}; \quad \frac{|\dot{w}|}{F_0} \doteqdot \frac{\omega}{k}; \quad \frac{|\ddot{w}|}{F_0} \doteqdot \frac{\omega^2}{k}$$

i.e. the amplitudes are controlled by the *stiffness, k,* and the displacement tends to be independent of frequency.

At very high frequencies (above ω_n),

$$\frac{|w|}{F_0} = \frac{1}{m\omega^2}; \quad \frac{|\dot{w}|}{F_0} = \frac{1}{m\omega}; \quad \frac{|\ddot{w}|}{F_0} = \frac{1}{m}$$

i.e. the amplitudes are controlled by the *mass, m,* and the acceleration tends to be independent of frequency.

3.3.5 Alternative representation of damping

So far, we have used the concept of a viscous damper, which exerts a damping force which is in counterphase with and proportional to the velocity. In practical structures, this type of damping is seldom realized, and experimental evidence has suggested that a more realistic form, for harmonically vibrating systems, is the 'complex stiffness' type of damping. We represent the stiffness and damping forces together in the expression:

$$(k + ih)w. \tag{3.18}$$

This can only be used if we use the complex exponential form of solution and *applies, strictly, only to harmonic motion.* However, in the absence of any other satisfactory method of representation, it is usually used (with care) when dealing with random motion. The imaginary part of the stiffness implies the existence of a resistive force, which has an amplitude proportional to the displacement w, but which (by virtue of the i) is in counterphase with the velocity $\dot{w}$.

If such 'structural' or 'hysteretic' damping replaces the viscous damping of equation (3.1), the expression for w (equation (3.7)) becomes:

$$w(t) = \frac{F_0 e^{i\omega t}}{(-\omega^2 m + k + ih)}; \tag{3.19}$$

also

$$|\alpha(i\omega)| = \frac{1}{\sqrt{[(k - \omega^2 m)^2 + h^2]}} \tag{3.20}$$

and

$$\tan \phi = \frac{h}{(k - \omega^2 m)}. \tag{3.21}$$

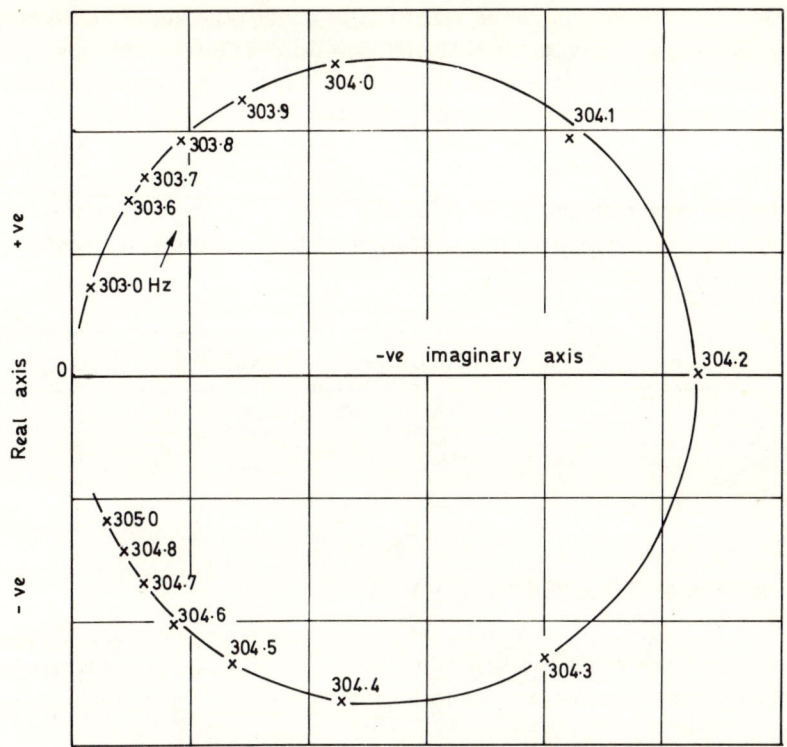

Fig. 3.6 – Steady-state response of a simply supported beam in its 2nd mode.

The undamped resonant frequency, ω_n, now coincides with the frequency for maximum displacement, at which

$$\left.\frac{|w|}{F_0}\right|_{max} = \frac{1}{h}. \tag{3.22}$$

This is independent of frequency.

In all the expressions we have derived which contain $(\omega/\omega_n)\zeta$, we may replace this term by

$$\zeta_H = \frac{h}{2k}. \tag{3.23}$$

We may call ζ_H the **hysteretic damping factor**. The principal characteristic of the 'hysteretic' damping is that the energy dissipated per cycle of harmonic displacement is independent of the frequency, and is proportional to the square of the displacement amplitude. The energy dissipated by a viscous damper is proportional to the product of the frequency and to the square of the displacement amplitude. It should be emphasized that these are two different mathematical models of what really goes on in practice. One or the other might be the best approximation to make in specific applications. For ease of manipulation the complex stiffness representation is often used. The complex stiffness representation (expression (3.18)) can be rewritten as

$$(1 + i\eta)\,kw. \tag{3.24}$$

We have now mentioned several different ways of defining or measuring the damping present in a system. The comparison of these quantities for *harmonic vibration* is as follows.

Loss factor	Viscous damping ratio	Log decrement	Q factor
$\eta \quad = $	$2\zeta \quad = $	$\dfrac{\delta}{\pi} \quad = $	$\dfrac{1}{Q}$

3.4 RESPONSE TO A PERIODIC FORCE

Next, we consider the response of the system to a force which is no longer harmonic, but is periodic, repeating itself after every T secs. The force may be analysed into its Fourier (harmonic) components, and may be expressed in the form

$$F(t) = \sum_{n=1}^{\infty} (a_n \cos \omega_n t + b_n \sin \omega_n t) \tag{3.25}$$

where $\omega_n = 2n\pi/T$, i.e. the frequency having n cycles in the interval T. This equation can be rewritten in complex notation as:

$$F(t) = \mathrm{Re} \sum_{n=1}^{\infty} (a_n - ib_n)\exp(i\omega_n t) \tag{3.26}$$

The coefficients a_n and b_n are found by the usual methods to be

$$a_n = \frac{2}{T}\int_0^T F(t)\cos \omega_n t\, \mathrm{d}t; \quad b_n = \frac{2}{T}\int_0^T F(t)\sin \omega_n t\, \mathrm{d}t \tag{3.27}$$

or

$$a_n = ib_n = \frac{2}{T}\int_0^F F(t)\exp(-i\omega_n t)\, \mathrm{d}t.$$

The total response of the system may now be found by superimposing the harmonic responses corresponding to the indivudual components of the analysed force. Using the form of equation (3.26), the nth component gives rise to the component of displacement (after transients have decayed):

$$w_n = \mathrm{Re}\,\alpha(\omega_n)\,(a_n - ib_n)\exp(i\omega_n t) \tag{3.28}$$

$\alpha(\omega_n)$ is the value of the complex receptance at the frequency ω_n. The total displacement at any instant is therefore

$$w(t) = \sum_{n=1}^{\infty} w_n = \mathrm{Re} \sum_{n=1}^{\infty} \alpha(\omega_n)(a_n - ib_n)\exp(i\omega_n t) \qquad (3.29)$$

This represents a series of frequency components, and since the relative phase angle ϕ is a function of frequency, each component of displacement lags behind its corresponding component of force by a different phase angle. Also, since $\alpha(\omega_n)$ has its minimum value when ω_n is close to ω_r, those components of the force closest to ω_r will cause the largest components of displacement.

3.5 RESPONSE OF MECHANICAL SYSTEMS TO TRANSIENT FORCES

In addition to excitation by periodic forces arising from drive motors, gears, periodic motions of sliders, etc., many types of factory machines and power units are excited by transient forces. These transient forces arise from such processes as combustion (e.g. diesel engines), impacts (presses and stamping machines), or contact forces (gear meshing). The impact or impulsive forces are usually of short duration compared with the time which the subsequent vibration takes to decay away to a negligible level. The time taken for the vibration to decay is directly related to the damping in the system as shown in Fig. 3.2. This generally means that the noise radiated from a lightly damped system subjected to transient forces is much greater than the noise from a heavily damped system. Where the impulsive forces are well separated in time such as in press tools, one effective means of noise control is to artificially increase the damping of the machine.

The theory of the response of systems to transient forces can be built up from a knowledge of the response of a single degree of freedom system to an impulse as follows:

Using equation (3.1) for the simple degree of freedom system we can write the forcing function $F(t)$ as

$$F(t) = I\delta(t)$$

where I is the magnitude of the impulse and $\delta(t)$ is the Dirac delta function which has the properties:

$$\int_{-\infty}^{\infty} \delta(t)\,dt = 1 \qquad \delta(t) = 0 \quad \text{for} \quad t \neq 0.$$

The solution to equation (3.1) is

$$w(t) = I \frac{\omega_n}{k\sqrt{(1-\zeta^2)}} \exp(-\zeta\omega_n t)\sin \omega_n \sqrt{(1-\zeta^2 t)}. \qquad (3.30)$$

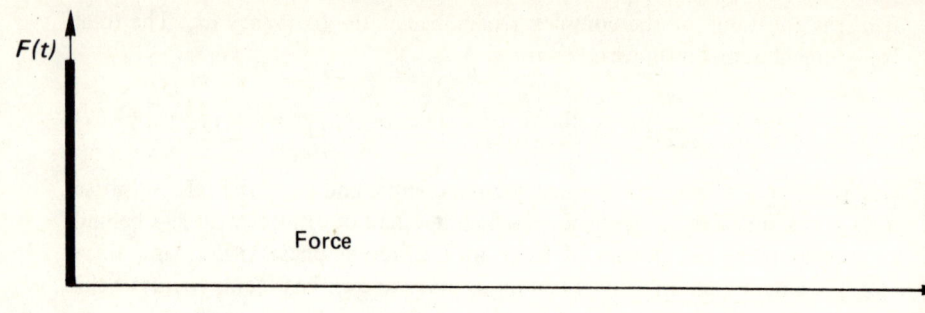

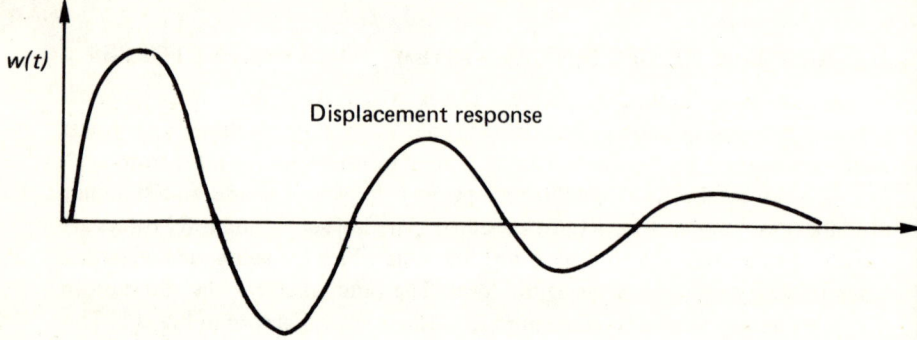

Fig. 3.7 – Response to an impulse.

In general we can write this as

$$w(t) \;=\; I h(t) \quad \text{where } h(t) \text{ is the response of the system to a unit}$$
impulse.

This form has the advantage that $h(t)$ can be used to represent the response of any system (not just the simple single degree of freedom system) to a unit impulse. For a multi-degree of freedom system the actual form of $h(t)$ will of course not be as simple as that given in equation (3.30).

We can now build up the expression for the response of the system to an arbitrary transient force, such as that shown in Fig. 3.8, by considering the force to be made up of a sequence of impulses. The nth impulse has a magnitude of $I = F(\tau_n)\,\Delta\tau_n$. The response of the system to this incremental impulse is

$$\Delta w(t) \;=\; F(\tau_n)\,\Delta\tau_n\,h(t - \tau_n)$$

where the impulse response function $h(t)$ has a time delay to allow for the fact that the impulse is not occurring at the time origin.

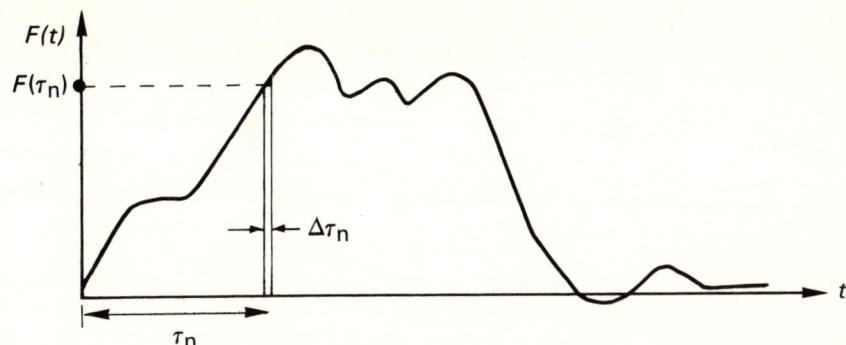

Fig. 3.8 — Representation of a Transient Force.

Now for a *linear* system we can superimpose the effect of a sequence of impulses to obtain the complete expression for the response at time t;

$$w(t) = \Sigma\, F(\tau_n)\, \Delta\tau_n\, h(t - \tau_n).$$

In the limit as $\tau_n \to 0$ the summation becomes the convolution integral

$$w(t) = \int_0^t F(\tau)\, h(t - \tau)\, \mathrm{d}\tau. \qquad (3.31)$$

In general the excitation may have been continuing from before the time origin, and therefore we can write:

$$w(t) = \int_{-\infty}^t F(\tau)\, h(t - \tau)\, \mathrm{d}\tau.$$

It is often more convenient to change the time variable and finally write

$$w(t) = \int_0^\infty F(t - \tau)\, h(\tau)\, \mathrm{d}\tau. \qquad (3.32)$$

Exact solutions can be obtained for simple transients such as step or ramp functions, and for more complicated practical forms a numerical integration procedure can be used.

3.6 TRANSMISSIBILITY

In equipment installation problems we are not concerned with forces acting directly upon the mass but rather with a mass connected to a moving base by a spring and damper as illustrated overleaf.

The equation of motion of the mass is now

$$m\ddot{x} + c(\dot{x} - \dot{y}) + k(x - y) = 0,$$

For harmonic motion of the support $y(t) = A \cos \omega t$ say.

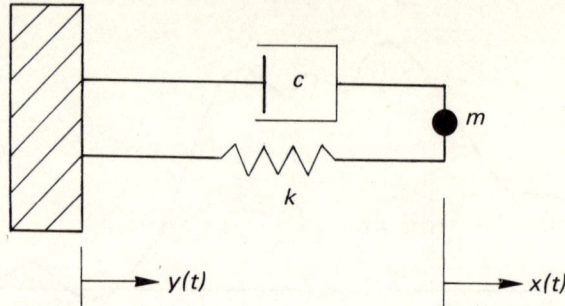

The solution is then

$$x(t) = X \cos(\omega t - \phi)$$

where

$$X = A \left\{ \frac{1 + (2\zeta\omega/\omega_n)^2}{[1 - (\omega/\omega_n)^2]^2 + (2\zeta\omega/\omega_n)^2} \right\}^{1/2} = A \left[1 + \left(\frac{2\zeta\omega}{\omega_n}\right)^2 \right]^{1/2} |\alpha(\omega)|$$

$$\tan\phi = \frac{2\zeta(\omega/\omega_n)^2}{1 - (\omega/\omega_n)^2 + (2\zeta\omega/\omega_n)^2}.$$

(3.33)

The non-dimensional ratio X/A is known as **transmissibility**. This is shown in Fig. 3.9. It can be seen from this that an increase in damping *increases* the transmissibility at frequencies *above* $\sqrt{2}\,\omega_n$.

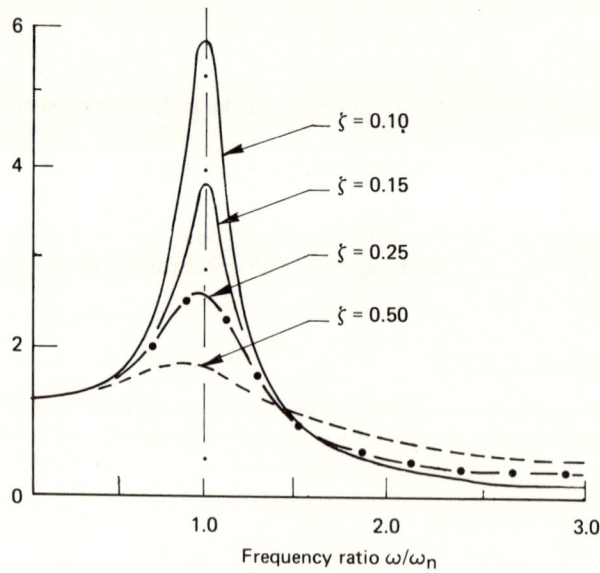

Fig. 3.9 – Transmissibility.

3.7 LUMPED PARAMETER SYSTEMS

In most practical situations it is not possible to make a realistic model of the vibration characteristics on the basis of a single degree of freedom system. Two methods for achieving more realism are in common use. In the first case the system is represented by a series of interconnected masses and springs, and the equations of motion, such as equation (3.1), for each mass are written down. There are now as many equations as there are masses, and each equation contains at least one term which couples it to another equation. To find analytical solutions which satisfy the coupled set of equations is difficult if there are more than three or four degrees of freedom. However, if a computer is available the coupled equations can be written in matrix form and a numerical solution obtained. This is often known as the lumped parameter method. Consider for example the transverse vibration of a non-uniform cantilevered beam. The distributed mass can be lumped together at a finite number of points. These points can then be connected together by springs which represent the average bending stiffness of the cantilever at that section. It is assumed that the springs are massless and that the mass is concentrated at infinitesimal points. Any damping can be represented by (massless) dashpots attached between the masses and some imaginary rigid support. Good approximate solutions can be obtained in this way, and intuitively we would expect that an increase in the number of elements would lead to a better approximation.

An alternative approach to a practical vibration problem is to treat it as a continuous system. Analytical methods for this are well developed. The most common method, known as the normal mode method, is described next.

3.8 THE FORCED VIBRATIONS OF CONTINUOUS SYSTEMS

A 'continuous system' implies that the mass and flexibility are distributed throughout the system. It is not a system of discrete masses connected by massless springs and links. As an example, a uniform simply supported beam will be considered.

The manner in which such a beam can vibrate *freely*, in the absence of damping, is well known. The fundamental mode of vibration is a simple sine wave, and the overtones are sine waves with different integral numbers of half-waves along the beam length. Now any deflected shape of the beam can be resolved into spatial harmonic components, by the methods of *Fourier* analysis. Symbolically, it follows that if the arbitrary deflected shape is $w(x)$, we may write

$$w(x) = \sum_{r=1}^{\infty} q_r \sin \frac{r\pi x}{\ell} \qquad (3.34)$$

where the q_rs are the Fourier coefficients.

In this case, the terms of the Fourier series are identical with the natural modes of vibration of the beam. In the more general case, when the beam is non-uniform or is not simply-supported, the modes of vibration are not sine waves, but are more awkward functions of x. Denote these modes by $f_1(x), f_2(x), \ldots,$ $f_r(x), \ldots,$ etc., $f_1(x)$ being the fundamental mode, and the others being the overtones. Now it is characteristic of the natural modes of vibration that they may be readily used in a 'generalized' type of Fourier analysis, i.e. for a given beam, any possible deflected from of the beam may be represented by

$$w(x) = \sum_{r=1}^{\infty} q_r f_r(x). \tag{3.35}$$

This suggests an alternative form of describing the deflection of the beam. Instead of specifying the continuous function $w(x)$ for all points in the beam, we may specify the q_rs. If it was necessary to specify an infinite set of q_rs, there would be no advantage gained by this alternative method, but as with an ordinary Fourier series, a good approximation to $w(x)$ may be obtained by using the first few terms only of the infinite series.

This feature effects a very great simplification in the calculations of the motion of a beam or plate as follows:

If w is now a function of both space and time, equation (3.35) may be written

$$w(x, t) = \sum_{r=1}^{\infty} q_r(t) f_r(x). \tag{3.36}$$

The time-dependence is contained entirely within the q_rs, and spatial dependence entirely within the f_rs. We call the q_rs **generalized coordinates** corresponding to the modes of displacement f_r. The problem of finding the deflection $w(x, t)$ at any point of the beam or plate is now resolved into the problem of finding the q_rs. Now it may be shown that when the system is subjected to a time-dependent load, the displacement q_r of any one of the natural modes of vibration is governed by the same sort of equation as that of a simple mass-spring-damper oscillator.

For most purposes we may write

$$M_r \ddot{q}_r + C_r \dot{q}_r + K_r q_r = L_r(t). \tag{3.37}$$

In this the terms M_r, C_r, K_r and $L(t)$ are known respectively as the **generalized mass**, the **generalized damping coefficient**, the **generalized stiffness**, and the **generalized force**, each corresponding to the rth natural (or normal) mode. The fact that the equation does not contain any other general coordinates q_s is due to the characteristic 'normal' property of the natural modes and the neglect of any damping coupling between the modes.

The generalized mass, M_r, has the same magnitude of that mass which when moving with velocity $\dot{q}_r$ has the same kinetic energy as the whole system when moving with the velocity $\dot{q}_r f_r(x)$, i.e.

$$M_r \frac{\dot{q}_r^2}{2} = \frac{\dot{q}_r^2}{2} \times \int_A \mu f_r^2(x)\,\mathrm{d}A.$$

(3.38)

Hence $M_r = \int_A \mu f_r^2(x)\,\mathrm{d}A.$

μ is the mass per unit area of the system (plate), and the integration is taken over the whole vibrating surface A.

The generalised stiffness, K_r, is the stiffness of the linear spring which, when displaced by q_r from its unstrained position, has the same potential energy stored within it as the actual system when displaced by $q_r f_r(x)$. For a beam, this is expressed by

$$K_r \frac{q_r^2}{2} = \frac{q_r^2}{2} \int_\ell EI \left\{ \frac{\mathrm{d}^2 f_r(x)}{\mathrm{d}x^2} \right\}^2 \mathrm{d}x,$$

(3.39)

i.e.

$$K_r = \int_\ell EI \left\{ \frac{\mathrm{d}^2 f_r(x)}{\mathrm{d}x^2} \right\}^2 \mathrm{d}x.$$

EI is the flexural stiffness of the beam, and ℓ is the length.

The generalized force, $L_r(t)$, is that single force which when moved through the 'virtual' displacement δq_r, does the same amount of work as all the externally applied forces and pressures acting on the system when the system is moved through the virtual displacement $\delta q_r \cdot f_r(x)$, i.e.

$$L_r(t)\delta q_r = \int_A p(x, t)\,\delta q_r \cdot f_r(x)\,\mathrm{d}A.$$

or

(3.40)

$$L_r(t) = \int_A p(x, t) f_r(x)\,\mathrm{d}A.$$

Here, $p(x, t)$ is the instantaneous pressure at the point x and time t. The integration again covers the whole surface.

The generalized damping coefficient C_r is the rate of that damper which, when extended at velocity $\dot{q}_r$, dissipates energy at the same rate as the whole system of damping forces and pressures on and within the system, when moving with velocity $\dot{q}_r f_r(x)$, i.e.

$$C_r \frac{\dot{q}_r^2}{2} = \int_A p_{dr}(x) \frac{\dot{q}_r^2}{2} f_r^2(x)\,\mathrm{d}A$$

or

(3.41)

$$C_r = \int_A p_{dr}(x) f_r^2(x)\,\mathrm{d}A$$

$p_{dr}(x)$ is the local viscous damping pressure per unit velocity in counterphase with the velocity $\dot{q}_r f_r(x)$. If hysteretic damping stresses exist within the system, we may include a hysteretic damping force, $iH_r.q_r$, in equation (3.37). H_r is then the complex part of the stiffness of the linear spring which dissipates the same amount of energy in the harmonic displacement cycle of amplitude q_r as the actual system when displaced in the mode $f_r(x)$ through the cycle of amplitude q_r.

There is an equation of the type (3.37) for each of the normal modes of vibration of the structure. When the generalized force is periodic, q_r is given by an expression of the same form as equation (3.29), i.e.

$$q_r = R \sum_{n=1} \alpha_r(\omega_n)(a_n - ib_n) \exp(i\omega_n t). \tag{3.42}$$

$\alpha_r(\omega_n)$ is the generalized complex receptance of the rth mode at the frequency ω_n, and is given by

$$\alpha_r(\omega_n) = \frac{1}{(K_r - M_r\omega_n^2) + i\omega_n C_n}. \tag{3.43}$$

The term $\alpha_r(\omega)$ varies in exactly the same way with frequency as the term for the simple system. In particular, at or near the frequency

$$\omega_r = \sqrt{K_r/M_r}$$

resonance will occur in the mode $f_r(x)$. We may also speak of the 'damping factor' corresponding to the rth mode, i.e.

$$C_r/2\sqrt{M_r K_r} = \zeta_r.$$

This term will govern the width and height of the response curve corresponding to the rth mode.

When the continuous system is excited by a harmonically varying pressure distribution giving rise to the rth generalized force, $L_r \cos \omega t$, the response in the rth mode is (in the real form)

$$q_r = |\alpha_r| L_r \cos(\omega t - \phi_r). \tag{3.44}$$

The total displacement at any point, from equation (3.36), is clearly

$$w(x, t) = \sum_{r=1}^{\infty} |\alpha_r| L_r f_r(x) \cos(\omega t - \phi_r), \tag{3.45}$$

and if only the first s modes are contributing appreciably,

$$w(x, t) \doteq \sum_{r=1}^{s} |\alpha_r| L_r f_r(x) \cos(\omega t - \phi_r). \tag{3.46}$$

If we are now concerned with the stress at some point in the system, it is necessary to specify the stress at that point due to unit displacement in each of the significant modes. Let unit displacement in the rth mode cause a stress at the point x of $\sigma_r(x)$. Then the total stress is obviously

$$\sigma(x, t) = \sum_{r=1}^{\infty} |\alpha_r| L_r \sigma_r(x) \cos(\omega t - \phi_r).$$ (3.47)

If the exciting pressure is periodic, but not harmonic, then the total stress is the sum of all such expressions as equation (3.47), each one corresponding to a different harmonic component, ω_n, of the exciting pressure. Hence

$$\sigma(x, t) = \sum_{n=1}^{\infty} \sum_{r=1}^{\infty} |\alpha_r(\omega_n)| L_r \sigma_n(x) \cos(\omega t - \phi_{rn}).$$ (3.48)

We write 'ϕ_{rn}' since ϕ_r depends on the frequency, ω_n. Further, if L_r derives from an acoustic pressure field, it too will depend on the frequency.

3.9 PARTICULAR FORMS OF THE GENERALIZED FORCE

We shall now examine the magnitude of the generalized force corresponding to different modes and pressure distributions, $p(x, t)$. Consider the normal modes of flexural vibrations of a uniform rectangular plate which is simply supported along each edge. The modes are represented by:

$$f_r(x) = \sin\frac{r \pi x}{\ell} \sin\frac{s \pi y}{b}.$$ (3.49)

(r and s are integers, ℓ and b are the plate length and breadth respectively). For simplicity, put $s = \ell$. There are then r half-waves along the length, and one across the width of the plate. If the plate is excited by a uniform harmonic pressure over the surface (i.e. in phase at all points, and of uniform amplitude), $p(x, t)$ is of the form $p_0 \cos \omega t$. The generalized force (from equation (3.40)) is then:

$$L_r(t) = \int_A p_0 \cos \omega t \sin\frac{r \pi x}{\ell} \sin\frac{\pi y}{b} \, dx \, dy$$

i.e. $$L_r(t) = p_0 \cos \omega t . \frac{4}{\pi^2} \frac{\text{plate area}}{r} \quad (r \text{ odd})$$

and $$L_r(t) = 0 \quad (r \text{ even}).$$

That this should be zero for even r may be seen from Fig. 3.10.

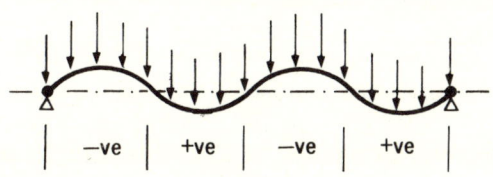

When the pressure $p_0 \cos \omega t$ is moved through a small displacement $\delta q_r \cdot f_r(x)$ as shown, the pressure does positive work over two of the half-wave sections, but an equal amount of negative work over the other two half-wave sections. Obviously, whenever r is even, the net work done is zero, and the corresponding generalized force is zero. When r is odd, a non-zero amount of work (negative or positive) is always done, and the generalized force is non-zero, but it gets progressively smaller as r increases since the area over which work is done without being cancelled gets smaller.

When the rth generalized force is zero, it follows from equation (3.37) that the rth coordinate and mode cannot be excited — always provided there is no significant damping coupling with other modes.

If the plate is excited by a harmonic pressure the amplitude of which varies sinusoidally over the plate (i.e. $p(x, t) = p_0 \cos \omega t \sin r \, \pi x / \ell$) we find that the generalized force is zero for every mode except the rth, provided that the half-wavelength of the pressure on the plate (ℓ/r) is an integral fraction of the plate length, i.e. r is an integer. When this is not so, an interesting and important effect occurs which is described later.

3.10 THE FORCED VIBRATION OF A PLATE IN A SIMPLE SOUND FIELD

The simplest sound field we can consider is a field of plane harmonic waves. Suppose such a field impinges upon a plate, the wavefronts making an angle θ with the plate surface. Consider only the case when the sound wave motion has no component of velocity across the width of the plate, i.e. the wave travels in the direction of the *length* of the plate (Fig. 3.11).

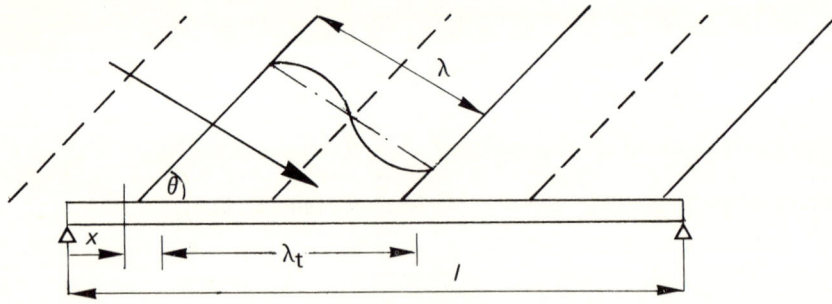

Fig. 3.11 – Diagram illustrating a sound field impinging on the plate.

The wavefronts make an intercept of λ_t on the plate. We call λ_t the 'trace wavelength', and the speed at which the wavefront moves along the plate the 'trace velocity' (a_t). Clearly, $\lambda_t = \lambda \, \text{cosec} \, \theta$, and $a_t = a \, \text{cosec} \, \theta$ (a = speed of sound in air).

The instantaneous pressure at any point along the plate is

$$p(x, t) = p_0 \cos\left(\omega t - \frac{2\pi x}{\lambda_t} + \phi\right). \tag{3.50}$$

The generalized force (see equation (3.41)) is given by:

$$L_r(t) = \int_0^b \int_0^\ell p_0 \cos\left(\omega t - \frac{2\pi x}{\lambda_t}\right) \sin \frac{r\pi x}{\ell} \sin \frac{\pi y}{b} \, dx \, dy$$

$$= p_0 \frac{2b}{\pi} \int_0^\ell \cos\left(\omega t - \frac{2\pi x}{\lambda_t}\right) \sin \frac{r\pi x}{\ell} \, dx.$$

Evaluating the integral, and replacing ℓ/r by $\lambda_m/2$ (the half-wavelength of the mode of vibration) we find:

$$L_r(t) = p_0 \ell b \frac{4}{\pi^2} \left\{ \frac{\cos(\frac{\pi}{2} r \lambda_m/\lambda_t)}{r\{(\lambda_m/\lambda_t)^2 - 1\}} \right\} \cos(\omega t + \epsilon) \quad (r \text{ odd})$$

$$\tag{3.51}$$

or

$$= p_0 \ell b \frac{4}{\pi^2} \left\{ \frac{\sin(\frac{\pi}{2} r \lambda_m/\lambda_t)}{r|(\lambda_m/\lambda_t)^2 - 1|} \right\} \cos(\omega t + \epsilon) \quad (r \text{ even})$$

3.10.1   The joint acceptance function

Equation (3.51) may be written in the form

$$L_r(t) = p_0 \ell b j_r \cos(\omega t + \epsilon). \tag{3.52}$$

The total force acting on the plate at any instant when the pressure is distributed uniformly over the plate is $p_0 \ell b \cos(\omega t + \epsilon)$. The joint acceptance function j is a factor which describes the proportion of this force which a particular mode of distortion can 'accept' and convert into the corresponding generalized force. Since it is a function of both the modal wavelength, λ_m, and the trace-wavelength, λ_t, we may call it the **joint acceptance** of the mode and pressure field. Fig. 3.12 shows how j_r varies with both r and λ_m/λ_t. Apart from the curve for $r = 1$, it is characteristic of each curve that its maximum value occurs at $(\lambda_m/\lambda_t) = 1$, i.e. when the modal wavelength is the same as the trace-wavelength. The response of the plate in any of its modes to this pressure field is given (as before) by equation (3.42) or (3.44). Substituting from equation (3.32) into equation (3.44) we have:

$$q_r = p_0 \ell b |\alpha_r| j_r \cos(\omega t + \epsilon - \phi_r). \tag{3.53}$$

It is now clear that a large response in this mode will occur if, at the resonant frequency ω_r (when $|\alpha_r|$ is a maximum) j_r is at the maximum value corresponding to $\lambda_m/\lambda_t = 1$. A dual coincidence then occurs, the sound field frequency being equal to the resonant frequency and the trace-wavelength being equal to the modal wavelength. For this reason the term **coincidence effect** is often used to describe this particular phenomenon.

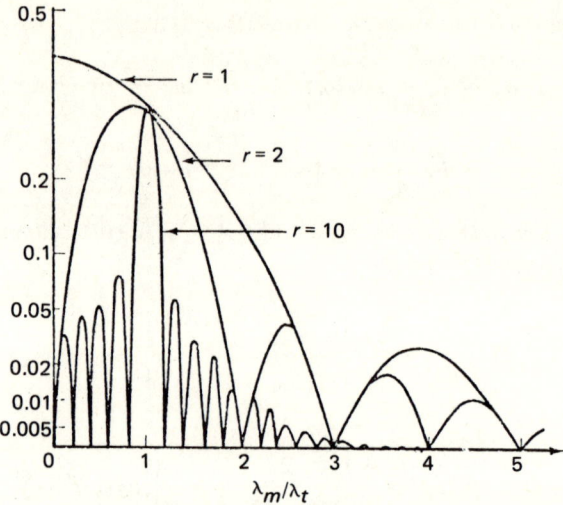

Fig. 3.12 – The variation of the joint acceptance, j_r, with the wave-length ratio λ_m/λ_t.

CHAPTER 4

Random processes

B. L. Clarkson
Institute of Sound and Vibration Research, University of Southampton

4.1 INTRODUCTION

In many of the phenomena in acoustics and vibration the quantity of interest such as sound pressure or acceleration varies in a way which is not completely predictable. It is not possible to state the exact value of the quantity at any instant of time, but instead we have to fall back on more general descriptions such as average value, etc. For example, the sound pressure at a point some distance from a jet exhaust or the force in the suspension of a car travelling over a rough road cannot be specified precisely for each instant of time. However, the quantity of interest will possess certain general properties which can be used to describe the process. The statistical theory which has been built up to describe these processes is known as random process theory. In the majority of our applications we are primarily concerned with quantities which are continuous functions of time, and so our task is to describe in a meaningful way a complex time history.

In this chapter I shall try to outline some of the relevant basic concepts of probability theory and then consider in some detail the quantities which can be used to describe a continuous stationary random record of infinite duration. The simplest property which can be used to describe a random time history is the mean square or root mean square value. Generally, the mean value is zero, or if, in a particular case, it has a finite value, this can be treated separately by relatively simple methods. The root mean square value gives an order of magnitude measure of the amplitude of the signal, but does not give sufficient information for more detailed investigation of the characteristics. Probably the two most important properties are the amplitude probability distribution and the power spectrum. The auto-correlation and cross-correlation functions are also used in special cases, but usually as an alternative method of obtaining the power spectrum. Another property of interest is the expected number of zero crossings per second.

4.2 SOME BASIC CONCEPTS OF PROBABILITY THEORY

Probability theory attempts to put a mathematical model to the outcome of a random experiment. In this case, a random experiment implies that the outcome of a particular performance of the experiment cannot be predicted. Nevertheless, it is often possible to say that in the *long run* a particular outcome will occur in a given proportion of the total number of repetitions of the experiment.

For example, if one considers the random experiment of throwing a die. The outcome of a particular throw is unpredictable and may take on any of the values 1, 2, 3, 4, 5, or 6. If, however, the experiment is carried out a large number of times, we should expect that about one sixth of the throws would result in the number 1 because any of the numbers 1 to 6 are equally likely to occur. The probability of the number 1 is therefore said to 1/6. If now we perform the experiment n times, we can count the number of times f_1 when a 1 is obtained. In many types of experiments it is found that as n increases the ratio f_1/n settles down to a constant value equal to the probability of the 1 being obtained. Thus we have the proposition or operational rule

$$P(1) = \lim_{n \to \infty} f_1/n \qquad\qquad (4.1)$$

$$= 1/6 \text{ in this case,}$$

where f_1/n is known as the **frequency ratio** or **relative frequency**. It cannot be emphasized too strongly that the '*a priori*' evaluation of probability depends on postulating a certain number of equally likely outcomes of the experiment, whereas the frequency ratio is the result of counting the number of occurrences of the desired outcome. The probability of a certain event is, therefore, unity and the probability of an impossible event is zero. The converse of these statements is not true in all cases as an event with probability zero is not necessarily impossible. Consider, for example, the probability that the room temperature is exactly 20°C. There are an infinite number of possible values of temperature in the continuous range t_1 to t_2 and therefore the probability of getting one particular value is zero. But it is clearly not impossible to get a value of exactly 20°C.

4.2.1 Probability distribution for a discrete random variable

A random variable, ξ, can now be defined by the probability that it takes on different admissible values, x, denoted by the probability function $P_\xi(x)$ or $P(\xi = x)$. Here x is the range variable and has a *finite* number of admissible values (e.g. 1, 2, 3, 4, 5, or 6 in the case of the die).

If the number of possible values of x is very large, a large number of probability functions are required to describe the random variable. In this case it is often more convenient to use the **probability distribution function** which is defined as the probability that ξ takes on a value less than or equal to x.

i.e. $F_\xi(x) = P(\xi \leqslant x) = \sum_i P_\xi(x_i) \quad x_i \leqslant x.$ (4.2)

From this it is clear that

$$F_\xi(-\infty) = 0$$
$$F_\xi(+\infty) = 1 \ .$$ (4.3)

Examples of the distribution function and the probability function for the probability of the number of heads obtained when six coins are tossed are shown in Fig. 4.1.

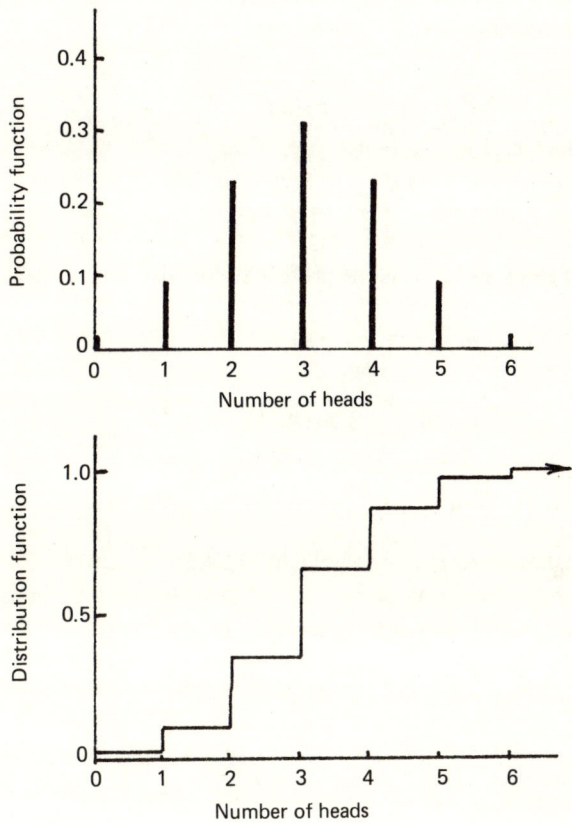

Fig. 4.1 – Probability distribution for a discrete random process (number of heads obtained when 6 coins are thrown).

4.2.2 Probability distribution for a continuous random variable

If the random variable can take on any value within the continuous range x_1 to x_2, say, then because there are an infinite number of possible values of ξ in this range the '*a priori*' probability that ξ takes on one particular value x is zero. Thus the probability function $P_\xi(x)$ is no longer a useful parameter of the process. The distribution function $F_\xi(x)$, however, is still a valid and useful description.

Now the probability that ξ lies in the range x to $x + \Delta x$ is given in terms of the distribution function as

$$P(x \leqslant \xi \leqslant x + \Delta x) = F_\xi(x + \Delta x) - F_\xi(x)$$
$$= \Delta F_\xi(x)$$

which can be considered as the probability increment ΔP at x. The rate of change of this increment with x can be written

$$\lim_{\Delta x \to 0} \frac{\Delta P}{\Delta x} = \lim_{\Delta x \to 0} \frac{\Delta F_\xi(x)}{\Delta x}$$

$$= \frac{\mathrm{d} F_\xi(x)}{\mathrm{d} x} = p_\xi(x). \tag{4.4}$$

The function $p(x)$ is known as the **probability density**, and we have

$$P(x_1 \leqslant \xi \leqslant x_2) = \int_{x_1}^{x_2} p_\xi(x) \, \mathrm{d} x$$

$$F_\xi(x) = \int_{-\infty}^{x} p_\xi(x) \, \mathrm{d} x \tag{4.5}$$

and $\quad \int_{-\infty}^{\infty} p_\xi(x) \, \mathrm{d} x = 1.$

Examples of these functions are shown in Fig. 4.2.

If a discrete random variable is mixed in with the continuous variable, the probability density of the mixed process can be represented in the following way

$$p_\xi(x) = p_\xi^*(x) + \sum_i P_\xi(x_i) \, \delta(x - x_i) \tag{4.6}$$

where $p_\xi^*(x)$ is the probability density of the continuous variable and $P_\xi(x_i)$ is the probability that ξ takes on the discrete value x_i. $\delta(x - x_i)$ is the Dirac delta function

$$= 0 \text{ when } x \neq x_i$$

and $\quad \int_{-\infty}^{\infty} \delta(x - x_i) \, \mathrm{d} x = 1.$

In the case of the mixed process the probability density diagram will have discontinuities at the points x_i, but the distribution function will be continuous as shown in Fig. 4.3.

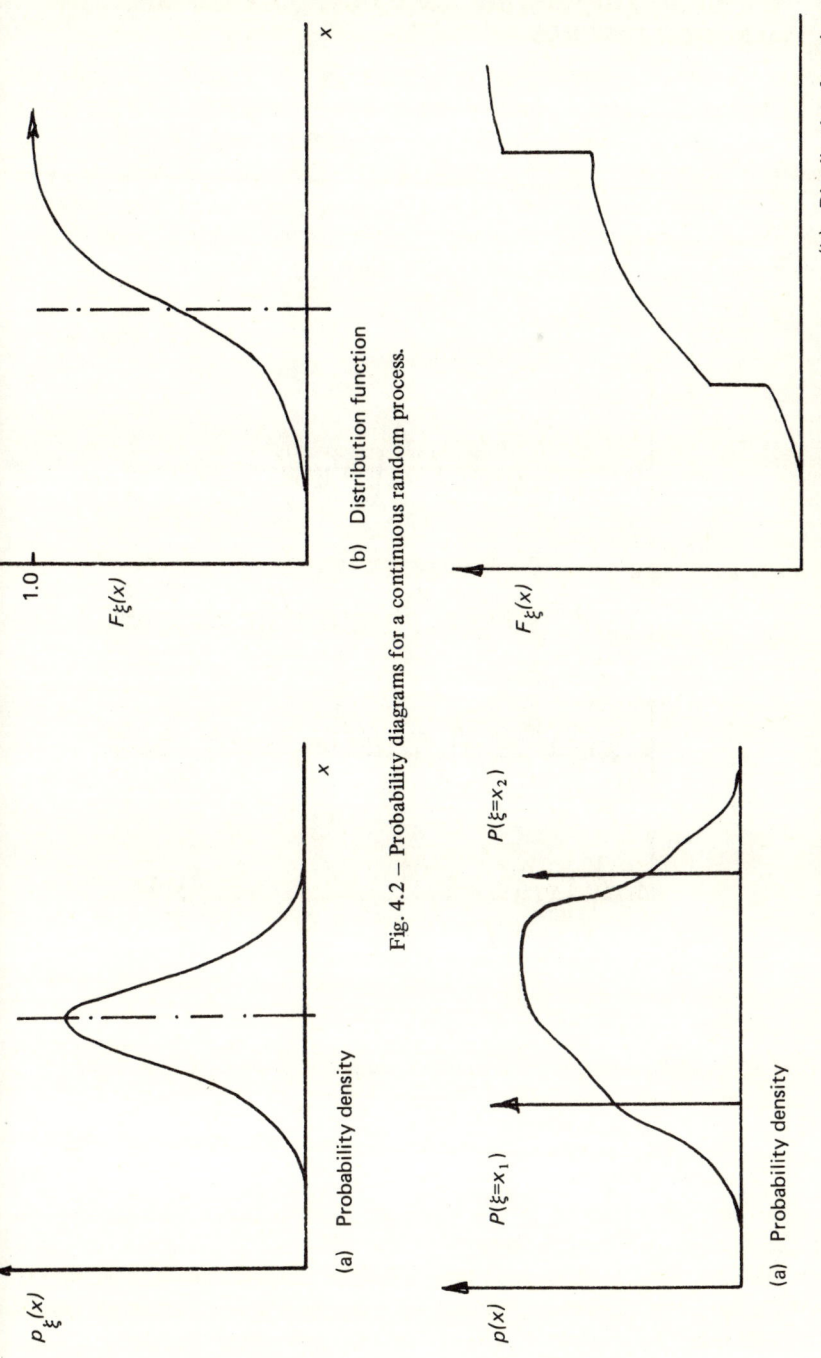

(a) Probability density

(b) Distribution function

Fig. 4.2 – Probability diagrams for a continuous random process.

(a) Probability density

(b) Distribution function

Fig. 4.3 – Probability diagrams for a mixed discrete and continuous process.

4.3 DESCRIPTION OF CONTINUOUS ACOUSTIC OR RANDOM VIBRATION RECORDS

Consider now the acceleration at the guidance bay of a guided missile or the acceleration at the centre of gravity of a passenger in the rear seat of a fast sports car. If we take continuous recordings of the acceleration from each of many missiles or many sports cars travelling over the same route, we obtain a series of records as shown in Fig. 4.4. The acceleration represented by these series of traces forms a random process in which we try to describe the whole family (or ensemble) of traces not just one specific trace.

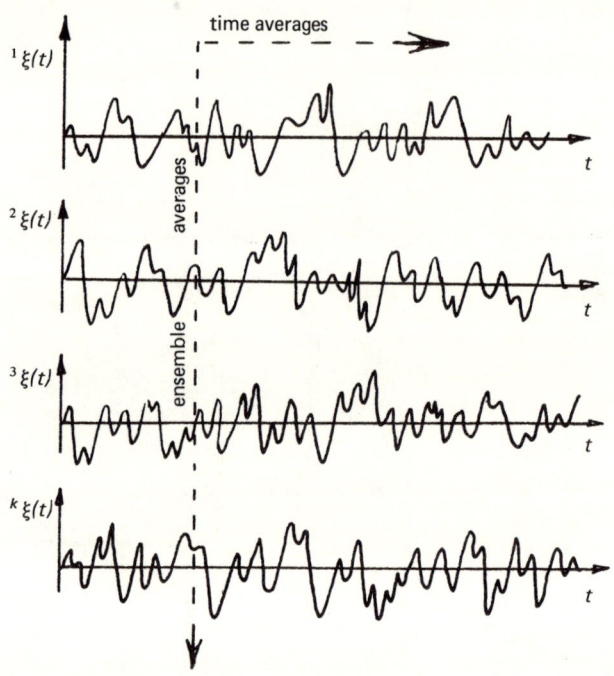

Fig. 4.4 – Averages of a family (or ensemble) of random vibration records.

4.3.1 Ensemble averages

Take the case of the sports car, for example. If a number of identical models left the factory at 5-minute intervals in the middle of the morning to travel down the same road at the same average speed for embarkation on board ship at Southampton, the acceleration record for each would vary. Although details of the individual records would vary we should expect them to take on similar characteristics. We can write down the acceleration for one car (say the kth) as $^k\xi(t)$ which in general is a function of time and the particular car being considered. Statistically we can now consider two kinds of averages. In the first place, we can consider the acceleration at time t from the start for all the cars and write down the **ensemble average** which is the average over all the n cars at the particular time t:

$$\langle^k\xi(t)\rangle = \frac{1}{n}\sum_{k=1}^{n}{}^k\xi(t) \tag{4.7}$$

where the angular brackets denote an ensemble average.

If we now imagine that we have a very large number of samples at the particular time t we can consider the probability of obtaining a particular value x at time t. The variation in ξ can then be described by the probability density function $p_\xi(x)$. In terms of this function the mean value or expected value of acceleration (at time t implied throughout) is:

$$E[\xi] = \int_{-\infty}^{\infty} x\,p_\xi(x)\,\mathrm{d}x. \tag{4.8}$$

Similarly the mean square value is

$$E[\xi^2] = \int_{-\infty}^{\infty} x^2\,p_\xi(x)\,\mathrm{d}x. \tag{4.9}$$

The variance of ξ is given by

$$\begin{aligned}
\sigma_\xi^2 &= E[(\xi - E[\xi])^2] \\
&= E[\xi^2] - (E[\xi])^2.
\end{aligned} \tag{4.10}$$

4.3.2 Time averages

The ensemble averages discussed above relate to conditions at one particular time t. In many natural processes it is found that if these probability distributions are measured across the ensemble for different times the results are independent of the particular time. If this independence of time can be established, the process is said to be **stationary**. Conversely, a **non-stationary** process is one in which the probability distributions across the ensemble are dependent on time. In many practical situations a sufficiently close approximation can be obtained by the assumption of stationarity. For example, if one considers the vertical component of the turbulent velocity in the atmosphere, the 'average' value and the probability distributions clearly vary from day to day. But if one is considering the response of an aircraft or missile to this turbulence, it may be adequate to assume that during one flight or a part of one flight the turbulence is stationary. In other words, the rate at which the properties are changing is very much slower than the typical velocities induced in the aircraft or missile.

In the stationary case it is now possible to consider just one sample $^k\xi(t)$ and write down averages and probability distributions of this variable over all time. Equations (4.8), (4.9) and (4.10) are still used, but the average is over all t for one k rather than over all k for one t. These are known as time or temporal averages. In the stationary process all the properties are independent of time, and therefore the time origin is of no significance.

4.3.3 Ergodic processes

There exist in practice many random processes in which the time averages are approximately equal to ensemble averages. These processes are known as **ergodic processes**. The very great practical significance of these conditions is that if we take just one sample record $^k\xi(t)$ and work out the time averages we can then equate these directly with the ensemble averages. Thus from the measurement of one sample we can predict the statistical properties of the behaviour of *all* other similar systems in a similar environment.

It is a very lengthy procedure to verify experimentally that a particular process is ergodic. But if it can be assumed that it is stationary or measurements from a long sample show stationarity, then it may be reasonable from an engineering point of view to assume ergodicity.

4.4 AMPLITUDE DESCRIPTION OF A RANDOM WAVEFORM

It is convenient to separate out the steady or mean value from the signal before dealing with the fluctuating part. The steady component can be treated by the standard static analyses. Thus the initial step in any data processing procedure is to determine the mean value and then separate out this 'DC' component before subsequent processing.

The gross measure of amplitude of a fluctuating signal whose mean value is zero is the mean square value. Rigorously the mean square value of a time varying signal $x(t)$ is defined as follows:

$$\overline{x^2(t)} = \lim_{T \to \infty} \frac{1}{T} \int_0^T x^2(t)\, dt = \sigma_x^2. \tag{4.11}$$

The square root of this quantity, the r.m.s. value of $x(t)$, is usually used as an amplitude measure. In practice the integration time T must be long compared with the period of the lowest frequency component in the signal.

Signals having the same root mean square value can have very different waveforms and peak to r.m.s. values. A measure of these differences can be obtained from the probability density distribution function $p(x)$.

4.4.1 Gaussian distribution

The probability density distribution of sound pressure close to a jet exhaust, for example, can be obtained by the use of the expressions (4.4) and (4.5). The accuracy of the measurements is discussed in the section 'ensemble averages'. In general, each random process will have a particular probability density distribution associated with it. However, it is found that there is one theoretical random process, known as the **Gaussian or normal random process**, which is a good approximation to many practical processes. The advantage of the theoretical distribution is that the response of a system to a random input which is Gaussian can be obtained analytically in many cases. Unless the process is Gaussian it is not generally possible to get analytical solutions.

In practice many random processes result from the action of a very large number of independent random inputs or sources. For example, the pressure close to a jet exhaust is made up of components due to radiation from a very large number of sources in the turbulent mixing region of a jet. In this type of situation the resulting random process is Gaussian, whatever the distribution of one of the particular independent components. A general statement of this idea is known as the **Central limit theorem.** Thus it can be expected that many practical processes will approximate closely to the Gaussian process.

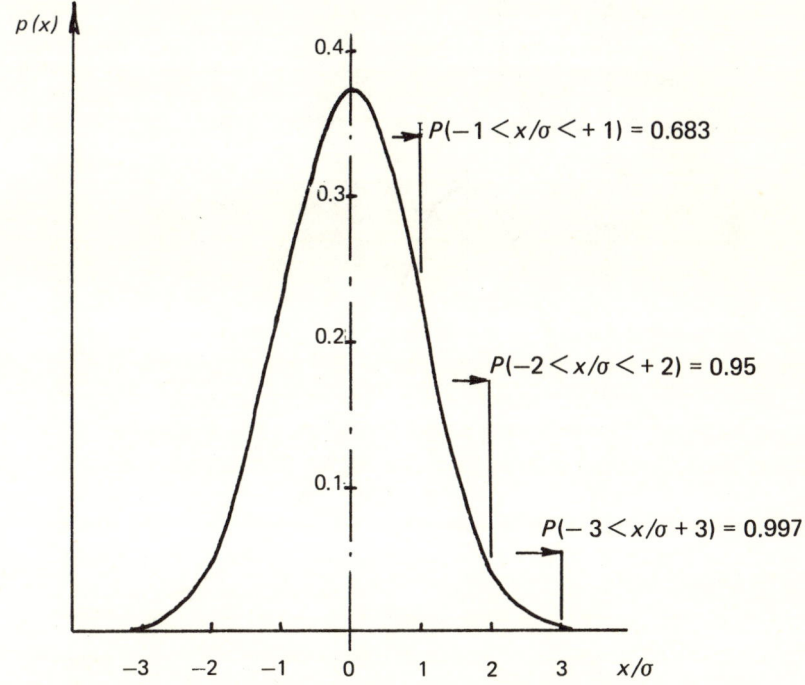

Fig. 4.5 – Gaussian probability density distribution.

The probability density distribution of the Gaussian process is given by

$$p_\xi(x) = \frac{1}{\sqrt{(2\pi\sigma_\xi)}} \exp\left(\frac{-x^2}{2\sigma_\xi^2}\right) \tag{4.12}$$

for zero mean value. This is shown in Fig. 4.5.

4.4.2 Rayleigh distribution

One specific application of probability density distribution is to fatigue life extimates. It is often assumed that the amplitudes of the peaks in the waveform of stress are the determining factors in the fatigue life of a component. The number of peaks in each amplitude band can be counted or computed and a probability density distribution of peaks determined. An estimate of fatigue life can be obtained from a cumulative damage theory and the S-N curve for the material. If it can be assumed that the amplitude distribution is Gaussian and that the random stress process is narrow band, then it can be shown that the probability density of peaks is the Rayleigh distribution

$$p_{\text{peaks}}(x) = \frac{x}{\sigma_x^2} \exp\left(-\frac{x^2}{2\sigma_x^2}\right). \tag{4.13}$$

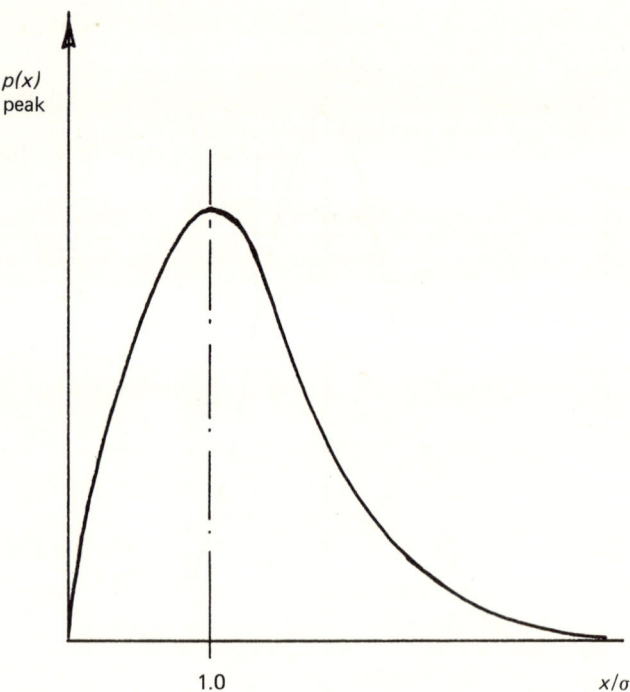

Fig. 4.6 – Probability distribution of peaks for narrow band random process (Rayleigh Distribution).

This is illustrated in Fig. 4.6. (In a process having zero mean value the standard deviation σ_x becomes the r.m.s. value.) In many processes which cannot strictly be termed narrow band the peak distribution is approximately Rayleigh and can therefore be determined solely from a knowledge of the mean square value σ_x^2. Thus a fatigue estimate can be made from a measure of the r.m.s. value of the stress and the $S-N$ curve of the material.

4.5 TIME DOMAIN DESCRIPTION

The averages and probability distributions discussed above lose all information on typical periods or frequencies present in the record. Whilst these measurements may be adequate for the consideration of fatigue life and reliability in general, they are not adequate if one is considering the response of various components mounted in the vibration environment. One measure which does take into account the typical periods present in the vibration record is the auto-correlation function or covariance of the random variable $\xi(t)$ at two times t_1 and t_2. This covariance is

$$E[(\xi(t_1) - E[\xi]) \, (\xi(t_2) - E[\xi])]$$

which is generally written for zero mean values as

$$R(t_1 \tau) = E[\xi(t_1)\, \xi(t_1 + \tau)] \tag{4.14}$$

where $t_2 = t_1 + \tau$.

If the record is stationary and ergodic we can replace the ensemble average by the time average.

$$R(\tau) = \lim_{T \to \infty} \frac{1}{T} \int_0^T \xi(t)\, \xi(t + \tau) \, \mathrm{d}t. \tag{4.15}$$

Thus the auto-correlation function is only a function of the time difference or delay τ between the two sample points on the record.

In all that follows we shall restrict ourselves to a consideration of processes which have a zero mean value. The results for the non-zero mean conditions can easily be obtained but they are more lengthy to write out, and therefore to avoid undue complexity in the expressions used only the case of a zero mean value is quoted.

If $\xi(t)$ is a periodic function, say $\xi(t) = \sin \omega t$ then the auto-correlation function is given by

$$R(\tau) = \lim_{T \to \infty} \frac{1}{T} \int_0^T \sin \omega t \, \sin \omega(t + \tau) \, \mathrm{d}t$$

$$= \lim_{T \to \infty} \frac{1}{T} \int_0^T (\sin^2 \omega t \cos \omega \tau + \tfrac{1}{2} \sin 2\omega t \sin \omega \tau) \, \mathrm{d}t$$

$$= \tfrac{1}{2} \cos \omega \tau. \tag{4.16}$$

For a non-periodic function a time delay will be reached when $R(\tau)$ reaches a steady zero value as shown diagrammatically in Fig. 4.7. If a variable has a random and a periodic component then the auto-correlation function will take on a form similar to that shown in Fig. 4.7(c). Thus the auto-correlation function forms a means of identifying hidden periodicities in random noise and has been used extensively in radar tracking systems.

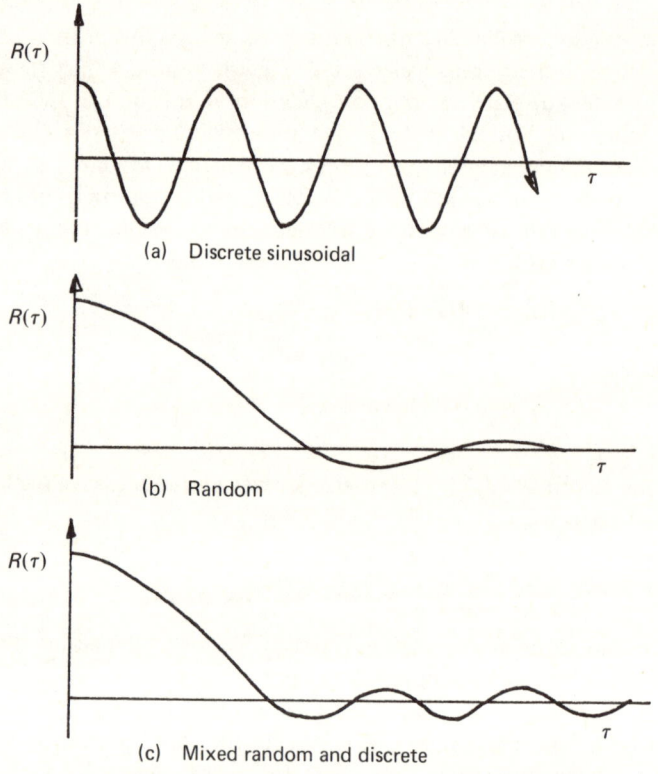

(a) Discrete sinusoidal

(b) Random

(c) Mixed random and discrete

Fig. 4.7 – Examples of correlation functions.

4.6 FREQUENCY DOMAIN DESCRIPTION

From an engineer's point of view many so-called random signals appear to have many different frequency components within them. Thus it would seem reasonable to try to express the signal as some sort of assembly of individual frequency components. This concept is carried out rather more rigorously by representing the signal in the following way:

$$\overline{x^2(t)} = \int_0^{\infty} G(f)\, df \qquad (4.17)$$

where $G(f)$ is known as the **spectral density function** and is the frequency domain representation of the signal. Typical examples are shown in Fig. 4.8. A sine wave appears as a delta function or line in the spectrum. Thus we can say that from a physical point of view the spectrum indicates the frequency range or frequency content of the signal.

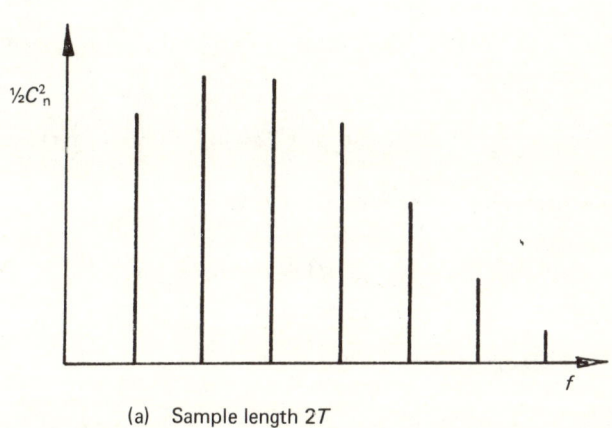

(a)   Sample length $2T$

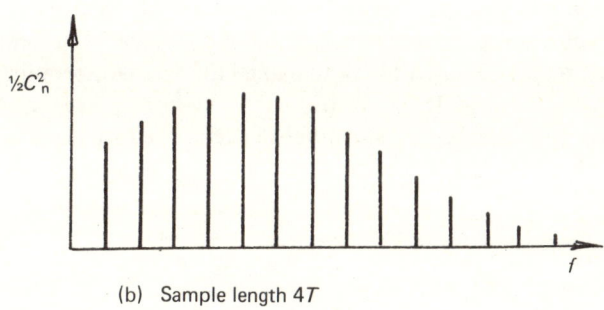

(b)   Sample length $4T$

Fig. 4.8 – Fourier spectra for a periodic function.

A more rigorous development can be obtained by the use of Fourier series analysis methods. Unfortunately, the Fourier series method can only be applied to periodic functions, but it is possible to extend the ideas to apply to random variables.

Consider a *periodic* function $x(t)$ of period $2T$

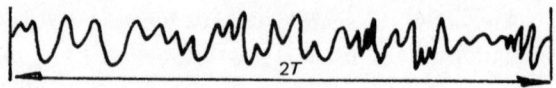

The function can be expanded in a Fourier series

$$x(t) = \frac{a_0}{2} + \sum_{n=1}^{\infty} (a_n \cos n \, \omega_0 t + b_n \sin n \, \omega_0 t) \qquad (4.18)$$

where in our case of zero mean value $a_0 = 0$, and the fundamental frequency $\omega_0 = \pi/T$, also

$$a_n = \frac{1}{T} \int_{-T}^{T} x(t) \cos n \, \omega_0 t \, dt, \quad b_n = \frac{1}{T} \int_{-T}^{T} x(t) \sin n \, \omega_0 t \, dt. \qquad (4.19)$$

The mean square value of the function is given by

$$E[x^2(t)] = \frac{1}{2T} \int_{-T}^{T} x^2(t) \, dt = \frac{1}{2} \sum_{n=1}^{\infty} (a_n^2 + b_n^2)$$

or

$$\frac{1}{2} \sum_{n=1}^{\infty} c_n^2. \qquad (4.20)$$

Thus the mean square value of $x(t)$ is equal to the sum of the squares of the Fourier coefficients. A Fourier spectrum could then be drawn as shown in Fig. 4.9(a) to represent the frequency composition of the signal. The spectrum is made up of a series of lines of amplitude $\frac{1}{2} c_n^2$ separated in frequency by an amount ω_0. We are tempted to try to extend this to a non-periodic function by letting T go to infinity. If we do this the fundamental frequency approaches zero, but the coefficients $\frac{1}{2} c_n^2$ also approach zero. The first stages of this process can be seen in Fig. 4.9(b) where the period has been increased to $4T$.

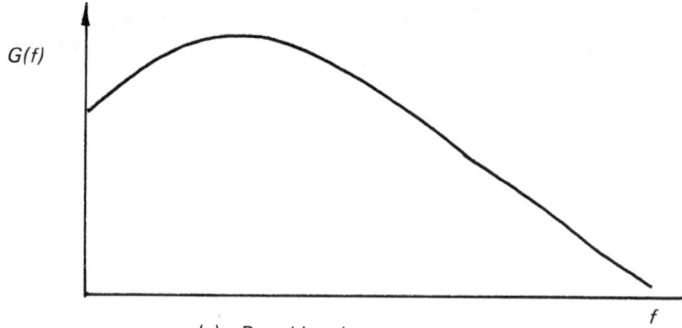

(a)   Broad-band process

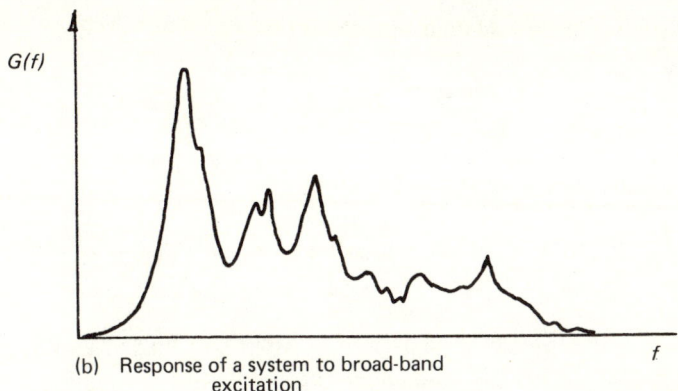

(b)   Response of a system to broad-band
      excitation

Fig. 4.9 – Typical spectral density diagrams for random processes.

To avoid the mathematical difficulty of letting the limits of equation (4.19) go to infinity a synthetic function is constructed as shown below. The synthetic periodic function $x(t)$ of period $2T$ is equal to the random variable $\xi(t)$ over the time interval $-t_0 \leqslant t \leqslant +t_0$ and zero in the ranges

$$-T \leqslant t \leqslant -t_0, \quad t_0 \leqslant t \leqslant T.$$

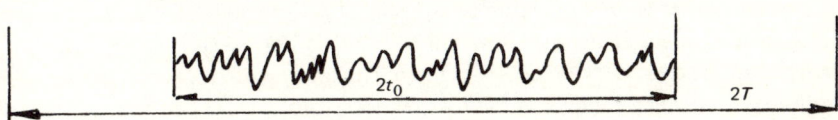

We can now let $T \to \infty$ whilst t_0 remains finite and therefore the restrictions on the Fourier transforms do not apply.

Using the complex Fourier series representation

$$x(t) = \sum_{n=-\infty}^{\infty} A_n \exp(in\omega_0 t) \tag{4.21}$$

where $\qquad A_n = \dfrac{1}{2T} \int_{-T}^{T} x(t) \exp(-in\omega_0 t)\,\mathrm{d}t$

as $\quad T \to \infty \quad A_n \to 0$

but $\quad \int_{-T}^{T} x(t) \exp(-in\omega_0 t)\,\mathrm{d}t \quad$ remains finite. $\omega_0 = \pi/T = \Delta\omega \to 0$.

We can therefore use $A_n \times 2T$ or $\dfrac{A_n}{\Delta\omega}$ as a measure of relative amplitude of frequency components.

$$\frac{A_n}{\Delta\omega} = \frac{1}{2\pi} \int_{-\infty}^{\infty} x(t)\,\mathrm{e}^{-i\omega t}\,\mathrm{d}t = F(i\omega) \quad \text{say,} \tag{4.22}$$

$F(i\omega)$ is thus the Fourier transform of $x(t)$

and $\qquad x(t) = \int\limits_{-\infty}^{\infty} F(i\omega)\, e^{i\omega t}\, d\omega.$ $\qquad\qquad\qquad\qquad$ (4.23)

The mean square value of $\xi(t)$ is given by

$$E[\xi^2(t)] = \lim_{t_0 \to \infty} \frac{1}{2t_0} \int\limits_{t_0}^{t_0} x^2(t)\, dt = \int\limits_{-\infty}^{\infty} \lim_{t_0 \to \infty} \frac{\pi|F(i\omega)|^2}{t_0}\, d\omega.$$
$\qquad\qquad\qquad\qquad\qquad\qquad\qquad\qquad\qquad\qquad\qquad\qquad$ (4.24)

The mean square value is again defined by a series of 'coefficients' as in (4.20), but in this case the series is continuous. The 'coefficients'

$$\lim_{t_0 \to \infty} \frac{\pi|F(i\omega)|^2}{t_0}$$

are known as the **spectral density** $S(\omega)$. Therefore

$$E[\xi^2(t)] = \int\limits_{-\infty}^{\infty} S(\omega)\, d\omega \qquad\qquad\qquad\qquad (4.25)$$

and in this case $S(\omega)\, d\omega$ is equivalent to $\frac{1}{2} c_n^2$ in the periodic case.

Mathematically it is often more convenient (as above) to use the alternative double-sided spectral density function $S(\omega)$. This is an even function which is defined in the frequency range $-\infty$ to $+\infty$. The integral of this function over the complete frequency range is equal to the mean square value of the signal (as in (4.17)) and therefore we have the following relationship between the single and double-sided functions

$$G(\omega) = 2\, S(\omega). \qquad\qquad\qquad\qquad (4.26)$$

If a random variable has a constant power spectral density over all frequencies it is often referred to as 'white' noise from the analogy with white light which contains all frequencies of light waves. In practice, the term 'white noise' is used to describe a forcing function which has a constant power spectral density over the frequency range which is being considered.

4.7 RELATIONSHIP BETWEEN SPECTRAL DENSITY AND CORRELATION FUNCTION

As the spectral density function $G(f)$ and the auto-correlation function $R(\tau)$ both describe the same waveform we would expect there to be a relationship between them. This is indeed so, and the two are found to be related through the Fourier transform:

$$R(\tau) = \int_0^\infty G(f) \cos 2\pi f\tau \, \mathrm{d}f$$

$$G(f) = 4 \int_0^\infty R(\tau) \cos 2\pi f\tau \, \mathrm{d}\tau. \tag{4.27}$$

These relationships are often used to determine $G(f)$, the more physically meaningful quantity, from $R(\tau)$ which is easier to measure digitally.

The equivalent relationships for the double-sided function are:

$$R(\tau) = \int_{-\infty}^\infty S(\omega) \, \mathrm{e}^{i\omega\tau} \, \mathrm{d}\omega$$

$$S(\omega) = \frac{1}{2\pi} \int_{-\infty}^\infty R(\tau) \, \mathrm{e}^{-i\omega\tau} \, \mathrm{d}\tau. \tag{4.28}$$

These relationships are generally known as the **Wiener-Khintchine relations.**

4.8 CROSS-CORRELATION AND CROSS-SPECTRAL DENSITY

It will be realized from the way in which the auto-correlation function has been introduced that a similar function can be defined for two variables $f(x_1 t)$, $f(x_2 t)$ where for example $f(x_1 t)$ is the random input to the system and $f(x_2 t)$ is the output. The cross-correlation function is defined as

$$R(x_1 x_2 \tau) = \lim_{T \to \infty} \frac{1}{T} \int_0^T f(x_1 t) f(x_2 t + \tau) \, \mathrm{d}t. \tag{4.29}$$

If the process is stationary this function is independent of t.

The cross-spectral density can now be defined as the Fourier transform of the cross-correlation function. As the cross-correlation is not now an even function the Fourier transformation has both real and imaginary components.

and

$$S_{x_1 x_2}(\omega) = \frac{1}{2\pi} \int_{-\infty}^\infty R_{x_1 x_2}(\tau) \, \mathrm{e}^{-i\omega\tau} \, \mathrm{d}\tau$$

$$R_{x_1 x_2}(\tau) = \int_{-\infty}^\infty S_{x_1 x_2}(\tau) \, \mathrm{e}^{i\omega\tau} \, \mathrm{d}\omega. \tag{4.30}$$

The application of cross-spectral density is usually in studying the response of systems to a random input.

4.9 ANALYSIS AND INTERPRETATION OF MEASUREMENTS

This section gives an introduction to the theoretical aspects of estimation. The topic is covered in more detail in Chapter 6.

It has been shown that all the commonly used descriptions of a random process such as the mean values, probability density distributions, autocorrelation function, and power spectrum are all defined as a limit in which the sample length becomes infinite in time. In practical situations we are restricted to samples of finite length and we are therefore interested to know how close our measured value is to the limiting value. Thus for any particular measurement we should like to know the error involved. Unfortunately, the error itself is a random process, and therefore as a measure of accuracy it is usual to use the variance of the quantity being considered. If the process is approximately Gaussian, it is then possible to state the probability of a sample value lying within a certain percentage of the true value. The assessment of the accuracy of each of the quantities described is now considered for a staionary time series.

4.9.1 Ensemble averages

The error in ensemble averages applies equally well to stationary and non-stationary processes. It can be shown that the variance of the mean of the independent sample values at time t is equal to the variance of the original random process divided by the number of samples, N.

$$\sigma^2_{\text{mean}}(t) = \frac{1}{N} \sigma^2_\xi(t) \qquad (4.31)$$

Thus if the process is Gaussian and one takes, for example, only 5 samples, then there is a 95% probability that the measured mean of 5 samples lies within the range

$$\mu(t) - \frac{2}{\sqrt{5}} \sigma_\xi(t) \text{ to } \mu(t) + \frac{2}{\sqrt{5}} \sigma_\xi(t)$$

where $\mu(t)$ is the true mean value at time t.

The variance of the mean square value measured from N independent samples at time t is given by

$$\sigma^2_{\text{var}}(t) = \frac{2}{N} \{\sigma^4_\xi(t) + 2 \sigma^2_\xi(t) \mu t\}. \qquad (4.32)$$

It can be seen from these results that unless a very large number of samples is taken it is not possible to get a high accuracy for non-stationary processes. For example, if a hundred samples were taken, there would be 95% probability that the measured mean of the samples fell in the range $0.4 \sigma_\xi(t)$ wide, which is still a relatively wide band.

4.9.2 Time averages for a stationary random variable

In the case of a stationary random variable, measurements of mean and mean square values are taken from a single continuous record of length T. The accuracy of the measurement can be determined from equations (4.31) and (4.32) if one substitutes for N the equivalent number of statistical degrees of freedom n where

$$n = 2 BT. \qquad (4.33)$$

Here B is the effective frequency bandwidth and T is the sample length or integration time if this is less than the sample length. If $S(f)$ is the power spectrum of the signal, then the effective bandwidth for use in equation (4.33) is

$$B = \int_0^\infty \frac{G(f)}{G(f)_{max}}\, df. \tag{4.34}$$

4.9.3 Probability density distributions

Theoretical and experimental work carried out by Bendat and his co-workers have shown that the variance of the *measured* value of probability density $p(x)$ from a sample of length T is given approximately by

$$\sigma_{p(x)}^2 \simeq \frac{0.045\, p(x)}{\Delta x\, BT}. \tag{4.35}$$

In this expression $p(x)$ is the true probability density of the variable, Δx is the amplitude bandwidth, B and T are as in (4.33).

4.9.4 Spectral density

The discussion in this section is more appropriate to the filtering method of interpretation of the spectral density function.

In considering estimates of spectral density, there are two types of errors involved. The first type is the inherent statistical error due to the finite length of the sample and the theoretical bandwidth of the filter. The second type of error is the experimental or practical error arising from the fact that the shape of the filter is different from the theoretical shape.

It has been shown by various workers that the variance of the estimated power spectral density is given approximately by

$$\sigma_{G(f)}^2 = \frac{1}{BT}\, G^2(f). \tag{4.36}$$

For example, if a sample record is analysed with a product BT equal to 100, then the estimated power spectral density $G(f)$ at the frequency f will have a standard deviation of 10% of the true power spectral density. Or, alternatively, if the estimates can be assumed to have a Gaussian distribution, then there is a 95% probability that the estimate will lie within $\pm 20\%$ of the true value.

Turning now to the experimental errors due to the type of analysing system being used, we consider first the use of electronic filters to measure directly the energy present in a given bandwidth. With normal electronic analysers the time T in equation (4.36) now becomes the integration time of the system, rather than the sample length. Thus a longer integration time is needed for narrow-band analysis.

The quantity we measure is not the quantity we require, but the integral of that quantity over a certain interval of frequency. If we assume the spectrum to be flat the correction is quite simple, for the reading is

$$X = \int_{\omega_2}^{\omega_1} G(\omega)\,d\omega = G_0(\omega)\int_{\omega_1}^{\omega_2} d\omega = G_0(\omega)\,(\omega_2 - \omega_1).$$

Ideally the electronic filter should have a rectangular response characteristic. In practice, however, it is not possible to construct a filter with a sharp cut-off, and a smoother curve such as shown in Fig. 4.10 generally results. Another difficulty is that in the practical filter there is a relative phase shift between the input and output. For a wide filter, such as one-third octave, the relative phase only changes slowly across the passband, reaching perhaps 30° at the edge. Outside the cut-off frequencies it changes more rapidly. In a narrow-band filter the change across the effective passband is very large. This is of great importance in the measurement of cross-spectra, but does not affect measurements of direct spectral density.

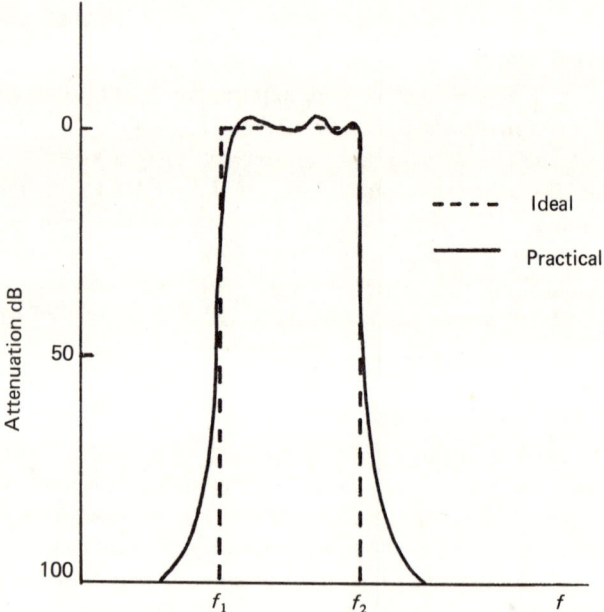

Fig. 4.10 – Filter response characteristics.

The effective cut-off frequencies of a practical filter may be defined as the cut-off frequencies of an ideal filter which will pass the same power, for a given noise signal, as the practical filter.

An alternative method for estimating the spectral density of a signal is to measure the correlation function and then transform this to obtain the spectral density using the relationships given by equation (4.27).

The limitations of the transformation process for obtaining direct and cross-spectra, are that the transformation involves an infinite integral which in turn requires a knowledge of $R(\tau)$ for all values of τ up to infinity. In practice the auto-correlation function must be truncated at some point τ_{max}. If the correlegram has reached steady zero values at this point, as in the case of broad-band noise, then the transformation can be carried out. However, for the type of spectrum which has narrow peaks (such as the strain in a panel excited by jet noise) the correlegram continues to oscillate for a considerable time, and it may be necessary for practical reasons to truncate it before steady zero values have been reached. If this truncated auto-correlegram is transformed, a mathematical filter or spectral window has been effectively introduced owing to the fact that a weighting function has been applied to the correlegram.

4.9.5   Choice of sample length

Each of the quantities discussed above will require a different sample length to get a constant overall accuracy. Generally, the mean square value estimates require the smallest sample length. The same length applies to the auto-correlation at zero time delay, but as the delay is increased the sample length must be increased, as discussed by Blackman & Tukey, to keep the error constant. Where spectral densities of response, rather than inputs, are required, it is generally necessary to use relatively narrow frequency bands, in order to pick out sharp peaks in the spectrum. With a narrow bandwidth it is necessary to increase the integration time to maintain a constant error. Finally, the probability density distribution requires the longest sample, because generally we are interested in estimating the probability of obtaining very high values or amplitudes in the range 3σ to 4σ.

STANDARD REFERENCE TEXTS

Bendat, J. S. & Piersol, A. G. (1971) *Random data: analysis and measurement procedures* Wiley-Interscience.

Blackman, R. B. & Tukey, J. W. (1958) *The measurement of power spectra* Dover.

Newland, D. E. (1975) *An introduction to random vibrations and spectral analysis* Longman.

CHAPTER 5

Random vibration

B. L. Clarkson and **J. K. Hammond**

Institute of Sound and Vibration Research, University of Southampton

5.1 INTRODUCTION

When a mechanical (or electrical, or fluid) system is subjected to continuous random disturbances, the response of the system will also be random. Examples are the motion of an aircraft as it flies through atmospheric turbulence, a car travelling over a rough road, a ship in rough sea, etc. The forces on the system can only be described in terms of their statistical properties, and similarly the response of the system can only be estimated in statistical terms. The statistical descriptors of a continuous random process will be used. These include amplitude distributions, correlation functions, power spectral density, etc. We will consider how these quantities can be applied in the analysis of a mechanical system to quantify its response to forces which are random in time. The power spectral density function will be shown to be of primary importance in this scheme, for although it describes a property of a non-deterministic or random variable, it can still be used within the conventional framework of analysis which usually involves deterministic quantities. The inclusion of random forces in the analysis does, however, preclude a description of the response in deterministic terms. For these cases, the response can only be given in terms of time-averaged quantities which for most engineering purposes is sufficient information.

In considering the response of a linear system to random excitation it is first necessary to summarize a few of the conventional methods of analysis which involve deterministic rather than random variables. For simplicity each of these methods will be developed in connection with the familiar equation of motion governing the response of a damped simple oscillator in the form:

$$m\ddot{x} + c\dot{x} + kx = F(t)$$

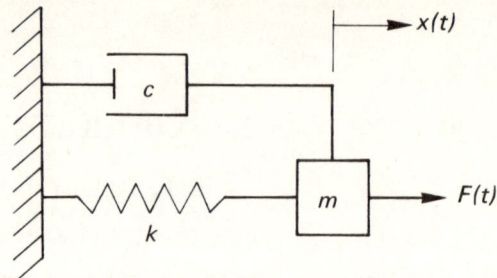

where m = mass, c = viscous damping coefficient, k = stiffness constant, F = forcing function.

5.2 SUMMARY OF DETERMINISTIC FORCE-RESPONSE RELATIONSHIP

5.2.1 Frequency domain

If $F(t) = F_0 e^{i\omega t}$

Response $x(t) = \alpha(i\omega) F_0 e^{i\omega t}$ (5.1)

Receptance $\alpha(i\omega) = \dfrac{1}{(k - m\omega^2 + i\omega c)}$ (5.2)

for a single degree of freedom system.

5.2.2 Time domain

Response $x(t) = \displaystyle\int_0^\infty h(\tau) F(t - \tau) \, d\tau$ (5.3)

where unit impulse response function

$$h(\tau) = \frac{1}{m\omega_n \sqrt{(1 - \zeta^2)}} \exp(-\zeta\omega_n\tau) \sin \omega_n \sqrt{(1 - \zeta^2)}\tau \qquad (5.4)$$

for a single degree of freedom system.

5.3 THE FOURIER INTEGRAL AND SPECTRAL DENSITIES

Another technique used in the analysis of the response of linear systems to transient forces is the Fourier integral method. This is the more general case of the Fourier series method which is used to describe a periodic function in the frequency domain in terms of an infinite series of components with amplitudes determined by the time domain description of the function. Transient or aperiodic functions can be described in a similar manner, but their aperiodicity requires that their frequency domain descriptions be in terms of continuous rather than discrete spectra. Thus, for example, a truncated random signal could be defined in the frequency domain by an appropriate spectrum. Before applying this technique to a particular problem a few preliminary definitions are required. Consider first the pair of Fourier integrals used to relate a real-time function, e.g. $x(t)$, with its corresponding frequency domain representation $X(i\omega)$:

$$
\left.
\begin{aligned}
\text{Fourier transform} \quad X(i\omega) &= \frac{1}{2\pi} \int_{-\infty}^{\infty} x(t)\, e^{-i\omega t}\, dt \\[2em]
\text{Inverse transform} \quad x(t) &= \int_{-\infty}^{\infty} X(i\omega)\, e^{i\omega t}\, d\omega.
\end{aligned}
\right\}
\qquad (5.5)
$$

Any time history of $x(t)$ of *finite* length can be described in terms of a Fourier spectrum. The second integral is finite, however, only if $x(t)$ is zero at $\pm\infty$, so for applications to random variables it is necessary to first consider a finite sample of length T and then allow T to become very large.

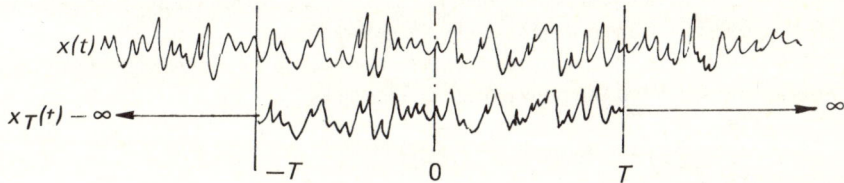

The Fourier transform equation can be written in the form:

$$
X(i\omega) = \lim_{T\to\infty} \frac{1}{2\pi} \int_{-T}^{T} x_T(t)\, e^{-i\omega t}\, dt.
$$

To develop a relationship between the Fourier transform of a random variable and its power spectral density representation, it is convenient to begin with the definition for the mean square value of a particular random variable, $x_T(t)$ as

$$
\overline{x_T^2(t)} = \frac{1}{2T} \int_{-T}^{T} x_T^2(t)\, dt = \frac{1}{2T} \int_{-\infty}^{\infty} x_T(t) x_T(t)\, dt.
$$

Substitution of one of the $x(t)$'s from the inverse transform equation gives:

$$
\overline{x_T^2(t)} = \frac{1}{2T} \int_{-\infty}^{\infty} x_T(t) \int_{-\infty}^{\infty} X_T(i\omega)\, e^{i\omega t}\, d\omega\, dt
$$

$$
\overline{x_T^2(t)} = \frac{\pi}{T} \int_{-\infty}^{\infty} X_T(i\omega)\, X_T^*(i\omega)\, d\omega.
$$

In the limit as $T \to \infty$,

$$\overline{x^2(t)} = \int_{\infty}^{\infty} \lim_{T \to \infty} \frac{\pi}{T} |X_T(i\omega)|^2 \, d\omega. \tag{5.6}$$

Keeping the previous equation in mind, consider next the expression relating the mean square value of a random variable to its power spectral density function:

$$\overline{x^2(t)} = \int_{-\infty}^{\infty} S_x(\omega) \, d\omega. \tag{5.7}$$

A simplified interpretation is as follows. If a random variable is thought of as a summation of an infinite number of infinitesimally small, randomly phased, sinusoidal components of continually distributed frequencies, $S(\omega)$ can then be interpreted as the mean square value of all of these components having angular frequencies within one radian/second bandwidth centred on the angular frequency ω.

Thus for a continuous frequency distribution from $-\infty$ to $+\infty$, the mean square value of the random variable $x(t)$ is equal to the integral of $S(\omega)$ over the entire frequency range. It should be noted that the units of power spectral density are the square of a linear time variable per unit radian.

A comparison of the equations gives the desired relationship between the Fourier spectrum and the power spectral density as

$$S_x(\omega) = \lim_{T \to \infty} \frac{\pi}{T} |X_T(i\omega)|^2. \tag{5.8}$$

This relationship will be used to determine the spectral density of the response of a system to random forces.

5.4 RESPONSE OF A SINGLE DEGREE OF FREEDOM SYSTEM TO A RANDOM FORCE

To apply the Fourier integral method to the case of the randomly excited linear system, consider again the equation which describes the forced motion of a damped simple oscillator:

$$m\ddot{x} + c\dot{x} + kx = F(t) = \int_{-\infty}^{\infty} F(i\omega) \, e^{i\omega t} \, d\omega.$$

The Fourier transform pairs relating the displacement $x(t)$ in the time and frequency domains can be written as

$$x(t) = \int_{-\infty}^{\infty} X(i\omega) \, e^{i\omega t} \, d\omega$$

and

$$X(i\omega) = \frac{1}{2\pi} \int_{-\infty}^{\infty} x(t)\, e^{-i\omega t}\, \mathrm{d}t.$$

Taking the necessary derivatives for the first two terms in the equation $m\ddot{x} + c\dot{x} + kx = F(t)$ gives,

$$\frac{\mathrm{d}x(t)}{\mathrm{d}t} = \int_{-\infty}^{\infty} (i\omega)\, X(i\omega)\, e^{i\omega t}\, \mathrm{d}\omega = i \int_{-\infty}^{\infty} \omega\, X(i\omega)\, e^{i\omega t}\, \mathrm{d}\omega,$$

and

$$\frac{\mathrm{d}^2 x(t)}{\mathrm{d}t^2} = \int_{-\infty}^{\infty} (i\omega)^2\, X(i\omega)\, e^{i\omega t}\, \mathrm{d}\omega = - \int_{-\infty}^{\infty} \omega^2\, X(i\omega)\, e^{i\omega t}\, \mathrm{d}\omega.$$

A transform pair for the force $F(t)$ can also be written as

$$F(t) = \int_{-\infty}^{\infty} F(i\omega)\, e^{i\omega t}\, \mathrm{d}\omega$$

and

$$F(i\omega) = \frac{1}{2\pi} \int_{-\infty}^{\infty} F(t)\, e^{-i\omega t}\, \mathrm{d}t$$

where $F(t)$ could, for example, define a truncated random signal of time duration T.

Rewriting the equation in terms of the transform representations gives

$$\int_{-\infty}^{\infty} X(i\omega)\, \{-m\omega^2 + ic\omega + k\}\, e^{i\omega t}\, \mathrm{d}\omega = \int_{-\infty}^{\infty} F(i\omega)\, e^{i\omega t}\, \mathrm{d}\omega.$$

Since the equality must hold for all values of t, the integrands must be equal, so that

$$X(i\omega)\, \{-m\omega^2 + ic\omega + k\} = F(i\omega)$$

or

$$X(i\omega) = \frac{F(i\omega)}{\{-m\omega^2 + ic\omega + k\}}$$

$$= F(i\omega)\, \underline{\alpha(i\omega)}$$

$$X(i\omega) = F(i\omega)\, \alpha(i\omega). \tag{5.9}$$

It is interesting to note at this point that transforming to the frequency domain has simplified the mathematics to some extent. In the time domain $x(t)$ and $F(t)$ were related through a second-order differential equation, whereas in the frequency domain, $X(i\omega)$ and $F(i\omega)$ are related in a simple algebraic equation. The response of the system in the time domain is obtained by writing the Fourier transform as:

$$x(t) = \int_{-\infty}^{\infty} F(i\omega)\, \alpha(i\omega)\, e^{i\omega t}\, \mathrm{d}\omega. \tag{5.10}$$

If $f(t)$ is a unit impulse, i.e. $F(t) = \delta(t)$,

$$F(i\omega) = \frac{1}{2\pi} \int_{-\infty}^{\infty} \delta(t)\, e^{-i\omega t}\, dt = \frac{1}{2\pi}$$

and

$$x(t) = h(t) = \frac{1}{2\pi} \int_{-\infty}^{\infty} \alpha(i\omega)\, e^{i\omega t}\, d\omega.$$

In the inverse form   (5.11)

$$\alpha(i\omega) = \int_{-\infty}^{\infty} h(t)\, e^{-i\omega t}\, dt$$

and thus it is evident that the unit impulse response function and the complex receptance are Fourier transforms of each other.

To develop a similar expression which relates the power spectral densities of the displacement and force through the receptance, recall

$$S_x(\omega) = \lim_{T\to\infty} \frac{\pi}{T} |X(i\omega)|^2$$

where $S_x(\omega)$ defines the power spectral density of the displacement function $x(t)$. Squaring the expression for $X(i\omega)$ and substituting gives

$$S_x(\omega) = \lim_{T\to\infty} \frac{\pi}{T} |F(i\omega)|^2 |\alpha(i\omega)|^2$$

therefore $S_x(\omega) = S_f(\omega) |\alpha(i\omega)|^2.$ (5.12)

This states that the power spectral density of the displacement is equal to the power spectral density of the force multiplied by the square of the modulus of the receptance. This is the most important result in random vibration theory. Examples of its application are shown in Fig. 5.1. The expression for the spectral density of response is often used to determine the modulus of the receptance of a system by recording the ratio $S_x(\omega)/S_f(\omega)$ over a given frequency range. For the particular case where $S_f(\omega) =$ constant, i.e. for white noise excitation, $|\alpha(i\omega)|^2$ is the directly proportional to $S_x(\omega)$. If receptance and power spectral density of the force are known for a given system, the mean square response can be written as

$$\overline{x^2(t)} = \int_{-\infty}^{\infty} |\alpha(i\omega)|^2\, S_f(\omega)\, d\omega.$$ (5.13)

When both phase and magnitude of the receptance are required, an expression using similar arguments can be derived in the form

$$S_{fx}(\omega) = \alpha(i\omega)\, S_f(\omega)$$ (5.14)

where $S_{fx}(\omega)$ is defined as the cross-power spectral density of the force and displacement. Fig.5.2 shows an example of the use of this result. The importance of this result is that both amplitude and phase of the receptance can be determined, whereas the simpler form of equation (5.13) only contains the square of the modulus of the receptance.

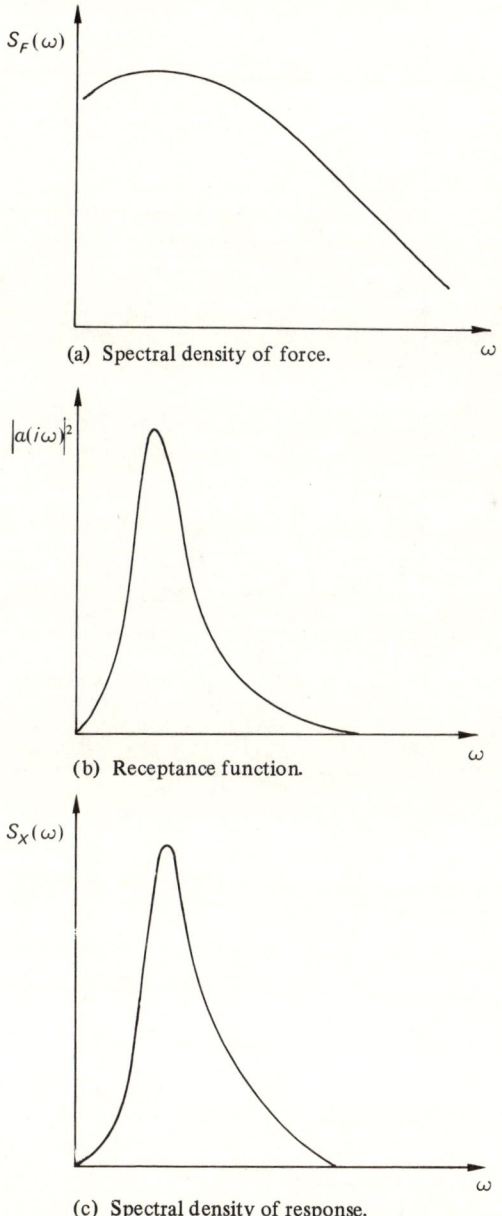

(a) Spectral density of force.

(b) Receptance function.

(c) Spectral density of response.

Fig. 5.1(i) – Spectral density relationships $S_x(\omega) = |\alpha(i\omega)|^2 S_F(\omega)$.

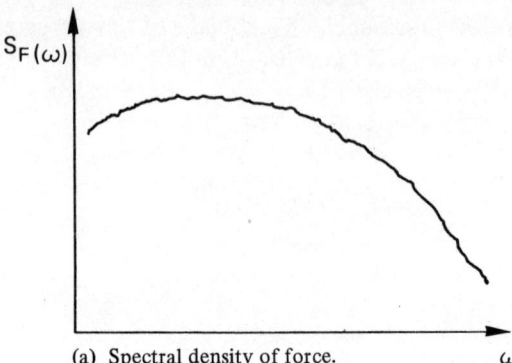

(a) Spectral density of force.

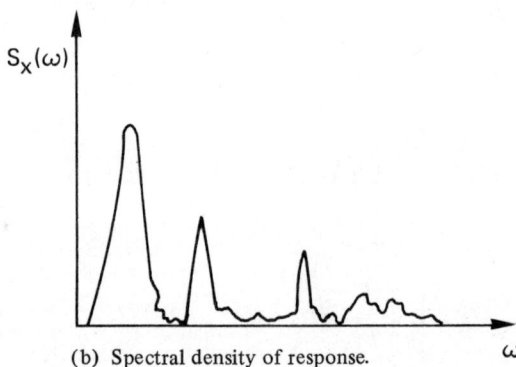

(b) Spectral density of response.

Fig. 5.1(ii) – Response of lightly damped structure to broad band force.

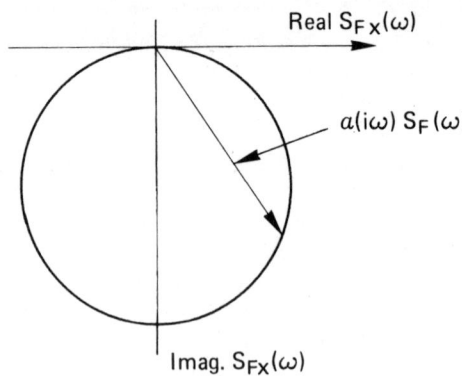

$$S_{FX}(\omega) = a(i\omega) S_F(\omega)$$

Fig. 5.2 – Cross spectral density between the force and displacement of single degree of freedom system.

5.5   RESPONSE OF A SINGLE DEGREE OF FREEDOM SYSTEM TO WHITE NOISE

The examples shown in Fig. 5.1 indicate that for typical lightly damped systems the response spectra contains narrow peaks at the natural frequencies of the system. If these peaks are well separated in frequency a good approximation to the overall response can be obtained by computing the response at each natural frequency separately and then adding the results together. The basic 'building brick' for this process is the response of one degree of freedom to a force which has a flat spectral density in the region of the natural frequency, as indicated in Fig. 5.3.

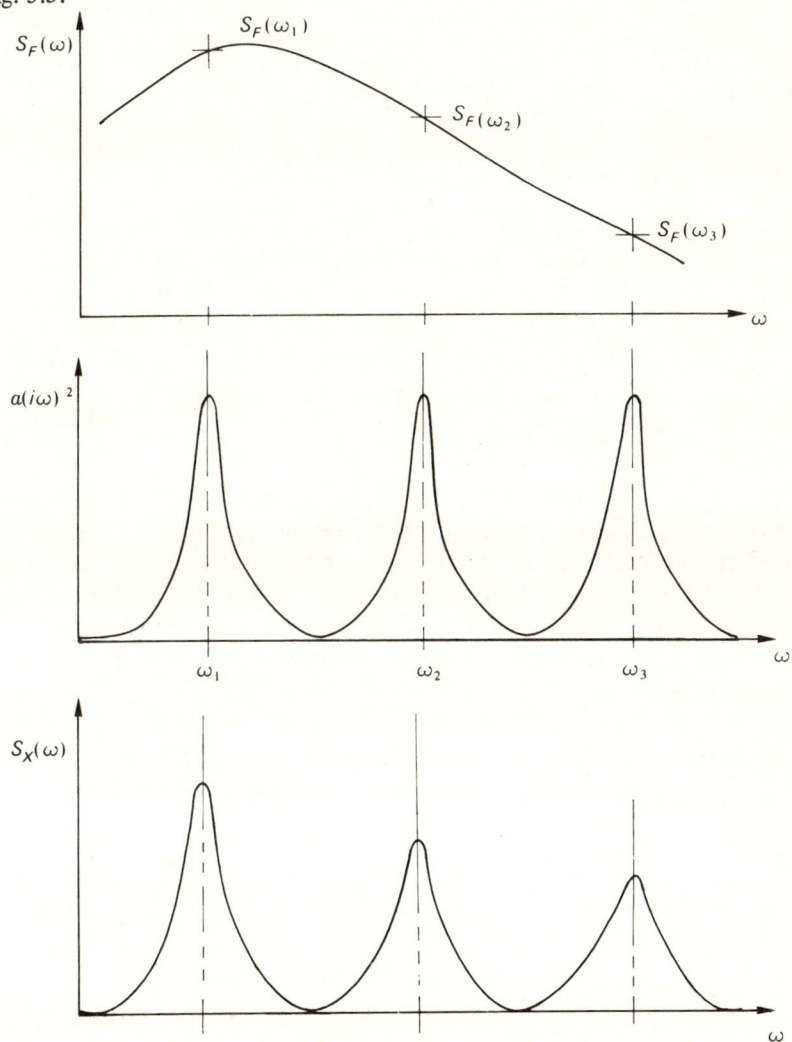

Fig. 5.3 – Response of lightly damped multi degree of freedom system to a broad band random force.

The mean square response to a force having a constant spectral density ('white noise') can be obtained as follows:

$$\overline{x^2(t)} = \int_{-\infty}^{\infty} S_x(\omega)\, d\omega$$

$$= \int_{-\infty}^{\infty} |\alpha(i\omega)|^2 \, S_F(\omega)\, d\omega$$

$$\alpha(i\omega) = \frac{1}{(k - m\omega^2) + i\omega c} \quad \text{and} \quad S_F(\omega) = S_{FW} \quad \text{say.}$$

Using the notation $\omega_n = \dfrac{k}{m} \quad c = 2\zeta\sqrt{(km)}$

$$\overline{x^2(t)} = \frac{S_{FW}}{m^2} \int_{-\infty}^{\infty} \frac{d\omega}{(\omega_n^2 - \omega^2) + 4\zeta^2\,\omega_n^2\,\omega^2}$$

$$= \frac{S_{FW}}{m^2} \frac{\pi}{2\omega_n^3\zeta} \tag{5.15a}$$

$$= \frac{\pi}{2} S_{FW} \frac{\omega_n}{k^2\zeta} \qquad \left.\right\} \text{ alternative forms} \tag{5.15b}$$

$$= \frac{\pi}{2} \frac{\omega_n}{\zeta} S_{FW}\, \delta_0^2 \tag{5.15c}$$

where δ_0 is the static deflection for unit magnitude force.

This analytical result has been derived on the assumption that the spectral density of the force is constant from frequency $-\infty$ to ∞. Crandall has shown that this 'infinite' result is a close approximation to practical situations providing that the bandwidth of the force is wide in comparison with the bandwidth of the system $(2\zeta\omega_n)$.

The time history of the response takes the form of a randomly modulated 'sine' wave.

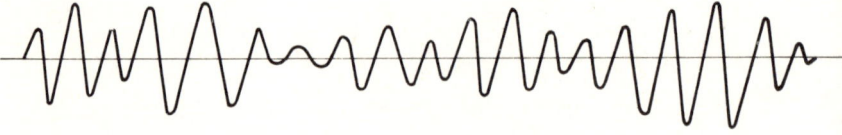

Period is $2\pi/\omega_0$ and the modulation 'period' is proportional to $1/\xi\omega_0$. Therefore for small damping the typical periods in the modulation are very long.

5.6 RESPONSE OF CONTINUOUS SYSTEMS TO RANDOM EXCITATION

The approach used to examine the response of a single degree of freedom system to random excitation can be extended to include the more complex behaviour of real systems. As discussed in Chapter 3 on periodic vibrations, any system having a continuous distribution of mass and stiffness can be represented by an infinite number of free vibration modes called normal modes. Each of these modes can be treated as a single degree of freedom system with mass and stiffness chosen such that the same general vibration characteristics as the continuous system are simulated. The overall response of the continuous system can be approximated by summing the responses of each mode.

Recall that the equation of motion for n^{th} mode was given by

$$M_n \ddot{q}_n(t) + C_n \dot{q}_n(t) + K_n q_n(t) = F_n(t)$$

where $q_n(t)$ is the time varying amplitude of the mode called the normal or generalized coordinates. The quantities M_n, C_n, K_n and F_n are the generalized mass, damping, stiffness, and force respectively, and are all related to their counterparts in the continuous system through the mode shape, ϕ, which is a function of the spatial variable, y.

The single degree of freedom results can now be generalized to apply to any system which has any number of random forces acting upon it. The basic results for the direct and cross-spectral density form the starting point:

Direct spectral density of response $S_x(\omega) = |\alpha(i\omega)|^2 S_F(\omega)$.

Cross spectral density of force and response $S_{Fx}(\omega) = \alpha(i\omega) S_F(\omega)$.

Consider, for example, a beam

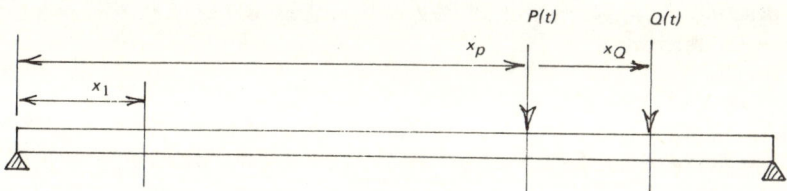

The receptance for the displacement at x_1 due to unit amplitude harmonic force at x_p can be given the notation α_{1P}. Similarly α_{1Q} relates the response at x_1 to a force at x_Q. The actual form of α_{1P} or α_{1Q} may be as shown in the figure below where the peaks represent the response at each of the natural frequencies.

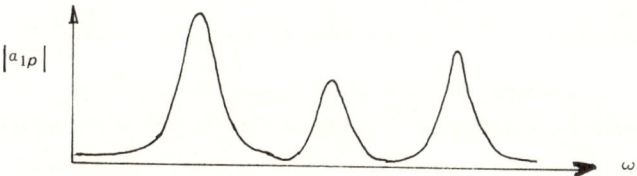

Thus in the normal mode representation α_{1P} will be expressed as the sum of the responses in each normal mode.

It can be shown that the spectral density of the response to two random forces is given by:

$$S_x(\omega) = |\alpha_{xP}(i\omega)|^2 S_p(\omega) + |\alpha_{xQ}(i\omega)|^2 S_Q(\omega)$$
$$+ \alpha_{xP}^*(i\omega)\, \alpha_{xQ}(i\omega)\, S_{PQ}(\omega) + \alpha_{xP}(i\omega)\, \alpha_{xQ}^*(i\omega)\, S_{QP}(\omega).$$
$$(5.16)$$

The first two terms in this sum are those which would be expected from an independent consideration of the effect of each force. The third and fourth term contain the cross-spectral densities of the two forces and effectively take account of any correlation or phase relationship which exists between the two forces. If the forces are statistically independent then $S_{PQ}(\omega) = S_{QP}(\omega) = 0$.

If there are several random forces $P_1, P_2, \ldots, P_n$, the expression for the spectral density of response can be expressed in matrix form as:

$$S_x(\omega) = \begin{bmatrix} \alpha_{xP_1}^*, \alpha_{xP_2}^* \ldots \alpha_{xP_n}^* \end{bmatrix} \times \begin{bmatrix} S_{P_1P_1}(\omega) & S_{P_1P_2}(\omega) & \ldots & S_{P_1P_n}(\omega) \\ S_{P_2P_1}(\omega) & S_{P_2P_2}(\omega) & \ldots & S_{P_2P_n}(\omega) \\ \vdots & \vdots & & \vdots \\ S_{P_nP_1}(\omega) & S_{P_nP_2}(\omega) & \ldots & S_{P_nP_n}(\omega) \end{bmatrix} \times \begin{bmatrix} \alpha_{xP_1} \\ \alpha_{xP_2} \\ \vdots \\ \alpha_{xP_n} \end{bmatrix}$$
$$(5.17)$$

The $[S_{PP}(\omega)]$ matrix contains direct spectral densities of forces $S_{P_2P_2}$ etc. and all the cross-spectral densities $S_{P_3P_2}$ etc. If the forces are all statistically independent, $S_{P_3P_2}$ terms will all be zero, i.e. there will only be terms on the diagonal of the matrix.

5.7 RESPONSE OF A CONTINUOUS SYSTEM TO RANDOM PRESSURES

If the structure is subjected to random pressure loads such as those produced by a high-intensity acoustic field close to a jet engine, beneath a turbulent boundary layer or turbulent pipe flow, it is more convenient to reformulate the problem rather than extend the matrix form given above. The result for the spectral density of response at point $\bar{x}_1$ can be simply expressed in the form:

$$S_{x_1}(\omega) = \int_{\text{Area}} \int_{\text{Area}} \alpha_{1A}(i\omega)\, \alpha_{1B}^* \, S_P(\bar{x}_A \bar{x}_B\, \omega)\, d\bar{x}_A\, d\bar{x}_B \qquad (5.18)$$

$\bar{x}_A\, \bar{x}_B$ are vectors defining points on the surface of the structure.
$S_P(\bar{x}_A \bar{x}_B\, \omega)$ is the cross-spectral density of the pressure at the points $\bar{x}_A$ and $\bar{x}_B$.

$\alpha_{1A}(i\omega)$, $\alpha_{1B}(i\omega)$ are the receptances as before. They will be expressed in the form of sums over the normal modes of the structure as follows:

$$\alpha_{1A}(i\omega) = \sum_r \frac{\phi_r(\overline{x}_1)\,\phi_r(\overline{x}_A)}{M_r} \frac{1}{(\omega_r^2 - \omega^2 + i\eta_r\omega_r^2)} \tag{5.19}$$

where M_r is the generalized mass
$\quad\quad\ \omega_r$ is the natural frequency $\left.\vphantom{\begin{array}{c}1\\1\\1\\1\end{array}}\right\}$ of the rth mode.
$\quad\quad\ \eta_r$ is the loss factor
$\quad\quad\ \phi_r(x)$ is the mode shape

An alternative formulation can be made using the joint acceptance function which was introduced in Chapter 3. This is obtained by substituting (5.19) into (5.18) and rearranging as follows:

$$S_{x_1}(\omega) = S_P(\omega)\,A^2 \sum_{r=1}^{\infty} \sum_{s=1}^{\infty} \frac{\phi_r(\overline{x}_1)\,\phi_s(\overline{x}_1)}{M_r M_s} \frac{j_{rs}^2}{(\omega_r^2 - \omega^2 + i\eta_r\omega_r^2)\,(\omega_s^2 - \omega^2 - i\eta_s\omega_s^2)}$$

$$\tag{5.20}$$

where the joint acceptance function is given by

$$j_{rs}^2 = \frac{1}{S_P(\omega)\,A^2} \int_{\text{Area}} \int_{\text{Area}} \phi_r(\overline{x}_A)\,\phi_s(\overline{x}_B)\,S_P(\overline{x}_A\overline{x}_B\omega)\,\mathrm{d}\overline{x}_A\,\mathrm{d}\overline{x}_B. \tag{5.21}$$

5.8 APPROXIMATE RESULT FOR RESPONSE IN ONE MODE TO RANDOM ACOUSTIC PRESSURES

If we assume that one mode only is providing the majority of the response then we can substituted just one term of equation (5.19) in equation (5.18) to give:

$$S_{x_1}(\omega) = \frac{\phi_r^2(x_1)}{M_r^2} \int_{\text{Area}} \int_{\text{Area}} \phi_r(x_A)\,\phi_r(x_B)\, \frac{S_P(x_A\,x_B\,\omega)}{[(\omega_r^2 - \omega^2) + \eta_r^2\,\omega_r^4]}\,\mathrm{d}x_A\,\mathrm{d}x_B. \tag{5.22}$$

The mean square displacement response y_{x_1} can then be obtained by integrating over the frequency domain

$$\overline{y_{x_1}^2(t)} = \int_{-\infty}^{\infty} S_{x_1}(\omega)\,\mathrm{d}\omega.$$

$$\overline{y^2(t)} = \frac{\pi}{\omega_r^3\,\eta_r}\frac{\phi_r^2(x_1)}{M_r^2} \int_A \int_A \phi_r(x_A)\phi_r(x_B)S_P(x_A\,x_B\,\omega_r)\,\mathrm{d}A\,\mathrm{d}A \tag{5.23}$$

where r is the predominant mode – usually the first mode. This now eliminates the mode summation and simplifies the structure to a set of independent plates responding in their fundamental mode only.

The second major simplification is to assume that the pressures are exactly in phase over the whole plate. The equation then becomes:

$$\overline{y^2(t)} = \frac{\pi}{\omega_r^3 \eta_r} \frac{\phi_r^2(x_1)}{M_r^2} \left[\int_A \phi_r(x_A) \, dA\right]^2 S_P(\omega_r). \tag{5.24}$$

This can be written in terms of the displacement response of the plate to a uniform static pressure of unit magnitude. The static displacement y_0 at x_1 is given by

$$y_0 = \frac{\int_A \phi_r(x_A) \, dA}{\omega_r^2 M_r} \cdot \phi_r(x_1).$$

Thus equation (5.24) can be written as:

$$\overline{y^2(t)} = \frac{\pi}{\eta_r} \omega_r S_P(\omega_r) y_0^2. \tag{5.25}$$

The more usual expression for the mean square stress can be written in terms of the viscous damping ratio ζ and frequency in sycles per second f, as:

$$\overline{\sigma^2(t)} = \frac{\Pi}{2\zeta} f_r S_P(f_r) \sigma_0^2 \tag{5.26}$$

where σ_0 is the stress at the point of interest due to a uniform static pressure of unit magnitude. This expression was first derived from the consideration of a single degree of freedom system by Miles [5.1], as in equation (5.4).

5.9 MULTIPLE INPUT SYSTEMS, RESIDUAL SPECTRA AND PARTIAL COHERENCE

Recently interest has been shown in the application of the concepts of residual spectra and partial and multiple coherence for systems with more than one input. The objective of the remainder of this chapter is to explain the significance and use of some of these concepts, drawing extensively on the work reported by Dodds & Robson [5.13] and Bendat [5.7–5.10] and Bendat & Piersol [5.12]. It is appropriate, however, to begin by noting the definition and significance of the ordinary coherence function for a single input–single output system.

5.9.1 The ordinary coherence function and single input systems

The ordinary coherence function between two stationary random processes $x(t)$ and $y(t)$ is defined as

$$\gamma_{xy}^2(\omega) = \frac{|S_{xy}(\omega)|^2}{S_{xx}(\omega) S_{yy}(\omega)} \tag{5.27}$$

and can be shown to satisfy the inequality

$$0 \leqslant \gamma_{xy}^2(\omega) \leqslant 1.$$

If x and y are completely unrelated $\gamma_{xy}^2(\omega) = 0$, whilst if x and y are linearly related $\gamma_{xy}^2(\omega) = 1$.

As an example consider the linear system below where x denotes the input and y the (noise-contaminated) measurement

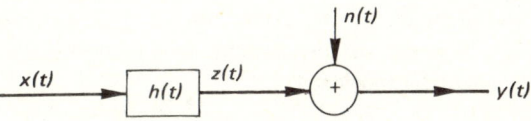

Assuming n to be uncorrelated with x, then

$$\gamma_{xy}^2(\omega) = \frac{1}{1 + S_{nn}(\omega)/S_{zz}(\omega)}$$

showing how the coherence between input and output depends on the output noise/signal ratio.

The 'coherent output power' is that part of the output that is linearly related to x and is given by $S_{zz}(\omega) = \gamma_{xy}^2(\omega) S_{yy}(\omega)$, whilst the noise or 'uncoherent output power' is $S_{nn}(\omega) = [1 - \gamma_{xy}^2(\omega)] S_{yy}(\omega)$. Furthermore, the coherence function may be written $\gamma_{xy}^2(\omega) = S_{zz}(\omega)/S_{yy}(\omega)$, showing that it is a measure of the proportion of the power of y that may be accounted for by linear operation on the input.

Least squares analysis

Before leaving the example above it is useful to look at it rather differently, treating it as the starting point in a system identification problem. More specifically, suppose two measurements, x and y, are available and one wishes to establish a linear transfer characteristic linking the two that account for $y(t)$. Let $y_0(t)$ denote a stationary random process produced by operating on $x(t)$ linearly, i.e.

$$y_0(t) = \int_0^\infty h(\tau) x(t - \tau) \, d\tau$$

and $e(t) = y(t) - y_0(t)$ denote the error. The problem of system identification now becomes that of finding the transfer function $h(t)$ that minimizes $E[e^2(t)]$. This problem can be solved in the time domain using variational methods and results in the equation

$$E[e(t) x(t')] = 0 \quad \text{for } t' < t$$

or, in words, the estimation error is uncorrelated with past observations (the condition $t' < t$ results from the imposed causality condition on $h(t)$). This is called the **Wiener–Hopf** equation and is often written

$$R_{xy}(\tau_1) = \int_0^\infty h(\tau_2) R_{xx}(\tau_1 - \tau_2)\, d\tau_2 \quad \tau_1 > 0,$$

and the solution of this equation for $h(\tau)$ yields the required transfer function. Owing to the added condition $\tau_1 > 0$ the solution cannot be accomplished simply by Fourier transforming the equation, but requires the method of spectral factorization. If, however, the requirement of causality is relaxed, then Fourier transforming the equation yields

$$H(\omega) = \frac{S_{xy}(\omega)}{S_{xx}(\omega)}.$$

The figure above is now a useful conceptual model where $y_0(t)$ is identified with $z(t)$ and the error $e(t)$ is identified with $n(t)$. One can also offer the alternative diagram below (following Dodds & Robson [5.13]) and regard $y(t)$ as being composed of two parts y_2 and y_3. y_2 is fully coherent with y_1, and y_3 is uncoherent with y_1. Furthermore it is easily verified that

$$S_{y_2 y_2}(\omega) = \gamma_{xy}^2(\omega) S_{yy}(\omega)$$

and

$$S_{y_3 y_3}(\omega) = (1 - \gamma_{xy}^2(\omega)) S_{yy}(\omega).$$

$$(5.28)$$

It will be seen later that it is useful to think of y_3 as a residual random variable, i.e., the result of y (or x_2) having had the (linear) effects of x_1 removed in an optimal (least squares) way, and so one might write $y_3 = x_{2.1}$ for this residual variable.

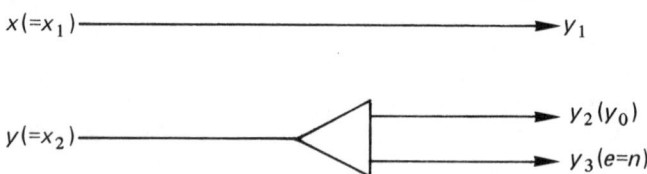

5.9.2 Multiple-input, multiple-output systems

Multi-input multi-output systems may be treated in the same manner as single-input single-output systems, and the results look similar so long as the appropriate vector-matrix interpretation is adopted. (In the figure below a system characterized by the matrix of impulse response functions $H_I(t)$ has m inputs (vector $\mathbf{g}(t)$) and n outputs (vector $\mathbf{x}(t)$), so that

$$x(t) = \int_0^\infty H_I(\tau)\, g(t - \tau)\, d\tau$$

resulting in the spectral input–output relationship

$$S_{xx}(\omega) = H_F^*(\omega)\, S_{gg}(\omega)\, H_F^T(\omega) \qquad (5.29)$$

$S_{xx}(\omega)$, $S_{gg}(\omega)$ are the output and input spectral density matrices respectively (which are Hermitian), and $H_F(\omega)$ denotes the matrix of system frequency response functions. The input–output cross-spectral relationships are

$$S_{gx}(\omega) = S_{gg}(\omega)\, H_F^T(\omega).$$

The equations above are easily derived, but the results themselves require further consideration if we are to use them to reveal details about the system; for example it may be of interest to know the relative importance of the inputs with reference to one particular output, or indeed whether the set of inputs considered is 'adequate'. These and other problems associated with systems with more than one input will now be discussed, but not starting with the full equation (5.29) but rather using the concepts introduced in section 5.9.1. To explain the basic ideas a simple two-input single-output system will be discussed. The equations developed are easily extended to describe more general situations.

5.10 A TWO-INPUT SINGLE-OUTPUT SYSTEM

Suppose three random processes x_1, x_2 and x_3 are measured (where one might choose to interpret one of these as an 'output' y, for example, x_3) and suppose the objective is to identify some form of excitation–response relationship between them. As noted in [5.11], when a single component of a random process is influenced by (many) others a distinct response relationship connecting it with any one of them is likely to be obscured by the action of the remainder. It is only by removing the effects of the remainder in some way that the connection can be established. Suppose x_1, x_2, x_3 are measured (see the figure below) and it is suspected that x_1 and x_2 cause x_3, i.e. (linear) systems S_1, S_2 are believed to link x_1, x_2 to x_3. Suppose further that in reality link S_2 is not present (but this is unknown) and also that both x_1 and x_2 are themselves 'caused' by some fundamental (unmeasured) process θ ensuring that x_1 and x_2 are highly coherent. If the ordinary coherence function $\gamma_{x_2 y}^2(\omega)$ is formed this will be large, tending to 'confirm' the belief in the presence of path S_2. But if the effect that x_1 may have on both x_2 and y is removed before trying to establish a link between x_2 and y, then it would soon be realized that no such link existed. The procedure for effecting this 'removal' is now explained.

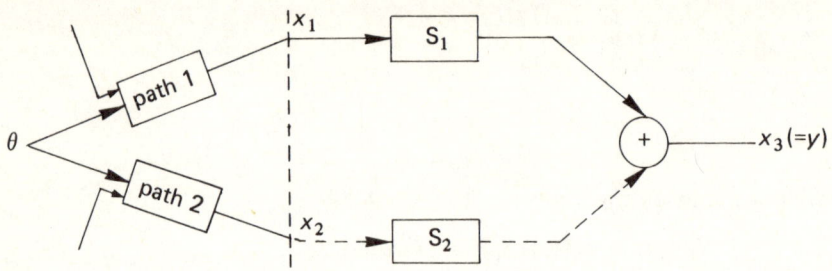

Consider a two-input single-output situation as below (n may be measurement noise or introduced to model non-linearities) and the objective might be to (i) establish the relative magnitudes of noise to 'linear effects' at the output, (ii) to find the relative importance of inputs x_1, x_2, (iii) to estimate the transfer functions H_1, H_2. (It is assumed that the ordinary coherence function $\gamma^2_{x_1 x_2}(\omega)$ is not unity since in this case it is only a single input problem.) (It is noted at the outset that one can easily write down vector matrix expressions as in equation (5.29) involving cross-spectra to estimate H_1, H_2, but this approach will not be followed directly.)

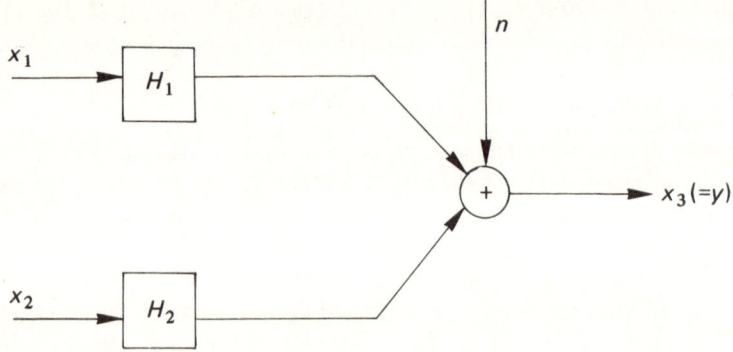

As a starting point it is convenient to decompose the processes involved as shown in the next diagram (following the method of section 5.9.1).

x_2 and x_3 are split into two components each where y_2 and y_4 are fully coherent with y_1 (i.e., those parts of x_2 and x_3 accounted for by optimal linear operations on x_1 through filters L_1, L_2). y_3 is uncoherent with y_1, and is a residual random variable denoted $x_{2.1}$ (similarly for $x_{3.1}$). From these residual random variables one can define residual spectra which are ordinary spectra formed from residual variables, and partial coherences which are ordinary coherences formed from residual variables.

Stage 1   Stage 2

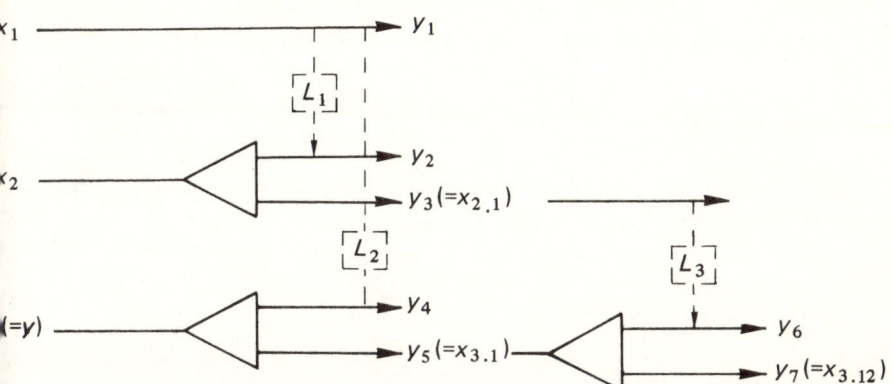

Referring in particular to 'stage 1' in the figure it is noted that x_2 and x_3 have been conditioned with respect to x_1 (this choice is arbitrary), and from the earlier work one can write down the following:

(i) The 'coherent output power', i.e., that proportion of the power of x_2 accounted for by linear operations on x_1 is $\gamma_{12}^2 S_{22}$ [N.B. $S_{22} = S_{x_2 x_2}$, etc.].

(ii) The 'noise' power $S_{y_3 y_3}$ is a residual power spectral density written $S_{22.1}$ and is

$$S_{22.1} = (1 - \gamma_{12}^2) S_{22}.$$

(iii) The optimal linear filter L_1 is S_{12}/S_{11}.

Identical expressions hold for processes x_1, x_3 with 3 replacing 2 above.

It is important to realize that L_1 and L_2 are not H_1 and H_2. In fact since $x_{2.1}$ and $x_{3.1}$ denote what is left after the linear effect of x_1 is removed from x_2 and y it should be clear that

$$H_2 = \frac{S_{23.1}}{S_{22.1}} \qquad \text{(to find } H_1 \text{ reverse the roles of 1 and 2).}$$

It only remains to obtain an expression for the residual cross-spectrum $S_{23.1}$ in terms of x_1, x_2, x_3.

Since

$$S_{23.1} = \lim_{T \to \infty} E\left[\frac{\pi}{T}(X_{2.1}^*(\omega) X_{3.1}(\omega))\right],$$

then using the signal flow algebra of the figure and expressions for L_1, L_2 gives

$$S_{23.1} = S_{23} - \frac{S_{21} S_{13}}{S_{11}}$$

i.e., the residual spectral density is evaluated in terms of ordinary spectra.

These concepts introduced at 'stage 1' can be extended in similar fashion to 'stage 2' where x_3 can be further conditioned, now with respect to x_2. y_6 is that part of $x_{3.1}$ coherent with $x_{2.1}$, and y_7 is that part of x_3 that is uncoherent with both x_1 and x_2 or in other words that part of x_3 not accounted for by linear operations on x_1 and x_2. This means y_7 is the residual random variable formed by conditioning x_3 with respect to both x_1 and x_2 and is written $x_{3.12}$.

At stage 2 (by analogy with stage 1) one can write down the following:

(i) The output noise power spectral density is $S_{33.12} = (1 - \gamma_{23.1}^2) S_{33.1}$ where
(ii) The partial coherence function $\gamma_{23.1}^2$ is

$$\gamma_{23.1}^2 = \frac{S_{23.1} S_{32.1}}{S_{22.1} S_{33.1}}$$

and

(iii) The optimal filter L_3 is $\dfrac{S_{23.1}}{S_{22.1}}$.

The total output power spectral density can be decomposed as the sum of three terms (and this can be done directly from the diagram):

$$S_{33} \quad = \quad \gamma_{13}^2 S_{33} \quad + \quad \gamma_{23.1}^2 S_{33.1} \quad + \quad S_{33.12}$$

| part fully coherent with x_1 | part fully coherent with x_2 after x_1 has been removed from x_2 and x_3 | uncoherent with both x_1 and x_2 |

The multiple coherence function (written $\gamma_{y.x}^2$) is defined by analogy with the ordinary coherence function as for a single-input system, i.e. it is that fraction of power accounted for in the output via linear relationships between input and output, i.e.

$$\gamma_{y.x}^2 = \frac{S_{33} - S_{33.12}}{S_{33}}$$

and by the expressions above can be written

$$\gamma_{y.x}^2 = 1 - (1 - \gamma_{13}^2)(1 - \gamma_{23.1}^2).$$

It is appropriate to emphasize that the multiple coherence function is a measure of how well the two inputs x_1 and x_2 account for the measured response of the system, whilst the partial coherence is a measure of how well an additional signal (in this case x_2) improves the predicted output.

All the formulae above can be related back to the direct and cross-spectra computed from the original three signals and can be applied directly to higher order systems to recover the general formulae given by Bendat [5.8] by simply adding on more 'stages' to the figure for appropriately more inputs and designating higher order residual processes in the obvious way (e.g., $x_{4.123}$).

Finally, with reference to the computation of the transfer functions H_1, H_2, an expression for H_2 has been written down earlier, but the H_i are easily expressed in terms of the L_j (themselves easily written down). By simply following the signal flow algebra in the figure one can write $H_2 = L_3$ and $H_1 = L_2 - L_3 L_1$.

It is appropriate to note (see the discussion following [5.10]) that the manipulations to evaluate the transfer functions amount to a Gauss elimination solution of the input–output cross-spectral equations, but that the approach as outlined here at least offers some physical interpretation in the unravelling of complex problems.

Finally it is pointed out that Bendat [5.9] and Bendat & Piersol [5.12] have given a comprehensive summary of the statistical properties of residual processes.

5.11 AN EXAMPLE

Fig. 5.4 depicts a simple 'synthetic' example constructed to demonstrate some of the above concepts. In order to keep the discussion short only the coherence function will be considered. From Fig.5.4 it is clear that x_1 and x_2 are correlated (the broken lines indicate the way in which the processes are made up), and in Fig. 5.5 this is confirmed with a plot of the ordinary coherence function $\gamma_{12}^2(\omega)$.

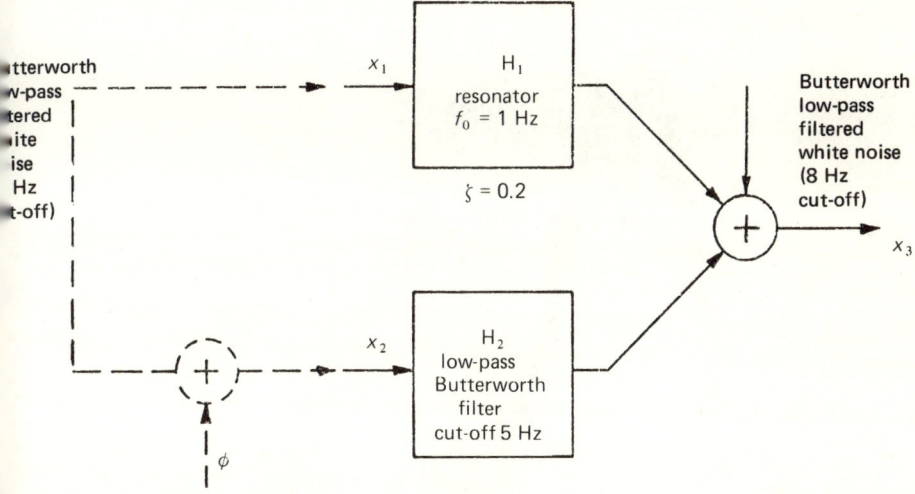

Fig. 5.4 – Two input single output system.

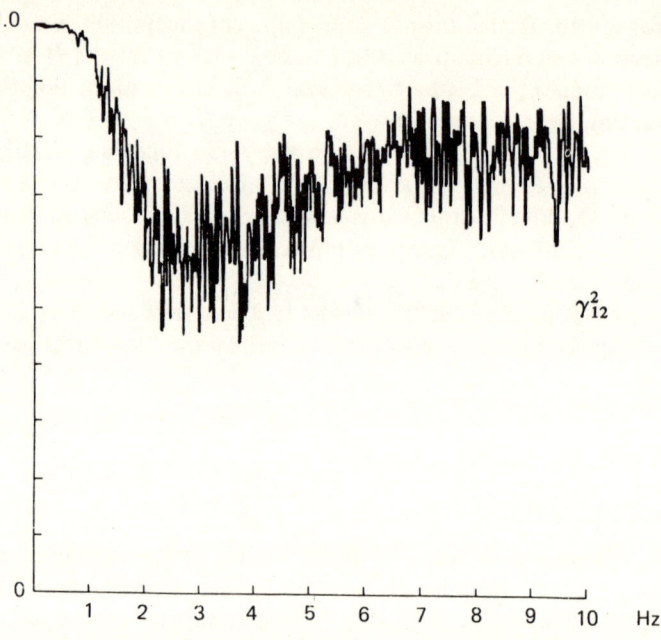

Fig. 5.5 – Ordinary coherence function.

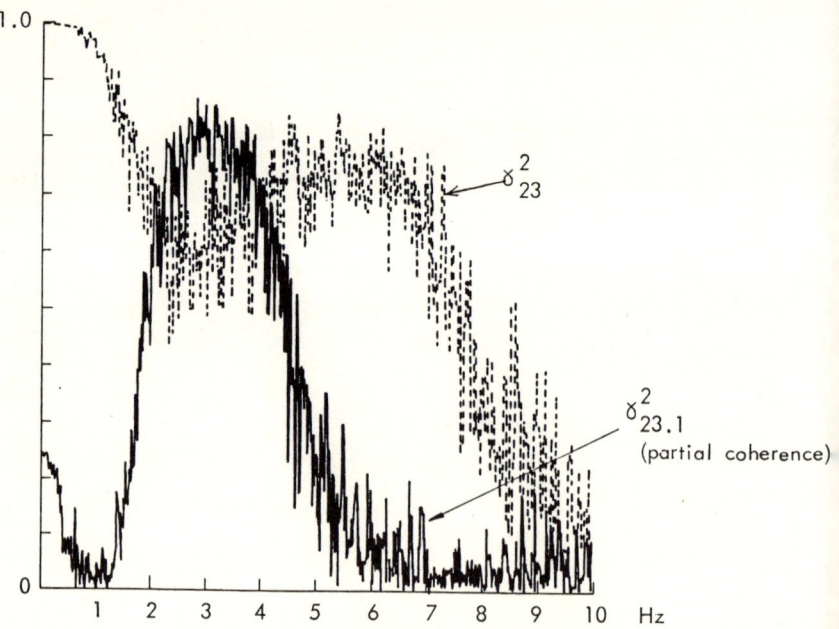

Fig. 5.6 – Ordinary and partial coherence functions.

In Fig. 5.6 the ordinary coherence function $\gamma_{23}^2(\omega)$ (dotted) shows that it is largest at low frequency and near 6 Hz, which might be interpreted as indicating that disturbances at these frequencies originating at x_2 are connected via a significant transmission path through H_2. However, when the effects of x_1 are removed from x_2 and x_3 and the partial coherence function $\gamma_{23.1}^2$ (shown in full line) found, it is immediately apparent that after conditioning with respect to x_1 the frequency range of any importance in the link x_2 to x_3 for disturbances originating at x_2 is from about 2 Hz to 4 Hz, confirming that the high coherence in $\gamma_{23}^2(\omega)$ is due to the fundamental source θ common to both x_1 and x_2, and not due to a source originating at x_2. This example serves to show that residual spectra can very simply help clarify situations that might otherwise be misleading.

5.12 REFERENCES AND SUGGESTED FURTHER READING

[5.1] Miles, J. W. (1954), On Structural Fatigue under Random Loading, *J. Aero Sci.* **21**, 753-762.

[5.2] Robson, J. D. (1963), *An introduction to random vibration.* Edinburgh University Press.

[5.3] Crandall, S. H. & Mark, W. D. (1963), *Random vibration in mechanical systems.* Academic Press.

[5.4] Newland, D. E. (1975), *An introduction to random vibrations and spectral analysis.* Longman.

[5.5] Richards, E. J. & Mead, D. J. (1968), *Noise and acoustic fatigue in Aeronautics.* J. Wiley.

[5.6] Price, W. G. & Bishop, R. E. D. (1974), *Probabilistic theory of ship dynamics.* Chapman & Hall.

[5.7] Bendat, J. S. (1976) *Journal of Sound and Vibration* **49(3)**. System identification from multiple input/output data.

[5.8] Bendat, J. S. (1976) *Journal of Sound and Vibration* **44(3)**. Solutions for the multiple input/output problem.

[5.9] Bendat, J. S. (1978) *Journal of Sound and Vibration* **59(3)**. Statistical errors in measurement of coherence functions and input/output quantities.

[5.10] Bendat, J. S. (1977), *Stochastic problems in dynamics,* Ed. Clarkson, B. L. 'Procedures for frequency decomposition of multiple input/output relationships'. Pitman Pub. Ltd.

[5.11] Bendat, J. S. & Piersol, A. G. (1971), *Random data: analysis and measurement procedures.* J. Wiley.

[5.12] Bendat, J. S. & Piersol, A. G. (1980), *Engineering applications of correlation and spectral analysis.* J. Wiley.

[5.13] Dodds, C. J. & Robson, J. D. (1975) *Journal of Sound and Vibration* **42(2)**. Partial coherence in multivariate random processes.

[5.14] Otnes, R. K. & Enochson, L. (1978), *Applied time series analysis,* Vol. 1, Basic Techniques. J. Wiley.

CHAPTER 6

Data analysis

J. K. Hammond

Institute of Sound and Vibration Research, University of Southampton

6.1 INTRODUCTION

Data analysis involves the three phases of data acquisition, processing, and interpretation (of the results of the processing), and of course all three are linked in any application. However, the last phase is naturally very much related to the subject under investigation, but the first two may be discussed independent of specific applications. A vast body of theory and methodology has been built up as a consequence of the problems raised by the need for data analysis often referred to as 'signal processing' or 'time series analysis' depending on the context. This chapter is concerned with methods of analysis of time histories which have usually been obtained as measurements in the course of conducting an experiment or when recording the behaviour of a physical system. We shall limit our considerations to 'basic procedures', and inevitably the treatment will be superficial owing to the restricted scope a single chapter can afford, but the references listed should serve most requirements.

Types of data to be analysed

Fig. 6.1 broadly categorizes the time histories that may arise [6.3]. A basic distinction is the designation of a signal as 'random' or 'deterministic', where by 'random' we mean one that is not predictable exactly. Very often processes are mixed and the demarcations of Fig. 6.1 are not easily applied, and consequently the analysis procedure to be used may not be apparent. Even if we restrict data to the categories listed, the two classes of data called 'nonstationary random' and 'almost periodic deterministic' are often difficult to handle, and we shall comment briefly on these topics later in the chapter.

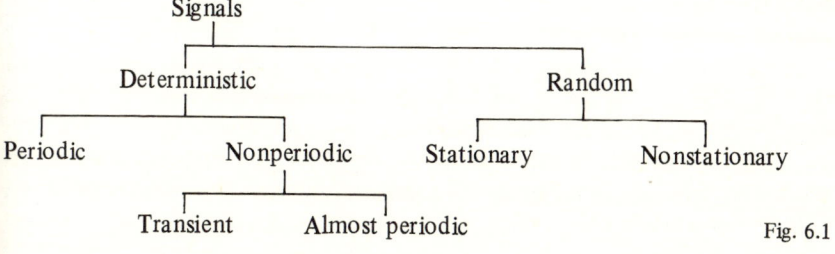

Fig. 6.1

Objectives of data analysis

Broadly speaking, the objective of data analysis is to highlight/extract information contained in a signal that direct observation may not reveal. Sometimes only 'gross' characteristics of a signal may be required (e.g., mean value, mean square level, total energy) in which case some simple processing will produce the required quantities whilst ignoring other information contained in the signal which might only unnecessarily confuse matters. At other times it may be necessary to probe more deeply in an attempt to extract features of interest.

Methods of data analysis

The signals that are processed are time histories (often electrical outputs from transducers), or equivalent (e.g., rough ground height profiles where a space variable is the equivalent of time) and useful information extraction procedures may be performed whilst still in the 'time domain'. However, very often transform methods are employed since they offer simplifications both in theory and physical interpretation, and 'frequency domain methods' in particular are commonly used.

The signals recorded from physical phenomena are generally 'continuous time' in nature (sometimes called analogue data), but developments in digital computation methods have now ensured that most signal processing is achieved either on special purpose analysers or on larger (e.g., time-shared, interactive) digital systems. This means the data is sampled, and the analysis of discrete (in time) data requires that the analyst should have a good grasp of the fundamentals of digital signal processing, which is now a vast area of study (see, for example [6.10, 6.11]), and in this chapter we shall emphasize digital signal processing methods. It is appropriate to begin by briefly discussing the digitization of continuous data.

6.2 ANALOGUE-TO-DIGITAL CONVERSION

An analogue-to-digital converter (ADC) is a device that operates on a continuous time history (input) and produces a sequence of numbers (output) that are sample values of the input. It is convenient to regard the process as consisting of two stages, namely 'sampling' and 'quantization'. In Fig. 6.2 $x(n\Delta)$ is the exact value attained by time history $x(t)$ at time $t = n\Delta$ (Δ is the sample interval for uniform rate sampling). $\hat{x}(n\Delta)$ is the representation of $x(n\Delta)$ in a computer and differs from $x(n\Delta)$ since a finite number of bits are used to represent each number. Actual ADCs do not consist of two separate components (as in the 'conceptual' (Fig. 6.2)), and a variety of types are available. One widely used type of ADC is that based on successive approximations (see Fig. 6.3). The digital word in the register is converted to an analogue signal and compared with $x(t)$. The error (which should be as small as possible) is used to update the digital word. The most significant bit in the register is adjusted first, working on down to the least significant. See refs [6.1, 6.9] for additional details on ADCs.

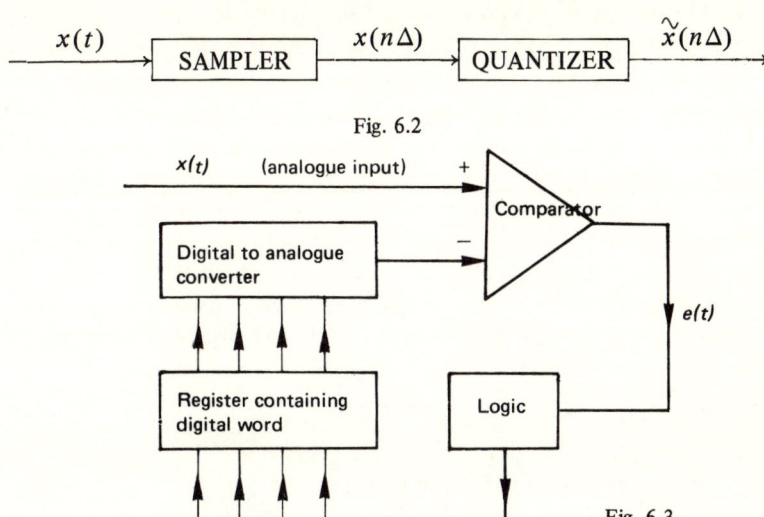

Fig. 6.2

Fig. 6.3

The problem of quantization error is treated in $[6.10, 6.11, 6.4, 6.9]$. Referring to Fig. 6.2, the output $\tilde{x}(n\Delta)$ can be written as $\tilde{x}(n\Delta) = x(n\Delta) + e(n\Delta)$, and finite word length effects are described by treating $e(n\Delta)$ as 'noise'. The probability distributions describing the error depend on the particular way in which the quantization is done. If the ADC introduces rounding (as opposed to truncation) error, then the error is ascribed a uniform distribution (with zero mean) over one quantization step and furthermore is assumed to be stationary and 'white'. For a b-bit word length (excluding sign), and if the full range of the ADC corresponds to X volts, say, then the variance of the error $\sigma_e^2 = \left(\frac{X}{2^b}\right)^2 / 12$.

Further, if x is assumed to be a random signal (see $[6.10]$) and σ_x^2 denotes the variance of $x(n\Delta)$, then a measure of signal-to-noise ratio is $S/N = 10 \log_{10}\left(\frac{\sigma_x^2}{\sigma_e^2}\right)$.

If $\sigma_x = X/4$ (to 'avoid' clipping) then $S/N = 6b - 1.24$ dB, giving a value of about 70 dB for a 12-bit ADC. This is reduced further by practical considerations (see $[6.4]$).

It is appropriate to note that the rate at which $x(t)$ is sampled should be high enough to avoid aliasing (see section 6.3). This usually means that $x(t)$ should be low-pass filtered using analogue filters preceding the ADC. The cut-off frequency and cut-off rate of these 'anti-alias' filters should be chosen with particular applications in mind, but very roughly speaking if the 3 dB point is a quarter of the sampling frequency and the cut-off rate better than 48 dB/octave, then this at least ensures a 40–50 dB reduction at the folding frequency (half the sampling frequency), and so probably an acceptable level of aliasing (though we must emphasize that this may not be adequate for some applications).

6.3 DETERMINISTIC DATA – FOURIER METHODS

In Fig. 6.1 we defined three types of deterministic signal, namely periodic, transient, and almost periodic. Periodic data is analysed using the Fourier series, transient data uses the Fourier integral, and some classes of almost periodic signals (and also damped signals) may be studied using the Prony series [6.24]. The basic difference between the Prony series and the Fourier series is that in the former case the frequencies are not assumed known, and so both amplitude and frequency are computed, while only amplitudes (including phase) are calculated at particular frequencies using the Fourier method. The Prony method is computationally more 'difficult' but can be very attractive under certain circumstances (e.g., where data lengths are short). We shall concentrate on Fourier methods in this section.

6.3.1 The Fourier series and Fourier integral

If a signal repeats itself exactly every T_p seconds then $x(t)$ may be represented as a sum of sines and cosines written in complex form as

$$x(t) = \sum_{n=-\infty}^{\infty} C_n \, e^{j2\pi nt/T_p} \tag{6.1}$$

whilst if $x(t)$ is a transient then the discrete set of sines and cosines becomes a continuum and we write

$$x(t) = \int_{-\infty}^{\infty} X(f) \, e^{j2\pi ft} \, df \tag{6.2}$$

The specification of C_n or $X(f)$ is equivalent to the original time history. For example, $X(f)$ is given by

$$X(f) = \int_{-\infty}^{\infty} x(t) \, e^{-j2\pi ft} \, dt. \tag{6.3}$$

An important consideration when calculating the transform $X(f)$ is that of data truncation (or windowing).

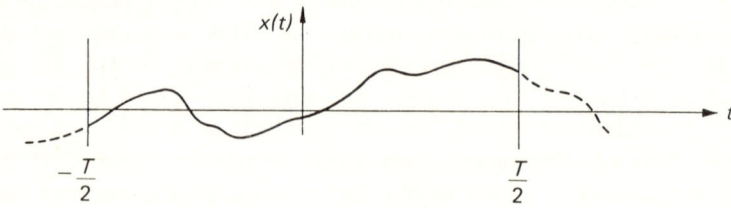

Fig. 6.4

The infinite integral (6.3) may be replaced by a finite one because, for instance, $x(t)$ may only be known (measured) for a limited duration, say for T seconds, and we might define

$$x_T(t) = x(t), \quad |t| \leqslant T/2$$
$$= 0, \quad \text{elsewhere.}$$

x_T is a truncated version of $x(t)$, and from Fig. 6.4 we can think of this cutting off the tails of the data as though we are looking at the data through a 'data window' $w(t)$ where

$$w(t) = 1 \quad |t| \leqslant T/2$$
$$= 0 \quad \text{elsewhere}$$

so that $x_T(t) = x(t)w(t)$.

Fourier transforming $x_T(t)$ gives $X_T(f)$ which is related to the Fourier transforms of x and w by

$$X_T(f) = \int_{-\infty}^{\infty} X(g)W(f-g)\,dg \tag{6.4}$$

where in turn $W(f)$ is sometimes called a 'spectral window'.

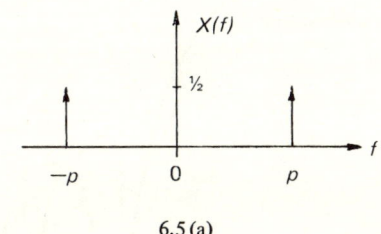

6.5 (a)

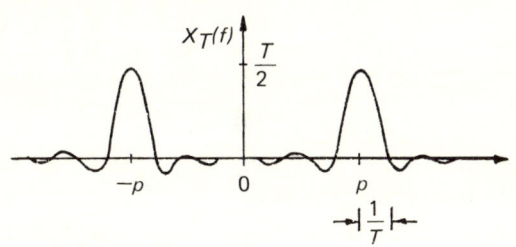

6.5 (b)

Fig. 6.5

The convolution integral (6.4) indicates that $X_T(f)$ is a distorted version of $X(f)$ (unless $W(f)$ is a delta function). This effect is simply demonstrated with a truncated cosine wave. If $x(t) = \cos 2\pi p t$ then

$$X(f) = \tfrac{1}{2}[\delta(f + p) + \delta(f - p)]$$

(Fig. 6.5(a)) and

$$X_T(f) = \tfrac{1}{2}[W(f + p) + W(f - p)]$$

(Fig. 6.5(b), for a rectangular data window). The appearance of energy at frequencies other than p is sometimes called leakage. If two frequencies are present, say p_1, p_2, then in order to be able to get two distinguishable peaks it is necessary for the data length T to (approximately) satisfy $T \geqslant 2/(p_2 - p_1)$ for a rectangular window. The rectangular window is a poor window since the side lobes are large and do not decay rapidly. Numerous alternatives have been suggested to reduce leakage in Fourier transform calculations. By tapering the windows to zero, the side lobes are reduced but at the expense of widening the main lobe (increasing smearing). The frequency domain behaviour of windows is often given in decibels (dB) where the characteristics have been normalized to unity gain (0 dB) at zero frequency, and some terms are defined in Fig. 6.6 which relates to the rectangular window.

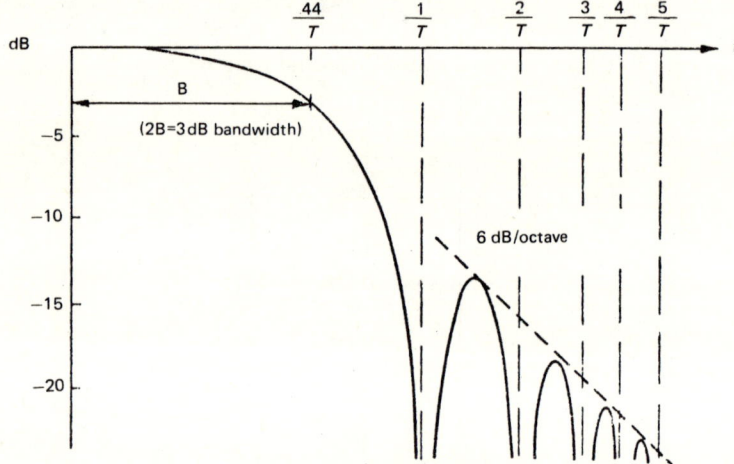

Fig. 6.6

The properties of a few commonly used windows are summarized below (see [6.9] for the appropriate formulae).

Window (length T)	Highest side lobe (dB)	3 dB bandwidth	Asymptotic roll off (dB octave)	First zero crossing
Rectangle	-13	$0.88\ 1/T$	6	$1/T$
Bartlett	-26	$1.25\ 1/T$	12	$2/T$
Hann(ing) or Tukey	-32	$1.4\ \ 1/T$	18	$2/T$
Hamming	-42	$1.3\ \ 1/T$	6	$3/T$

The paper by Harris [6.18] contains a comprehensive treatment of windows and their effects.

6.3.2 The Fourier transform of a sequence

Since the calculations of Fourier coefficients and transforms are usually carried out on a digital computer it is necessary to re-examine the preceding work for data in discrete form. We might treat a discrete version of (6.3), for example, as an approximation to the 'correct' result. However, completely rigorous, exact theories can be developed for discrete data which are, of course, closely related to the continuous time counterparts, but there are fundamental differences, and it is often more convenient to consider the problems of discrete data as self-contained and relate the results to those of the continuous domain only as and when they are required, accounting at that stage for any errors or approximations that might have been incurred in using discrete methods in place of the continuous operations. Having said this we will, nevertheless, attempt to relate the Fourier analysis of continuous and discrete data.

Impulse train modulation

One way of introducing the Fourier transform of a sequence involves the use of the mathematical notion of 'ideal sampling' of a continuous wave. If an analogue signal $x(t)$ is to be sampled every Δ seconds it is convenient to model the sampled signal as the product of the continuous signal with a 'train' of delta functions $i(t)$ where

$$i(t) = \sum_{n=-\infty}^{\infty} \delta(t - n\Delta), \quad \text{i.e.,} \quad x_s(t) = x(t)i(t).$$

The Fourier transform of $x_s(t)$ is $X_s(f)$, and using properties of the delta function, it is written

$$X_s(f) = \sum_{n=-\infty}^{\infty} x(n\Delta) e^{-j2\pi f n\Delta}. \tag{6.5}$$

Multiplying both sides by $e^{j2\pi f r\Delta}$ and integrating with respect to f from $-1/2\Delta$ to $1/2\Delta$ gives the inverse

$$x(n\Delta) = \Delta \int_{-1/2\Delta}^{1/2\Delta} X_s(f) e^{j2\pi f n\Delta} \, df. \tag{6.6}$$

Equations (6.5) and (6.6) relate the *sequence of numbers* $x(n\Delta)$ to the quantity $X_s(f)$ which is termed the Fourier transform of the sequence.

A natural question to ask at this stage is 'How is $X_s(f)$ (or as it is often written, $X(e^{j2\pi f\Delta})$) related to $X(f)$?'

The answer to this can be obtained formally as follows. Since $i(t)$ is periodic we might represent it with a Fourier series, i.e.,

$$i(t) = \frac{1}{\Delta} \sum_{n=-\infty}^{\infty} e^{(2\pi j n t)/\Delta}.$$

Now substituting this into $x_s(t) = x(t)i(t)$ and Fourier transforming gives an alternative right-hand side to (6.5), namely,

$$X(e^{j2\pi f\Delta}) = \frac{1}{\Delta} \sum_{n=-\infty}^{\infty} X(f - (n/\Delta)). \tag{6.7}$$

This important equation relates the Fourier transform of a continuous signal and the Fourier transform of the sequence formed by sampling the signal at equispaced intervals. It will be discussed further in the explanation of 'aliasing'.

The z-transform

An alternative route to equation (6.5) is to use the notion of the z-transform of a sequence (widely used in the solution of difference equations). The z-transform of a sequence of numbers $x(n)$, say, (where the notion of time is not made explicit) is

$$X(z) = \sum_{n=-\infty}^{\infty} x(n)z^{-n} \tag{6.8}$$

z is interpreted as a complex number and $X(z)$ a function of a complex variable. If z is allowed to take values on the unit circle in the z plane, i.e., $z = e^{j\omega} = e^{j2\pi f}$, then (6.8) is

$$X(e^{j2\pi f}) = \sum_{n=-\infty}^{\infty} x(n)e^{-j2\pi fn}$$

which is seen to correspond identically with (6.5) if the sample interval Δ is 1.

6.4 ALIASING

Equation (6.7) describes how the frequency components of the sampled waveform relate to that of the continuous waveform, and Fig. 6.7 explains this pictorially. In Fig. 6.7(a) it is assumed that $X(f) = 0$ for $|f| > f_0$ (say), and Fig. 6.7(b) is a plot of $\Delta X(e^{j2\pi f\Delta})$ assuming that the sampling rate $f_s = 1/\Delta$ is such that $f_0 < 1/2\Delta$. Some commonly used terms are defined on the diagram.

Suppose now $f_0 > f_s/2$, then there is an overlapping of the shifted versions of $X(f)$ resulting in a distortion of the frequency description for $|f| < 1/2\Delta$ as in Fig. 6.7(c). This 'distortion' is due to the fact that high frequencies in the data are indistinguishable from lower frequencies owing to the sampling rate not being fast enough. More specifically, consider $x(t) = \cos 2\pi pt$ sampled at $f_s = 1/\Delta$, i.e., at $t = n\Delta$ (with, say, $p < 1/2\Delta$). This may be written

$$x(n\Delta) = \frac{1}{2}(e^{j2\pi n\Delta p} + e^{-j2\pi n\Delta p}).$$

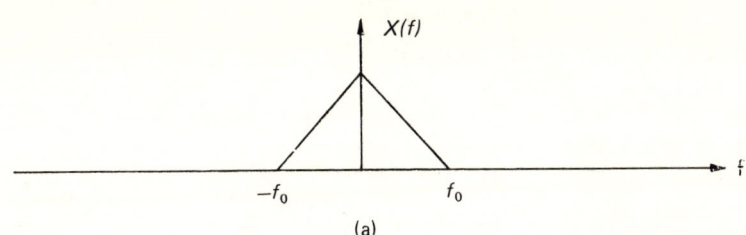

(a)

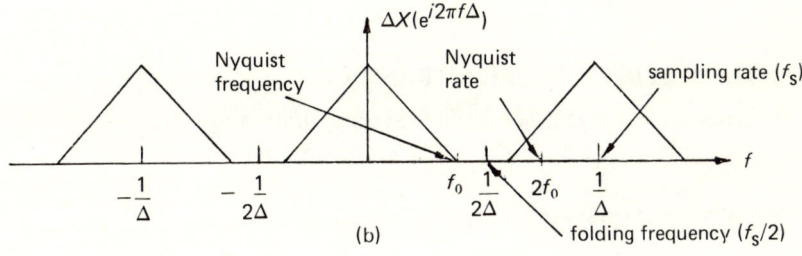

(b)

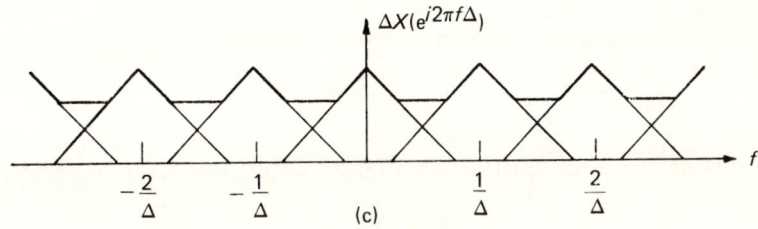

(c)

Fig. 6.7

Now letting frequency p above be replaced by $\pm p + (k/\Delta)$ $(k = 1, 2, \ldots,)$ results in an identical sequence. So if a frequency component were detected at p Hz, any one of these higher frequencies might have been responsible for this rather than a 'true' component at p Hz. This phenomenon is called **aliasing**. The values $\pm p + (k/\Delta)$ can be seen graphically for some p between 0 and $1/2\Delta$ by 'pleating' the frequency axis as in Fig. 6.8.

To avoid aliasing, the sample rate must be chosen to be greater than twice the highest frequency contained in the signal. (See the discussion in section 6.2 regarding 'anti-alias' filters.)

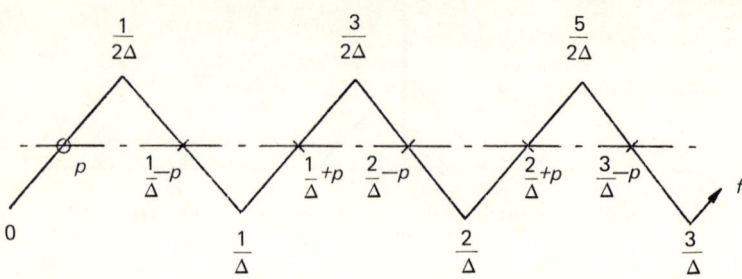

Fig. 6.8

6.5 THE DISCRETE FOURIER TRANSFORM

The equation (6.5) is still not a practical proposition, and a finite sum would be more appropriate which, for an N-point sequence, could be written

$$\sum_{n=0}^{N-1} x(n\Delta)e^{-j2\pi fn\Delta}.$$

Obviously, if this involved truncating the data there would be some distortion explained earlier by the discussion on windowing. Assuming for the moment that we can ignore this, we can think of simplifying our task by only evaluating f at specific points, say at $k/N\Delta$ Hz, $k = 0, ..., N-1$. Then

$$X(e^{j2\pi k/N}) = \sum_{n=0}^{N-1} x(n\Delta)e^{-j(2\pi/N)nk}.$$

Or, adopting the more usual notation (and omitting the Δ for convenience),

$$X(k) = \sum_{n=0}^{N-1} x(n)e^{-j(2\pi/N)nk}. \tag{6.9}$$

Multiplying both sides by $e^{j(2\pi/N)mk}$ and summing over k gives

$$x(n) = \frac{1}{N}\sum_{k=0}^{N-1} X(k)e^{j(2\pi/N)nk}. \tag{6.10}$$

The pair (6.9) and (6.10) constitute the Discrete Fourier Transform (DFT) and are the form that are suitable for machine computation.

It is important to realize that even though the original sequence $x(n\Delta)$ may have been zero for n outside the range $0 \to N-1$, the act of 'sampling in frequency' has imposed a *periodic structure* to the sequence, i.e., it follows from (6.10) that $x(n + N) = x(n)$.

We are now in a position to summarize the various Fourier transforms we have defined and their differences, and this is done in Fig. 6.9 where we follow the method of display given in [6.5].

Fourier series

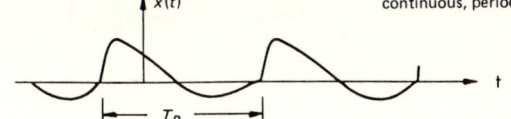

$$x(t) = \sum_{n=-\infty}^{\infty} C_n e^{j2\pi nt/T_P}$$

$$C_n = \frac{1}{T_p} \int_0^{T_P} x(t) e^{-j2\pi nt/T_P} \, dt$$

Fourier integral

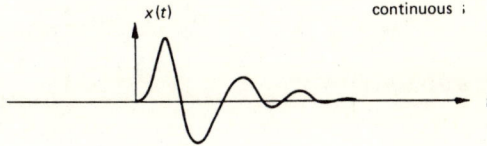

$$x(t) = \int_{-\infty}^{\infty} X(f) e^{j2\pi ft} \, df$$

$$X(f) = \int_{-\infty}^{\infty} x(t) e^{-j2\pi ft} \, dt$$

Fourier transform of a sequence

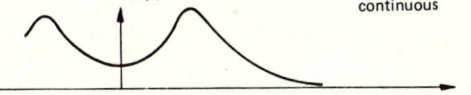

$$x(n\Delta) = \Delta \int_{-1/2\Delta}^{1/2\Delta} X(e^{j2\pi f\Delta}) e^{j2\pi fn\Delta} \, df$$

$$X(e^{j2\pi f\Delta}) = \sum_{n=-\infty}^{\infty} x(n\Delta) e^{-j2\pi fn\Delta}$$

Discrete Fourier transform

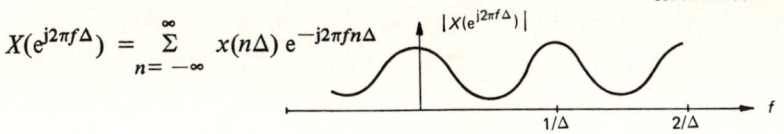

$$x(n) = \frac{1}{N} \sum_{k=0}^{N-1} X(k) e^{j(2\pi/N)nk}$$

$$X(k) = \sum_{n=0}^{N-1} x(n) e^{-j(2\pi/N)nk}$$

Fig. 6.9

Fourier transform of original data

data $x(t)$

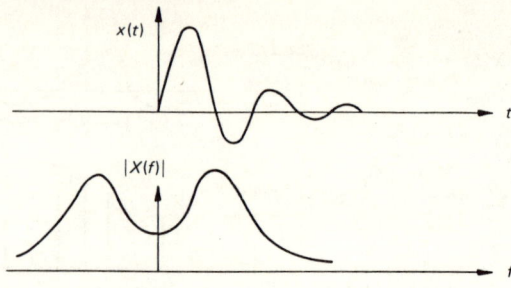

$$X(f) = \mathcal{F}[x(t)]$$

Fourier transform of truncated data

$$x_T(t) = x(t)w(t)$$

$w(t)$ is a data window

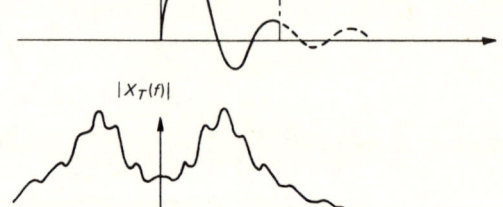

$$X_T(f) = \mathcal{F}[x_T(t)]$$

Fourier transform of sampled, truncated data

$x_T(t)$ is sampled
every Δ seconds

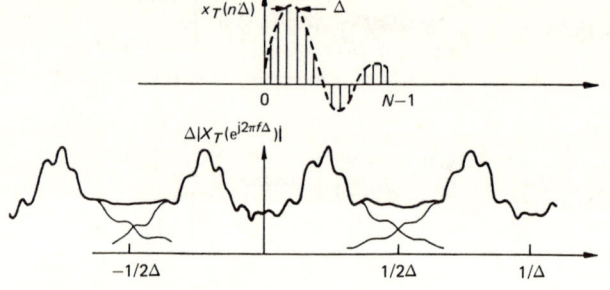

DFT of sampled, truncated data

$X_T(e^{j2\pi f\Delta})$ is sampled
every $1/N\Delta$ Hz

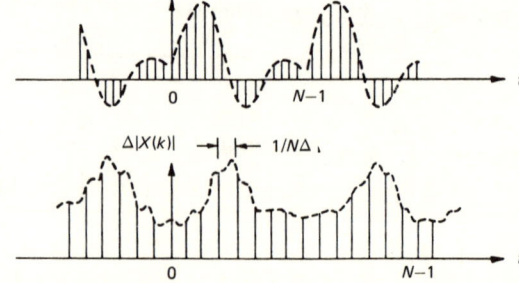

Fig. 6.10

In addition it is instructive to follow (diagramatically) the interpretation of a DFT of an aperiodic function that is truncated, as in Fig. 6.10 (see also ref [6.8]).

6.6 THE FAST FOURIER TRANSFORM

The evaluation of the DFT can be accomplished efficiently by a set of algorithms known as the fast Fourier transform (FFT). This section briefly outlines the basic 'decimation in time' method for a radix 2 FFT and follows references [6.10] and [6.11].

The radix 2 FFT

Writing $e^{-j(2\pi/N)}$ as W_N then the DFT is

$$X(k) = \sum_{n=0}^{N-1} x(n) W_N^{nk}.$$

The FFT algorithms essentially exploit the periodicity and symmetry properties of W_N^{nk}, and so reduce the number of multiply and add opperations needed to calculate $X(k)$ from about N^2 to approximately $N \log_2 N$ which for $N = 1024$ is a reduction by a factor of about 100.

We shall take N to be a power of two $(= 2^\nu)$ and so obtain a base-2 or radix-2 algorithm. As the name 'decimation in time' suggests, the sequence $x(n)$ is successively decomposed into smaller subsequences. In fact $N/2$ point sequences $x_1(n)$, $x_2(n)$ can be defined as the even and odd members of $x(n)$, so that

$$X(k) = \sum_{n=0}^{\frac{N}{2}-1} x(2n) W_N^{2nk} + \sum_{n=0}^{\frac{N}{2}-1} x(2n+1) W_N^{(2n+1)k}.$$

Then noting that $W_N^2 = W_{N/2}$ this can be written

$$X(k) = \sum_{n=0}^{\frac{N}{2}-1} x_1(n) W_{N/2}^{nk} + W_N^k \sum_{n=0}^{\frac{N}{2}-1} x_2(n) W_{N/2}^{nk}$$

or $$X(k) = X_1(k) + W_N^k X_2(k),$$

where $X_1(k)$, $X_2(k)$ are $N/2$ point DFTs of $x_1(n)$, $x_2(n)$ whose 'usual' properties are used to calculate $X(k)$ for $0 \leqslant k \leqslant N-1$.

Some signal flow graph terminology is defined in Fig. 6.11(a), and Fig. 6.12 shows the first decomposition for the case $N = 8$, showing the two 4-point transforms making up the full 8-point transform.

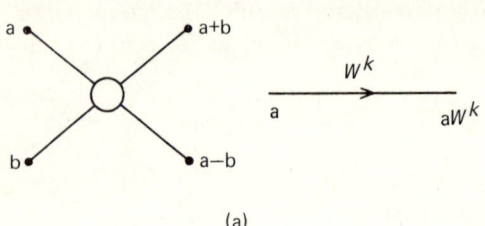

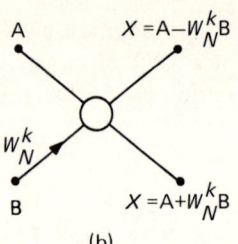

(a)   (b)

Fig. 6.11

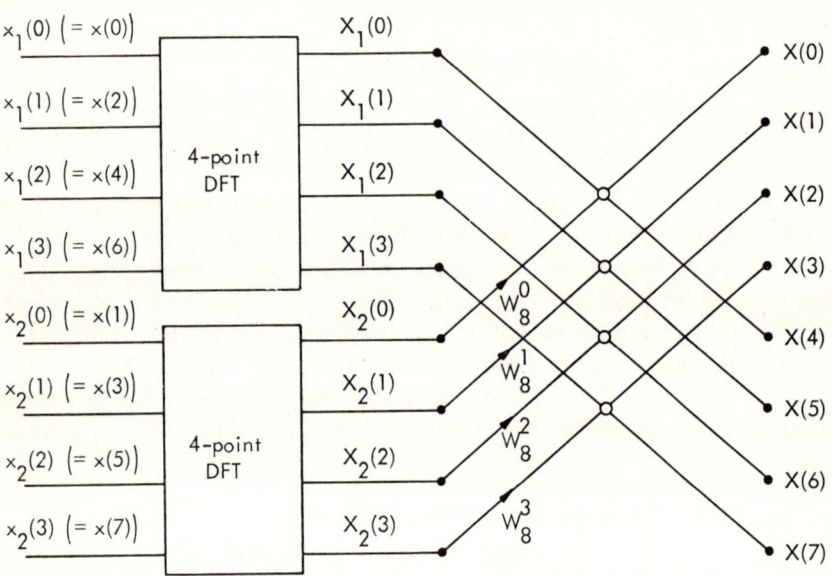

Fig. 6.12

The same now is done to the 4-point transforms by dividing each of $x_1(n)$, $x_2(n)$ into two sequences of even and odd members, resulting in four 2-point transforms. Following reference [6.11] the overall scheme is shown in Fig. 6.13, where the basic operation is the 'decimation in time butterfly' (Fig. 6.11(b)).

There are $N/2$ butterflies at each stage, and if $N = 2^\nu$ there are $\nu = \log_2 N$ stages, and with approximately 2 multiply/add operations 'per butterfly' we have of the order of $N \log_2 N$ operations in all.

(Note that the output array $X(k)$ is in the correct order, but the input array is shuffled. In fact it is in 'bit reversed' order (see Fig. 6.13).)

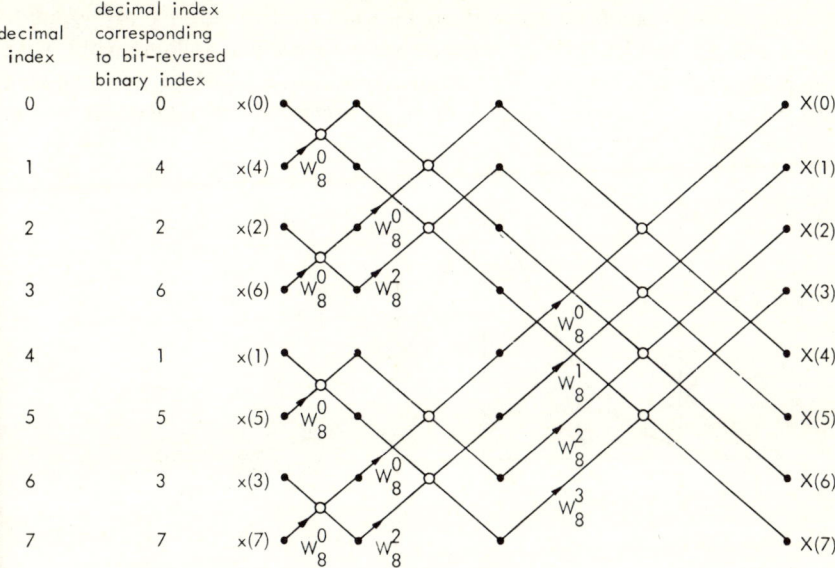

Fig. 6.13

6.7 RANDOM DATA

In Chapter 4 random processes were described in terms of probability functions, correlation functions, and spectral densities. The definition of these 'theoretical' concepts serves as the basis upon which measured data may be interpreted. In this section we introduce and discuss estimation methods for random signals, based on a single recording of the process.

6.7.1 Estimator errors and accuracy

Suppose the mean value of a random process is μ_x (unknown to us) and we only have a record length T of the signal. A reasonable estimate $\bar{x}$ of μ_x is

$$\bar{x} = \frac{1}{T} \int_0^T x(t)\, \mathrm{d}t.$$

The value obtained is a sample value of a random variable $\bar{X}$, having its own probability distribution (called as sample distribution), and $\bar{x}$ is a single realization. Each $\bar{x}$ computed from a (different) record of length T will, in general, be different. If the estimation procedure is to be useful we would hope that (i) the scatter of values of $\bar{x}$ is not 'too great' and should lie 'close' to μ_x; (ii) the more data we take (the larger T is) the 'better' the estimate. We can formalize these ideas as follows.

Let ϕ be a parameter we wish to estimate (in the introductory example above ϕ is μ_x) and let $\hat{\Phi}$ be an estimator for ϕ, then $\hat{\Phi}$ is a random variable with the probability distribution, for example, as in Fig. 6.14 where it can be seen that estimates $\hat{\phi}$ of random variable $\hat{\Phi}$ would predominantly take values near α.

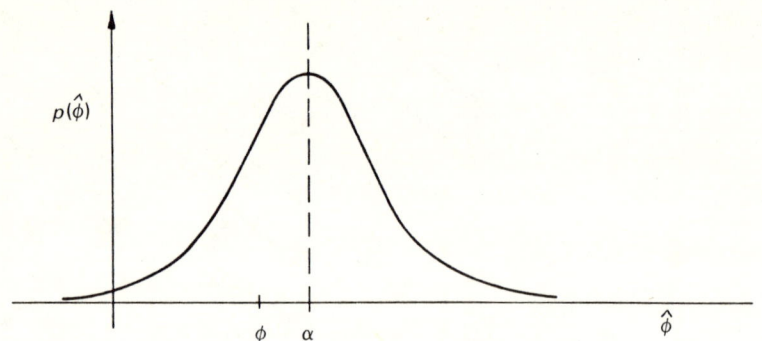

Fig. 6.14

The sampling distribution $p(\hat{\phi})$ is often difficult to obtain, and so we shall settle for a few summarizing properties.

(1) *Bias error*. The bias of an estimator is defined

$$b(\hat{\Phi}) = E[\hat{\Phi}] - \phi. \tag{6.11}$$

This is the difference between the average of the estimator and the true value. If $b(\hat{\Phi}) = 0$, then $\hat{\Phi}$ is 'unbiased'. Whilst it might seem desirable to use an unbiased estimator, it is sometimes prudent to tolerate some bias if it means that the variability of the estimate is reduced (relative to that of an unbiased estimator).

(2) *Variance*. The variance of an estimator is a measure of the spread of $\hat{\Phi}$ about its own mean value

$$\text{Variance}[\hat{\Phi}] = E[(\hat{\Phi} - E(\hat{\Phi}))^2]. \tag{6.12}$$

The square root of the variance is the standard deviation ($\sigma(\hat{\Phi})$) of the estimator. We would hope that this would be 'small', i.e., the probability density function should be 'peaky', but this requirement often increases the bias. A measure of the relative importance of bias and variance is the mean square error.

(3) *Mean square error* (m.s.e.). The m.s.e. of an estimator is a measure of the spread of values of $\hat{\Phi}$ about the value ϕ, i.e.,

$$\text{m.s.e.}[\hat{\Phi}] = E[(\hat{\Phi} - \phi)^2]. \tag{6.13}$$

The right-hand side of (6.13) can be rewritten as

$$\text{m.s.e.}[\hat{\Phi}] = \text{Variance}[\hat{\Phi}] + b^2[\hat{\Phi}] \tag{6.14}$$

showing that m.s.e. reflects both variance and bias.

If the m.s.e. reduces as the sample size (amount of data used) increases then the estimator is *consistent*.

Confidence intervals

The estimates $\hat{\phi}$ referred to so far are *point estimates*, i.e., single values. It is frequently desirable to know that a parameter is likely to fall within a certain *interval of values*. For example, if we estimate a mean value $\bar{x}$ as 5, perhaps it is 'likely' that μ_x lies in the interval 4.5–5.5. These estimates are *interval estimates*, and are called *confidence intervals* when we attach a number describing the likelihood of the parameter falling within the interval.

If, for example, we state 'a 95% confidence interval for μ_x is $(4.5, 5.5)$', then this means we are 95% confident that μ_x lies in the range $(4.5, 5.5)$. *N. B.* This does *not* mean that the probability that μ_x lies in the interval $(4.5, 5.5)$ is 0.95 (μ_x is not a random variable and so we do not assign probabilities to it). We mean that if we could realize a large number of samples and find a confidence interval for μ_x for *each* sample, then approximately 95% of these intervals would contain μ_x.

In order to calculate confidence intervals we need to know the sampling distribution of the estimator. We shall return to this problem for a specific estimator.

6.8 ESTIMATORS FOR STOCHASTIC PROCESSES

The problem of the study of estimators for stochastic processes is a wide one, and some idea of the extent of the field can be obtained from [6.7]. Several other texts are not as detailed and give useful summaries; these are [6.1, 6.2, 6.3, 6.4, 6.6].

We shall pick out the autocorrelation function and power spectral density as examples, and try to indicate some of the features that should be considered when analysing a sample time history. Following this we briefly discuss cross spectra and system identification. Unless otherwise stated we shall assume we are dealing with realizations of a continuous, stationary random process. The data is sampled so that the computation may be done and is therefore subject to the attendant 'aliasing' problems. It is assumed throughout that the sampling rates are 'sufficiently high'.

6.8.1 The autocorrelation function

(a) *Definition.* If we have a signal defined for $0 \leqslant t \leqslant T$, then there are two (ref. [6.7]) commonly used estimators of the function $R_{xx}(\tau)$. Following [6.7] we write them as $\hat{R}_{XX}(\tau)$ and $\hat{R}'_{XX}(\tau)$ where (assuming a zero mean process)

$$\hat{R}_{XX}(\tau) = \frac{1}{T} \int_0^{T-|\tau|} X(t)X(t+|\tau|)\mathrm{d}t \qquad 0 \leqslant |\tau| \leqslant T \qquad (6.15)$$
$$= 0 \qquad\qquad\qquad |\tau| > T$$

and

$$\hat{R}'_{XX}(\tau) = \frac{1}{T-|\tau|} \int_0^{T-|\tau|} X(t)X(t+|\tau|)\mathrm{d}t \qquad 0 \leqslant |\tau| \leqslant T \qquad (6.16)$$
$$= 0. \qquad\qquad\qquad |\tau| > T$$

Often the expression (6.16) is used since it is unbiased, but it is the m.s.e. that should be the criterion for a choice between the two.

(b) *Statistical considerations.* For τ in the range $0 \leqslant |\tau| \leqslant T$, $\hat{R}'_{XX}$ is unbiased, and $E[\hat{R}_{XX}(\tau)] = R_{xx}(\tau)[1 - \frac{|\tau|}{T}]$ showing (6.15) to be 'asymptotically unbiased'.

An approximate expression for the variance of $\hat{R}_{XX}(\tau)$ is

$$\mathrm{Var}[\hat{R}_{XX}(\tau)] \simeq \frac{1}{T} \int_{-\infty}^{\infty} [R_{xx}^2(r) + R_{xx}(r+\tau)R_{xx}(r-\tau)]\mathrm{d}r \quad (6.17)$$

showing this to be a consistent estimator. When $\tau \ll T$ there is little to choose between the two estimators, but as τ approaches T the variance of the unbiased estimator diverges (hardly surprising considering the divisor in (6.16)). Another important feature of the estimators is that values computed for neighbouring values of τ are sample values of (in general) strongly correlated random variables, meaning that $\hat{R}_{XX}(\tau)$ is more strongly correlated than the original time series $x(t)$, so the estimate may not decay as rapidly as one might expect (see section 5.3 of ref. [6.7]).

(c) *Computation.* Computation of (6.15) is achieved with sampled data with the formula

$$\hat{R}_{xx}(m) = \frac{1}{N} \sum_{n=0}^{N-|m|-1} x(n)x(n+|m|). \qquad (6.18)$$

The divisor is $N - |m|$ for $\hat{R}'_{xx}(m)$, i.e., (6.16); m is the lag and may take values $0 \leqslant m \leqslant M-1$ (where the maximum allowable value of M is N).

The calculations to be done in (6.18) are time-consuming if M is not small, and FFT methods can be employed with considerable speed advantage. A method due to Rader [6.23] is particularly effective, the procedure being based on the idea of 'fast' convolution (see [6.11]).

6.8.2 The power spectral density

Estimation methods for spectra considered here will relate to the 'traditional' methods rather than the recently developed 'parametric' methods. The standard methods are (i) the Fourier transform of the autocorrelation function (indirect method); (ii) the direct method of averaging periodograms; (iii) filtering (either bandpass or complex demodulation (heterodyning) and low-pass filtering). We shall outline (i) and (ii). (An interesting and readable paper [6.14] considers various aspects (computational and statistical) of spectral estimation, though recently [6.29] some criticisms of the tendency to favour method (ii) have been voiced.)

(i) The Fourier transform of the autocorrelation function

(a) *Definition.* An estimate of the power spectral density from a length of data T (here assumed to be defined for $|t| < T/2$) is

$$\hat{S}_{XX}(f) = \frac{|X_T(f)|^2}{T} = \int_{-T}^{T} \hat{R}_{XX}(\tau)\, e^{-j2\pi f\tau}\, d\tau. \qquad (6.19)$$

This is termed the 'raw' spectral density since it turns out (see (b) which follows) that the variability of this estimator is independent of the data length T and so is no use as an estimator.

To reduce the sampling fluctuations the estimate must be smoothed in the frequency domain, and this can be achieved by multiplying the autocorrelation function estimate by a 'lag window' $w(\tau)$ which has a Fourier transform $W(f)$ (a 'spectral window'), and Fourier transforming the product. This defines the estimator $\overset{\vee}{S}_{XX}(f)$ where

$$\overset{\vee}{S}_{XX}(f) = \int_{-T}^{T} \hat{R}_{XX}(\tau)w(\tau)e^{-j2\pi f\tau}\, d\tau \qquad (6.20)$$

which by the work of section 6.3 is the convolution of the 'raw' spectral density with $W(f)$, i.e., $\hat{S}_{XX}(f)*W(f)$. This estimation procedure may be viewed as a smoothing operation in the frequency domain or, in the time domain, the lag window can be regarded as reducing the 'importance' of values of $\hat{R}_{XX}$ as τ increases (since the variability of the estimator R_{XX} is increased as τ increases).

(b) *Statistical properties.* In order to discuss the properties of estimators $\hat{S}_{XX}$ and $\overset{\vee}{S}_{XX}$ it is necessary to first outline a few preliminary ideas in statistics [6.7]. The *chi-squared distribution.* If $X_1, X_2, ..., X_n$ are n independent random variables each having a normal distribution with zero mean and unit standard deviation, and if

$$\chi_n^2 = X_1^2 + X_2^2 + ... + X_n^2 \qquad (6.21)$$

then the distribution of χ_n^2 is called the chi-squared distribution with 'n degrees of freedom' (the number of degrees of freedom representing the number of independent random variables X_i entering the expression). The first two moments of χ_n^2 are

$$E[\chi_n^2] = n \quad \text{and} \quad \text{Var}[\chi_n^2] = 2n. \tag{6.22}$$

If a set of independent χ^2 variables are added together, i.e.,

$$\chi_\nu^2 = \chi_{n_1}^2 + \chi_{n_2}^2 + \dots + \chi_{n_k}^2 \tag{6.23}$$

then the sum is χ_ν^2 (as indicated), and $\nu = n_1 + n_2 + \dots + n_k$.

These properties of the chi-squared distribution are used in the development of some of the following results (see [6.7]).

Bias. $\hat{S}_{XX}(f)$ is an asymptotically unbiased estimator.

Variance. Since $\hat{S}_{XX}(f)$ is a squared quantity (and if we assume $X(t)$ is Gaussian) then it can be argued that $2\hat{S}_{XX}(f)/S_{xx}(f)$ is distributed as a chi-squared random variable with two degrees of freedom, i.e., χ_2^2 (for all values of data length T). From (6.22) it follows therefore that the variance of $\hat{S}_{XX}(f)$ is equal to $S_{xx}^2(f)$. So $\hat{S}_{XX}(f)$ is an inconsistent estimator of $S_{xx}(f)$, and the standard deviation of the estimate is as large as the quantity being estimated. It is these unsatisfactory properties that lead to the consideration of another estimate.

Smoothed spectral estimators. Since $\overset{\vee}{S}_{XX}(f)$ is a convolution of $\hat{S}_{XX}(f)$ with $W(f)$ it turns out that $\overset{\vee}{S}_{XX}(f)$ is biased. In fact (for large T)

$$E[\overset{\vee}{S}_{XX}(f)] \simeq \int_{-\infty}^{\infty} S_{xx}(g)W(f-g)dg \tag{6.24}$$

and

$$\text{bias}[\overset{\vee}{S}_{XX}(f)] \simeq \int_{-\infty}^{\infty} [w(\tau) - 1] R_{xx}(\tau)e^{-j2\pi f\tau}d\tau. \tag{6.25}$$

Equation (6.24) indicates how the average of the estimate is distorted by smearing and 'leakage' of the spectral density owing to the spectral window width and shape (side lobes), and the window discussions earlier relate directly to this situation. Owing to the complicated shape of the $W(f)$ functions we can only comment broadly that the general effect is to reduce the dynamic range of the spectra, i.e., peaks are underestimated and troughs are overestimated, and the effect is reduced as the spectral window gets narrower. However, the windows cannot be made too narrow since then there is little smoothing and the random errors increase. Consequently there is always the problem of trading bias errors against random errors. (The bias problem is sometimes referred to under 'bandwidth considerations'.)

Since $\tilde{S}_{XX}$ is a weighted sum of values of $\hat{S}_{XX}$ it is argued [6.7] that $\dfrac{n\tilde{S}_{XX}(f)}{S_{xx}(f)}$ is approximately distributed as χ_n^2 where n (the number of degrees of freedom) $= 2BT$ and B is the resolution bandwidth. It follows that

$$\mathrm{Var}[\tilde{S}_{XX}(f)] = \frac{S_{xx}^2(f)}{n/2} = \frac{S_{xx}^2(f)}{BT},$$

and in Chapter 6 of ref. [6.7] details of estimator accuracy are tabulated for various lag windows.

The choice of window should depend on whether the concern is for statistical *stability* (low variance) or small bias. If the bias is small for all f, then we have high *fidelity*. Of course in a general situation one must be careful with both. For example, if the spectral function has narrow peaks of importance we may willingly tolerate some loss of stability to resolve the peaks properly, whilst if the spectral function is smooth then bias errors are not likely to be so important.

Confidence intervals. We now give 'interval estimates' based on the 'point estimates' for the smoothed spectral density. Since $n\tilde{S}_{XX}(f)/S_{xx}(f)$ is distributed as a χ_n^2 random variable then the probability density function has the appearance as in the Fig. 6.15.

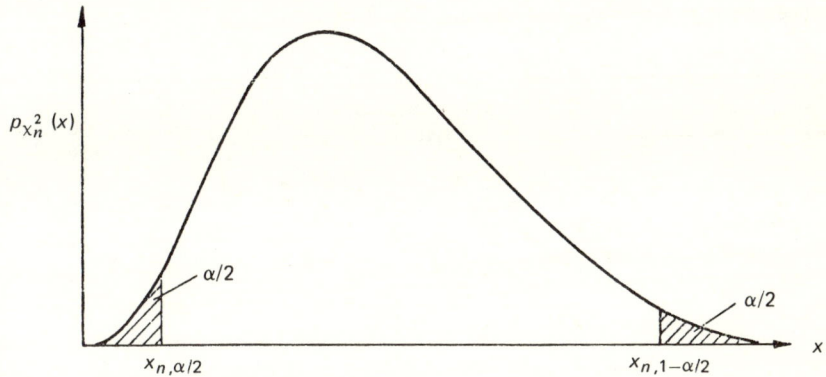

Fig. 6.15

If we choose a number $\alpha\,(0 < \alpha < 1)$ such that sections of area $\alpha/2$ are marked off as shown by the points $x_{n,\alpha/2}$ and $x_{n,1-\alpha/2}$ then we can say

$$\mathrm{Prob}\,[x_{n,\alpha/2} \leqslant \frac{n\tilde{S}_{XX}(f)}{S_{xx}(f)} \leqslant x_{n,1-\alpha/2}] = 1 - \alpha.$$

The points $x_{n,\alpha/2}, x_{n,1-\alpha/2}$ can be obtained from tables of χ_n^2 for different α and the inequality 'solved' for the true spectral density $S_{xx}(f)$ to yield the equivalent inequality

$$\frac{n\tilde{S}_{xx}}{x_{n,1-\alpha/2}} \leqslant S_{xx}(f) \leqslant \frac{n\tilde{S}_{xx}}{x_{n,\alpha/2}}. \tag{6.26}$$

So for a particular point estimate $\tilde{S}_{xx}$ the $100(1-\alpha)\%$ confidence limits for S_{xx} are

$$\frac{n}{x_{n,1-\alpha/2}}\tilde{S}_{xx}, \quad \frac{n}{x_{n,\alpha/2}}\tilde{S}_{xx}.$$

The confidence *interval* is the difference between them.

Note that on a linear scale the confidence interval depends on the estimate, but on a log scale the interval is $\log(n/x_{n,\alpha/2}) - \log(n/x_{n,1-\alpha/2})$ which is a constant independent of $\tilde{S}_{xx}$.

(c) *Computation* of $\tilde{S}_{xx}(f)$ is achieved by implementing a discrete form of (6.20). Reference [6.4] contains details.

(ii) Segment averaging

A method of spectral estimation that has become very popular (owing mainly to its speed of computation) is that of segment averaging discussed in Jenkins & Watts' book (section 6.3.4) as Bartlett's procedure, in some detail by Welch [6.26] and discussed by Yuen [6.29]. The books by Bendat & Piersol [6.3] Otnes & Enochson [6.4] and Beauchamp [6.2] also discuss it in detail. The basic procedure is outlined with reference to Fig. 6.16.

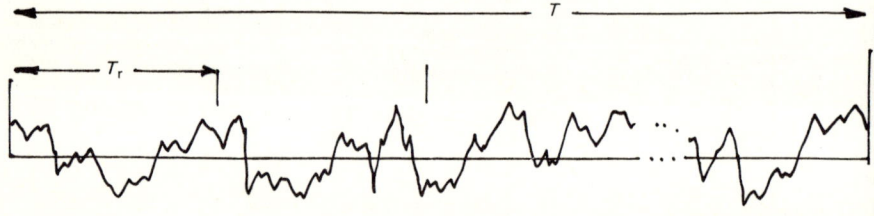

Fig. 6.16

A data length T is split up into q seqments of length T_r (in this case non-overlapping) and the raw periodogram formed for each segment, i.e.,

$$\hat{S}_{xx_i}(f) = \frac{1}{T_r}|X_{T_{ri}}(f)|^2 \quad \text{for } i = 1, \ldots, q$$

and to reduce the fluctuations the average

$$\tilde{S}_{xx}(f) = \frac{1}{q}\sum_{i=1}^{q}\hat{S}_{xx_i}(f)$$

is formed and it can be argued (very approximately) that since each $2\hat{S}_{XX_i}/S_{xx}$ is a χ_2^2 random variable then $2\overset{\vee}{S}_{XX}q/S_{xx}$ is χ_{2q}^2 from which it follows that

$$\text{Var}[\overset{\vee}{S}_{XX}] \simeq \frac{S_{xx}^2(f)}{q}$$

and since the resolution bandwidth B is approximately $1/T_r$ then

$$\text{Var}[\overset{\vee}{S}_{XX}] \simeq \frac{S_{xx}^2(f)}{BT} \qquad \text{(as before)}.$$

Whilst this summarizes the essential features of the method, references [6.26] and [6.14] give more elaborate procedures and insight. An important aspect is the use of data windows (or linear tapering as it is sometimes called) that are assigned to each segment (it is important to realize that the word linear here does not refer to the window shape but to the fact that it operates on the data directly and not on the autocorrelation function where it is sometimes called quadratic tapering). Furthermore, the segments may be chosen to overlap each other. A criticism of this method [6.29] is that linear windowing in effect ignores some data because of the tapered shape of the window. Intuitively the over-lapping of segments compensates for this in some way, and Welch [6.26] has results for this case though it is not easy to relate these results to the indirect method of Fourier transforming the autocorrelation function.

The use of a data window on a segment before transformation reduces leak-age but has the disadvantage of also reducing the total energy of that segment, and this must be compensated. This results in the calculation of modified periodograms for each segment of the form

$$\hat{S}_{xx_i}(f) = \frac{\dfrac{1}{T_r}\left| \displaystyle\int_{i\text{th interval}} x(t)w(t)e^{-j2\pi ft}\,dt \right|^2}{\dfrac{1}{T_r}\displaystyle\int_{-T_r/2}^{T_r/2} w^2(t)\,dt} \qquad (6.27)$$

where the denominator compensates for the 'energy reduction'.

As far as computation is concerned the DFT (FFT) is used to compute the required modified periodograms and the appropriate summations performed.

6.8.3 Cross-spectra, coherence and transfer functions

(a) *Cross-spectra*

The basic considerations above also relate to cross-spectral estimation together with some additional features. Detailed results can be found in

Chapter 9, ref. [6.7]. If the smoothed cross-spectral density $\tilde{S}_{xy}(f)$ is written as $|\tilde{S}_{xy}(f)|\exp(j\,\arg\,\tilde{S}_{xy}(f))$ then the variances of the amplitude ($|\tilde{S}_{XY}|$) and phase ($\arg\,\tilde{S}_{XY}$) estimators are proportional to $1/BT$ but are also strongly dependent on the coherency $\gamma^2_{xy}(f)$. As Jenkins & Watts emphasize ([6.7] p. 379), the sampling properties of the cross-amplitude and phase estimators may be dominated by the (uncontrollable) influence of the coherency spectrum rather than by the (controllable) influence of the smoothing factor $1/BT$.

(b) *The coherence function*
The estimate for the coherence function is made up from estimates of smoothed power and cross-spectra, i.e.,

$$\overset{\sim}{\gamma}^2_{xy} = \frac{|\tilde{S}_{xy}(f)|^2}{\tilde{S}_{xx}(f)\tilde{S}_{yy}(f)}.$$

(It should be noted that if 'raw' spectra are used in this estimate it is a simple matter to verify that the sample coherence function is unity for all frequencies for any signals x and y, ref. [6.4].) Statistical errors and confidence limits are given in [6.1, 6.3, 6.4, 6.6] and a recent paper [6.16] discusses the bias of the estimator. Results derived by Jenkins & Watts show that the bias is proportional

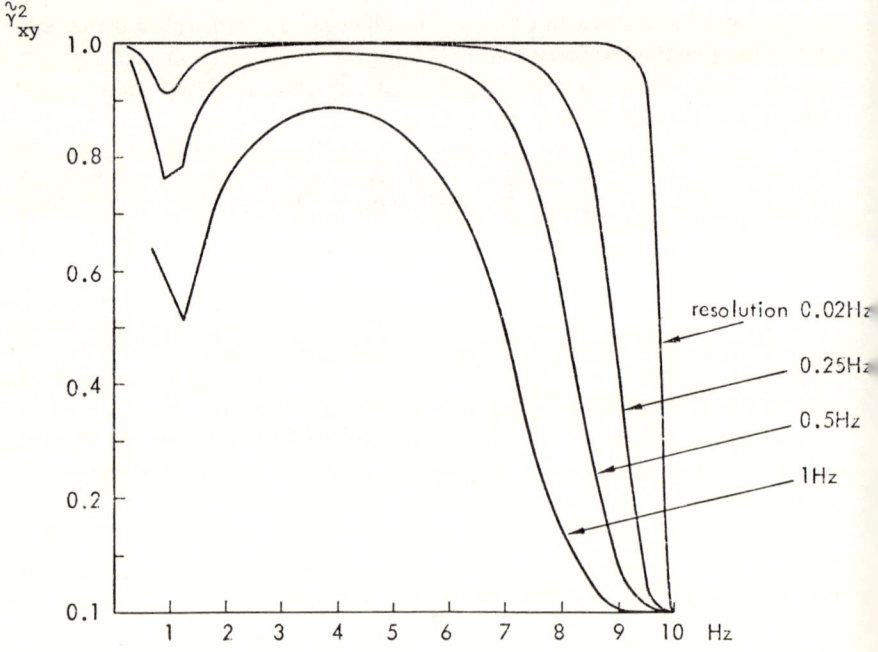

Fig. 6.17

to the square of the derivative of the phase spectrum between x and y. This means that the estimator is sensitive to delays between x and y and can also be expected to be inaccurate when, for example, x and y are input and output for a lightly damped oscillator. The bias error can be reduced by improving the resolution of the estimate. In Fig. 6.17 the coherence function is estimated for simulated input/output results for an oscillator with centre frequency $f_0 = 1$ Hz. The theoretical value of coherence is unity and the resolutions used are shown on the figure.

(c) *Frequency response functions* are estimated using smoothed spectra from the expression $\hat{H}(f) = \hat{S}_{xy}(f)/\hat{S}_{xx}(f)$, and results for errors and confidence limits can be found in [6.3, 6.4, 6.6]. The use of this nonparametric approach for systems with feedback is treated in [6.27]. (If feedback loops are not correctly accounted for, the estimate of the frequency response function of an element in the loop can be biased.)

6.9 DIGITAL FILTERING

The computer offers enormous flexibility in the processing of data, and some of the vast array of procedures that may be carried out on a sequence of numbers is referred to as 'digital filtering'. Several extremely good books exist [6.10, 6.11, 6.9] that have become standard texts, and all we shall do in this short section is introduce some terminology and direct the reader toward some of the methods that may prove relevant. We shall restrict our discussion to linear operations on input data with time (shift) invariant filters.

If $x(n)$ denotes an input sequence (time increment Δ is dropped for convenience) and $y(n)$ the output sequence, then a difference equation relating the sequences can be written

$$y(n) = -b_1 y(n-1) - b_2 y(n-2) \ldots -b_N y(n-N) +$$
$$a_0 x(n) + a_1 x(n-1) + \ldots + a_M x(n-M). \qquad (6.28)$$

This is a general digital filter which can be programmed to produce an output sequence given an input sequence and some starting conditions. The z-transform of section 6.3 may also be used to solve the equation, and this leads to a definition of a transfer function. By suitable selection of the coefficients b_i, a_j (and the orders N, M) the transfer characteristics can be adjusted to approximate some desired form. It is important to note that finite word length effects mean that coefficients cannot be represented exactly, and arithmetic round-off in the multiply operations constitute another source of error. These effects are not discussed further here (see Chapter 9 of ref. [6.10], for example, for more details).

Returning to equation (6.28), if at least one coefficient b_i is not zero the filter is recursive, whilst if all the b_i are zero then the filter is nonrecursive.

If the filter has a finite memory (i.e., its impulse response sequence is zero after a certain 'time') then it is called a Finite Impulse Response or FIR filter, in contrast to Infinite Impulse Response (IIR) filters which have infinite memory. We note that the terms recursive and nonrecursive describe how a filter is realized and not whether an impulse response is finite or infinite, though FIR filters are usually nonrecursive and IIR filters recursive in implementation.

Numerous design procedures are available for both types of filter. A popular procedure for IIR filters is to discretize some well understood analogue filter (bearing in mind that any digital counterpart is bound to have a frequency response that is periodic in frequency). The 'impulse invariant' method of design, as its name suggests, creates a filter whose impulse response sequence matches the impulse response function of an analogue filter at equi-spaced time instants. This procedure is simple to understand but limited in effectiveness since high-pass and bandstop filters cannot be designed this way. The 'bilinear transform' method is a procedure that overcomes the 'aliasing' problems of the impulse invariance procedure at the expense of introducing some frequency distortion when transforming from the analogue to the digital domain. However, this can be compensated for, and this is an effective and widely used procedure. Finally, optimal IIR filters can be designed to meet certain criteria, and listings of Fortran programs are available [6.25] which produce optimal filters.

FIR filters have certain advantages over IIR filters, for example that they are always stable and can have linear phase characteristics; but a major disadvantage is that they must be long enough to achieve adequate cutoff. A very easily implemented Fortran FIR optimal filter design program [6.20] is available that works very effectively, and such a filter could then be implemented using a (fast) convolution algorithm. Recent technological developments have exploited the use of FIR filters (or tapped delay lines as they are often called), and filtering, correlation, etc., can be implemented cheaply and simply using CCDs (Charged Couple Devices). A basic text covering much of this subject is [6.13].

We note that linear phase FIR filters can easily be designed (thereby ensuring no phase distortion), but, if data files can be stored, then zero phase filtering can easily be achieved even with IIR filters. This is done by simply processing the data 'forwards' and 'backwards' with the same filter. The block diagram below shows a possible scheme (see [6.11]). The only things to watch for are the filter 'starting transients' at both ends of the data, but this apart it is a simple and effective procedure.

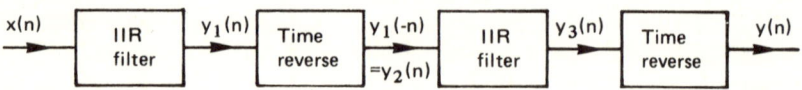

Fig. 6.18

6.10 'ADVANCED' METHODS

A host of signal analysis methods now exist, some of which are at a very advanced state of development and widely used in certain fields but which are perhaps still not in general use. The objective of this section is to list some of these methods and areas of application with the hope that the suggestions and references may prove useful.

Cepstral analysis (sometimes called homomorphic filtering) is essentially aimed at separating convolved signals. The name arises since it is a spectrum of a (log) spectrum, and some sort of new name seems in order to avoid confusing the spectra. The original paper [6.15] (which coined the terminology) still remains one of the most illuminating and easy to read descriptions, though Chapter 10 of reference [6.10] is the current way of treating the topic. Applications include echo detection, system identification, and pattern recognition, and the procedures are sufficiently well developed to be used regularly in speech processing and seismology. Chapter 11 of [6.1] offers an introduction to cepstral methods and also a useful set of references.

Nonparametric *multi-input single output* systems involving the concepts of partial and multiple coherence functions have been discussed for some time, but some recent publications by Bendat [6.12, 6.6] have reawakened interest in this area. The theoretical ideas underlying these notions are explained in the notes in Chapter 5 of this volume.

Parametric system identification methods occupy a dominant position in control applications and speech processing and have been used in many other fields. The various procedures that may be employed are too numerous to mention, but Chapter 15 of [6.1] gives an overview of some of the methods. It is possible to include the Prony series approach (mentioned earlier) and 'maximum entropy' spectral analysis within this area. A readable tutorial account of theory and application is given in [6.19].

The Chirp-Z transform (ref. [6.11]) has interesting possibilities allowing efficient evaluation of transforms of time histories other than along the unit circle in the z-plane (or the imaginary axis in the 's-plane') and can 'sharpen' resonances permitting more accurate evaluation of natural frequencies. Another application is in the so-called 'zoom' transform (though there are other methods of 'zooming' [6.5]).

Adaptive noise cancellation [6.27] is a procedure whereby a contaminating noise superimposed on a data signal can be 'subtracted off' as long as a separate, linearly related measure of the contamination is available. This may be achieved (subject to certain conditions) without *a priori* knowledge of the statistical character of either signal or noise. The method is based on minimizing a squared error with respect to the filter weights of a tapped delay line (FIR) adaptive filter.

It was noted in the introduction that *nonstationary phenomena* are 'difficult' to deal with. However, the 'evolutionary spectral density' [6.22] offers an approach to the description of such processes that still retains much of the physical interpretation of the usual spectral density for stationary processes. Attempts have been made [6.17] to use this in the study of vibrating systems.

For *nonlinear* systems procedures are much more application dependent. Chapters 18-21 of [6.1] provide an idea of some of the methods used, but this is still a fertile research area. For studies of input-output response relations 'equivalent linearization' and 'stochastic averaging' (Chapter 21, ref. [6.1]) are conceptually simple and hence quite attractive, whilst the bispectrum ([6.1, 6.21]) offers an approach to treating nongaussian data, which could possibly have resulted from a nonlinear process.

We conclude by noting that conspicuously absent from this chapter has been any discussion of hardware. With the pace of technological developments the scene is a rapidly changing one, and any overview of available equipment would soon be outdated. However, Chapters 12 and 14 in ref. [6.1] give summaries of some aspects of hardware architecture.

REFERENCES/BIBLIOGRAPHY

(A)   General texts covering time series analysis

   'Engineering' approach

[6.1] Applications of Time Series Analysis – Course Notes available from ISVR, University of Southampton.

[6.2] Beauchamp, K. G. (1973) *Signal processing*. George Allen & Unwin.

[6.3] Bendat, J. S. & Piersol, A. G. (1971) *Random data: analysis and measurement procedures*. Wiley Interscience.

[6.4] Otnes, R. K. & Enochson, L. (1978) *Applied time series analysis* Vol. 1. *Basic techniques*. Wiley & Sons.

[6.5] Randall, R. B. (1977) *Frequency analysis*. Bruel & Kjaer Publication.

[6.6] Bendat, J. S. & Piersol, A. G. (1980) *Engineering applications of correlation and spectral analysis*. Wiley & Sons.

   Statistical/mathematical approach

[6.7] Jenkins, G. M. & Watts, D. G. (1968) *Spectral analysis and its applications*. Holden Day.

(B)   General texts covering digital signal processing

[6.8] Brigham, E. O. (1974) *The fast Fourier transform*. Prentice Hall.

[6.9] Childers, D. & Durling, A. (1975) *Digital filtering and signal processing*. West Publishing Co.

[6.10] Oppenheim, A. V. & Schafer, R. W. (1975) *Digital signal processing*. Prentice Hall.

[6.11] Rabiner, L. R. & Gold, B. (1975) *Theory and application of digital signal processing*. Prentice Hall.

(C) Other references

[6.12] Bendat, J. S. (1978) *Journal of Sound and Vibration* **59** (3). Statistical errors in measurement of coherence functions and input/output quantities.

[6.13] Beynon, J. D. E. & Lamb, D. R. (1980) *Charge-coupled devices and their applications*. McGraw-Hill.

[6.14] Bingham, C., Godrey, M. & Tukey, J. W. (1967) *IEEE Trans. Audio and Electroacoustics* **AU-15**, No. 2. Modern techniques of power spectrum estimation.

[6.15] Bogert, B. P., Healy, W. J. R. & Tukey, J. W. (1963). *Time series analysis*, ed. Rosenblatt, M. The quefrency alanysis of time series for echoes. Wiley & Son.

[6.16] Carter, G. C. (1980) *IEEE Trans. on Acoustics, Speech and Signal Processing*. **ASSP-28**, No. 1. Bias in magnitude-squared coherence estimation due to misalignment.

[6.17] Hammond, J. K. (1973) *Journal of the Royal Statistical Society Series B*, **35**, No. 2. Evolutionary spectra in random vibrations.

[6.18] Harris, F. J. (1978) *Proc. IEEE* **66**, No. 1. On the use of windows for harmonic analysis with the discrete Fourier transform.

[6.19] Makhoul, J. (1975) *Proc. IEEE* **63**, No. 4. Linear prediction: A tutorial review.

[6.20] McClellan, J. H., Parks, T. W. & Rabiner, L. R. (1973) *IEEE Trans. Audio and Electroacoustics*. **AU-21**, No. 6. A computer program for designing optimum FIR linear phase digital filters.

[6.21] Perrochaud, J. B. (1982) *Bispectral analysis of nonlinear systems*. MSc Thesis ISVR, University of Southampton.

[6.22] Priestley, M. B. (1967) *Journal of Sound and Vibration* **6** (1). Power spectral analysis of nonstationary random processes.

[6.23] Rader, C. M. (1970) *IEEE Trans. Audio and Electroacoustics* **AU-18**. An improved algorithm for high-speed autocorrelation with application to spectral estimation.

[6.24] Spitznogle, F. R. & Quazi, A. H. (1970) *Journal of the Acoustical Society of America* **47**, No. 5. Representation and analysis of time-limited signals using a complex exponential algorithm.

[6.25] Steglitz, K. (1970) *IEEE Trans. Audio and Electroacoustics* **AU-18**. Computer aided design of recursive digital filters.

[6.26] Welch, P. D. (1967) *IEEE Trans. Audio and Electroacustics* **AU-15**. The use of the fast Fourier transform for the estimation of power spectra: a method based on time averaging over short, modified periodograms.

[6.27] Wellstead, P. E. (1979) *Nonparametric methods of system identification*. 5th IFAC Symposium on Identification and System Parameter Estimation. Ed. R. Isermann, held at Darmstadt. Pergamon Press.

[6.28] Widrow, B., *et al.* (1975) *Proc. IEEE* **63**, No. 12. Adaptive noise cancellation; principles and applications.

[6.29] Yuen, C. K. (1979) *IEEE Trans. Acoustics, Speech and Signal Processing* **ASSP–27**, No. 3. Comments on modern methods for spectrum estimators.

CHAPTER 7

Statistical energy analysis

F. J. Fahy

Institute of Sound and Vibration Research, University of Southampton

7.1 RATIONALE OF STATISTICAL ENERGY ANALYSIS (SEA)

We have become accustomed to applying statistical analyses to problems which involve the response of mechanical structures and fluids to force fields which are random in time and space, and which are represented by various probabilistic models. It is common experience that all but the simplest and most carefully constructed structures exhibit vibration characteristics which differ, sometimes considerably, in natural frequencies, damping, and mode shapes from the predictions of apparently relevant theoretical calculations based upon idealized models having well-defined characteristics. The uncertainty of prediction increases with increase of frequency above the fundamental mode of the system (or as the vibration wavelengths become rather small compared with the system dimensions); the difficulty of prediction increases with the degree of mechanical and geometrical complexity of the system. Room acousticians have been faced with this problem for many years, and the most widely used theoretical analyses of sound fields in rooms use probabilistic models (e.g. the diffuse field) and provide statistical measures of behaviour. It is not surprising then that the past two decades have seen the development of an approach to the analysis of random vibrations of complex structural and fluid systems which represent the systems as having parameters drawn from populations with random distributions.

R. H. Lyon, the main originator of Statistical Energy Analysis (SEA) says that "SEA has been described as a point of view in dealing with vibration of complex structures, and as such it employs a series of analytical and experimental 'methods', most of which pre-date the identification of SEA. The viewpoint is *statistical* because the systems under analysis are presumed drawn from populations with random parameters; *energy* is the independent dynamical variable chosen because, by using it, distinctions between acoustical and mechanical systems disappear; and *analysis* emphasises that SEA is an approach to problems rather than a set of techniques as such".

SEA has most commonly been applied to situations in which two or more simple identifiable, but often complex, mechanical or acoustical systems are coupled together and vibrate under the action of broadband stationary random forces applied to one or more of them, e.g. sound fields and shell structures, beams and plates. The primary aim of the analysis is to estimate the distribution of vibration energy among the coupled 'subsystems', and for this purpose energy balance equations are set up which involve expressions for power flowing from one subsystem to another. Because of the probabilistic descriptions of the forces and the system parameters, the equations involve energies and power flows which are averaged over time and also over the population of systems, or what is more practical, over the modes of vibration having natural frequencies in a band of frequency which is large enough to contain a statistically usable modal population. Because their will nearly always be some frequency dependence of interaction behaviour (e.g. the coincidence phenomenon is structural-acoustic interactions), and because practical force spectra are not usually flat, such frequency bands must be narrow enough not to obscure such frequency trends. Hence SEA is most useful and reliable in applications to systems which have rather close natural frequencies (high modal densities).

The subsystems are, commonly, finite linear elastic systems which can, in principle, be described in terms of their uncoupled natural frequencies, mode shapes, and dampings when uncoupled from contiguous subsystems, although these are not usually exactly known. Each normal mode can be modelled by a simple oscillator, and the interaction between two coupled subsystems may be represented by that between two oscillator sets. Hence the study of random excitation of coupled oscillators forms an extremely important part of SEA. An advantage of having the scalar, energy, as a primary variable is that the total energy of an *uncoupled* subsystem equals the sum of the energies of the individual oscillators in that subsystem. The power flow-energy relationships which have been derived for randomly excited systems *are not directly applicable to single frequency, or very narrow band, excitation.* However, some progress has been made in the statistical analysis of response to single frequency and narrow band excitation [7.1, 7.2, 7.3].

SEA has so far been used mainly in the analysis of vibration by stationary random forces. However, some progress has been made in the analysis of power flow between coupled oscillators excited by non-stationary (transient or slowly time varying) forces, although applications have not yet been widely reported [7.4, 7.5, 7.6]. A critical review of SEA is available in reference [7.7].

7.2 ENERGY AS AN INDEPENDENT VARIABLE

The relationship between the 'energy of vibration' of a system and the measureable, or practically significant, system variables depends upon the particular systems, and also upon the type of motion considered. The 'energy of vibration' is here taken to mean the long-time averaged sum of the kinetic and potential energies under stationary random excitation. For a simple, linear, damped oscillator the two components of the sum are equal, and the sum equals twice either one: $\overline{T} + \overline{U} = \overline{E}; \overline{T} = \overline{U}$. A measure of the distribution of energy within a system is the energy density. Expressions for energy density in terms of commonly measured quantities for different types of vibration are given in Table 7.1.

Table 7.1

Energy Densities

System	Type of motion	Energy density
		($\overline{}$ denotes time average)
Uniform fluid	Acoustic wave	$\overline{p^2}/\rho c^2$
Uniform bar	Quasi-longitudinal wave	$\rho A \overline{v_x^2} = EA\overline{\epsilon^2}$/unit length
Uniform bar (mass m/unit length)	Flexural wave	$m\overline{v^2}$/unit length
Uniform plate (mass m/unit area)	Flexural wave	$m\overline{v^2}$/unit area
Uniform bar (polar moment of inertia I/unit length)	Torsional wave	$I\overline{\dot{\theta}^2}$/unit length

v = transverse particle velocity; v_x = longitudinal particle velocity

In general the more resonant modes which are simultaneously excited, the more uniform is the energy density. Formulae for distribution of energy density are given in references [7.8] and [7.9]. SEA does not give any information about concentrations of energy in particular parts of the system, e.g. dynamic strain concentrations such as might lead to structural failure. The analysis of travelling wave field models, as in [7.10], is best suited to this problem in cases of high-frequency vibration. Of course, statistics of vibration variables other than that which is directly proportional to the time averaged energy, such as probability distribution, are needed for fatigue life estimates (see Chapter 20). These are related to the statistics of the excitation and to the number and density of excited modes.

7.3 LOSS FACTOR

Damping in linear oscillatory systems is usually represented by viscous or by hysteretic models. The equation of motion of a viscously damped simple oscillator is $M\ddot{x} + R\dot{x} + Kx = F(t)$. The dissipated power is related to the time-averaged oscillator energy $\bar{E} = M\overline{\dot{x}^2}$ by $\bar{P}_d = \dfrac{R}{M}\bar{E}$ where R is termed the resistance. The equation of motion of a hysterically damped oscillator is $M\ddot{x} + K(1 + i\eta)x = F(t)$. The power dissipated is given by $\bar{P}_d = \dfrac{\eta K}{M\omega_o}\bar{E} = \omega_o\eta\bar{E}$.

Hence the dissipation loss factor, hereinafter simply termed 'loss factor' η, is given by $\eta = R/\omega_o M$, where ω_o is the undamped natural frequency $= 2\pi f_o = (K/M)^{\frac{1}{2}}$. The energy of a simple oscillator, when averaged over an integral number of half cycles, decays approximately as $\bar{E} = E_o e^{-2\pi\eta f_o t}$ during free decay after the cessation of excitation, where E_o is the initial, long-time averaged energy. The corresponding 'reverberation time' (time to decay to $E_o/10^6$ or by 60 dB) is related to η by $T = 13.8/\eta\omega_o$. This expression is valid for small damping, i.e. $(\eta/2)^2 \ll 1$. The half-power bandwidth of the oscillator is given by $\Delta\omega = \eta\omega_o$.

7.4 ENERGY FLOW BETWEEN TWO LINEARLY COUPLED OSCILLATORS

7.4.1 Power flow-energy difference relationship

The most general system of two linearly coupled oscillators is shown in Fig. 7.1.

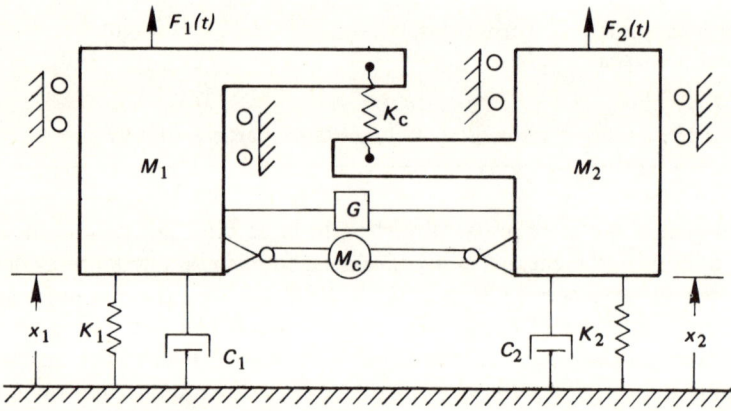

Fig. 7.1 – Two mass, spring, and gyrostatically coupled linear oscillators (ref. 7.11).

Non-conservative, or dissipative, couplings are excluded. The equations of motion are as follows;

$$(M_1 + M_c/4)\ddot{x}_1 + C_1\dot{x}_1 + (K_1 + K_2)x_1 + (M_c/4)\ddot{x}_2 - G\dot{x}_2 - K_c x_2$$
$$= F_1(t) \qquad (7.1a)$$

$$(M_2 + M_c/4)\ddot{x}_2 + C_2\dot{x}_2 + (K_1 + K_2)x_2 + (M_c/4)\ddot{x}_1 + G\dot{x}_1 - K_c x_1$$
$$= F_2(t) \qquad (7.1b)$$

K_c represents linear stiffness coupling, G represents linear gyrostatic coupling [7.12], and M_c represents linear inertial coupling: acoustic coupling between a fluid and a structure takes a gyrostatic form. The input forces $F_1(t)$ and $F_2(t)$ are *statistically independent* forces and are assumed to have spectra which are uniform (flat). The time-averaged power flow from oscillator 1 to oscillator 2 is given *exactly*, for any coupling strength, by

$$\bar{P}_{12} = g(\bar{E}_1 - \bar{E}_2) \tag{7.2}$$

where $\bar{E}_1$ is the time-averaged total vibration energy of oscillator 1 and $\bar{E}_2$ is the total time-averaged vibration energy of oscillator 2, where the total energy is defined as that associated with the relevant displacement (potential) and velocity (kinetic) of one oscillator when the other oscillator is held motionless (or strain free).

$$g = \frac{\mu^2 [\beta_1\omega_2^4 + \beta_2\omega_1^4 + \beta_1\beta_2(\beta_1\omega_2^2 + \beta_2\omega_1^2)^2] + (\gamma^2 + 2\mu\kappa)(\beta_1\omega_2^2 + \beta_2\omega_1^2) + \kappa^2(\beta_1 + \beta_2)}{(1-\mu^2)[(\omega_1^2 - \omega_2^2)^2 + (\beta_1 + \beta_2)(\beta_1\omega_2^2 + \beta_2\omega_1^2)]}$$

where $\omega_1^2 = (K_1 + K_c)/(M_1 + M_c/4)$, $\omega_2^2 = (K_2 + K_c)/(M_2 + M_c/4)$,

$\beta_1 = \eta_1\omega_1$, $\beta_2 = \eta_2\omega_2$, $\eta_1 = C_1/(M_1 + M_c/4)$,

$\mu = (M_c/4)/[(M_1 + M_c/4)(M_2 + M_c/4)]^{\frac{1}{2}}$,

$\gamma = G/[(M_1 + M_c/4)(M_2 + M_c/4)]^{\frac{1}{2}}$

and $\kappa = K_c/[(M_1 + M_c/4)(M_2 + M_c/4)]^{\frac{1}{2}}$.

The important features of equation (7.2) are as follows:

(a) Energy flows on average from an oscillator of higher energy to one of lower energy (analogues to heat and temperature).
(b) The rate of flow is proportional to the time-averaged energy difference.
(c) The constant of proportionally is a function of the oscillator parameters and is independent of the excitation source strength.
(d) The constant of proportionality, hereinafter called the rate factor, is very sensitive to the difference in 'blocked' natural frequencies, $(\omega_1^2 - \omega_2^2)$. All other things being equal, oscillators having close blocked natural frequencies exchange energy more easily than those having remote natural frequencies. Note that for purely gyrostatic coupling ($K_c = M_c = 0$) the blocked and uncoupled natural frequencies are identical and that the rate factor becomes independent of blocked frequencies as $(\omega_1^2 - \omega_2^2) \to 0$.

7.4.2 Coupling loss factor

By analogy with the dissipation loss factor η of section 7.3 we can define coupling loss factors η_{12} and η_{21} such that the transmission of energy from one oscillator to another to which it is coupled is related to the first-oscillator energy. However, because the flow depends upon energy *difference* we must define it for zero energy in the second oscillator, hence

$$[P_{12}]_{\bar{E}_2=0} = g\bar{E}_1 = \eta_{12}\omega_1\bar{E}_1 . \qquad (7.3)$$

Since g is symmetric in the oscillator parameters, we can also write

$$[P_{21}]_{\bar{E}_1=0} = g\bar{E}_2 = \eta_{21}\omega_2\bar{E}_2 . \qquad (7.4)$$

Hence $\eta_{21}\omega_2 = \eta_{12}\omega_1 .$ \qquad (7.5)

A discussion of the definition of coupling loss factor is found in reference [7.13].

7.4.3 Energy balance in the two-oscillator system

The power balance equations for the two oscillators can now be written as

$$\bar{P}_1 = \omega_1\eta_1\bar{E}_1 + \eta_{12}\omega_1(\bar{E}_1 - \bar{E}_2) \qquad (7.6)$$

$$\bar{P}_2 = \omega_2\eta_2\bar{E}_2 + \eta_{21}\omega_2(\bar{E}_2 - \bar{E}_1) \qquad (7.7)$$

where $\bar{P}_1$ and $\bar{P}_2$ are the time averaged power inputs from the applied forces. If $\bar{P}_1$, $\bar{P}_2$, η_1, η_2, η_{21}, ω_1 and ω_2 are known, these equations can be solved for the oscillator energies $\bar{E}_1$ and $\bar{E}_2$.

A frequently encountered situation is when one oscillator is directly driven by external forces and the other oscillator is driven only through the coupling, e.g. isolated load system. The equations then reduce to

$$\bar{P}_1 = \omega_1\eta_1\bar{E}_1 + \eta_{12}\omega_1(\bar{E}_1 - \bar{E}_2) \qquad (7.8)$$

$$0 = \omega_2\eta_2\bar{E}_2 + \eta_{21}\omega_2(\bar{E}_2 - \bar{E}_1) \qquad (7.9)$$

The solution for energy ratio is

$$\frac{\bar{E}_2}{\bar{E}_1} = \frac{\eta_{21}}{\eta_2 + \eta_{21}} \qquad (7.10)$$

and of course

$$\bar{P}_1 = \omega_1\eta_1\bar{E}_1 + \omega_2\eta_2\bar{E}_2 . \qquad (7.11)$$

Equation (7.10) is very revealing. It says

(a) If $\eta_2 \ll \eta_{21}$, $\bar{E}_2/\bar{E}_1 \to 1$ (equipartition of energy) irrespective of the actual magnitude of the coupling. The practical significance of this result is that the application of additional damping will not be effective unless η_2 can be increased at least to a value similar to η_{21}.

(b) $\bar{E}_2 < \bar{E}_1$ because η_{21} is always positive.

The conditions which favour equipartition are thus (i) close natural frequencies, (ii) small damping in the indirectly driven oscillator, (iii) large coupling between the oscillators.

The results are good approximations if the input spectra are fairly uniform over a frequency range which includes *both the blocked natural frequencies*. In the case of strong coupling this range can be significantly larger than that based upon the uncoupled natural frequencies. The spectra of oscillator energies naturally peak at both the coupled natural frequencies so that with strong coupling it should not be assumed that measurements of vibration in finite frequency bands in which the *uncoupled* frequencies lie, necessarily include the peaks in the coupled-response spectra.

Newland [7.14] calculates the spectral density of power flow between two oscillators; he finds peaks at $(\omega_1 - \omega_2)$ and $(\omega_1 + \omega_2)$, nowhere near ω_1 and ω_2. Note carefully that the spectral density of power flow is a measure of the frequency distribution of the square of the instantaneous rate of energy flow, and hence is nothing to do with the time-averaged or mean value of the power flow discussed above. Indeed the spectral density of power flow is non-zero even if the mean power flow is zero. The frequency-dependent quantity of more interest is the cross spectral density of coupling force and velocity of which the integral is the mean power flow. This does peak at ω_1 and ω_2.

7.5 ENERGY FLOW BETWEEN MORE THAN TWO OSCILLATORS

7.5.1 Exact analyses
Kakar [7.15] and, more recently, Woodhouse [7.16] have shown that the exact two-oscillator result cannot be extended to more than two oscillators, although Woodhouse shows that the power flow-energy difference relationship is true even when the two oscillators are connected through a third *undamped* oscillator.

7.5.2 Special case of weak coupling
In a series of perturbation analyses [7.14, 7.17, 7.18] Newland has investigated power flow between many oscillators for the special case of 'weak coupling'. The concept and definition of weak coupling have been argued over by many authors, and no universally agreed conditions have been established. Intuitively one would expect the strength of coupling between oscillators to be related to the relative magnitudes of coupling forces and oscillator internal forces, or impedances [7.19, 7.20, 7.21]. However, it appears that a more correct criterion is based upon the relative magnitude of the coupling loss factors and the internal loss factors of the coupled systems [7.16, 7.21, 7.22]. Coupling is weak if the ratio of coupling loss factor to internal loss factor of *each* oscillator is substantially less than unity. This condition ensures that the actual oscillator energies are much closer to their uncoupled values than to those which obtain

under the state of equipartition of energy. Note that the energies of directly-driven oscillators fall when coupled to an undriven set, and the energy of the undriven set naturally rises.

Weak coupling perturbation analysis shows that the power flow between a mode of one set and a mode of the coupled set can be evaluated without including the interaction of any other mode. *This is the vital simplification that weak coupling affords and which is not true in general.* Under conditions of weak coupling there is a power flow-energy difference relationship between two sets of oscillators (which represent the uncoupled modes of two subsystems) which is analogous to the two-oscillator result, *with the restriction that the external (modal) forces applied to the individual oscillators (modes) of each set are statistically independent.* This condition is not satisfied by single point excitation of a distributed elastic system [7.23].

A related assumption, central to SEA, is that the modal vibrations of coupled sets are *uncorrelated.* This assumption is necessary to justify the linear dependence on modal energies of power flow expressions. Smith [7.24] discusses the hypothesis in terms of the correlation between waves incident upon a coupling element, or junction, from its two sides. By considering the generation of waves in distributed systems by broadband random forces, their incidence upon junctions, and their return to the junction after reflection at boundaries of contiguous subsystems (i.e. after some time delay and, perhaps, dispersive transformation process), it would seem that the degree of correlation would be related to frequency bandwidth, physical size, wave speed, and damping of the subsystems (via the autocorrelation of the force-time history, and the time delay). Of course these three factors are all related to the number of resonant modes in the band considered, through the bandwidth and modal density, and to the modal bandwidths (damping).

7.5.3 Two-coupled oscillator sets

Extension of the two-oscillator analysis to energy flow between *weakly coupled* sets of oscillators (modes) yields the following energy balance equations:

$$\bar{P}_\alpha = \eta_\alpha \omega \bar{E}_\alpha + \eta_{\alpha\beta} \omega N_\alpha \left(\frac{\overline{E_\alpha^*}}{N_\alpha} - \frac{\overline{E_\beta^*}}{N_\beta} \right) \tag{7.12a}$$

$$\bar{P}_\beta = \eta_\beta \omega \bar{E}_\beta + \eta_{\beta\alpha} \omega N_\beta \left(\frac{\overline{E_\beta^*}}{N_\beta} - \frac{\overline{E_\alpha^*}}{N_\alpha} \right). \tag{7.12b}$$

Sets α and β contain N_α and N_β uncoupled modes respectively which are resonant in the frequency band of interest. The modal average loss factors η_α and η_β, and the modal average coupling loss factors $\eta_{\alpha\beta}$ and $\eta_{\beta\alpha}$, are defined below.

The energies $\overline{E^*}$ in equations (7.12) are the energies of each of the two oscillator sets *when uncoupled from the other set* but driven by the same external forces (the uncoupled modal energies) [7.21, 7.25]. They are given by

$$\overline{E_\alpha^*} = \sum_i (\pi s_{\alpha_i}/M_{\alpha_i}\eta_{\alpha_i}\omega_{\alpha_i}) \tag{7.13a}$$

$$\overline{E_\beta^*} = \sum_j (\pi s_{\beta_j}/M_{\beta_j}\eta_{\beta_j}\omega_{\beta_j}) \tag{7.13b}$$

where s_α and s_β are the spectral densities of the external generalized modal forces, and M symbolizes modal mass.

The average number of modes in a band N is more conveniently defined in terms of the average density of resonance frequencies, which is termed the modal density n; hence $N = n(\omega)\Delta\omega = n(f)\Delta f$. Formulae for the modal densities of a number of idealized subsystem elements are presented in Table 7.2. The modal densities of one-, two- and three-dimensional elastic systems are proportional to length, area, and volume respectively.

The modal average internal loss factors are defined as

$$\eta_\alpha = \sum_i \eta_{\alpha_i}/N_\alpha; \quad \eta_\beta = \sum_j \eta_{\beta_j}/N_\beta \;. \tag{7.14}$$

The modal average coupling loss factors are defined as

$$\eta_{\alpha\beta} = \sum_{i,j} \eta_{\alpha_i\beta_j}/N_\alpha; \quad \eta_{\beta\alpha} = \sum_{i,j} \eta_{\alpha_i\beta_j}/N_\beta \tag{7.15}$$

where $\eta_{\alpha_i\beta_j} = g_{\alpha_i\beta_j}/\omega$ by analogy with equation (7.3). Hence

$$n_\alpha n_{\alpha\beta} = n_\beta n_{\beta\alpha} \;. \tag{7.16}$$

This relationship can also be derived from reciprocity [7.21].

In equations (7.12) the uncoupled set energies *per mode*, rather than the total energies, appear. These modal energies generally have more practical significance than the total energy because for a given bandwidth, N is proportional to the physical size of a subsystem, and therefore $\overline{E^*}/N$ is proportional to the energy density. Quantities of concern to the engineer, such as mean square vibration velocities, strains, and sound pressures, are proportional to the local energy density, rather than the total energy of a subsytem.

Note that the energy flow between oscillator sets is proportional to the difference of time average uncoupled energy per mode, but the energy dissipated by each set is proportional to its actual energy.

With zero power fed from external sources into set β, $\overline{P_\beta} = 0$; therefore the uncoupled energy $\overline{E_\beta^*}$ of set β is zero. Hence,

$$\overline{E_\beta}/\overline{E_\alpha^*} = \eta_{\alpha\beta}/\eta_\beta = (n_\beta/n_\alpha)(\eta_{\beta\alpha}/\eta_\beta) \;. \tag{7.17}$$

By definition of weak coupling $\eta_{\beta\alpha}/\eta_\beta \ll 1$; hence the modal energy ratio

$$(\overline{E_\beta}/n_\beta)/(\overline{E_\alpha^*}/n_\alpha) \ll 1 \;. \tag{7.18}$$

Equipartition of *actual* (not uncoupled) modal energy may be assumed as a limiting case in cases of strong coupling [7.21].

7.5.4 Many-coupled oscillator sets

In many practical cases it will be necessary to model a total system as an assemblage of many physical sybsystems, in which case each subsystem may be physically coupled to more than one other subsystem. Each subsystem will be modelled by one, *or more,* oscillator sets. For instance, flexural and longitudinal wave modes of individual plates or beams would need to be modelled as distinct oscillator sets because of their vastly different characteristic impedances, modal densities and, probably, internal loss factors.

Strictly speaking, equation (7.12) cannot be extended to more than two coupled oscillator sets because a simple consideration [7.21] of a linear chain of three oscillator sets, with external power supplied only to set 1, indicates that the uncoupled modal energies of both sets 2 and 3 are zero, and hence no power will flow from set 2 to set 3; therefore set 3 does not respond. This conclusion is clearly absurd. A disturbing implication of this argument is that the energies of each oscillator set in a multiple-set system depend upon the power inputs to all sets, or to the uncoupled energies of all sets, not just those directly physically coupled to it.

In order to proceeed with the analysis of multiple set systems, many analysts have assumed that equations (7.12) hold for *actual* coupled modal energies, an assumption which may lead to significant error which has not, as yet, been specifically evaluated. Extension of equations (7.12) on the basis of this assumption, to the general case of k oscillator sets, yields k simultaneous energy balance equations. These can be written in matrix form:

$$\omega \begin{bmatrix} (\eta_1 + \sum_{i \neq 1}^{k} \eta_{1i})n_1 & -\eta_{12}n_1 & -\eta_{13}n_1 \ldots \ldots -\eta_{1k}n_1 \\ -\eta_{21}n_2 & (\eta_2 + \sum_{i \neq 2}^{k} \eta_{2i})n_2 & -\eta_{23}n_2 \ldots \ldots -\eta_{2k}n_2 \\ \vdots & & \vdots \\ \vdots & & \vdots \\ -\eta_{k1}n_k & & (n_k + \sum_{i \neq k}^{k} \eta_{ki})n_k \end{bmatrix} \begin{bmatrix} \bar{E}_1/n_1 \\ \bar{E}_2/n_2 \\ \vdots \\ \vdots \\ \bar{E}_k/n_k \end{bmatrix} = \begin{bmatrix} \bar{P}_1 \\ \bar{P}_2 \\ \vdots \\ \vdots \\ \bar{P}_k \end{bmatrix}$$

$$(7.19)$$

The matrix of loss factors is symmetric because of the reciprocal relationship of equation (7.16). In most applications of equation (7.19) it is assumed that the input powers $\bar{P}_i$ are known, or can be estimated. However, analysis based upon this equation may also be used as the basis for parametric studies of power flow distribution for the purpose of optimizing noise control design.

In most SEA applications it is assumed that the greater part of the power flow is between uncoupled modes of each subsystem which are resonant in the band of interest and have proximate natural frequencies (resonant power flow). Woodhouse [7.16] suggests that in the general case of coupling by an *undamped* single degree of freedom element, it is the proximity of modal frequencies to that of the coupling, rather than to each other, which determines the coupling loss factor. In the special case of coupling through a pure mass (zero natural frequency), the lower the frequency of the subsystem mode, the greater the power flow; in this case, non-resonant power flow may dominate. A recent example of non-resonant power flow calculation is given by Wilby & Pope [7.45].

7.6 SYSTEM MODELLING FOR SEA

The initial step in the application of SEA to any system is the choice of sub-systems, and of the oscillator set (or sets) which represent the subsystems. A subsystem is most simply characterized by its geometric boundaries and by its material properties; it is also necessary to specify the associated kinematic and dynamic boundary conditions. There are no well-established criteria for the choice of subsystem boundaries. It is hoped that the major part of the energy flow can be associated with resonant motion in modes which bear some resemblance to the uncoupled subsystem modes (but see the previous section).

From the dynamic point of view it would seem reasonable to suggest that an appropriate choice of boundaries would be at points, or on surfaces, where freely propagating incident waves are substantially reflected, i.e. a boundary may be associated with a large change of wave impedance. This choice is necessary, but not sufficient, to satisfy the weak coupling criterion $\eta_{\alpha\beta} \ll \eta_\alpha, \eta_\beta$. There are two problems associated with such a choice. First, many elements which form discontinuities in structural systems, such as ribs, frames, etc., are highly selective in frequency and angle of incident waves in their reflective properties; in particular they may exhibit coincidence effects. Second, a physical discontinuity in a system may produce more or less reflection depending upon the *type* of wave incident upon it (e.g. a line mass on a plate carrying both longitudinal and flexural waves: see ref. [7.26]). In such cases it is necessary to define different subsystem boundaries, and of course oscillator sets, for each type of wave. Wave type transformation at discontinuities will necessitate the assumption of coupling between these sets of oscillators (e.g. ref. [7.27]). A related factor is that modal densities of different types of waves may have very different values. For example, the ratio of longitudinal mode density to flexural mode density in a flat plate is given by $n_L/n_F = k_L h/\sqrt{3}$ which is typically of order 10^{-2}. Hence, although energy per mode is normally independent of sub-system size (for a plate in flexure $\bar{E}/n = \rho h A \langle\bar{v^2}\rangle/\sqrt{3}A/hc_L$, which is independent of A), consideration of resonant/non-resonant power flow, and modal overlap factor (see 7.7.3) involve absolute modal densities; also the power flow $\sim$ energy difference equations involve both total energies and energies per mode. Ref. [7.49] warns of longitudinal wave problems.

7.7 ESTIMATION OF COUPLING LOSS FACTORS

There is no universal method of estimating modal average coupling loss factors.

7.7.1 Equation of motion

If coupled equations of modal motion can be formed, then the rate factors g_{ij} can be formed by analogy with equation (7.2) and (7.3) for the two-oscillator model. The modal average coupling loss factors are then found from equation (7.15), for which integration using statistical models of modal frequency distribution, rather than summation, may be used, e.g. references [7.28, 7.29, 7.30].

7.7.2 Natural frequency shift

Another method is based upon the average shift of modal natural frequencies of each of the coupled subsystems from the completely uncoupled configuration to the coupled, blocked configuration, with the motion of the other system constrained to be zero [7.17, 7.19]. The result is applicable to lightly damped distributed systems coupled at discrete points, but is considered to have somewhat limited usefulness [7.31, 7.25, 7.16].

7.7.3 Wave transmission analyses

A third method of estimating coupling loss factors is based upon the fact that many examples of bounded, multimode systems behave like the corresponding infinitely extended systems as far as frequency (or modal) averaged dynamic characteristics are concerned, e.g. point mobility (admittance) of a plate, acoustic power radiated by a point monopole source into a reverberation room (cf. (7.1)). Many examples of this approach exist in the literature, e.g. references (7.11, 7.19, 7.31, 7.32, 7.27, 7.33, 7.34]. On this basis it is assumed that the modal average coupling loss factor is closely related to the frequency average wave energy transmission coefficient across the subsystem coupling element(s), which may, itself, be derived in terms of characteristic impedance of the corresponding semi-infinite systems. The specific relationship depends upon the types of wave field and coupling concerned; some examples follow.

(i) Line junctions between homogeneous plates [7.35, 7.36, 7.26].

$$\eta_{\alpha\beta} = c_{g_\alpha} L \tau_{\alpha\beta} / \pi \omega S_\alpha \tag{7.20}$$

where $\eta_{\alpha\beta}$ is the diffuse field wave energy flux transmission coefficient, c_{g_α} is the group velocity of waves in subsystem α, L is the junction length, and S_α is the surface area of subsystem α.

(ii) Plate-cantilever beam junctions [7.36].

$$\eta_{bp} = (2\rho_b c_b \kappa_b S_b)^2 (\omega M_b)^{-1} \operatorname{Re}(z_p^{-1}) \, | \, z_p/(z_p + z_b)|^2 \tag{7.21}$$

where κ_b is the radius of the beam, z_p and z_b are the appropriate moment impedances of the plate and the beam respectively, c_b is the longitudinal wave speed in the beam, κ_b is its radius of gyration, S_b is the cross-sectional area of the beam, and M_b is the mass of the beam.

(iii) Stiff point bridges between plates [7.26].

$$\eta_{12} = \frac{2}{\pi \omega n_1} \frac{z_1 z_2}{(z_1 + z_2)^2} \tag{7.22}$$

where z_1 and z_2 are the two plate point impedances and n_1 is the modal density of plate 1.

(iv) Non-resonant sound transmission through a panel from a room [7.48].

$$\eta_{nr} = (cS/4\omega V)\tau_{nr} \text{ where}$$

$$\tau_{nr} = \frac{\pi^9}{2^{13}} \frac{M^2}{\rho_0^2 S} \left[1 - \left(\frac{10\omega}{\omega_{cr}}\right)^2 \right] + \left(\frac{\omega M}{2\rho_0 c}\right)^2 \; ; \; \omega_o < \omega < \omega_{cr}/10$$

$$= \left(\frac{\omega M}{2\rho_0 c}\right)^2 \; ; \; \omega > \omega_{cr}/10$$

where ω_{cr} = critical frequency, M = panel mass per unit area, S = panel area, and V = room volume.

(v) Plate-cavity acoustic coupling [7.28].

$$\eta_{\alpha\beta} = \rho_0 c\sigma/\omega m \dagger \tag{7.23}$$

where σ is the radiation efficiency and m is the mass per unit area of the plate.

(vi) Cylindrical shell-cavity acoustic coupling [7.29].

$$\eta_{\alpha\beta} \text{ as equation } (7.23)^\dagger$$

(vii) Acoustic cavity–cavity coupling [7.38].

$$\eta_{\alpha\beta} = cS_w\tau_{\alpha\beta}/8\pi f V_\alpha \tag{7.24}$$

where S_w is the coupling aperture area, V_α is the volume of cavity α and $\tau_{\alpha\beta}$ is the sound intensity transmission coefficient (transmission loss $R = 10 \log_{10}(1/\tau)$). This coupling loss factor holds for open apertures and for common partitions.

(viii) Beam–beam coupling [7.30].

†Valid above a lower limiting frequency.

Davies *et al.* [7.30] compare estimates of modal average coupling loss factor derived on the basis of summation (integration) of mode-to-mode couplings (see section 7.7.1) with the wave transmission approach. They find that there is a lower limiting frequency below which the wave transmission estimate exceeds the true value, the difference increasing rapidly with decrease of frequency. This conclusion confirms earlier findings by Fahy [7.28, 7.29] and Scharton [7.39].

The accuracy of this approach to estimating the coupling loss factor is dependent upon the degree of correlation between the vibrations of the two subsystems on either side of the junction [7.24]. Clearly, if waves returning to a junction after travelling around a subsystem are not correlated with the waves simultaneously incident upon the junction from the contiguous system, there is no time-average interaction between the two sets of waves, and they transmit power independently of one another, as though into non-reflective contiguous systems. The factors favouring low correlation are: large subsystems, low group velocities, and high wave damping. The first two factors produce high modal densities (see Table 7.2) and the third produces large modal bandwidths. It transpires [7.30, 7.24] that the relevant parameter is the Modal Overlap Factor $M = [\eta\omega] \, n(\omega)$, where η is the subsystem dissipation loss factor and $n(\omega)$ is the modal density (modes/rad/s), and that the wave transmission method of estimating coupling loss factor is only reliable if M exceeds unity. Particular care should be taken with one-dimensional systems in which the bending wave modal density decreases with frequency.

7.8 MODAL DENSITY

Table 7.2 provides formulae for various idealized forms of subsystem.

7.9 EXPERIMENTAL ASPECTS OF THE APPLICATION OF SEA

7.9.1 Modal density

Although satisfactory results of attempts to estimate modal density by experiment have been reported [7.41] it is the author's experience that the method of 'counting peaks' in a frequency response curve is very unreliable since no criteria for the identification of a resonance peak are available. The magnitudes of the various peaks depend upon the positions of the driving point and of the response measurement transducer and, at relatively high modal densities, modal overlap completely obscures the individual peaks. Some improvements in resolution can be obtained by summing the squared output of many transducers. Indeed, at sufficiently high modal densities the frequency of peaks in a frequency response curve is not determined by modal density at all [7.2]. Phase response techniques are useful at low frequencies but these also fail under conditions of high modal overlap.

Attempts have been made recently [7.42] to take advantage of the theoretical relationship between the frequency-averaged value of the real part of the point input mobility (admittance) of a uniformly distributed multi-mode system and its modal density [7.26]. It is found that the measured values agree well with theory, but unfortunately it does not seem possible to extend this technique to non-uniformly distributed systems such as plates carrying stiffeners or concentrated masses because of uncertainties regarding suitable points of application and total effective masses. Consequently reliance must currently be placed upon theoretical estimates.

7.9.2 Estimate of energy of vibration from measurement

The time-averaged vibration energy of multi-mode, lightly damped elastic systems subject to broadband excitation can be related to certain field variables as described in section 7.2. The major problem in obtaining estimates of the total vibration energy is that there is inevitably a spatial variation of the time-averaged quantity over a field and consequently there is an associated uncertainty in using measurements at a finite number of points in specific directions. This is less of a problem in an acoustic field in which the energy can be simply related to the scalar quantity, pressure, and for which a considerable body of theory concerning distribution statistics exists, e.g. [7.8]. But in a structural vibration field many different types of waves can exist, and practical structures are generally not homogeneous and uniform like fluids; in addition, relatively few investigations have been made of the statistics of structural fields [7.9]. Consequently, provided the type(s) of vibration fields are known (e.g. flexural, torsional, transverse shear, longitudinal waves), empirical measures of uncertainty, or confidence, based upon measures of spatial variance of the measured quantities, must be employed, so that estimates of the uncertainties in the quantities ultimately derived from SEA theory can be established. Reference [7.27] presents some interesting data in this respect, in which variances of 6 dB or less were considered to be acceptable. Separation of in-plane longitudinal and transverse shear waves in plates from flexural waves was also accomplished in the research by the use of tri-axial accelerometers, with cross-sensitivity corrections. Estimates of structural stiffener or beam energies are particularly difficult to obtain.

7.9.3 Measurement of power flow

Measurements have been made of power flow through discrete coupling elements [7.31, 7.43] but in general, with distributed coupling regions it is unpractical to attempt such a measurement.

7.9.4 Internal loss factor

The concept of modal average internal loss factor is an essential element of SEA. Unfortunately, insufficient attention has been paid to the question of

Table 7.2 – Asymptotic Modal Density Formulae

System	Wave Type	Modal Density n(f) Modes/Hz	Notation
Beam, bar or rod	Longitudinal	$2L/C_l'' = 2L/C_g$	L = length of beam. C_l'' = phase speed = $\sqrt{E/\rho}$ for homogeneous beam; C_g = group speed.
Beam, bar or rod	Flexural	$(2\pi)^{-\frac{1}{2}} L (m'/B)^{\frac{1}{4}} f^{-\frac{1}{2}} = 2L/C_g$	L = length of beam, m' = mass per unit length; B = bending stiffness = EI for homogeneous beam; I = second moment of area of cross-section.
Beam, bar or rod	Torsional	$2L/(Z/I)^{\frac{1}{2}} = 2L/C_g$	L = length of beam; Z = torsional rigidity. I = polar moment of inertia.
Flat plate	Flexural	$\frac{1}{2} A (m''/B)^{\frac{1}{2}} = 4\pi A f/C_g^2$	A = area of plate; m'' = mass per unit area of plate; B = bending stiffness per unit width = $Eh^3/12(1-\nu^2)$ for homogeneous plate; h = plate thickness.
Flat honeycomb panel (rectangular)		$\dfrac{\pi^2}{2\omega_0} \cdot \dfrac{a}{b} \cdot r\!\left(\dfrac{\omega}{\omega_0}\right) \cdot \left\{1 + \dfrac{1 + \frac{1}{2} r^2(\omega/\omega_0)}{r(\omega/\omega_0)\left[1 + \frac{1}{4} r^2(\omega/\omega_0)\right]^{\frac{1}{2}}}\right\}$	a,b = side length; $\omega_0 = (\pi/b)^2 (D/MA)^{\frac{1}{2}}$; $r(\omega/\omega_0) = 2\pi f [DMA/D_{Qx} D_{Qy}]^{\frac{1}{2}}$; $D = E_f t_f t_c^2 (1 + t_f/t_c)^2/2(1-\nu^2)$ = bending stiffness, $D_{Qx} = G_{cx} t_c (1 + t_f/t_c)^2$ = shear stiffness; E_f = elastic modulus of face plate of thickness t_f; G_{cx} = shear modulus of core of thickness t_c in the x direction G_{cy} = shear modulus of core of thickness t_c in the y direction

circular cylin- (radial) drical shell	$f/f_R \leq 0.48$, $\dfrac{}{\pi}\left(\dfrac{f}{f_R}\right) \Big/ tC_i'$	$C_i' = [E/\rho(1-\nu^2)]^{\frac{1}{2}}$; f_R = ring frequency = $C_i'/2\pi R$; R = cylinder radius; F = bandwidth factor = (upper frequency/lower frequency)$^{\frac{1}{2}}$. NB: These formulae are designed specifically to yield estimates of average modal densities in one third octave bands.
	$0.48 \leq f/f_R \leq 0.83$; $\dfrac{7.2}{\pi}\left(\dfrac{f}{f_R}\right)\cdot\dfrac{A}{tc_l}$	
	$f/f_R > 0.83$	
	$\left\{2 + \dfrac{0.6}{F-F^{-1}}\left[F\cos\left(\dfrac{1.75}{F^2}\left(\dfrac{f}{f_R}\right)\right) - \dfrac{1}{F}\cos\left(1.75 F^2\left(\dfrac{f}{f_R}\right)\right)\right]\right\}$	
Fluid volume Acoustic		L = length of tube; A = area of shallow cavity; V = volume of enclosure; c = speed of sound = C_g; S = total wall area of enclosure.
One-dimensional (tube)	$2L/c$	
Two-dimensional (shallow cavity)	$\pi Af/c^2$	
Three-dimensional (enclosure)	$4\pi Vf^2/c^3 + \pi Sf/2c^2$	

accuracy of estimation from measurement of decay rates or equilibrium energy level under given conditions of power input. The fundamental problem is discussed in reference [7.44] in which it is stated that, to provide a suitable measure of modal average loss factor, the distribution of modal energies which holds under the operational equilibrium conditions should be simulated during measurement of the internal loss factor of the *uncoupled* subsystems. This is not normally achieved by impact loading, especially of inhomogeneous mechanical structures, and very inaccurate estimates can result from the application of impact excitations, unless averaging over many loading positions is achieved. Even then, the decay results are often very difficult to interpret. Frequently it is impossible to decouple the subsystems, and then, of course, the measured 'loss factor' is a function of both internal and coupling losses. One example would be that of an element of an erected building structure. Comparison of loss factors estimated by measuring input power, and those derived from decay measurements, suggests that the steady-state equilibrium technique may be more reliable, but it is often more difficult to implement than the decay method, especially outside the laboratory.

It is probably fair to say that the estimation of internal loss factors is the major source of uncertainty in the application of SEA, as it is in most calculations of dynamic response of mechanical systems. The magnitude of this uncertainty is often such as to make quibbles regarding the finer points of analysis pale into insignificance – a fact which is too frequently overlooked by enthusiastic 'pursuers of the truth'. Upper and lower bounds to account for this factor should always be quoted in the presentation of theoretical predictions.

7.9.5 Coupling loss factor

In principle, the coupling loss factor between two coupled subsystems can be estimated by first measuring the apparent loss factor of one of them with the other attached and again with it detached. In the former configuration the damping of the attached system should be artificially increased as much as possible to reduce feedback of energy, in which case the apparent loss factor of the driven system equals the sum of its dissipation loss factor and the coupling loss factor. Unfortunately, this technique does not usually yield accurate answers because the coupling loss factor is usually very much smaller than the dissipation loss factor, and is given by the small difference between two large numbers. An alternative technique [7.44] is to measure the dissipation loss factors of both uncoupled subsystems, and the equilibrium energies when one is directly driven. The resulting equation for coupling loss factor is:

$$\eta_{\alpha\beta} = \eta_\beta n_\beta \bar{E}_\beta/(n_\beta \bar{E}_\alpha - n_\alpha \bar{E}_\beta) \tag{7.25}$$

where the modal densities n_α and n_β are theoretically estimated. Of course, this technique is best suited to laboratory situations in which subsystems can easily be isolated and uncoupled. Attempts to infer coupling loss factors from matrix equations such as (7.19), involving many subsystems, are almost certainly doomed to failure.

7.10 CONCLUSION

SEA has frequently been found to provide an approach to the estimation of high-frequency vibration levels in complex systems where no other rational technique has been feasible. The basis on power balance equations is eminently sensible and fundamental. The approach is very attractive to the engineer who is faced with design decisions because the results display the influence of the system parameters in an explicit and readily manipulated form. Perhaps because of this very attraction it has unfortunately been abused by people who attempt to apply it without understanding its *basis* and its *limitations*. What is now sorely needed is a number of systematic and carefully worked experiments which will establish the conditions under which SEA can provide acceptably accurate estimates of vibration levels in coupled systems, together with quantitative descriptions of the sensitivity of confidence levels to variation in system parameters such as coupling strength, modal overlap factor and perturbations of geometry and dynamic properties.

REFERENCES

[7.1]   Lyon, R. H. (1969). *J. Acoust. Soc. Amer.*, **45**(3), 545, Statistical analysis of power injection and response in structures and rooms.

[7.2]   Shroeder, M. (1954). *Acustica*, **4**, 594. Die statistischen Parameter der Frequenz kurven von grossen Räumen.

[7.3]   Fahy, F. J. (1971), *Proc. of the 7th International Congress on Acoustics*, Budapest, Paper 19V5. Statistics of acoustically induced vibration.

[7.4]   Mercer, C. A., Rees, P. L. & Fahy, F. J. (1971), *J. Sound Vib.*, **15**(3), 373. Energy flow between two weakly coupled oscillators subject to transient excitation.

[7.5]   Hammond, J. K. (1971). PhD Thesis, Southampton University. Frequency-time methods in vibrations.

[7.6]   Manning, J. E. & Lee, K. (1968). *Shock Vib. Bull.*, **37**(4), 65. Predicting mechanical shock transmission.

[7.7]   Fahy, F. J. (1974). *The Shock and Vibration Digest*. Statistical energy Analysis: A Critical Review.

[7.8]   Schroeder, M. (1969). *J. Acoust. Soc. Amer.*, **46**(3), Pt. 1, 277. Effect of frequency and space averaging on the transmission responses of multi-mode media.

[7.9]   Stearn, S. M. (1970). *J. Sound Vib.*, **12**(1), 85. Spatial variation of stress, strain and acceleration in structures subject to broad frequency band acceleration.

[7.10] Stearn, S. M. (1971), *J. Sound Vib.*, **15**(3), 193. The concentration of dynamic stress in a plate at a sharp change of section.

[7.11] Scharton, T. D. & Lyon, R. H. (1968). *J. Acoust. Soc. Amer.*, **43**(6), 1332. Power flow and energy sharing in random vibration.

[7.12] Rayleigh, Lord (1894). *The theory of sound.* Vol. 1, Section 82. Dover Publications, New York.

[7.13] Newland, D. E. (1966). *J. Acoust. Soc. Amer.*, **39**(4), 755. Comment on "Vibration energy transmission in a three-element structure".

[7.14] Newland, D. E. (1966). *J. Sound Vib.*, **3**(3), 262. Calculation of power flow between coupled oscillators.

[7.15] Kakar, M. P. (1969). *Proc. of the British Acoustical Society meeting 'Sonically induced vibration of structures'*. Power flow between linearly coupled oscillators. (Also Sheffield University PhD thesis, Dept. of Mechanical Engineering.)

[7.16] Woodhouse, J. (1981). *J. Acoust. Soc. Amer.*, **69**(6), 1695-1709. An approach to the theoretical background of statistical energy analysis applied to structural vibration.

[7.17] Newland, D. E. (1969). *J. Acoust Soc. Amer.*, **43**(3), 553. Power flow between a class of coupled oscillators.

[7.18] Newland, D. E. (1965). *J. Inst. Maths. Applics.*, **1**, 199. Energy sharing in the random vibration of non-linearly coupled modes.

[7.19] Crandall, S. H. & Lotz, R. (1971). *J. Acoust. Soc. Amer.*, **49**(1), Pt. 1, 352. On the coupling loss factor in statistical energy analysis.

[7.20] Gulizia, C. & Price, A. J. (1977). *J. Acoust. Soc. Amer.*, **61**(6), 1511-1515. Power flow between strongly coupled oscillators.

[7.21] Smith, P. W. Jr. (1979). *J. Acoust. Soc. Amer.*, **65**(3), 695-698. Statistical models of coupled dynamical systems and the transition from weak to strong coupling.

[7.22] Fahy, F. J. (1970). PhD Thesis, Southampton University. Acoustically induced vibration of containing structures.

[7.23] Fahy, F. J. (1970). *J. Sound Vib.*, **11**(4), 481. Energy flow between oscillators: special case of point excitation.

[7.24] Smith, P. W. Jr. (1980). *J. Sound Vib.*, **70**(3), 343-353. Random response of identical one-dimensional coupled systems.

[7.25] Maidanik, G. (1976). *J. Sound Vib.*, **46**(4), 561-584. Response of coupled dynamic systems.

[7.26] Cremer, L., Heckl, M. & Ungar, E. (1973). *Structure-borne sound.* Springer-Verlag, Berlin.

[7.27] Plunt, Juha (1980). PhD Thesis, Chalmers University of Technology, Goteborg, Sweden. Methods for predicting noise levels in ships.

[7.28] Fahy, F. J. (1969). *J. Sound Vib.*, **10**(3), 490. Vibration of containing structures by sound in the contained fluid.

[7.29] Fahy, F. J. (1970). *J. Sound Vib.*, **13**(2), 171. Response of a cylinder to random sound in the contained fluid.

[7.30] Davies, H. G. (1981). *J. Sound Vib.*, **77**(3), 311-321. Ensemble averages of power flow in randomly excited coupled beams.

[7.31] Lotz, R. & Crandall, S. H. (1973). *J. Acoust. Soc. Amer.*, **54**(2), 516-524. Prediction and measurement of the proportionality constant in SEA of structures.

[7.32] Remington, P. J. & Manning, J. F. (1975). *J. Acoust. Soc. Amer.*, **57**(2), 374-379. Comparison of Statistical Energy Analysis power flow predictions with an 'exact' calculation.

[7.33] Garrelick, J. M. (1972). Cambridge Acoustical Associates, Cambridge, Massachusetts. Technical report U-392-213. Dynamic response of coupled systems: a comparison between statistical energy and deterministic techniques.

[7.34] Gibbs, B. M. & Gilford, C. L. S. (1976). *J. Sound Vib.*, **49**(2), 267-286. The use of power flow methods for the assessment of sound transmission in building structures.

[7.35] Budrin, S. V. & Nikiforov, A. J. (1965). *Sov. Phys. Acoust.*, **9**, 33. Wave transmission through assorted plate joints.

[7.36] Lyon, R. H. (1975). *Statistical energy analysis of dynamical systems: theory and applications*. M.I.T. Press.

[7.37] Maidanik, G. (1978). *J. Sound Vib.*, **60**(3), 313-318. Influence of fluid loading and compliant coating on the coupling loss factor across a rib.

[7.38] Beranek, L. L. (ed.) (1971). *Noise and vibration control*. McGraw-Hill Book Co. Inc., New York.

[7.39] Scharton, T. D. (1971). *J. Acoust. Soc. Amer.*, **50**(1), Pt. 2, 373. Frequency averaged power flow into a one-dimensional system.

[7.40] Hart, F. D. & Shah, K. C. (1971). NASA-CR-1773. *Compendium of modal densities of structures* (obtainable from the National Technical Information Service, Springfield, Virginia 22151, USA).

[7.41] Crocker, M. J. & Price, A. J. (1969). *J. Sound Vib.*, **9**(3), 469. Sound transmission using statistical energy analysis.

[7.42] Clarkson, B. L. & Pope, R. (1980). *Proc. of a Conference on 'Recent Advances in Structural Dynamics'*, I.S.V.R., Southampton University. Experimental determination of structural modal densities and average loss factors.

[7.43] Pinnington, R. J. & White, R. G. (1981). *J. Sound Vib.*, **75**(2), 179-197. Power flow through machine isolators to resonant and non-resonant beams.

[7.44] Clarkson, B. L., Cummins, R. J., Eaton, D. C. G. & Vessaz, J. P. (1981). *European Space Agency Journal*, **5**, 137-150. Prediction of high-frequency structural vibrations using Statistical Energy Analysis (SEA).

[7.45] Wilby, J. F. & Pope, L. D. (1980). *Journal of Spacecraft and Rockets,* **17**(3), 232--239. Prediction of the acoustic environment in the space shuttle.

[7.46] Fahy, F. J. (1975). *Revue d'Acoustique.* L'Analyse Statistique Energetique: Une Revue Critique.

[7.47] Chintsun Hwang & Pi, W. S. (1973). NASA-CR-1224450. Investigation of vibrational energy transfer in connected structures.

[7.48] Guvovich, Yu. A. (1980). *Sov. Phys. Acoust.,* **26**(3), 221–223. Nonresonant sound transmission through a rectangular panel.

[7.49] de Vries, D., van Bakel, J. G. & Berkhout, A. J. (1981). *Proc. Internoise, 81,* 465–469. Application of S.A.E. in building acoustics: a critical note.

CHAPTER 8

Nonlinear acoustics

C. L. Morfey

Institute of Sound & Vibration Research, University of Southampton

8.1 INTRODUCTION

Propagating sound waves of finite amplitude exhibit a phenomenon known as nonlinear steepening. Fig. 8.1 shows the spatial waveform (p versus x at one instant) of a plane progressive pressure wave. Nonlinear steepening means that sections of the waveform like ABC, which slope downwards, become steeper in slope as time advances. Upward sloping sections like CDE simultaneously become less steep. These effects are due to the fact that positive pressure disturbances travel faster than negative pressure disturbances.

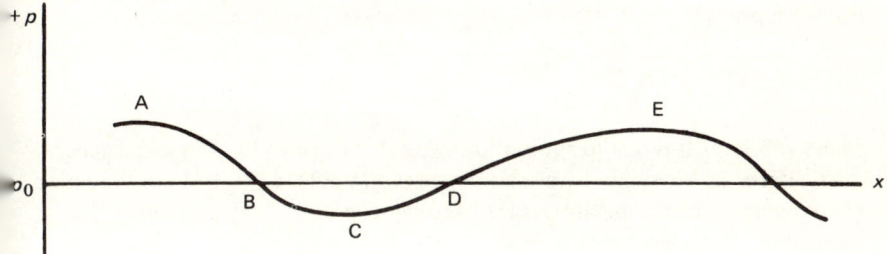

Fig. 8.1 – Pressure waveform of a one-dimensional travelling wave.

Nonlinear steepening is opposed, in real fluids, by dissipative effects (e.g. viscosity). These work in the reverse direction so as to smooth out any steep gradients in the waveform [8.1]. The two opposing effects lead to a number of phenomena of practical interest in acoustics, ranging from shock formation to anomalous atmospheric propagation. Examples of these are presented in later sections.

8.2 THEORY OF NONLINEAR PROPAGATION

The simplest description of nonlinear distortion is obtained by neglecting dissipative effects entirely; this is the first model discussed below. The non-dissipative model becomes unrealistic once large slopes appear in the waveform, and the second step towards a realistic model is to introduce shocks of zero thickness, while still neglecting dissipation between the shocks. Finally, waves in which nonlinear steepening is relatively weak require a more complete model still, in which both nonlinearity and dissipation are retained throughout the waveform.

8.2.1 Plane waves: shock-free distortion

In an ideal fluid whose undisturbed state is uniform, Riemann [8.2] showed that plane waves propagate with the variable velocity

$$u(x,t) = v \pm c \ . \tag{8.1}$$

The two signs in equation (8.1) represent two independent solutions; each solution on its own is a progressive wave, in the sense that values of the flow variables (p, v, c, ρ) are propagated in the x direction at velocity u. The thermodynamic variables (p, c, ρ) are isentropically related, while the fluid velocity v is given by

$$v = \pm b(p) \quad (\pm \text{ progressive waves}); \ b(p) = \int_{p_0}^{p} \frac{\mathrm{d}p'}{\rho(p')\,c(p')} \ . \tag{8.2}$$

For wave propagation in a perfect gas, equations (8.1) and (8.2) give

$$u = \pm c_0 \pm \frac{\gamma + 1}{2}v ; \quad c = c_0 \pm \frac{\gamma - 1}{2}v \tag{8.3}$$

where subscript 0 refers to the stationary undisturbed fluid. The signal speed u thus differs from c_0 by a variable amount $\frac{1}{2}(\gamma + 1) \, v$. Part of this variation $\frac{1}{2}(\gamma - 1) \, v$ is due to the difference between c and c_0, and part (v) is due to convection of the wave by the fluid.

In what follows we consider the $+x$ progressive solution. The waveform distortion due to u variation is conveniently described in moving coordinates

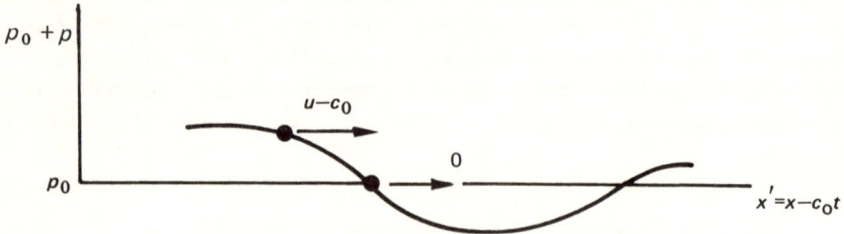

Fig. 8.2 – Waveform propagation in moving coordinates.

(Fig. 8.2). In the $p-x'$ plane ($x' = x - c_0 t$), each point of the waveform advances to the right at velocity

$$u - c_0 = \frac{\gamma + 1}{2} v .$$ (8.4)

Since v is positive when p is positive, the downward-sloping parts of the waveform become steeper with time. Eventually the waveform acquires an infinite slope, and subsequently becomes multivalued, according to the Riemann

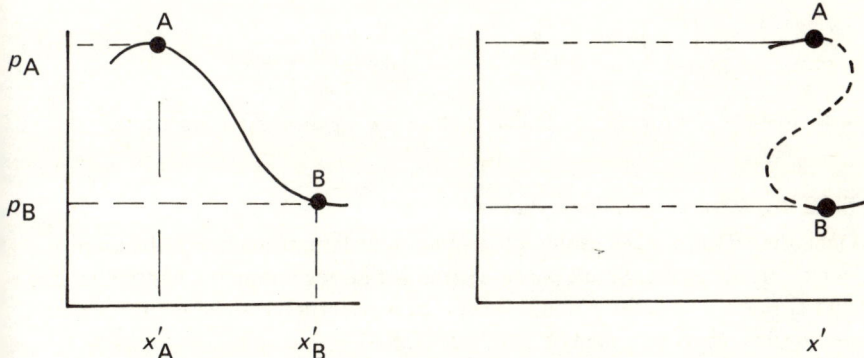

Fig. 8.3 – Sketch to illustrate distortion of a finite-amplitude travelling wave.

solution (Fig. 8.3). Dissipation must be allowed for in the model as soon as this occurs, by introducing shocks as outlined below.

During the pre-shock phase, waveform distortion is described by two equivalent equations, either of which describes plane waves exactly:

$$\frac{1}{c_0}\left(\frac{\partial x'}{\partial t}\right)_\sigma = \sigma \quad \text{or} \quad c_0\left(\frac{\partial t'}{\partial x}\right)_\sigma = \frac{-\sigma}{1+\sigma} .$$ (8.5 a, b)

Here σ is the non-dimensional excess wavespeed,

$$\sigma = \frac{u}{c_0} - 1 = \frac{\gamma + 1}{2}\frac{v}{c_0} .$$ (8.6)

Equation (8.5a) treats the wave variable σ as a function of (x', t), and corresponds to Figs. 8.1 to 8.3 (note that p is a function of σ, so the pressure waveform may be deduced from the σ waveform). Equation (8.5b) treats σ as a function of (t', x), where $t' = t - x/c_0$ is the retarded time. The first equation is more convenient if the spatial waveform is prescribed at $t = t_0$ and its evolution with time is required. More commonly, the time-waveform is prescribed at $x = x_0$; the second equation is then appropriate, and describes the waveform evolution with distance.

The choice of σ as wave variable is natural for plane isentropic waves, since it leads to simple exact equations. For spherical waves, or waves containing shocks, this advantage disappears. In what follows we prefer either p, or the non-dimensional wave pressure

$$P = \frac{\gamma + 1}{2} \frac{p}{\rho_0 c_0^2} \quad . \tag{8.7}\dagger$$

Equations (8.5 a,b) may be written in terms of P as

$$\frac{1}{c_0}\left(\frac{\partial x'}{\partial t}\right)_P \cong \frac{P}{1 + \frac{1}{2}P} \quad \text{or} \quad c_0\left(\frac{\partial t'}{\partial x}\right)_P \cong \frac{-P}{1 + \frac{3}{2}P} \quad , \tag{8.8 a,b}$$

with an error of order P^3. Note that σ and P are equal, to first order in P.‡

8.2.2 Plane waves containing shocks

Once the waveform steepening process leads to infinite gradients, the isentropic model breaks down. Shock waves appear — thin regions across which the pressure changes by a finite amount — and the waveform is divided up into separate isentropic regions separated by shocks (Fig. 8.4).

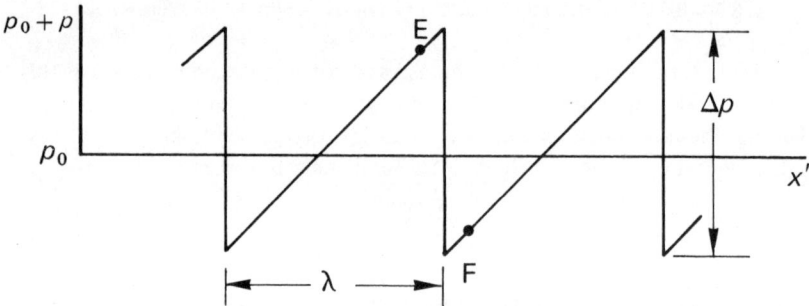

Fig. 8.4 — Waveform containing shocks.

The propagation of each shock is described by the Rankine-Hugoniot relations [8.3], which connect the flow properties on the two sides of the shock. For weak shocks (pressure jump $p_s \ll \rho_0 c_0^2$), the entropy change is of order p_s^3. We describe here a weak-shock theory, in which such terms are neglec-

---

†In general, $\dfrac{\gamma + 1}{2}$ is replaced by the nonlinearity coefficient $\beta = \dfrac{1}{c}\left(\dfrac{\partial \rho c}{\partial \rho}\right)_s$.

‡The exact $\sigma - P$ relation follows from equations (8.1) and (8.2).

ted. This also means that in the regions between shocks, reflected waves may be neglected. The motion of each shock is given by

$$\frac{1}{c_0}\frac{dx_s'}{dt} \cong \frac{P_m}{1+\frac{1}{2}P_m} \quad \text{or} \quad c_0\frac{dt_s'}{dx} \cong \frac{-P_m}{1+\frac{3}{2}P_m} , \qquad (8.9\ a,b)$$

with an error of order P^3; P_m is the average value of P on the two sides of the shock.

The simplicity of equations (8.9 a,b), and their similarity to the corresponding shock-free waveform relations (8.8 a,b), follow from the use of P as the wave variable. Using σ or any other variable would lead to more complicated expressions, at this level of accuracy.

It follows from a comparison of equations (8.8) and (8.9) that each shock moves slower than the point on the waveform just behind it (E in Fig. 8.4). Likewise the shock overtakes the point just ahead (F in Fig. 8.4). The shock therefore evolves as it propagates, at a rate determined by the slopes on either side. The entire pressure waveform, including shocks, is determined to second-order accuracy by these equations.

8.2.3 Accuracy of the second-order waveform relations

Numerical calculations have been carried out to check the accuracy of the waveform distortion equation (8.8b), and the shock overtaking rates given by (8.8b) and (8.9b). For $\gamma = 1.4$, the accuracy is better than 1% provided the magnitude of P, and the shock strength P_s, remain less than 0.2.

8.2.4 First-order nonlinear theory

Values of P encountered in high-intensity sound propagation are typically less than 0.05; this is the peak value in a harmonic sound wave at 166 dB in air (at 1 atm pressure). It then becomes reasonable to replace (8.8) and (8.9) by simpler first-order approximations. Thus (8.8b) and (8.9b) become

$$c_0\left(\frac{\partial t'}{\partial x}\right)_P \cong -P , \quad c_0\frac{dt_s'}{dx} \cong -P_m , \qquad (8.10)$$

to better than 10% accuracy for $|P| < 0.05$.

The first-order relations (8.10) are sufficiently simple to yield analytical predictions, some of which are discussed below. First, however, we note a useful generalization of (8.10) to spherical waves.

8.2.5 Nonlinear distortion of spherical waves

In the limit $P \to 0$, equation (8.10) gives $(\partial t'/\partial x)_P = 0$, which describes linearly-propagating plane waves in an ideal fluid. The corresponding linear equation for outgoing spherical waves is $(\partial t'/\partial r)_{rP} = 0$, where r is the radius; this suggests that the first of equations (8.10) may be generalized to

$$c_0 \left(\frac{\partial t'}{\partial r} \right)_{rP} \cong -P ,$$
(8.11)

for nonlinear outgoing waves [8.4].

Equation (8.11) uses the same nonlinear term for spherical wave distortion as for plane waves, which is justified at sufficiently large distances from the origin $(r \gg \lambda)$ where the spherical wavefront appears locally plane. Similarly the second of equations (8.10), describing the motion of shocks, becomes

$$c_0 \frac{dt'_s}{dr} \cong -P_m .$$
(8.12)

Note that by introducing the transformed variables

$$y = r_0 \ln \frac{r}{r_0} , \quad Q = rP/r_0$$
(8.13)

the spherical-wave equations (8.11) and (8.12) become formally identical with the plane-wave equations, with y in place of x and Q in place of P.

The accuracy of using Q in spherical waves as the counterpart of P in plane waves may be tested in two ways: (i) is *energy conserved* prior to shock formation? (ii) does a *shock wave* (with uniform pressure on either side) actually propagate with rP_s = constant? Calculations show that an energy-conserving outgoing wave would have rP constant within 1% for $|P| < 0.2$ $(\gamma = 1.4)$. For shocks, the work of Chester [8.5] gives the error in assuming rP_s constant as less than about 5%, under the same conditions.

8.2.6 Initially-sinusoidal waves: shock formation distance

A sinusoidal plane wave, with peak-to-peak amplitude $(\Delta P)_0$, starts to form shocks after travelling a distance

$$\bar{x} = \frac{\lambda}{2(\Delta P)_0} \quad (\lambda = \text{wavelength}) .$$
(8.14)

This result follows from the first-order distortion equation (8.10). The corresponding result for spherical waves, starting from radius r_0, is that shocks form at a radius

$$\bar{r} = r_0 \exp \frac{\lambda}{2(\Delta P)_0 r_0} .$$
(8.15)

It follows that $(\bar{r} - r_0)$ is always greater than $\bar{x}$ for the same value of $(\Delta P)_0$. This is to be expected, since the spherical wave weakens as it spreads out.

8.2.7 Periodic waveforms of arbitrary shape

For arbitrary periodic waveforms of wavelength λ and peak-to-peak amplitude ΔP, some rather general settlements are possible on the basis of first-order weak shock theory.

(i) The characteristic propagation distances, over which significant non-linear waveform distortion occurs, are $\bar{x}$ (plane waves) or $\bar{r} - r_0$ (spherical waves) as defined above.

(ii) The waveform eventually takes on a sawtooth shape as in Fig. 8.4. The amplitude is then given asymptotically by

$$\Delta P \cong \lambda/x \quad or \quad \lambda \bigg/ \left(r \ln \frac{r}{r_0} \right) , \qquad (8.16)$$

for plane or spherical waves respectively at large distances.

In the special case of a *sawtooth* initial waveform, with initial amplitude $(\Delta P)_0$ at $x = 0$, the exact plane-wave result is

$$\frac{1}{\Delta P} = \frac{1}{(\Delta P)_0} + \frac{x}{\lambda} . \qquad (8.17)^{\dagger}$$

This reduces to the asymptotic expression (8.16), at distances sufficiently large that

$$x \gg \frac{\lambda}{(\Delta P)_0} \quad (\text{i.e. } x \gg 2\bar{x}) . \qquad (8.18)$$

The fact that ΔP becomes independent of $(\Delta P)_0$ at large initial amplitudes is referred to as 'saturation'. Whatever value $(\Delta P)_0$ takes, the value at ΔP cannot exceed the asymptotic limit (8.16).

8.2.8 Basic limitation on weak-shock theory

The preceding analysis is based on an ideal-fluid model; no dissipation is allowed for except (indirectly) within any shocks present. The results must therefore eventually break down, at propagation distances over which significant attenuation would occur for linear waves.

The implications for periodic waves are as follows.[‡] Let the linear attenuation coefficient be α at the fundamental frequency (implying $e^{-\alpha x}$ decay). Then the preceding weak-shock analysis will give the *fundamental* component of the waveform when αx is small, but not otherwise.

†The corresponding spherical-wave result follows if (y, Q) are substituted for (x, P).
‡Transient signals are discussed in reference [8.1].

For the higher harmonic components of the waveform, the distance limitation on weak-shock theory is more severe, since the attenuation coefficient α_n of the nth harmonic is generally larger than α.

8.3 CRITERIA FOR SIGNIFICANT NONLINEARITY

In order for significant distortion to take place in a propagating waveform, two conditions must be met. First, the nonlinear steepening effect must be powerful enough to overcome the opposing effects of dissipation. Second, the distance propagated must also be large enough for the cumulative distortion to become apparent. These two criteria are discussed further below.

8.3.1 Effects of dissipation in limiting nonlinear distortion

The balance between nonlinear and dissipative effects, in propagating periodic waves of arbitrary shape, is expressed by the non-dimensional ratio

$$\frac{(\Delta P)}{\lambda \alpha} = \hat{R} \quad \text{(acoustic Reynolds number)} \ . \tag{8.19}$$

When $\hat{R}$ is large, nonlinear steepening is powerful enough to counteract the smoothing-out tendency of dissipation; any shocks formed are relatively thin (compared with λ), and account for most of the energy dissipation in the waveform. The weak-shock model is then realistic.

For small values of $\hat{R}$, on the other hand, energy dissipation occurs throughout the waveform, and the decay of energy with distance closely follows linear theory. Thus a necessary condition for strong nonlinear distortion is that the initial value $\hat{R}_0$ be large. The evolution of the waveform may then be followed using weak-shock theory, *until* its amplitude is attenuated (by spreading and/or shock dissipation) to the point where $\hat{R}$ is of order 1.

In fact by changing from weak-shock theory to linear theory at $\hat{R} = 2$, a good approximation to the fundamental component of the waveform is obtained at *all* distances for plane waves in a thermoviscous fluid (α proportional to f^2). In the saturation limit ($\hat{R}_0 \to \infty$), the weak-shock value $\hat{R} = 2$ corresponds to $\alpha x = \frac{1}{2}$ (for plane waves), and the corresponding waveform at the change-over point is shown in Fig. 8.5.

The evolution of periodic plane waves (for different values of $\hat{R}_0$) is illustrated in Fig. 8.6, based on Blackstock's [8.6] solution of Burgers' equation for initially-sinusoidal waves in the thermoviscous fluid. The fundamental component is plotted as a function of αx, in such a way that signals which obey linear theory appear as a horizontal line. For values of $\hat{R}_0$ below about 1, the effect of nonlinearity on the fundamental is evidently quite small. The same applies at distances from $\alpha x = 1$ onwards, regardless of $\hat{R}_0$.

The simple picture above does not apply, however, to higher harmonic components of the waveform. These are continuously regenerated by nonlinear

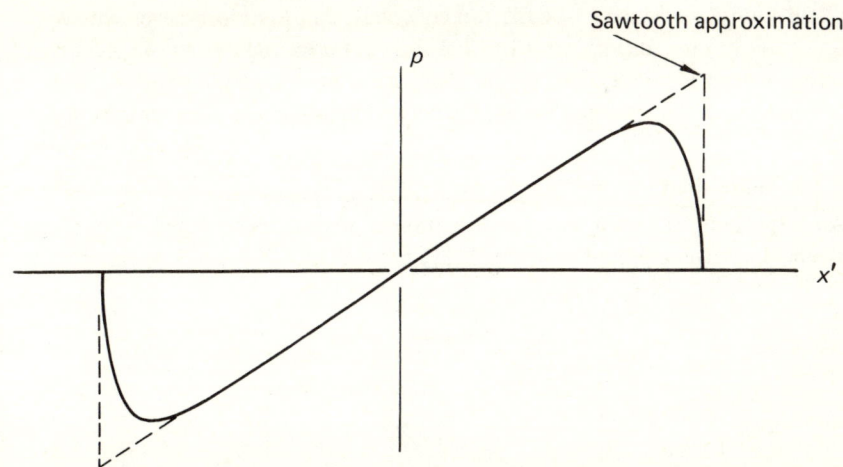

Fig. 8.5 – Asymptotic waveform for plane periodic waves at $\alpha x = 0.5$, in the saturation limit. The solid line is the exact solution from Burgers' equation; the dashed line is the approximate solution given by weak-shock theory.

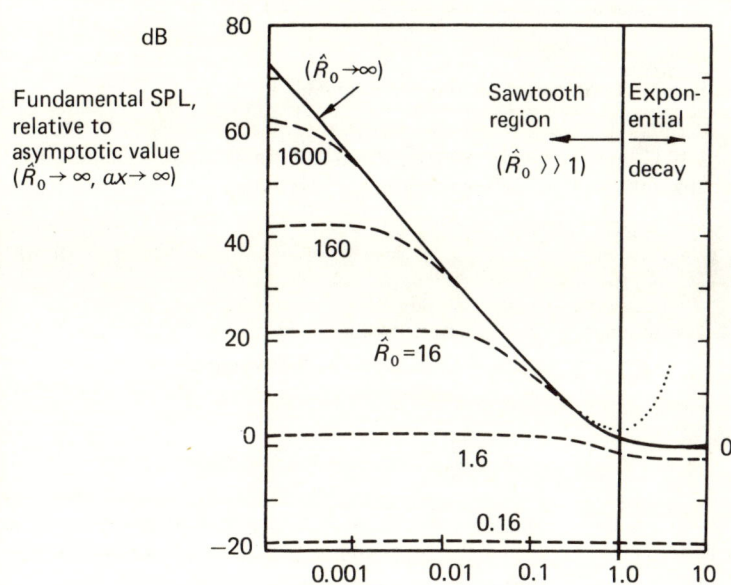

Fig. 8.6 – Fundamental component of an initially sinusoidal plane wave, showing the onset of nonlinearity as $\hat{R}_0$ exceeds 1 and saturation as $\hat{R}_0$ approaches infinity. The ordinate is the nondimensional pressure $\dfrac{\pi\beta}{2\alpha\rho\,c^2\,\lambda}p_1 e^{\alpha x}$ (squared, in dB).

distortion (at the expense of the fundamental), and therefore never achieve the linear decay rate. Blackstock [8.7] has shown, for example, that the second and higher harmonics decay *less rapidly* than linear theory predicts for *all* distances beyond $1/\alpha$ in a thermoviscous fluid, whether the waves are plane or spherical.

8.3.2 Finite rise time of shock waves

An important consequence of dissipation is that shocks formed by nonlinear steepening have a finite thickness [8.1]. Equivalently, the shock pressure p_s has

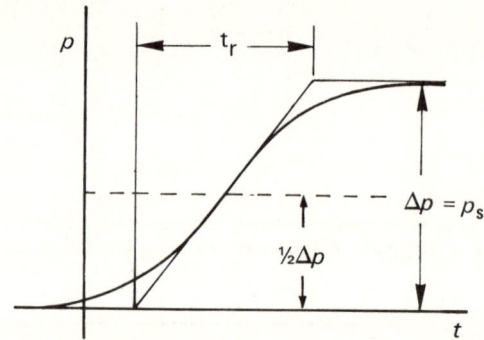

Fig. 8.7 – Definition of rise time based on slope at mid-height.

a finite rise time (Fig. 8.7). In the simple case where the dissipation mechanism is thermoviscous, the rise time defined in Fig. 8.7 is given by

$$t_r = \frac{4}{\beta} \frac{\bar{\mu}}{p_s} . \tag{8.20}$$

Here β is the nonlinearity coefficient introduced in section 8.2.1 (see Table 8.1 for typical values). The equivalent longitudinal viscosity $\bar{\mu}$ is

$$\bar{\mu} = \mu_v + \left(\frac{4}{3} + \frac{\gamma - 1}{Pr} \right) \mu , \tag{8.21}$$

where μ_v, μ are the bulk and shear viscosities, Pr is the Prandtl number and γ is the specific-heat ratio as previously. Note that Lighthill [8.1] defines the rise time by the 5% and 95% points on the pressure-rise curve of Fig. 8.7; this definition gives a value 1.5 times greater than equation (8.20).

Dissipative processes whose relaxation times are not short compared with t_r lead to complicated shock profiles; examples are discussed in references [8.8] and [8.9]. Weak shocks propagating in the atmosphere, with $p_s < 100$ Pa, fall in this category.

Table 8.1

Values of nonlinearity parameter β for various gases $\left(\doteq \dfrac{\gamma + 1}{2}\right)$ and liquids [8.16].

Fluid (pressure 1 atm)	Temperature °C	β
Air	−100 to 150	1.20
	150 to 250	1.19
Steam (low pressure)	0 to 150	1.16
	150 to 250	1.15
	250 to 400	1.14
	400 to 550	1.13
Distilled water	10	3.3
	30	3.6
Sea water, 35°/$_{oo}$ salinity	10	3.55
	30	3.7
Carbon tetrachloride	30	6.8
Mercury	30	4.9
Sodium	110	2.35

8.3.3 Initial-amplitude criterion for significant waveform distortion

A periodic signal is regarded as undergoing significant distortion, if it reaches the shock formation stage (as described by weak-shock theory) before $\hat{R}$ falls below 2. This implies, for plane waves, simply

$$\hat{R}_0 > 2 \ . \tag{8.22}$$

Fig. 8.8 shows the boundary $\hat{R} = 2$ for sinusoidal waves in air, and also the boundary $\hat{R} = \frac{1}{4}$ below which nonlinear distortion has virtually no influence on the signal energy.

For spherical waves, the amplitude is progressively reduced by spherical spreading. If a value $\hat{R} > 2$ is to be sustained out to $r = \bar{r}$ (as defined by equation (8.15)), we require that

$$\hat{R}(\bar{r}) = R_0 r_0 / \bar{r} > 2 \ . \tag{8.23}$$

This implies that

$$\frac{2(\Delta P)_0 r_0}{\lambda} \exp \frac{-\lambda}{2(\Delta P)_0 r_0} > 4 \, \alpha r_0 \ . \tag{8.24}$$

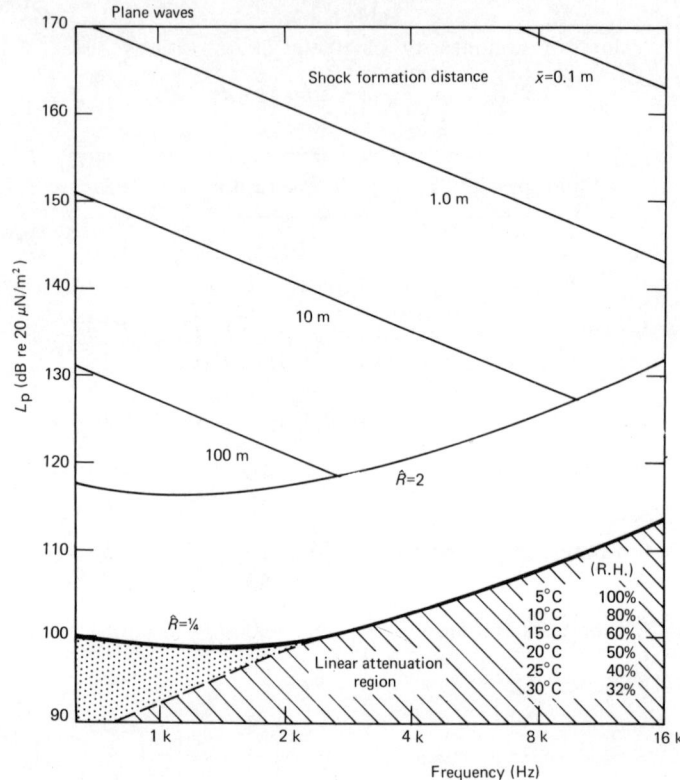

Fig. 8.8 – Amplitude-frequency map for plane waves in air, showing the boundary of the linear propagation region (where the signal component at the original frequency is not significantly affected by nonlinearity). In the nonlinear region, distances to shock formation are shown as contours; these are based on weak-shock theory, which becomes inaccurate when $\hat{R} < 2$.

Thus when αr_0 is small, $\hat{R}_0$ has to be considerably greater than 2 produce significant distortion in a spherically spreading wave.

The corresponding power output, for a 200 Hz spherical source radiating in air ($15°C$, 1 atm) which just meets criterion (8.24), is indicated by the following table.

αr_0	10^{-6}	10^{-4}	10^{-2}	10^{0}
W(dB re 1 pW)	170	174	182	204

Similar values apply at a frequency of 100 kHz in water ($15°C$, 1 atm). As a rough guide, a source diameter of 1 m corresponds to $\alpha r_0 \approx 10^{-4}$ in both cases.

8.4 APPLICATIONS OF NONLINEAR THEORY

8.4.1 Shock-wave radiation from a supersonic fan rotor

An axial-flow fan rotor operating supersonically (i.e. with supersonic inflow velocity relative to the fan blades) produces a set of rotating shock waves, which may be regarded as a finite-amplitude propagating sound field. If the fan operates in an inlet duct (for example the front fan on a high bypass ratio turbofan engine), the propagating shock pattern is constrained to follow a spiral path within the duct as in Fig. 8.9 [8.10].

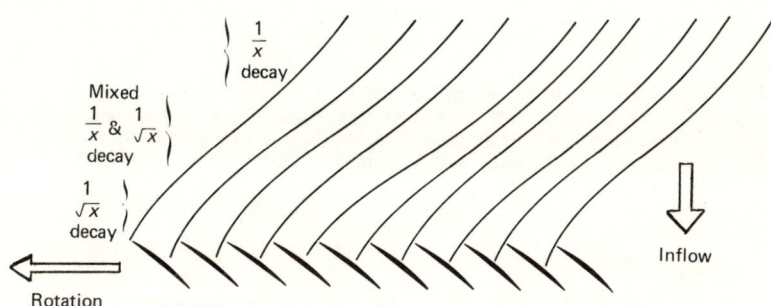

Fig. 8.9 – Sketch showing shock pattern upstream of a supersonic fan rotor. The decay laws indicated are for the shock strengths produced by an array of identical blades. In practice, slight differences between individual blades cause an irregular shock pattern. From Sofrin & Pickett [8.10].

Under these conditions the propagation is approximately one-dimensional, in the direction normal to the wavefronts. If the fan blades produce identical equally-spaced shocks, their amplitude ahead of the rotor may be estimated from equation (8.17), with x and λ interpreted appropriately [8.11]. The shock strength in the $1/x$ decay region is independent of the initial strength, and is controlled by the inflow and rotational Mach numbers.

In practice, slight irregularities in blade geometry produce an irregular shock pattern (Fig. 8.10). As such a pattern propagates in the inlet duct, the unevenness becomes more pronounced (Figs. 8.11 and 8.12). The signal energy is spread over a wide range of rotational harmonics, instead of being concentrated at the blade-passing frequency. This is the origin of the 'buzz-saw' noise which is produced by turbofan aircraft engines during take-off [8.10, 8.12].

8.4.2 Nonlinear distortion of random noise spectra

The propagation of high-intensity random noise and periodic signals is described in principle by the same theory, although in the case of noise it is not the waveform itself which is usually of interest, but rather its statistical properties. We

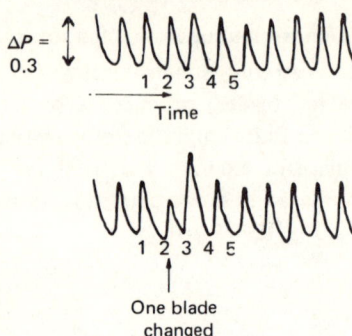

$\Delta P = 0.3$

Time

One blade
changed

Fig. 8.10 – The pressure-time history measured 0.5 blade spaces upstream of a supersonic fan rotor, at the outer duct wall. Trace (a) is for the unmodified rotor with nominally identical blades. In trace (b), the stagger angle of one blade has been changed, causing unequal shock strengths. (Stratford & Newby [8.12]).

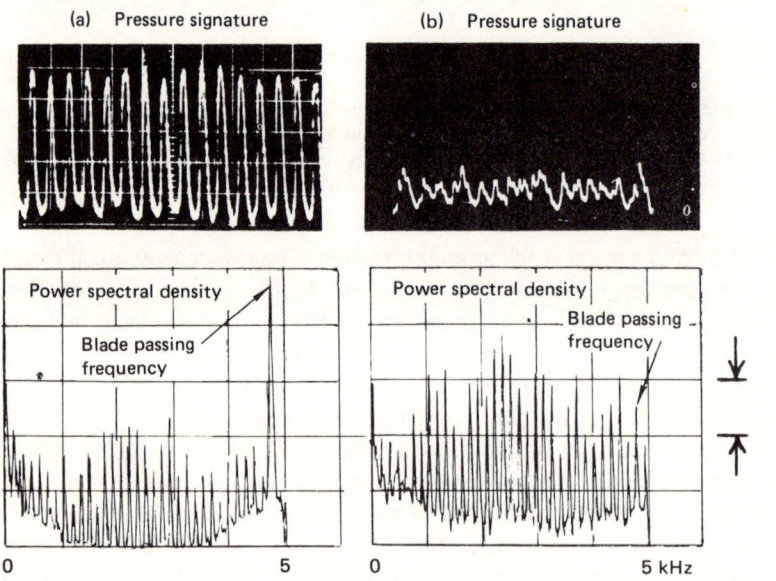

(a) Pressure signature   (b) Pressure signature

Power spectral density   Power spectral density

Blade passing
frequency   Blade passing
frequency

0   5   0   5 kHz

Fig. 8.11 – Comparison of supersonic fan pressure-time histories and corresponding spectra, measured at different distances from the fan rotor.
(a) Just upstream of rotor face.
(b) Several blade spaces upstream, showing the irregular waveform produced by slight differences between blades.
(Sofrin & Pickett [8.10]).

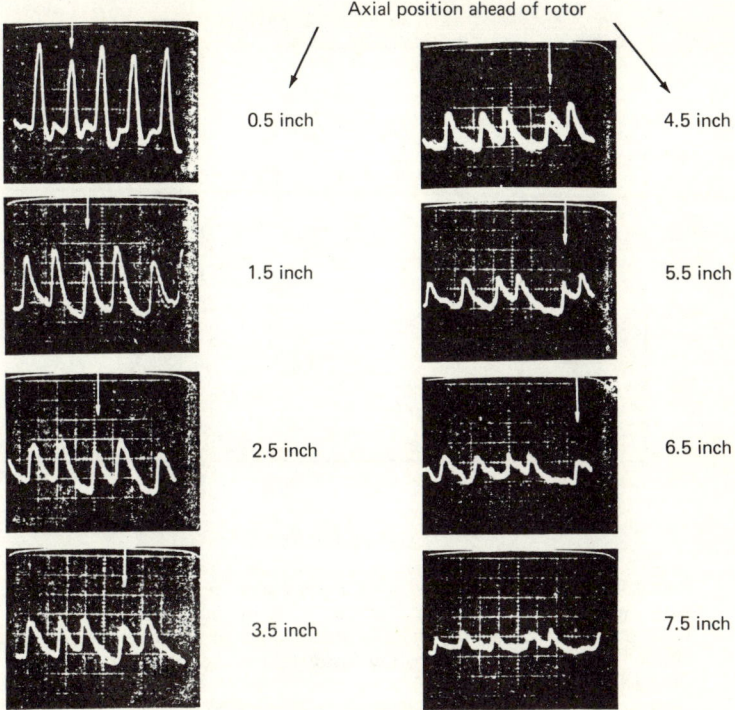

Fig. 8.12 – Evolution of pressure-time histories from a supersonic fan rotor, measured along outer wall of annular inlet duct. Blade spacing at tip is 2.75 inches. (Sofrin & Pickett [8.10]).

can therefore expect that over short distances, weak-shock theory will be appropriate; while for longer distances (αx of order 1 or more), attenuation and possibly dispersion will have to be applied to the whole waveform, instead of being confined to shocks.

Fig. 8.13 shows measurements of high-intensity noise propagation in a hard-walled pipe [8.13]. The spectrum after propagating 15 m is significantly broadened, indicating the formation of sum and difference frequency signals. Results of this type have been explained theoretically, by applying a combination of weak-shock theory and linear attenuation (plus dispersion) to the initial waveform [8.13]. Note that dispersion, which has no influence on the signal spectrum according to linear theory, can have a significant effect at finite amplitudes since nonlinear distortion depends on the waveform shape.

Spectral broadening similar to that in Fig. 8.13 has also been observed during the propagation of aircraft noise over distances up to 1 km [8.14, 8.15]. Figs. 8.14 and 8.15 show results from a controlled flyover experiment, during which a significant shortfall was observed in high-frequency sound attenuation

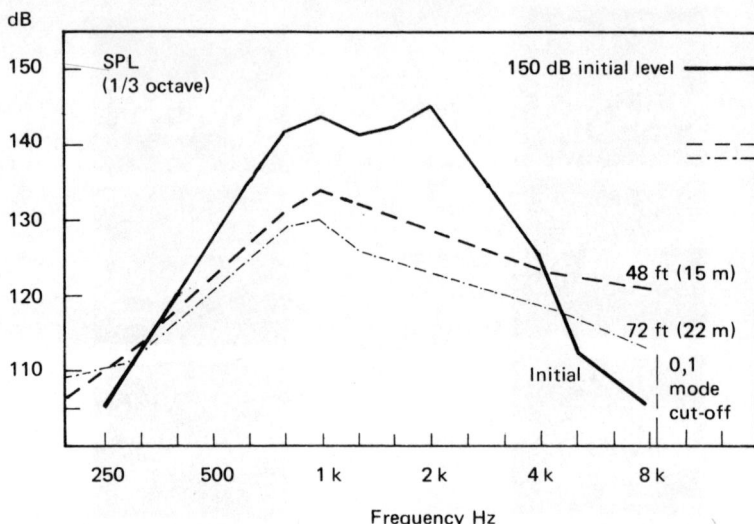

Fig. 8.13 – Nonlinear distortion of a plane-wave noise signal propagating in a long pipe of 50 mm internal diameter. Most of the nonlinear distortion takes place in the first 15 m; beyond this distance, the main effect is a loss of energy above 1 kHz, due to thermoviscous dissipation at the pipe wall. Data from Pestorius [8.13].

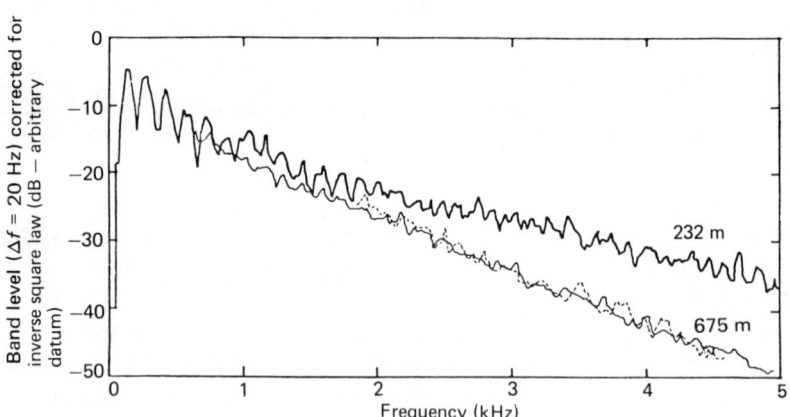

Fig. 8.14 – Aircraft noise spectra measured at the same angle and two different distances, showing apparent attenuation due to atmospheric propagation. From Morfey & Howell [8.15].

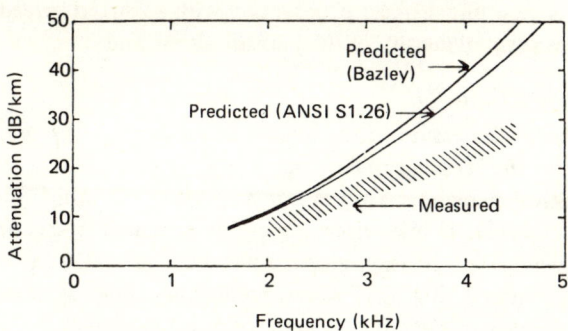

Fig. 8.15 – Atmospheric attenuation rate deduced from Fig. 8.14, compared with standard prediction procedures.

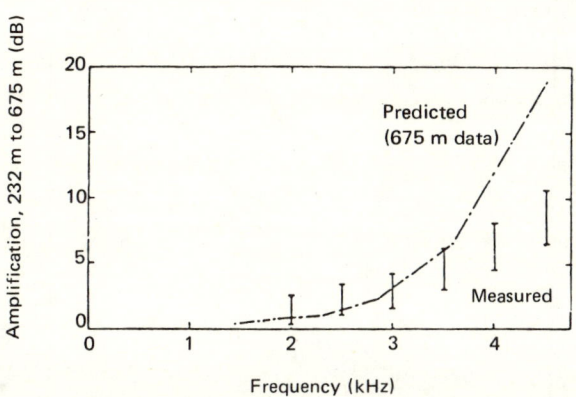

Fig. 8.16 – Shortfall in high-frequency attenuation (from Fig. 8.15), compared with spectral transfer prediction from nonlinear theory (using statistical data measured at 675 m).

(2 kHz upwards). Subsequent analysis of the noise signals showed the shortfall to be consistent with the nonlinear transfer of energy from low to high frequencies (Fig. 8.16). The transfer process is described by the equation

$$\frac{\partial L}{\partial (\ln r/r_0)} = -kr \cdot P_{\mathrm{rms}} \cdot \Omega \qquad (8.25)$$

[8.14], where L is a measure of nonlinear spectral distortion. L is defined at each frequency by

$$L = \ln[(r^2 e^{2\alpha r} S_p)/(r^2 e^{2\alpha r} S_p)_0] \ . \qquad (8.26)$$

(Note that L is zero under linear propagation with spherical spreading.) S_p denotes the power spectral density of the pressure signal, and

$$\Omega = Q_{pp^2}/(\mathscr{P}_{rms}S_p) \tag{8.27}$$

is a non-dimensional measure of the quad-spectral density of p and p^2 (i.e. the imaginary part of the cross-spectral density).

The statistical parameter $\Omega(f)$ is not a quantity for which measurements are normally available. If one wishes to estimate nonlinear spectral distortion from power spectral information alone, some kind of statistical model of the noise signal is required. Fig. 8.17 shows preliminary progress in this direction, for the particular case of aircraft (jet) noise [8.15].

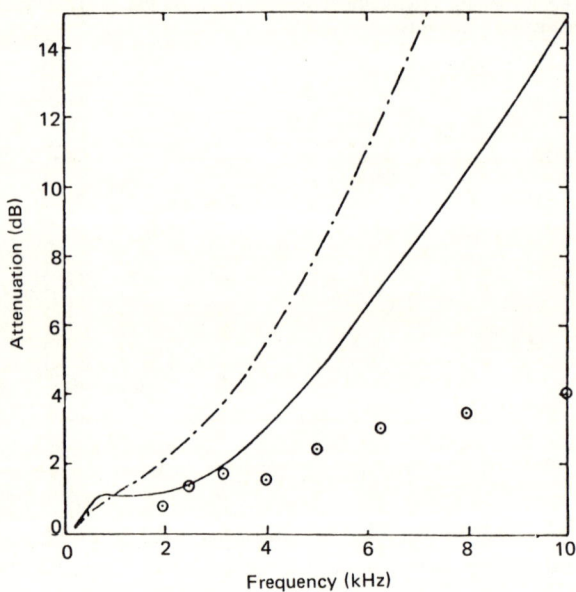

Fig. 8.17 – Measured attenuation of aircraft noise between 262 m and 501 m (⊙), in 1/3-octave bands as a function of centre frequency. Broken curve (— · —) shows linear prediction. Solid curve (——) shows current nonlinear prediction, without allowing for actual response of 1/3-octave filter. (Morfey & Howell [8.15]).

REFERENCES

[8.1]   Lighthill, M. J. (1956) in *Surveys in Mechanics,* (ed. G. K. Batchelor & R. M. Davies), 250–351. London: Cambridge University Press.

[8.2]   Riemann, B. (1860) *Abhandlungen Ges. Wiss. Göttingen,* **8,** 43–65.

[8.3]   Landau, L. D. & Lifshitz, E. M. (1959) *Fluid Mechanics.* Oxford: Pergamon Press.

[8.4]   Blackstock, D. T. (1964) *J. Acoust. Soc. Am.*, **36**, 217–219.

[8.5]   Chester, W. (1953) *Quart. J. Appl. Math.*, **6**, 440–452.

[8.6]   Blackstock, D. T. (1964) *J. Acoust. Soc. Am.*, **36**, 534–542.

[8.7]   Blackstock, D. T. (1979) *Applied Research Laboratories, University of Texas at Austin*, ARL-TR-79-36.

[8.8]   Morfey, C. L. (1979) in *Special course on acoustic wave propagation*, Chapter 9. AGARD report No. 686.

[8.9]   Johannesen, N. H. & Hodgson, J. P. (1979) *Reports on Progress in Physics*, **42**, 629–677.

[8.10] Sofrin, T. G. & Pickett, G. F. (1974) in *NASA SP-304*, 435–460.

[8.11] Morfey, C. L. & Fisher, M. J. (1970) *Aeronautical Journal R.Ae.S.*, **74**, 579–585.

[8.12] Stratford, B. S. & Newby, D. R. (1977) *AIAA Paper 77-1343*.

[8.13] Pestorius, F. M. (1973) *Applied Research Laboratories, University of Texas at Austin*, ARL-TR-73-23.

[8.14] Morfey, C. L. (1980) *Royal Aircraft Establishment, Farnborough, U.K.*, TR 80 004.

[8.15] Morfey, C. L. & Howell, G. P. (1981) *AIAA Journal*, **19**, 986–992.

[8.16] L. Bjørnø (1976) *Nonlinear Acoustics* (Vol. 2 of *Acoustics and Vibration Progress*, ed. R. W. B. Stephens & H. G. Leventhall). London: Chapman & Hall.

CHAPTER 9

Structural wave motion

D. J. Mead

Department of Aeronautics & Astronautics, University of Southampton

9.1 INTRODUCTION

A noise field impinging on a structural surface generates vibration. This has been analysed in previous chapters in terms of the normal modes of vibration of the structure. An acoustic pressure field consists of acoustic pressure waves, and when these impinge on, and propagate across the surface, wave motion is induced in the surface. The waves propagate in the same direction as the acoustic waves which generate them, but on reaching a boundary or discontinuity in the surface they are partly reflected, partly transmitted across the boundary, and also possibly partly transformed into different types of waves.

When the reflections interact with the incident structural waves, standing waves are established, and at particular frequencies these constitute the normal modes of vibration of the surface. It is particularly helpful in problems of acoustically excited vibration to see the vibration from the wave rather than the mode viewpoint. A better understanding of the problem may often thus be obtained, and certain specific problems are more easily solved using wave concepts. Accordingly, this chapter is an introduction to the type of wave-motion most commonly encountered in acoustically excited structures, i.e. flexural wave motion.

9.2 PLANE FLEXURAL WAVE MOTION IN UNIFORM PLATES

A plane wave is one in which the wave displacement at any instant is identical at all points along a line perpendicular to the direction of propagation. In this section we consider such waves propagating parallel to the x-axis of an infinite uniform plate.

When a flat plate is subjected to a transverse, time-dependent pressure loading, $q(x, y, t)$, the transverse deflection $\bar{w}$ is governed by the fourth-order differential equation.

$$D \, \nabla^4 \bar{w} = q \tag{9.1}$$

where
$$\nabla^4 = \frac{\partial^4}{\partial x^4} + 2\frac{\partial^4}{\partial x^2 \partial y^2} + \frac{\partial^4}{\partial y^4}$$

and
$$D = Eh^3/12(1 - \nu^2) \ . \tag{9.2}$$

D is the *flexural rigidity* of the plate, E is its Young's Modulus, h its thickness, and ν the material Poisson's Ratio.

The transverse loading under free wave conditions stems entirely from inertia loading, $-\rho h \partial^2\bar{w}/\partial t^2$, so equation (9.1) becomes the flexural wave equation

$$D \nabla^4 \bar{w} + \rho h \, \partial^2\bar{w}/\partial t^2 = 0 \tag{9.3}$$

where ρ is the plate density. When the wave motion is harmonic we express $\bar{w}$ in the form

$$\bar{w} = w \, e^{i\omega t} \tag{9.4}$$

in which w is a function of x and y only. The differential equation for w, from (9.3), now becomes

$$\nabla^4 w - k^4 w = 0 \tag{9.5}$$

where
$$k^4 = \rho h \omega^2 / D \ . \tag{9.6}$$

This is a general equation, applicable to all forms of the plate flexural wave motion, and not just to the plane waves parallel to the x axis. However, the latter wave motion, being independent of the y co-ordinate, is represented by the solution of (9.5) in the form

$$w = A \, \exp(k_n x) \tag{9.7}$$

where k_n can have any one of the four roots of

$$k_n = \{\rho h \omega^2 / D\}^{\frac{1}{4}} \ . \tag{9.8}$$

Let the real positive root of this be denoted by k. The four different k_n's are

$$k_1 = k, k_2 = -k, k_3 = ik, k_4 = -ik \ . \tag{9.9}$$

The complete solution of the wave equation includes all these k's in the form

$$\bar{w} = A_1 e^{kx} e^{i\omega t} + A_2 e^{-kx} e^{i\omega t} + A_3 e^{i(\omega t + kx)} + A_4 e^{i(\omega t - kx)} \ . \tag{9.10}$$

The last term of this expression represents a true flexural wave, sinusoidal in spatial waveform, constant in amplitude (A_4) and propagating to the right (in the positive x direction). In this term k represents the difference in phase between deflections at unit x-wise distance apart, and is the *flexural wave number*. The wavelength of the flexural wave is given by

$$\lambda = 2\pi/k = 2\pi \, \{D/\rho h \omega^2\}^{\frac{1}{4}} \ . \tag{9.11}$$

The flexural wave velocity, a_f, is given by $\lambda \times$ frequency (Hz) so

$$a_f = \omega^{\frac{1}{2}} \, \{D/\rho h\}^{\frac{1}{4}} \, . \tag{9.12}$$

Now the longitudinal wave velocity in the flat plate, a_l, is given by $\{E/\rho(1-\nu^2)\}^{\frac{1}{2}}$. Also, $h^2/12 = \kappa^2$, where $\kappa = $ radius of gyration of the plate section. Hence

$$a_f = \{\omega a_l \kappa\}^{\frac{1}{2}} \, . \tag{9.13}$$

This increases with (frequency)$^{\frac{1}{2}}$, so the waves of different frequency travel at different speeds. The wave motion is therefore *dispersive*.

The third term of equation (9.10) represents a similar wave which propagates to the left. Each of these waves (A_3 and A_4) transmits energy along the beam (in opposite directions in each case). The energy flow across a given plate section per unit time is found from the average value of (Shear force, S_x, on the section due to the wave $\times$ transverse plate velocity) + (Bending moment, M_x, on the section $\times$ angular plate velocity).

These two component energies are found to be equal in a free wave, and the total power transmitted by a single propagating flexural wave of amplitude $|A|$ is

$$\Pi_f = Dk^3\omega |A^2| = \omega^2 \rho \, ha_f |A^2| \tag{9.14}$$

per unit width of plate, normal to the wave direction.

The first two terms of equation (9.10) do not represent propagating waves. The motion represented by $A_1 \, e^{kx} \, e^{i\omega t}$ is in phase at all points along the beam, and is a deflection which decays exponentially in space from right to left. $A_2 \, e^{-kx}$ represents a similar deflection which decays from left to right. No energy is transmitted by these 'waves', so we refer to them as 'non-propagating' waves. They always decay with distance away from the point at which they are generated, and will not be noticed (owing to the exponential factor) at large distances from the point. The propagating waves, on the other hand, have the same amplitude at large or small distances, provided there is no damping present to attenuate them. Since the non-propagating waves are only noticed near to the point of generation, they may be called 'near field' waves, whereas the propagating waves spreading outwards without attentuation, may be called 'far-field' waves.

9.3 WAVE MOTION GENERATED BY LINE FORCES OR MOMENTS

Suppose an infinite plate is excited by a harmonic force

$$\bar{P} = P \, e^{i\omega t} \tag{9.15}$$

per unit length along the line $x = 0$, from $y = \infty$ to $-\infty$. Propagating and near-field waves are sent out in either direction from the line of excitation. If there are no discontinuities, reflecting boundaries, or other external forces at any point in the infinite plate, the deflections in the right-hand and left-hand regions (on either side of $x = 0$) will be given by

RH　　　　$w_+ = A_2 e^{-kx} + A_4 e^{-ikx}$　　　　　　　(9.16a)

LH　　　　$w_- = A_1 e^{kx} + A_3 e^{ikx}$.　　　　　　(9.16b)

The A's are found by satisfying the continuity and equilibrium conditions at $x = 0$. These yield

$$A_1 = A_2 = -iA_3 = -iA_4 = -P/4Dk^3 \ . \tag{9.17}$$

Notice that the amplitudes of the near-field and propagating waves are equal, that the near-field wave is in anti-phase with the applied force, but that the propagating wave is in quadrature with the applied force. The total deflection in the right-hand part of the plate is

$$w_+ = -\{P/4Dk^3\} \ \{e^{-kx} + ie^{-ikx}\} \ . \tag{9.18}$$

At $x = 0$, this lags in phase behind the force by $135°$.

Notice further, that when $kx = 4.6$ (i.e. $x = 0.732\lambda$), e^{-kx} is $1/100$ i.e. over the distance of approximately $3/4$ of the propagating wavelength, the near-field wave has decayed to $1/100$th of its initial value at $x = 0$. This serves to show that the effect of such near-field waves can usually be neglected at points further than $3\lambda/4$ from the source of the wave.

A similar expression for w_+ or $w-$ may be derived for a plate which is excited along the line $x = 0$ by the distributed moment,

$$M_x = M_x e^{i\omega t} \text{ per unit length } . \tag{9.19}$$

We find　　　$w_+ = \{M_x/4Dk^2\} \{e^{-ikx} - e^{-kx}\} \ . \tag{9.20}$

Again, the amplitudes of the near-field and propagating waves at $x = 0$ are equal. The deflection at $x = 0$ is zero, but the rotation, $w'_+(0)$ is given by

$$w'_+(0) = M_x(1-i)/4Dk \ . \tag{9.21}$$

These deflection-force relationships may be used to obtain expressions for the plate impedance, Z. The impedance is defined as the complex harmonic force (or moment) required to produce unit harmonic transverse (or angular) velocity. A *direct* impedance is the force required to produce unit velocity in the same direction as the force at the driving point. A *transfer* impedance is the force required to produce unit velocity at another point. The term 'cross' impedance is often used to indicate the relationship between force and velocity in different planes. For the plate with the transverse line force, we see that

$$Z_{\text{direct}} = \bar{P}e^{i\omega t}/\bar{\dot{w}}(0) = -4Dk^3/i\omega(1+i) = 2a_f\rho h(1+i) \ , \tag{9.22}$$

which has the dimensions of force per unit length per unit velocity. Notice that this impedance increases with frequency in the same way as a_f. Table 9.1 shows

Table 9.1

Component wave amplitudes and impedances for plates with line forces or moments

Beam and Excitation	A_4	A_2	Z_{direct}	Z_{cross}
Infinite plate; force ex^n:	$\dfrac{-iP}{4Dk^3}$	$-iA_4$	$\dfrac{2a_f\rho h(1+i)}{\omega^{\frac{1}{2}}(a_f\kappa)^{\frac{1}{2}}} = 2\rho h(1+i)$	—
Infinite plate; moment ex^n:	$\dfrac{M_x}{4Dk^2}$	A_4	$\dfrac{2a_f\rho h(1-i)/k^2}{\omega^{-\frac{1}{2}}(a_f\kappa)^{3/2}} = 2\rho h(1-i)$	—
Semi-inf. plate; force ex^n: free end	$\dfrac{-P(1+i)}{2Dk^3}$	A_4	$a_f\rho h(1+i)/2$	$-a_f\rho h/k$
Semi-inf. plate; moment ex^n: free end	$\dfrac{-M_x(1+i)}{2Dk^2}$	iA_4	$a_f\rho h(1-i)/2k^2$	$a_f\rho h/k$

the direct and cross impedances of infinite and semi-infinite plates under different conditions of excitation. The cross impedance here is the force per unit angular velocity at the driving point or moment per unit transverse velocity at the driving point.

Notice that the direct force impedance is proportional to $\omega^{\frac{1}{2}}$, whereas the direct moment impedance is proportional to $\omega^{-\frac{1}{2}}$.

9.4 PLANE WAVES INCIDENT UPON LINE BOUNDARIES AND DISCONTINUITIES; NORMAL INCIDENCE

When a propagating wave encounters a boundary, line support, or line discontinuity, it will be partially or wholly reflected, and at the line of reflection a near-field wave will be generated. If the plate extends on both sides of the support or discontinuity, a propagating wave will be transmitted across it, and a near-field wave will spread out (decaying as it does so) on either side of it.

Suppose the uniform plate has the discontinuity at $x = 0$, and is excited far to the left by a harmonic line force which generates the incident flexural wave $\bar{A}_i$ in the region of the plate $x < 0$. Then

$$\bar{A}_i = A_i e^{i(\omega t - kx)} \ . \tag{9.23}$$

Reflected back into this region is the wave

$$\bar{A}_r = A_r e^{i(\omega k + kx)} \tag{9.24}$$

in addition to the near-field wave

$$\bar{A}_{nl} = A_{nl} e^{kx} e^{i\omega t} \ . \tag{9.25}$$

On the other side of the discontinuity, $x > 0$, are the transmitted and right-hand near-field waves

$$\bar{A}_t = A_t e^{i(\omega t - kx)} \tag{9.26}$$

and $$\bar{A}_{nr} = A_{nr} e^{-kx} e^{i\omega t} \ . \tag{9.27}$$

The total motion in the left-hand and right-hand sections of the plate is now

LH: $$\bar{w}_- = w_- e^{i\omega t} = \left\{ A_i e^{-ikx} + A_r e^{ikx} + A_{nl} e^{kx} \right\} e^{i\omega t} \tag{9.28}$$

RH: $$\bar{w}_+ = w_+ e^{i\omega t} = \left\{ A_t e^{-ikx} + A_{nr} e^{-kx} e^{i\omega t} \right\}. \tag{9.29}$$

The four unknown wave amplitudes, A_r, A_{nl}, A_t, A_{nr} are found from four equations expressing continuity and equilibrium at the discontinuity. Table 9.2. shows expressions for these, in terms of A_i, for several different types of discontinuity or end conditions.

Also shown are the expressions for the *reflection and transmission coefficients* which are defined (as in acoustics theory) by

$$\text{Reflection coefficient} \quad \alpha_r = \frac{\text{Reflected wave energy}}{\text{Incident wave energy}} = \left| \frac{A_r}{A_i} \right|^2 \qquad (9.30)$$

$$\text{Transmission coefficient} \ \alpha_t = \frac{\text{Transmitted wave energy}}{\text{Incident wave energy}} = \left| \frac{A_t}{A_i} \right|^2 \qquad (9.31)$$

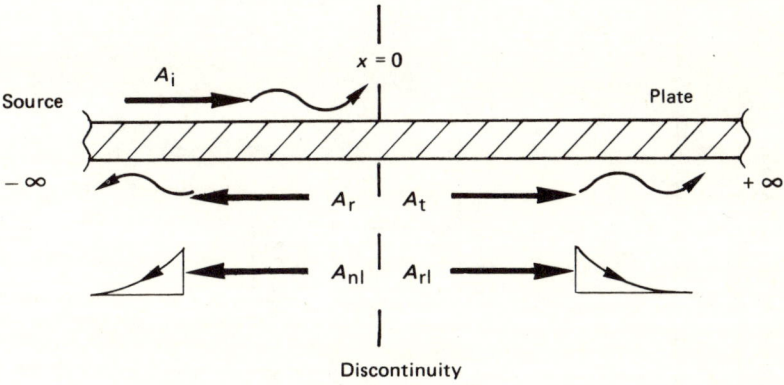

Fig. 9.1 – Plate with a single discontinuity and the component wave motions.

The following features of the first four discontinuities should be noticed from Table 9.2.

(a) The amplitudes of the two near-field waves are equal to one another, and to the amplitude of the reflected wave.

(b) The amplitude of the reflected wave never exceeds one half of the amplitude of the incident wave. However, if the mass in (2) had rotational inertia, and the spring in (3) provided rotational constraint the factor of one half could be exceeded.

(c) With simple support (1), and the simple hinge (2), one half of the incident wave energy is transmitted and one half reflected. In (1), all the energy transmitted by the moment and rotation of the incident wave motion is transmitted into the right-hand region, but none of the energy transmitted by shear force and deflection. In (2) all of the energy transmitted by the shear force and deflection in the incident wave is transmitted into the right-hand region but none transmitted by the moment.

Table 9.2

Component waves, reflection and transmission ratios for different infinite plates with discontinuities;
$\mu = Mk/\rho h$, $\epsilon = (K/\omega^2)\,k/\rho h$

Type of Discontinuity	A_r	A_t	A_{n1}	A_{nr}	α_r	α_t
1) Simple support	$\dfrac{A_i}{1-i}$	$\dfrac{i \cdot A_i}{1-i}$	$-iA_r$	$-iA_r$	$\dfrac{1}{2}$	$\dfrac{1}{2}$
2) Line mass M per unit length	$\dfrac{\mu A_i}{\mu - i(4+\mu)}$	$\dfrac{i(4+\mu)A_i}{\mu - i(4+\mu)}$	$-iA_r$	$-iA_r$	$\dfrac{\mu^2}{(4+\mu)^2 + \mu^2}$	$\dfrac{(4+\mu)^2}{(4+\mu)^2 + \mu^2}$
3) Line spring K per unit length	$\dfrac{\epsilon A_i}{\epsilon + i(4-\epsilon)}$	$\dfrac{i(4-\epsilon)A_i}{\epsilon + i(4-\epsilon)}$	$-iA_r$	$-iA_r$	$\dfrac{\epsilon^2}{(4-\epsilon)^2 + \epsilon^2}$	$\dfrac{(4-\epsilon)^2}{(4-\epsilon)^2 + \epsilon^2}$
4) Simple hinge $(w''_+ = 0 = w''_-)$	$\dfrac{(1+i)A_i}{2}$	$\dfrac{(1-i)A_i}{2}$	A_t	A_t	$\dfrac{1}{2}$	$\dfrac{1}{2}$
5) Simply-supported end $(w_- = 0;\, w''_- = 0)$	$-A_i$	—	0	—	1	—
6) Fully-fixed end $(w_- = 0,\, w'_- = 0)$	$-iA_i$	—	$-(1-i)A_i$	—	1	—
7) Free-end $(w''_- = 0,\, w'''_- = 0)$	$-iA_i$	—	$(1-i)A_i$	—	1	—

The reflected wave amplitudes in the beams (5), (6) and (7) are equal to the incidence wave amplitudes, but there is a phase difference of π or $-\pi/2$ between the two waves. Where the near-field wave exists (6 and 7) its amplitude is $\sqrt{2}$ times that of the reflected wave.

At large distances to the right of the discontinuities, the amplitude of transverse deflection is A_t. Close to the discontinuity the near-field wave modifies this. At large distances to the left of the discontinuity the total motion consists of the incident and reflected wave, and the amplitude of total deflection varies along the length. The maximum amplitude may be found from the value of $|A_i e^{-ikx} + A_r e^{ikx}| = A_{max}$. Similarly, we can find amplitudes of *curvature* at and far from the discontinuity. These are important as being proportional to the vibrational bending stresses. Tables 9.3 and 9.4 compare some of these amplitudes, and show (a) the influence of the discontinuity in reducing vibration levels in the right-hand region or at the point of discontinuity, (b) the build-up (or otherwise) of vibration levels at a boundary. For instance, notice how the curvature (and hence the bending stresses) builds up at the fixed end of the semi-infinite beam, and how the displacement builds up at the free end of the semi-infinite beam.

9.5 PLANE WAVES AT ANY INCIDENCE IMPINGING ON A BOUNDARY

Let the direction of propagation of the wave be inclined at α to the x axis. The more general solution for the wave equation (9.5) must now be used. The incident plane wave of frequency ω and wave number k in the direction of propagation is represented by

$$\bar{w}_i = w'_i e^{i\omega t} = A_i e^{i(\omega t - (k\cos\alpha)x - (k\sin\alpha)y)} \; . \tag{9.32}$$

The intercepts of the wavefronts on the x axis give an x-wise 'trace wavelength' of $\lambda_x = 2\pi/k_x$, where $k_x = k\cos\alpha$. Likewise, the intercepts on the y axis yield y-wise trace wavelengths of $\lambda_y = 2\pi/k_y$ where $k_y = k\sin\alpha$. Putting w from equation (9.32) into equation (9.5) we find

$$(k_x^2 + k_y^2)^2 = k^4 \quad \text{or} \quad k_y^2 = k^2 - k_x^2 \; . \tag{9.33}$$

Suppose this wave impinges on a fully-fixed boundary at $x = 0, y = -\infty$ to $+\infty$. A propagating wave, $\bar{w}_r$, is reflected from the boundary with an angle of reflection equal to the angle of incidence. Its y-wise wave motion is in the same direction as that of the incident wave, but its x-wise wave motion has been reversed. Hence

$$\bar{w}_r = w_r e^{i\omega t} = A_r e^{i(\omega t + (k\cos\alpha)x - (k\sin\alpha)y)} \; . \tag{9.34}$$

In addition to this propagating wave, a near-field wave is generated. This must have the same y-wise wave-number, k_y, as the incident and reflected waves (for

Table 9.3

Displacement amplitudes at different points in the infinite plates with discontinuities

Type of discontinuity	Amplitude at		
	Far Left	$x = 0$	Far Right
1. $x=0$	$A_i\left\{1+\dfrac{1}{\sqrt{2}}\right\}$	0	$A_i\dfrac{1}{\sqrt{2}}$
2. **M** $x=0$	$A_i\dfrac{\mu+\sqrt{\mu^2+(4+\mu)^2}}{\sqrt{\mu^2+(4+\mu)^2}}$	$\dfrac{4A_i}{\sqrt{\mu^2+(4+\mu)^2}}$	$\dfrac{(4+\mu)A_i}{\sqrt{\mu^2+(4+\mu)^2}}$
3. $x=0$	$A_i\dfrac{\epsilon+\sqrt{\epsilon^2+(4-\epsilon)^2}}{\sqrt{\epsilon^2+(4-\epsilon)^2}}$	$\dfrac{4A_i}{\sqrt{\epsilon^2+(4-\epsilon)^2}}$	$\dfrac{(4-\epsilon)A_i}{\sqrt{\epsilon^2+(4-\epsilon)^2}}$
4. $x=0$ HINGE	$A_i\left(1+\dfrac{1}{\sqrt{2}}\right)$	$A_i\sqrt{2}$	$A_i\dfrac{1}{\sqrt{2}}$
7. $x=0$ free end	$2A_i$	$A_i2\sqrt{2}$	—

Table 9.4

Curvature amplitudes at different points in the infinite plates with discontinuities

Type of Discontinuity	Curvature Amplitude at		
	Far Left	$x = 0$	Far Right
1. $x=0$ (pinned support)	$A_i k^2 \left(1 + \dfrac{1}{\sqrt{2}}\right)$	$A_i k^2 \sqrt{2}$	$A_i k^2 \dfrac{1}{\sqrt{2}}$
2. $x=0$, M (mass)	$A_i k^2 \dfrac{\mu + \sqrt{\mu^2 + (4+\mu)^2}}{\sqrt{\mu^2 + (4+\mu)^2}}$	$A_i k^2 \dfrac{(4 + 2\mu)}{\sqrt{\mu^2 + (4+\mu)^2}}$	$A_i k^2 \dfrac{(4 + \mu)}{\sqrt{\mu^2 + (4+\mu)^2}}$
3. (spring)	$A_i k^2 \dfrac{\epsilon + \sqrt{\epsilon^2 + (4-\epsilon)^2}}{\sqrt{\epsilon^2 + (4-\epsilon)^2}}$	$A_i k^2 \dfrac{(4 - 2\epsilon)}{\sqrt{\epsilon^2 + (4-\epsilon)^2}}$	$A_i k^2 \dfrac{(4 - \epsilon)}{\sqrt{\epsilon^2 + (4-\epsilon)^2}}$
4. HINGE	$A_i k^2 \left(1 + \dfrac{1}{\sqrt{2}}\right)$	0	$A_i k^2 \dfrac{1}{\sqrt{2}}$
6.	$A_i\, 2k^2$	$A_i\, 2\sqrt{2}\, k^2$	—

continuity and equality of wavelength along the boundary) but its decay rate is no longer numerically equal to k_x, as in the previous section. We represent the near-field wave by

$$w_{ne}e^{i\omega t} = A_{ne}\exp(k_d x)\exp(i(\omega t - k_y y)) \qquad (9.35)$$

and put this into equation (9.5), to yield $(k_d^2 - k_y^2)^2 = k^4$.

Hence, $\quad k_d = \pm\sqrt{(+ k_y^2 \pm k^2)}$.

The positive sign is required in the square root, so

$$k_d = \pm k\sqrt{(1 + \sin^2\alpha)} \ . \qquad (9.36)$$

Hence, the rate of decay of the near-field wave, generated by an incident wave of wave number k, depends on the inclination of the incident wave to the boundary.

The total motion generated by the incident wave is

$$\bar{w}_{total} = \bar{w}_i + \bar{w}_r + \bar{w}_{ne}$$

$$= \{A_i\exp(-ik_x x) + A_r\exp(ik_x x) + A_{ne}\exp(k_d x)\}\exp(i(\omega t - k_y y)) \ . (9.37)$$

At the fully fixed boundary at $x = 0$, both $\bar{w}$ and $d\bar{w}/dx$ must vanish. From these two conditions we find

$$A_{ne} = -A_i\left(\cos^2\alpha - i\cos\alpha\sqrt{(1 + \sin^2\alpha)}\right) \qquad (9.38)$$

and $\quad A_r = -A_i\left(\sin^2\alpha + i\cos\alpha\sqrt{(1 + \sin^2\alpha)}\right)$. $\qquad (9.39)$

The amplitudes A_r and A_i are the same, but the phase difference between them depends upon α. It becomes $90°$ when $\alpha = 0$.

At a simply-supported boundary, no near-field wave is generated, and we find $A_r = -A_i$. There is a reflection with phase change of π irrespective of angle of incidence α. The total motion is then

$$\bar{w}_{total} = A_i\{\exp(-ik_x x) - \exp(ik_x x)\}\exp(i(\omega t - k_y y)) \ .$$

$$= -2iA_i\sin k_x x \exp(i(\omega t - k_y y)) \ . \qquad (9.40)$$

There is now a standing wave in the x direction, and nodal lines exist at distances $n\pi/k_x$ from the boundary. The motion has the form of an x-wise sine-standing-wave propagating in the y direction with wave number k_y.

Since a nodal line exists parallel to the simply supported boundary, the plate to the left of the line could be removed and the line itself made into a simply-supported edge. We should now have a plate of finite width, b, simply supported along two edges. Flexural waves can propagate along the infinite length with x-wise standing-wave-form $\sin k_x x = \sin(m\pi x/b)$ (m is any integer) and y-wise wave number $k_y = \sqrt{(k^2 - k_x^2)}$ (from equation 10.33). This wave motion consists, basically, of propagating waves incident at α to the x axis, of wave number k in the α direction and being reflected backwards and forwards across the plate. When $k^2 = k_x^2 = m^2\pi^2/b$, we have $k_y = 0$, so the y-wise wavelength is infinite and there is no effective propagation along the plate. When $k^2 > m^2\pi^2/b^2$, the x-wise deflection mode $\sin(m\pi x/b)$ cannot be propagated down the plate. The critical frequency is $(m^2\pi^2/b^2)\,(D/\rho h)^{\frac{1}{2}}$, which is the natural frequency of a simply-supported strip of length b, thickness h, vibrating in its mth mode. This is the cut-off frequency for this mode of propagation, for below it, no propagation can occur. The lowest cut-off frequency is with $m = 1$, i.e. equal to the natural frequency of the strip vibrating in its fundamental mode.

If the boundary at $x = 0$ of the semi-infinite plate is rotationally elastic, but transversely rigid, then further expressions can be derived for the amplitudes of the reflected and near-field waves. If the rotational stiffness of the boundary is k_r per unit length, then

$$A_{\text{ne}} = A_i\, 2i \cos \alpha/(\sqrt{(1 + \sin^2\alpha)} - 2Dk/k_r - i \cos \alpha) \ , \qquad (9.41)$$

$$A_r = -A_i\, (\sqrt{(1 + \sin^2\alpha)} - 2Dk/k_r + i \cos \alpha)/ \qquad (9.42)$$

$$(\sqrt{(1 + \sin^2\alpha)} - 2Dk/k_r - i \cos \alpha) \ .$$

Ungar [9.1] has derived further expressions of this type for the infinite plate with a single stiffener which has both torsional and transverse flexibility and inertia. Propagating and near-field waves exist on both sides of the stiffener, and transmission and reflection coefficients have been derived. The expressions he derived, together with (9.41) and (9.42), apply to very ideal situations. They have practical usefulness and real situations where a plate has multiple stiffeners only if the wavelength of the incident wave is small compared with the stiffener spacing. Otherwise the inter-acting influence of all the stiffeners must be taken into account.

9.6 THE NATURAL MODES OF VIBRATION OF A RECTANGULAR PLATE

Suppose the plate is simply-supported along two opposite edges. It has been shown above that flexural waves can propagate in the direction parallel to these edges with a standing-sinusoidal wave-form between the edges, and a propagating or decaying wave-form parallel to the edges. Let the propagating wave be

$$w_+ = A_1 \exp(i(\omega t - k_y y)) \sin (m\pi x/a) \ . \qquad (9.43)$$

Suppose the x-wise edges at $y = 0$, b are also simply-supported. The wave w_+ will be completely reflected with a phase change of π at the edge $y = b$, and the reflected wave will itself be reflected with the phase change of π at the edge $y = 0$. Hence, as the incident wave traverses the length b of the plate, is reflected from the top to the bottom boundary, back to an arbitrary starting point, the total phase change of the wave will be $2b.k_y + 2\pi$. If this is an exact multiple of 2π then the wave on returning will reinforce a later wave just departing, and a resonant condition will build up. The frequency of this is given by $2b.k_y + 2\pi$

$= 2(n + 1)\pi$ where n is an integer > 0. Hence, $k_y = n\pi/b$. But $k_x^2 + k_y^2 = k^2 = (\omega^2 \rho h/D)^{\frac{1}{2}}$, and $k_x = m\pi/a$. Hence,

$$\omega = \omega_{nm} = \pi^2 \{m^2/a^2 + n^2/b^2\} \{D/\rho h\}^{\frac{1}{2}}. \tag{9.44}$$

This is the well-known expression for the natural frequency for the simply supported rectangular plate.

Now suppose the two edges at $y = 0$ and b are fully-fixed. The propagating wave (9.43) is reflected from the boundary at $y = b$, and a propagating and near-field wave are returned. It can be shown that the reflected wave lags behind the incident wave by $2 \tan^{-1} \{(k^2 - k_x^2)/(k^2 + k_x^2)\}^{\frac{1}{2}}$ at the boundary. The total phase change as the propagating wave travels backwards and forwards once, and is reflected twice, is

$$2k_y . b + 4 \tan^{-1} \{(k^2 - k_x^2)/(k^2 + k_x^2)\}^{\frac{1}{2}}. \tag{9.45}$$

If we ignore the presence of near-field waves (which is justifiable if they decay away rapidly from the boundaries) we can say that resonant frequencies occur when this total phase change is $2(n + 1)\pi$. The resulting equation for k, and hence for ω, is more complicated than that for the simply-supported plate. However, if the frequency is high, k is very large and the phase change at a boundary becomes $\pi/2$. The total phase change is $2k_y b + \pi$. Putting this equal to $2(n + 1)\pi$ we obtain $k_y = (n + \frac{1}{2})\pi$. Hence

$$\omega_{nm} = \pi^2 \{m^2/a^2 + (n + \frac{1}{2})^2/b^2\} \{D/\rho h\}^{\frac{1}{2}}. \tag{9.46}$$

This is a good approximation to the frequency of a high-order mode of the plate with two simply-supported and two fully-fixed edges.

It is now evident, from this and the preceding section, that a natural mode of vibration can be regarded as the superposition of propagating waves and their reflections from the boundaries. This section shows them to consist of propagating waves with wave-numbers k_x and k_y, while the previous section shows these waves to consist of propagating waves of wave number k and incidence α to the x axis. ($\alpha = \tan^{-1} k_y/k_x = \tan^{-1} \lambda_x/\lambda_y$, where the λ's are the wavelengths in the two directions). This direction is perpendicular to the diagonals of the nodal rectangles. Acoustic plane waves of wave number k which convect across the plate in direction α can readily excite these waves and some of the corresponding natural modes.

9.7 DIFFUSE FLEXURAL WAVE FIELDS

Natural modes have been seen to consist of incident and reflected waves inclined at particular angles of incidence to the boundaries. When a plate is excited by a high-frequency random source (distributed or multiple point forces) flexural waves in many different directions will be generated. The limiting case of this occurs when waves are incident in all directions, but with no waves from any discrete direction. If the wave intensity per unit angle of incidence is constant for all incidence, the flexural wave field is said to be 'diffuse'.

It is sometimes convenient to analyse high-frequency plate vibrations using this diffuse wave field concept. It can be shown that if there are more than about six or seven natural modes participating in a forced vibration, and their contributions are more-or-less equal, then the total vibration field can quite accurately be described and analysed as a diffuse field. The techniques developed for diffuse fields in room acoustics may be adopted.

The detailed methods will not be elaborated here, but certain useful results will be quoted. These relate to the stress concentration factors at the boundaries of a plate in which the diffuse field exists. We have shown already that the curvature at the fixed boundary of a plate with flexural waves of given wave number and at normal incidence is $\sqrt{2}$ times the curvature at distances remote from the boundary. This factor varies as the incidence of the wave on the boundary varies. The r.m.s. curvature remote from the boundary is simply related to the local r.m.s. displacement through the relationship: r.m.s. curvature $= k^2 . w_{rms}$. The r.m.s. curvature at the boundary is therefore $\sqrt{2}\, k^2$ times the r.m.s. displacement remote from the boundary. When the wave is inclined by α to the straight boundary, we find: r.m.s. curvature at boundary $= \sqrt{2} . k^2 \cos\alpha\, w_{rms}$.

When waves are incident on to the boundary with uniform intensity from all directions, the r.m.s. curvature is found by appropriate integration over the incident range $-\pi/2 \leqslant \alpha \leqslant \pi/2$. When all these waves have the same wave-number, the curvature at the boundary is still found to be equal to $\sqrt{2}\, k^2$ times the r.m.s. displacement remote from the boundary. The maximum r.m.s. bending stress, at and normal to the boundary, is therefore $Ehk^2 w_{rms}/\sqrt{2}(1 - \nu^2)$. This can be shown to be about 2.2 times the surface bending stress remote from the boundary.

These results can be obtained either by assuming a true diffuse field, or by assuming that a large number of normal modes of vibration contribute to the vibration level. Stearn [9.2] has studied such 'stress concentration factors' for different types of boundary and discontinuity. They are particularly useful when it is required to estimate approximately the bending stress in a plate when the total vibrational energy and its spectrum are known.

9.8 THE FORCED VIBRATIONS OF UNIFORM FLAT PLATES; THE WAVE APPROACH

Consider a plate which is simply-supported along the two x-wise edges, and which is subjected to a propagating pressure wave on the surface of the form

$$p(x, y, t) = p_0 \exp(i(\omega t - k_p x)) \qquad (9.47)$$

where k_p is the wave-number of the pressure wave on the plate surface. This represents a plane harmonic pressure wave propagating in the x direction only at a velocity of

$$a_p = \omega/k_p . \tag{9.48}$$

At a given instant and x position, p is constant in the y direction. It is convenient, however, to analyse this constant value into sinusoidal y-wise components, thus:

$$p_0 = \sum_{r=1}^{\infty} P_r \sin(2r-1)\,\pi y/l_y = \sum_{r=1}^{\infty} P_r \sin(k_{ry}y) . \tag{9.49}$$

For simplicity in the remainder of this analysis, we shall consider only the first term of this, i.e.

$$p(x, y, t) = P_1 \sin (k_{1y}y) \exp(i(\omega t - k_p x) . \tag{9.50}$$

Now the response of the plate, $\bar{w}(x, y, t)$, to this pressure is governed by

$$D\nabla^4\bar{w} - \omega^2\rho h\bar{w} = P_1 \sin(k_{1y}y) \exp(i(\omega t - k_p x) . \tag{9.51}$$

Damping may now be allowed in the plate by assigning the complex form $D(1 + i\eta)$ to the flexural rigidity. The solution to equation (9.51) has two parts — the particular integral and the complementary function. The particular integral is identical to the response of the plate if it was infinite in the x direction, and represents a forced flexural wave travelling in the x direction. It is given by

$$w_{PI} = P_1 \sin(k_{1y}y) \exp(i(\omega t - k_p)/(D(k_{1y}^2 + k_p^2)^2 - \omega^2\rho h) . \tag{9.52}$$

Since the plate is in fact finite, this wave will be reflected when it encounters the right-hand boundary of the plate. Likewise, reflection will occur from the left-hand boundary. The reflections, however, will be *free* flexural waves and there will be a propagating wave and a near-field wave reflected from each boundary. These free waves constitute the complementary function of the solution to equation (9.51) and are represented by

$$w_{CF} = \{A_1\exp(k_n x) + A_2\exp(-k_n x) + A_3\exp(ik_x x) +$$

$$A_4\exp(-ik_x x)\} \exp(i\omega t) \sin (k_{1y}y)$$

where $\tag{9.53}$

$$k_n^2 = \{k^2 + k_{1y}^2\}, \ k_x^2 = k^2 - k_{1y}^2, \ k^4 = \omega^2\rho h/D .$$

The total displacement at any point (x, y) is therefore given by

$$w(x, y, t) = \{A_1\exp(k_n x) + A_2\exp(-k_n x) + A_3\exp(+ik_x) + \tag{9.54}$$

$$A_4\exp(-ik_x x) + P_1\exp(-ik_p x)/\{D(k_{1y}^2 + k_p^2) - \omega^2\rho h\}\} \sin(k_{1y}y) \exp(i\omega t) .$$

This solution contains four unknown constants of integration which can be found by ensuring that the solution satisfies the four boundary conditions of the plate at $x = 0$ and $x = l_x$. Suppose these two boundaries are encastré, i.e.

$$w = 0, \quad w' = 0, \quad \text{at } x = 0 \text{ and } x = l_x \ . \tag{9.55}$$

These yield the following four equations for the A's.

$$
\begin{bmatrix}
1 & , & 1 & , & 1 & , & 1 & \times \\
k_n & , & -k_n & , & ik_x & , & -ik_x & \times \\
\exp(k_n l_x) & , & \exp(-k_n l_x) & , & \exp(ik_n l_x) & , & \exp(-ik_x l_x) & \times \\
k_n\exp(k_n l_x), & -k_n\exp(-k_n l_x), & ik_x\exp(ik_x l_x) & , & -ik_x\exp(-ik_x l_x)
\end{bmatrix}
$$

$$
\times
\begin{Bmatrix}
A_1 \\
A_2 \\
A_3 \\
A_4
\end{Bmatrix}
=
\begin{Bmatrix}
1 \\
-ik_p \\
\exp(-ik_p l_x) \\
-ik_p\exp(-ik_p l_x)
\end{Bmatrix}
\times \frac{P_1}{D(k_{1y}^2 + k_p^2) - \omega^2 \rho h} \tag{9.56}
$$

Notice that the A's will have very large values under two different conditions:

(a) When the real part of the denominator of the right-hand side approaches zero, i.e. when $\mathrm{Re}\,(D(k_{1y}^2 + k_p^2) - \omega^2 \rho h) \to 0$

To a first approximation this means $k_p \to \sqrt{(k^2 - k_{1y}^2)} = \mathrm{Re}\,(k_x)$.

Hence, when the wave number of the exciting pressure field is equal to the natural wave number of free flexural wave motion, a large response is generated. These two wave numbers will be equal if the propagation velocity of the pressure field is equal to the free wave velocity of the corresponding flexural waves in the plate at that frequency.

(b) When the determinant of the matrix on the left-hand side has a minimum value. This occurs whenever the frequency is equal to one of the natural frequencies of the finite plate.

The response expression (9.54) will therefore give rise to resonant type peaks at frequencies corresponding to each of these conditions. Condition (a) is the 'COINCIDENCE CONDITION', for the propagation speed of the pressure wave *coincides with* the natural propagation velocity of the flexural wave being generated. Condition (b) is a simple 'resonance' condition.

Now a set of equations of identical form to (9.56) exists for each value of k_{ry} belonging to the pressure field. The equations can be solved for each of these values and for given values of p_0, k_p and ω. Hence, the displacement at any point on the plate can be found for a given value of k_p and ω. We can therefore write

$$w(x, y, t) = Y(x, y, k_p, \omega) P_0 e^{i\omega t} \tag{9.57}$$

where $Y(x, y, k_p, \omega)$ is the 'wave receptance function'. It is the complex harmonic plate response at point (x, y) to a harmonic pressure wave of given y-wise form of unit amplitude and given frequency and wave-number.

9.9 THE RANDOM VIBRATION OF UNIFORM FLAT PLATES; THE WAVE ANALYSIS METHOD: THE WAVE–NUMBER/FREQUENCY SPECTRUM

The normal mode method (as outlined in a previous chapter) can be applied to any type of plate, whether flat or curved, rectangular or irregular in shape, uniform or variable in thickness. The general expression obtained for the response spectral density applies to all such conditions. It involves the calculation of the joint acceptances, and this can be an extremely tedious task; so also can be the summing of the modal responses allowing for the correlation between different modes.

Some simple types of structure can be more easily analysed by making use of a wave approach developed from the method described above. The uniform rectangular flat plate is one such case.

A random pressure field can be regarded as a continuous assembly of harmonic pressure waves of all different frequencies and wavelengths (or wave numbers). A random pressure field which is convected along at a uniform velocity a_p and which maintains its precise waveform as it moves along, can be analysed into its continuum of spectral components of all frequencies. Each frequency can be associated with a unique wavelength of pressure wave (given by $\lambda = a_p/f = 2\pi a_p/\omega$) or with a unique wave number (given by $k_p = \omega/a_p$). Thus, the spectral density $S_p(\omega)$ of the pressure fluctuation corresponds to a wave-number $k_p = \omega/a_p$.

If the random pressure field does not maintain its precise waveform as it convects along (and this is the more realistic case) it can be analysed into a continuous assembly of harmonic pressure waves of all frequencies and all wave numbers at each frequency. Physically, this means that we have pressure waves of all different frequencies travelling along at all different velocities. The spectral information about such pressure waves can be described by the two-dimensional wave number/frequency spectrum. For any given frequency, a spectrum is drawn in the wave number domain showing how the pressure 'power' is distributed over the different wave numbers.

Now the power spectrum of the pressure at a single point can be found from the Fourier transform of the pressure auto-correlation function, i.e.

$$S_p(\omega) = (1/2\pi) \int_{-\infty}^{+\infty} R_p(\tau) e^{-i\omega t} \, d\tau \qquad (9.58)$$

where $R_p(\tau)$ is the pressure auto-correlation function with time-delay τ and is defined by

$$R_p(\tau) = \lim_{T \to \infty} (1/T) \int_{-T}^{+T} p(t) p(t + \tau) \, dt \ . \tag{9.59}$$

We can also define the space time correlation function as the correlation between the pressure, $p(x, t)$ at one point and the pressure $p(x + \xi, t + \tau)$ at another point distance ξ (in the x direction) from the first point. This is

$$R_p(\xi, \tau) = \lim_{T \to \infty} (1/T) \int_{-T}^{+T} p(x, t) p(x + \xi, t + \tau), \, dt \ . \tag{9.60}$$

The wave number/frequency spectrum $S_p(k_p, \omega)$ is then defined as the double Fourier Transform of the space-time correlation function, i.e.

$$S_p(k_p, \omega) = 1/(2\pi)^2 \int_{-\infty}^{+\infty} \exp(-ik_p\xi) \int_{-\infty}^{+\infty} R_p(\xi, \tau) \, e^{-i\omega t} \, d\tau d\xi \ . \tag{9.61}$$

Now the cross-power spectral density of the pressures at any two points distance ξ apart is

$$S_p(\xi, \omega) = (1/2\pi) \int_{-\infty}^{+\infty} R_p(\xi, \tau) \, e^{-i\omega t} \, d\tau \ . \tag{9.62}$$

This can be expressed in the form

$$S_p(\xi, \omega) = S_p(\omega) \, \rho(\xi, \omega) \tag{9.63}$$

where $\rho(\xi, \omega)$ is the narrow-band correlation coefficient between pressures distance ξ apart.

Hence $\qquad S_p(k_p, \omega) = S_p(\omega) \dfrac{1}{2\pi} \int_{-\infty}^{+\infty} \rho(\xi, \omega) \exp(-ik_p\xi) \, d\xi \ . \tag{9.64}$

The total power spectral density of the pressure at frequency ω is found from this by integrating over the whole range of k_p. Thus

$$S_p(\omega) = \int_{-\infty}^{+\infty} S_p(k_p, \omega) \, dk_p \ , \tag{9.65}$$

and the mean square value of the pressure at a point is found by integrating this over the whole of the frequency range, hence

$$<p^2> = \int_{-\infty}^{+\infty} S_p(\omega) \, d\omega \ . \tag{9.66}$$

Now the spectral density of the plate displacement is related to the spectral density of the pressure by

$$S_w(k_p, \omega) = |Y(x, y, k_p, \omega)|^2 S_p(k_p, \omega) \tag{9.67}$$

where $Y(x, y, k_p, \omega)$ is the wave receptance function of equation (9.57). The frequency spectrum of the displacement is obtained by integrating this over the whole wave number domain, so

$$S_w(\omega) = \int\limits_{-\infty}^{+\infty} |Y(x, y, k_p, \omega)|^2 S_p(k_p, \omega) \, dk_p \qquad (9.68)$$

and the mean square displacement is

$$<w^2> = \int\limits_{-\infty}^{+\infty} S_w(\omega) \, d\omega \ . \qquad (9.69)$$

REFERENCES AND FURTHER READING

[9.1] Ungar, E. E. (1961). *J. Acoust. Soc. Am.*, **33**, 5, 633–639. Transmission of plate flexural waves through reinforcing beams; dynamic stress concentrations.

[9.2] Stearn, S. M. (1971). *J. Sound Vib.*, **15**, 3, 353–365. The concentration of dynamic stress in a plate at a sharp change of section.

[9.3] Snowdon, J. C. (1968). *Vibration and shock in damped mechanical systems.* John Wiley & Sons, Inc., New York.

[9.4] Cremer, L., Heckl, M. & Ungar, E. E. (1973). *Structure-borne sound.* Springer-Verlag, New York, Heidelburg, Berlin.

[9.5] Skudrzyk, E. (1968). *Simple and complex vibratory systems.* Pennsylvania State University Press, University Park and London.

CHAPTER 10

Structure of turbulence

P. O. A. L. Davies

Institute of Sound and Vibration Research, University of Southampton

10.1 INTRODUCTION

Turbulent flow is the most common form of fluid motion and may be regarded as a complex assembly of locally organised but usteady velocity patterns which interact strongly with each other as they move with the flow. An explanation of how sound is produced by turbulent motion in the absence of solid boundaries was first provided by Lighthill some thirty years ago, with his classical papers on sound generated aerodynamically. He showed that aerodynamic noise is generated by the interactions between unsteady (fluctuating) fluid motions in neighbouring volumes of fluid, a proportion of the fluctuating stresses providing acoustic sources.

We are concerned here with a study of the characteristics of turbulence which allow us to estimate the size, strength, and spatial distribution of the sources as they move downstream. Such a study raises serious experimental problems since measurements normally consist of a velocity time history record obtained at fixed points and are difficult to interpret in terms of moving unsteady complex patterns of motion. When one examines such a record, which is obtained from a succession of disturbances passing the measuring stations, information from those parts which are relatively remote from each other in space or time does seem to be statistically independent. Thus the measured signal obtained at any fixed point is normally regarded as random in statistical terms.

The adoption of a statistical approach represents an enormous experimental simplification both in the type and number of measurements required and in their subsequent analysis. Even so, an adequate statistical description of a developing turbulent flow remains a formidable experimental task. Unfortunately, statistical descriptions tend to be inappropriate for evaluating the detailed mechanisms of sound generation, since they average out many of the significant features of the motion. On the other hand, they provide a concise, systematic and convenient method of presenting large quantities of experimental observation. Such results can then be used to indicate trends and relative orders of magnitude of source strength in different regions of a flow field.

An alternative description is to represent the moving velocity patterns mathematically by waves. The original mathematical formulation was set out by Rayleigh about 1880 [10.1] during a study of the development of instabilities in vortex sheets. The aim of such studies is to develop techniques for predicting which wave components are most strongly amplified in a particular flow configuration and calculate the rate at which this occurs. The calculations employ measured descriptions of mean velocity field as a starting point. To maintain a finite balance of turbulent energy, the extent of the waves must lie within the boundaries of the flow (i.e., they must be dispersive) with a mechanism established to limit growth once an appropriate amplitude has been reached. Great care is necessary in applying the results of such calculations for prediction purposes.

Though avoiding many of the limitations created by time-averaging in the purely statistical descriptions, wave models involve other practical difficulties. With all free turbulent flows the scale of the largest motions grows monotonically in the downstream direction. Though this is helpful as it tends to provide limits to growth of the wave components, it also restricts useful modelling to short distances in the streamwise direction.

Yet another alternative is to develop a mathematical model of the flow field with statistical properties that correspond to those of the measured ones. Such models can then be employed to estimate the source characteristics of the turbulent flow [10.2]. One such model which consists of a three-dimensional solution of the complete unsteady Navier-Stokes equation, is being studied by Orsag [10.3]. This model also requires a measured mean velocity field as a starting point and is again severely limited in streamwise extent. Another kinematic model [10.4] calculates the velocity field of an impulsively started flow; representing shear layers by vortex sheets and using source distributions to satisfy boundary conditions. This model provides much useful insight into the detailed structures of the velocity patterns and of the way in which they interact. However, practical considerations of computer running time restrict the representation to two-dimensional flows or those with axial symmetry, while all turbulent motions have a three-dimensional structure. Furthermore, the temporal length of the record for spatially extensive flows is similarly restricted to provide only a modest number of statistical degress of freedom at time scales of practical interest.

The different approaches to modelling the structure of turbulence are to some extent complementary, and taken together provide some of the insight needed for a proper understanding and evaluation of the sound source mechanisms. The statistical description of the structure of a turbulent jet will be presented first since it provides the most concise and comprehensive description of the flow structure within the limits of present knowledge. An outline of the other approaches to models of turbulent flow structure will follow to provide further insight into the interpretation of the statistical descriptions in terms of noise-producing mechanisms.

10.2 STATISTICAL RELATIONS AND FRAMES OF REFERENCE

Observations in a turbulent flow normally consist of a time history of the velocity or some other property at one or more fixed points. The structure of the flow can then be described in terms of a time averaged or mean motion and a fluctuating motion by separating the record of velocity components, pressure, temperature, density, etc., into mean and fluctuating parts. Thus a velocity component v_i can be represented as

$$v_i = U_i + u_i \tag{10.1}$$

where
$$U_i = \frac{1}{T} \int_0^T v_i \mathrm{d}t = \bar{v}_i \tag{10.2}$$

and
$$\bar{u}_i = \frac{1}{T} \int_0^T u_i \mathrm{d}t = 0 \ . \tag{10.3}$$

Similar expressions can be written out for pressure or density. Provided the records are long enough they will be statistically stationary, for a steady mean flow, so that mean values obey the simple rules

$$\overline{v_i + v_j} = U_i + U_j \tag{10.4}$$

and
$$\overline{v_i v_j} = U_i U_j + \overline{u_i u_j} \tag{10.5}$$

after making use of (10.3).

In most hydrodynamic studies the basic equations, which provide a mathematical model of the fluid, are those of conservation of mass and momentum. With reference to the measurements they relate to a unit volume of fluid at some fixed reference point in space. Provided it contains no sources of mass, conservation of mass for an elementary volume surrounding the reference point is expressed by

$$\frac{\partial \rho}{\partial t} + \frac{\partial}{\partial x_i} (\rho v_i) = 0 \ , \tag{10.6}$$

where ρ is the density and v_i the velocity in the i direction. For the mean motion (10.6) becomes simply

$$\bar{\rho} \frac{\partial}{\partial x_i} U_i = 0 \ , \tag{10.6a}$$

while for the fluctuating part, (10.6) becomes

$$\frac{\partial \rho}{\partial t} + \frac{\partial}{\partial x_i} (\rho u_i) = 0 \ . \tag{10.7}$$

Since we are concerned generally with moving eddy patterns the fluctuating parts of the motion are of more direct interest. Thus the distributions of the mean and fluctuating parts of the motion are normally described separately in statistical models of the flow structure.

Conservation of momentum, for a similar elementary volume in a continuous medium, under no external forces can be expressed [10.5] as

$$\frac{\partial}{\partial t}(\rho v_i) + \frac{\partial}{\partial x_i}(\rho v_i v_j + p_{ij}) = 0. \tag{10.8}$$

Here p_{ij} is the compressive stress tensor, representing the force in the x_i direction acting on a portion of the fluid with inward normal in the x_j direction and for real fluids includes the viscous stresses. The term ρv_i represents the fluid momentum per unit volume, while $\rho v_i v_j$ is the stress tensor arising from the momentum exchanges associated with turbulent mixing in the flow. This stress is the result of transport of momentum components v_i by velocity components v_j. The momentum stress tensor is generally dominant in turbulent motion and its components are normally referred to as the Reynolds' stresses.

One can again separate equation (10.8) into a pair of equations describing conservation of momentum for the time averaged or mean motion and the fluctuating motion. We note, however, that in this case the expression for the mean motion must include the mean Reynolds stresses $\overline{\rho v_i v_j}$ in accordance with (10.5). This fact has proved a major problem in attempts to provide analytical models of turbulent flow structure, since the values of the correlation $\overline{u_i u_j}$ must be obtained or inferred from observations.

So far we have considered observations and their interpretation in terms of a fixed (Eulerian) frame which is appropriate for measurements made at fixed points. Since aerodynamic sound arises from interactions between moving eddies, it is more revealing if the observations are resolved into those component fluctuations that can be assigned to the spatial variation of a convected eddy structure and to those components that can be assigned to the purely temporal fluctuations within the moving eddies, as their structure is modified by interactions. An observed flow component v_1, say, is given in terms of displacement ξ, by

$$v_1 = \frac{\partial x}{\partial t} + U_\phi \frac{\partial x}{\partial \xi}, \tag{10.9}$$

where the first term represents the contribution to v_1 from the temporal variation and the second the contribution from the spatial variation of the moving velocity pattern. This resolution thus requires an independent estimate of the relative contribution of these two components and, in particular, the choice of an appropriate value of the local phase velocity U_ϕ of the eddy pattern.

10.2.1 Correlations and spectra

We have already encountered one correlation $\overline{u_i u_j}$ which describes the mean Reynolds stress components at a point. More generally we can investigate the spatial properties of the velocity pattern in terms of the correlations between fluctuating velocity components measured at neighbouring points. Since the velocity field at a point must be continuous, one might expect a high value of the correlation between velocity components at nearby points and that its magnitude will decay as the separation increases. In a moving eddy pattern, however, if we increase the separation in a streamwise direction, a given eddy will pass the successive measuring points at time intervals fixed by its average velocity of travel over the separation distance. In this case we will expect to recover a higher value of the correlation if we delay the upstream signal by the same time interval. It is convenient in evaluating such correlations to normalize the result by the standard deviation σ_i of the velocity fluctuations u_i, thus producing a coefficient whose value can never exceed unity.

We can describe the statistical properties of a convecting pattern of eddies in terms of space-time or cross-correlation functions. In its most general form the normalized cross-correlation function is expressed as the tensor

$$R_{ij}(\xi, \tau) = \overline{u_i(x, t) \cdot u_j(x + \xi, t + \tau)}/\sigma_i \sigma_j, \qquad (10.10)$$

where ξ is the separation vector between the two points and τ the time delay between the two recorded time histories. Experimentally determined components of the tensor R_{ij} are estimates whose quality depends on the magnitude of the integration time T (see eqn. (10.2)) compared to the longest significant fluctuation period in the velocity u_i.

Correlations performed with zero time delay describe the spatial characteristics of the velocity field, while those performed with zero separation define the spectral or time-dependent characteristics of the velocity patterns. Thus correlations can be employed to perform the resolution into spatial and temporal fluctuations defined in equation (10.9). Typical examples of such spatial and temporal correlation functions are given in Fig. 10.1.

$R_{11}(\xi, 0)$

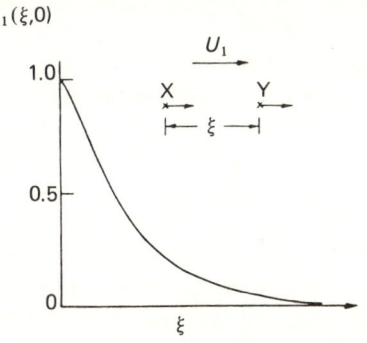

Fig. 10.1a – Space correlation.

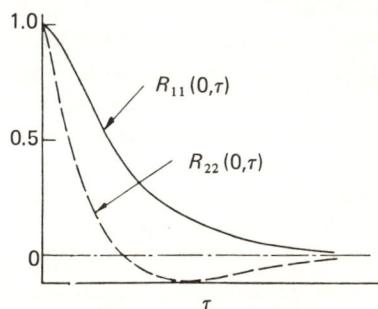

Fig. 10.1b – Autocorrelation.

The interpretation of the auto-correlation function is straightforward. Since the velocity field is continuous the auto-correlation function at a point will be a function only of the time delay τ, so must be symmetric in time, or an even function of τ. This can be transformed to yield the frequency power spectrum of the signal or

$$G_{11}^{\tau}(0, \omega) = \int_{-\infty}^{\infty} R_{11}(0, \tau) \exp(-i\omega\tau) \, d\tau \ , \tag{10.11}$$

where ω is the radian frequency. If the moving pattern is frozen, the space correlation function is a function only of the separation ξ, so must also be symmetric in the displacement. In this case the wave number spectrum can be obtained directly from the transformation

$$G_{11}^{\xi}(k, 0) = \int_{-\infty}^{\infty} R_{11}(\xi, 0) \exp(-ik\xi) \, d\xi \ , \tag{10.12}$$

where k is the wave number $2\pi/\lambda$, if λ is the wavelength.

Measurements of the type shown in Fig. 10.1 can be used to derive mean space and time scales of the eddy pattern. Thus the length scale L_{ξ} is normally defined as

$$L_{\xi} = \int_{0}^{\infty} R_{11}(\xi, 0) \, d\xi \ . \tag{10.13}$$

with an equivalent time scale L_{τ} defined as

$$L_{\tau} = \int_{0}^{\infty} R_{11}(0, \tau) \, d\tau \ . \tag{10.14}$$

Comparison of (10.13) with (10.12) and (10.14) with (10.11) shows that these integral scales are equal to the spectral energy level of the normalized spectral components at zero wave number and zero frequency, respectively.

Figure 10.1b includes a typical representation of the autocorrelation function for the cross-stream velocity components, $R_{22}(0, \tau)$ which clearly differs from the streamwise component correlation. If one performs the integration in (10.14) the resultant time scale is nearly zero. Such a result can be interpreted as the existence of almost zero energy in the zero component of the spectrum, which is compatible with a zero transverse mean velocity. These results indicate the convenience of spectral methods for presenting large quantities of experimental data in a concise form which can be clearly interpreted.

Frozen turbulence patterns are rare, if they exist at all. Such turbulence is not of direct interest in the present context since it cannot generate aerodynamic sound. In developing turbulence the length and time scales always increase in the streamwise direction so that the space or space-time correlations are no longer even functions. The transformation of such measurements yields a complex wave number or frequency spectrum, which is more difficult to interpret [10.6].

This can be best illustrated by first examining the relationship between space and time scales in a frozen turbulence pattern and then in a more practical non-frozen pattern. A single spectral component with wave number k moving with a phase velocity c will produce, at a fixed measuring point, a signal of frequency ω, where

$$\omega = ck . \tag{10.15}$$

With a frozen pattern, all the phase velocity components c are equal to the mean streamwise velocity U_1. Thus the wave number and frequency are related by

$$G_{11}^{\tau}(0, \omega) = G_{11}^{\tau}(0, U_1 k) ,$$

so that $U_1 G_{11}^{\tau}(0, \omega) = G_{11}^{\xi} \left(\frac{\omega}{U_1}, 0 \right) . \tag{10.16}$

However, if the wave number spectrum is complex, then, since the frequency ω is real, the phase velocity c must also be complex to satisfy (10.15). This shows that the wave number components of developing or reacting turbulence must be dispersive, a point noted earlier. This result is thus compatible with the concept of a turbulent motion with finite energy and of limited spatial extent.

10.2.2 Correlations and spectra in shear flows and developing turbulence

Strongly sheared turbulence is generated in boundary layers, jets and wakes. The turbulence intensity is high, that is the ratio of the root mean square velocity fluctuations (their standard deviation) to the mean local velocity is of order 0.1 or more. In free shear layers, i.e., those found in jets and wakes, the scale of the turbulence increases in the streamwise direction. Typical cross-correlation measurements of the velocity fluctuations in a jet mixing region are shown in Fig. 10.2. The rapid fall in peak correlation with distance shows that the structure is changing relatively rapidly, showing that this turbulent flow provides a significant source of aerodynamic noise. The changing structure is also indicated by the broadening of the individual curves as the displacement increases, since with a frozen pattern the shape of each curve would remain the same.

We noted earlier that the phase velocity of the spectral components of a distorting turbulence pattern varies with wave number k. However, one can estimate the convection velocity of the eddy patterns as a whole by finding the characteristic time τ_c at which $\partial R(\xi, \tau)/\partial \xi = 0$. This can be determined experimentally from the values of time delay τ at which the correlation curves of constant separation ξ are tangent to the envelope in Fig. 10.2. It can be shown that the ratio ξ/τ_c, which we shall call the convection velocity U_c, is equal to the mean rate of convection of turbulent energy [10.7], and this is analogous to a group velocity of the spectral components. In addition the envelope to the correlation curves on Fig. 10.2 represents the autocorrelation in a reference frame travelling with the eddies at the convection velocity U_c.

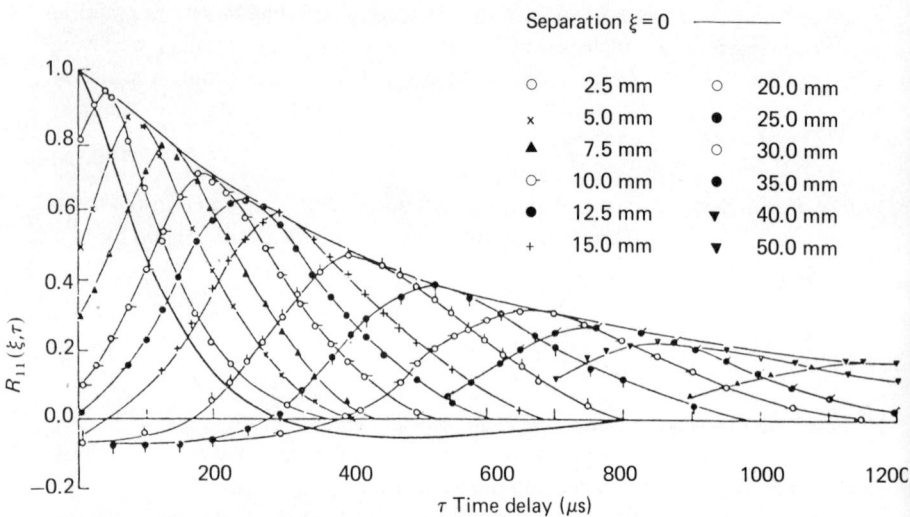

Fig. 10.2 – Cross-correlation measurements in jet mixing region.

One can obtain an estimate of the phase velocities $c(\omega)$ of the spectral component waves making up a moving eddy pattern averaged over a flight path ξ from the cross-power spectrum $G_\xi(\omega)$ of the velocity fluctuations at x_1 and x_2 (where $\xi = x_1 - x_2$). Alternatively one can calculate $G_\xi(\omega)$ by finding the Fourier transform of the appropriate cross-correlation curve illustrated in Fig. 10.2. We noted earlier that such a spectrum will be complex and can be expressed as

$$G_\xi(\omega) = |G_\xi(\omega)| \, \exp(i\theta(\omega)) \, . \tag{10.17}$$

For a component at frequency ω, the time of flight over the separation ξ is given by $\tau(\omega) = \partial/\omega$. Thus the average velocity $u(\omega)$ for each component over the path will be ξ/τ, or

$$u(\omega) = \omega\xi/\theta \, . \tag{10.18}$$

Also, from (10.15) we have

$$d\omega = c\,dk + k\,dc \, ,$$

and as $\xi \to 0$, the change in phase velocity will also vanish. Thus, as this limit is approached, then

$$\mathrm{d}\omega = c\,\mathrm{d}k = \frac{\omega\xi}{\theta}\,\mathrm{d}k \ . \tag{10.19}$$

Furthermore, we see that $u \to c$ as $\xi \to 0$. Thus one can estimate the spectral component phase velocities from the measured cross-power spectrum or the cross-correlation function.

10.3 STATISTICAL MEASUREMENTS OF TURBULENT JET STRUCTURE

A statistical description of the structure of a turbulent jet includes the distribution of the mean velocity, pressure, and density fields as well as the statistical properties of their fluctuating (time-dependent) components. Fig. 10.3 illustrates an extensive set of such measurements made on a model jet with a steady mean flow which have been reproduced from reference [10.8]. In the figure it can be seen that the width of the turbulent flow field increases linearly with distance downstream; this width represents roughly the size of the largest eddies at that point. For the first 5 diameters, the turbulent flow consists of a conically shaped annulus surrounding a relatively undisturbed core, which is moving at the initial jet velocity U_j. Beyond this the mean structure changes, and similarity of the mean motion is finally achieved beyond 20 or 30 initial jet diameters from the orifice. From here on the centre line velocity falls as X^{-1}. Note also that the plot of the mean radial velocity V has been inverted on the figure and its scale increased 50 times relative to the axial component U.

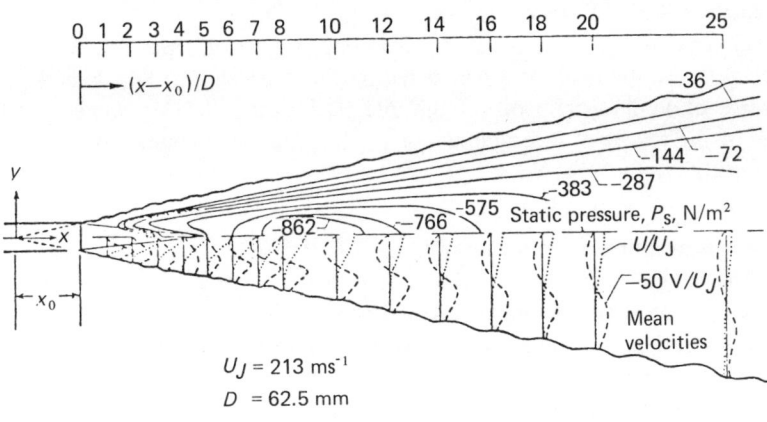

Fig. 10.3 – Mean flow field of a jet.

In terms of mean velocity measurements, the jet can be subdivided into three regions. Firstly there is an initial region where a zone of intense turbulence surrounds a core of substantially potential motion; secondly, an adjustment region, where the turbulent flow changes over from an annular to cylindrical distribution. Here again the turbulence is relatively intense and the eddies are still convecting at a little over 0.6 of the jet exit velocity at least as far as the first 10 to 15 jet diameters. The last region, called the fully developed region, extends beyond the adjustment region.

The first two regions, where the turbulence is most intense and its structure changes most rapidly, provide the major source of jet noise, in terms of the most intense sources. For this reason detailed studies have been concentrated on the first ten or so diameters of the flow. To estimate the noise source strength, one requires information on the distribution of turbulent length and time scales, the eddy convection speed, and the rates at which the patterns change under the influence of the Reynolds' stresses. Over most of these two regions, similarity in radial distributions of mean and fluctuating velocity component can be expressed in terms of radial coordinates which subtend constant angles of magnitude $\tan^{-1}\eta$ with the jet lip. The mean velocity remains constant over the conical surfaces $\eta = $ constant for from 0 to 9 diameters from the orifice for negative and as much as 15 diameters or more for zero or positive values of η. The cylindrical surface $\eta = 0$ represents a boundary between the inner and outer parts of the jet mixing region.

10.3.1 Structure of the initial region

In the initial region the inner edge of the mixing region is bounded by the conical surface $\eta = -0.1$, while the outer edge lies just inside the conical surface $\eta = 0.2$ [10.8, 10.9]. Thus, the average width of the turbulent annulus and the integral length scales of the turbulence grow linearly with distance x downstream. Thus the value of the mean shear, dU/dY, which represents the average effects of the Reynolds stress in redistributing momentum, falls linearly with increase of x. Measurements of the integral length scale vary somewhat, but most of the results of measuremnts of the longitudinal components lie close to the value given by

$$L_\xi = 0.12x \; , \tag{10.20}$$

where x is measured from the jet orifice.

The integral time scales of the turbulence will provide a measure of the rate at which turbulence is being distorted. Since the eddies are moving, one requires this information in a frame moving with the eddies rather than measurements of the time scale made at fixed points. We noted already that the moving frame time scale could be calculated from the envelope of the correlation curves shown in Fig. 10.2. This represents the time scale of eddies in a frame moving at the mean eddy convection speed U_c. Estimates of the convection speed can be made in a number of ways, and typical results are shown in Fig. 10.4. The value of the convection speed U_c remains almost constant across the mixing layer and is that of the most energetic eddies which seem to stretch completely across the mixing layer [10.10 to 10.13].

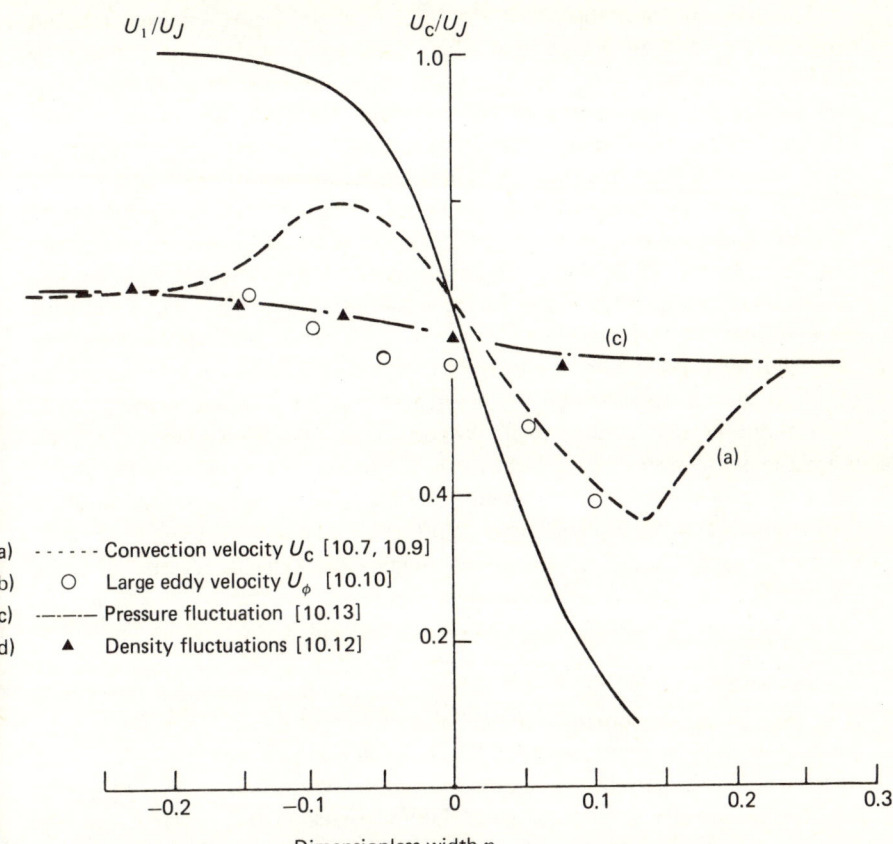

Fig. 10.4 — Convection and phase velocity measurements in a jet mixing layer.

A constant value of the convection velocity and a characteristic spatial scale equal to the width of the mixing layer provides a simple generalization of the mixing layer structure. Measurements [10.9] show that the moving frame time scale L_{TC} is proportional to the inverse of the local mean shear $\partial U / \partial Y$. This result seems plausible since the time scale should be inversely proportional to the rate of distortion of the eddy patterns, which in turn should be related to the mean shear. The moving axis time scale L_{TC} in the initial region is given approximately by

$$L_{TC} = \tfrac{3}{4} \frac{x}{U_j} \, . \tag{10.21}$$

This means that typical frequencies of the velocity fluctuations fall like $1/x$.

Measurements also show that the local axial turbulence intensity is about equal to the ratio of the integral length scale to the moving axis time scale [10.9]. This ratio, $L_{\xi 1}/L_{Tc}$ remains constant, as does the turbulent intensity, over surfaces $\eta = $ constant in the initial region. The magnitude of intensity of the radial and circumferential fluctuating velocity components are of the same order, though a little less than this [10.8]. Measurement of radial and circumferential scales by integration of measured space correlations, is subject to considerable uncertainty, since the zero wave number components have energies that lie near zero. Since the larger eddies appear to span the mixing layer, one would expect radial eddy scales of the same order of magnitude as the axial ones. Recent observations [10.10] support this view. They also suggest that circumferential scales may be of similar order of magnitude as well, though this is not always supported by circumferential correlation measurements.

Conditions in the adjustment region appear more complex. The outer mixing region continues to exhibit the characteristic behaviour outlined above, but the inner region behaves rather differently once the core has vanished. Characteristic frequencies still vary as $1/x$ though the mean structure of the flow is now changing. One can assume that, roughly speaking, the volume of active turbulence creating noise will be proportional to x per unit length of the annular mixing zone of the initial region and to x^2 per unit length beyond the adjustment region.

The description of the flow structure given so far is somewhat oversimplified. More details are obtained if the behaviour of the different spectral components is studied. However, attempts to do so are accompanied by an increasing uncertainty in the physical interpretation of the observations. Cross-correlation measurements with 1/3 octave band filtered signals [10.7] suggest that the low-frequency components are convected more slowly than the high-frequency ones (Fig. 10.5a). The moving axis integral time scale (Fig. 10.5b) remained constant at lower frequencies but fell rapidly at the higher ones.

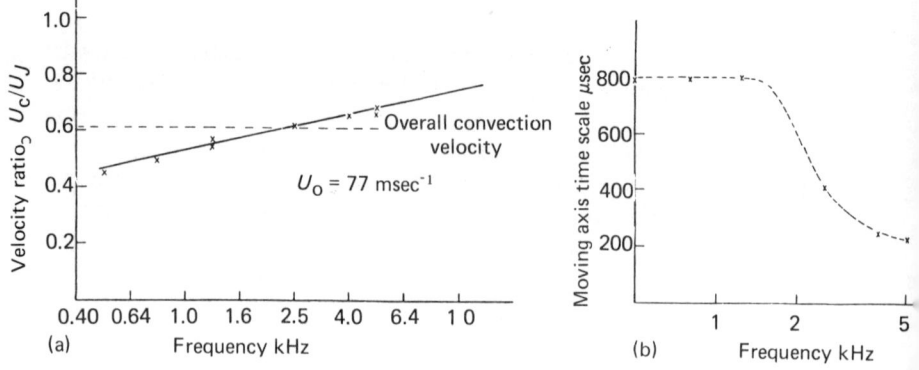

Fig. 10.5 – Variation of (a) convection velocity and (b) moving axis time scale with centre band frequency.

These results suggest that the smallest eddies are being strongly distorted while the larger ones are all being distorted at the same rate. If they were all being stretched at equal rates by the shear, the time scale would vary inversely with their characteristic length scale, but not more rapidly than this! This warns us that we cannot regard the structure of turbulent motion as being similar at different spectral scales. Analysis of the way the phase velocity (see (10.18)) of the higher frequency components varies with path length suggests that the eddy patterns may only be regarded as frozen over flight paths of around half a wavelength or so. This explains the rapid fall in integral time scale at the higher frequencies, since these were deduced from a set of fixed length flight paths.

10.4 ALTERNATIVE MODELS OF JET STRUCTURE

Observations of the initial development of circular jets show that the mixing layer has its origins in the nozzle boundary layer that is cast off as a thin cylindrical vortex sheet. This sheet first rolls up into a train of vortex rings whose development has been described in much detail by Wille [10.14]. These rings grow by coalescence and the migration of vorticity from the sheet towards the vorticity concentrations represented by the rings. At the same time these rings develop circumferential waves, so the motion rapidly becomes three-dimensional [10.10]. The rapid development of such three-dimensionality means that the flow structure is difficult to observe or describe, except in qualitative terms.

10.4.1 Wave models of turbulence
Considerable attention has been paid recently to the similarity between the large-scale ordered motions in a high Reynolds number turbulent flow and the instabilities of a laminar flow [10.15]. The application of stability analysis to describe the development of instability in the vortex sheet in the terms of amplifying waves was first suggested by Rayleigh. There have been a number of interesting developments in this approach which has been an active interest of many contributors over the past 20 years [10.15]. One might summarize the progress achieved so far by noting that there has been some marked success in predicting the most likely modes and frequencies of the initial flow instabilitoes. Furthermore, the observed flow behaviour is most appropriately modelled by spatially growing disturbances [10.16].

A representative example of the approach and its application to noise predictions has been presented recently by Morris [10.17]. The velocity and pressure in the jet are divided into three parts; the first being the mean components based on the measurements, the second a time (or space)-dependent organized fluctuation, and the third representing the background disorganized turbulence whose effects are approximated by introducing an appropriate eddy viscosity. The equations for the organized motion are Fourier decomposed and linearized, while the mean flow is assumed locally parallel. The structure of the organized motion is seen to be dominated by the spatially unstable modes which are Eigen solutions to the stability equations. The downstream growth of the organized large-scale motion, though initially highly amplified, is then damped by nonlinear interactions and flow divergence.

Though they have not yet provided the information required for a quantitative prediction of the noise, the potential of wave models can be illustrated by an example. Suppose we have a distribution of acoustic source components that can be described by travelling waves and expressed as $A \cos (\omega t - kx)$. Suppose also there is a distant observer at r_0 (with coordinate r_0, ϑ) as shown in Fig. 10.6. Components of the sound will arrive at the observer at times $t - r(x)/a_0$ after they are emitted. If the angle θ is effectively constant then $r(x) = r - x \cos \theta$, so that the apparent source strength seen by the observer at r_0 will be given by

$$\int_{-\infty}^{\infty} \frac{A}{r} \cos \left[\omega (t - \frac{r_0}{a_0} + \frac{x \cos \theta}{a_0}) - kx \right] dx$$

and setting $t' = t - \dfrac{r_0}{a_0}$, and $\omega = ck$ this becomes

$$\int_{-\infty}^{\infty} \frac{A}{r} \cos(\omega t' + \left(\frac{c \cos \theta}{a_0} - 1 \right) kx) \, dx \ .$$

As r_0 tends to infinity this integral is made up of cosine and sine terms like

$$\frac{A}{r_0} \cos \omega t' \int_{-\infty}^{\infty} \cos \left[\left(\frac{c \cos \theta}{a_0} - 1 \right) kx \right] dx ,$$

while the integral is non-zero only if $(c \cos \theta/a_0) - 1$ is zero. That is if

$$c = \frac{a_0}{\cos \theta} \ . \tag{10.22}$$

This result shows that only those wave number components of a distributed moving pressure source field with supersonic phase velocities make any contribution to the far field. It is clear that for calculation of the noise one requires a model or description that provides both the amplitude and relative phase information for these noise-producing disturbances.

This example illustrates the problems that arise when attempting to estimate the noise experienced at an observer from statistical or wave models of the turbulence, since these provide the group velocities U_c rather than the phase velocities c of the wave components in the turbulence. One can also see that supersonic phase velocities can arise only as a result of relatively rapid changes in the local structure of the moving eddy patterns.

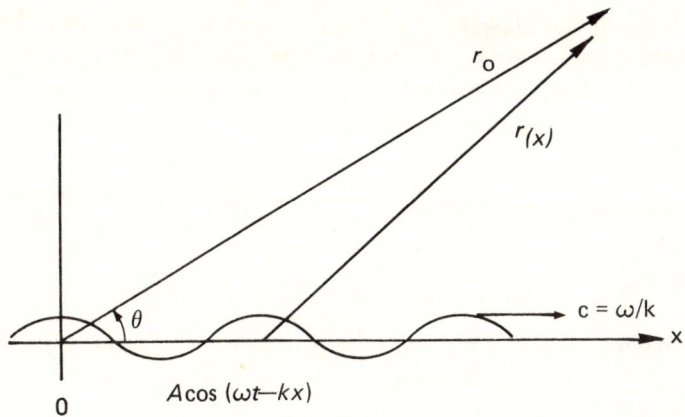

Fig. 10.6

10.4.2 Averaged eddy model

A second approach to modelling the flow is to solve the complete Navier-Stokes equation in three dimensions. This is a formidable task and requires substantial computational power. Nevertheless, encouraging progress has been made recently by Orsag and his colleagues [10.3]. So far, their model requires measured mean velocity profiles as boundary conditions, and is restricted to strictly cylindrical segments of the flow extending for two or three diameters axially. These restrictions and requirements are similar to those employed in the wave-guide models mentioned above [10.17].

It appears that with these constraints with a zero entrainment restriction such a solution of the Navier-Stokes equations is a practical proposition. It has been used to describe the complete distribution of quadruple moments, so that the Lighthill noise terms can be evaluated over a limited though substantial volume of the jet flow fluid. By repeating the procedure for a succession of sections of the jet and superimposing the results, a model of the noise generation of a complete jet can be achieved. How realistic and useful this model will prove must depend on how successfully it can reproduce those rapid changes in local flow structure in the moving eddy patterns which represent the more significant noise sources.

An alternative approach to modelling the flow would be a numerical simulation of Kelvin-Helmholtz waves of finite amplitude based on the Boussinesq equations. A recent study of two-dimensional wave development has been published by Corcos and his colleagues [10.18, 10.19]. So far the solution represents a limited extent of the layer and is confined to two dimensions. Again, too, this model exhibits many of the limitations of other wave models, though with care both the rapid local changes in structure and the associated phase relation ships can be preserved during the calculations.

10.4.3 Inviscid Flow Models

A different approach again is to first calculate the velocity field by first establishing all the conditions needed to specify the complete jet flow field. The velocity at any point v can be specified by

$$v = v_e + v_v + v_b , \qquad (10.23)$$

where v_e represents an isentropic expansion, with a rate of increase per unit volume Δ, so that

$$\nabla . v_e = \Delta; \quad \nabla \wedge v_e = 0 . \qquad (10.24)$$

The component v_v represents a rigid body notation with angular velocity $\Omega/2$, so that

$$\nabla . v_v = 0; \quad \nabla \wedge v_v = \Omega . \qquad (10.25)$$

Finally, the component v_b is due to the motion of the solid boundaries, so is given by

$$\nabla . v_b = 0; \quad \nabla \wedge v_b = 0 . \qquad (10.26)$$

In a free jet flow this component may be ignored as no boundaries are present apart from the nozzle exit.

The simplest realization of the flow is to start it impulsively from rest. Thus at zero time the jet outflow can be represented by a distributed source of strength $2U_J$ per unit area, while the velocity everywhere else is zero. The model then proceeds to calculate flow at later times using a time marching procedure. Observations of real jet flows show that the flow separates at the lip. For short time steps the velocity discontinuity introduced by the separated flow may be closely approximated by a short cylindrical vortex sheet. This represents the local rotational component of rhe velocity field. It is easy to show both by observation and analysis that this vortex sheet begins to roll up at its free end. Since the computation of the evolution of doubly curved sheets of vorticity is very laborious, the model is further simplified by replacing the elementary sheet by an equivalent elementary vortex ring [10.4].

Though this avoids one difficulty, it introduces others which arise as a consequence of the representation of a continuous sheet by isolated concentrations of vorticity. The generators of such vortices represent singularities since, as the spacing between them vanishes, the induced velocities tend to infinity. This is spurious since in real flows such motions are reduced to local solid rotations by the action of viscosity. In developing a working model, some form of smoothing must be introduced to avoid the singular behaviour and reproduce the effects of viscosity [10.4].

Continuous operation of the model requires the generation of a succession of elementary vortex rings to represent the continuous generation of vorticity by the flow separation at the jet lip. At each time step the velocity at the centre of each vortex ring is calculated as the sum of the contributions of the induced velocities of all the other elementary vortices and of the source distribution representing the jet outflow. The resultant flow field is unsteady, so mass and momentum are conserved separately at each element of the field, represented by the elementary vortices and the components of the source.

Some consideration of the way each element moves during each time step is also required, since the velocity at each elementary vortex position is the sum of contributions from a large number of other elementary vortices which are also moving. Since, at any instant, the position of any vortex in the model is calculated by integration of their total previous motion, care is needed to avoid the accumulation of errors during the integration process. Thus successful operation of a realistic model of this type requires careful housekeeping, much of which is discussed in more detail in [10.4]. The results are, however, encouraging as can be seen in Fig. 10.7.

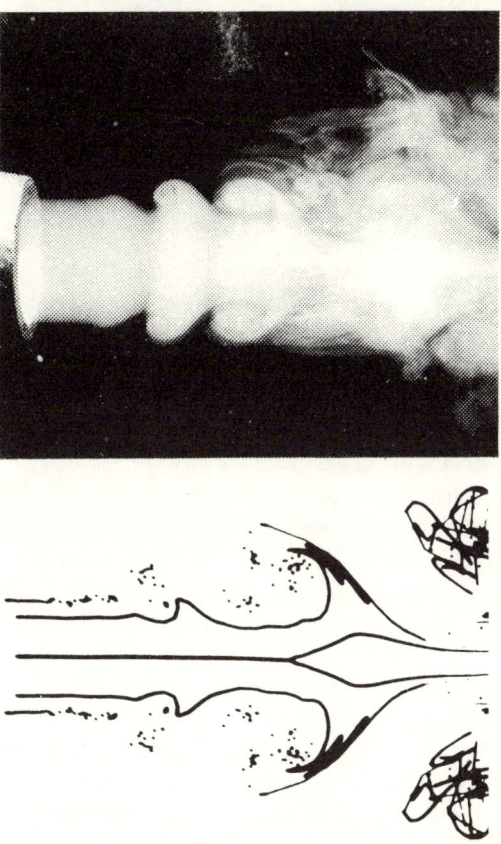

Fig. 10.7 − A jet flow and its computed realization.

Assuming a new elementary vortex is created at each time step, each time step involves the evaluations of $N(N - 1)$ interactions between the vortices. Thus computer running time increases as N^3 approximately, while the extent of the flow is proportional to $N\delta t$, where δt is the length of a time step. For a fixed computational cost there must be a compromise between the representation of finer detail ($\delta t \rightarrow 0$) and the extent of time and flow that can be modelled. For similar reasons of computational economy the model flow is constrained to remain axisymmetric, though this is hardly realistic as is shown by the flow cross-section illustrated in Fig. 10.8.

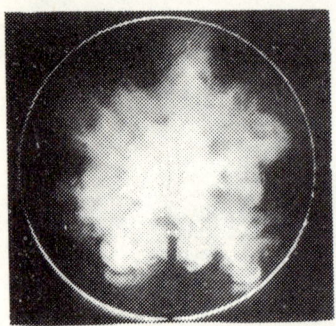 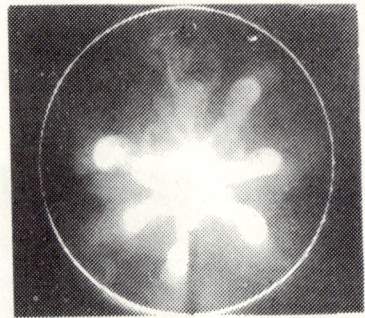

Fig. 10.8 – Typical cross-section of a smoke-filled jet.

The far-field noise can be calculated using methods described by Hardin [10.2]. In this case the phase relations between the components and the rapid changes in flow structure are properly represented, so the noise calculations are, in principle, straightforward. Since the velocity field is known, the unsteady pressure field can also be calculated by solving the Poisson equation, to yield the near-field pressure as well. The calculated mean velocity distributions agree closely with observations, as do the time history and spatial statistics of the flow. Amplitude statistics exhibit similar trends within the statistical uncertainties that result from limited length of the record available within a practical computation budget, and the constraint of the model to axial symmetry.

10.5 CONCLUSIONS

Attempts to calculate the noise radiated by jet flows from measurements of the turbulence or from models of the flow have not been successful, to date, in a strictly quantitative sense. The work has been of great value, however, in developing insight into the details of processes that are responsible for the generation of aerodynamic sound. This has been of considerable value in the development of practical noise prediction schemes since such insight is necessary for the establishment of scaling laws in empirically based prediction schemes [10.20].

REFERENCES
[10.1] Rayleigh, Lord (1896). *The theory of sound,* 2nd ed., Chapter XXI, MacMillan (republished by Dover, 1945).
[10.2] Hardin, J. C. (1973). *NASA T.N. D-7242.* Analysis of noise produced by an orderly structure of turbulent jets.
[10.3] Metcalfe, R. W. & Orsag, S. A. (1974). *Flow Research Inc. Report No. 53.* Numerical simulation of turbulent jet noise I.
[10.4] Davies, P. O. A. L., Hardin, J. C., Edwards, A. V. J. & Mason, J. P. (1975). *AIAA Paper 75-441.* A potential flow model for calculation of jet noise.
[10.5] Landau, L. D. & Lifshitz, E. M. (1959); *Fluid mechanics.* Pergamon Press.
[10.6] Davies, P. O. A. L. (1973). *Journal of Sound and Vibration,* **28,** 513-526. Structure of turbulence.
[10.7] Fisher, M. J. and Davies, P. O. A. L. (1964). *J. Fluid Mech.,* **18,** 97-116. Correlation measurements in a non-frozen pattern of turbulence.
[10.8] Maestrello, L. (1976). PhD Thesis, University of Southampton. Acoustic energy flow from subsonic jets and their mean and turbulent flow structures.
[10.9] Davies, P. O. A. L., Fisher, M. J. & Barrett, M. J. (1963). *J. Fluid Mech.,* **15,** 337-367. Turbulence in the mixing region of a round jet.
[10.10] Yule, A. J. (1978). *J. Fluid Mech.,* **89,** 413-432. Large scale structure in the mixing layer of a round jet.
[10.11] Lau, J. C. & Fisher, M. J. (1975). *J. Fluid Mech.,* **67,** 299-377. The vortex-street structure of 'turbulent' jets. Part 1.
[10.12] Harper-Bourne, M. (1970). *ISVR Memo No. 398.* Optical Measurements of jet turbulence.
[10.13] Ko, N. W. M. & Davies, P. O. A. L. (1971). *J. Fluid Mech.,* **50,** 49-78. The near-field within the potential cone of subsonic cold jets.
[10.14] Wille, R. (1963). *AFOSR Technical Report,* Hermann Föttinger Institut, Berlin. Growth of velocity fluctuations leading to turbulence in a free shear layer.
[10.15] Davies, P. O. A. L. & Yule, A. J. (1975). *J. Fluid Mech.,* **69,** 513-537. Coherent structures in turbulence.
[10.16] Michalke, A. (1971). *Z. Flugwiss,* **19,** 319-328. Instabilität eines kompressiblen runden Freistrahls.
[10.17] Morris, P. J. (1976). *J. Fluid Mech.,* **77,** 511-529. The spatial viscous instability of axisymmetirc jets.
[10.18] Patnaik, P. C., Sherman, F. S. and Corcos, G. M. (1976). *J. Fluid Mech.,* **73,** 215-240. A numerical simulation of Kelvin-Helmholtz waves of infinite amplitude.
[10.19] Corcos, G. M. & Shermann, F. S. (1976). *J. Fluid Mech.,* **73,** 241-264. Vorticity concentration and the dynamics of unstable free shear layers.

[10.20] Tester, B. J. & Morfey, C. L. (1975). *AIAA Paper No. 75-477.* Developments in jet noise modelling – theoretical predictions and comparisons with measured data.

Additional References

[10.21] Davis, M. R. (1971). *J. Fluid Mech.,* **46**, 631-656. Measurements in a subsonic jet using quantitative Schlieren.

[10.22] Moore, C. J. (1977). *J. Fluid Mech.,* **80**, 321-367. The role of shear-layer instability waves in jet exhaust noise.

[10.23] Townsend, A. A. (1976). *The structure of turbulent shear flow.* Cambridge University Press.

[10.24] Davis, M. R. & Davies, P. O. A. L. (1979). *J. Fluid Mech.,* **93**, 281-303. Shear fluctuations in a turbulent jet shear layer.

[10.25] Davies, P. O. A. L. & Baxter, D. R. J. (1978). *Lecture notes in Physics* **75**, 125-135. Springer-Verlag, Berlin. Transition in free shear layers.

[10.26] Muller, E. A. (Ed.) (1979). *Mechanism of sound generation in flows,* Springer-Verlag, Berlin.

[10.27] Browand, F. K. & Weidman, P. D. (1976). *J. Fluid Mech.,* **76**, 127-144, Large scales in the developing mixing layer.

[10.28] Lau, J. C., Morris, P. S. & Fisher, M. J. (1979). *J. Fluid Mech.,* **93**, 1-27. Measurements in subsonic and supersonic free jets using a laser velocimeter.

[10.29] Orszag, S. A. & Kells, L. C. (1980). *J. Fluid Mech.,* **96**, 159-205. Transition to turbulence in plane Poiseuille and plane Couette flow.

CHAPTER 11

Structure–fluid interaction

F. J. Fahy

Institute of Sound and Vibration Research, University of Southampton

11.1 INTRODUCTION

This chapter concerns the complementary problems of flexural vibration induced in thin plate and shell structures by sound in an adjacent fluid, and sound radiation into fluid by such motion when otherwise generated. It is the aim of this chapter not so much to present mathematical analyses of particular problems as to give the reader some 'feel' for the physical vibration behaviour of coupled structural-fluid systems, and to draw attention to references which may be studied by those who wish to obtain a more detailed knowledge, and a deeper understanding, of the subject. There is not room in this chapter to provide the comprehensive formulae, tables, and figures which are necessary to a reference work, but it is hoped that the guidance provided to the most usable data will be a satisfactory substitute. Reference [11.1] provides an introduction to the forms of mathematical analysis central to the theoretical treatment of the coupling problem.

11.2 THE RECIPROCITY PRINCIPLE: AN ELEMENTARY CASE

Consider an elastically suspended, but otherwise rigid, circular plate element, mounted flush with an infinite rigid baffle, as shown in Fig. 11.1; we shall call the element a piston. The piston has a mass m and radius a, the elastic suspension has a stiffness k, and the viscous damper has a damping coefficient c. We consider only the fluid to the right of the piston.

(i) Mechanical excitation of the piston
Let a simple harmonic force $F\,e^{i\omega t}$ act on the piston.
The equation of motion is

$$m\,\ddot{\xi} + b\,\dot{\xi} + k\,\xi = F\,e^{i\omega t} - p(r_s, t)\,\pi a^2 \tag{11.1}$$

where $p(r_s, t)$ is the pressure in the fluid at the piston surface. Elementary analysis of the acoustic field generated by motion of the piston shows that, if the non-dimensional size of the piston as indicated by the parameter $\omega a/c$, or ka, is small compared with unity, then the pressure p, when averaged over the face of the piston, is

$$p = P\,e^{i\omega t} = \rho_0 c\left(\frac{(ka)^2}{2} + i\,\frac{8ka}{3\pi}\right)\dot{\xi} = Z_{\text{rad}}\dot{\xi} \ . \tag{11.2}$$

The term in brackets, which is complex, is the specific radiation impedance of the piston; it represents the fluid loading due to piston motion. The imaginary part represents an inertial, or *mass-like*, loading, and the real part represents *damping-like* loading. If Z_{rad} is written as $R_{\text{rad}} + i\,X_{\text{rad}}$, and ξ as $\xi\,e^{i\omega t}$, equation (11.1) can be written

$$(-\omega^2 m - \omega\pi a^2 X_{\text{rad}})\,\xi + i\omega(b + \pi a^2 R_{\text{rad}})\,\xi + k\xi = F \ . \tag{11.3}$$

Because X_{rad} is linearly proportional to frequency through k, the effective mass of the piston/fluid combination becomes

$$m' = m + 8\,\rho_0 a^3/3 \tag{11.4}$$

and the damping coefficient becomes

$$b' = b + \rho_0 c\pi a^2(ka)^2/2 \ .$$

Hence it is seen that the undamped natural frequency of the fluid-loaded piston system differs from the *in vacuo* values by an amount which depends upon the density of the fluid and radius of the piston, but not the speed of sound in the fluid. Lord Rayleigh [11.7] first obtained this result.

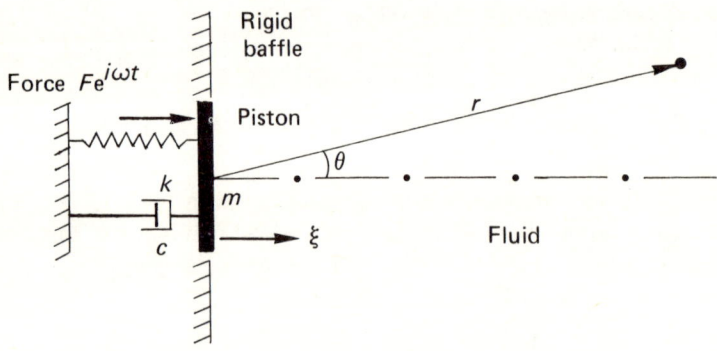

Fig. 11.1 – Vibrating piston in a rigid baffle.

Elementary textbooks show that the acoustic pressure at a distance r from the piston is virtually independent of an angle θ, if $ka \ll 1$, and is given by

$$P(r) = -\frac{\rho_0 \omega^2 a^2}{2r}\,\xi \ . \tag{11.5}$$

Hence the application of force $F\,e^{i\omega t}$ to the piston creates an pressure at r given by

$$P(r) = -\frac{\rho_0\omega^2 a^2 F}{2r(-\omega^2 m' + i\omega b' + k)} \ . \tag{11.6}$$

(ii) Acoustic excitation of the piston

A point monopole source of volume velocity $Q\,e^{i\omega t}$, located in the fluid at position r, generates sound waves which impinge on the piston surface. The total acoustic pressure on the piston can be exactly decomposed into two components: (a) the *blocked* pressure which would exist if the piston were held stationary; (b) the *radiated* pressure caused by piston motion. Such a decomposition is valid for all problems of acoustic excitation of flexible structures. The equation of motion of the piston is hence

$$m\,\ddot{\xi} + b\,\dot{\xi} + k\,\xi = -(p_{bl} + p_{rad})\,\pi a^2 \ . \tag{11.7}$$

Since the piston is assumed to be small compared with an acoustic wavelength, the blocked pressure can be considered to be uniform over its face. It is given by

$$p_{bl} = \frac{i\omega\rho_0}{2\pi r}Q(r) \ . \tag{11.8}$$

The relationship between p_{rad} and the motion of the piston is independent of whether the motion is caused by a mechanical force or acoustic excitation. Hence equation (11.7) can be written

$$(-\omega^2 m' + b' + k)\,\xi = -\frac{i\rho_0\pi a^2\omega}{2\pi r}Q(r) \ . \tag{11.9}$$

The velocity of the piston $U = U\,e^{i\omega t} = i\omega\xi\,e^{i\omega t}$. Hence

$$U = \frac{\rho_0\omega^2 a^2}{2\pi r(-\omega^2 m' + i\omega b' + k)}Q(r) \ . \tag{11.10}$$

Comparison of equations (11.6) and (11.10) shows that

$$U/Q(r) = -P(r)/F \ . \tag{11.11}$$

This result is not confined for the $ka \ll 1$ case analysed, which is simply a special case of a general reciprocity principle which applies to *all acoustically coupled, linear, elastic solid and fluid systems*. Lyamshev [11.2] derives the general result which states that the sound pressure produced at any point in a fluid by the motion of a structure subjected to a mechanical point force can be determined from a knowledge of the velocity of the free structure at the point of force application when excited by the acoustic field of a point *monopole* situated at the point of observation of the pressure, and vice versa. A useful review paper on this subject has been written by Belousov & Rimskii-Korsakov [11.3]. Advantage has been taken of this principle in experimental investigations of noise radiated from ship structures excited by internal machinery [11.4, 11.5]. A consequence of considerable practical significance in the choice of mounting points for machinery in vehicles, buildings, etc. is that measurements with an accelerometer of vibration distributions on a structure when acoustically excited. will indicate optimum mounting positions for minimum noise radiation due to machinery excitation.

11.3 RELATIONSHIP BETWEEN SOUND RADIATION AND RESPONSE TO ACOUSTIC EXCITATION

Another useful form of reciprocity relationship concerns the acoustic pressure in a fluid produced by the motion of a surface and the blocked pressure produced at the surface when it is held stationary. The general mathematical expression of the dependence of the acoustic field on the harmonic source distribution over the volume of the fluid, and on the surface boundary conditions, is the Kirchoff–Helmholtz integral equation [11.6]:

$$p(r) = -\frac{i\omega\rho}{4\pi} \int_{vol} q(r_0)\frac{e^{-ikR}}{R} dV -$$

$$\frac{1}{4\pi} \int_{surf} \left[p(r_0^s) \frac{\partial}{\partial n}\left(\frac{e^{-ikR}}{R}\right) - \frac{\partial p(r_0^s)}{\partial n}\frac{e^{-ikR}}{R} \right] dS$$

where r is the observation point, $q(r_0)$ is the distribution of volume velocity source strength density throughout the fluid volume V, $p(r_0^s)$ is the pressure at the boundary S, and $\partial p/\partial n$ is the pressure gradient normal to the boundary, n is the unit normal vector, and $R = |r - r_0|$ or $|r - r_0^s|$ is the distance from source (or boundary) point to observation point. Note that the pressure at r is in general a function of *both* surface pressure and surface velocity distributions, since $\partial p/\partial n = -i\omega\rho\, u_n$, where u_n is the normal surface velocity, and $k = \omega/c$. For infinite *plane* surfaces the equation can be expressed purely in terms of the normal velocity and volume source distribution (the Rayleigh integral); for surfaces which have dimensions and radii of curvature large compared with a wavelength the Rayleigh integral gives a good approximation to the correct pressure field. This is useful when mechanically induced surface vibration distributions are known and it is wished to estimate the radiated field, i.e. $q(r_0) = 0$. Then

$$p(r) = -\frac{i\omega\rho}{4\pi} \int_{surf} \frac{e^{-ikR}}{R} 2u_n dS \ .$$

In considering the radiation of sound by surface motion, the Kirchoff-Helmholtz integral shows that an element of surface of area δS, which has a normal particle velocity u_n, does not, in general, radiate sound like a simple isolated source of strength $u_n \delta S$, because of the presence of the rest of the surface. Only when the surface is an infinite plane does an element radiate like an isolated simple source, and then with a strength $2u_n \delta S$ (cf. the Rayleigh integral: reference [11.7], section 278). However, there is a simple reciprocal relationship between the acoustic pressure produced at an observation point in the radiation field of any vibrating body by an elemental volume velocity $u_n \delta S$, and the pressure produced by a simple source at the observation point on the *rigid body* at the location of the element (cf. ref. [11.8], section VII.29, equations 29.6 and 29.12). This example of the general principle of reciprocity forms the basis of a method of analysis of acoustically induced vibration due to Smith [11.9].

This analysis is particularly useful when fluid loading is light, i.e. the natural frequencies and mode shapes of the structure when immersed in the fluid are not significantly different from those *in vacuo*. (In practice heavy fluid loading usually affects natural frequencies to a far greater extent than it affects mode shapes, which may usually be assumed to remain unchanged.) In general it may be assumed that fluid loading is light if the parameter $(\omega m/\rho_o c)$ is small compared with unity; here m is the mass per unit area of the structure which is assumed to be shell- or plate-like. Some general results of Smith's analysis follow.

A coupling factor Γ between the blocked pressure transverse displacement frequency ω and direction Ω, and an individual transverse displacement mode of shape $\psi_m(r)$ is defined in terms of the blocked generalized force on the mode

$$F_{m,bl} = - \int_{\text{surface}} p_{bl}(r)\, \psi_m(r)\, dr = P_o \Gamma_m(\omega, \Omega) . \qquad (11.12)$$

Reciprocity gives the following results:

$$|P_{rad}(\Omega)/V_m| = (\rho\omega/4\pi R_o) |\Gamma_m(\omega, \Omega)| \qquad (11.13)$$

where $P_{rad}(\Omega)$ is the amplitude of the pressure radiated to a point at a far distance R_o from the structure in a direction Ω by modal vibration of amplitude V_m. The intensity of the radiated sound wave is $|P_{rad}(\Omega)|^2/2\rho c$. The total power radiated is given by

$$\frac{1}{2\rho c} \int_{\text{all } \Omega} |P_{rad}(\Omega)|^2 R_o^2 d\Omega .$$

Writing this also as $R_{m,\ rad} V_m^2/2$, where R_{rad} is the real part of the radiation impedance, Z_{rad}, and substituting for $|P_{rad}|$ from equation (11.13), we obtain a relationship between the radiation resistance of a mode and its blocked force coupling factor

$$R_{m, rad} = \frac{\rho c k^2}{4\pi} <|\Gamma_m(\ ,\Omega)|^2>_{\text{all } \Omega} \tag{11.14}$$

where $<>$ means 'averaged over'. R_{rad} is related to the radiation efficiency σ of a surface of area A, vibrating with a space-time averaged mean square velocity $<\overline{v^2}>$, by $R_{rad} = \rho c A \sigma$. The radiation loss factor $\eta_{rad} = R_{rad}/\omega M$, where M is the total mass of the vibrating structure.

Now, just as pressure on the elementary piston could be decomposed into a blocked and radiated component, the pressure field on the surface of any vibrating body can be decomposed into a blocked *modal* pressure distribution (equation 11.12)) and a radiated *modal* pressure distribution. The latter corresponds to the pressure field generated by isolated motion in the mode considered which can be accounted for by a complex modal radiation impedance Z_{rad}, as for the piston. This impedance has two components: an imaginary component which normally corresponds to fluid inertial loading on the mode, and a real part corresponding to modal acoustic damping, or energy radiation. With heavy fluid loading the *in vacuo* modes become coupled by mutual impedance mechanisms which, except in the case of a totally enclosed fluid space, render the fluid loaded modes non-orthogonal [11.10].

By adding the mechanical modal impedance and the fluid-loading impedance, equation (11.15) is obtained, which is equivalent to equation (11.9) for the piston.

$$V_m(\omega, \Omega)(Z_{int} + Z_{rad}) = \text{blocked modal force} = P_o \Gamma(\omega, \Omega) \tag{11.15}$$

where Z_{int} is the modal impedance of the *in vacuo* structure. Equations (11.14) and (11.15) together provide a modal response equation in terms of the radiative properties of a mode. If we ignore the imaginary (reactive) part of Z_{rad}, the *pure tone* response to *diffuse field* acoustic excitation, i.e. waves incident from all directions with equal amplitude and random phase, becomes

$$|V_m(\omega)|^2[R_{m, int} + R_{m, rad}]^2 = |P_o|^2 <|\Gamma m(\omega, \Omega)|^2>_{\text{all } \Omega}$$

$$= \frac{4\pi |P_o|^2 R_{m, rad}}{\rho c k^2}$$

Hence　　　$$|V_m(\omega)|^2 = \frac{4\pi c |P_o|^2 R_{m, rad}}{\rho \omega^2 (R_{m, tot})^2} \tag{11.16}$$

where $R_{m, tot} = R_{m, int} + R_{m, rad}$: $R_{m, int} = \omega M_m \eta_{m, int}$. It should be noted that pure tones rarely produce 'diffuse' fields except in large enclosures and at high frequencies. The corresponding result for response to a *diffuse field* of *uniform pressure spectral density* $S_{p_o}(\omega_m)$ is

$$|V_m|^2 = \frac{S_{p_o}(\omega_m) 2\pi^2 c}{M \rho \omega^2} \mu_m \tag{11.17}$$

where $\mu_m = (R_{m, \text{rad}})/(R_{m, \text{rad}} + R_{m, \text{int}})$. μ is an important parameter because it has an upper limit of unity when $R_{m, \text{rad}} \gg R_{m, \text{int}}$, i.e. the radiation damping is far greater than the internal damping. Then $\mu = 1$ gives an upper limit on acoustically induced response, and added mechanical damping will produce very little reduction in amplitude unless $R_{m, \text{int}}$ can be made comparable with $R_{m, \text{rad}}$. For metal structures in water, and even in pressurized gases, μ frequently approaches unity.

An approximate expression for the response energy of a *multi-modal* structure in a particular frequency band to a *diffuse, broad band noise field* is obtained by simply adding the energies in each mode:

$$\frac{<\overline{v^2}>}{\overline{P_0^2}} = 2\pi^2 c \, \frac{n(\omega)}{M\rho\omega^2} \, \frac{<R_{\text{rad}}>_m}{<R_{\text{rad}} + R_{\text{mech}}>_m} \, . \tag{11.18}$$

$\overline{P_0^2}$ is the mean square acoustic pressure in the frequency band of interest, $<\overline{v^2}>$ is the spaced averaged mean square velocity, and $n(\omega)$ is the average modal density of the structure in the band of interest: $<>_m$ means averaged over all modes.

The practical implications of these results is that *good radiators are good receivers,* and that measurements and calculations of modal radiation can be used directly in acoustically induced response analyses. Great care would have to be used in attempts to mechanically induce 'pure modal' vibration in heavily damped, or fluid-loaded, plates by the application of localized forces, because the near field of the forces can dominate the radiation. However, Lyamshev's reciprocity result for forces could be validly applied.

11.4 BENDING WAVE PROPAGATION IN FLUID-LOADED PLATES

11.4.1 Plates in vacuo

Free bending, or flexural, waves in homogeneous thin *flat* plates *in vacuo* travel at a speed given by

$$c_B = (\omega^2 D/M)^{1/4} \tag{11.19}$$

where m is the mass per unit area and D is the bending stiffness. The corresponding wavenumber is given by

$$k_B = (\omega^2 m/D)^{1/4} \, . \tag{11.20}$$

Such waves are dispersive, and therefore a wave pulse will change shape as it travels. The frequency at which the *in vacuo* bending wave speed equals the speed of sound in a fluid with which a plate is in contact is termed the CRITICAL FREQUENCY, which is indicated in Fig. 11.2: $\omega_c = c^2(m/D)^{\frac{1}{2}}$.

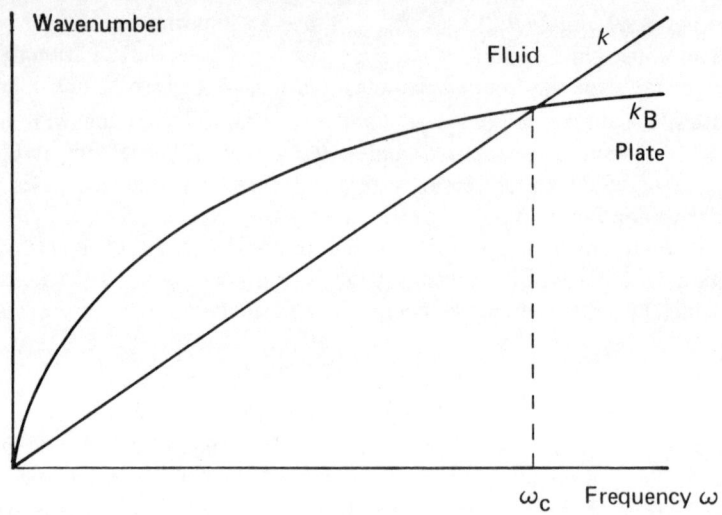

Fig. 11.2 – Dispersion curves for acoustic fluid and plate bending waves.

11.4.2 Plates in a fluid

The bending wave equation for a plate in contact with a fluid is

$$D\nabla^4 w + m\ddot{w} = -p\,|_{z=0} \tag{11.21}$$

subject to the boundary condition

$$\rho\ddot{w} = -\partial p/\partial z\,|_{z=0} \ . \tag{11.22}$$

A simple harmonic plane wave takes the form $w = W\exp(i(\omega t - k_f x)$. Using equations (11.21) and (11.22), together with the fluid wave equation, yields the dispersion equation

$$\frac{\rho\omega}{\sqrt{(k_f^2 - k^2)}} + \omega m\left(1 - \frac{Dk_f^4}{m\omega^2}\right) = 0 \ . \tag{11.23}$$

The phase velocity $c_f = \omega/k_f$ therefore satisfies the equation

$$\frac{\rho c}{\omega m} + \sqrt{\left(\left(\frac{c}{c_f}\right)^2 - 1\right)\left(1 - \left(\frac{\omega_c^2}{\omega_c c_f^2}\right)^2\right)} = 0 \ . \tag{11.24}$$

There is *only one purely real root* of this equation which represents a wave which has a phase velocity always *less* than the speed of sound, c, in the fluid [11.11, 11.12]. At frequencies well below ω_c, ($\omega/\omega_c \ll 1$), the wavenumber k_f is given approximately by

$$k_f \cong (\rho\omega^2/D)^{1/5}$$

which, by comparison with equation (11.20), shows that the fluid density, rather than the plate mass, controls the wave speed. Under these circumstances the fluid loading can be said to be heavy. Strictly, the condition for an assumption of heavy fluid loading is $\tilde{\omega} = \omega(\omega_c m^2/\rho^2 c^2) \ll 1$, but the less restrictive condition $M = k_0/k_B = c_B/c_0 \ll 1$ appears to be adequate [11.13]. These subsonic, free bending waves are purely surface waves, involving motion of the plate and of the fluid only in the near vicinity of the plare.

About the other physically significant roots there is considerable controversy and a number of different opinions [11.12, 11.14, 11.15, 11.16]. However, there appears to be general agreement that fluid-loaded, free plate waves cannot travel at exactly the same speed as sound in the fluid (i.e. coincidence cannot occur at the critical frequency), it also appears that a supersonic branch of the dispersion curve can exist above the critical frequency, but it is highly damped by radiation and ultimately becomes a shear wave at frequencies well above the critical. Since, in general, critical frequencies in water are so high (frequency $\times$ plate thickness $= 250$ Hz–m for steel or aluminium), behaviour above f_c is not generally of great importance in audio-frequency problems.

It should be noted that homogeneous thin-plate theory is adequate only for frequencies below $\omega_l = 0.3\, c_s/h$ where c_s is the material shear wave speed $= (E/2\rho_s(1+\nu))^{\frac{1}{2}}$; ρ_s is the material density, and h is the plate thickness. For metal plates in water the critical frequency ω_c given by thin-plate theory is about 40% greater than ω_l. Hence, in order to analyse fluid-loaded plate behaviour near coincidence, Timoshenko–Mindlin thick-plate theory, which accounts for transverse shear and rotary inertia, must be used [11.1]. It is concluded in ref. [11.15] that the combined effect of water loading and finite thickness is to increase the actual critical frequencies of aluminium and steel plates by 52% and 34% respectively.

11.5 SOUND RADIATION FROM FLUID-LOADED PLATES

Consider an infinite plate carrying a siusoidal transverse displacement wave of frequency ω and phase velocity v; the wavenumber $\kappa = \omega/v$. The displacement normal to the plane of the plate may be written as $w(x, t) = W \exp i(\omega t - \kappa x)$. Solution of the fluid wave equation, subject to this boundary condition, yields the following expression for acoustic pressure in a fluid in contact with the plate.

$$p(x, z, t) = \frac{i\omega^2 \rho W}{(k^2 - \kappa^2)^{\frac{1}{2}}} \exp(i(\omega t - \kappa x)) \exp(-i(k^2 - \kappa^2)^{\frac{1}{2}}z) , \quad (11.25)$$

where z is directed from the boundary into the fluid. If the plate wave speed v exceeds the speed of sound c in the fluid, $\kappa < k$ and plane acoustic waves are radiated away from the plate at an angle $\theta = \cos^{-1}(\kappa/k)$, thereby carrying energy away from the plate. Bending waves having subsonic phase velocities, which can propagate both below and above the critical frequency, generate only near fields in the fluid, which impose a mass-like loading on the plate. Supersonic waves, which can propagate only just below, and at any frequency above, the critical frequency [11.15] are damped by radiation and hence have complex wavenumbers. Radiation damping is proportional to ρc and can be very high in liquids.

11.5.1 Infinite-plate radiation

How then could a uniform infinite plate radiate sound energy at frequencies well below critical if not driven by a distributed field which has a supersonic wavenumber component parallel to the plate, such as an incident acoustic field? The answer lies in the nature of wave fields induced in plates by *locally* acting forces and moments. In addition to the freely propagating waves already discussed, such locally acting forces generate inhomogeneous *near* fields in the plate which decay exponentially with distance from the point of action. Just as a signal which decays exponentially with time contains components of all frequencies, so a displacement which decays exponentially with distance contains components of all wavenumbers. Those components which are of wavenumber less than $k = \omega/c$ can radiate sound energy.

The presence of local discontinuities or constraints, such as masses, holes, stiffness, and supports can cause otherwise uniform plates to radiate, owing to the near fields created by their presence. One approach to estimating the radiating effect of such features is to analyse the sound radiation from plates subjected to point and line distributions of forces and moments, since reaction forces are generated by the scattering of incident waves. Analysis of the scattering process can be formulated in terms of the impedance of the plate and of the scattering element. A brief example is presented to illustrate the general procedure.

Consider a subsonic free wave in a plate to be incident upon a line constraint as shown in Fig. 11.3. A wave of transverse velocity amplitude v_i is normally incident upon a line constraint which is assumed to provide only a transverse reaction force given by $F = -z_c v_0$. The response of the plate to the force is given by $v = F/Z_p$, where Z_p is the fluid-loaded line impedance. Scattered v_r and v_t are reflected and transmitted according to this relationship. The total velocity of the plate at the constraint is given by $v_0 = v_i + v_r = v_t$. Combining these equations yields and expression for the reaction force F.

$$F = -v_i[Z_p Z_c/(Z_p + Z_c)] . \tag{11.26}$$

In terms of mobilities (or admittances), $Y = 1/Z$,

$$F = -v_i(Y_p + Y_c)^{-1} . \tag{11.27}$$

The action of fluid loading affects the resulting radiation in two ways: (i) it influences the plate impedance Z_p; (ii) it controls the strength and directivity of the radiated and transmitted fields. Expressions for fluid-loaded plate admittances are given in references [11.17], [11.13], and [11.18] and summarized in Table 11.1. Fluid loading effects are greatest at low frequencies, but they generally only make a significant change to the imaginary part of the line force admittance which is substantially reduced below its *in vacuo* value. A general analysis of sound radiation from periodically stiffened, fluid-loaded plates is presented in reference [11.19].

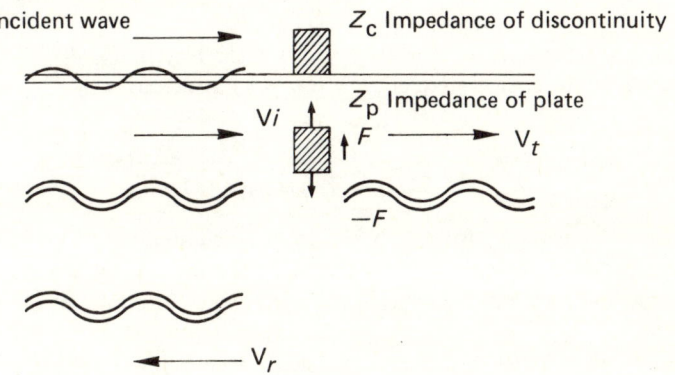

Fig. 11.3 – Reflection and transmission of bending waves at a discontinuity.

Table 11.1
Low-frequency mobility admittance formulae (Refs. [11.13] and [11.18]).

Line force:
$$A_F = \frac{\omega}{5Dk_B^3\sigma^3}\left(1 + i\tan\frac{\pi}{10}\right)$$

$$A_{F_0} = \frac{\omega}{4Dk_B^3}(1 + i)$$

$$: \sigma = (\rho/mk_B)^{1/5}$$

Line moment:
$$A_M = \frac{k_B^3}{5m\omega\sigma}\left(1 - i\cot\frac{\pi}{5}\right)$$

$$A_{M_0} = \frac{k_B^3}{4m\omega}(1 - i) \ .$$

Point force:
$$A_F = \frac{1}{8(Dm)^{\frac{1}{2}}}\frac{4}{5}\left(\frac{k}{\nu k_B}\right)^{2/5}\left(1 - i\tan\frac{\pi}{10}\right)$$

$$: \nu = \rho\, k/mk_B^2$$

$$A_{F_0} = \frac{1}{8(Dm)^{\frac{1}{2}}}$$

The subscript o indicates *in vacuo* values. Frequency condition: $\sigma \gg 1$. *In vacuo* bending wavenumber $k_B = (\omega^2 m/D)^{1/4}$

11.5.2 Point-force excitation

Reference [11.1] gives the following expression for the acoustic pressure at a radius R from the point of application of the force and at an angle θ to the line of action:

$$p(R, \theta, t) = -\frac{ikFe^{ikR}}{2\pi R} \frac{\cos\theta}{1 - ikh(\rho_s/\rho)\cos(1 - (\omega/\omega_c)^2 \sin^4\theta)} e^{i\omega t} \tag{11.28}$$

where ρ_s is the plate material density and h is the plate thickness. For frequencies well below the critical frequency ($\omega/\omega_c \ll 1$) this reduces to

$$p(R, \theta, t) \cong -\frac{ikFe^{ikR}}{2\pi R} \left[\frac{\cos\theta}{1 - ikh(\rho_s/\rho)\cos\theta} \right] e^{i\omega t} . \tag{11.29}$$

At very low frequencies, for which $kh\rho_s/\rho \ll 1$ equation (11.29) reduces to the expression for the acoustic field generated by the direct application of a point force to a fluid, i.e. *the presence of the plate does not influence the radiation behaviour*.

The total radiated power at frequencies well below $\omega_c((\omega/\omega_c)^2 \ll 1)$ is given by

$$W = \frac{\rho F^2}{4\pi c\rho_s^2 h^2} \left(1 - \frac{\rho}{kh\rho_s} \tan^{-1} \frac{kh\rho_s}{\rho} \right) . \tag{11.30}$$

In air, the bracketed term is normally very close to unity. At low frequencies, for typical marine plate structures, the fluid loading is so high as to reduce the radiated power almost to zero. For instance, at 500 Hz, with a 12.5 mm thick steel plate in water, the bracketed term equals 0.015: at 1 kHz it is 0.04, and at 5 kHz it is 0.44. The radiation from point-exited plates under conditions of light fluid loading ($kh\rho_s/\rho \gg 1$) is almost omni-directional, except for $\theta \to \pi/2$; with heavy fluid loading when $kh\rho_s/\rho \ll 1$ the directionality approaches that of a dipole source in a fluid [11.21].

According to Lyamshev's reciprocity relationship [11.2], referred to in section 11.2, equations (11.28) and (11.29) should also indicate the characteristics of plate response to an acoustic point source of volume velocity source strength Q located at a distance R and angle θ to the normal. Application of the relevant equation gives the following expression for the velocity response of a plate which separates a dense fluid from vacuum (or a lightweight fluid):

$$V = -\frac{ikQ\,e^{ikR}}{2\pi R} \left[\frac{\cos\theta}{1 - ikh(\rho_s/\rho)\cos\theta} \right] . \tag{11.31}$$

When specialized to response to an incident plant acoustic wave of pressure amplitude P, reciprocity gives

$$V = \frac{-(2/\rho c)\, P \cos\theta}{[1 - ikh(\rho_s/\rho)\cos\theta]} \qquad (11.32)$$

which, in the limit of heavy fluid loading, or large angle of incidence, tends to the free surface particle velocity $-2P\cos\theta/\rho c$ in the absence of the plate. These expressions can be used in conjunction with equations (11.26) or (11.27) to estimate the scattering of incident sound energy by discontinuities in an otherwise uniform plate [11.20].

11.5.3 Line-force excitation

Reference [11.21] gives the expression for the pressure field as

$$p(R, \theta) = \frac{1}{2} k^{\frac{1}{2}} F'\beta [R^{-\frac{1}{2}}\exp(-ikR + i\pi/4)]\, f(\theta) \qquad (11.33)$$

where $f(\theta) = \cos\theta/(\cos\theta + \beta)$ and $\beta = -i\rho/kh\rho_s$. F' is the force/unit length. In the case of light fluid loading, $|\beta| \ll 1$, $f(\theta) \cong 1$ and the field is almost independent of θ, except as $\theta \to \pi/2$. The total power radiated per unit length is given by

$$W' = \pi k (F')^2 |\beta|^2/4\rho c = \pi(F')^2 \rho/4\omega h^2 \rho_s^2 . \qquad (11.34)$$

In the case of heavy fluid loading $f(\theta) \to \cos\theta/\beta$, and the radiation field becomes like that of a line dipole. The radiated power is given by

$$W' = \pi k (F')^2/8\rho c$$

which is independent of the plate parameters, as with the point force.

11.5.4 Point-moment excitation

Reference [11.22] presents expressions for the radiated power for frequencies well below the critical frequency,

$$W = \frac{\rho k^2 M^2}{12\pi c(\rho_s h)^2} \left\{ 1 - \frac{3}{2}\beta\tan^{-1}(1/\beta) + \frac{3}{2}\beta^2[1 - \beta\tan^{-1}(1/\beta)] \right\} \qquad (11.35)$$

where M is the applied moment. The power varies as the square of the frequency, which suggests that moment-excited radiation should be considered at high frequencies. Heavy fluid loading ($|\beta| > 1$), greatly reduces the radiated power, as it does with a point force. For $|\beta| \gg 1$, $W = \frac{1}{60\pi}(k^4 M^2)/\rho_c$, which is independent of the plate parameters; in this case the radiation is as from a quadrupole, which renders it very inefficient.

11.5.5 Line-moment excitation

Reference [11.21] gives the following expressions:

$$W = (\pi/2)k^2\beta^2(M')^2/4\rho c : \text{light fluid loading } (|\beta| \ll 1) \qquad (11.36a)$$

$$W = \pi k^2(M')^2/32\rho c : \text{heavy fluid loading } (|\beta| \gg 1) . \qquad (11.36b)$$

11.5.6 Bounded-plate radiation

The mechanisms of radiation by infinite plates discussed above also apply to the case of vibration of a bounded plate. The latter exhibits the phenomenon of characteristic modes and frequencies because of interference between free bending waves reflected from the boundaries. The characteristic modes take the form of standing waves which dominate the vibration pattern when the plate is excited into resonance at one of its characteristic, or natural, frequencies. The modes of an isolated bounded plate form an orthonormal set *in vacuo*, but fluid loading effects couple these modes so that no such set exists (except when the fluid itself is bounded). Reference [11.23] discusses the conditions under which modal interaction is of practical significance.

The fact that the standing waves exist only within the boundaries of a plate produces another mechanism of sound radiation by subsonic-flexural waves, in addition to those described above. Just as a 'sinusoidal' signal which has a finite duration can be synthesized from, or decomposed into, an infinity of infinite duration sinusoids by Fourier Integral transformation, so a 'sinusoidal' displacement pattern in space can be analysed into an infinity of infinitely extended sinusoidal waves, by wavenumber transformation (see Fig. 11.4).

Hence there are wavenumber components of a standing wave formed from subsonic components which satisfy $k_p < k_n$, where $k_n = \omega_n/c$, ω_n being the natural frequency of the mode concerned. These components lead to sound radiation (equation (11.25)), even though the primary component $k_o > k_n$. The components for which $k_p > k_n$ give rise to mass loading by a near acoustic field. A single exception to the general picture illustrated by Fig. 11.5 is the fundamental mode of a baffled panel of which the wavenumber spectrum peaks at $k_p = 0$. This mode is relatively efficient at radiating compared with the high-order modes immediately above it in natural frequency, and it can be heavily damped by radiation.

An alternative approach to the analysis of radiation by bounded-plate modes represents the modal displacement by a distribution of adjacent, inter-nodal regions of positive and negative volume velocity (e.g. [11.24], pp. 489–492). It is evident that, at sub-critical frequencies, radiation originates primarily from regions adjacent to the boundaries of a baffled plate, because the inner regions suffer mutual cancellation. For plates of a given material and thickness, the radiation efficiency increases as the plate area decreases. The effect of the addition of dividing elements, such as stiffeners, to a plate is to increase its radiation efficiency. Such addition may also decrease its transmission loss at frequencies in the range one or two octaves below the plate critical frequency. The edges of unbaffled plates are not good radiators, and hence such plates exhibit very low sub-critical radiation efficiencies. As shown in [11.25], there is a limit to the amount by which the addition of damping treatment to a plate can reduce sub-critical radiation from resonant plates. If the plates are excited by localized inputs, such as vibrating edge-mounting structures, damping will suppress the vibration of non-radiating interior regions, but may not significantly reduce the strength of the boundary source regions.

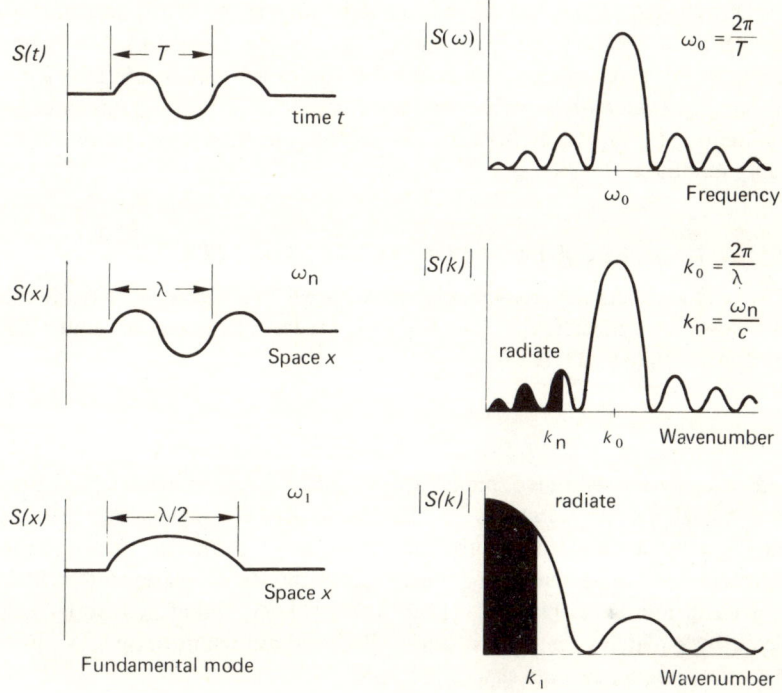

Fig. 11.4 – Equivalence of frequency and wavenumber decomposition: identification of radiating wave-number components.

The radiation damping of a bounded plate can be expressed in terms of its radiation efficiency σ as $\eta_{rad} = (\rho c/\omega m)\sigma$, where m is the mass per unit area and s is the plate area. Reference [11.24] shows the sub-critical radiation efficiency of a baffled plate to be approximately equal to $(U \lambda_c/\pi^2 s) (f/f_c)^{\frac{1}{2}}$, where U is the length of the plate perimeter and λ_c is the wavelength at the critical frequency f_c. For a plate of given material, dimensions, and mass it can be shown that $\eta_{rad}(f) \propto (\rho/c)$. Hence η_{rad} for a plate in water at a given frequency would be about 190 times η_{rad} in air at the same frequency. Fig. VI/2 of reference [11.24] corresponds to a ratio of about 40. Data presented in another reference on plate radiation into water [11.35] do not show such a large ratio, but the plates tested were unbaffled. Indeed, the impedance of water is so great that the *rigid* baffle model may never be valid, and hence estimates of radiation efficiency based upon such a model may always yield excessively large values.

A complete presentation of sound radiation behaviour of plate modes is not possible within the confines of this chapter; the reader is directed to reference [11.25] for information on radiation from resonant vibration of plates in air. If modal coupling is neglected the method of this reference can be applied to heavily loaded plates, but the intermodal coupling terms can be significant, as shown in reference [11.10]. It must, however, be realized that the natural frequencies of plate modes, especially the low-order modes, are affected by fluid loading. Hence the *in vacuo* modal wavenumber lattice diagrams employed in reference [11.25] to evaluate modal average radiation efficiencies must be suitably modified.

11.6 NATURAL FREQUENCIES OF FLUID-LOADED PLATES

Fluid loading affects the free bending wave speeds most strongly at frequencies well below the critical frequency. The fluid-loaded plate mass per unit area is given approximately [11.10] by

$$m' = m(1 + \rho/mk_{mn})$$
(11.37)

where k_{mn} is the *in vacuo* modal wavenumber (ω_{mn}/c) and ω_{mn} is the *in vacuo* modal natural frequency. Fluid-loaded natural frequencies will be less than ω_{mn} by a factor of approximately $(m/m')^{\frac{1}{2}}$. Formulae for the natural frequencies of fundamental and higher-order modes of plates and cylinders are given in references [11.26], [11.27] and [11.36]. Fluid in a closed cavity covered by a plate can exert a stiffness effect and can significantly affect lower-order mode frequencies and shapes [11.33].

11.7 TRANSMISSION OF BENDING WAVES ACROSS STRUCTURAL DISCONTINUITIES

Many structures of engineering interest consist of plates or shells stiffened by beam or rib members, and therefore the scattering properties of such features are of interest. In Chapter 13 methods of analysing wave propagation in periodically constrained plates are presented. Heavy fluid loading can have a profound effect on the reflection, transmission, and sound radiation characteristics of localized constraints. In reference [11.28] the bending-wave energy transmission coefficient across a clamped line constraint on a steel plate in water is calculated: *in vacuo* this coefficient is, of course, zero. At normal incidence the coefficient is between −10 dB and −15 dB. A considerable proportion of the incident energy is radiated into the fluid by the scattering process. The influence of fluid loading on the energy flow and coupling loss factor between plates joined by a rib is analysed in reference [11.29]. This factor is important in the application of Statistical Energy Analysis to power flow in stiffened, fluid-loaded panel structures.

The practical implication of these analyses is that localized constraints are less effective in reflecting bending wave energy in fluid-loaded plates than the *in vacuo* analyses would suggest.

11.8 WAVE PROPAGATION IN FLUID-FILLED PIPES AND CAVITY-BACKED PLATES

A problem of considerable practical interest is that of wave propagation in thin-walled, circular shells which are filled with fluid. Structural wave motion in cylindrical shells *in vacuo* are considerably more complicated than those in plates; in particular, four dispersion curves exist for every value of the circumferential wavenumber (n/a): n is an integer, a is the pipe radius. The four types of waves differ in their relative proportions of radial, axial, and tangential displacements. Complex and imaginary axial wavenumbers exist in addition to real propagating wavenumbers.

The presence of fluid within the shell introduces extra branches of the dispersion curves which are associated with the acoustic modes of the fluid waveguide [11.31]. Fluid 'short-circuiting' of discontinuities, similar to that described in section 11.7, above, is expected to render the reflection effects of structural discontinuities and constraints in a pipe less effective than *in vacuo* analysis would suggest, especially at frequencies greater than the ring frequency. It has been established that measurements of radial acceleration levels on either side of structural discontinuities cannot be relied upon to indicate their energy reflection and transmission coefficients, because of mode conversion at the discontinuities [11.30]. Constraints which are not axisymmetric transform energy in a wave of given mode order n into waves of various other orders.

Similar coupled fluid waveguide and structural wave dispersion curves occur in the case of a flat plate coupled to a fluid layer [11.32]. One of the interesting features of the behaviour of this system is that of multiple coincidence frequencies for a sound wave incident at a given angle on the side of the plate not adjoining the layer, because of the waveguide characteristic of the dispersion curve. In particular a coincidence condition can occur well below the critical frequency. However, in general, the response of a cavity-backed panel is considerably less than that of a panel in space; hence, by reciprocity, far less sound is radiated by a cavity-backed panel excited by mechanical forces.

11.9 NUMERICAL ANALYSIS TECHNIQUES

Many practical fluid-structure coupling problems do not admit of even approximate analytical solution, and various numerical calculation procedures have been developed. Some of the more widely used and successful analyses are referred to in reference [11.34] which is a general review of the coupling problem. Reference [11.37] is of particular interest for Finite Element analyses.

REFERENCES

[11.1] Junger, M. C. & Feit, D. (1972). *Sound, Structures and Their Interaction.* Cambridge, Massachusetts: M.I.T. Press.

[11.2] Lyamshev, L. M. (1960). *Soviet Physics Acoustics,* **5**(4), 431. Theory of sound radiation by thin elastic plates and shells.

[11.3] Belousov, Yu. I. & Rimskii-Korsakov, A. V. *Soviet Physics Acoustics,* **21**(2), 103. The reciprocity principle in acoustics and its application to the calculation of sound fields of bodies.

[11.4] Steenhoek, H. F. & Ten Wolde, T. (1970). *Acustica,* **23,** 301. The reciprocal measurement of mechano-acoustical transfer functions.

[11.5] Ten Wolde, T., Verheij, J. W. & Steenhoek, H. F. (1975). *Journal of Sound and Vibration* **42**(1), 49-56. Reciprocity method for the measurement of mechano-acoustical transfer functions.

[11.6] Pierce, A. D. (1981). *Acoustics: an Introduction to its Physical Principles and Applications.* New York: McGraw-Hill Inc.

[11.7] Rayleigh, Lord (1894). *Theory of Sound.* New York: Dover Publications.

[11.8] Morse, P. M. (1948). *Vibration and Sound.* New York: McGraw-Hill Book Co. Inc., second edition.

[11.9] Smith, P. W. Jr. (1962). *J. Acous. Soc. Amer.,* **34,** 640. Response and radiation of structural modes excited by sound.

[11.10] Davies, H. G. (1971). *Journal of Sound and Vibration,* **15**(1), 107-120. Low frequency random excitation of water-loaded rectangular plates.

[11.11] Crighton, D. G. (1979). *Journal of Sound and Vibration,* **63**(2), 225-235. The free and forced waves in a fluid-loaded elastic plate.

[11.12] Strawderman, W. A., Ko, S.-H. & Nuttall, A. N. (1979). *J. Acous. Soc. Amer.,* **66**(2), 579-585. The real roots of the fluid-loaded plate.

[11.13] Crighton, D. G. (1972). *Journal of Sound and Vibration,* **20**(2), 209-218. Force and moment admittance of plates under arbitrary fluid loading.

[11.14] Pierucci, M. & Graham, T. S. (1979). *J. Acous. Soc. Amer.,* **65**(5), 1190-1197. A study of bending waves in thick fluid-loaded plates.

[11.15] Guicking, D. & Boisch, R. (1980). *Acustica,* **44**(1), 41-45. Zur Grenzfrequenz ebener Platten in dichten Medien.

[11.16] Schoch, A. & Feher, K. (1952). *Acustica* **2**(5), 189-204. The mechanism of sound transmission through single leaf partitions, investigated using small scale models.

[11.17] Crighton, D. G. (1980). *Journal of Sound and Vibration,* **68**(1), 15-33. Approximations to the admittances and free wavenumbers of fluid-loaded panels.

[11.18] Crighton, D. G. (1977). *Journal of Sound and Vibration,* **54**(3), 389-391. Point admittance of an infinite thin elastic plate under fluid loading.

[11.19] Mace, B. R. (1980). *Journal of Sound and Vibration,* **73**(4), 473-504. Periodically stiffened fluid-loaded plates.

[11.20] Fahy, F. J. (1973). *Admiralty Underwater Weapons Establishment Publication* 35333. Vibration and sound radiation of fluid-loaded marine structures. Part I. Fundamental concepts and review of the open literature. (Unclassified).

[11.21] Maidanik, G. & Kerwin, E. M. Jr. (1966). *J. Acous. Soc. Amer.,* **40**(5), 1034. Influence of fluid loading on radiation from infinite plates below the critical frequency.

[11.22] Thompson, W. Jr. & Rattaya, J. V. (1964). *J. Acous. Soc. Amer.,* **36**(8), 1488-1490. Acoustic power radiated by an infinite plate excited by a concentrated moment.

[11.23] Mkhitarov, R. A. (1972). *Soviet Physics Acoustics,* **18**(1), 123. Interaction of the vibrational modes of a thin bounded plate in a liquid.

[11.24] Cremer, L., Heckl, M. & Ungar, E. (1973). *Structure-borne Sound.* Berlin: Springer-Verlag.

[11.25] Beranek, L. L. (Ed.) (1971). *Noise and Vibration Control.* New York: McGraw-Hill Book Inc. Co.

[11.26] Greenspon, J. E. (1961). *J. Acous. Soc. Amer.,* **33**(11), 1485. Vibrations of cross-stiffened and sandwich plates with application to underwater sound radiation.

[11.27] Au-Yang, M. K. (1978). *Journal of Sound Vibration,* **57**(3), 341-356. Natural frequencies of cylindrical shells and panels in vacuum and in a fluid.

[11.28] Lyapunov, V. T. (1969). *Soviet Physics Acoustics,* **14**(3), 352. Flexural wave propagation in a liquid loaded plate with an obstruction.

[11.29] Maidanik, G. (1978). *Journal of Sound and Vibration,* **60**(3), 313-318. Influence of fluid loading and compliant loading on the coupling loss factor across a rib.

[11.30] Fuller, C. R. (1981). *Journal of Sound and Vibration,* **75**(2), 207-228. The effects of wall discontinuities on the propagation of flexural waves in cylindrical shells.

[11.31] Fuller, C. R. & Fahy, F. J. (1981). *Journal of Sound and Vibration,* **81**(3). Characteristics of wave propagation and energy distributions in cylindrical elastic shells filled with fluid.

[11.32] Schroter, V. & Fahy, F. J. (1981). *Journal of Sound and Vibration,* **74**(4), 465-476, Point-force excited vibrations of a thin, infinite panel separating a fluid layer from a fluid half-space.

[11.33] Pretlove, A. J. (1965). *Journal of Sound and Vibration*, **2**(3), 197–209. Free vibrations of a rectangular panel backed by a closed rectangular cavity.

[11.34] Chen, L. H. & Peirrucci, M. (1977). *The Shock and Vibration Digest*, **9**(4), 23-24; **9**(5), 17–24; **9**(6), 13-18; **9**(7), 29-37. Underwater fluid-structure interaction. Parts I-IV.

[11.35] Teubner, V. (1974). *Acustica*, **31**(4), 203-214. Wasserschallabstrahlung von Platten mit und ohne Versteifungen (Plate radiation in water with and without stiffening).

[11.36] Guicking, D. & Boisch, R. (1979). *Acustica*, **42**(2), 89-96. Vereinfachte Berechnung der Eigenfrequenzen dickwandiger Zylinder in Luft und Wasser.

[11.37] Everstine, G. C. (1981). *Journal of Sound and Vibration*, **79**(1), 157-160. A symmetric potential formulation for fluid-structure interaction (Letter to the Editor).

CHAPTER 12

Fundamental duct acoustics

P. O. A. L. Davies

Institute of Sound and Vibration Research, University of Southampton

12.1 INTRODUCTION

Experimental and theoretical studies of flow duct acoustics are best approached on the basis of transmission line equations. In their simplest form, they describe the propagation of plane waves in a tube with acoustically hard walls, where the diameter is small compared with the wavelength. More complex relations are needed to describe more complex wave motions, involving higher-order modes of wave propagation. Any quantitative description of energy propagation also involves a careful specification of the boundary conditions. Nonlinear behaviour exists when flow is present, for certain classes of discontinuity or boundary conditions, or when the pressure amplitude is high (of order 0.1 bar or more).

Plane-wave analysis is appropriate for many practical examples of sound propagation in ducts where the wavelength is large compared with the transverse dimensions. This is true, for example, when the source arises from the inlet and exhaust flow of internal combustion engines, reciprocating or rotating compressors, and pumps, as well as some classes of furnaces, boilers, and ventilating systems. Examples where the transverse dimensions may be large compared with the wavelength included ducted fans, compressors, and turbines, and in such cases sound propagation takes place in one or more higher-order modes.

Prediction of sound propagation in ducts requires a basic understanding of both the generating mechanism of the noise sources and of the way that sound propagates along a given duct. In many cases the analysis must include the convective and scattering effects of the mean flow, with refraction by velocity gradients and temperature gradients.

Having set out the general relationships for sound propagation in ducts we shall consider first plane mode propagation, to gain a general physical understanding, before considering the more general case of energy propagation in the higher-order modes.

12.2 THE WAVE EQUATION FOR FLOW DUCTS WITH FLOW AND THERMAL BOUNDARY LAYERS

In a number of recent papers [12.1-12.9], wave equations governing the sound propagation in ducts of various cross-sectional shape have been derived with regard to various conditions of mean flow and transverse gradients of mean flow and temperature. The basic equations underlying all these derivations are the equation for (a) conservation of momentum in inviscid media (Euler's equation), (b) the equation of continuity (conservation of mass), and (c) the energy equation (conservation of entropy). The acoustic fluctuations are treated as small perturbations of velocity, pressure, density, and temperature superimposed on the steady-state values. Combining the linearized expressions of the above equations leads to a wave equation describing propagation of acoustic pressure.

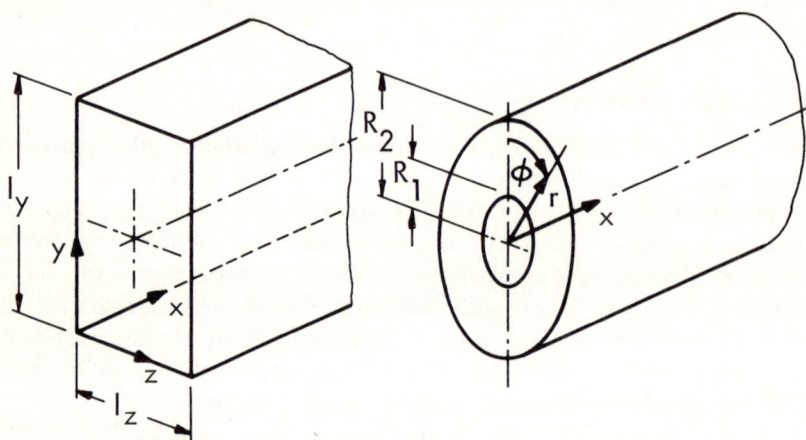

Fig. 12.1 – Coordinate systems.

The propagation of sound in a rectangular duct (see Fig. 12.1) with a steady flow in the X direction $U(y, z)$ is described by

$$\left(\frac{\partial^2 p}{\partial x^2} + \frac{\partial^2 p}{\partial y^2} + \frac{\partial^2 p}{\partial z^2}\right) - \frac{1}{c^2}\left(\frac{\partial^2 p}{\partial t^2} + 2U\frac{\partial^2 p}{\partial t \partial x} + U^2\frac{\partial^2 p}{\partial x^2}\right) +$$

$$2\bar{\rho}\left(\frac{\partial U}{\partial y}\frac{\partial v}{\partial x} + \frac{\partial U}{\partial z}\frac{\partial w}{\partial x}\right) - \frac{1}{\bar{\rho}}\left(\frac{\partial\bar{\rho}}{\partial y}\frac{\partial p}{\partial y} + \frac{\partial\bar{\rho}}{\partial z}\frac{\partial p}{\partial z}\right) = 0 , \qquad (12.1)$$

where U, c and $\bar{\rho}$ are the time.average values of flow velocity, sound speed and density which may vary with both transverse coordinates, but remain constant along the duct. The dependence of $\bar{\rho}$ and c on y and z is due to the assumed temperature profile, while u, v and w are the components of the acoustic particle velocity.

For a circular or annular duct, the wave equation, expressed in cylindrical coordinates (see Fig. 12.1), becomes

$$\left(\frac{\partial^2 p}{\partial x^2} + \frac{\partial^2 p}{\partial r^2} + \frac{1}{r}\frac{\partial p}{\partial r} + \frac{1}{r^2}\frac{\partial^2 p}{\partial \phi^2}\right) - \frac{1}{c^2}\left(\frac{\partial^2 p}{\partial t^2} + 2U\frac{\partial^2 p}{\partial t \partial x} + U^2\frac{\partial^2 p}{\partial x^2}\right) +$$

$$2\bar{\rho}\frac{dU}{dr}\frac{\partial v_r}{\partial x} - \frac{1}{\bar{\rho}}\frac{\partial\bar{\rho}}{\partial r}\frac{\partial p}{\partial r} = 0 \ . \tag{12.2}$$

In this equation the time average values U, c and $\bar{\rho}$, will vary only along the radial coordinates, while v_r is the acoustic particle velocity in the radial direction.

The first term on the left hand side (l.h.s.) of both equations (12.1) and (12.2) can be written using the Laplace operator: $\nabla^2 p$. The second term is the material derivative of the sound pressure which includes the convective effect of superimposed mean flow. The last two terms represent the refraction of sound owing to the presence of transverse gradients of mean flow and temperature.

The effective phase velocity, in terms of displacement along the duct, is the algebraic sum of the local acoustic and mean flow velocity. For downstream propagation, when flow is present, the wave travels faster on the duct axis than near the duct walls, so the sound wave is bent towards the wall. For an equivalent reason the wave is bent towards the centre of the duct when the sound propagates against the flow.

The refractive effect of transverse temperature gradients can be explained in a similar manner. In the case of a cooled flow the sound wave is bent towards the wall owing to the higher sound speed on the duct axis, and if the temperature of the wall is higher than that of the fluid, the energy of sound is concentrated in the centre of the duct.

For an isothermal medium at rest ($\bar{\rho} = $ const, $c = $ const, and $U = 0$) equations (12.1) and (12.2) reduce to the familiar form of the wave equation presented in previous chapters:

$$\nabla^2 p - \frac{1}{c^2}\frac{\partial^2 p}{\partial t^2} = 0 \ . \tag{12.3}$$

The sound field in the duct is not completely determined by the wave equation alone; the acoustic pressure must also satisfy the boundary conditions at the duct walls, at the duct termination, and at the sound source. To illustrate this, we shall consider first zero-order mode, or plane-wave energy propagation in ducts before considering the more general case of propagation in other modes.

12.3 PLANE-WAVE PROPAGATION IN FLOW DUCTS

With one-dimensional plane wave propagation the Euler equation becomes

$$\left(\frac{\partial}{\partial t} + U\frac{\partial}{\partial x}\right) u = \frac{1}{\bar{\rho}}\frac{\partial p}{\partial x} \ . \tag{12.4}$$

If p_0 and u_0 are the pressure and particle velocity at the origin for x, for a component of frequency ω, equation (12.4) is satisfied by the wave system defined by

$$p(x, t) = (p_x^+ + p_x^-)\, e^{i\omega t}\,,$$

$$= p_0^+ \exp(i(\omega t - \beta^+ x)) + p_0^- \exp(i(\omega t + \beta^- x)) \qquad (12.5)$$

and $\qquad u(x, t) = (u_0^+ + u_0^-)\, e^{i\omega t}\,,$

$$= u_0^+ \exp(i(\omega t - \beta^+ x)) + u_0^- \exp(i(\omega t + \beta^- x))\,. \qquad (12.6)$$

With hard walls, the complex wave-numbers are given by (see Rayleigh [12.10]),

$$\beta = \omega/c + \alpha(1-i); \quad \beta^+ = \beta/(1+M); \quad \beta^- = \beta/(1-M)\,, \qquad (12.7)$$

with M the Mach number U/c and α the viscothermal dissipation, which can be evaluated [12.10] from

$$\alpha = \frac{1}{ac}\sqrt{\frac{\overline{\nu\omega}}{2}}\left[1 + \left(\sqrt{\gamma} - \frac{1}{\sqrt{\gamma}}\right)\sqrt{\frac{1}{Pr}}\right]\,, \qquad (12.8)$$

where a is the pipe radius (or duct hydraulic mean radius), ν the kinematic viscosity, γ the ratio of the specific heats, and Pr the Prandtl number for the gas in the pipe.

Each positively or negatively travelling component wave must satisfy equation (12.4) independently. For the positive wave, substitution for u_x and p_x in equation (12.4) from (12.6) and (12.5) gives, in the most general case with hard walls and axial temperature gradients and hence mean axial velocity gradients etc.,

$$u_x^+ = \frac{1}{\overline{\rho_x c_x}}\; \frac{\beta_x^+}{k_x - \beta_x^+ M_x}\; p_x^+ = \frac{\delta_x^+ p_x^+}{\overline{\rho_x c_x}}\,, \qquad (12.9)$$

where $k_x = \omega/c_x$, $M_x = U_x/c_x$ and $\delta_x^+ = \beta_x/[k_x - M_x\,\alpha_x\,(1-i)]$. Similarly, for the negatively travelling wave we find

$$u_x^- = \frac{-\delta_x^- p_x^-}{\overline{\rho_x c_x}}\,, \qquad (12.10)$$

with $\delta_x^- = \beta_x/[k_x + M_x\alpha_x(1-i)]$.

12.3.1 Acoustic energy flux
The mean acoustic energy flux per unit area, in a hard-walled duct with flow, can be calculated from the general expression

$$I = (1 + M^2)\,\overline{p.u} + M\left[\frac{\overline{p^2}}{\overline{\rho c}} + \overline{\rho c\, \overline{u}^2}\right]\,, \qquad (12.11)$$

where the overbars represent time averages (see [12.7]). The first term represents the energy flux due to the convecting wave, and the second the convection of the mean acoustic energy density with the flow. In terms of the wave components p^+ and p^- the energy flux becomes

$$I = \frac{1}{\bar{\rho c}} \left[(1 + M)^2 \, |\overline{p^+}\,|^2 - (1 - M)^2 \, |\overline{p^-}\,|^2 \right] , \tag{12.12}$$

using the mean square values of the wave components. The first term in the square brackets represents energy flux with the flow (positive) and the second, flux against the flow carried by the reflected waves.

12.3.2 Boundary conditions at open duct terminations
As well as the equations governing plane-wave propagation in a uniform length of duct given in the last two sections, a quantitative description of acoustic conditions in the duct requires a specification of the boundary conditions at each end. At any discontinuity in duct geometry, or wall impedance, including the terminations, some of the incident wave energy will be reflected, some will be transmitted, while losses in acoustic energy will also occur. With plane waves the boundary condition at each discontinuity can be represented by the local pressure reflection coefficient $r_p = p^-/p^+$. Alternatively, the boundary condition can be specified as an impedance $Z_d = \bar{\rho c}(1 + r_p)/(1 - r_p)$. See for example references [12.8] and [12.9].

For the reflection coefficient one has

$$r_p = R \exp i \theta = - R \exp ik2l , \tag{12.13}$$

where R is the modulus, θ the phase, and l is an end correction. For circular ducts R and θ are both functions of the dimensionless wave number ka (Helmholtz number) and of the mean flow Mach number M.

With an open end for a circular duct, a theoretical determination of both R and l has been made by Levine & Schwinger [12.11] for $M = 0$. Their predictions are in good agreement with observation. Expressions which provide a close fit to their predictions can be found in reference [12.12]. More recently, Munt [12.13] has calculated the value of R with a cold subsonic outflow, where he showed that the effect of flow is to increase the relative magnitude of R as can be seen in Fig. 12.2, taken from reference [12.12]. These predictions are also in good agreement with observation. Corresponding results for a baffled opening [12.14] are given in Fig. 12.3. Relations for calculating the sound radiated from an open end can be found in references [12.15] and [12.16].

12.3.3 Conditions at area and other discontinuities
Boundary conditions for other discontinuities in duct geometry, for example an abrupt change in cross-section area from S_1 to S_2, can be specified once the corresponding local complex values of the wave components p^+ and p^- have

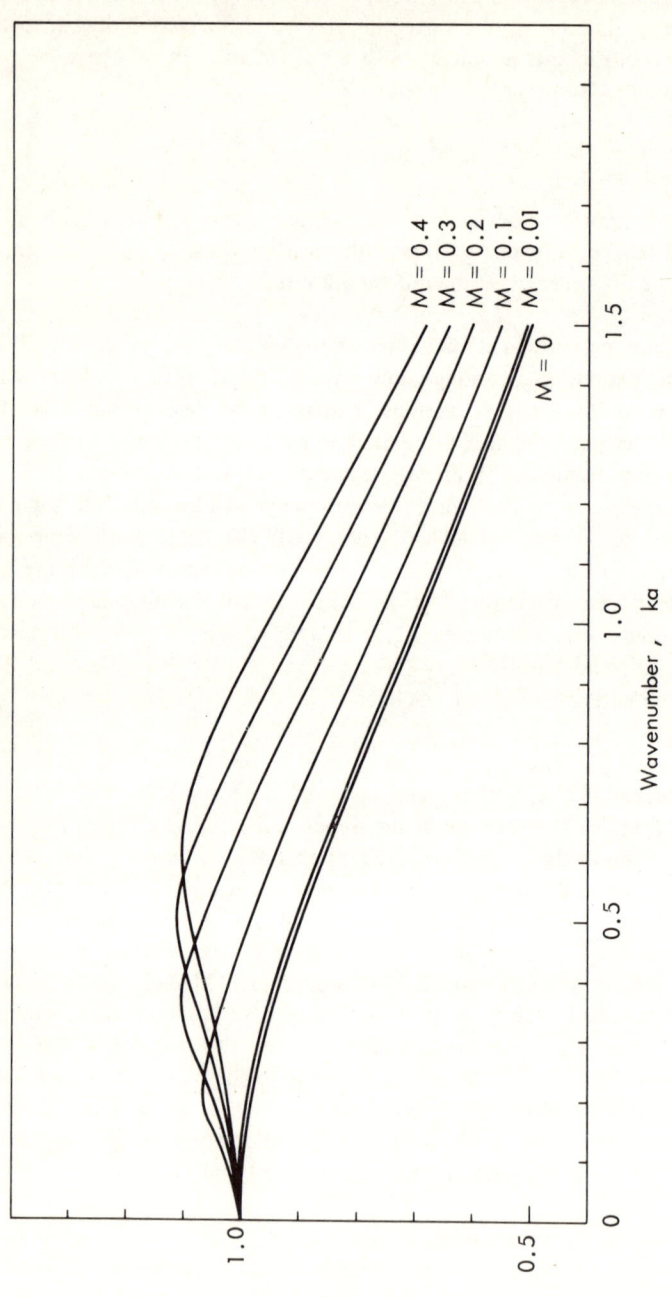

Fig. 12.2 — Reflection coefficient modulus R as predicted by Munt [12.13], and for $M = 0$ by Levine & Schwinger [12.11].

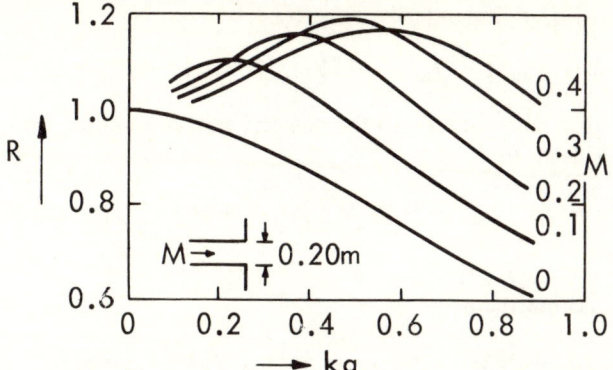

Fig. 12.3 — Pressure reflection coefficient of flanged circular pipe end for the plane-wave mode with mean flow (after Ref. [12.14].

been derived. A satisfactory practical method for doing so for plane waves which is summarized here can be found in reference [12.17]. Assuming that the values of p_1^+ and p_1^- at one side of the discontinuity are known, the related values p_2^+ and p_2^- can be found from the equations for conservation of energy, mass, and momentum. To complete the analysis it is necessary to estimate or predict the value of the end correction l for the particular geometry of each case.

Conservation of energy is satisfied if the stagnation enthalpy h_o is constant, or

$$0 = \mathrm{d}h_o = \mathrm{d}h + u\,\mathrm{d}u = \theta\,\mathrm{d}\sigma + \frac{\mathrm{d}(p)}{(\rho)} + u\,\mathrm{d}u \ , \qquad (12.14)$$

with θ the absolute temperature, h the enthalpy, $\mathrm{d}\sigma$ the entropy change, while u, (p), and (ρ) respectively represent the total instantaneous values of velocity, pressure, and density. These equations can be linearized by replacing the perturbation variables by the appropriate acoustic variables, for example, $\bar{\rho}c\,\mathrm{d}u = \bar{\rho}cu = p^+ - p^-$.

Abrupt changes in cross-section involve losses and flow separations which introduce both entropy and phase changes. Following reference [12.2] appropriate acoustic expressions for representing the entropy and density perturbations are

$$\mathrm{d}\sigma = \frac{1}{\bar{\rho}\theta}\phi \ \text{ and } \ \rho = \frac{p}{c^2} - \frac{\gamma-1}{c^2}\phi \ ,$$

where ϕ is a rather complicated function representing the irreversable exchanges. Defining the loss factor δ as $\delta = -(\gamma-1)\phi$, the perturbation variables in equation (12.14) can be replaced by their acoustic equivalents. Doing so, conservation of energy in an expansion or contraction from area S_1 to an area S_2 downstream is satisfied, to first order [12.17], by

$$p_2^+ (1 + M_2) + p_2^- (1 - M_2) = p_1^+ (1 + M_1) + p_1^- (1 - M_1) - \delta(\gamma - 1) \quad .(12.15)$$

Making a similar substitution into the one-dimensional equation for the conservation of mass leads to

$$S_2[p_2^+ (1 + M_2) - p_2^-(1 - M_2)] = S_1[p_1^+ (1 + M_1) - p_1^-(1 - M_1) + \delta M_1] \quad .$$
$$(12.16)$$

Momentum is conserved if

$$p_2^+[S_1 + S_2(M_2^2 + 2M_2)] + p_2^-[S_1 + S_2(M_2^2 - 2M_2)]$$
$$= p_1^+[S_1 (1 + M_1)^2] + p_1^-[S_1 (1 - M_1)^2] + \delta S_1 M_1^2 \quad . \tag{12.17}$$

The phase changes that occur across the discontinuity can be determined from a zero flow analysis which includes the non-propagating higher modes to satisfy the boundary conditions explicitly [12.18]. Of particular practical importance are special acoustical effects associated with separated flow discussed in [12.19].

The usefulness of this two-step approach can be illustrated by an example where predictions of the acoustic transfer characteristics of a contraction, which includes a sidebranch, are compared with measurements [12.17]. Figure 12.4a presents the measurements made at three mean flow Mach numbers, predictions using equations (12.15) to (12.17) and a higher-order mode analysis for zero flow. The exact analysis predicts the frequency correctly, while the three equations predict the amplitude characteristics quite well.

12.4 SOUND PROPAGATION IN RECTANGULAR DUCTS IN THE ABSENCE OF TEMPERATURE GRADIENTS AND MEAN FLOW

We turn now to the more general case when the transverse duct dimensions are no longer a fraction of the acoustic wavelengths so that the restriction to zero order mode energy propagation is now relaxed. Initially, we shall again consider the simplest situation of zero flow and uniform temperature. For this case, the wave equation becomes

$$\frac{\partial^2 p}{\partial x^2} + \frac{\partial^2 p}{\partial y^2} + \frac{\partial^2 p}{\partial z^2} - \frac{1}{c^2} \frac{\partial^2 p}{\partial t^2} = 0 \quad . \tag{12.18}$$

Equation (12.18) can be solved by the method of separation of variables which leads to the result

$$p(x, y, z, t) = [A_1 \exp(ik_x x) + B_1 \exp(-ik_x x)] \times$$
$$[A_2 \exp(ik_y y) + B_2 \exp(-ik_y y)] \times$$
$$[A_3 \exp(ik_z z) + B_3 \exp(-ik_z z)] \times$$
$$[A \exp(i\omega t) + B \exp(-i\omega t)] \ . \tag{12.19}$$

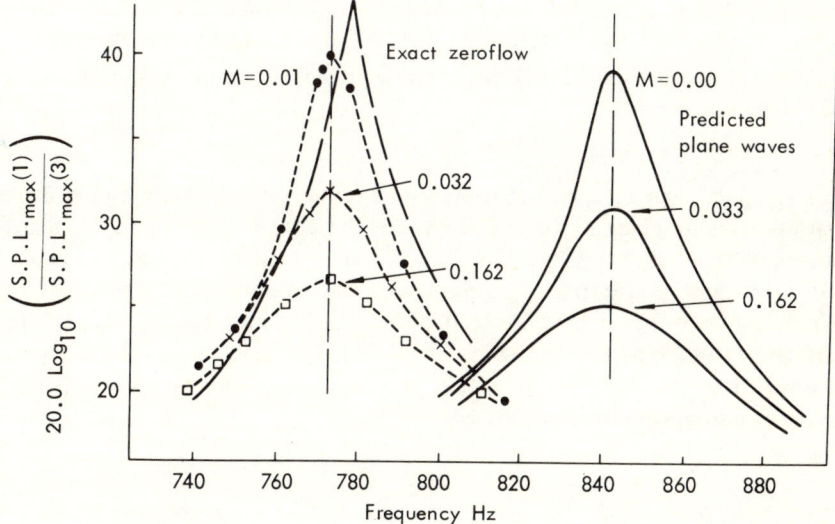

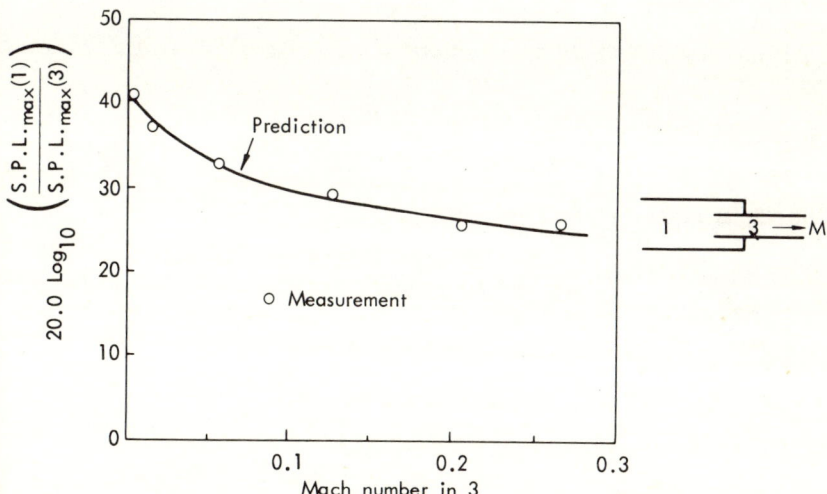

Fig. 12.4 − Acoustic energy transport at a contraction with sidebranch.

Each of the four terms in brackets [ ] on the r.h.s. of equation (12.19) represents the mathematical description of a wave motion, each depending on only one space or time variable. The variation of the acoustic pressure with time is determined by the last term where A and B are arbitrary constants and $\omega = kc$ is related to the angular frequency of the source fluctuation. The phase of this fluctuation can be chosen such that $B = 0$. The remaining three terms represent the pressure field in the duct as a triple product of waves going in the positive and negative $x-$, $y-$, and $z-$ directions, the corresponding wave numbers being k_x, k_y, and k_z, respectively. Inserting the solution given by equation (12.19) into equation (12.18) yields the so-called dispersion relation between the wave numbers:

$$k^2 = (\omega/c)^2 = k_x^2 + k_y^2 + k_z^2 \;, \tag{12.20}$$

where k_x, k_y, and k_z appear as components of the wave-number vector k. Each set of waves travelling in the x, y and z directions can combine to give resultant waves of wave-number k, propagating at an angle to the duct axis. The reflections of these waves from the duct walls generate an interference pattern ('standing waves') over the cross-section of the duct. These interference patterns are termed the 'modes of the duct'. The fluctuating velocities u, v and w can be derived from the linearized condition of momentum conservation in a homogeneous medium at rest expressed as

$$\frac{\partial u}{\partial t} = -\frac{1}{\rho}\frac{\partial p}{\partial x}, \quad \frac{\partial y}{\partial t} = -\frac{1}{\rho}\frac{\partial p}{\partial y}, \quad \frac{\partial w}{\partial t} = -\frac{1}{\rho}\frac{\partial p}{\partial z} \;.$$

Together with equation (12.20) then one obtains

$$u(x,y,z,t) = \frac{-k_x}{k\rho c}\,[A_1 \exp(ik_x x - B_1 \exp(-ik_x x] \times$$

$$[A_2 \exp(ik_y y) + B_2 \exp(-ik_y y)] \times$$

$$[A_3 \exp(ik_z z) + B_3 \exp(-ik_z z)]\; A\,\exp(i\omega t)\,, \tag{12.21}$$

with similar expressions for v and w.

12.4.1 Rigid duct walls and steady acoustic excitation
If the walls of the duct are rigid, the velocity fluctuations normal to the walls must vanish; the consequences of this boundary condition are

$v = 0$ at $y = 0$ and $y = l_y$ gives $A_2 = B_2$ and $\sin k_y l_y = 0$,

$w = 0$ at $z = 0$ and $z = l_z$ gives $A_3 = B_3$ and $\sin k_z l_z = 0$,

which in turn leads to

$$k_y = m\pi/l_y \quad \text{and} \quad k_z = n\pi/l_z \;, \tag{12.22}$$

where m and n are integers including zero. The expressions for the pressure and velocity field from equations (12.19) and (12.21) now become

$$p_{mn} = \cos(m\pi y/l_y)\cos(n\pi z/l_z) \times$$

$$[A_{mn}\exp(i(\omega t + k_{mn}x)) + B_{mn}\exp(i(\omega t - k_{mn}x))] \quad,$$

$$u_{mn} = \frac{-k_{mn}}{\bar{\rho}ck}\cos(m\pi y/l_y)\cos(n\pi z/l_z) \times$$

$$[A_{mn}\exp(i(\omega t + k_{mn}x)) - B_{mn}\exp(i(\omega t - k_{mn}x))] \quad,(12.23)$$

where $A_{mn} \doteq AA_1 B_2 B_3$ and $B_{mn} = AB_1 B_2 B_3$; the axial wave number k_x is now denoted by k_{mn}. From equations (12.20) and (12.22) one obtains

$$k_x^2 = k_{mn}^2 = (\omega/c)^2 - (m\pi/l_y)^2 - (n\pi/l_z)^2 \quad . \tag{12.24}$$

For each integer value of m and n, equation (12.23) represents one possible solution of the wave equation (12.18) together with the boundary conditions specified, or one possible mode. The complete solution is given by the super-position of all the linearly independent solutions. Thus

$$p = \sum_{m=0}^{\infty} \sum_{n=0}^{\infty} p_{mn} \quad, \tag{12.25}$$

with similar expressions, for u, v and w.

The terms on the r.h.s. of equation (12.23) represent waves travelling in the negative and positive x-directions respectively, the amplitudes of which vary over the duct cross-section owing to the two cosine terms. This front modulated wave is of frequency ω and travels at a phase velocity, $c_{mn} = \omega/k_{mn}$, which is greater than the speed of sound. As was explained in connection with equation (12.20) the front modulation is the result of waves travelling at an angle to the duct axis which are reflected from the duct walls. Hence, the waves propagating along the duct are a superposition of acoustic waves each of which travels at the speed of sound. When m and n are equal to zero, the axial phase velocity is equal to the speed of sound, and the mode is a plane acoustic wave (fundamental mode) as we have already seen.

Wave propagation in the axial direction is possible so long as the axial wavenumber $k_x = k_{mn}$ is real, see equation (12.23). According to equation (12.24) this is true for all frequencies given by

$$\omega > c\,[(m\pi/l_y)^2 + (n\pi/l_z)^2]^{1/2} \quad . \tag{12.26}$$

For frequencies below this limiting value, known as the cut-off frequency, the axial wave number k_{mn} becomes imaginary, and the propagation factors in equation (12.23) become $\exp(i\omega t - k_{mn}x)$. Thus the amplitudes of these modes decay exponentially with axial distance from the sound source or refer-ence discontinuity.

The previous discussion has shown that the effect of the boundary conditions at the duct walls is to permit only certain types of fluctuations to propagate along the duct. Each of these modes is characterized by its amplitude pattern over the duct cross-section, depending only on the duct dimensions. Energy in any given mode, described by fixed values for m and n, can only propagate along the duct if the excitation frequency is above the cut-off, or the mode is cut-on. Thus at a given frequency only a limited number of modes actually propagate; the others decay exponentially with distance from the source. The two arbitrary constants in equation (12.23), A_{mn} and B_{mn}, are determined by the boundary conditions at the sound source and at the duct termination.

12.4.2 Boundary conditions at the source and duct termination

We saw earlier that reflections will occur at the duct end. For a given duct termination impedance, $z_{mn} = p_{mn}/u_{mn}$, one may express A_{mn} in terms of B_{mn}. In the absence of reflections (i.e. $A_{mn} = 0$), the termination impedance is equal to $\bar{\rho}ck/k_{mn}$; this is usually called the 'modal impedance', which is not a function of x, y or z. For the plane wave modal, $k_{mn} = k$, the modal impedance becomes equal to the characteristic impedance $\bar{\rho}c$, as we saw earlier.

If a pressure wave in the form of a single acoustic duct mode is incident on the duct termination, the reflected and transmitted pressures and acoustic velocities are not confined to that mode alone. In general, other modes will be 'generated' at the termination. Hence, complete specification of the reflection and transmission process requires that the complete expressions for the sound fields inside and outside the duct are matched at the termination by applying usual boundary conditions of continuity of pressure and particle velocity (or displacement). For details see for example references [12.6] or [12.20] to [12.23]. An interesting result is that any well cut-on mode higher-order suffers very little reflection; it carries on past the termination as if it were continuing down an infinitely long duct. Another important feature is that modes that would be cut-off in an infinite duct can carry appreciable acoustic power down to frequencies an octave or so below their cut-off frequency, when the axial distance between source and duct end is only of the order of a representative duct width.

A sound source may be represented by a distribution of fluctuating velocity or force. Again the basic boundary condition is that, for example, the sum of all component velocities over all modes must equal the source velocity. Again further details may be found in reference [12.6] and [12.20] to [12.23].

12.5 SOUND PROPAGATION IN DUCTS WITH COMPLIANT WALLS

When the duct walls are not rigid or are lined with some absorbing material, the sound field in the duct must still satisfy the wave equation (12.18), and is described by the general solution given in equations (12.19)–(12.21). The boundary condition to be satisfied at the wall requires that the fluctuating pressure and the normal component of the velocity (or displacement) of the fluid near the sound absorbent wall must be equal to those on the wall. The absorptive properties of the wall are characterized by the normal acoustic impedance Z_w which is defined as the ratio of acoustic pressure to normal velocity [12.25].

If the duct walls are symmetrically lined with respect to the duct axis, it is more convenient to have the coordinate system on the duct axis. The boundary conditions then become

at $y = \pm \dfrac{l_y}{2}$;

$$Z_{wy} = \frac{p}{v} = - \bar{\rho}c \, \frac{k}{k_y} \, \frac{A_2\exp(ik_yl_y/2) + B_2\exp(-ik_yl_y/2)}{A_2\exp(ik_yl_y/2) - B_2\exp(-ik_yl_y/2)} \,,$$

at $z = \pm \dfrac{l_z}{2}$;

$$Z_{wz} = \frac{p}{w} - \bar{\rho}c \, \frac{k}{k_z} \, \frac{A_3\exp(ik_zl_z/2) + B_3\exp(-ik_zl_z/2)}{A_3\exp(ik_zl_z/2) - B_3\exp(-ik_zl_z/2)} \,. \tag{12.27}$$

Owing to the symmetry of the duct lining, the sound field will be either symmetric or antisymmetric with respect to the duct axis. Hence on the duct axis, $y = z = 0$, the following conditions apply:

$$\frac{\partial p}{\partial y} = 0 \text{ and } \frac{\partial p}{\partial z} = 0 \text{ for symmetric modes which gives } A_2 = B_2 \text{ and } A_3 = B_3$$

$$\tag{12.28}$$

$p = 0$ for antisymmetric modes which gives $A_2 = -B_2$ and $A_3 = -B_3$.

Inserting equation (12.28) into equation (12.27) leads to

$$\frac{Z_{wy}}{\bar{\rho}c} \, \frac{1}{kl_y/2} = i \frac{\cot(k_yl_y/2)}{k_yl_y/2} \text{ for symmetric modes and}$$

$$\frac{Z_{wy}}{\bar{\rho}c} \, \frac{1}{kl_y/2} = i \frac{\tan(k_yl_y/2)}{k_yl_y/2} \text{ for antisymmetric modes .} \tag{12.29}$$

These transcendental equations are known as 'eigen-equations'. The corresponding expression for the z-component are obtained by replacing the subscript y by z. As in the hard-walled case, an infinite number of eigen-solutions for the complex numbers k_{ym} and k_{zn} exists. Once these eigen-solutions are found. the axial wave number k_{xmn}, which in general will also be complex, can be derived using equation (12.20) as

$$k_{xmn}^2 = k^2 - k_{ym}^2 - k_{zn}^2 \,. \tag{12.30}$$

Denoting the axial wave number by

$$k_{xmn}^2 = \beta_{mn} - i\alpha_{mn} \; , \qquad (12.31)$$

one may express the acoustic pressure given by (12.19) for the conditions specified by (12.28) as

$$P_{mn} = {\textstyle\frac{\cos}{\mathrm{isin}}}(k_{ym}y) \, {\textstyle\frac{\cos}{\mathrm{isin}}}(k_{zn}z) \times$$

$$[A_{mn}\exp(\alpha_{mn}x + i(\omega t + \beta_{mn}x) + B_{mn}\exp(-\alpha_{mn}x + i(\omega t - \beta_{mn}x)] \; , \quad (12.32)$$

where the cosine and the sine terms correspond respectively to the symmetric and antisymmetric modes. The real part of the axial wave number, β_{mn}, represents the phase distribution of the mode, and the imaginary part, α_{mn}, describes the attenuation of the mode amplitude in the acoustically lined duct. This mode attenuation must not be confused with the decay of modes in rigid-walled ducts below their cut-off frequency. As in rigid-walled ducts, however, wave propagation in lined ducts is possible only so long as the axial wave number has a real component. Consequently, the cut-off frequency for ducts with sound-absorbent walls may be derived from the condition $\beta_{mn} = 0$.

12.6 SOUND PROPAGATION IN CYLINDRICAL OR ANNULAR DUCTS IN THE ABSENCE OF TEMPERATURE GRADIENT AND MEAN FLOW

The wave equation for an isothermal medium at rest, expressed in cylindrical coordinates, reads

$$\frac{\partial^2 p}{\partial x^2} + \frac{\partial^2 p}{\partial r^2} + \frac{1}{r}\frac{\partial p}{\partial r} + \frac{1}{r}\frac{\partial^2 p}{\partial \phi^2} - \frac{1}{c^2}\frac{\partial^2 p}{\partial t^2} = 0 \; . \qquad (12.33)$$

Again the solution is derived by the method of separation of variables which for a cylindrical or annular duct without radial splitters leads to

$$p(x, r, \phi, t) = (A_1\exp(ik_x x) + B_1\exp(-ik_x x)) (A_2\exp(im\phi) + B_2\exp(-im\phi))$$

$$\times [A_3 J_m(k_r r) + B_3 N_m(k_r r)] \, Ae^{i\omega t} \; , \qquad (12.34)$$

where J_m and N_m are the Bessel and Neumann functions of integer order m. (in the case of a duct with lined radial splitters, the azimuthal wave number $k_\phi = m$ can be fractional and complex). The wave-numbers are

$$k^2 = (\omega/c)^2 = k_x^2 + k_r^2 \; . \qquad (12.35)$$

The first term in the r.h.s. of equation (12.34) again represents waves travelling in the positive and negative x-directions. The second term describes waves going in both circumferential directions. The superposition of both wave types represents waves propagating along the duct and spinning at the same time. The third term on the r.h.s. expresses the wave amplitudes as functions of the radial coordinate.

Such spiralling modes can be exicited by the rotating blades of ducted fans or compressors; they can be related to the modes of a rectangular duct that, as was pointed out in reference [12.24], would be formed by cutting the annulus along a radius and 'unrolling' it until it becomes a rectangular cross-section.

As in case of the rectangular duct, the various constants A_i, B_i, and k_i are determined by the boundary conditions at the duct walls, the duct termination, and the sound source. For a circular duct, the condition of finite amplitude of the pressure on the axis requires that $B_3 = 0$.

A comprehensive study of the sound field in a rigid-walled annular duct in the absence of flow is given in reference [12.20] and [12.24]. Sound transmission in hollow and annular ducts with non-rigid walls is discussed in references [12.3] and [12.25] where sound fields in a more general environment of flow and shear are also considered.

The radial component of the fluctuating velocity can be derived from equation (12.34), with the momentum equation expressed as

$$\frac{\partial v_r}{\partial t} = i\omega v_r = -\frac{1}{\bar{\rho}} \frac{\partial p}{\partial r} . \tag{12.36}$$

The eigen-equation for the lined annular duct is then obtained from the condition that the ratio of pressure to normal velocity near the boundaries must be equal to the acoustic wall impedance, so that

$$\frac{ik \dfrac{\bar{\rho}c}{Z_{\mathrm{w}1}} J_m(k_r R_1) + k_r J_m'(k_r R_1)}{ik \dfrac{\bar{\rho}c}{Z_{\mathrm{w}1}} N_m(k_r R_1) + k_r N_m'(k_r R_1)}$$

$$-\frac{ik \dfrac{\bar{\rho}c}{Z_{\mathrm{w}2}} J_m(k_r R_2) + k_r J_m'(k_r R_2)}{ik \dfrac{\bar{\rho}c}{Z_{\mathrm{w}2}} N_m(k_r R_2) + k_r N_m'(k_r R_2)} = 0 . \tag{12.37}$$

$Z_{\mathrm{w}1}$ and $Z_{\mathrm{w}2}$ are the values of wall impedances at the inner radius R_1 and at the outer radius R_2 of the annulus. Unlike the rectangular duct problem where k_y and k_z could be evaluated independently, the radial wave number k_r has to be calculated separately for each value of the azimuthal wave number $k_\phi = m$; for each value of m, k_r is multi-valued and therefore is denoted by k_{rmn}. The axial wave-number follows from equation (12.35) or

$$k_{xmn}^2 = k^2 - k_{rmn}^2 . \tag{12.38}$$

The eigen-equation for the hard-walled annular duct is easily obtained from equation (12.37) by applying the condition $Z_{\mathrm{w}1} = Z_{\mathrm{w}2} = \infty$.

12.7 CONCLUSIONS

There are several other relevant topics, for example higher-order mode sound propagation with flow present, which are not pursued here. The relevant theory and, where available, experimental results, can be found among the references cited. Further information and references to material describing applications to practical noise control problems can be found elsewhere in Chapter 21.

The author gratefully acknowledges the material kindly provided by W. Neise and the helpful discussions with C. L. Morfey in the preparation of this chapter.

REFERENCES

[12.1] Tack, D. H. & Lambert, R. R. (1965). *Journal of the Acoustical Society of America*, **38**, 655. Influence of shear flow on sound attenuation in lined ducts.

[12.2] Mungur, P. & Gladwell, G. M. L. (1969). *Journal of Sound and Vibration*, **9**, 28–48. Acoustic wave propagation in a sheared fluid contained in a duct.

[12.3] Mungur, P. & Plumblee, H. E. Jr. (1969). *Basic Aerodynamic Noise Research NASA*, **SP-207**, 305–327. Propagation and attenuation of sound in a soft-walled annular duct containing a sheared flow.

[12.4] Ko, S. H. (1972). *Journal of Sound and Vibration*, **22**, 193–210. Sound attenuation in acoustically lined circular ducts in the presence of uniform flow and shear flow.

[12.5] Kapur, A. & Mungur, P. (1972). *Journal of Sound Vibration*, **23**, 401–404. On the propagation of sound in a rectangular duct with gradients. of mean flow and temperature in both transverse directions.

[12.6] Doak, P. E. (1973). *Journal of Sound Vibration*, **31**, 1–72 and 137–174. Excitation, transmission and radiation of sound from source distributions in hard-walled ducts of finite length. Part I: The effects of duct cross-section geometry and source distribution space-time pattern. Part II: The effects of duct length.

[12.7] Morfey, C. L. (1971). *Journal of Sound and Vibration*, **14**, 37–55. Sound generation and transmission in ducts with flow.

[12.8] Cummings, A. J. (1974). *Journal of Sound and Vibration*, **35**, 451–477. Sound transmission in curved duct bends.

[12.9] Cummings, A. J. (1975). *Journal of Sound and Vibration*, **41**, 375–379. Sound transmission in a folded annular duct.

[12.10] Rayleigh, Lord (1894). *The theory of sound*. London: MacMillan. Second edition. See Articles 346–350.

[12.11] Levine, H. & Schwinger, J. (1948). *Physical Review*, **73**, 383. On the radiation of sound from an unflanged circular pipe.

[12.12] Davies, P. O. A. L., Bento-Coelho, J. L. & Bhattochanga, M. (1980). *Journal of Sound and Vibration*, 72, 543-546. Reflection coefficients for an unflanged open pipe with flow.

[12.13] Munt, R. M. (1981). (Submitted to the *Journal of Sound and Vibration*) Acoustic transmission properties of a jet pipe with subsonic jet flow. 1: the cold jet reflection coefficient.

[12.14] Meyer, E. & Neumann, E. G. (1972). *Physical acoustics,* **Chapter 11.** Academic Press.

[12.15] Davies, P. O. A. L. & Halliday, R. F. (1981). *Journal of Sound and Vibration,* 76, 591-594. Radiation of sound by a hot exhaust.

[12.16] Cargill, A. M. (1979). *Mechanics of sound generation in flows.* **Ed.** Muller, E. A. See p. 19-25.

[12.17] Davies, P. O. A. L. (1978). United States Environmental Protection Agency 550/9, 78-206, 5-47. Bench test procedures and exhaust system performance prediction.

[12.18] Cummings, A. J. (1975). *Journal of Sound and Vibration,* 38, 149-155. Sound transmission at a sudden expansion in circular ducts with superimposed mean flow.

[12.19] Davies, P. O. A. L. (1981). *Journal of Sound and Vibration,* 77, 191-209. Flow-Acoustic Coupling in Ducts.

[12.20] Tyler, J. M. & Sofrin, T. G. (1962). *SAE Transactions,* 70, 309. Axial flow compressor noise studies.

[12.21] Morfey, C. L. (1969). *Journal of Sound and Vibration,* 9, 367-372. A note on the radiation efficiency of acoustic duct modes.

[12.22] Wright, S. E. (1972). *Journal of Sound and Vibration,* 25, 163-178. Waveguides and rotating sources.

[12.23] Mungur, P., Plumblee, H. E. & Doak, P. E. (1974). *Journal of Sound and Vibration,* 36, 21-52. Analysis of acoustic radiation in a jet flow environment.

[12.24] Morfey, C. L. (1964). *Journal of Sound and Vibration,* 1, 60-87. Rotating pressure patterns in ducts; their generations and transmission.

[12.25] Plumblee, H. E., Dean, P. D., Wynne, G. A. & Burrin, R. H. (1973). *NASA CR-2306.* Sound propagation in the radiation from acoustically lined flow ducts; a comparison of experiment and theory.

CHAPTER 13

Response of periodic structures to noise fields

D. J. Mead

Department of Aeronautics and Astronautics, University of Southampton

13.1 INTRODUCTION

A periodic structure is one which consists of a number of identical elements, connected in an identical manner to one another, end-to-end and/or side-by-side. It may be constructed either by actually assembling together the identical elements (as in a multi-storey building of modular construction) or by subdividing a uniform structure into identical elements, as in a stiffened plate with regular and identical stiffening members attached, or in a turbine disc with regularly-spaced blades around the periphery.

The vibrations of such structures are most conveniently analysed by taking advantage of the spatially periodic property. In the simpler cases, it is possible to predict vibration levels or natural frequencies by considering just one of the periodic elements together with the boundary conditions at the extreme ends of the periodic structure. This can effect an enormous reduction in the magnitude of the computational problems involved. For instance, if there are N elements in the whole structure, and each element has j degrees of freedom, a conventional vibration analysis requires the solution of up to $N \times j$ equations involving all the degrees of freedom. Using periodic structure theory, no more than j equations have to be solved, and in some cases substantially fewer.

The analysis of damped periodic structures is no more difficult than that of undamped structures. It is well known that analyses of highly damped general systems often become more difficult when the natural frequencies of the system are close, and the damping is large enough to create problems of 'modal overlap'. This problem does not arise when periodic structure theory is used in the appropriate manner.

The forced vibration of periodic structures is best understood by first considering the nature of the free waves which can exist in infinite periodic structures. These waves can be reflected if a boundary or discontinuity is added to the otherwise uniform structure. Understanding this mechanism gives insight into the nature of the natural modes of finite periodic structures, and leads to a simple method for predicting natural frequencies. These natural free waves also assist in the understanding of vibration caused by moving loads or pressure fields. The emphasis is therefore on a *wave-motion* approach to the subject.

13.2 DIFFERENT TYPES OF PERIODIC STRUCTURE

One-dimensional periodic systems consist of a 'line' of periodic elements joined together end-to-end. The simplest example of this is the rail of a railway track, supported at regular intervals in an identical manner on the sleepers. These provide large transverse and small bending-rotational constraint at the supports. Flexural wave motion can propagate along the rail in either vertical or horizontal planes. Each element of this system is connected to its neighbour through two coordinates – the transverse deflection and bending rotation at the support. We call this a *bi-coupled* system. If transverse deflection is prevented by a vertically rigid sleeper, the system becomes *mono-coupled* [13.1]. More complicated structures may have more coupling coordinates between adjacent elements, and so are called *multi-coupled* systems [13.2].

Two-dimensional periodic systems consist of an array of periodic elements, coupled together end-to-end *and* side-by-side. A simple example is a flat plate reinforced by an orthogonal set of equi-spaced, identical stiffeners. The element consists of a single plate 'bay' surrounded by a 'half'-stiffener attached to each edge. This constitutes a multi-coupled periodic system, since an infinite number of coordinates (displacement and rotation) couple one element to its neighbour along an edge.

Some periodic plate structures can be regarded as one-dimensional, e.g. a plate of finite width, stiffened in the direction of the width by equi-spaced, identical stiffeners. In general, this must be regarded as multi-coupled, but it effectively reduces to a bi-coupled system if the plate edges perpendicular to the stiffeners are simply-supported. The transverse deflection and the twisting of the stiffeners constitute the two coupling co-ordinates.

Three-dimensional periodic systems consist of 'layers' of two-dimensional systems coupled one-on-top of another. A multi-storey building of modular construction is an example. Simpler, and better, examples are found in physics, where the atomic lattices of crystal structures form almost ideal periodic structures. It is in this context that many of the ideas and analytical techniques for periodic structure analysis have been developed (see Brillouin [13.3]).

13.3 FREE HARMONIC WAVE MOTION THROUGH A MONO-COUPLED PERIODIC BEAM

Consider an infinite beam on regularly-spaced simple supports. One of the bays is subjected to a transverse harmonic force. As the frequency varies, the response at the forcing point varies as shown in Fig. 13.1. The resonant peaks occur at the natural frequencies of a single beam element with its ends either simply-supported or fully-fixed. At higher frequencies, resonances are found to occur in similar pairs.

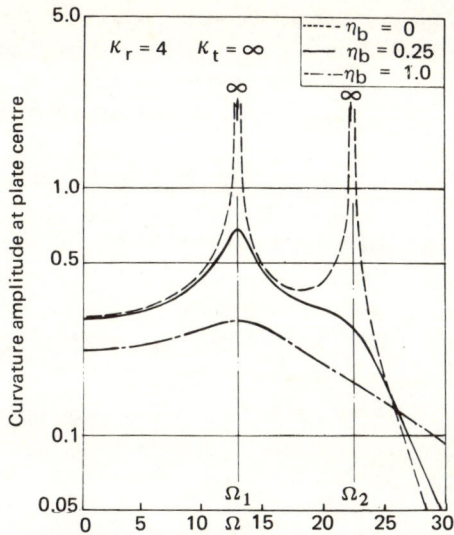

Fig. 13.1 – Variation with frequency and plate damping of the curvature amplitude at the centre of the loaded bay of an infinite stiffened plate. Single point
load.

At very low frequencies, the displacement response of an undamped beam
varies as shown in Fig. 13.2a. Each bay vibrates in counter-phase with its neighbour and with a reduced amplitude, but otherwise in an identical mode. The
amplitude of any point in one bay is $e^{-\mu}$ times the amplitude of the corresponding point in the previous bay. μ depends only on frequency, and not on the
particular pair of points chosen. It is clearly a measure of the rate of decay of
the particular type of wave motion being generated and is known as the *attenuation constant* for this periodic system. If there is no damping in the system,
no net energy is being propagated along the beam under these conditions.

As the frequency increases, the decay rate drops while the amplitude at
the point of excitation increases. At the frequency Ω_1 the decay rate is zero;
each bay vibrates at the same (infinite) amplitude as its neighbour but in counterphase with it (Fig. 13.2b). Above this frequency, each bay vibrates with the
same finite amplitude and in the same mode as its neighbour, but a phase difference ϵ (between 0 and $\pm \pi$) exists between motions at all pairs of corresponding
points in adjacent bays. The response (i.e. displacement, slope, bending moment,
or shear force) at a point in one unloaded bay is then $e^{-i\epsilon}$ times the response at
the corresponding point in the preceding unloaded bay. Energy now propagates
along the beam, from bay to bay, even in the absence of damping. ϵ is called
the phase constant.

There is now a phase difference between displacements at different points *within* a single bay, so the complex flexural motion cannot be represented by a single curve. When $\epsilon = \pi/2$, the total motion may appear at a particular instant as shown in Fig. 13.2c. At a later instant the pattern will have moved along by one or more bays.

At the frequency Ω_2, ϵ has become 0. μ is also zero, so adjacent bays vibrate identically, in phase, in the mode shown in Fig. 13.2d. In the absence of damping, the amplitude generated by the finite force on the loaded bay is infinite, as at Ω_1.

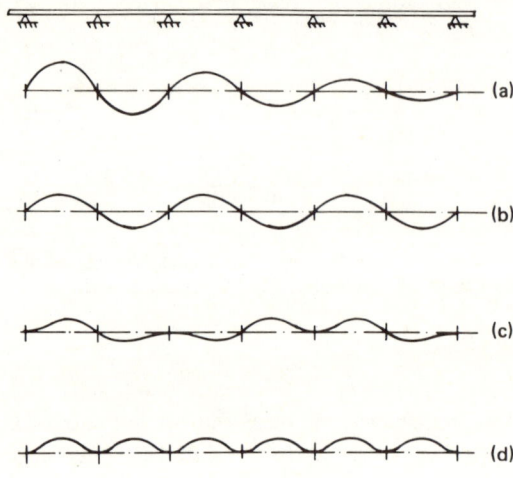

Fig. 13.2 – Modes of displacement of a periodic beam at different frequencies.

Figure 13.3 shows how μ and ϵ vary over an extended frequency range for a particular beam which has rotational constraints at its otherwise 'simple' supports. When μ is non-zero, ϵ is constant and all wave motions decay. These frequency regions are known as *attenuation zones*. On the other hand when μ is zero and constant, ϵ varies. Energy is then propagated and the frequency regions are known as *propagation zones*. Figure 13.3 shows clearly that energy-propagating wave motion can only occur in particular frequency bands. Outside these bands wave motions decay as they attempt to spread out. This 'frequency passband' characteristic is one of the principal features of all periodic structures. It is in these bands that the greatest vibration response can be expected to occur. Also, the natural frequencies of many finite periodic structures tend to occur within these bands. Both the vibration response and natural frequencies depend significantly on the phase constant within the band.

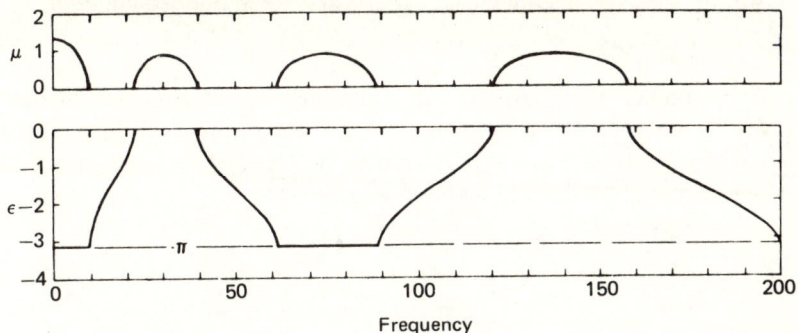

Fig. 13.3 — The real and imaginary parts of the propagation constant for a beam on periodic simple supports.

The complex quantity, $\mu + i\epsilon$, is *the complex propagation constant* and will be denoted by δ. It depends only on the dynamic characteristics of a single element of the whole periodic system and not upon the nature or distribution of loading in the loaded bay. Moreover, the 'mode' of wave motion which is being generated at a particular frequency is a characteristic of the elements of the periodic system, and not of the forcing in the loaded bay. In addition, in all but the loaded bay the wave motion is *free*. Accordingly, the waves may be called 'characteristic free waves'. The motion is freely transmitted from one bay to the next through the forces or moments which are internal to the periodic beam. The propagation constants may therefore be called 'free' propagation constants. An expression for these constants for the simply-supported periodic system is now sought.

Consider two adjacent unloaded bays (Fig. 13.4) of the beam through which a free harmonic wave motion is passing. Harmonic moments, M, act on each end of each bay from the adjacent bays, as shown below. Harmonic rotations θ are caused. From the principle already stated, the moment and rotation at the end B' of bay II must be equal to $e^{-\delta}$ times the moment and rotation at the end A of bay I.

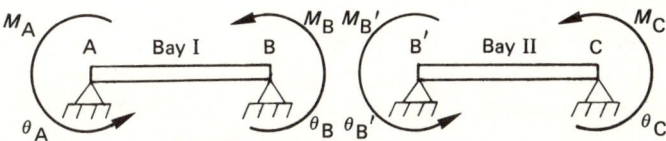

Fig. 13.4 — Moments and rotations of adjacent beam elements.

Now $\theta_B = \theta_{B'}$ for continuity at B–B' and $M_B = -M_{B'}$ for equilibrium.

Hence $\theta_B = \theta_A \, e^{-\delta}$ and $M_B = - M_A \, e^{-\delta}$.   (13.1a, b)

The harmonic rotations, θ_A, θ_B, are related to the harmonic moments M_A, M_B, through the frequency-dependent receptance functions α_{ll}, α_{rr}, (α_{rl} = rotation at r due to moment at l, etc. Suffix l refers to left hand end, r to right-hand end. See ref [13.4] for beam receptances.) Then

$$\begin{Bmatrix} \theta_A \\ \theta_B \end{Bmatrix} = \begin{bmatrix} \alpha_{ll} & \alpha_{lr} \\ \alpha_{rl} & \alpha_{rr} \end{bmatrix} \begin{Bmatrix} M_A \\ M_B \end{Bmatrix} . \qquad (13.2)$$

From equations (13.1) and (13.2) it is easily shown that

$$\cosh \delta = (\alpha_{ll} + \alpha_{rr})/2\alpha_{rl} . \qquad (13.3)$$

If the beam element is symmetrical, $\alpha_{ll} = \alpha_{rr}$, and this equation becomes

$$\cosh \delta = \alpha_{ll}/\alpha_{rl} . \qquad (13.4a)$$

This is clearly satisfied by $\pm\delta$, which shows that for every free wave which propagates or decays in the positive direction, there is a negative-going counter-part. A *pair* of free waves therefore exists at each frequency. If the beam element is undamped, the α's are real and $\cosh \delta$ is real. δ is purely real if $\cosh \delta > 1$, is real $+ i\pi$ if $\cosh \delta < -1$, and is purely imaginary ($= i\epsilon$) if $-1 < \cosh \delta < 1$. In the latter case, which constitutes a propagation zone,

$$\cos \epsilon = (\alpha_{ll} + \alpha_{rr})/2\alpha_{lr} . \qquad (13.4b)$$

Equation (13.4b) for the phase constant may be expressed in the alternative form

$$\epsilon = \cos^{-1} [(\alpha_{ll} + \alpha_{rr})/2\alpha_{lr}] \qquad (13.4c)$$

which emphasizes the multi-valued nature of the solution. If ϵ is the solution between 0 and π, then $\epsilon' = 2n\pi + \epsilon$ is also a solution. n is any integer between $\pm\infty$. Now ϵ or ϵ' is the change of phase of the propagating wave motion over the periodic distance l_x (the length of the periodic element). The corresponding 'effective wave number' is the phase change per unit length and is

$$K_n = (2n\pi + \epsilon)/l_x . \qquad (13.5a)$$

The beam free wave motion at a particular frequency with phase constant ϵ can actually be analysed into a group of sinusoidal travelling waves of all these wave-numbers. Since each component wave has the same frequency ω as the total wave motion, the phase velocity of the component, a_n, is given by ω/K_n, i.e.

$$a_n = \omega l_x/(2n\pi + \epsilon) , \qquad (n = -\infty \text{ to } +\infty) . \qquad (13.6a)$$

Using the values of ϵ from Fig. 13.3 (from the first propagation band) we can calculate a_n for different values of n, and these are shown in Fig. 13.5. Since n can have negative as well as positive values, it is evident that some of the component waves in the group may be travelling in the negative direction, while others are travelling in the positive direction.

Since equations (13.4 b, c) are satisfied by $\pm \epsilon$, the negative-going free wave motion has wave-number components

$$K_n = (2n\pi - \epsilon)/l_x \tag{13.5b}$$

with phase velocities

$$a_n = \omega l_x / (2n\pi - \epsilon) \tag{13.6b}$$

which yields a similar curve to that of Fig. 13.5, but in the negative a_n domain.

The magnitude of the contribution of the nth component wave to the total motion has been considered elsewhere [13.5]. The phase velocity curves are very important in understanding and predicting the response of periodic beams or plates to convected pressure fields (i.e. sound fields), or in explaining the sound radiation properties of periodically-stiffened plates [13.6]. We shall return to this at a later stage.

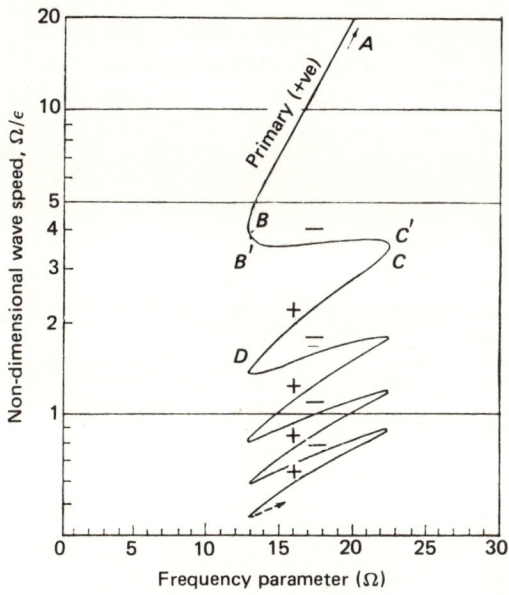

Fig. 13.5 – Variation of component wave speed frequency; for first propagation band of beam of Fig. 13.3.

13.4 FREE HARMONIC WAVE MOTION THROUGH A MULTI–COUPLED BEAM

A beam on periodic flexible supports has two coupling coordinates between adjacent periodic elements. At any frequency there are now two pairs of complex propagation constants, $\pm\delta_1$, $\pm\delta_2$, which are obtained from the quadratic equation

$$\cosh^2 \delta - X \cosh \delta + Y = 0 . \tag{13.7}$$

X and Y are functions of the eight receptance functions for the ends of the beam element. Section 13.5 gives explicit forms for X and Y. δ_1 and δ_2 depend upon the frequency, and also upon the degree of rotational and transverse stiffness at each support. If the system is undamped, δ_1 and δ_2 may both be real (or real $+ i\pi$), both purely imaginary, one real and one imaginary, or one the complex conjugate of the other. The purely imaginary value corresponds to an energy-carrying propagating wave, adjacent elements vibrating with equal amplitudes but different phase. The real or complex conjugate values correspond to spatially-decaying waves.

Figure 13.6 shows how δ_1 and δ_2 vary with frequency for a beam on periodic elastic supports. Propagation zones, attenuation zones and a complex-conjugate zone are seen to exist. The bounding frequencies of the propagation zones vary with the rotational and transverse stiffnesses at the supports. If the periodic element is symmetrical, these bounding frequencies can be identified with certain natural frequencies of a single element, with the coupling coordinates either fixed or free [13.2]. The corresponding wave motion in an element at these frequencies is real, and identical with the corresponding mode of natural vibration.

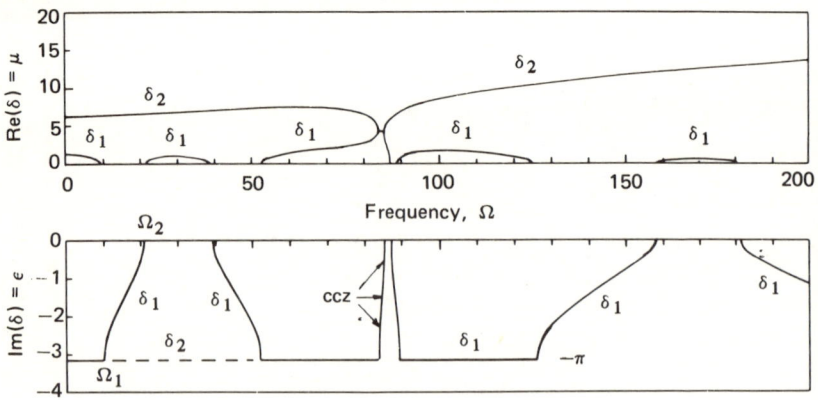

Fig. 13.6 – Propagation constants δ_1 and δ_2 for a beam on deflecting supports.

13.5 FREE HARMONIC WAVE MOTION THROUGH A STIFFENED, SIMPLY-SUPPORTED PLATE OF FINITE WIDTH

Figure 13.7 illustrates the type of plate considered here. The x-wise edges are simply-supported. The equispaced, identical stiffeners span the plate in the y-direction and are simply-supported at their ends. The periodic element is one plate bay with a 'half-stiffener' along each y-wise edge. Flexural plate motion only will be considered, coupled with flexure and torsion of the stiffener.

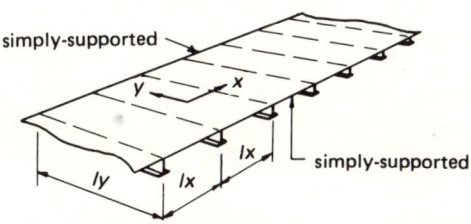

Fig. 13.7 – The stiffened plate.

Although this is strictly a two-dimensional system, the periodic elements are assembled side-by-side in a one-dimensional array. One-dimensional periodic structure theory may therefore be applied. Further, although each element is coupled to its neighbour through an infinite number of coordinates (the coupling is *continuous* along the edge, through both translation and rotation), the wave motion can be studied as though only two coupling coordinates exist – one translational and one rotational. This is accomplished simply by allowing the plate and stiffener motion to vary in a single sinusoidal mode across the plate in proportion to sin $m\pi y/l_y$ (m is an integer). The whole system is then effectively reduced to a one-dimensional bi-coupled periodic system.

Wave motion can propagate in the x-direction in such a way that the complex motions in any pair of adjacent bays are identical apart from either a particular phase difference, or a particular exponential decay factor. These, of course, are the phase constant or attenuation constant respectively of the wave motion. Being effectively a bi-coupled system, there are two pairs of different propagation constants at each frequency, for each chosen value of m. The values of cosh δ are obtained from equation (13.7) in which X and Y are given by

$$X = (\cosh \Lambda_1 + \cosh \Lambda_2) - C_r (\Lambda_2 \sinh \Lambda_2 - \Lambda_1 \sinh \Lambda_1)$$

$$- C_t (\sinh \Lambda_1/\Lambda_1 - \Lambda_1 - \sinh \Lambda_2/\Lambda_2) \qquad (13.8a)$$

$$Y = \cosh \Lambda_1 \cosh \Lambda_2 - C_r \left(-\Lambda_1 \sinh \Lambda_1 \cosh \Lambda_2 + \Lambda_2 \cosh \Lambda_1 \sinh \Lambda_2\right)$$

$$- C_t \left(\sinh \Lambda_1 \cosh \Lambda_2 / \Lambda_1 - \cosh \Lambda_1 \sinh \Lambda_2 / \Lambda_2\right) \qquad (13.8\text{b})$$

$$+ C_r C_t \left\{(\Lambda_1/\Lambda_2 + \Lambda_2/\Lambda_1) \sinh \Lambda_1 \sinh \Lambda_2 + \right.$$

$$\left. + 2 - 2 \cosh \Lambda_1 \cosh \Lambda_2\right\}$$

where $\quad \Lambda_1 = (P^2 + \Lambda^2)^{\frac{1}{2}}; \qquad\qquad \Lambda_2 = (P^2 - \Lambda^2)^{\frac{1}{2}} \qquad (13.8\text{c, d})$

$$\Lambda = \omega^{\frac{1}{2}} \{\mu l_x^4 / D(1 + i\eta)\}^{\frac{1}{4}}; \qquad P = m \pi l_x / l_y \qquad (13.8\text{e,f})$$

$$C_r = P^2 \{GJ/l_x + P^2 E\Gamma/l_x^3\}/4D\Lambda^2 \qquad (13.8\text{g})$$

$$C_t = P^4 E I_s / 4D\Lambda^2 . \qquad (13.8\text{h})$$

D is the plate flexural rigidity; GJ and $E\Gamma$ are the stiffener torsional stiffnesses (St Venant and torsion bending respectively); EI_s is the stiffener flexural stiffness. Torsional and flexural inertia of the stiffener can be included by multiplying C_r by $\{1 - (\omega/\omega_t)^2\}$ and C_t by $\{1 - (\omega/\omega_f)^2\}$, where ω_t and ω_f are the uncoupled torsional and flexural natural frequencies respectively. (Equations (13.8 a-h) may be modified, with care, to apply to a narrow beam resting on point elastic supports.)

13.6 ENERGY SOLUTIONS FOR THE PROPAGATION CONSTANTS

A great deal of tedious algebra is required to derive the above equation for the flat stiffened plate. There are alternative computer-based methods of determining the propagation constants, and these include transfer matrix [13.7] and finite element methods [13.8]. The transfer matrix method is based on the fact that e^δ is an eigenvalue of the transfer matrix relating the state vector at one end of the periodic element to the state vector at the other. Finite element methods will not be discussed here, but an important underlying principle will now be presented which leads to a simple but approximate method of determining the *phase* constants of *propagating* waves.

It can be shown that when a true harmonic propagating wave progresses through the system, the sum of the maximum kinetic energies of all particles of mass is equal to the sum of the maximum strain energies of all 'elements' of stiffness. If the system is a uniform beam, each element of which vibrates in the complex flexural mode $w = e^{i\omega t} f(x)$, then this energy equality is expressed in the familiar form

$$\frac{\omega^2}{2} \int_0^{l_x} \mu |f|^2 \, \mathrm{d}x = \frac{1}{2} \int_0^{l_x} EI \, |(\mathrm{d}^2 f/\mathrm{d}x^2)|^2 \, \mathrm{d}x \qquad (13.9)$$

where $|\,|$ implies the modulus of the complex function. If the correct complex mode of vibration of the beam element, $f(x)$, corresponding to a particular phase constant ϵ, is put into this equation, it yields the correct frequency at which this wave propagates. If an approximate mode is used in this equation, it yields a *good approximation* to the frequency at which the wave propagates. This is clearly Rayleigh's Principle expressed in a more general form to deal with waves and their propagation frequencies, rather than with natural modes and their natural frequencies.

The successful use of this method of finding curves of phase constant vs. frequency depends on a wise choice of the complex mode of vibration. For beam elements, especially in the lowest propagation band, this is not difficult, and is discussed in detail in Ref. [13.9]. The simplest polynomial function for a beam element on simple supports with no rotational constraints is found to be

$$f(\xi) = i(\xi^3 - \xi) - \frac{1}{4}\left[(3 \cot \frac{\epsilon}{2} + 5 \tan \frac{\epsilon}{2}) - \right.$$

$$\left. - (6 \cot \frac{\epsilon}{2} + 6 \tan \frac{\epsilon}{2}) \xi^2 + (3 \cot \frac{\epsilon}{2} + \tan \frac{\epsilon}{2}) \xi^4 \right] \qquad (13.10a)$$

where $\xi = 2x/l_x$ \qquad\qquad (13.10b)

This looks complicated, but is obtained from a systematic application of four boundary conditions to the 4th order polynomial function. Using this function in equation (13.9), and computing ω for a range of values of ϵ, yields the phase constant curve shown in Fig. 13.8 for the first propagation band of a uniform periodic beam on simple supports. The agreement with the curve from the exact method is striking.

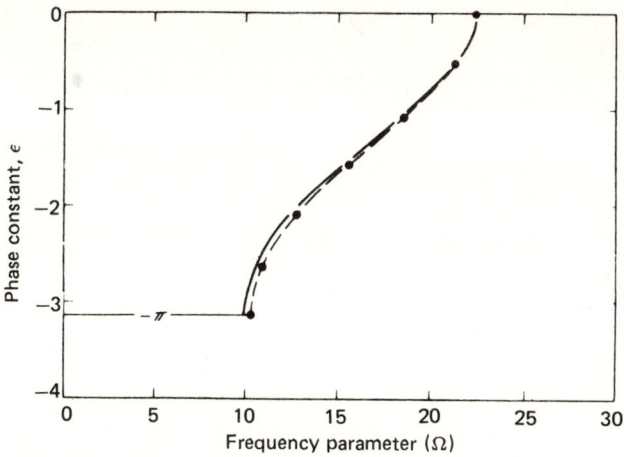

Fig. 13.8 – Approximate ($\bullet - - - \bullet$) and exact (———) values of ϵ *vs.* Ω for a uniform beam on periodic simple supports.

Just as Rayleigh's Method for natural frequencies can be developed into the Rayleigh-Ritz method by using several preassigned modes in the analysis and minimizing the frequency, so can the above method be extended to incorporate several modes. By minimizing the frequency or the appropriate energy functions, much improved values can be obtained for the frequencies for given ϵ's. In addition, ϵ vs. frequency curves are thereby found for the higher propagation zones. This has been investigated in Ref. [13.10].

This energy method is readily adaptable to finding the propagation constants and frequencies of 'plane' wave motion across plates which are periodically stiffened in both x and y directions [13.15].

13.7 SOME APPLICATIONS OF THE PROPAGATION CONSTANT CURVES

13.7.1 The coincidence frequency of a periodically supported plate subjected to a convected harmonic sound field

Suppose the periodic system is subjected to a convected form of loading (e.g. a harmonic sound pressure field) which convects over the system at velocity c. At frequency ω, the difference in phase between the pressures in adjacent bays is $\omega l_x/c = \epsilon_p$. The system responds at this frequency and *with this phase difference between adjacent bays*. In fact the pressure field forces wave motion along the system at its own convection velocity c. It is physically intuitive (and mathematically verifiable) that if this phase difference ϵ_p is equal to the phase constant ϵ of the free harmonic waves of the periodic system at the frequency, then the beam will respond readily to the pressure field. In the absence of any damping, the response would be infinite.

The frequency at which this occurs is obtained from the phase-constant curve by the intersection of this curve and the phase-angle 'line' of the pressure field, $\epsilon_p = \omega l_x/c$. A simple graphical construction on the phase constant curve then yields this 'coincidence frequency', for at this frequency the free wave velocity and pressure field velocity coincide. The phase constant curve must include all the multiple values of $\epsilon \pm 2n\pi$ as described before. Such coincidence frequencies exist in each propagation zone, and not just at one particular frequency as in the simplest acoustic coincidence theory. The lowest frequency at which such coincidence can exist is usually much lower than the 'simple' coincidence frequency.

13.7.2 The natural frequencies of finite periodic beams on simple supports

Let the finite beam have N bays. Provided there are no external masses attached at the extreme ends of the beam, and that its elements are symmetrical, it can be shown [13.1] that all the natural frequencies of the beam fall within the propagation zones. Now a general free motion of the beam at a given frequency can be represented by the superposition of all the possible free wave motions of the beam at the frequency. The periodic beam on simple-supports has two possible free waves, a positive and negative-going pair. As the positive-going wave progresses, its phase changes by ϵ per bay traversed. On encountering the right-hand boundary B, it is totally reflected (if the boundary is undamped) with a phase change ϵ_B. It returns as the negative-going wave, changing phase by ϵ per bay to the left-hand boundary A. Here it is totally reflected with phase change ϵ_A and then proceeds as the positive-going wave. The total phase change in one complete excursion through the whole system is $2N\epsilon + \epsilon_A + \epsilon_B$. If this is an integral multiple of 2π, a resonance of the system can occur. Thus resonant frequencies occur at those frequencies for which

$$\epsilon = (2n\pi + \epsilon_A + \epsilon_B)/2N, \quad (0 \leqslant n \leqslant N-1) \; . \tag{13.11}$$

If the beam is fully-fixed at one end (A, say), $\epsilon_A = 0$. If the beam is free at that end, $\epsilon_A = \pi$. The natural frequencies of periodic beams with two fully-fixed extreme ends are therefore those frequencies at which $\epsilon = n\pi N$, whereas for beams with two 'free' extreme ends $\epsilon = (n+1)\pi/N$. These frequencies can be found from a simple graphical construction [13.11] on the ϵ vs. frequency curves, by drawing in the lines $\epsilon = (n+1)\pi/N$ and locating the frequencies of intersection.

13.8 THE RESPONSE OF PERIODIC SYSTEMS TO CONVECTED HARMONIC PRESSURE FIELDS

13.8.1 Closed-form solutions and results

Suppose an harmonic pressure field convects at uniform velocity over a uniform, infinite, periodic beam or plate with simply-supported (long) sides. It is possible to find closed form solutions to the response of the system. Details of the method are given in Refs. [13.12] and [13.13]. Since each periodic element is identical and is subjected to the same pressure field, the response of each element must be identical apart from phase. The phase difference between responses at corresponding points in adjacent elements is equal to the phase difference between the pressures acting at those two points. This fact may be used to reduce the analysis of the whole infinite beam to the analysis of a single element.

Let the phase difference between the pressures in adjacent elements be ϵ_p when the non-dimensional frequency of the pressure is Ω. A convenient non-dimensional convection velocity, CV, can now be defined as

$$CV = \Omega/\epsilon_p \; . \tag{13.12}$$

Figure 13.9 shows how the response at a point in a periodic beam varies both with CV and Ω of a convecting harmonic pressure field of unit amplitude. The response quantity considered is the amplitude of the beam curvature, $|d^2w/dx^2|$, at the centre of a beam element, i.e. mid-way between two beam supports. The beam has supports of infinite transverse stiffness but of rotational stiffness $k_r = 4EI/l_x$ (i.e. $\kappa_r = k_r l_x/EI = 4.0$). Internal damping in the beam was allowed for by assigning to the hysteretic loss factor, η, the value 0.25.

The curves of Fig. 13.9 extend only over the frequency ranges of the lowest attenuation and propagation zones of the beam. The peaks in the response curves occur at the coincidence frequencies described in section 13.7.1. Notice that the highest peak occurs when the convection velocity is 4, which is the free wave velocity of the characteristic wave at the lower bounding frequency, $\Omega \cong 12.8$ (See Fig. 13.5). Lower, but broader peaks occur if the convection velocity is higher than this.

Response curves for higher propagation zones have also been computed (results unpublished). They are generally of more complicated appearance, and often have more than one peak in a given propagation zone. These can always be explained, however, on the grounds of section 13.7.1.

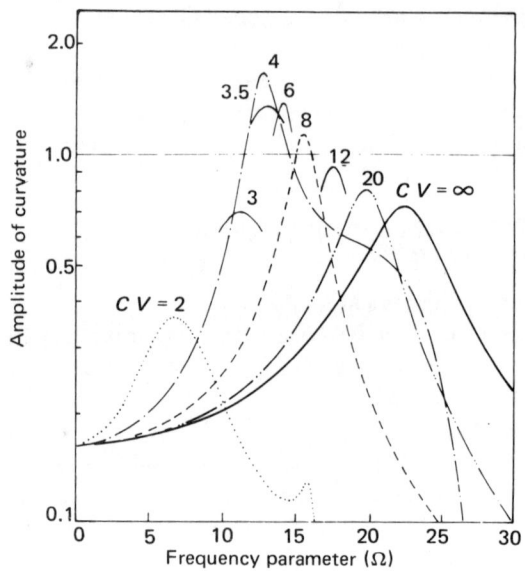

Fig. 13.9 — Amplitides of curvature at bay centre of an infinite beam; acoustic plane waves of different convection velocity $\eta_b = 0.25, \kappa_r = 4.0, \kappa_t = \infty$.

13.8.2 Approximate methods of response prediction

The likelihood of correctly predicting the precise response at a point in a large periodic system is remote, and the validity of attempting such a prediction is dubious. The space-average response, however, should be amenable to more accurate and valid prediction. This, of course, is part of the underlying philosophy of statistical energy methods of structural dynamic analysis (SEA), and is applicable to the periodic structure analysis.

We start from the following feature first observed by Abrahamson [13.14]. When a plane harmonic pressure field convects over a periodic structure, imposing the phase difference ϵ_p upon the responses in adjacent bays, the total response may be regarded as the sum of responses in all the free characteristic waves of the periodic structure which have the phase constant ϵ_p.

There is a free wave of phase constant ϵ_p in each of the propagation zones, and its 'natural' frequency of propagation in the rth zone may be denoted by ω_r. However, when the structure is forced by the pressure wave at frequency ω, each of these free waves is forced to propagate at ω and not at ω_r. Let the complex modal displacement function of the rth characteristic wave be $f_r(x)$, and the corresponding generalised coordinate describing the wave amplitude be q_r. The actual displacement of the system is given by

$$w(x, t) = e^{i\omega t} \sum_{r=1}^{\infty} q_r f_r(x) \ . \tag{13.13}$$

q_r can be shown to be given by

$$q_r = \{Q_r/M_r\}/\{\omega_r^2(1+i\eta_r) - \omega^2\} \tag{13.14}$$

where $\quad Q_r = p_0 \int_0^{l_x} f_r^*(x) \exp(-i\epsilon_p x/l_x)\, dx \quad$ and $\quad M_r = \int_0^{l_x} \mu\,|f_r^2(x)|\, dx$ (13.15a, b)

η_r is the loss factor of the wave and is defined by

$$\eta_r = \text{Energy dissipated in one element per cycle} \div \pi\omega_r^2 M_r |q_r^2| \ . \tag{13.16}$$

Equation (13.14) has the familiar appearance of an expression for the harmonic response of a natural mode of an elastic system. Q_r is the complex generalized force exciting the rth wave, and M_r is the (real) generalized mass of the rth wave. The corresponding (real) generalized stiffness is $K_r = \omega_r^2 M_r$. It can be shown that $\frac{1}{2}|q_r^2|M_r = T_{max}$ is equal to the sum of the maximum kinetic energies possessed by all mass particles of a periodic element when vibrating with harmonic velocity $q_r e^{i\omega t}$. Likewise, $\frac{1}{2}|q_r^2|K_r = U_{max}$ is the sum of the maximum strain energies possessed by all stiffness 'particles' in the periodic element when vibrating with harmonic displacement $q_r\, e^{i\omega t}$.

It follcws from the above equations that when just one characteristic wave is being excited by the pressure wave

$$U_{max} = Q_r^2/2M_r\omega_r^2\{[1 - (\omega/\omega_r)^2]^2 + \eta_r^2\} \ . \tag{13.17}$$

The time-averaged strain energy is one-half of this. Its value will be obtained exactly if M_r and Q_r are calculated from the exact, known complex wave motion, $f_r(x)$. If $f_r(x)$ is only known approximately, the corresponding value of ω_r found from an energy analysis (see section 13.6) is a good approximation, We cannot say whether Q_r^2/M_r will be a good approximation, but calculations for a special case show that it can, in fact, be a very good approximation. This is shown by Fig. 13.10 which compares the values of U_{max} calculated by this method with exact values of U_{max} calculated from the closed-form solution for a periodic beam on simple supports. The approximate mode, $f_r(x)$, used in the approximate method was similar to that of equation (13.10a), but allowed for rotational constraint at the supports of $\kappa_r = 4.0$.

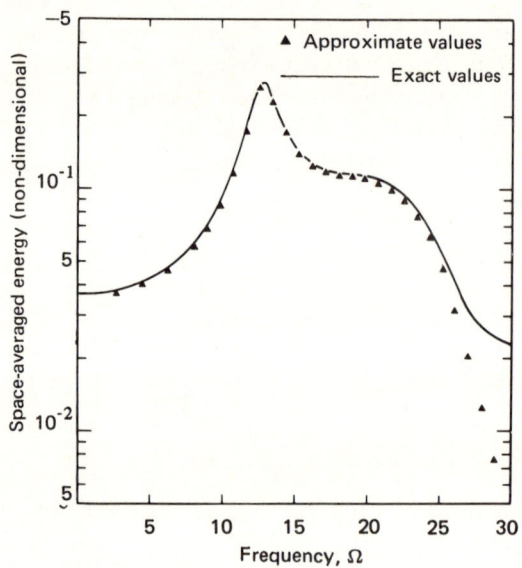

Fig. 13.10 – U_{max} calculated by approximate and exact methods: $\kappa_r = 4$, $CV = 4$.

Equation (13.17) shows that U_{max} must have a maximum ('resonant') value when the exciting frequency, ω, is equal to ω_r. This is the condition for coincidence, previously explained. The pressure wave phase difference ϵ_p is then equal to the phase constant ϵ_r of the free characteristic wave, so ϵ_p can be replaced by ϵ_r in equations (13.15a and 13.17) in order to find the peak value of U_{max}. This phase constant is also required in the modal function, $f_r(x)$ (equation 13.10a).) It is then found that Q_r^2/M_r only varies slowly and relatively slightly, as ϵ_r varies from 0 to $\pm \pi$, i.e. as the coincidence frequency varies from one end of a propagation zone to the other. This feature can be used to obtain 'rule of thumb' values of the space-averaged energy at coincidence, and from this can be found approximate space-averaged and space-peak values of the internal stresses. Current research is directed to this end.

13.9 THE RESPONSE OF PERIODIC SYSTEMS TO CONVECTED RANDOM PRESSURE FIELDS

'Frozen' convected pressure fields only will be considered here. The power spectral density of the pressure, $S_p(\omega)$, will be assumed to have a constant value through a propagation zone of the system. The convection velocity V of the pressure field is constant for all spectral components, so a component of frequency ω is associated with a wave number of ω/V and a phase-difference between adjacent periodic elements of $\epsilon_p = \omega l_x/V = \omega/k_p$. Suppose the response of a system element to a harmonic pressure field of amplitude p_0, frequency ω, and wave-number k_p is $p_0 e^{i\omega t} Y(i\omega, k_p)$. The spectral density of the response to the random convected field is given by

$$S_r(\omega) = S_p(\omega) | Y(i\omega, k_p) |^2 . \tag{13.18}$$

If $S_p(\omega)$ does not vary with frequency (as assumed) the response spectral density is proportional to the square of the system response to a convected harmonic pressure field, i.e. it is proportional to the square of such curves as those of Fig. 13.9. The random response spectrum will have a peak at a coincidence frequency in each propagation zone.

The mean square response of the system is given by $\int\limits_0^\infty S_r(\omega)\, d\omega$. If only one propagation zone contributes significantly to the response, the limits in this integral can be set just outside that zone. Such integrations have been carried out for periodic beams subjected to random pressure fields of different convection velocities, with the significant response confined to the first propagation zone. This response depends on the stiffness of constraint at the supports, and Fig. 13.11 shows how the mean square curvature of the beam at a support varies with convection velocity, CV, and with support rotational stiffness, κ_r. Similar curves have been plotted (but are not shown here) to show how the space-time averaged energy in the beam element varies with these parameters. The following conclusions have been drawn:

(a) As the convection velocity increases (the pressure spectral density remaining the same) the r.m.s. response of the periodic system approaches a constant value.

(b) With CV greater than 6, an increase of rotational constraint at the supports increases the curvature in the beam at the supports.

(c) As the damping of the beam, η, increases, the response decreases in proportion to $1/\sqrt{\eta}$, provided the convection velocity exceeds the minimum value for coincidence excitation.

(d) The greatest response of a lightly-damped beam, either at the supports or at the beam centre, occurs when the convection velocity is equal to the lowest value for coincidence excitation.

(e) The energy method for predicting the space-time averaged response gives results which are almost indistinguishable from the exact method, when a single wave-mode is allowed which satisfies both geometric and natural wave boundary conditions.

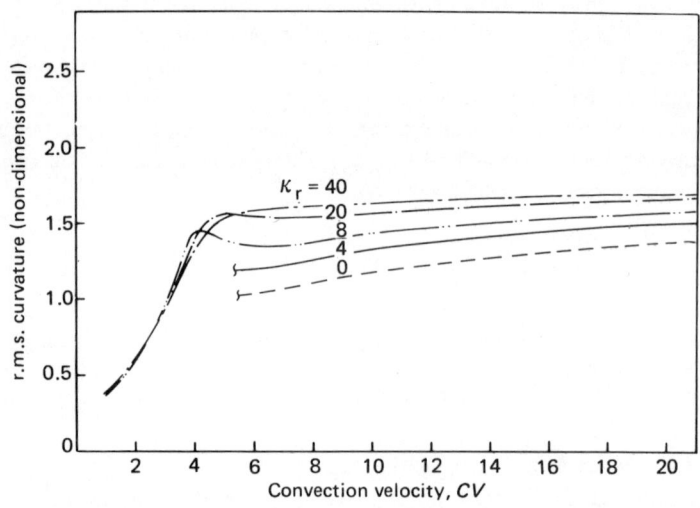

Fig. 13.11 – R.M.S. curvature at a support of a periodic beam; the effect of changing CV and κ_r.

13.10 THE EFFECTS OF NON-PERIODICITY

Real structures can never be exactly periodic. Even were they designed to be so, manufacturing inaccuracy would introduce slight departures from the periodicity. An aeroplane fuselage structure, in some areas, may have stiffening frames and stringers at nominally identical intervals, but it cannot be manufactured with precisely identical intervals. Moreover, practical considerations usually require that extra stiffening or mass be added at some locations, and this destroys the local periodicity. It makes the system either locally discontinuous or more generally disordered.

Some effort has been directed to examining these effects on response levels and free wave propagation, as described below.

13.10.1 Local discontinuities

A 'local discontinuity' is one which occurs only once in the whole periodic structure; e.g. one bay of a periodic beam may be of non-periodic length, or a rotational constraint (elastic or inertial) may be added at a single support of the beam [13.16].

The propagation constants of such a system are unchanged by the local discontinuities. In the regions on either side of the discontinuity, the structure is periodic, and wave motion propagates there as in a normal periodic structure. However, when a free wave which propagates in one region encounters the discontinuity, part of it is reflected and part of it is transmitted across the discontinuity. Both reflected and transmitted waves are free waves, propagating with the usual propagation constants.

Both waves have amplitudes less than that of the initially-incident wave, but the reflected wave interferes with the incident wave as it travels backwards. In some parts of the system the total wave motion is then greater than that of the incident wave, and in other parts it is less. This is always true, irrespective of the type of discontinuity.

It is possible for a local discontinuity to resonate against the adjacent periodic systems at discrete natural frequencies, but only in the frequency attenuation zones. This means that a source of harmonic excitation can excite large amplitudes of motion in the attenuation zones. The motion is that of an attenuating wave, not a propagating wave, so the large amplitudes are confined to the region close to, and on either side of, the discontinuity.

When forced wave motion occurs in the locally disordered system (due, say, to a convected harmonic pressure field [13.17]) free waves are scattered from each side of the discontinuity when the forced wave impinges on it. Interference between the forced and free waves causes local increases and decreases of the total response relative to the forced wave motion on its own. Furthermore, a forced resonant-type response will occur in those attenuation zones where the disorder can resonate against the adjacent periodic system. In the forced motion due to a convected pressure field there can therefore be large peaks at these resonant frequencies *and* at the coincidence frequencies which are always associated with convected pressure field excitation. The presence of these additional resonant peaks means that the overall response due to a random pressure field is greater when the disorder is present than when it is removed.

13.10.2 Distributed disorders

This term describes nominally periodic systems, but each periodic element differs in a small way from its neighbours. Bansal [13.18, 13.19] has considered the special type of disordered system in which N adjacent bays are disordered in a random way (e.g. the span of each bay differs from the others) but the next N adjacent bays are disordered identically, and so on throughout the whole infinite system. The system is now periodic, but the periodic element is N-bays long, and is internally-disordered.

The propagation constants of this system differ from those of the simpler, ordered system. Each frequency propagation zone of the simpler system now contains $N-1$ attenuation zones and N propagation zones. In the attenuation zones of the former ordered system, the attenuation constants of the disordered system are increased approximately by a factor of N. (This constant now describes the attenuation over N bays, whereas formerly it applied to just one bay.)

If the system is excited by a convected random pressure field, there will be a coincidence frequency and response peak in each propagation zone. Since the disordered system has N times as many propagation zones as the ordered system, there will be N times as many coincidence peaks. However, not all of these are highly excited, but the overall response level of the disordered system is found to be greater than that of the ordered system. The greater the degree of disorder, the greater is the response [13.20].

13.10.3 General remarks on non-periodicity

The forced response of a disordered or discontinuous periodic system will always be greater at some points than the response anywhere in a periodic system. The effect of disorder or discontinuity is always to produce a local, if not a general, increase in response level. It is therefore advantageous to design and manufacture a system as closely as possible to a truly periodic condition. This is particularly true for a turbine disc assembly. The slightest irregularity in the array can cause a large local vibration level.

13.11 CONCLUDING REMARKS

Periodic structure theory may be used to great advantage to reduce the size of the computational problem when a large periodic system is to be analysed and its response levels predicted. Closed form solutions may be obtained for beam-type systems. Energy methods using approximate series form solutions can yield very accurate results for forced beam vibrations, especially when the space-averaged vibration level is sought. The approximate methods which have been described for beams have also been applied to two-dimensional plates [13.21, 13.22], and finite element methods are easily adaptable to periodic structure analysis.

When a periodic beam- or plate-like structure is excited by a convected pressure field, the response is minimized by having the least possible rotational (or torsional) restraint along the edges of the periodic element. Any non-periodicity (local or distributed) tends to increase the local (or general) response level. The smallest response will be obtained with the most periodic structure.

REFERENCES

[13.1] Mead, D. J. (1975) *J. Sound & Vib.*, **40(1)**, 1–18. Wave propagation and natural modes in periodic systems: I. Mono-coupled systems.

[13.2] Mead, D. J. (1975) *J. Sound & Vib.*, **40(1)**, 19–39. Wave propagation and natural modes in periodic systems: II. Multi-coupled systems.

[13.3] Brillouin, L. (1953) *Wave propagation in periodic structures*. Dover Publications, Inc., New York.

[13.4] Bishop, R. E. D. & Johnson, D. C. (1960) *The mechanics of vibration*. Cambridge University Press.

[13.5] Mead, D. J. (1970) *J. Sound & Vib.*, **11(2)**, 181–197. Free wave propagation in periodically-supported infinite beams.

[13.6] Pujara, K. K. (1970) PhD Thesis, ISVR, University of Southampton. Vibration of and sound radiation from some periodic structures under convected loadings.

[13.7] De Espindola, J. J. (1974) PhD Thesis, ISVR, University of Southampton. Numerical methods in wave propagation in periodic structures.

[13.8] Orris, R. M. & Petyt, M. (1974) *J. Sound & Vib.*, **33(2)**, 233–236. A finite element study of harmonic wave propagation in periodic structures.

[13.9] Mead, D. J. (1973) *J. Sound & Vib.*, **27(2)**, 235–260. A general theory of harmonic wave propagation in linear periodic systems with multiple coupling.

[13.10] Mead, D. J. & Mallik, A. K. (1976) *J. Sound & Vib.*, **47(4)**, 457–472. An approximate method of predicting the response of periodically supported beams subjected to random convected loading.

[13.11] Sen Gupta, G. (1970) *J. Sound & Vib.*, **13(1)**, 89–101. Natural flexural waves and the normal modes of periodically supported beams and plates.

[13.12] Mead, D. J. (1971) *J. Engineering for Industry Trans. ASME*, **93**, Ser. B, 3, 783–792. Vibration response and wave propagation in periodic structures.

[13.13] O'Keefe, J. M. (1972) MSc Thesis, University of Southampton; A study of the forced response of a highly-damped periodic structure.

[13.14] Abrahamson, A. L. (1973) PhD Thesis, University of Southampton. The response of periodic structures to aero-acoustic pressures, with special reference to aircraft skin-rib-spar structures.

[13.15] Mead, D. J. & Parthan, S. (1979) *J. Sound & Vib.*, **64(1)**, 325–348. Free wave propagation in two-dimensional periodic plates.

[13.16] Mead, D. J. & Bansal, A. S. (1978) *J. Sound & Vib.*, **61(4)**, 481–496. Monocoupled periodic systems with a single disorder: free wave propagation.

[13.17] Mead, D. J. & Bansal, A. S. (1978) *J. Sound & Vib.*, **61(4)**, 497–515. Mono-coupled periodic systems with a single disorder: response to convected loadings.

[13.18] Bansal, A. S. (1978) *J. Sound & Vib.*, **60(3)**, 389–400. Free wave motion in periodic systems with multiple disorders.

[13.19] Bansal, A. S. (1979) *J. Sound & Vib.*, **62(1)**, 39–49. Flexural wave motion in beam-type disordered periodic systems: coincidence phenomenon and sound radiation.

[13.20] Bansal, A. S. (1977) PhD Thesis, University of Southampton. Dynamic response of disordered periodic systems.

[13.21] Mead, D. J. & Parthan, S. (1982) (Paper in preparation). The analysis of the forced response of periodic plates using simple approximate modes.

[13.22] Abdel-Rahman, A. Y. A. (1980) PhD Thesis, University of Southampton. Matrix analysis of wave propagation in periodic systems.

Other important literature on periodic structures

Cremer, L. & Leilich, H. O. (1953) *Arch. elekt. Ubertr.*, **7**, 261. Zur Theorie der Biegekettenleiter.

Miles, J. W. (1956) *Proc. Am. Soc. Civil Engineers*, **82**, EMI, 1. Vibrations of beams on many supports.

Heckl, M. (1961) *J. Acoust. Soc. Am.*, **33**, 640. Wave propagation in beam-plate systems.

Heckl, M. (1964) *J. Acoust. Soc. Am.*, **36**, 1335. Investigations on the vibrations of grillages and other simple beam structures.

Ungar, E. E. (1966) *J. Acoust. Soc. Am.*, **39**, 887. Steady state response of one-dimensional periodic flexural systems.

Cremer, L., Heckl, M. & Ungar, E. E. (1973) Structure-borne Sound, (Chapter V.5), Springer–Verlag.

CHAPTER 14

Jet noise

M. J. Fisher and C. L. Morfey

Institute of Sound and Vibration Reasearch, University of Southampton

14.1 INTRODUCTION

The term *jet noise* is frequently used to describe the total noise emanating from an aircraft exhaust system. However, studies have shown that this *total noise* is composed of several components which should, whenever possible, be considered separately. The most fundamental of these, and certainly the component which is in principle the most difficult to eliminate is that due to the turbulent mixing of the jet exhaust with the ambient fluid downstream of the nozzle exit plane. We term the resulting sound *jet mixing noise*. In incorrectly expanded jet exhaust flows, the presence of the resulting shock waves leads to a further source of noise which we shall term *shock associated noise*. In general, this source gives rise to two components, one a set of discrete tones, often referred to as *screech*, together with more broadband radiation, which we shall term *broadband shock-associated noise*.

These sources are, to the best of our knowledge, the only significant contributors to the 'total noise' which exist downstream of the nozzle exit plane. However the advent, in particular, of high bypass ratio, relatively low jet efflux velocity engines has brought to attention additional sources of noise. These are variously referred to as *excess noise, tailpipe noise,* or *core noise*. Historically the term excess noise was coined to account for measurements of total noise which were in excess of that anticipated from available predictions of the downstream noise sources introduced above. As the understanding of mixing noise developed in the 1970s, and prediction schemes consequently improved, the discrepancies at low jet velocities were significantly reduced. Much of what was once termed excess noise (in aircraft exhaust noise data) is now recognized as mixing noise, enhanced by density fluctuations in the turbulent mixing region.

Extraction of the mixing noise component from aircraft engine exhaust noise measurements is nevertheless found to leave some sources unaccounted for. These all appear – from the far field – to be located at the nozzle exit, but clearly originate in part from further upstream (e.g. core engine noise). The general description *tailpipe noise* covers such sources; they fall outside the scope of the present chapter.

In the course of this chapter we shall review jet mixing noise and shock cell noise respectively, and attempt to outline the extent to which current fundamental understanding will stand the test of prediction. We shall not consider purely empirical prediction methods, which may in some cases be more satisfactory in practice at the present time.

14.2 JET MIXING NOISE

The source of jet noise which has historically received the most attention, both theoretically and experimentally, is the jet mixing noise component. Theoretical work due to Lighthill [14.1, 14.2] showed how this noise could be generated in a freely exhausting jet flow as a result of the fluctuating Reynolds shear stress. These concepts have since dominated the study of jet noise and offer a strong foundation for both the study and prediction of jet mixing noise. A basic attraction of the Lighthill formulation is undoubtedly the relative simplicity with which an expression for the strength of the contributing 'noise sources' is obtained. It is equally true, however, that proper evaluation of that expression from available fluid mechanics (i.e. turbulence) data is by no means straightforward.

Certain very useful predictions do emerge fairly readily, notably that the noise output should vary as the eighth power of the jet efflux velocity. However, in 1971 Lush [14.3] published a series of carefully conducted jet noise measurements which highlighted certain significant and systematic discrepancies between measurement and the predictions available from the Lighthill formulation. These observations have since been amply confirmed by independent measurements, among which we would note in particular those of Tanna [14.4] which cover arguably the largest envelope of test conditions available in a single systematic study of jet mixing noise.

It is to be emphasized that the now established existence of such discrepancies does not represent errors in the basic Lighthill theory *per se*. It is the knowledge of the quantities required for evaluation of the source term which is inadequate. A portion of that source term represents as equivalent acoustic sources the processes of refraction and scattering of acoustic radiation by the jet flow. This, together with the nature of the discrepancies reported by Lush, led Lilley *et al.* [14.5] to undertake a reformulation of the governing equations in a manner which separated more explicitly the generation of acoustic energy and its subsequent transmission through the jet flow field. This work has led to a new area of jet noise study termed *flow-acoustic interaction*, as represented for example in the papers of Mani [14.6], Tester & Morfey [14.7] and Howe [14.8].

In the following sections we begin by outlining the predictions available from the original Lighthill formulation and indicate where discrepancies are experienced. We then attempt to trace, in outline, developments during the 1970s which have served to bring prediction and measurement into significantly closer agreement.

14.2.1 Predictions from the acoustic analogy

The essence of the Lighthill theory of aerodynamic noise is the formulation of an acoustic analogy, in which the complicated process of sound generation by turbulence is modelled in terms of an *acoustically equivalent* set of acoustic sources embedded in an otherwise uniform medium at rest. By a simple re-arrangement of the equations of fluid motion it was shown that this process was described by a wave equation for the fluid density ρ:

$$\frac{\partial^2 \rho}{\partial t^2} - c_0^2 \frac{\partial^2 \rho}{\partial x_i^2} = \frac{\partial^2 T_{ij}}{\partial x_i \, \partial x_j} \tag{14.1}$$

where $T_{ij} \equiv \rho v_i v_j + (p - c_0^2 \rho) \delta_{ij} + \sigma_{ij}.$ (14.1a)

The left side of this equation therefore describes the propagation of acoustic disturbances through the ambient medium at the speed of sound c_0. The right side is then interpreted as a forcing function or source term.

The solution of (14.1) for an observer at $\mathbf{x}$, at a large distance, r, from a typical element of the source at $\mathbf{y}$, (such that $p' \cong c_0^2 \, \rho'$), is

$$p(\mathbf{x}, t) - p_0 = \frac{1}{4\pi c_0^2 r} \int_V \frac{\partial^2 T_{rr}}{\partial t^2} (\mathbf{y}, t - r/c_0) \, \mathrm{d}V(\mathbf{y}) \, , \tag{14.2}$$

where $T_{rr} = \rho v_r^2 + (p - c_0^2 \rho)$ (14.2a)

and v_r is the fluid velocity in the direction of the observer. (We have omitted the viscous stress tensor in view of the high Reynolds numbers in jet flows of practical interest.) This result shows the relationship between the far-field pressure amplitude and the turbulent flow field, while the appearance of the retarded time, $t - r/c_0$, emphasizes the fact that, in general, phase differences must be accounted for during the integration process over the source region.

The latter may, however, be neglected for compact source regions; that is, whenever the acoustic wavelength is long compared with the extent of the source region over which coherent source fluctuations exist. A scaling law for the radiated intensity is then derived in [14.2] with the aid of the following assumptions:

(i) Pressure and density variations are related as in the ambient medium (i.e. $p' - c_0^2 \, \rho' = 0$). As discussed below, this is not appropriate for hot jets, or in general for any flow containing gradients of entropy or composition.

(ii) All velocities vary in direct proportion to the jet efflux velocity U_J so that

$$\rho v_r^2 \sim \rho_s U_J^2 \; ,$$

where ρ_s is a density appropriate to the most intense source region.

(iii) The second time derivative is assumed to be equivalent to a frequency squared weighting, and frequency is assumed to scale on a Strouhal number based on jet diameter and jet efflux velocity; thus

$$\frac{\partial^2}{\partial t^2} \sim \omega^2 \sim \left(\frac{U_J}{D}\right)^2 .$$

(iv) The noise-producing volume scales on the cube of the nozzle diameter, i.e. $\int dV(\mathbf{y}) \sim D^3$.

Combining these assumptions with equation (14.2) yields the following predicted dependence of far-field intensity, for similar jet flows:

$$I(r, \theta) \sim \frac{\rho_s^2 U_J^8 D^2}{\rho_0 c_0^5 r^2} . \tag{14.3}$$

The observer angle θ is the polar angle measured from the jet downstream axis.

Of particular significance is the prediction that the intensity varies in proportion to the eighth power of jet efflux velocity, implying a 24 dB noise increase for each doubling of velocity.

The compact source restriction appropriate to equation (14.2) is, however, valid only for jet velocities which are small compared to the speed of sound, a situation seldom experienced in aero-engine applications. Lighthill argued that this restriction could be eased if the estimates of T_{rr} were based on a set of sources convecting at speed appropriate to the convection velocity of the intense turbulence. With such a reformulation equation (14.3) above becomes [14.9]

$$I(r, \theta) \sim \frac{\rho_s^2 U_J^8 D^2}{\rho_0 c_0^5 r^2} (1 - M_c \cos\theta)^{-5} \tag{14.4}$$

where M_c is normally taken to be of order 0.7 U_J/c_0. Note that equation (14.4) reduces to equation (14.3) as the jet Mach number tends to zero.

For $\theta = 90°$, therefore, the dimensional variation remains unaltered, but at angles closer to the jet axis, i.e. $\theta < 90°$, augmentation in accordance with five powers of the Doppler factor $(1 - M_c \cos\theta)$ is predicted. This reflects the enhanced efficiency of the convecting quadrupole sources for radiation directions close to their direction of motion.

Equation (14.4) thus represents the anticipated scaling of overall intensity as a function of jet velocity, angle of observation, etc. It was, however, the extension by Lush [14.3] of this type of scaling argument to the variation of intensity in proportional frequency bands (1/3 octave for example) which

conclusively demonstrated the limitations of the Lighthill approach for prediction purposes (see the following section). Lush argued that the intensity
in proportional frequency bands, centre frequency f, should vary at a given angle
θ in accordance with

$$I(f, r, \theta) \sim \frac{\rho_s^2 U_J^8 D^2}{\rho_o c_o^5 r^2} (1 - M_c \cos\theta)^{-5} F \left\{ \frac{fD}{U_J} (1 - M_c \cos\theta) \right\} . \quad (14.5)$$

The variation is identical to that for overall intensity, except for the addi-
tion of the spectrum function $F \left\{ \frac{fD}{U_J} (1 - M_c \cos\theta) \right\}$. The appearance of the
Doppler factor emphasizes again that the equivalent sources are assumed to be in
motion, the frequency factor $(1 - M_c \cos\theta)$ ensuring that the same source
frequency is considered irrespective of the angle from which it is observed.

We close this section with the reminder that at jet velocities such that
$M_c \cos\theta$ can approach unity, the more complete Doppler factor

$$\{(1 - M_c \cos\theta)^2 + (\alpha_c^2 \sin^2\theta + \beta_c^2 \cos^2\theta) M_c^2\}^{\frac{1}{2}}$$

should be employed [14.9]. The parameters α_c and β_c are a measure of the
lateral and axial non-compactness of the sources respectively.

14.2.2 Comparison with experiment

In this section we shall explore the measure of agreement between the predictions of equation (14.4) and (14.5) and jet noise measurements, utilizing the
results of [14.4] in view of the large range of parameters avalable therein. We
shall also restrict ourselves initially to isothermal jet flows; that is, flows where
the jet static temperature at the nozzle exit is equal to the ambient temperature.

The comparison between equation (14.4) and experimental observation is
shown in Fig. 14.1, for four angles of observation. At 90° to the jet axis, very
acceptable agreement is observed. However, at angles less than 90° the predicted
convective amplification clearly overestimates the measured levels, while in the
forward arc, $\theta > 90°$, the converse applies; that is, the anticipated convective
attenuation is not observed. The influence of these discrepancies on the directivity of the overall sound is shown in Fig. 14.2, where it is clear that the dimensional reasoning, equation (14.4), significantly overestimates the degree of
directionality observed in practice.

The origin of these differences becomes clearer when data and prediction
are compared on a spectral basis as suggested in equation (14.5). Figure 14.3
a-d shows comparisons of the predicted and measured directivities for four
values of the reduced frequency

$$\frac{fD}{U_J} (1 - M_c \cos\theta) = \frac{f_s D}{U_J} ,$$

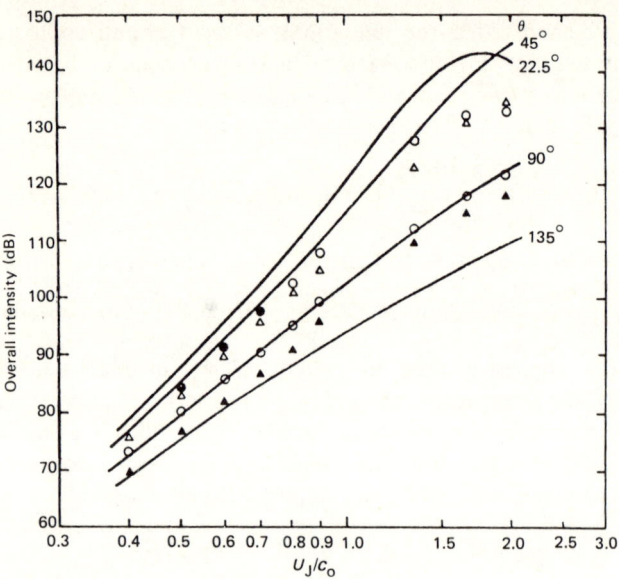

Fig. 14.1 – Velocity dependence of overall intensity: $T_J/T_O = 1$. ——— Freely convecting quadrupole theory. θ: ▲ 135°; O 90°; △ 45°; ● 22.5°.

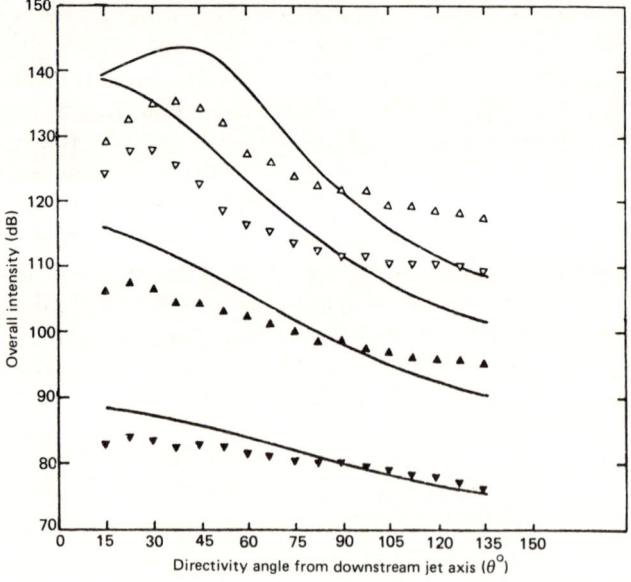

Fig. 14.2 – Directivity of overall intensity: $T_J/T_O = 1$. ——— Freely convecting quadrupole theory. U_J/c_O; ▼ 0.5; ▲ 0.9; ▽ 1.33; △ 1.95.

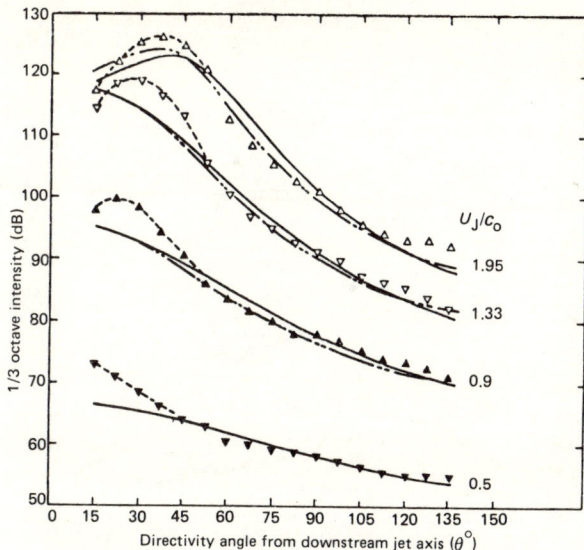

Fig. 14.3a – Directivity of $\frac{1}{3}$ octave intensity at $f_s D/U_J = 0.1$: $T_J/T_0 = 1$. ——— freely convecting quadrupole theory; ———·——— same theory with modified θ (source at $x/D = 10$); ‑‑‑‑‑‑‑‑ experiment, $r_m/D = 120$ (r_m = distance of microphone from nozzle exit).

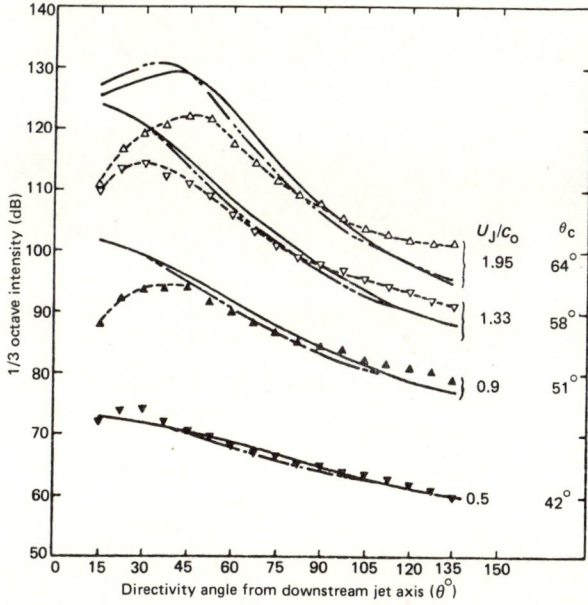

Fig. 14.3b – Directivity of $\frac{1}{3}$ octave intensity at $f_s D/U_J = 0.3$: $T_J/T_0 = 1$. ——— freely convecting quadrupole theory; ‑‑‑‑ same theory with modified θ (source at $x/D = 10$); ‑‑‑‑‑‑‑‑ experiment.

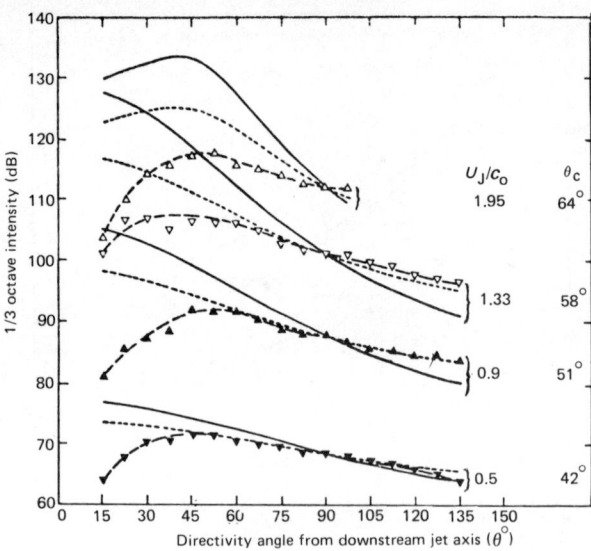

Fig. 14.3c – Directivity of $\frac{1}{3}$ octave intensity at $f_s D/U_J = 1.0$: $T_J/T_0 = 1$.
——— Freely convecting quadrupole theory (5 powers of Doppler factor); - - - - - -
3 powers of Doppler factor; – – – experiment.

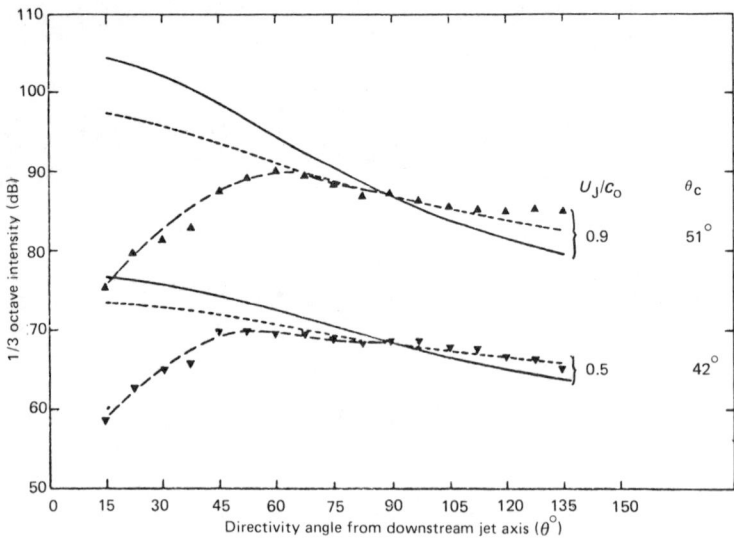

Fig. 14.3d – Directivity of $\frac{1}{3}$ octave intensity at $f_s D/U_J = 3.0$: $T_J/T_0 = 1$.
——— Freely convecting quadrupole theory (5 powers of Doppler factor); - - - - - - -
3 powers of Doppler factor: – – – – experiment.

equal to 0.1, 0.3, 1.0, and 3.0 respectively. We note that constant values of $f_s D/U_J$ are required to keep the spectrum function $F(f_s D/U_J)$ constant as dictated by equation (14.5), with the result that the observed frequency increases with decreasing angle of observation. Inspection of Fig. 14.3 indicates the following:

(a) At the lowest Strouhal number, reasonable agreement is obtained except at angles close to the jet axis, where measurement significantly exceeds prediction. Also demonstrated in Fig. 14.3a is the change of prediction involved by assuming that the source is ten diameters downstream of the nozzle exit plane, and allowing for the increase of observation angle relative to the source thus incurred. For the present measurement arrangement, $r/D = 72$, the correction is relatively small, but does offer some improvement in the comparison. The influence of such a correction is particularly significant where experimental limitations impose the use of small r/D values.

(b) The remaining comparisons, Figs. 14.3b, c and d, all exhibit the same general feature that the predicted degree of directivity exceeds that observed experimentally, the magnitude of the discrepancy becoming progressively larger as the reduced frequency is increased. We can summarize these observations as follows.

For the majority of frequencies the theory of freely convecting quadrupole sources overestimates the observed directivity, the magnitude of the discrepancies increasing as

(i) The frequency is increased;

(ii) The jet efflux velocity is increased;

(iii) The angle of observation is decreased.

We shall return to offer some degree of explanation for these observations in section 14.2.4 below.

14.2.3 Effect of temperature on jet mixing noise

In the review of jet mixing noise above we have avoided, as, strictly, does the Lighthill theory, the question of the effect of jet temperature on the radiated noise field. Early considerations appear to have concentrated on the concept that the principal effect of the elevated temperatures used in practice would be to reduce noise, as a result of the reduced density in the source region being reflected in the Lighthill stress tensor $\rho v_i v_j$.

However, carefully controlled experiments at the NGTE in England and by SNECMA in France showed that such a picture was too simple. These results [14.10], subsequently confirmed by other studies [14.11], [14.12], show that at low jet efflux velocities the use of high jet temperatures increases the noise radiation; while at higher velocities the converse is true, as shown in Fig. 14.4. Hoch *et al.* [14.10] chose to characterize this variation empirically by the parameter $(\rho_J/\rho_o)^\omega$ where ω varies from -0.75 at a velocity of 500 ft/sec to a value approaching $+2$ at 1500 ft/sec.

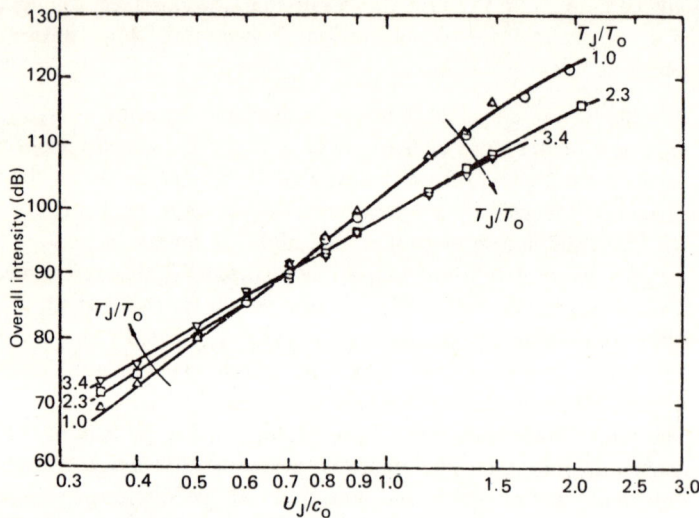

Fig. 14.4 – Effect of T_J/T_0 on velocity dependence of overall intensity at $\theta = 90°$, T_J/T_0 (nominal): $\triangle$ cold; O 1.0; $\Box$ 2.3; ∇ 3.4.

It appears to have been Lush [14.11] who first attempted to provide a rational explanation for these observations. He identified the fact that in a source region in which the speed of sound differed appreciably from that in the ambient sound field, it was no longer permissible to ignore the second (i.e. $p - c_o^2 \rho$) term of the Lighthill source term in equation (14.2a).

A detailed investigation of this additional source term was undertaken by Morfey [14.13]. The starting point of that analysis is the Lighthill wave equation, equation (14.1), written in terms of pressure instead of density as above, i.e.

$$\frac{1}{c_o^2} \frac{\partial^2 p}{\partial t^2} - \nabla^2 p = \frac{\partial^2 \rho v_i v_j}{\partial x_i \, \partial x_j} - \frac{\partial^2}{\partial t^2} (\rho - c_o^{-2} p) \, , \tag{14.6}$$

and the use of the energy equation in the form

$$\frac{D\rho}{Dt} = \frac{1}{c^2} \frac{Dp}{Dt} \tag{14.7}$$

to relate density and pressure. That is, instead of assuming that density and pressure fluctuations at a fixed point are approximately related by a factor $1/c^2$ (where c is the local speed of sound), the more general relationship (equation (14.7)) is applied which remains valid for flows containing either hot spots or pockets of differing composition, as long as molecular diffusion effects (i.e. heat conduction and viscosity) can be ignored.

The resulting analysis revealed two source terms relevant to heated jets *in the limit of low Mach number,* namely

(i) a quadrupole source

$$\frac{\partial^2(\rho_o v_i v_j)}{\partial x_i \, \partial x_j}$$

which we note depends on the ambient density, *not* that in the source region;

(ii) a dipole source

$$-\frac{\partial}{\partial x_i}\left(\frac{\rho - \rho_o}{\rho} \frac{\partial p}{\partial x_i}\right).$$

At low Mach numbers, where the acoustic wavelength is long compared to the flow dimensions, it is reasonable to assume that these two sources radiate directly into the ambient fluid. Their contributions to the far-field intensity would then scale according to

$$I_q \sim \frac{\rho_o U_j^8 D^2}{c_o^5 r^2} \tag{14.8}$$

and

$$I_d \sim \frac{\rho_s^2 U_j^6 D^2}{\rho_o c_o^3 r^2}\left(\frac{\Delta T}{T_o}\right)^2 \quad \text{(hot air jets)} \tag{14.9}$$

respectively, where, in obtaining the latter, it has been assumed that pressure fluctuations in the source region scale as $\rho_s U_j^2$.

We see therefore that the latter term constitutes an additional source whenever a temperature difference (ΔT) exists between the jet and ambient fluid, and furthermore that the weaker velocity dependence is liable to make this term progressively more dominant as the jet velocity is reduced.

However, while this could account for the observed increase of noise with temperature at low jet velocities, it clearly cannot account for the converse behaviour at the higher velocities. The quadrupole term, which might then be expected to be dominant, is predicted to be independent of temperature.

A probable explanation [14.13] lies in the fact that at higher Mach numbers the acoustic wavelength becomes comparable to the shear layer thickness. Thus instead of assuming that the sources radiate directly into the ambient fluid, as above, it might be equally reasonable to consider them radiating into a fluid whose properties are close to those of the source region. The effect of such a hypothesis on the scaling laws above is to replace ambient values of density and speed of sound (ρ_o, c_o), wherever they occur, by the source region values (ρ_s, c_s). Thus the temperature dependence for the quadrupole term, when the intensity is measured in the flow, is $(T_s/T_o)^{-7/2}$. This factor reduces to $(T_s/T_o)^{-3}$ for positions outside the flow when refraction across the flow boundary is taken into account. In a similar manner the dipole term is predicted to vary as $(T_s/T_O)^{-2}$. By this modification of the Lighthill model the observed reduction of noise with increased temperature at high Mach number is predicted, and, as we shall see below, found to be in quantitative agreement with measurements.

These latter observations do, however, raise the question, at Mach numbers of relevance to jet noise, of the shielding of the 'sources' by the surrounding jet flow field. This is considered in the next section.

14.2.4 Acoustic mean flow interaction

As reviewed in section 14.2.2 above, significant discrepancies are observed between the measured directional properties of jet noise and those predicted by the Lighthill acoustic analogy of freely convecting quadrupole sources. The nature of these discrepancies led Lilley [14.5] to formulate an alternative acoustic analogy in which the sources are embedded in an otherwise steady sheared flow with, in principle, arbitrary mean velocity and temperature profiles. The principal disadvantage of this approach is that the general Lilley equation requires numerical solution, although both low- and high-frequency asymptotic solutions are available as reported by Tester [14.7].

In this chapter we shall concentrate on the asymptotic, geometric acoustics, approach which has been used with some success in developing a physically based prediction scheme for jet noise [14.14], [14.15].

We begin with the very simple two-dimensional situation, shown in Fig. 14.5, in which a disturbance $P_s(x, y, t)$ is approaching an interface which divides fluid of velocity U_s and sound speed c_s from an ambient medium with sound speed c_0. The disturbance approaches the boundary in the form of waves whose normal makes an angle θ_s with the interface, and is refracted to emerge at angle θ_0 into ambient fluid.

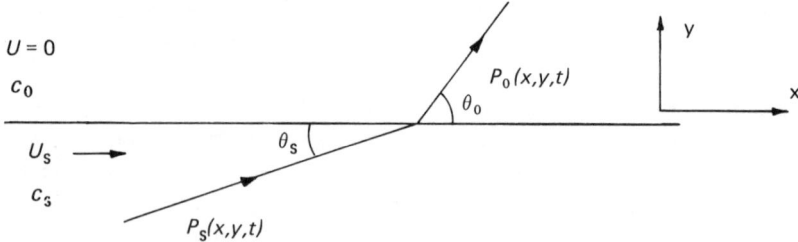

Fig. 14.5 – Transmission of wavelike pressure disturbances across a plane fluid interface (two-dimensional model).

We may write descriptions of these disturbances in the form

$$P_s(x, y, t) = A_s \exp j(\omega t - k_s x \cos\theta_s - k_s y \sin\theta_s) \qquad (14.10)$$

$$P_0(x, y, t) = A_0 \exp j(\omega t - k_0 x \cos\theta_0 - k_0 y \sin\theta_0) \qquad (14.11)$$

in the flow and ambient regions respectively, although it must be clearly understood that the descriptions relate to a space-fixed observer in both cases.

Matching the x-component wave-numbers across the boundary,

i.e. $k_s\cos\theta_s = k_o\cos\theta_o$ (14.12)

and using Snell's law of refraction, i.e.

$$\frac{\cos\theta_s}{\cos\theta_o} = \frac{c_s + U_s\cos\theta_s}{c_o}$$ (14.13)

(which amounts to matching phase speeds across the interface), is sufficient to yield an expression for $k_s\sin\theta_s$, which as the defining equations above demonstrate, describes propagation in the y direction within the flow region. We find

$$K_y \equiv k_s\sin\theta_s = \frac{\omega}{c_o}\left\{(1-M\cos\theta_o)^2\left(\frac{c_s}{c_o}\right)^{-2} - \cos^2\theta_o\right\}^{\frac{1}{2}},\quad (14.14)$$

where $M = U_s/c_o$.

Of particular relevance to our present considerations is the fact that K_y may be either real or imaginary, depending on the sign of the term in the square brackets. It is in fact imaginary for all values of the emergence angle θ_o less than some critical value θ_c, where[†]

$$\cos\theta_c = \frac{1}{c_s/c_o + M}\ .$$

The physical significance of θ_c is that it corresponds to the angle of emergence of a sound wave travelling in the flow region in a direction parallel to the x axis, as can be seen by putting $\theta_s = 0$ in the Snell's law relation above. Thus all genuine sound waves in the flow field (that is waves whose phase speed relative to the moving fluid is c_s) must emerge at angles greater than θ_c. For this reason the angular range $0 < \theta_o < \theta_c$ is commonly referred to as the *cone of silence*, since sound waves in the flow cannot penetrate into that angular range. Consideration of the value of θ_c, even for the relatively modest flow conditions $c_s/c_o = M = 1$, indicates that it will be of order $60°$. We thus obtain the apparently contradictory result that in many practical cases the angle of peak noise radiation from jet flows, circa $45°$, is contained within the cone of silence. It is necessary, therefore, to consider the nature of disturbances within the flow region which can emerge into the cone of silence.

For cases in which K_y is imaginary (i.e. $\theta_o < \theta_c$), the defining equation for disturbances in the flow region, equation (14.10), becomes

$$P_s(x,y,t) = A_s\exp[j(\omega t - k_o x\cos\theta_o)]\ \exp[-|K_y|y]\ ,\quad (14.15)$$

[†]We note also a maximum value of θ_o for which K_y is real, but shall not consider this further here.

with $\quad |K_y| \equiv \dfrac{\omega}{c_o} \left\{ \cos^2\theta_o - (1 - M\cos\theta_o)^2 \left(\dfrac{c_s}{c_o}\right)^{-2} \right\}^{\frac{1}{2}} .$ $\qquad$ (14.16)

It appears, therefore, that a disturbance originating a distance $\bar{y}$ below the interface, which subsequently radiates into the cone of silence, will suffer an exponential decay, $\exp[-|K_y|\bar{y}]$, prior to its arrival at the interface. The analogy between this process and the behaviour of cut-off duct modes is of course obvious.

Some idea of the physical processes involved can be obtained by considering the wave-number/frequency decomposition of a source located a distance $\bar{y}$ below the interface. It is well known that for such a source located in an infinite uniform medium only those portions of the wave-number/frequency spectrum, $S(\mathbf{k}, \omega)$ having axial phase speeds in excess of the effective sound speed can contribute to the far-field radiation. Further, the radiation angle to which a particular value of $S(\mathbf{k}, \omega)$ contributes is that for which its phase speed, resolved in that direction, is equal to the appropriate speed of sound. In similar vein we can identify the wave-number components which radiate outside and inside the cone of silence respectively. Those for which the axial phase speed is greater than $(c_s + U_s)$ can generate an acoustic field in the moving stream which would radiate to the boundary even in the limit that the distance $\bar{y} \to \infty$. However, on arrival at the boundary the minimum phase speed of this set is $(c_s + U_s)$ so that in radiating into the ambient fluid, the minimum angle of emergence is given by

$$(c_s + U_s) \cos\theta_{min} = c_o,$$

which again defines the cone of silence angle. We see, therefore, that those wave-number components, whose axial phase speeds are supersonic relative to the moving fluid, all yield radiation outside the cone of silence.

However, let us now consider a further subset of wave-number/frequency components whose axial phase speeds, U_x, are in the range

$$c_o < U_x < (U_s + c_s) .$$

That is, they are subsonic relative to the moving fluid, but supersonic relative to the ambient fluid. They cannot therefore produce an acoustic field in the normal sense within the moving fluid. In fact a destructive interference occurs leading to the exponential decay specified above. However, if $|K_y|\bar{y}$ is not too large some residual of these disturbances arrives at the interface. Their axial phase speed is now supersonic relative to the ambient medium so that, in essence, the boundary can be regarded as a new source which can radiate an acoustic field to fill the angular range between the flow axis and the cone of silence angle. The amplitude of the acoustic field in this region will, of course, now depend, among other things, on the severity of the exponential decay and hence on the distance of the source region from the flow boundary, $\bar{y}$. We see, therefore, that the behaviour of source fluctuations which radiate inside and outside the cone of silence respectively is likely to be rather different, and it is useful to use the cone of silence angle as a boundary for future discussion.

(a) *Radiation outside the cone of silence* $(\theta_o > \theta_c)$
The early comparisons due to Lush [14.3], with their relatively limited range of variables, suggested that outside the cone of silence the predictions of directivity based on the Lighthill acoustic analogy were reasonably acceptable. However, the extended range due to Tanna [14.4] does, as we have seen in Figs. 14.3b, c and d, suggest that even at relatively large angles to the jet axis the measured directivity is less than that predicted by five powers of Doppler factor.

A detailed remodelling of that angular range is reported by Morfey, Szewczyk & Tester [14.14], based on a geometric acoustics approximation of the Lilley equation. The essentials of the model are a combination of quadrupole and dipole sources whose radiated intensity would be isotropic[†] in the absence of convection or flow. These are then considered to convect with and radiate into a moving fluid. Finally, acoustic energy conservation principles are employed to allow for the transmission of the resulting acoustic energy out of this flow field into the ambient medium.

The resulting scaling laws as summarized in [14.14] are

$$I_q \propto \left(\frac{\rho_s}{\rho_o}\right)\left(\frac{c_s}{c_o}\right)^{-4}\left(\frac{U_J}{c_o}\right)^8 \frac{D_s^6}{D_m^9} \tag{14.17}$$

$$I_d \propto \left(\frac{\rho_s}{\rho_o}\right)\left(\frac{c_s}{c_o}\right)^{-2}\left(\frac{T_s - T_o}{T_s}\right)^2\left(\frac{U_J}{c_o}\right)^6 \frac{D_s^4}{D_m^7} \tag{14.18}$$

for the quadrupole and dipole source respectively. The directivity is contained in the ratio of Doppler factors where

$$D_m^2 = \{1 - D_s U_c \cos\theta_s/c_s\}^2 + (U_J D_s/c_s)^2 \{\alpha^2 \sin^2\theta_s + \beta^2 \cos^2\theta_s\}$$

and $D_s = (1 + U_s \cos\theta_s/c_s)^{-1}$.

The modified Doppler factor D_m represents convective amplification due to source motion, for radiation within the flow at an angle θ_s to the flow direction. The acoustic influence of the mean flow itself is contained in the Doppler factors D_s.

Using the relationships given above in equations (14.12) and (14.13), these Doppler factors can be written explicitly in terms of the observer angle θ_o *outside* the flow and become

---

[†] In its final form, the model allows for non-isotropic sources by including axially-oriented quadrupole and dipole contributions.

$$D_m^2 = \{1 - U_c\cos\theta_0/c_0\}^2 + \left(\frac{U_J}{c_0}\right)^2 \{\alpha^2(D_s^2 \frac{c_0^2}{c_s^2} - \cos^2\theta_0) + \beta^2\cos^2\theta_0\}$$

$$D_s = (1 - U_s\cos\theta_0/c_0)$$

respectively.

Finally we note that for conditions where $U_c\cos\theta_0/c_0$ does not approach unity the modified Doppler factor approximates to $(1 - U_c\cos\theta_0/c_0)$. If further the source-region flow velocity U_s equals the convection velocity U_c, D_s reduces to the same value.

The far-field directivity factors associated with both the quadrupole and dipole sources then become identical and are given by $(1 - U_c\cos\theta_0/c_0)^{-3}$. The improved prediction offered by this directivity factor, as opposed to the five powers associated with the freely convecting quadrupole model, is demonstrated in Fig. 14.3c, d. It is also clear, however, that this agreement breaks down as one enters the cone of silence ($\theta_0 < \theta_c$).

(b) *Radiation inside the cone of silence*

We have already noted above that a principal difference between radiation inside and outside the cone of silence is that the former suffers exponential decay during its passage through the flow field. The result of this 'tunnelling' process, for a source a distance $\bar{y}$ below the fluid interface, is to reduce the level of radiation by Δ dB where

$$\Delta\,(\text{dB}) = 20\log_{10}\{\exp(-|K_y|y)\}$$

which using the expression for $|K_y|$, equation (14.16), becomes

$$\Delta\,(\text{dB}) = 55\frac{f\bar{y}}{c_0}\left\{\cos^2\theta_0 - (1 - M\cos\theta_0)^2\left(\frac{c_s}{c_0}\right)^{-2}\right\}^{\frac{1}{2}}.$$

A first, very crude, feasibility study of this process as an explanation for the reduced high-frequency levels within the cone of silence was carried out by Fisher & Szewczyk [14.16]. They equated the observed discrepancy between prediction and experiment to the quantity Δ (dB) above and then determined the required 'effective depth' $\bar{y}$. This work demonstrated that the effective depths so determined were of a physically realistic order of magnitude, commensurate with the jet shear layer dimensions.

Subsequently a more sophisticated model of this process was developed by Morfey & Szewczyk [14.15]. Using the combination of quadrupole and dipole sources which had yielded a satisfactory fit to measurements outside the cone of silence, they calculated the radiation expected from these at smaller angles to the jet axis in the absence of any flow shrouding. The discrepancy between these levels and those actually measured within the cone of silence was again attributed to the exponential decay process. However, due account was taken of the fact that the rapidity of the decay, as reflected in $|K_y|$, would diminish as one proceeded from the source region towards the lower velocity and temperature regions in the outer portion of the shear layer. Incorporating an assumed velocity and temperature profile, it was then possible to determine the shear layer thickness required to account for the difference of levels defined above.

The success of this modelling for subsonic jet efflux velocities is demonstrated in Fig. 14.6. Here the shear layer thickness inferred from the acoustic data in the manner described above has been compared with that measured at locations which the source location techniques [14.17] show to be the major noise-producing region for each Strouhal number. The agreement is remarkably close and does indicate both the relevance and importance of this flow-shrouding mechanism in determining the directional properties of jet noise over a majority of the frequencies of practical interest.

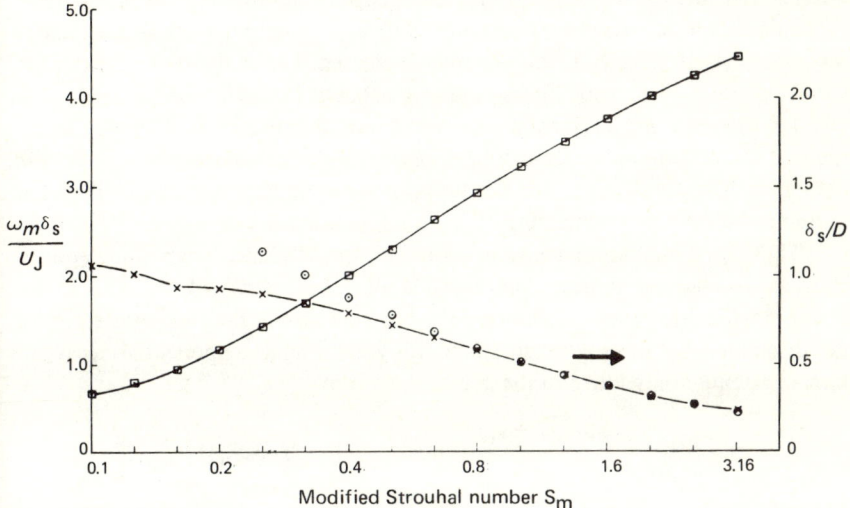

Fig. 14.6 – Optimum values of the shear layer parameters $\omega_m \delta_s / U_J$ and δ_s / D at fixed values of S_m; $T_J / T_0 = 1.0$.
Code: $\square$ $\omega_m \delta_s / U_J$; $\times$ δ_s / D; O δ_s / D from flow and source location measurements.

However, an attempt to extend this same modelling procedure to supersonic jet velocities was not entirely successful. Using the shear layer parameters derived from the subsonic data leads to a significant under-prediction of the measured levels. Several possibilities for such discrepancies are given by Morfey & Szewczyk [14.15], but the need to resolve this difficulty remains. In the meantime, ongoing work at ISVR is extending these concepts into the areas of co-axial jets and jets in flight.

(c) *Flow-acoustic interaction and jet noise suppressors*

In this final section on the topic of flow/acoustic interaction we shall explore the possible role of such phenomena in the operation of jet noise suppression devices. A number of jet noise suppressors appear to operate either by causing the jet flow to spread more rapidly than that for the datum conical nozzle (e.g. fishtailing jets [14.18]), or by surrounding certain noise-producing regions with flow (e.g. multi-tube or chuted nozzles).

Fisher & Szewczyk [14.16] suggest that a rational explanation for the observed characteristics of these suppressors is possible if one assumes that the principal role of the mean flow modification is to increase the depth of flow separating the noise-producing region from the ambient atmosphere (i.e an increase of the effective depth $\bar{y}$). This differs from more conventional explanations, based on the Lighthill analogy, in which attenuations are attributed to factors such as reduced shear, reduced source volumes, etc. However, the latter do not adequately explain two commonly observed features of suppressor nozzles: first the lack of benefit obtained at large angles to the jet axis, and more particularly the increased attenuation, relative to a datum conical nozzle, observed as the jet velocity is progressively increased. Let us therefore explore the way in which flow-acoustic interaction may account for such observations.

To this end we assume that the only modification created by the introduction of the suppressor flow is to increase the effective depth from $\bar{y}_c$ for the datum circular nozzle to $\bar{y}_s$ for the suppressor nozzle, the character of the main noise-producing region remaining otherwise unmodified.

Thus, for observation angles outside the cone of silence, where the depth of flow separating the source from the boundary has no effect, no attenuation is anticipated. However on entering the cone of silence the increased effective depth created by the suppressor flow will yield a larger exponential decay, to give an attenuation relative to the conical nozzle of

$$A \text{ (dB)} = \frac{55f(\bar{y}_s - \bar{y}_c)}{c_0} \left\{ \cos^2\theta_0 - (1 - M\cos\theta_0)^2 \left(\frac{c_s}{c_0}\right)^{-2} \right\}^{\frac{1}{2}} .$$

For interpretive purposes it is useful to write this expression in the expanded form

$$A \text{ (dB)} = \frac{55fD}{U_J} \frac{(\bar{y}_s - \bar{y}_c)}{D} \frac{U_J}{c_0} \left[\cos^2\theta_0 - (1 - M\cos\theta_0)^2 \left(\frac{c_s}{c_0}\right)^{-2} \right]^{\frac{1}{2}}$$

and to assume on the basis of the data in Fig. 14.6 that both $\bar{y}_s/D$ and $\bar{y}_c/D$ are constant for a given value of Strouhal number fD/U_J.

The characteristics of the observed attenuation are then expected to be as follows:

(a) Attenuation will occur as one enters the cone of silence and will progressively increase with further reduction of observation angle, as a result of the increase of value of the [ ] term above.

(b) An increase in jet efflux velocity will increase the angular range over which attenuations are observed, as a result of the increased cone of silence angle.

(c) An increase of jet efflux velocity will increase the magnitude of the attenuations, as a result of the U_J/c_0 term in the expression above.

Some confirmation of these ideas is presented in Fig. 14.7, which shows a comparison of measured field shapes for a circular nozzle and rapidly spreading fishtailed jet. Outside the cone of silence angle the levels are quite similar, but once one enters the cone of silence the fishtailed jet shows increasing benefits, whose characteristics are qualitatively in agreement with the suggestions above.

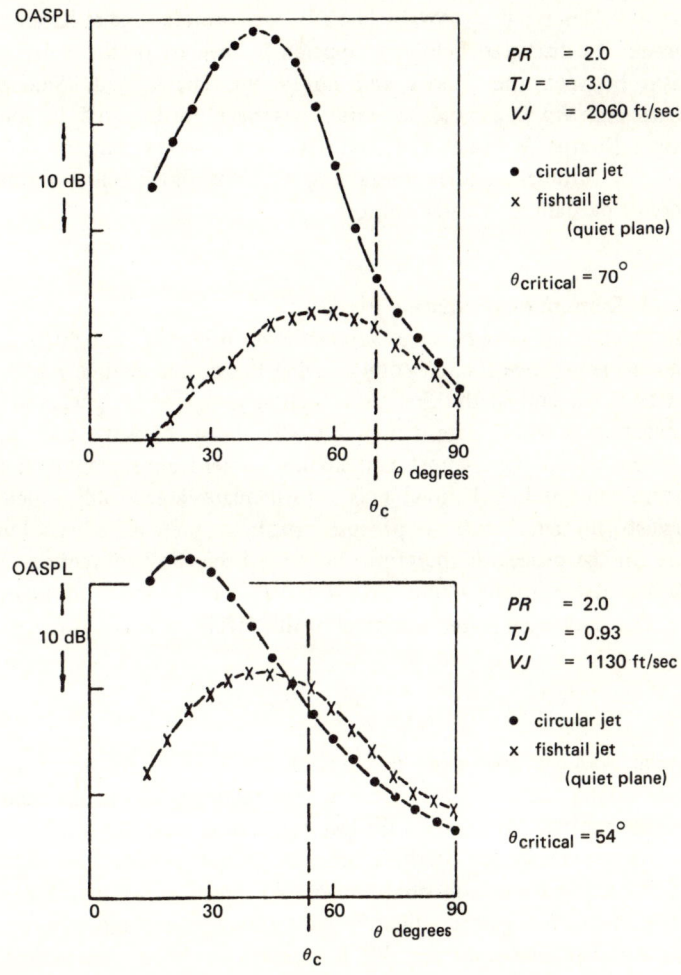

Fig. 14.7 – Comparative field shapes for circular and fishtail jets.

14.3 SHOCK-ASSOCIATED NOISE

As the pressure ratio of a convergent nozzle exceeds a certain critical value
(1.89 for air with $\gamma = 1.4$), a series of *shock cells*, sometimes also termed *shock
diamonds*, are observed to form in the jet exhaust flow. Further increase of
pressure serves to extend the length and spacing of the successive cells. The
spacing is given approximately by

$$L = 1.25\,\beta D \ .$$

Here $\beta \equiv \sqrt{(M_J^2 - 1)}$, and M_J is the fully expanded jet Mach number, a
function of pressure ratio alone.

The presence of these shock cells is known to give rise to two types of noise.
The first, described by Powell [14.19], is a discrete tone radiation often termed
screech. As discussed below, it appears to owe its origin to a feedback mechanism between the shocks and nozzle lip. The second component, termed
broadband shock associated noise, has been investigated in some detail by
Harper–Bourne & Fisher [14.20]. That work shows that the noise arises as a
result of turbulent eddies interacting with the shock cell structure to form an
array of partially correlated sources.

14.3.1 Discrete tone radiation (screech)

The mechanism giving rise to screech is, in principle, straightforward. We consider a disturbance (i.e. an eddy) leaving the nozzle at time $t = 0$. It therefore
arrives at the end of the first shock cell at time $t_1 = L/U_c$ where U_c is the eddy
convection velocity. Here it interacts with the shock wave, generating an acoustic wave which travels upstream in the ambient air surrounding the jet to re-
disturb the nozzle exit flow. This in turn creates a new eddy which travels away
downstream, and hence the process repeats to create a feedback loop. The cycle
time for the process is therefore the sum of the eddy convection time L/U_c and
the time taken for the sound wave to travel from the shock to the nozzle L/c_0.
The frequency is the reciprocal of this 'cycle' time and is therefore

$$f_s = \frac{U_c}{L(1 + M_c)} \ , \quad (M_c = U_c/c_0) \ .$$

Ample evidence does exist to confirm that discrete frequency radiation from
shock-containing flows does occur at this frequency and its harmonics.

By contrast, one can be far less categorical regarding the parameters which
control the amplitude of these tones. Common experience indicates that they
are often important contributors on cold model jets, but seldom occur significantly on aero-engine configurations, although some minor structural damage
to a VC10 tailplane, attributable to screech, has been reported by Hay & Rose
[14.21].

Experience on cold model jets at the ISVR indicates that for a normal nozzle configuration screech tone amplitudes are frequently non-stationary, varying by a factor of five while the jet is operated at ostensibly constant conditions. This phenomenon can be eliminated with the provision of a large reflector plate in the nozzle exit plane. The tone amplitudes are both stabilized and increased. Stable tones, but at a much lower level, are also obtained if the reflector plate is covered with a layer of acoustic foam. It was also found that the addition of a small projection on the nozzle lip, with this latter configuration, was effective in eliminating the tones at least to the extent where they were not detectable on 6% bandwidth spectral analysis.

From this experience it is concluded that the amplitude of screech tones is very dependent both on the presence of acoustically reflecting surfaces in the vicinity of the nozzle exit plane and on the state of the nozzle flow.

14.3.2 Broadband shock associated noise
The character and dependences of this second component of shock cell noise have been investigated in considerable detail by Harper–Bourne & Fisher [14.20], using the configuration outlined above to eliminate screech tones.

The variation of *the overall sound pressure level* for an unheated jet as a function of jet velocity is shown in Fig. 14.8a for three angles of observation. It is clear that once the nozzle chokes ($M_J = 1$) the OASPL's at 90° and 143° rise extremely rapidly and the field becomes virtually omni-directional. A more informative manner of plotting the data for $M_J > 1$ is shown in Fig. 14.8b. Also shown there is a linear extrapolation of the mixing noise data from Fig. 14.8a. It is apparent that once the shock noise dominates, the intensity varies as β^4 over the majority of the pressure ratio range. This in turn suggests that the

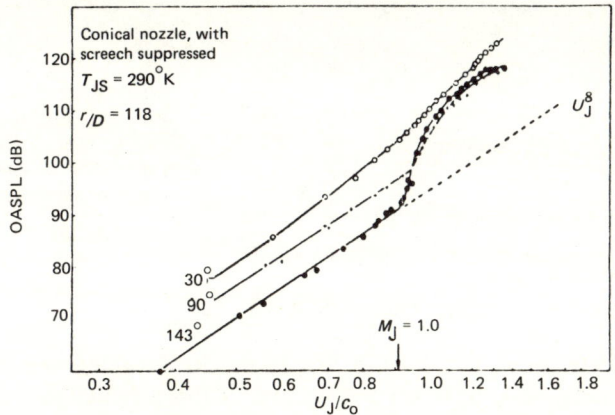

Velocity dependence of overall intensity of jet noise at several angles to the jet showing shock associated noise

Fig. 14.8(a) – Shock cell noise.

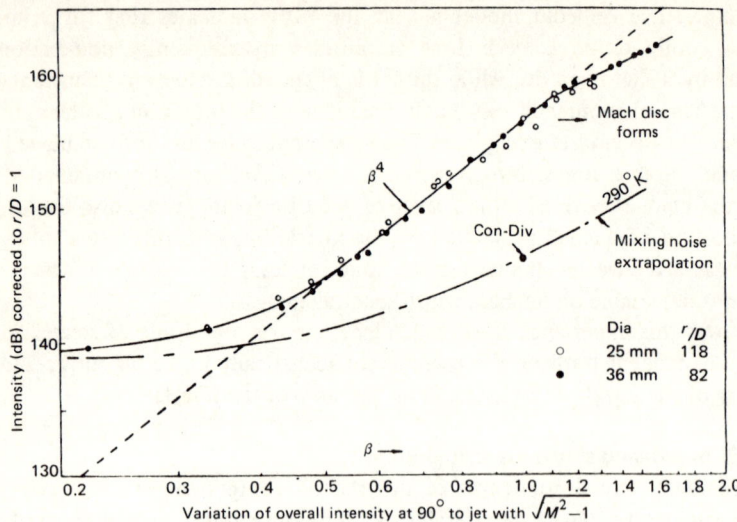

Fig. 14.8(b) – Shock cell noise.

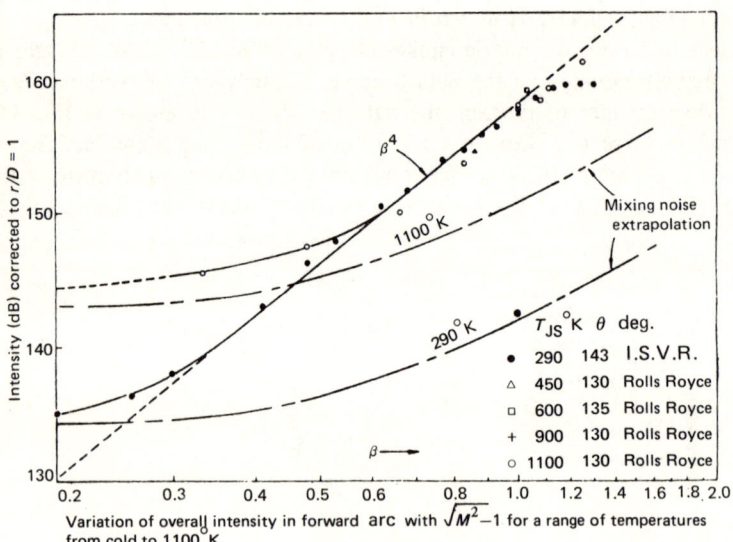

Fig. 14.8(c) – Shock cell noise.

source amplitude varies as β^2, which use of the Rankine-Hugoniot relations shows is in direct proportion to the pressure jump across the shock wave. Harper-Bourne & Fisher furthermore were able to confirm, using a crossed beam Schlieren system, that this was indeed the variation of fluctuation amplitude observed at the shock cell ends. Finally shown in Fig. 14.8c is the variation of OASPL as a function of pressure ratio (i.e. β) for a range of jet efflux temperatures, together with estimates of the mixing noise contribution at the two extremes of temperature considered. We note that the mixing noise contribution is larger at the higher temperature as a result of the higher associated jet efflux velocity at a fixed pressure ratio. However, again once the shock cell noise dominates, a β^4 dependence is observed at levels which are independent of jet temperature.

Thus with respect to the variation of overall intensity of broadband shock associated noise we conclude

(a) It is independent of observation angle.

(b) It is independent of jet temperature and hence jet efflux velocity, being solely a function of pressure ratio in accordance with the empirical relationship

$$OASPL\,(dB) \; = \; 158.5 \; + \; 10\log_{10}\left(\frac{D}{R}\right)^2\beta^4 \quad .$$

The spectral character of shock-associated noise is demonstrated in Fig. 14.9, where the spectrum of noise from a convergent under-expanded nozzle is compared with that from a convergent-divergent, shock-free nozzle, both of which were operated at the same pressure ratio and temperature. This shock-associated noise component is clearly reasonably broadband, but exhibits a distinct spectral peak. This peak arises owing to interference between radiation from the various shock cells as shown schematically in Fig. 14.10.

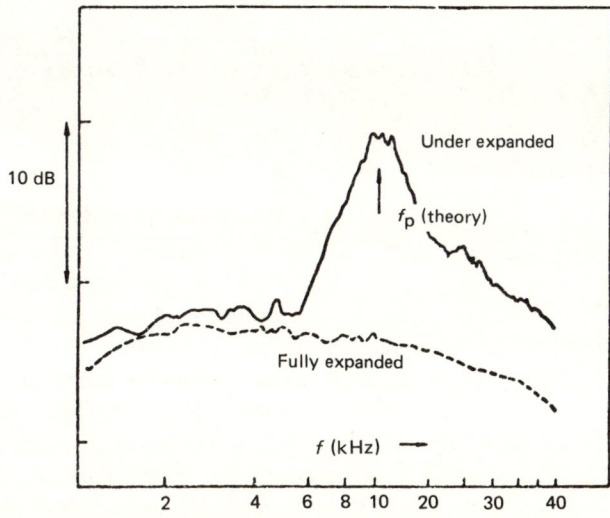

Fig. 14.9 – Comparison of supersonic jet noise spectra for a fully-expanded and under-expanded flow ($\theta = 90°$, $\beta = 1.0$).

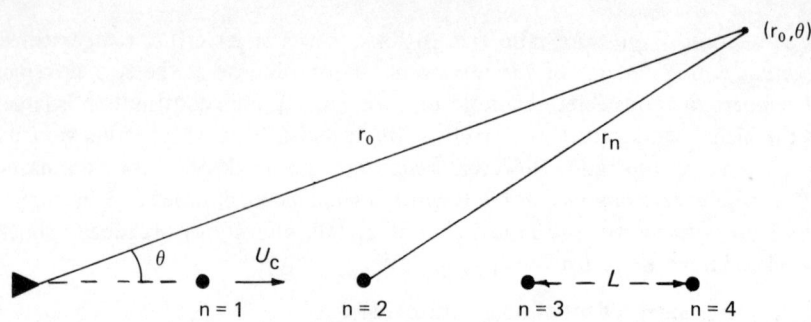

Fig. 14.10 – Definition sketch for shock cell noise model.

We suppose that an eddy convecting down the shock cell array causes each shock to emit a signal whose relative phasing is set by the eddy convection time. Hence, considering for simplicity a narrow band of frequencies, the source fluctuation at the nth source is of the form

$$A_n(\omega) \cos\omega \left(t - \frac{nL}{U_c} \right) .$$

We can next sum the contribution from all such sources, with due allowance for retarded time. The signal received by a far-field observer at distance r_0 and angle θ is therefore

$$p(r, \theta, t + r_0/c_0) = \frac{1}{r_0} \Sigma_n A_n(\omega)\cos\omega \left(t - \frac{nL}{U_c} + \frac{nL\cos\theta}{c_0} \right)$$

$$= \frac{1}{r_0} \Sigma_n A_n(\omega)\cos \omega(t - \frac{nL}{U_c}(1 - M_c\cos\theta))$$

where $M_c \equiv U_c/c_0$.

The mean square pressure at frequency ω is therefore

$$\overline{p^2(r, \theta, \omega)} = \frac{1}{r_0^2} \Sigma_n \Sigma_m A_n(\omega) A_m(\omega) \cos \left\{ \frac{(n - m)\, \omega L\, (1 - M_c\cos\theta)}{U_c} \right\} .$$

We note that the summation above is made up of two distinct types of term. First, those for which $n = m$ represents the sum of contributions from each shock cell acting alone. This is the mean square pressure which would be obtained if the sources were statistically independent or uncorrelated. However, since the sources are driven by the same eddy, some degree of correlation is anticipated and is represented by the remaining terms for which $n \neq m$. We note also that these remaining terms will yield a maximum contribution whenever the argument of the cosine is either zero or an integer multiple of 2π. The former condition corresponds to the Mach angle, $M_c\cos\theta = 1$, but we shall not consider this further as experience suggests that jet mixing noise is normally the major contributor at this angle. The second condition, however, implies that we should expect a peak shock cell noise contribution at frequencies given by

$$f_p = \frac{U_c}{L(1 - M_c\cos\theta)} \ .$$

That is for an observation angle of $\theta = 90°$ the peak frequency is U_c/L and then varies in the manner of an apparent Doppler shift at other angles. A comparison between this prediction and experiment has been included in Fig. 14.9. We also note finally that for $\theta = 180°$

$$f_p = \frac{U_c}{L(1 + M_c)} \ ,$$

which is identical to the screech frequency discussed previously. It appears therefore that at the screech frequency the shock associated noise from an array of shocks will all combine constructively to yield the strong forward radiation needed to maintain the feedback loop.

These basic concepts were refined, allowing for example for non-equal shock spacing and the loss of coherence of eddies as they travel down the jet shear layer, to form the semi-empirical prediction method for broadband shock associated noise presented in [14.20], which also forms the basis of the S.A.E. prediction method contained in ARP 876.

14.4 THE STATIC TO FLIGHT DILEMMA

A set of notes on jet noise would be incomplete without some brief mention of work on static to flight effects. Most of our knowledge on jet noise is clearly derived from static tests, but normally the practical requirement is to predict the noise field generated when the nozzle is in forward motion at speeds up to 250 knots ($M_A = 0.37$). Early estimates of the differences due to forward motion appear to have centred on the idea that the principal effect of flight would be to reduce the velocity difference between the jet efflux and the ambient fluid from U_J statically to ($U_J - U_A$) in flight (U_A = aircraft forward speed). Assuming that both turbulence intensity and the time scales of the turbulence scale on this velocity difference, then a static to flight noise reduction of

$$\Delta \text{ (dB)} = 10 \log_{10}\left(\frac{U_J}{U_J - U_A}\right)^8$$

might be expected for unheated jets.

However, a considerable volume of flight testing now shows such a prediction to be grossly over-optimistic. A comprehensive survey has been presented by Bushell [14.22], who chose to empiricize the static to flight difference in the form

$$\Delta \,(\text{dB}) \;=\; 10 \log_{10} \left[\left(\frac{U_{\text{J}}}{U_{\text{J}} - U_{\text{A}}} \right)^{m} (1 + M_{\text{A}} \cos\theta) \right] .$$

The variation of the empirical exponent m with angle of observation for a wide range of aircraft types is shown in Fig. 14.11. It is clear that m is a strong function of angle, reducing from about 5.5 at small angles to zero at $\theta = 90°$ and then going negative, implying that for the forward arc the noise actually *increases* in going from static to flight.

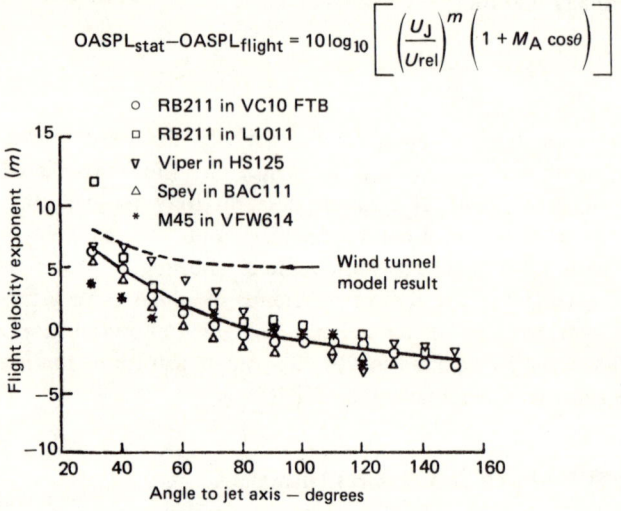

Fig. 14.11 − Comparison of flight velocity exponents.

These and other difficulties associated with accurate estimation of the flight performance of various types of jet noise suppressor nozzles have led to a search for methods of simulating forward motion, to avoid the expensive process of bringing these nozzles to a flight standard. Two principal methods of simulation are currently in vogue, namely the use of anechoically treated wind tunnels [14.23] and the use of large area ratio coaxial jet configurations. In both cases the principle is to submerge the jet under test in a large co-flowing stream.

The results from such simulations appear in general to be internally consistent and demonstrate a fairly uniform reduction of noise irrespective of angle of observation as indicated in Fig. 14.11. Clearly the results are quite significantly different. For example, the wind tunnel results would imply that for a jet velocity of 1000 ft/s and a flight speed of 200 ft/s (or 125 knots) a noise reduction at $\theta = 90°$ of order 5 dB might be expected, while actual flight experience would suggest no change.

The magnitude of this discrepancy has led, in the latter half of the 1970's, to numerous detailed studies of static to flight noise changes at both model and full scale. A particularly comprehensive series of tests is reported by Bashforth [14.24] in which a HS125 was flown over a noise-measuring station mounted on the top of the Severn Bridge. Static noise testing of the associated engine, a Viper 601, included simulation of the aircraft installation effects and employment of modern source location techniques [14.25] to determine residual core noise levels both with and without an acoustic lining in the engine tailpipe. The conclusion drawn was that to a large extent the type of anomaly revealed in Fig. 14.11 is accounted for by a combination of factors including reflections from the airframe, some residual core noise, and airframe noise, all of which contribute to the higher than anticipated flight noise levels. Thus, while provisionally it does appear that forward flight simulation facilities yield the correct static to flight changes of pure jet mixing noise, precise determination of the noise generated in flight by the full aircraft/engine combination is still not an entirely straightforward process.

14.5 SUMMARY

The purpose of this chapter has been principally to summarize recent developments in the understanding of jet noise and to examine the extent to which these developments will stand the test of prediction. We would note in particular:

(a) The identification of an additional source term in hot jets, associated with the previously ignored $(p - c_0^2\rho)$ term in the Lighthill stress tensor, equation (14.1a).

(b) An awareness of the practical importance of flow-shrouding effects in determining both the temperature dependence and directivity of jet noise. These include:

    (i) The existence and importance of the cone of silence within which flow-shrouding plays a frequently dominant role in determining the far-field levels;

    (ii) The existence of a $(1 - M\cos\theta)^{-3}$ directional factor outside the cone of silence as opposed to the $(1 - M\cos\theta)^{-5}$ dependence predicted by the original freely convecting quadrupole model of Refs. [14.1, 14.2].

Items (a) and (b) above have led to a physically based prediction scheme for jet mixing noise, described in [14.14] and [14.15], which shows typically ± 1 dB agreement with 1/3 octave data over an extensive range of both jet temperature and velocity. Examples of the agreement obtained with subsonic jet data are shown in Figs. 14.12 and 14.13, which refer respectively to $\theta = 90°$ and to radiation inside the cone of silence. Several further comparisons are shown in [14.15], [14.26]. It does remain, however, to extend this work into the cone of silence for supersonic jet velocities as discussed in section 14.2.4(b).

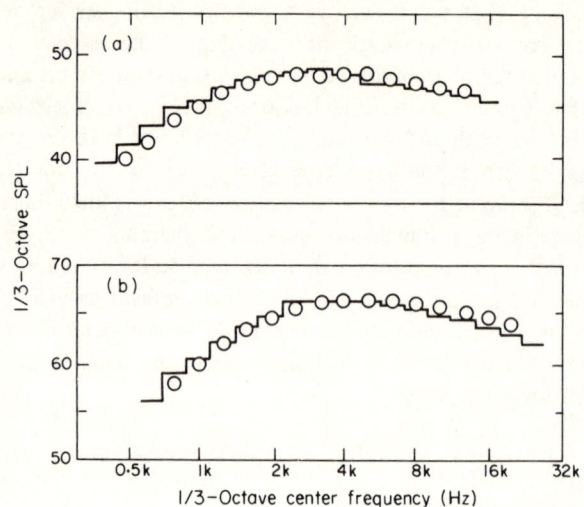

Fig. 14.12 – Direct comparison of spectra for low-velocity heated air jets at $\theta_m = 90°$ $(r_m/D = 120)$. (a) $T_J/T_0 = 1.18$, $U_J/c_0 = 0.308$; (b) $T_J/T_0 = 1.67$, $U_J/c_0 = 0.515$. Code: —— measurements; O prediction by using scheme [14.14].

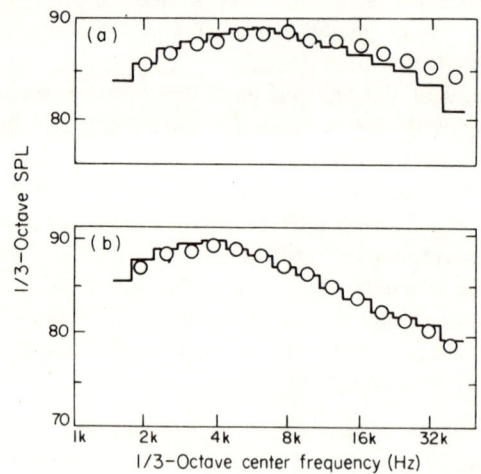

Fig. 14.13 – Measured and predicted spectra for subsonic jets near and within cone of silence boundary: $\theta_m = 45°$ $(r_m/D = 120)$. (a) $T_J/T_0 = 0.86$, $U_J/c_0 = 0.839$; (b) $T_J/T_0 = 1.57$, $U_J/c_0 = 0.848$. Code: —— measurements; O prediction by using extension of GA scheme [14.14].

(c) A prediction scheme for broadband shock-associated noise has been developed [14.20] which is satisfactory for most practical purposes.

(d) The static to flight behaviour of jet mixing noise is correctly simulated in wind tunnel/open jet facilities. The observed anomalous behaviour of aircraft in flight is probably associated with additional noise sources and installation effects. An improved physical understanding of the static to flight behaviour of the mixing noise component is, however, still required.

REFERENCES

[14.1]   Lighthill, M. J. (1952) *Proc. Roy. Soc. A.,* **211,** 564-587. On sound generated aerodynamically. I: General theory.

[14.2]   Lighthill, M. J. (1954) *Proc. Roy. Soc. A,* **222,** 1-32. On sound generated aerodynamically. II: Turbulence as a source of sound.

[14.3]   Lush, P. A. (1971) *J. Fluid Mech.,* **46,** 477-500. Measurements of subsonic jet noise and comparison with theory.

[14.4]   Tanna, H. K. (1976) *J. Sound & Vib.,* **50,** 405-428. An experimental study of jet noise. Part I: Turbulent mixing noise.

[14.5]   Lilley, G. M., Morris, P. J. & Tester, B. J. (1973) *AIAA Paper* No. 73-987. On the theory of jet noise and its applications.

[14.6]   Mani, R. (1972) *J Sound & Vib.,* **25,** 337-347. A moving source problem relevant to jet noise.

[14.7]   Tester, B. J. & Morfey, C. L. (1976) *J. Sound & Vib.,* **46,** 79-103. Developments in jet noise modelling — theoretical predictions and comparisons with measured data.

[14.8]   Howe, M. S. (1975) *J. Sound & Vib.,* **43,** 77-86. Application of energy conservation to the solution of radiation problems involving uniformly convected source distributions.

[14.9]   Ffowcs Williams, J. E. (1963) *Phil. Trans. Roy. Soc. A,* **255,** 469-503. The noise from turbulence convected at high speeds.

[14.10] Hoch, R. G. *et al.* (1973) *J. Sound & Vib.,* **28,** 649-668. Studies of the influence of density on jet noise.

[14.11] Lush, P. A., Fisher, M. J. & Ahuja, K. (1973) *Proc. British Acoustical Society Spring Meeting* (April 1973). Noise from hot jets. See also Fisher, M. J., Lush, P. A. and Harper-Bourne, M. (1973) *J. Sound & Vib.,* **28,** 563-585. Jet noise.

[14.12] Tanna, H. K., Dean, P. D. & Fisher, M. J. (1975) *J. Sound & Vib.,* **39,** 429-460. The influence of temperature on shock free supersonic jet noise.

[14.13] Morfey, C. L. (1973) *J. Sound & Vib.,* **31,** 391-397. Amplification of aerodynamic noise by convected flow inhomogeneities.

[14.14]   Morfey, C. L., Szewczyk, V. M. & Tester, B. J. (1978) *J. Sound & Vib.*, **61**, 255-292. New scaling laws for hot and cold jet mixing noise based on a geometric acoustics model.

[14.15]   Morfey, C. L. & Szewczyk, V. M. (1977) ISVR (University of Southampton) *Technical Report* No. 92. Jet noise modelling by geometric acoustics. Part II: Theory and prediction inside the cone of silence.

[14.16]   Fisher, M. J. and Szewczyk, V. M. (1974) *ARC Paper* No. 35, 212-N897. Flow-acoustic interaction effects in jet noise.

[14.17]   Fisher, M. J., Harper-Bourne, M. & Glegg, S. A. L. (1977) *J. Sound & Vib.*, **51**, 23-54. Jet engine noise source location: the polar correlation technique.

[14.18]   Hoch, R. & Hawkins, R. (1973) *AGARD Conference Preprint* No. 131 on Noise Mechanisms. Recent studies into Concorde noise reduction.

[14.19]   Powell, A. (1953) *Proc. Phys. Soc. B,* **66**, 1029-1056. On the mechanism of choked jet noise.

[14.20]   Harper-Bourne, M. & Fisher, M. J. (1973) *AGARD Conference Preprint* No. 131 on Noise Mechanisms. The noise from shock waves in supersonic jets.

[14.21]   Hay, J. E. & Rose, E. G. (1970) *J. Sound & Vib.*, **11**, 411-420. In flight shock cell noise.

[14.22]   Bushell, K. W. (1975) *AIAA Paper* No. 75-461. Measurement and prediction of jet noise in flight.

[14.23]   Cocking, B. J. & Bryce, W. D. (1975) *AIAA Paper* No. 75-462. Subsonic jet noise in flight based on some recent wind tunnel tests.

[14.24]   Bashforth, S. (1981) *Proc. AIAA 7th Aero-Acoustics Conference, AIAA Paper* No. 81-2029. The effect of flight and the presence of an airframe on exhaust noise.

[14.25]   Tester, B. J. & Fisher, M. J. (1981) *Proc. AIAA 7th Aero-Acoustics Conference, AIAA Paper* No. 81-2040. Engine noise source breakdown: theory, simulation and results.

[14.26]   Morfey, C. L. & Szewczyk, V. M. (1977) ISVR (University of Southampton), *Technical Report* No. 91. Jet noise modelling by geometric acoustics. Part I: Theory and prediction outside the cone of silence.

CHAPTER 15

Finite element techniques for structural vibration

M. Petyt

Institute of Sound and Vibration Research, University of Southampton

15.1 INTRODUCTION

The dynamic response of simple structures, such as uniform beams, plates and cylindrical shells, may be obtained by solving their equations of motion. However, in many practical situations either the geometrical or material properties vary, or the shape of the boundaries cannot be described in terms of known mathematical functions. Also, practical structures consist of an assemblage of components of different types, namely beams, plates, shells and solids. In these situations it is impossible to obtain analytical solutions to the equations of motion. This difficulty is overcome by seeking some form of approximate solution. There are a number of techniques available for determining approximate solutions, the most widely used one being the Rayleigh–Ritz method. An extension of this method, which is known as the 'Finite element displacement method', can be used to analyse any structure, however complex.

15.2 FORMULATION OF THE EQUATIONS OF MOTION

The first step in the analysis of any structural vibration problem is the formulation of the equations of motion. There are a number of ways of doing this, but the simplest, since it is less prone to error, is to use Hamilton's Principle. This principle states that 'among all the displacements which satisfy the prescribed (geometric) boundary conditions and the prescribed conditions at $t = t_1$, and $t = t_2$, the actual solution renders the integral $\int_{t_1}^{t_2} (T - U + W)\, \mathrm{d}t$ stationary'. In this integral T denotes the kinetic energy of the system, U the strain energy and W the work done by the non-conservative forces (e.g. external forces and friction forces). The principle can be stated briefly by the equation

$$\delta \int_{t_1}^{t_2} (T - U + W)\, \mathrm{d}t = 0 \qquad (15.1)$$

where δ denotes the first variation of the integral. It can be applied to both continuous and discrete dynamic systems.

In the case of a multi-degree of freedom system, the deformation of which is described by n independent displacements $q_1, q_2, \ldots q_n$, the condition that Hamilton's integral is stationary reduces to

$$\frac{d}{dt}\left(\frac{\partial T}{\partial \dot{q}_j}\right) + \frac{\partial D}{\partial \dot{q}_j} + \frac{\partial U}{\partial q_j} = f_j \quad j = 1, 2, \ldots n \qquad (15.2)$$

which are known as Lagrange's equations. In these equations D is a dissipation function which represents the instantaneous rate of energy dissipation caused by the frictional forces. f_j is a generalised force which is equal to the work done by the applied forces when the component q_j undergoes a unit displacement.

15.3 RAYLEIGH–RITZ METHOD

Consider a one-dimensional structure whose axial coordinate is x and displacement component $u(x, t)$. The Rayleigh–Ritz method approximates the solution by a finite series expansion of the form

$$u(x, t) = \sum_{j=1}^{n} \phi_j(x)\, q_j(t) \qquad (15.3)$$

where the $\phi_j(x)$ are prescribed functions of x and the $q_j(t)$ unknown functions of t.

Substituting (15.3) into the appropriate energy expressions T, D, U and W reduces the continuous structure to a multi-degree of freedom system with $q_1, q_2, \ldots q_n$ as degrees of freedom. Substituting into equation (15.2) gives the following equations of motion

$$[M]\,\{\ddot{q}\} + [C]\,\{\dot{q}\} + [K]\,\{q\} = \{f\} \qquad (15.4)$$

where $[M]$ is the inertia matrix, $[C]$ the damping matrix, and $[K]$ the stiffness matrix. $\{f\}$ is a column matrix of generalized forces corresponding to the generalised displacements $\{q\}$.

Equation (15.4) is solved for the unknowns $\{q\}$, which are then substituted into (15.3) to give the required approximate solution. The accuracy of the solution obtained can be increased by increasing the number of terms in (15.3), provided the prescribed functions, $\phi_j(x)$, satisfy certain criteria. If the integral in (15.1) involves derivatives up to order p, then the functions $\phi_j(x)$ must satisfy the following conditions:

(1) Be linearly independent.
(2) Be continuous and have continuous derivatives up to order $(p-1)$.
(3) Satisfy the geometric boundary conditions (these will involve derivatives up to order $(p-1)$).
(4) Form a complete series.

15.4 FINITE ELEMENT METHOD

The major drawback to the Rayleigh–Ritz method is the difficulty in constructing a set of prescribed functions, particularly for a built-up structure. This difficulty can be overcome by using the finite element displacement method which provides an automatic procedure for constructing such functions.

In the finite element displacement method the prescribed functions are constructed in the following manner:

(a) Select a set of reference or 'node points' on the structure.
(b) Associate with each node point a given number of degrees of freedom (displacement, slope, etc).
(c) Construct a set of functions such that each one gives a unit value for one degree of freedom and zero values for all the others.

This procedure is illustrated for the axial motion of a rod in Fig. 15.1, where five node points have been selected at equal intervals. The portion of the rod between two adjacent nodes is called an 'element'. The highest derivative appearing in the energy expressions for a rod is the first (see equation (15.8)). Therefore, only the prescribed functions themselves need to be continuous, and so the axial displacement, u, is the only degree of freedom required at each node point. In the figure, five prescribed functions are illustrated. They have been constructed by giving each node point in turn a unit displacement, whilst maintaining zero displacement at all other nodes. The geometric boundary conditions, for the particular problem to be analysed, can be satisfied by omitting any of the functions constructed which do not satisfy them. For example, if node 1 is clamped, then the function $\phi_1(x)$ is omitted.

It can be seen in Fig. 15.1 that each element deforms in only two deformation patterns and that the deformation patterns for each element are the same. Because of this, the emphasis is on determining deformation patterns for individual elements and not for the whole structure. It is also simpler to evaluate the energy expressions for each element and then add the contributions from the elements together. Approached in this way, there is no reason why the elements should be identical. For example, in Fig. 15.1, the elements could have different cross-sectional areas and densities as well as different lengths.

A typical element in Fig. 15.1 is illustrated in Fig. 15.2. The displacement variation for such an element, as shown in Fig. 15.1, is given by

$$u = \tfrac{1}{2}(1-\xi)u_1 + \tfrac{1}{2}(1+\xi)u_2 \tag{15.5}$$

where u_1, u_2 are axial displacements of nodes 1 and 2. This expression can be rewritten in the following matrix form:

$$u = \lfloor \tfrac{1}{2}(1-\xi) \ \tfrac{1}{2}(1+\xi) \rfloor \begin{bmatrix} u_1 \\ u_2 \end{bmatrix} = [N(\xi)] \ \{u\}_e \tag{15.6}$$

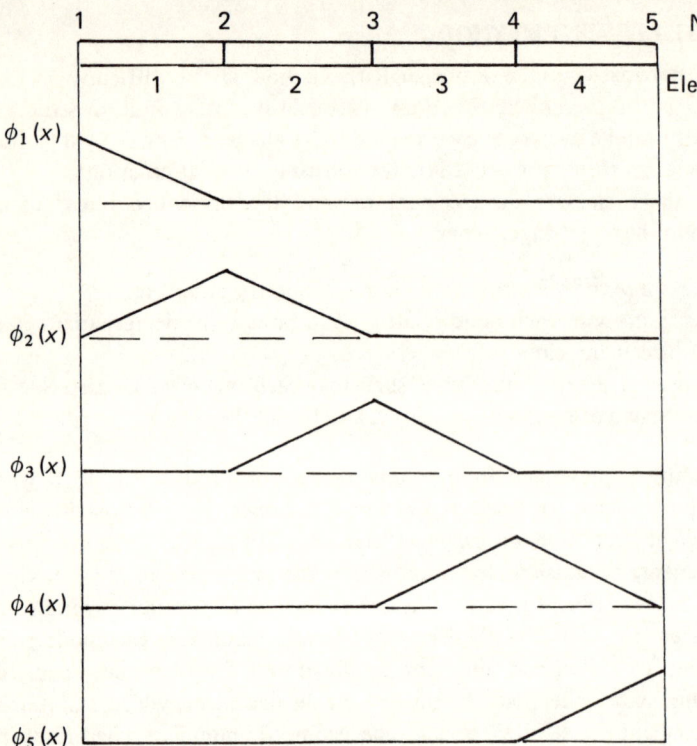

Fig. 15.1 – Prescribed functions for a rod.

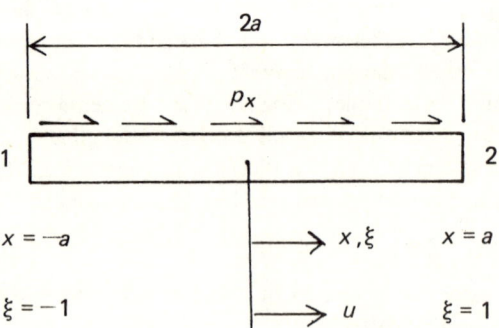

Fig. 15.2 – Geometry of a single axial element.

The energy expressions for a single element are

$$T_e = \tfrac{1}{2} \int\limits_{-1}^{+1} \rho A a \, \dot{u}^2 \, d\xi \tag{15.7}$$

$$U_e = \tfrac{1}{2} \int\limits_{-1}^{+1} \frac{EA}{a} \left(\frac{\partial u}{\partial \xi}\right)^2 d\xi \tag{15.8}$$

$$\delta W_e = \int\limits_{-1}^{+1} p_x a \, \delta u \, d\xi \tag{15.9}$$

where A is the cross-sectional area, E is Young's modulus, ρ the density, and p_x the distributed load per unit length. Substituting (15.6) into (15.7), (15.8) and (15.9) gives

$$T_e = \tfrac{1}{2} \{\dot{u}\}_e^T \, [m]_e \, \{\dot{u}\}_e$$
$$U_e = \tfrac{1}{2} \{u\}_e^T \, [k]_e \, \{u\}_e \tag{15.10}$$
$$\delta W_e = \{\delta u\}_e^T \, \{f\}_e$$

where $\quad [m]_e = \rho A a \int\limits_{-1}^{+1} \lfloor N(\xi) \rfloor^T \, \lfloor N(\xi) \rfloor \, d\xi$

$$= \rho A a \begin{bmatrix} 2/3 & 1/3 \\ 1/3 & 2/3 \end{bmatrix} \tag{15.11}$$

is the element inertia matrix,

$$[k]_e = \frac{EA}{a} \int\limits_{-1}^{+1} \lfloor N'(\xi) \rfloor^T \, \lfloor N'(\xi) \rfloor \, d\xi$$

$$= \frac{EA}{a} \begin{bmatrix} 1/2 & -1/2 \\ -1/2 & 1/2 \end{bmatrix} \tag{15.12}$$

is the element stiffness matrix, and

$$\{f\}_e = a \int\limits_{-1}^{+1} p_x \lfloor N(\xi) \rfloor^T \, d\xi$$

$$= a p_x^e \begin{bmatrix} 1 \\ 1 \end{bmatrix} \tag{15.13}$$

is the element load matrix, for $p_x = p_x^e$ (a constant) over the element.

The energy expressions for the complete rod are obtained by adding together the energies for all the individual elements. Before carrying this out it is necessary to relate the degrees of freedom of a single element, $\{u\}_e$, to the set of degrees of freedom for the complete rod, $\{u\}$. For the rod shown in Fig. 15.1 this is

$$\{u\}^T = \lfloor u_1 u_2 u_3 u_4 u_5 \rfloor \tag{15.14}$$

For element e the relationship is

$$\{u\}_e = [a]_e \{u\} \tag{15.15}$$

where for element 1, for example, $[a]_e$ takes the form

$$[a]_1 = \begin{bmatrix} 1 & 0 & 0 & 0 & 0 \\ 0 & 1 & 0 & 0 & 0 \end{bmatrix} \tag{15.16}$$

Substituting (15.15) into (15.10) and summing over all the elements gives

$$T = \tfrac{1}{2} \{\dot{u}\}^T [M] \{\dot{u}\} \tag{15.17}$$

where

$$[M] = \sum_{e=1}^{4} [a]_e^T [m]_e [a]_e \tag{15.18}$$

for the kinetic energy. Using (15.11) and (15.16) in (15.18) gives

$$[M] = \frac{\rho A a}{3} \begin{bmatrix} 2 & 1 & 0 & 0 & 0 \\ 1 & 4 & 1 & 0 & 0 \\ 0 & 1 & 4 & 1 & 0 \\ 0 & 0 & 1 & 4 & 1 \\ 0 & 0 & 0 & 1 & 2 \end{bmatrix}. \tag{15.19}$$

The strain energy and the work done by the applied loads can be treated similarly.

Notice that $[M]$ is symmetric and banded. These properties can be exploited when writing computer programs. All the information displayed in (15.19) can be stored in 9 locations, whereas 25 locations would be required to store the complete matrix. The stiffness matrix also has the same properties.

As in the Rayleigh–Ritz method, the total energy expressions (such as (15.17)) are substituted into Lagrange's equations (15.2) to give the equations of motion in the form (15.4), namely

$$[M] \{\ddot{u}\} + [C] \{\dot{u}\} + [K] \{u\} = \{f\} \tag{15.20}$$

where $\{u\}$ represents a column matrix of nodal degrees of freedom (axial displacements in this example, as shown in equation (15.14)).

It is not usual to derive the dissipation function for each element in the same manner as the kinetic and strain energies. This is because damping is not necessarily an inherent property of the vibrating structure. Damping forces depend not only on the structure itself, but also on the surrounding medium. Structural damping is caused by internal friction within the material and at joints between components. Viscous damping occurs when a structure is moving in air or a fluid. It is usual to use simplified damping models for the complete structure, rather than for individual elements. Suitable forms for the matrix $[C]$ will be introduced when discussing the solution of equation (15.20).

In the Rayleigh-Ritz method the accuracy of the solution is increased by increasing the number of prescribed functions in the assumed series. To increase the number of prescribed functions in the finite element method, the number of node points, and therefore the number of elements, is increased. The displacement function for each element should satisfy the Rayleigh-Ritz convergence criteria. In particular, the functions and their derivatives up to order $(p-1)$ should be continuous across element boundaries.

15.5 FINITE ELEMENT MODELS

The inertia, stiffness, and consistent load matrices can be derived for any finite element model once the energy and work expressions, T, U, W, and the displacement function $\lfloor N \rfloor$ are known. In this section a number of common finite element models are reviewed.

15.5.1 Beams

Although the beam is one of the simplest of structural elements, it is difficult to model mathematically owing to the large number of different forms it can take. It is used by engineers to describe the behaviour of everything from a uniform, rectangular section beam to the hull of a ship.

The simplest form is a uniform, slender beam which is vibrating in one of its principal planes. The energy expressions for a single element of length $2a$ are

$$T_e = \tfrac{1}{2} \int_{-a}^{+a} \rho A \dot{v}^2 \, dx \tag{15.21}$$

$$U_e = \tfrac{1}{2} \int_{-a}^{+a} EI_z \left(\frac{\partial^2 v}{\partial x^2} \right)^2 dx \tag{15.22}$$

$$\delta W_e = \int_{-a}^{+a} p_y \, \delta v \, dx \tag{15.23}$$

where A is the cross-sectional area, I_z is the second moment of area of the cross-section about the principal axis which is perpendicular to the plane of vibration, E is Young's modulus, ρ the density, p_y the distributed lateral load per unit length, and v the lateral displacement of the centroid. The highest derivative appearing in the energy expressions is the second. Hence, it is necessary to take v and θ_z $(= \partial v / \partial x)$ as degrees of freedom at each node to ensure that v and its first derivative are continuous between elements. With four degrees of freedom (2 at 2 nodes) it is possible to represent the displacement, v, by a cubic polynomial. Details of the element are gieven in reference [15.10].

When analysing three-dimensional frameworks, each member is capable of axial deformation, bending in two principal planes and torsion about ats axis. The axial and bending deformations are treated as described previously. The torsional deformation can be treated in a similar way to the axial deformation. After deriving the energy expressions for a single element with respect to local axes, the nodal components of displacement and rotation are transformed to global axes. The energies of each element are then added together as before.

When analysing deep beams at low frequencies and slender beams at high frequencies it is necessary to include the effect of shear deformation and rotary inertia in the energy expressions. A survey of methods of analysing such beams is given in reference [15.11]. Other factors to be considered are the off-set of the centroid from the node point and the off-set of the shear centre from the centroid. In the case of thin-walled, open section beams the effect of warping of the cross-section must also be included in the energy expressions [15.12]. Beams with variable cross-sections can also be analysed by treating A and I_z as functions of x in (15.21) and (15.22) (see, for example, reference [15.13]). Tapered and twisted beam elements have also been developed for analysing turbine blades [15.14]. Curved beam elements have been developed for analysing in-plane variations [15.15, 15.16] and out-of-plane vibrations [15.17].

15.5.2 Membranes
Membrane elements are used to analyse the low-frequency vibrations of complex shell type structures such as aircraft and ships. The highest derivative in the energy expressions for such elements is the first, and so only components of

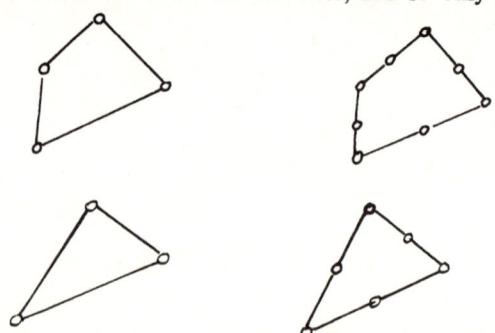

Fig. 15.3 – Typical shapes for two-dimensional elements.

in-plane displacements are used as nodal degrees of freedom. Some typical shapes are shown in Fig. 15.3, details of which are given in reference [15.3]. The accuracy of the two elements with nodes at the vertices only is not good, as the displacements vary linearly. The introduction of additional nodes at the mid-points of the sides allows parabolic variation of displacement. The displacement functions of these latter two elements can be used to transform the elements into curvilinear ones as shown in Fig. 15.4. These elements are known as iso-parametric elements [15.3]. The inertia, stiffness and load matrices of this type of element are obtained using numerical integration [15.3].

Fig. 15.4 – Two-dimensional isoparametric elements.

15.5.3 Plates

The highest derivative in the energy expressions for a thin, flat plate-bending element is the second. Therefore the displacement function for the lateral displacement, w, must ensure continuity of w and its derivatives between elements. Continuity of w and its tangential derivative, $\partial w/\partial s$, can easily be made continuous by ensuring that w varies cubically along an edge and taking w and $\partial w/\partial s$ as nodal degrees of freedom. Continuity of the normal derivative, $\partial w/\partial n$, has proved a little more difficult to achieve.

The first way of achieving continuity of normal slope between elements is to ensure that the normal slope varies linearly between node points. Reference [15.18] uses this method for a rectangular element. The element has nodes at the four corners only, and the degrees of freedom at each node are w, $\partial w/\partial x$, and $\partial w/\partial y$, x and y being parallel to the sides of the element. The displacement function is defined by dividing the rectangle into four triangles by means of two diagonals. Nine of the functions are defined over the complete rectangle and three over the individual triangles.

Another way of ensuring continuity of the normal slope is to introduce an additional node point at the mid-point of each edge. If w, $\partial w/\partial x$ and $\partial w/\partial y$ are taken as degrees of freedom at the corner nodes and $\partial w/\partial n$ as a degree of freedom at the mid-point nodes, then $\partial w/\partial n$ will vary quadratically along each edge. References [15.19, 15.20] give two alternative derivations of the same quadrilateral element which uses this technique. Again the quadrilateral is divided into four triangles by means of the two diagonals. Within each triangle w is represented by a complete cubic polynomial in x and y. Since each polynomial has 10 terms, there are a total of 40 parameters. Only 16 of these are independent because of continuity of displacement and slope between adjacent triangles. These 16 can be expressed in terms of the 16 nodal degrees of freedom ($4 \times 3 + 4 \times 1$).

The quadrilateral element is more useful than the rectangle because of its ability to model complex polygonal shapes. However, in some situations it is necessary to use triangular elements. These should be capable of maintaining the necessary continuity when used in conjunction with either rectangular or quadrilateral elements.

A triangular element with a linear variation of normal slope along its edges can be obtained by dividing it into three sub-triangles, by connecting the centroid to the three vertices [15.21, 15.22]. Within each sub-triangle, w is represented by an incomplete cubic to ensure that $\partial w/\partial n$ varies linearly over the external edge. Of the resulting 27 parameters (9 for each sub-triangle) only 9 are independent, because of the requirement of continuity between sub-triangles. These 9 can be expressed in terms of w, $\partial w/\partial x$, $\partial w/\partial y$ at the three vertices of the main triangle.

A triangular element with quadratic variation of normal slope can be obtained by using the complete cubic polynomial within each sub-triangle [15.23]. As in the case of the quadrilateral, $\partial w/\partial n$ is introduced as an additional degree of freedom at the mid-points of the edges.

For thick plates the effect of shear deformation and rotary inertia should be included in the energy expressions. This results in expressions which involve the lateral displacement, w, the cross-sectional rotations, θ_x and θ_y, and their first derivatives. If w, θ_x and θ_y are represented by independent functions, then it is only necessary to satisfy continuity of these quantities between elements. In this case the displacement functions developed for membrane elements can be used [15.24, 15.25], and typical element shapes are as shown in Fig. 15.3 and 15.4. It can be shown that these elements can also be used to model the behaviour of thin plates provided reduced integration techniques are used.

15.5.4 Solids

For very thick structures three-dimensional solid elements should be used. The energy expressions contain the first derivatives of the displacement components u, v, and w, therefore only continuity of displacements is required. Suitable displacement functions can be obtained by generalizing the membrane functions into three-dimensions [15.3]. Typical element shapes are shown in Fig. 15.5.

15.5.5 Shells

There are many theories available for analysing thin shells. A critical review of many of these can be found in reference [15.26]. One which is commonly used for doubly curved shells is given in reference [15.27].

The experience gained in constructing flat plate displacement functions can be used in developing suitable displacement functions for doubly curved shells. If the element is assumed shallow, then everything can be referred to the base plane. In this case, any of the shape functions described for thin, flat plate elements in section 15.5.3 can be used to represent the shell normal displacement. All that remains is to develop suitable displacement functions for the tengential components of displacement. Experience has shown that it is advantageous to represent all three components by polynomials of equal degree. Typical examples of such elements are given in references [15.28, 15.29]. A survey of user experience with various shell elements is given in reference [15.4].

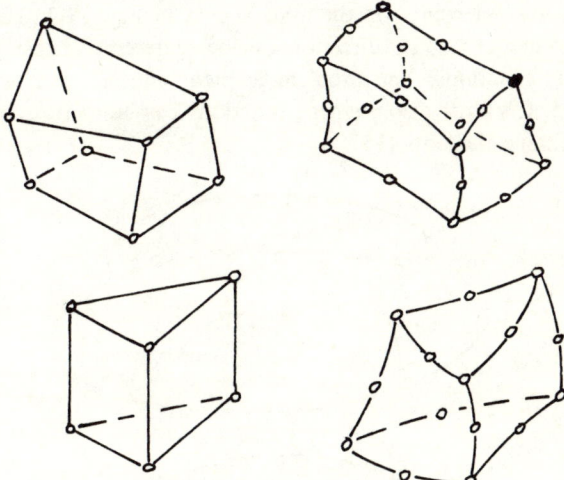

Fig. 15.5 – Three-dimensional elements.

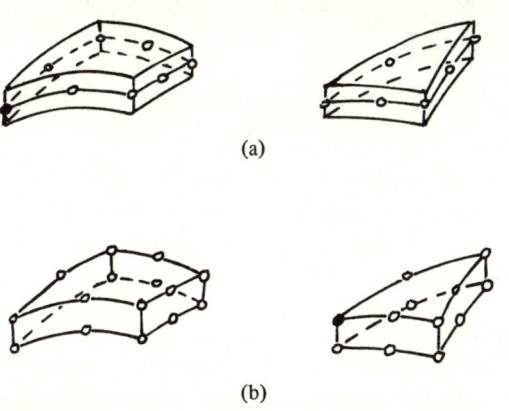

(a)

(b)

Fig. 15.6 – Shell elements: (a) middle type elements; (b) solid type elements.

Thick shells can be analysed using superparametric elements [15.3]. These are constructed by methods which are similar to the ones used for thick plate elements. Again, they can be used for thin shells provided reduced integration techniques are used. Solid type shell elements [15.30] can be constructed by modifying the solid elements described in section 15.5.4. Both types are shown in Fig. 15.6.

15.5.6 Axisymmetric elements
Many structures, such as cooling towers, pressure vessels, and parts of aerospace and marine vehicles, are axisymmetric. In this case they may be idealized by an assembly of ring elements of the type shown in Fig. 15.7. The variation of displacement in the angular direction can be represented exactly by means of trigonometric functions. The problem is then reduced to a two-dimensional one, and the necessary displacement functions can be constructed in a similar way to membrane elements [15.3].

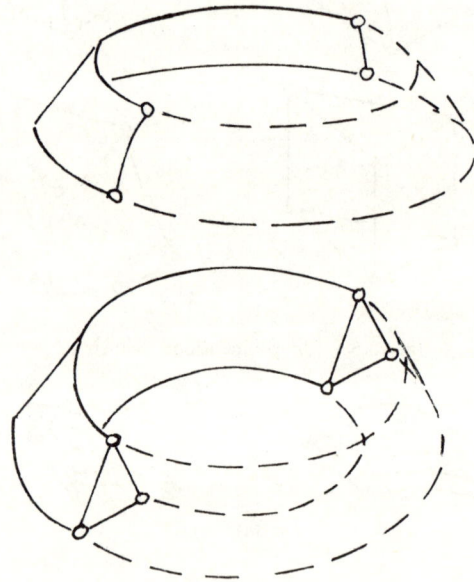

Fig. 15.7 – Axisymmetric ring elements.

15.5.7 Finite strip elements
In many problems the geometry and material properties do not vary in one coordinate direction. The analysis can be simplified by using orthogonal functions to represent the variation of displacement in this direction. Details of the method are given in reference [15.5].

15.5.8 Other types of elements

In the preceding sections the emphasis has been on displacement type elements which satisfy all the necessary convergence criteria. Not all elements to be found in the literature satisfy all of these criteria. In this case the convergence of the resulting solutions cannot be guaranteed. It has been suggested [15.32] that all elements should be tested by the recognized set of tests in order to indicate their accuracy. Other types of models exist such as hybrid elements and mixed variable elements. These will not be discussed here.

15.6 METHODS OF SOLUTION

In this section methods of solving the equation of motion (15.4) will be reviewed. Both free and forced vibration will be considered.

15.6.1 Free vibration

The free motion of an undamped structure is given by the solution of the equation

$$[M] \{\ddot{q}\} + [K] \{q\} = 0 \ . \tag{15.24}$$

Since the motion is harmonic then

$$\{q\} = \{A\} e^{i\omega t} \tag{15.25}$$

where ω is the frequency of vibration. Substituting (15.25) into (15.24) gives

$$[K - \omega^2 M] \{A\} = 0 \ . \tag{15.26}$$

This equation is a linear eigenvalue problem which can be solved by a number of methods. The eigenvalues (which are positive) represent the square of the natural frequencies, and the corresponding eigenvectors give the shape of the modes of vibration.

Methods of solving the eigenvalue problem (15.26) can be divided into four basic groups: vector iteration, transformation, polynomial iteration, and Sturm sequence methods. In fact some methods combine some of these techniques. The choice of method is mostly influenced by the order of the matrices **K** and **M** and the number of eigenvalues required (that is, all or a few). In finite element analysis the order of **K** and **M** is usually quite large (several hundred to a few thousand), and only a relatively few eigenvalues are required. In this case methods based upon sub-space iteration, simultaneous iteration, and Sturm sequences are to be recommended. Details of these methods can be found in references [15.7, 15.8] .

15.6.2 Reduction of degrees of freedom

The number of degrees of freedom in any vibration problem can be reduced in three ways: making use of symmetry, employing a direct reduction technique, and using component mode synthesis.

If a structure has a plane of symmetry, then a single eigenvalue problem for the complete structure can be replaced by two eigenvalue problems for the half-structure. These two eigenvalue problems will give the symmetric and anti-symmetric modes separately. This method greatly reduces both computer storage and computing time.

For more general structures it is necessary to employ a technique for reducing the number of degrees of freedom directly [15.33, 15.34]. In this method only a small proportion of the total number of degrees of freedom, referred to as 'masters', are retained. The remaining 'slave' degrees of freedom take the values giving least strain energy, regardless of what this does to the kinetic energy. In practice, large reductions in the numbers of degrees of freedom can be introduced without introducing significant errors in the lower natural frequencies. An automatic procedure for selecting the master degrees of freedom is presented in reference [15.35].

For very large structures, using the reduction technique above is not an efficient method of analysis. In this case a method known as 'component mode synthesis' should be used. This method is also known as 'substructure analysis'. The first step in the analysis is to divide the complete structure into a number of substructures. Each substructure is then represented by a finite element model. The next step is concerned with reducing the number of degrees of freedom for each substructure by model substitution. Finally, the substructures are coupled together and the complete structure analysed. There are various ways of carrying out this analysis. They can be classified as fixed interface methods, modal substitution, free interface, and hybrid methods. A survey of these techniques is given in reference [15.36].

In the case of cyclic structures the theory of periodic structures can be used [15.37]. In this method only one of the repeating sections is modelled by an assemblage of finite elements. Knowing the phase difference between the motion on the common boundaries with the adjacent sections on either side, for standing waves in the complete structure, the natural frequencies of the complete structure can be found.

15.6.3 Forced vibration
The equation of forced vibration of a structure takes the form

$$[M]\ \{\ddot{q}\} + [C]\ \{\dot{q}\} + [K]\ \{q\} = \{f\}\ . \qquad (15.27)$$

The method of solving this equation depends upon whether the exciting forces are harmonic, periodic, transient, or random. Whatever the nature of the exciting force, equation (15.27) can be solved either directly or after reduction by modal substitution. In finite element analysis the number of degrees of freedom is usually quite large, and so it is usual to use modal substitution. In this method use is made of the transformation

$$\{q\} = [\Phi]\ \{\xi\} \qquad (15.28)$$

where $[\Phi]$ is a matrix whose columns represent the modes of vibration of the structure. Usually only the lower modes of vibration contribute significantly to the response, and so the number of elements in $\{\xi\}$ is far fewer than the number in $\{q\}$. Substituting equation (15.28) into equation (15.27) and pre-multiplying by $[\Phi]^T$ gives

$$\{\ddot{\xi}\} + [\bar{c}]\{\dot{\xi}\} + \lceil \lambda \rfloor \{\xi\} = [\Phi]^T \{f\} \tag{15.29}$$

where $\lceil \lambda \rfloor$ is a diagonal matrix of eigenvalues (solutions of equation (15.26) with $\lambda_i = \omega_i^2$, $\omega_i = $ frequency of mode i) and

$$[\bar{c}] = [\Phi]^T [c] [\Phi] \ . \tag{15.30}$$

Inspection of equation (15.29) reveals that the equations of motion for the individual modes are coupled via the generalized damping matrix $[\bar{c}]$ only. These equations would be very much easier to solve if they were completely uncoupled. The simplest way of ensuring that $[\bar{c}]$ is diagonal is to assume $[c]$ takes the form

$$[c] = a_0 [M] + a_1 [K] \ . \tag{15.31}$$

In this case,

$$[\bar{c}] = a_0 \lceil I \rfloor + a_1 \lceil \lambda \rfloor \ . \tag{15.32}$$

a_0 and a_1 can be calculated from a knowledge (from test or experience) of the damping ratio in two modes. The damping ratio in the other modes can be calculated from (15.32).

The problem has now been reduced to solving a set of single degree of freedom equations. These can be solved by standard techniques which can be found in any book on vibration theory. The special problems associated with random response using finite element techniques are discussed in reference [15.38].

The above procedure for uncoupling the equations of motion can only be carried out if the damping is light. Fortunately, for most structural configurations this is the case. If the damping is heavy (e.g. if the structure is coated with a significant amount of viscoelastic material, or submerged in a fluid and there is significant radiation) then the equations of motion can only be uncoupled by using certain modes of the damped system which are complex.

15.6.4 Response to support motion

In some situations a structure is excited by support motion (e.g. effect of earthquakes on buildings, and vehicles passing over rough roads). To deal with this situation the degrees of freedom are partitioned into boundary degrees of freedom $\{q_B\}$, and internal degrees of freedom $\{q_I\}$. It can be shown [15.39] that the motion of the internal degrees of freedom is given by

$$\{q_I\} = - [K_{II}]^{-1} [K_{IB}] \{q_B\} + \{q_I\}_R \tag{15.33}$$

where $\{q_I\}_R$ represents the motion relative to the supports. This is given by the solution of the equation

$$[M_{II}] \{\ddot{q}_I\}_R + [C_{II}] \{\dot{q}_I\}_R + [K_{II}] \{q_I\}_R = \{f\}_{eff} \quad (15.34)$$

where the effective force $\{f\}_{eff}$ is given by

$$\{f\}_{eff} = [M_{II} K_{II}^{-1} K_{IB} - M_{IB}] \{\ddot{q}_B\} . \quad (15.35)$$

Equation (15.34) can be solved in the same way as equation (15.27). In equations (15.33) to (15.35) subscripts I, B denote the partitions of M, C and K which correspond to $\{q_I\}$ and $\{q_B\}$.

15.7 COMPUTER PROGRAMS

With the aid of the finite element method, the prediction of the dynamic response of structures is reduced to a routine procedure. It is therefore possible to write general computer codes for analysing any structure, the particular structure being specified by means of input data. Reference [15.6] gives an elementary introduction to the programming aspects of the method whilst references [15.7, 15.8] discuss more advanced aspects. Surveys of existing programs and their availability are given in references [15.9, 15.40].

REFERENCES

[15.1]   Desai, C. S. & Abel, J. F. (1972) *Introduction to the finite element method.* New York: Van Nostrand Reinhold.

[15.2]   Cook, R. D. (1981) *Concepts and applications of finite element analysis,* 2nd ed. New York: Wiley.

[15.3]   Zienkiewicz, O. C. (1977) *The finite element method,* 3rd ed. London McGraw Hill.

[15.4]   Ashwell, D. G. & Gallagher, R. H. (Editors) (1976) *Finite elements for thin shells and curved members.* London: Wiley

[15.5]   Cheung, Y. K. (1976) *Finite strip method in structural analysis.* Oxford: Pergamon.

[15.6]   Hinton, E. & Owen, D. R. (1977) *Finite element programming,* London: Academic Press.

[15.7]   Bathe, K. J. & Wilson, E. L. (1976). *Numerical methods in finite element analysis.* Eglewood Cliffs: Prentice-Hall.

[15.8]   Jennings, A. (1977) *Matrix computation for engineers and scientists.* London: Wiley.

[15.9]   Pilkey, W., Saczalski, K. & Schaeffer, H. (editors) (1974). *Structural mechanics computer programs, surveys, assessments and availablity.* Charlottesville, University Press of Virginia

[15.10] Leckie, F. A. & Lindberg, G. M. (1963) *Aeronautical Quarterly,* **14,** 224–240. The effect of lumped parameters on beam frequencies.

[15.11]   Thomas, D. L., Wilson, J. M. & Wilson, R. R. (1973) *J. Sound &*
          Vib., **31**, 315-330. Timoshenko beam finite elements.

[15.12]   Petyt, M. (1977) *J. Sound & Vib.*, **54**, 533-547. Finite strip analysis
          of flat skin-stringer structures.

[15.13]   Lindberg, G. M. (1963) *Aeronautical Quarterly*, **14**, 387-395. Vibra-
          tion of non-uniform beams.

[15.14]   Gupta, R. S. & Rao, S. S. (1978) *J. Sound & Vib.*, **56**, 187-200.
          Finite element eigenvalue analysis of tapered and twisted Timoshenko
          beams.

[15.15]   Petyt, M. & Fleischer, C. C. (1971) *J. Sound & Vib.*, **18**, 17-30. Free
          vibrations of a curved beam.

[15.16]   Davis, R., Henshall, R. D. & Warburton, G. B. (1972) *J. Sound & Vib.*,
          25, 561-576. Constant curvature beam finite elements for in-plane
          vibration.

[15.17]   Davis, R., Henshall, R. D. & Warburton, G. B. (1972) *International*
          Journal of Earthquake Engineering and Structural Dynamics, **1**,
          165-175. Curved beam finite elements for coupled bending and
          torsional vibration.

[15.18]   Deak, A. L. and Pian, T. H. H. (1967) *AIAA Journal*, **5**, 187-189.
          Application of the smooth surface interpolation to the finite element
          analysis.

[15.19]   Fraeijs de Veubeke, B. (1968) *International Journal of Solids and*
          Structures, **4**, 95-108. A conforming finite element for plate bending.

[15.20]   Orris, R. M. & Petyt, M. (1973) *J. Sound & Vib.*, **27**, 325-334. A
          finite element study of the vibration of trapezoidal plates.

[15.21]   Clough, R. W. & Tocher, J. L. (1965) *Proc. First Conference on*
          Matrix Methods in Structural Mechanics, AFF DL-TR-66-80. Finite
          element stiffness matrices for analysis of plate bending.

[15.22]   Dickinson, S. M. & Henshell, R. D. (1969) *AIAA Journal*, **7**, 560-561.
          Clough-Tocher triangular plare-bending element in vibration.

[15.23]   Clough, R. W. & Felippa, C. A. (1968) *Proc. Second Conference on*
          Matrix Methods in Structural Mechanics, AFF DL-TR-68-150. A
          refined quadrilateral element for analysis of plate bending.

[15.24]   Rock, T. A. & Hinton, E. (1974) *International Journal of Earthquake*
          Engineering and Structural Dynamics, **3**, 51-63. Free vibration and
          transient response of thick and thin plates using the finite element
          method.

[15.25]   Rock, T. A. & Hinton, E. (1976) *Computers and Structures*, **7**, 37-
          44. A finite element method for the free vibration of plates allowing
          for transverse shear deformation.

[15.26]   Kraus, H. (1967) *Thin elastic shells*. New York: Wiley.

[15.27]   Novozhilov, V. V. (1964) *The shell theory*. Groeningen: Noordhoff.

[15.28]   Cowper, C. R., Lindberg, G. M. & Olson, M. D. (1970) *International*

Journal of Solids and Structures, **8**, 1133-1156. A shallow shell finite element of triangular shape.

[15.29] Petyt, M. & Fleischer, C. C. (1972). *The mathematics of finite elements and applications* (Ed. by J. R. Whiteman) London: Academic Press, 367-378. Vibrations of curved structures using quadrilateral finite elements.

[15.30] Nicolas, V. T. & Citipitioglu, E. (1976) *Second National Symposium on Computerised Structural Analysis and Design*, Washington DC. A general isoparametric finite element program SDRC SUPERB.

[15.31] Robinson, J. (Editor) (1978) *Finite element methods in the commercial environment*. Wimborne: Robinson and Associates.

[15.32] Robinson, J. (1978) Reference [15.31], 217-248. Element evaluation: a set of assessment points and standard tests.

[15.33] Guyan, R. J. (1965) *AIAA Journal*, **3**, 380. Reduction of stiffness and mass matrices.

[15.34] Irons, B. M. (1965) *AIAA Journal*, **3**, 961-962. Structural eigenvalue problems − elimination of unwanted variables.

[15.35] Henshell, R. D. & Ong, J. H. (1975) *Earthquake Engineering and Structural Dynamics*, **3**, 375-383. Automatic masters for eigenvalue economization.

[15.36] Stavrinidis, C. (1978) Reference [15.31], 307-331. Theory and practice of modal synthesis techniques.

[15.37] Salama, A. M., Petyt, M. & Mota Soares, C. A. (1976). ASME Monograph, *Structural dynamic aspects of bladed disc assemblies*. Dynamic analysis of bladed discs by wave propagation and matrix difference techniques.

[15.38] Petyt, M. (1975) *Proceedings world congress on finite element methods in structural mechanics*, Bournemouth. Finite-element methods for the response of structures to random excitations.

[15.39] Clough, R. W. (1969) *Proceedings US-Japan Symposium on Recent Advances in Matrix Methods of Structural Analysis and Design*, Tokyo, 441-482. Analyses of structural vibrations and dynamic response.

[15.40] Imbert, J. F. (1978). Reference [15.31], 421-464. A survey of current capability for dynamic analysis of complex structures.

CHAPTER 16

Finite element techniques for acoustics

M. Petyt

Institute of Sound and Vibration Research, University of Southampton

16.1 INTRODUCTION

Finite element techniques were first developed for analysing complex engineering structures. Once the method had been given a firm mathematical foundation, it was only natural that it should be used for analysing other physical problems which could be represented by partial differential equations. The field of acoustics has been no exception. There are three main types of acoustic problems to which the method has been applied. The first is the determination of the acoustic pressures in a cavity which is enclosed by either rigid or flexible walls. The noise radiated by vibrating structures which are immersed in an infinite acoustic medium can also be predicted. Finally, the method has been used to predict the propagation of acoustic waves down variable ducts with mean flow.

16.2 IRREGULAR SHAPED CAVITIES WITH ACOUSTICALLY HARD WALLS

Consider a cavity of volume V enclosed by a surface S. Within V the acoustic pressure, p, must satisfy the Helmholtz equation

$$\nabla^2 p + (\omega^2/a_0^2) p = 0 \tag{16.1}$$

where ω is the frequency of vibration and a_0 the speed of sound. Over the bounding surface S, the fluid particle velocity normal to the surface is zero, giving

$$\partial p/\partial n = 0 \tag{16.2}$$

where n is the outward normal to the surface

The solution of the differential equation (16.1) subject to the boundary condition (16.2) can be replaced by an equivalent variational principle, namely,

$$\delta \int_V \tfrac{1}{2} [(\nabla p)^2 - (\omega^2/a_0^2) p^2] \, dV = 0 \ . \tag{16.3}$$

An approximate solution to this variational principle can be obtained using finite element techniques. The volume, V, is represented by an assemblage of three-dimensional finite elements. The highest derivative appearing in equation (16.3) is the first. Therefore, the guidelines given in Chapter 15 indicate that the pressure, p, should be continuous between elements. The pressure distribution within an element can, therefore, be approximated by an expression of the form

$$p = \lfloor N(x,y,z) \rfloor_e \{ p \}_e \qquad (16.4)$$

where $\lfloor N(x,y,z) \rfloor_e$ is the element shape function and $\{ p \}_e$ is a column matrix of nodal pressure-values for element e. Thus for each element

$$\int_{V_e} \tfrac{1}{2} (\nabla p)^2 \, dV = \tfrac{1}{2} \{ p \}_e^T [k]_e \{ p \}_e$$

and

$$\int_{V_e} (1/2\, a_0^2) p^2 dV = \tfrac{1}{2} \{ p \}_e^T [m]_e \{ p \}_e \qquad (16.5)$$

where

$$[k]_e = \int_{V_e} [B]_e^T [B]_e \, dV$$

$$[m]_e = \int_{V_e} (1/a_0^2) \lfloor N \rfloor_e^T \lfloor N \rfloor_e \, dV \qquad (16.6)$$

and

$$[B]_e = \begin{bmatrix} \dfrac{\partial}{\partial x} \\[2mm] \dfrac{\partial}{\partial y} \\[2mm] \dfrac{\partial}{\partial z} \end{bmatrix} \lfloor N \rfloor_e. \qquad (16.7)$$

The usual assembly procedures (see Chapter 15) can be used to produce the corresponding matrices $\mathbf{K_a}$ and $\mathbf{M_a}$ for the complete volume. The variational equation (16.3) then gives the equation

$$[\mathbf{K_a} - \omega^2 \mathbf{M_a}] \{ p \} = 0 \qquad (16.8)$$

where $\{ p \}$ is a column matrix of nodal pressures for the complete volume. Equation (16.8) is a linear eigenproblem. The eigenvalues yield the natural frequencies and the corresponding eigenvectors the shape of the acoustic modes of the cavity. Methods of solving equation (16.8) are indicated in Chapter 15.

16.2.1 Finite element models

Reference [16.3] derives both a tetrahedron and a cuboid element with the pressure as the only nodal degree of freedom. Reference [16.4] develops a higher order cuboid element. The nodal degrees of freedom are taken to be the velocity potential, ϕ, and its three first derivatives. The pressure can be obtained using the relationship $p = - \rho \partial \phi / \partial t$. A twenty-node, isoparametric hexahedron is presented in reference [16.5], the pressure being the only nodal degree of freedom. With this element it is possible to represent curved, bounding surfaces more closely.

If the cavity is axisymmetric then it is possible to model the cavity with axisymmetric ring elements having triangular, quadrilateral, or isoparametric cross-sections. In this case the pressure (or velocity potential) is represented by a Fourier series expansion in the angular coordinate. Finite element approximations are used in radial-axial coordinate plane. References [16.6-16.11] present various elements of this type. Reference [16.12] also presents an axisymmetric element but uses a displacement potential and its derivatives as nodal degrees of freedom.

The use of a Fourier series expansion in the angular coordinate effectively reduces the analysis of axisymmetric cavities to a two-dimensional finite element analysis. This is also the case if the cavity has one pair of parallel walls. Assuming that $z = 0$, l_z are the parallel walls, then it can be assumed that

$$p(x. y. z) = p(x, y) \cos \frac{n_z \pi z}{l_z} \quad n_z = 0, 1, \ldots \qquad (16.9)$$

Substituting (16.9) into (16.3) and integrating with respect to z gives

$$\delta \int_A \tfrac{1}{2} [(\nabla_1 p)^2 - k_1^2 p^2] \, \mathrm{d}A = 0 \qquad (16.10)$$

where A is the cross-sectional area of the cavity, ∇_1 is the two-dimensional form of the operator ∇, and

$$k_1^2 = (\omega/a_0)^2 - (n_z \pi / l_z)^2 . \qquad (16.11)$$

The variational principle (16.10) can now be solved using two-dimensional finite elements [16.13-16.16].

Reference [16.17] presents a number of one-, two-, and three-dimensional elements.

In some applications a large number of elements, and hence a large number of degrees of freedom, will be required. This results in a large amount of computer storage and long computing times. In structural analysis the problem is overcome by using component mode synthesis techniques. Reference [16.18] demonstrates that this method can also be used when analysing acoustic cavities.

16.3 IRREGULAR SHAPED CAVITIES WITH NON-RIGID WALLS

Consider a cavity of volume V enclosed by a non-rigid surface as shown in Fig. 16.1. Part of the surface S is acoustically hard, the part S' is flexible and vibrating with a normal velocity v_n, and the part S'' is covered with a sound-absorbing

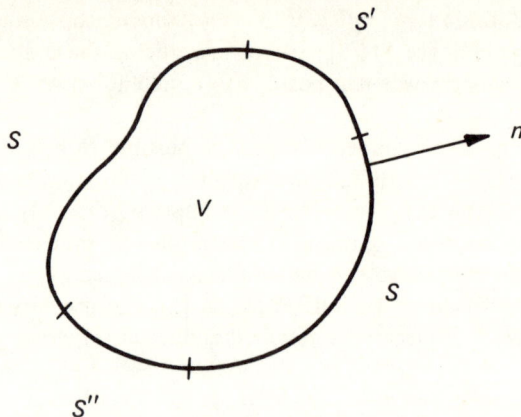

Fig. 16.1 — Irregular shaped cavity with non-rigid walls.

material, the specific acoustic impedance of which is Z_s. Within V the acoustic pressure, p, must satisfy the Helmholtz equation

$$\nabla^2 p + (\omega^2/a_0^2)\, p \;=\; 0 \tag{16.12}$$

where the notation is the same as in section 16.2. Over the bounding surface the fluid particle velocity normal to the surface is equal to the normal velocity of the surface. This gives rise to the following boundary conditions:

$$\frac{\partial p}{\partial n} = 0 \qquad\qquad \text{over } S$$

$$\frac{\partial p}{\partial n} = -\,i\rho\,\omega\,v_n \qquad\qquad \text{over } S' \tag{16.13}$$

and $\qquad \dfrac{\partial p}{\partial n} = -\,i\rho\,\omega\,\dfrac{p}{Z_s} \qquad\qquad \text{over } S''$

where ρ is the density of the acoustic medium.

The solution of equation (16.12) subject to the boundary conditions (16.13) can be replaced by an equivalent variational principle, namely:

$$\delta\,[\tfrac{1}{2}\int_V \{(\nabla p)^2 - (\omega^2/a_0^2)p^2\}\,\mathrm{d}V + \int_{S'} i\rho\omega p\, v_n\,\mathrm{d}S$$

$$+ \tfrac{1}{2}\int_{S''} i\rho\omega\frac{p^2}{Z_s}\,\mathrm{d}S] \;=\; 0 \ . \tag{16.14}$$

An approximate solution to this variational principle can be obtained using finite element techniques. The volume, V, is represented by an assemblage of three-dimensional finite elements. The pressure distribution within an element is again approximated by an expression of the form (16.4). If an element is in contact with either S' od S'' it will contribute to one of the surface integrals in equation (16.14). The variation of pressure over one surface of an element will be denoted by $\lfloor N_a \rfloor_e$. This is obtained from the volume shape function $\lfloor N \rfloor_e$. Also the normal velocity distribution over an element of S' (assumed identical to one face of an acoustic element) is approximated by

$$v_n = \lfloor N_s \rfloor_e \{v\}_e \tag{16.15}$$

where $\{v\}_e$ is a column matrix of nodal velocity values.

Substituting the pressure and velocity variations into the integrals in equation (16.14) and adding the contributions from each element leads to the following equation

$$[\mathbf{K}_a - \omega^2 \mathbf{M}_a + i\omega \mathbf{D}_a]\ \{p\} = -\ i\omega\ [S]\ \{v\}\ . \tag{16.16}$$

The matrices $\mathbf{K}_a$ and $\mathbf{M}_a$ are formed in the way described in section 16.2. The contributions from a single element to the matrices $\mathbf{D}_a$ and $\mathbf{S}$ are as follows:

$$[d]_e = \int_{S_e} (\rho/Z_s)\ \lfloor N_a \rfloor_e^T\ \lfloor N_a \rfloor_e\ \mathrm{d}S$$
$$\tag{16.17}$$
$$[s]_e = \int_{S_e} \rho\ \lfloor N_a \rfloor_e^T\ \lfloor N_s \rfloor_e\ \mathrm{d}S\ .$$

Equation (16.16) has been used to analyse the performance of mufflers [16.8, 16.20] and acoustic filters [16.9, 16.10]. The velocity distribution is specified at the inlet, and the acoustic impedance of the outlet is taken to be ρa_0. Absorbent linings can also be included by specifying their impedances. Reference [16.15] also analyses irregular shaped cavities with various absorbent boundaries.

In the above studies the absorbent surfaces were assumed to be locally reacting, and either empirical or measured values were used for the acoustic impedance. Although encouraging results were obtained, it has been shown that for sound transmission in ducts the liner is best considered as having a bulk reaction. As a first step towards including this effect, reference [16.21] investigates a finite element model for rigid porous absorbing materials. The coupling of this type of element to an acoustic element is treated in reference [16.22].

16.3.1 Structural-acoustic interaction

If S' represents the surface of a vibrating structure, then the velocity distribution will be given by the solution of the equations of motion of the structure. If the methods described in Chapter 15 are followed then the equation of motion of the structure takes the form:

$$[\mathbf{K} - \omega^2 \mathbf{M} + i\,\omega\,\mathbf{C}]\ \{q\} = \{f_m\} + \{f_a\} \tag{16.18}$$

for harmonic motion, where $\{q\}$ is a column matrix of nodal displacements. $\{f_m\}$ and $\{f_a\}$ are the mechanical forces and the forces exerted by the acoustic medium respectively. The acoustic forces acting on an individual element are

$$
\begin{aligned}
\{f_a\}_e &= - \int_{S_e} \lfloor N_s \rfloor_e^T \lfloor N_a \rfloor_e \, dS \, \{p\}_e \\
&= - \frac{1}{\rho} [s]_e^T \{p\}_e \; .
\end{aligned}
\tag{16.19}
$$

For the complete structure

$$
\{f_a\} = - \frac{1}{\rho} [S]^T \{p\} \; .
\tag{16.20}
$$

Substituting (16.20) into equation (16.18) gives

$$
[K - \omega^2 M + i \omega C] \{q\} + \frac{1}{\rho} [S]^T \{p\} = \{f_m\} \; .
\tag{16.21}
$$

Putting $\{v\} = i\omega \{q\}$ into equation (16.16) gives

$$
[K_a - \omega^2 M_a + i\omega D_a] \{p\} - \omega^2 [S] \{q\} = 0 \; .
\tag{16.22}
$$

Equations (16.21) and (16.22) are coupled equations for the structural displacements and acoustic pressures. Together they form an unsymmetric set of equations. They can, however, be manipulated into a symmetric form which is easier to solve. Reference [16.11] describes a modal method of solution. Reference [16.12] shows that using a displacement potential formulation for the cavity, the coupled equations of structural-acoustic interaction are symmetric. One or the other of these approaches has been used for interior noise studies of buildings and aircraft [16.4, 16.11], the dynamics of containers partially filled with liquid [16.12], and sound transmission between enclosures [16.14, 16.19].

In many applications the acoustic motion of the cavity has little effect on the response of the structure. In this case the term $(1/\rho) [S]^T \{p\}$ can be neglected in equation (16.21). The resulting equation represents the response of the structure *in vacuo*. The solution, $\{q\}$, of this equation can then be substituted into equation (16.22) which is then solved for the acoustic pressures $\{p\}$.

16.4 ACOUSTIC RADIATION

There are three ways of using finite element methods for predicting the acoustic radiation from arbitrary vibrating bodies immersed in an acoustic medium. The first approach is to use the finite element method described in section 16.3. This method can be made more efficient by using semi-infinite elements to model the fluid region. The third technique is to use the so called 'Boundary element' method.

16.4.1 Finite element analysis

Consider a vibrating surface S which is immersed in an acoustic medium as shown in Fig. 16.2. If S is completely enclosed by a spherical surface σ, then the

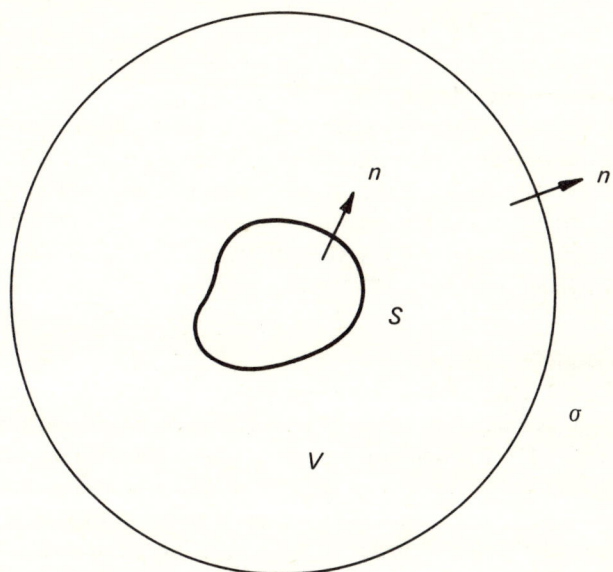

Fig. 16.2 – Radiation from a flexible body immersed in an acoustic medium: finite element solution.

acoustic pressures radiated by S will be given by the solution of the Helmholtz equation in V, the volume between S and σ, subject to the boundary conditions over both S and σ. Thus,

$$\nabla^2 p + k^2 p = 0 \qquad \text{in } V \qquad (16.23)$$

where $k = \omega/a_o$, and

$$\frac{\partial p}{\partial n} = i \rho \, a_o k \, v_n \qquad \text{over } S \qquad (16.24)$$

and

$$\frac{\partial p}{\partial n} = - i \rho \, a_o k \frac{p}{Z_\sigma} \qquad \text{over } \sigma \qquad (16.25)$$

where v_n is the normal velocity of the surface S, and Z_σ is the acoustic impedance over the surface σ.

If the radius of the sphere is large enough then the waves crossing σ can be assumed to be approximated by plane waves. In this case

$$Z_\sigma = \rho \, a_o \ . \qquad (16.26)$$

A smaller sphere can be used by assuming that the waves crossing σ are spherical waves; this gives

$$Z_\sigma = \rho\, a_0\, (1 + 1/ikR)^{-1} \qquad (16.27)$$

where R is the radius of σ. In order to reduce the radius of the sphere further it is necessary to obtain the exact solution for the acoustic pressures outside σ. This can be done using spherical harmonics [16.23].

The differential equation (16.23) and the boundary conditions (16.24) and (16.25) can be replaced by an equivalent variational principle which is similar to equation (16.14). An approximate solution can be obtained by representing the volume V between S and σ by an assemblage of finite elements. The resulting equation is similar to equation (16.16). Examples of the use of this method can be found in references [16.7, 16.23, 16.24].

16.4.2 Semi-infinite elements

The concept of using semi-infinite elements for analysing problems which involve unbounded media has already been applied to diffraction and refraction of surface waves [16.25] and viscous flow [16.26]. There is no reason why this approach should not be applied to acoustic radiation problems. The technique would be to model the volume bounded by S and σ in Fig. 16.2 using finite elements, and the volume outside σ using semi-infinite elements (references [16.25, 16.26] refer to these as infinite elements).

A two-dimensional, semi-infinite element can be formed by letting three nodes of a nine-node, Lagrangian element approach infinity. Thus the element has three finite boundaries and one at infinity. The shape function in the finite direction is a conventional Lagrange polynomial. In the semi-infinite direction a function which satisfies the Sommerfield radiation condition [16.27] is used. The function and its nodal values are complex. A more general treatment of such elements can be found in reference [16.28].

16.4.3 Boundary element method

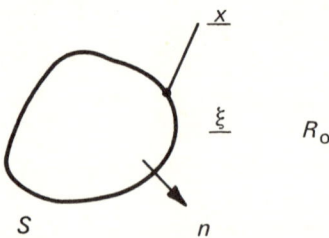

Fig. 16.3 – Radiation from a flexible body immersed in an acoustic medium: boundary element solution.

The general radiation problem can be stated to be the solution of the Helmholtz equation

$$\nabla^2 p + k^2 p = 0 \quad \text{in } R_0 \text{ (see Fig. 16.3)} \tag{16.28}$$

subject to the boundary condition

$$\frac{\partial p}{\partial n} = i \rho a_0 k v_n \qquad \text{over } S \tag{16.29}$$

and the Sommerfield radiation condition

$$\lim_{R \to \infty} R \left(\frac{\partial p}{\partial R} + ikp \right) = 0 \tag{16.30}$$

or equivalently the solution of the integral equation

$$\epsilon\, p(\underline{x}) = \int_S \{p(\underline{\xi}) \frac{\partial G}{\partial n}(\underline{x}, \underline{\xi}) + i\rho a_0 k\, v_n (\underline{\xi})\, G(\underline{x}, \underline{\xi})\}\, dS(\underline{\xi}) \tag{16.31}$$

$$\text{where} \qquad \epsilon = \begin{cases} 1 & \underline{x} \text{ in } R_0 \\ \frac{1}{2} & \underline{x} = \underline{\zeta} \text{ on } S \end{cases} \tag{16.32}$$

where $G(\underline{x}, \underline{\xi})$ is the free space Green's function which is given by

$$G(\underline{x}, \underline{\xi}) = \frac{\exp\left[-ik\, d(\underline{x}, \underline{\xi})\right]}{4\pi\, d(\underline{x}, \underline{\xi})} \tag{16.33}$$

$$\text{and} \qquad d(\underline{x}, \underline{\xi}) = |\underline{x} - \underline{\xi}| \,. \tag{16.34}$$

Evaluating equation (16.31) on the surface S (i.e. putting $\underline{x} = \underline{\zeta}$) gives an equation for the pressure distribution on the surface; that is

$$\tfrac{1}{2} p(\underline{\zeta}) = \int_S \{p(\underline{\xi}) \frac{\partial G}{\partial n}(\underline{\zeta}, \underline{\xi}) + i \rho a_0 k v_n(\underline{\xi})\, G(\underline{\zeta}, \underline{\xi})\}\, dS \tag{16.35}$$

Once the pressure distribution on S is known, the pressure anywhere in R_0 can be obtained using equation (16.31).

An approximate solution to equation (16.35) can be obtained by representing the surface S by an assemblage of finite elements. Over each element assume

$$v_n = \lfloor N_s \rfloor_e \{v\}_e \,, \qquad p = \lfloor N_a \rfloor_e \{p\}_e \tag{16.36}$$

Substituting (16.36) into (16.35) and adding the contributions from each element leads to an equation of the form

$$[A]\,\{p\} = [B]\,\{v\} \tag{16.37}$$

where $\{p\}$ and $\{v\}$ are column matrices of nodal pressures and velocities. The solution of equation (16.37) gives the nodal pressure values. Unfortunately, for certain wave-numbers (both $[A]$ and $[B]$ are functions of k), the matrix $[A]$ is singular, and so equation (16.37) does not have a unique solution. Several methods have been proposed for overcoming this problem (see the list of references in [16.29]), but probably the most successful one is the one given in reference [16.30].

Applications of this technique can be found in reference [16.31].

16.5 DUCT ACOUSTICS

16.5.1 Sound propagation in ducts without mean flow

The propagation of sound in ducts without mean flow can be analysed using the same techniques as described in section 16.3 for mufflers and filters. The advantage of using finite element techniques is that variable geometry and varying liner properties can easily be introduced in the analysis. References [16.32, 16.33] are typical examples of this approach. Acoustic radiation from open-ended ducts has been analysed using a combined finite element/analytical technique in references [16.34, 16.35].

16.5.2 Sound propagation in ducts with flow

A variational formulation of the governing equations of sound propagation in ducts with shear flow cannot easily be found. Therefore, the standard Rayleigh-Ritz/finite elemental analysis cannot be used. Instead an approximate solution of the governing differential equations is obtained using a Galerkin/finite element type of analysis. The technique will be illustrated by considering a two-dimensional duct as shown in Fig. 16.4.

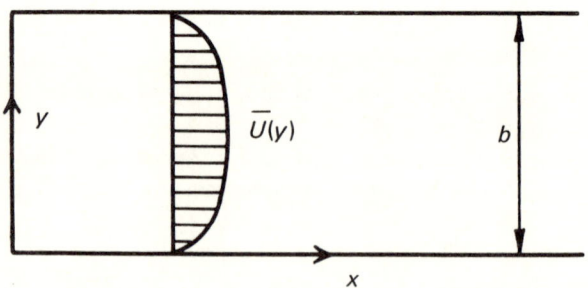

Fig. 16.4 – Two-dimensional duct with shear flow.

The linearized equations of motion for an inviscid, compressible fluid are

$$
\begin{bmatrix}
\left(\dfrac{1}{a_o}\dfrac{\partial}{\partial t}+M\dfrac{\partial}{\partial x}\right) & \dfrac{\mathrm{d}M}{\mathrm{d}y} & \dfrac{\partial}{\partial x} \\[2ex]
0 & \left(\dfrac{1}{a_o}\dfrac{\partial}{\partial t}+M\dfrac{\partial}{\partial x}\right) & \dfrac{\partial}{\partial y} \\[2ex]
\dfrac{\partial}{\partial x} & \dfrac{\partial}{\partial y} & \left(\dfrac{1}{a_o}\dfrac{\partial}{\partial t}+M\dfrac{\partial}{\partial x}\right)
\end{bmatrix}
\begin{bmatrix} u^* \\[2ex] v^* \\[2ex] p^* \end{bmatrix} = 0 \quad (16.38)
$$

where $u^*=u/a_o,\quad v^*=v/a_o,\quad M=\bar{U}/a_o,\quad p^*=p/\rho\,a_o^2$,   (16.39)

u,v being the components of the perturbation velocity in the x,y directions.

The boundary conditions are

$$v = 0\,(=v^*) \text{ at } y = 0,b \ . \tag{16.40}$$

If the duct is uniform in the x-direction the solution of (16.38) and (16.40) takes the form

$$
\begin{bmatrix} u^* \\ v^* \\ p^* \end{bmatrix} =
\begin{bmatrix} \bar{u} \\ \bar{v} \\ \bar{p} \end{bmatrix} \exp(ik(a_o t - \mu x)) \tag{16.41}
$$

where $k = \omega/a_o$. μ is a non-dimensional wave-number which may be complex.

Substituting (16.41) into equation (16.38) gives

$$
\begin{bmatrix}
ik(1-\mu M) & M' & -ik\mu \\
0 & ik(1-\mu M) & \mathrm{d}(\ )/\mathrm{d}y \\
-ik\mu & \mathrm{d}(\ )/\mathrm{d}y & ik(1-\mu M)
\end{bmatrix}
\begin{bmatrix} \bar{u} \\ \bar{v} \\ \bar{p} \end{bmatrix} = 0 \quad (16.42)
$$

or $[L]\,\{f\} = 0$.   (16.43)

The Galerkin method assumes a solution of the form

$$\{f\} = [N]\,\{\delta\} \tag{16.44}$$

where $[N]$ is a matrix of assumed functions which satisfy all the boundary conditions. In the present context, these functions are derived using finite element techniques. In fact, for two-dimensional ducts, these elements will be one-dimensional elements which span the width of the duct.

If the solution (16.44) is substituted into the left-hand side of (16.43) the result will not be a null matrix unless the assumed solution is the exact one, thus

$$[L]\,[N]\,\{\delta\} = \{R\} \tag{16.45}$$

where $\{R\}$ is a column matrix of residuals. The values of the parameters which give the 'best' approximation are the ones which make the weighted average of the residuals over the domain vanish. In the Galerkin method the weight functions are taken to be the same functions as the ones used in the trial solution, thus

$$\int_0^b [N]^T \{R\} \, dy = 0 \ . \tag{16.46}$$

Equations (16.45) and (16.46) give

$$[K] \{\delta\} = 0 \tag{16.47}$$

where

$$[K] = \int_0^b [N]^T [L] [N] \, dy \tag{16.48}$$

Equation (16.47) represents a linear eigenproblem in μ since the operator $[L]$ can be written in the form

$$[L] = [L_1] - \mu [L_2] \tag{16.49}$$

where

$$[L_1] = \begin{bmatrix} ik & M' & 0 \\ 0 & ik & d(\ )/dy \\ 0 & d(\ )/dy & ik \end{bmatrix} \tag{16.50}$$

and

$$[L_2] = \begin{bmatrix} ikM & 0 & ik \\ 0 & ikM & 0 \\ ik & 0 & ikM \end{bmatrix} \tag{16.51}$$

Further details can be obtained from references [16.36, 16.37]. Reference [16.38] also uses the same technique for axisymmetric ducts.

An alternative procedure is to eliminate the perturbation velocities $\bar{u}$ and $\bar{v}$ from equations (16.42). This gives

$$\frac{d^2\bar{p}}{dy^2} + \frac{2\mu M'}{(1 - \mu M)} \frac{d\bar{p}}{dy} + k^2 [(1 - \mu M)^2 - \mu^2] \bar{p} = 0 \ . \tag{16.52}$$

The boundary condition (16.40) now becomes

$$d\bar{p}/dy = 0 \quad \text{at } y = 0, b \ . \tag{16.53}$$

Equation (16.52) is now written in the alternative form

$$D(\bar{p}) = 0 \tag{16.54}$$

where

$$D(\ ) = \lambda^3 D_1(\ ) + \lambda^2 D_2(\ ) + \lambda D_3(\ ) + D_4(\ ) \tag{16.55}$$

$\lambda = 1/\mu$ and

$$D_1(\ ) = \frac{d^2}{dy^2} + k\ , \quad D_2(\ ) = -M\frac{d^2}{dy^2} + 2M'\frac{d}{dy} - 3k^2M$$

$$D_3(\ ) = -k^2(1 - 3M^2), \quad D_4(\ ) = k^2M(1 - M^2)\ . \tag{16.56}$$

Equation (16.54) is now solved using the Galerkin/finite element method as before. This leads to a cubic eigenproblem in λ which can be transformed to a linear eigenproblem and then solved by standard techniques. Further details are given in reference [16.39].

A Galerkin/finite element solution for non-uniform ducts with flow is given in references [16.40, 16.41].

REFERENCES

[16.1]   Huebner, K. H. (1975) *The finite element method for engineers.* New York: Wiley.

[16.2]   Zienkiewicz, O. C. (1977) *The finite element method in engineering science,* 3rd Edition. London: McGraw-Hill.

[16.3]   Craggs, A. (1972) *J. Sound Vib.,* **23**, 331-339. The use of simple three-dimensional acoustic finite elements for determining the natural modes and frequencies of complex shaped enclosures.

[16.4]   Craggs, A. (1971) *J. Sound Vib.,* **15**, 509-528. The transient response of a coupled plate-acoustic system using plate and acoustic finite elements.

[16.5]   Petyt, M., Lea, J. & Koopmann, G. (1976) *J. Sound Vib.,* **45**, 495-502. A finite element method for determining the acoustic modes of irregular shaped cavities.

[16.6]   Herting, D. N., Jospeh, J. A., Kuusinen, L. R. & MacNeal, R. H. (1971). NASA TM X-2378 NASTRAN: *Users' experiences* (ed. by P. J. Rainey), 285-324. Acoustic analysis of solid rocket motor cavities by a finite element method.

[16.7]   James, J. H. (1973) *Admiralty Research Laboratory Report* ARL/R/R4. Acoustic finite element analysis of axisymmetric fluid regions.

[16.8]   Craggs, A. (1976) *J. Sound Vib.,* **48**, 377-392. A finite element method for damped acoustic systems: an application to evaluate the performance of reactive mufflers.

[16.9]   Kagawa, Y. & Omote, T. (1976) *J. Acoust. Soc. America,* **60**, 1003-1013. Finite element simulation of acoustic filters of arbitrary profile with circular cross section.

[16.10] Kagawa, Y., Yamabuchi, T. & Mori, A. (1977) *J. Sound Vib.,* **53**, 357-374. Finite element simulation of axisymmetric acoustic transmission system with sound absorbing wall.

[16.11] Petyt, M. & Lim, S. P. (1980) *International Journal for Numerical Methods in Engineering,* **13**, 109-122. Finite element analysis of the noise inside a mechanically excited cylinder.

[16.12] Lim, S. P. & Petyt, M. (1980) *Recent advances in structural dynamics* (ed. by M. Petyt). Free vibration of a cylinder partially filled with a liquid.

[16.13] Shuku, T. & Ishihara, K. (1973) *J. Sound Vib.*, **29**, 67-76. The analysis of the acoustic field in irregularly shaped rooms by the finite element method.

[16.14] Craggs, A. & Stead, G. (1976) *Acustica*, **35**, 89-98. Sound transmission between enclosures – a study using plate and acoustic finite elements.

[16.15] Joppa, P. D. & Fyfe, I. M. (1978) *J. Sound Vib.*, **56**, 61-69. A finite element analysis of the impedance properties of irregular shaped cavities with absorptive boundaries.

[16.16] Richards, T. L. & Jha, S. K. (1979) *J. Sound Vib.*, **63**, 61-72. A simplified finite element method for studying acoustic characteristics inside a car cavity.

[16.17] McRae, G. J. (1974) *Proceedings Noise, Shock and Vibration Conference*, Melbourne 286-295. Finite element solutions of the wave equation.

[16.18] Petyt, M., Koopmann, G. H. & Pinnington, R. J. (1977) *J. Sound Vib.*, **53**, 71-82. The acoustic modes of a rectangular cavity containing a rigid, incomplete partition.

[16.19] Craggs, A. (1973) *J. Sound Vib.*, **30**, 343-357. An acoustic finite element approach for studying boundary flexibility and sound transmission between irregular enclosures.

[16.20] Craggs, A. (1977) *J. Sound Vib.*, **54**, 285-296. A finite element method for modelling dissipative mufflers with a locally reactive lining.

[16.21] Craggs, A. (1978) *J. Sound Vib.*, **61**, 101-111. A finite element model for rigid porous absorbing materials.

[16.22] Craggs, A. (1979) *J. Sound Vib.*, **66**, 605-614. Coupling of finite element acoustic absorption models.

[16.23] Hunt, J. T., Knittel, M. R. & Barach, D. (1974) *J. Acous. Soc. America*, **55**, 269-280. Finite element approach to acoustic radiation from elastic structures.

[16.24] Hunt, J. T., Knittel, M. R., Nichols, C. S. & Barach, D. (1975) *J. Acoust. Soc. America*, **57**, 287-299. Finite element approach to acoustic scattering from elastic structures.

[16.25] Bettess, P. & Zienkiewicz, O. C. (1977) *International Journal of Numerical Methods in Engineering*, **11**, 1271-1290. Diffraction and refraction of surface waves using finite and infinite elements.

[16.26] Bettess, P. (1977) *International Journal for Numerical Methods in Engineering*, **11**, 53-64. Infinite elements.

[16.27] Sommerfeld, A. (1949) *Partial differential equations in physics.*

New York: Academic Press.

[16.28] Bettess, P. (1980) *International Journal of Numerical Methods in Engineering*, **15**, 1613-1626. More on infinite elements.

[16.29] Wilton, D. T. (1978) *International Journal of Numerical Methods in Engineering*, **13**, 123-138. Acoustic radiation and scattering from elastic structures.

[16.30] Filippi, P. J. T. (1977) *J. Sound Vib.*, **54**, 473-500. Layer potentials and acoustic diffraction.

[16.31] Saylii, M. N., Ousset, Y. & Verchery, G. (1981) *J. Sound Vib.*, **74**, 187-204. Solution of radiation problems by collocation of integral formulations in terms of single and double layer potentials.

[16.32] Watson, W. & Lansing, D. L. (1976) *NASA* TN D-8186. A comparison of matrix methods for calculating eigenvalues in acoustically lined ducts.

[16.33] Astley, R. J. & Eversham, W. (1978) *J. Sound Vib.*, **57, 367-388**. A finite element method for transmission in non-uniform ducts without flow: comparison with the method of weighted residuals.

[16.34] Kagawa, Y., Yamabuchi, T., Yoshikawa, T., Ooie, S. & Kyouno, N. (1980) *J. Sound Vib.*, **69**, 207-228. Finite element approach to acoustic transmission-radiation systems and application to horn and silencer disign.

[16.35] Kagawa, Y., Yamabuchi, T., Sugihara, K. & Shindou, T. (1980) *J. Sound Vib.*, **69**, 229-244. A finite element approach to a coupled structural-acoustic radiation system with application to loudspeaker characteristic calculation.

[16.36] Astley, R. J. & Eversman, W. (1979) *J. Sound Vib.*, **65**, 61-74. A finite element formulation of the eigenvalue problem in lined ducts with flow.

[16.37] Astley, R. J. & Eversman, W. (1980) *J. Sound Vib.*, **69**, 13-25. The finite element duct eigenvalue problem: an improved formulation using Hermitian elements and no flow condensation.

[16.38] Abrahamson, A. L. (1977) *American Institute of Aeronautics and Astronautics. Paper* 77-1301. A finite element algorithm for sound propagation in axisymmetric ducts containing compressible mean flow.

[16.39] Fahmy, M. S. Y. (1979) University of Southampton. *ISVR Technical Report* No. 105. A finite element formulation of the eigenvalue problem of sound propagation in uniform ducts with shear flow.

[16.40] Ling, S-F. (1976) Purdue University, PhD Thesis. A finite element method for duct acoustic problems.

[16.41] Eversman, W. & Astley, R. J. (1981) *J. Sound Vib.*, **74**, 103-122. Acoustic transmission in non-uniform ducts with mean flow, Part II: The finite element method.

CHAPTER 17

Noise from industrial plant

The current state of the art and future prospects

A. H. Middleton

Institute of Sound and Vibration Research, University of Southampton

17.1 INTRODUCTION

Control of noise from large industrial plant has become an important aspect of the design and operation of such plant. Over the past ten years noise control requirements have become increasingly severe. Initially the stimulus for noise control was an appreciation of the possibility of permanent hearing damage to plant operators, and a growing awareness of noise amongst inhabitants of neighbouring communities produced complaints about plant noise. More recently, legislation has provided an even more persistent stimulus for noise control. In the design and operation of industrial plant the in-plant noise and the environmental noise must be separately assessed, although there is a relationship between them.

17.2 IN-PLANT NOISE

Company and legislative policies regarding in-plant noise vary, depending upon whether it is considered that 'noise emitted' is an important consideration, or whether 'noise received' should be the criterion.

If 'noise emitted' is considered important, noise control measures will be necessary for all plant items which are likely to produce an in-plant noise environment above the set limit. Current in-plant or work area noise limits are usually in the region of 85 to 90 dB(A), with occasional requirements down to 80 dB(A). Legislative requirements in many countries demand a level not exceeding 90 dB(A) in all areas normally accessible to personnel. Where no legislation exists some plant operators require 85 or 90 dB(A) in work areas and allow up to 95 dB(A) in areas where personnel are not permanently stationed. Any areas where noise is above 90 dB(A) are usually marked off as 'noise hazard zones', where ear protection must be used. Concawe report no. 3/81 reviews legislative requirements for in-plant noise [17.1].

Some special problems exist where 'noise received' is regarded as the criterion. Unlike operatives in mechanical manufacturing industries, many process plant operators have a roving commission around the plant, inspecting equipment and carrying out adjustment. Noise levels at specific points within an operating plant remain more or less constant with time, but the varying personal exposure patterns of the operators make it difficult to predict their daily noise dose. It is even more difficult for maintenance personnel, who sometimes have to work in places which are not normal access areas and which may be rather noisy. Some organizations have attempted to deal with this problem by posting notices defining the allowable daily exposure duration in noisy areas, and requiring personnel to regulate their own noise exposure. This no longer seems to be an acceptable way of managing the problem. Because exposure times can be so variable, and because the management cannot absolve itself from the responsibility of protecting the operatives from excessive noise exposure, the only reasonable course of action now is to designate all areas in which noise levels are over 90 db(A) (or other appropriate limit) as noise hazard zones, in which ear protection must be used at all times. Control of noise exposure by equipping all personnel with noise dosemeters is hardly practical, on the grounds of cost and inconvenience to the wearers.

17.3 SOURCES OF NOISE

Large industrial plants usually contain large machinery items and other non-mechanical noise producers such as furnaces and boilers. Each has its own particular characteristics and means of noise control.

17.3.1 Furnaces

The main sources of noise are airborne noise radiated from burners, noise radiated from furnace walls, and noise radiated from flue gas ducts and stacks. An indication of the relative importance of these sources is given in Table 7.1, taken from [17.2].

Table 7.1

Relative sound power emissions of furnace components
(dB re 10^{-12} watts)

	Furnace No.				
	A	B	C	D	E
Burner wall	112	114	115	105	104
Sidewalls	107	112	112	100	100
Stack	106	103	107	83	83
Total	114	116	117	106	105

Noise from burners is caused by turbulence generated in the air being drawn in through the air register and by the combustion process within the furnace. In an unsilenced burner the combustion space is directly connected to the outside air via the burner air inlets. Burner noise can be controlled by adding a plenum chamber around the burners which is itself supplied with air via silencers. These may be of the splitter type or the labyrinth type. The attenuation characteristics required of such a system can usually be estimated from burner vendors' data on noise from unsilenced burners in test furnaces. These data are usually in the form of sound pressure levels at one metre from the burner, or in the case of cylindrical vertical furnaces, are sometimes the sound pressure levels between the furnace legs. The sound power level emitted can be deduced by modelling the space beneath the furnace as a semi-reverberant room, with reflecting upper and lower surfaces and absorbent walls.

Where possible, the internal surfaces of the plenum chamber should be lined with mineral wool or glass wool acoustic absorption material. The lower surface of the plenum chamber should have sufficient transmission loss to reduce sound radiation into the space below it to an acceptable level, usually below 85 dB(A).

To achieve the full benefit from fitting a plenum chamber it is desirable that it should not be connected structurally to the burner. Controls should preferably be rotary action and should be led out through acoustically lined sleeves. Sight holes and lighting holes should be provided with spring-closed trap doors. Fuel pipes should be led out through lined sleeves.

An alternative and, usually, preferred solution to furnace burner noise is to convert to forced draught operation. This can produce side benefits in reduced furnace operating costs as a result of improved air to fuel mixture control. The forced draught ducting and fan casing need to be heavily lagged to prevent noise radiation, and a silencer must be provided which is able to deal with the fan sound power output (which should be obtainable from the fan manufacturers) and burner noise which will be transmitted down the ducting. Duct wall radiation can be effectively minimized by burying as much of the ducting as possible.

Furnace wall radiated noise is normally mainly low frequency in character. Furnace walls are of heavy construction and hence provide a considerable transmission loss at high frequencies. Wall-radiated sound power level is generally about 15 dB below unsilenced burner radiated power level. Reduction of wall-radiated noise is difficult and costly. Some double-wall constructions have been proposed, but it is not clear how much benefit can be obtained.

A recent development in furnace construction has been the use of low-density high-temperature ceramic fibre linings. These replace the conventional refractory lining. Whilst they are thermally effective the implications of this structural change on noise emission must be considered. The mass per unit area of the furnace wall is substantially reduced and hence the transmission loss is likely to be reduced, but the internal acoustic absorption is increased, thus pre-sumably reducing the reverberant sound level inside the furnace. The practical consequences of this have yet to be assessed by a direct comparison between otherwise similar furnaces.

Noise radiated from furnace stacks is usually difficult to quantify by simple noise measurements because of access problems on an operating plant. Directivity effects of stacks are substantial, so ground-level measurements are not helpful.

Where induced draught fans are used the fan sound power radiated from the stack and ducting should be assessed, and in-line silencers and lagging fitted as necessary. When operating in a hot gas, the performance of silencers will be modified by the change of wavelength for a given frequency of sound, and also by variations of the acoustic impedance of the silencer packing material in the case of absorption silencers.

Mechanical design of silencers in such situations can present problems in the fields of corrosion, mechanical integrity of packing materials and retainers and blockage of the pores of packing materials by combustion products. Split-ters which can be easily withdrawn for servicing may be desirable.

Stack noise problems at specific frequencies can be traced to low-frequency excitation of longitudinal or transverse standing waves in the ducting or stack. Fitting longitudinal dividing plates in ducts can cure transverse standing wave problems. Alterations of duct lengths by a quarter wavelength of sound at the problem frequency, or the introduction of reactive silencing elements in ducts, can eliminate longitudinal standing wave problems. In-line reactive silencers consisting of tuned length expansion chambers, or side branch resonators, can be used [17.3]. An expansion chamber can be effective if its length is a quarter wavelength of the sound in the duct.

17.3.2 Boilers

Noise problems with boilers are usually associated with the forced draught fan and ducting, which can be dealt with as described above, and with radiation from the burner front. Secondary enclosure of this area may be the only way of reducing the local noise to below 90 dB(A), but it is often acceptable to desig-nate the relatively small area around the boiler front as a noise hazard zone and avoid the need for noise control. Obvious sources of noise such as fuel feed pipes, valves, and pumps should be dealt with if possible. Proprietary package attenu-ator units for boilers are now available. These fit over the burner area and can be moved aside for servicing.

Low-frequency rumble problems from boilers are common. Analysis of the noise usually shows that a number of internal cavity resonances of the boiler and stack are excited intermittently. The cause can be instability in the fan airflow resulting from operating the fan out of its intended operating range.

The remedy is to select a fan which is operating at or near its design point under the damand of the boiler, and, if a large turndown ratio is necessary, to control the airflow by fan speed change rather than damper control.

Boiler noise can also be traced to pulsating combustion, sometimes related to rotation frequency of rotating cap burners, at other times caused by a forced oscillation of inlet air flow. The energy to keep the oscillation going is supplied by the fuel.

17.3.3 Compressors

Nearly all types of compressor produce excessive noise. The quietest, which produce around 85 to 96 dB(A) at one metre distance, are slow-speed reciprocating compressors. Other types, such as centrifugal, screw, or roots types, invariably require some form of enclosure to prevent substantial in-plant and environmental noise problems. The fitting of enclosures may cause explosion hazards. To overcome these, a highly efficient ventilation system will be required. This in turn presents noise control problems as it is necessary to provide a fan with an inlet silencer to suppress fan noise and duct-borne compressor noise and a similar ventilation air outlet silencer. It will normally be necessary to enclose the driving motor or turbine separately, possibly with a different enclosure ventilation pressure.

The reason for differential pressurization is to ensure that the motor enclosure pressure is always higher than the compressor enclosure pressure so that there can be no transfer of combustible gases into the motor enclosure.

It is essential with all rotary compressor installations to lag the inlet and outlet piping, and if possible to fit in-line silencers.

In-line silencers must have very high levels of mechanical integrity. Mechanical failure of a silencer upstream of a compressor could result in debris entering the compressor causing severe damage or the risk of explosion. As a safety measure 'debris catchers' consisting of perforated plates or honeycomb constructions may be fitted between the silencer and the compressor. Because of the high sound pressure levels, high flow velocities, and the danger of shedding fibres, normal fibrous silencer packing materials are not usually acceptable. Wire-wool packing or porous metal screens can be used instead. It may be necessary to lag intercoolers and aftercoolers and their associated water piping in some cases. Structure-borne vibration can cause these parts to be substantial noise radiators. Unsilenced rotary compressors cause particularly severe environmental problems because of the distinctive waveform of their noise emission.

The commonly used half-walled compressor house is of little value for community noise control, but it does increase the noise level around the compressor by a few decibels as a result of reverberation. Compressor houses with full-height walls on two or three sides can provide useful screening for community noise control.

Few compressor manufacturers appear to be willing to provide comprehensive information about the noise emitted by their machine, so site measurements of similar equipment must normally be used for prediction purposes.

17.3.4 Motors

In a modern noise-controlled plant low noise electric motors will usually be required. Motor noise is reduced by fitting more efficient or reduced size fans, fitting fan intake and exit silencers, and by increasing frame size and insulation grade to allow for a reduced cooling air flow. Sometimes complete enclosures are fitted. Most motor manufacturers now offer low-noise motors, often with an option of two stages of noise control. Off-load noise outputs are usually readily available, but on-load data very rarely. It is common to add 3 db(A) to the off-load data to allow for on-load conditions. Care should be taken to allow for the test shop and site environments when converting manufacturer's motor noise data to expected levels on site. The common position for motors, under a pipe rack, acts as a semi-reverberent space, thus causing increased noise levels close to motors.

The main noise source in totally enclosed fan cooling (TEFC), closed circuit air cooling (CACA), and water protected (WP) motors is the cooling fan. However, in some instances magnetic noise can be a problem. This is apparent as powerful tonal components in the noise spectrum. Total enclosure or a change of rotor design may be necessary to eliminate the problem. Most manufacturers are able to control magnetic noise at the design stage, and it should not be significant with any well-established designs.

Bearing noise is not normally significant unless the bearings are damaged.

Motor noise as a function of speed and power output may be estimated from information published by BEAMA and NEMA [17.4] and [17.5], although it is preferable to obtain manufacturer's data where possible.

17.3.5 Steam turbines

The steam turbine drivers used on process plants can be significant sources of tonal noise, predominantly at blade passing frequency and its harmonics. Control valves in the steam supply lines may create greater noise than the turbine, but this can be reduced by selection of appropriate valve noise control measures.

Thermal lagging, which is commonly applied to steam turbines, can be made to double as acoustic lagging by ensuring that it forms a complete enclosure round the turbine and that it is isolated from turbine casing vibrations by a fibrous resilient layer such as glass or mineral wool. Acoustic lagging is discussed separately.

17.3.6 Gas turbines

Large industrial gas turbines are used for power generation and compressor drivers. Power generation turbines are commonly sold as packaged units contained in an acoustic enclosure with silencers on the turbine inlet and outlet. Manufacturers of these units are usually prepared to quote guaranteed noise levels, often at distances of one metre from the casing and 120 m from the centre of the enclosure. The most difficult problem in the design of industrial gas turbine noise control is the design of the exhaust silencer. It is subjected to a harsh internal environment from the exhaust gas stream, and at the same time has to provide a high level of attenuation of sound in the exhaust gases. Methods for optimum design of hot-gas silencers need further investigation.

Gas turbine exhaust noise can be particularly troublesome at very low frequencies, in the 16 and 32 Hz octave bands. Accurate design procedures for low-frequency silencers do not exist. In some installations longitudinal duct modes can be excited by the turbine exhaust. These aggravate the low-frequency problem. A cure is often to be found in improvements in exhaust gas flow ducts.

When estimating the sound power level of a package gas turbine unit from 120 metre sound pressure level data it is advisable to bear in mind that the exhaust is normally ejected upward, and the directivity effects of the upward pointing stack will provide substantial reduction of high-frequency noise at 120 m distance at ground level. The directivity effects can be estimated from [17.6]. During conditions of atmospheric temperature inversion the upward radiated components of sound power could be important in the surrounding community.

Gas turbines sold for compressor driving and similar duties are not necessarily fitted with silencing packages, although most manufacturers are able to supply such items, or recommend suitable suppliers. It is usually preferable to buy a complete silenced unit from one source, if possible. This reduces the possibilities of problems arising over noise guarantees. Particular ventilation features may be necessary for gas turbine enclosures in potentially hazardous parts of a plant.

17.3.7 Pumps

Until recent legislation and operator requirements forced plant designers to specify low-noise electric motors, pump noise tended to be insignificant compared with motor noise. It is normally 10 to 15 dB(A) below unsilenced motor noise. Information on pump noise is still difficult to obtain. Few manufacturers have reliable data. Data from test beds usually relate to operation with water, and may have to be corrected to allow for use on process fluids.

Significant noise can be radiated from pipework associated with pumps if the dimensions of the pipework are such that standing waves are set up in the fluid column contained within the pipe.

High-pressure hydraulic power systems, with pressure supplied by gear or piston pumps, can be extremely noisy. Their noise is usually radiated from the pipework or reservoirs rather than the relatively massive pump casing. High-pressure pump noise can be reduced by detailed attention to pump design to minimize the rate of change of pressure during the pumping cycle and to prevent cavitation. In-line silencers are available which can reduce noise radiated from the pipework. Methods and positioning of pipework mountings are important to prevent noise radiation from supporting structures. A detailed discussion of causes of hydraulic system noise is given in [17.7].

17.3.8 Control valves

Unless control valve noise is considered at the design stage it can be one of the major sources of plant noise. Rectification is expensive and inconvenient. Normal control valves in gas service operating under choked flow conditions generate noise inside the pipe by a jet type phenomenon. The noise produced is radiated mainly from the pipe walls downstream of the valve, but there is also some radiation from upstream piping. Several noise control techniques are available:

(a) lagging or burying below ground of piping and valves
(b) fitting multihole orifice plates in the line downstream of the valve to split the required pressure drop into several stages, each stage preferably having less than critical pressure drop
(c) fitting in-line silencers downstream, and possibly upstream of the valve, in conjunction with lagging of the piping between valve and silencer,
(d) fitting low-noise control valves.

Lagging is simple in theory, but has problems in practice when it has to be fitted around pipe supports and flanges. It tends to be damaged rather easily and is often not refitted properly after maintenance operations have been carried out.

In-line silencers can offer 10 to 20 dB(A) reduction in piping-radiated noise. Most are absorptive types with some form of fibrous lining. The design of the lining retention means can be difficult. Regular lining replacement may be necessary. Some processes and machines cannot tolerate the passage of lining debris, thus prohibiting the use of absorptive silencers. Where low- or mid-frequency noise exists in a pipeline, particularly if it contains prominent tonal components such as may be emitted by a compressor, reactive silencers can only deal with plane-wave mode propagation within a duct, and thus the maximum frequency at which their performance can be predicted is inversely proportional to the pipe diameter.

Many types of low-noise control valve are available. They achieve noise reduction by one or more of three techniques:

(a) raising the frequency and lowering the overall level of the jet noise produced in the valve by using large numbers of small orifices rather than one large orifice. The increase in frequency is beneficial because the pipe transmission loss increases with frequency;
(b) using a valve plug or seat design which gives multiple expansions of the gas;
(c) passing the gas through a series of long narrow channels in which the loss mechanism is viscous, rather than turbulent.

Most manufacturers of control valves now have noise-prediction techniques for their particular valves. They normally predict noise at 1 metre from the pipe surface close to the valve. The more comprehensive of these are in the form of computer programmes which are able to indicate various alternative ways of achieving a given noise level, depending on the valve duty, pipe sizes, and noise levels required. The prediction techniques of the larger manufacturers are as accurate as could reasonably be expected of any prediction technique. 'Errors' which occur can often be traced to valves which are not operated in the manner which was envisaged at the time of design.

The biggest problem in the application of valve noise prediction data is in the estimation of the rate of decay of sound pressure level with distance within the pipe, and the estimation of the sound level which may be radiated from vessels fed by the pipe. These difficulties become particularly significant in plants with low machinery noise level because pipe-radiated noise is then likely to be dominant. At present there are no standardized ways of predicting or estimating noise from long pipes.

Commercial valve noise-prediction methods predict sound pressure levels outside the piping. They do not predict sound power level generated inside the pipe. Internal sound power level could be estimated using a theoretical pipe transmission loss. If internal PWL is calculated the methods of [17.8] could be used to predict the distribution of sound power level within branched piping systems, and hence to calculate the SPL outside branch pipes.

17.3.9 Gearboxes
Gearboxes on industrial machines usually produce noise levels of 85 dB(A) or above at 1 metre from their casing. Traditional methods of construction of gearboxes make little concession to noise control, by having casings which are efficient radiators of noise well coupled to the bearing housings through which gear-induced vibrations are fed. In consequence, it will usually be necessary to fit an enclosure over a gearbox. Apart from ensuring that there is adequate cooling of the lubricant by an oil/water heat exchanger or radiator outside the enclosure, there should be few problems in fitting enclosures to gearboxes which are not an integral part of the machine. Where a gearbox is built as an integral part of a machine, such as in unit construction pumps, or in rolling mills, the problem of containing gearbox noise is substantially more difficult because it is transmitted via structural paths to remote parts of the machine. In the design of such machines it would be desirable to isolate the driving units from the outer casing to prevent radiation of structure-borne noise, but this is rarely done.

Gearbox manufaturers are often able to supply noise level data for their products and should be consulted at an early stage of plant design.

17.3.10 Fans
Probably the most common source of industrial noise which causes community reaction is the fan. Such reaction tends to be strongest when the fan noise con-

tains tonal components, indicative of interaction between the moving fan blades and stationary features close to the blades. A large proportion of fan noise problems could be eliminated at the plant design stage if the noise emitted by the fans had been estimated, and suitable silencers fitted.

There are many noise-prediction methods available for fans, some being reviewed in [17.9]. These can give an indication at an early stage in the plant design process of the sound power level likely to be produced by a given fan. When a particular fan is selected the manufacturer can usually supply sound power level data. Be certain to correct these data for installation conditions which are different from test conditions. Reference to silencer manufacturers' catalogues will usually allow a suitable silencer to be selected to give the required insertion loss.

Fan noise prediction methods usually assume that the fan is operating on its design point. Many industrial installations depart substantially from this ideal situation. If the fan is operating in a stalled condition the low-frequency noise will generally increase and become irregular in its time history. These conditions can be particularly troublesome with such equipment as forced-draught fans for boiler combustion air, where the fan airflow is often controlled merely by a damper and the instability can react with the combustion process and the internal cavities of the boiler. The low-frequency spectral components can be difficult to deal with by silencing, and the provision of silencing may not be the best solution in such cases.

Two particular cases of fan noise which cause community complaint are cooling towers and air fin coolers. Both use large specially designed fans which have been the subject of intensive development programmes during the last ten years to enable their noise level to be reduced. Modern wide-bladed aerodynamically designed low tip speed fans can be 15 dB(A) or more quieter for the same airflow than the previous narrow-bladed high tip speed fans and are also more efficient. Individual fan manufacturers should be consulted for data.

Methods of measuring sound power level from air fin coolers and their fans are not standardized. Different measuring methods will indicate different apparent sound power levels for a given fan. This must be borne in mind when comparing data from different manufacturers. Concawe have published a standardized test procedure for fin fans, suitable for use in operating plant situation [17.10]. Results obtained by this method give estimates of sound power level which may differ by a few decibels from that estimated from more distant measurements (at say 10 m from the fan centre) on a test rig, but distant measurements are not usually possible in an operating plant.

Details of the design of the cooler bank and plenum chamber can affect the sound power level of a given fan when fitted in a range of installations. Small-scale features associated with fan supports or fan ring manufacture can cause variations of up to 5 dB in the higher frequency octave bands. Gross details of plenum chamber design can cause effects in the lower frequency octave bands, which are normally more critical. If a particularly low-noise installation is required a fan manufacturer should be asked to substantiate his predicted or guaranteed noise levels with a rig test on a typical fan and cooler installation.

Motor and belt noise may be significant in very quiet installations. Toothed belt drives must be assessed particularly carefully.

Fully ducted or heavily silenced industrial fan installations can emit undesirably high noise levels as a result of duct wall radiation. This is particularly troublesome with rectangular section ducts. Air supply ducts for forced-draught furnaces and boilers would normally need to be lagged. Stiffening and bracing of duct walls may be necessary.

17.3.11 Vents and stacks

Any plant in which steam is used will sometimes need to vent steam to atmosphere. This may happen for periods of several hours or days during commissioning of the plant, and may be done regularly for the purpose of cleaning or testing of relief valves. Vents will normally need to be fitted with silencers of adequate capacity to control the noise of the maximum steam flow which could predictably be emitted. The sound power level likely to be emitted from stacks and vents can be estimated using jet noise theory, as indicated in [17.11] and [17.12]. Note that noise produced by flow through a valve located upstream of the vent may also be radiated by the vent. In-plant or community noise requirements will dictate the performance required of a vent silencer. Allowance may be made for directivity effects [17.6].

Intense noise sources located high above ground level (for instance the stack of a cat cracker) can cause unusually unpleasant environmental conditions. They suffer no attenuation due to screening by other plant items and no ground attenuation, and in gusting wind conditions are exposed to high levels of atmospheric turbulence. This can result in up to 15 dB(A) variation of noise level received by a community over a period of a few seconds. Even if the noise emitted is of a broadband character, which would normally cause the minimum community annoyance, the variability caused by atmospheric turbulence immediately makes the noise very easily distinguishable and hence annoying. Measurements of average noise levels may not give a good indication of community reaction.

17.3.12 Flares

Noise sources in elevated flares are steam noise and combustion noise. Steam noise is usually predominant. As the noise source is elevated it suffers the same problem of atmospheric turbulence variations as the high stacks described above. In the case of a flare, attention is drawn to it by the flame itself.

During normal operation steam flows should be small and atmospheric attenuation may be sufficient to minimize noise at the community. At higher steam flows steam injection nozzle design becomes an important consideration.

Multihole nozzles are normally quieter than single-hole nozzles. The erection of silenced plenum chambers around the air entrainment region of the steam injectors can reduce directly radiated noise, but a considerable quantity of steam noise is inevitably conveyed with the entrained air to the flare tip and thence radiated to the community.

When very stringent noise limits are enforced it is desirable to consider the use of a non-steam-injected ground flare for normal flaring, if necessary backed up by an elevated flare for emergencies.

Noise levels close to ground flares are not outstandingly low, but being at a low elevation they have the benefit of ground attenuation effects before the noise reaches the community.

Flare manufacturers can usually provide information on noise levels.

17.3.13 Mechanical sources
Some plants contain mechanical sources of noise, such as conveyors, vibrating screens, and hoppers into which solids are discharged. There is little generalized information available which would allow predictions to be made of noise from such sources. It will usually be necessary to take measurements of similar equipment in operation.

Noise control can often be achieved by cushioning the impact of solids on metal surfaces by fitting resilient linings to hoppers and chutes. Proprietary linings are available for this. It is sometimes possible to enclose noisy mechanical equipment inside buildings. Conventional industrial building construction can provide substantial noise reductions. The limitation on the noise reduction achievable is usually that imposed by leakage beneath eaves, through ventilators, and through open doors and broken windows.

Vibrating screens and conveyors can produce particular problems at low frequencies, particularly if they are enclosed in buildings or fitted with inlet or outlet ducts one of whose major dimensions is a half wavelength of sound at the oscillation frequency, or multiples of that.

Handling of structural materials, particularly metallic ones, can produce high in-plant and community noise levels. Every effort should be made to avoid the application of impulsive force inputs.

17.3.4 Vehicle noise source
Many plants require the use of vehicles for the transport of materials around the plant, or the transport of products to or away from the plant. These can be a cause of substantial community annoyance, and in many cases hearing hazard to their operators. Wherever possible on-site equipment should be fitted with silencers and operated in such a way that noise is minimized.

The routeing and timing of regularly operating delivery vehicles should be regulated to minimize community annoyance.

17.4 LAGGING

Many types of lagging are used to obtain thermal insulation. Many of these are unsuitable for acoustic insulation. An essential requirement of acoustic insulation is that a heavy, damped impervious outer layer should be fitted over a vibration-isolating layer wrapped around the pipe. The vibration-isolating layer may be any unbonded fibrous material such as mineral or glass wool, usually of 65 to 80 kg-m^{-3} density, capable of supporting the weight of the outer layer without crushing to a degree where it ceases to be resilient. The outer layer may consist of steel or aluminium coated with damping material, or lead or loaded plastic sheeting. Surface weight should be 5 to 10 kg-m^{-2}. In extreme cases double lagging may be desirable and may be necessary in cases where combined thermal and acoustic lagging is required.

The actual performance of lagging achievable in the field tends to be controlled by acoustic 'short circuits' between the outer covering and the pipe [17.13]. The performance achieved under ideal laboratory conditions may not be representative.

17.5 PREDICTION OF COMMUNITY NOISE

The prediction of community noise from large plants can be simplified to a routine procedure. For the sake of convenience and compatibility between predictions by various bodies it is desirable that it should be standardized into a set of rules. One such set of rules is OCMA guides NWG-1, 2 and 3 [17.14]. The problem is that such a set of rules can deal only in the most general way with each individual case. The rules are generalized in so far as they assume one set of values for atmospheric attenuation and ascribe particular values to plant screening and ground absorption effects. In actuality atmospheric variations cause wide variations in noise levels received by a community [17.15] over a period of months or years as well as wide variations over periods of seconds or minutes. It has been shown that plant screening effects vary very substantially with source position and direction of noise radiation from a plant [17.16], and the OCMA specification cannot take this fully into account. Concawe have published a prediction model which takes account of meteorological conditions [17.17].

Writers of legislation and codes of practice like to imagine that noise levels can be specified by a single number. An average noise level can be specified by a single number, but it may not turn out to be a good measure of relative community annoyance. Even to check whether a plant complies on average with a noise specification, it is likely that measurements will have to be made over a period of months. Plant design to achieve a given environmental noise level is therefore fraught with difficulty, and can be tackled sensibly only if those setting noise limits and those designing the plant can agree at the outset on a set of rules by which the game is to be played, and to recognize that when the plant is completed it will not be easy to decide precisely whether the required noise level has been achieved. As rules for the game, specifications such as OCMA's are very valuable. Strict regulations which are enforced blindly do not necessarily bring any great benefits to the community.

Noise control may only cost an additional few per cent of the capital cost of the project, but because the initial cost is so high the noise control costs are also high ([17.12] and [17.18]). Thus accurate prediction of the sound power emitted by equipment at the design stage and accurate assessment of the reduction which can be achieved by noise control features are most important. Neither of these tasks is easy. The relatively simple relationship between plant sound power and community sound pressure level contains hidden pitfalls.

$$L_p = L_W - 20 \log r - 8 - (A + B + C + D)$$

where L_p = community SPL, received at a distance r(m) from the plant centre

       L_W = total sound power level of plant
       A = atmospheric attenuation
       B = attenuation due to ground
       C = attenuation, negative or positive, due to plant screening
       D = other attenuation.

17.5.1 Assessment of plant sound power levels

Rarely are measurement conditions suitable for precise measurement of the noise emitted from each particular piece of equipment on a plant. Many noisy items are interdependent in their operation and each provides a high background noise level for measurements of the noise of the others. Because of this background noise problem, measurement positions can rarely be far enough from the noise source to allow a proper hemispherical or spherical measurement envelope in the direct field to be used. The density of packing of equipment also tends to prevent ideal conditions being achieved. To overcome the problems, 'engineering approximations', such as these recommended in the OCMA guide, have to be made by measuring close to equipment, and correcting for equipment surface area. Fortunately such measurements tend to overestimate sound power levels, giving a built-in safety factor. Reference [17.19] discusses in detail the errors involved in taking measurements too close to sources.

At the design stage of a plant, predictions of sound power level must be made from previously measured data, manufacturers' data, or estimates. Whilst it is necessary to base a plant design on this evidence one should be aware of its limitations. Measurement accuracy of the original data was certainly no better than ± 1 dB, even using precision grade measuring equipment. Repeatability of the measurements may be no better than ± 2 dB. The grade of instrumentation may be unknown, so a greater error may be possible. Many new plants use equipment whose design is scaled up from previous experience to give greater performance. Scaling laws may not be very precise. Precise mechanical, thermodynamic, and aerodynamic scaling does not always occur in the equipment.

The difficulties involved in estimation of the sound power level of piping systems, one of the major plant noise sources, have already been discussed.

The greatest scope for improvement in the assessment of the sound power levels of items of equipment lies in the gathering and publishing of as much noise level data as possible on all classes of equipment, coupled with appropriate performance information. It is essential that uniform methods of measurement are used and that high-grade measurement equipment is competently handled.

17.5.2 Assessment of attenuation

The OCMA guide gives precise values of still air atmospheric attenuation and plant screening and ground absorption. Numerous papers ([17.20] and others) discuss atmospheric attenuation for particular conditions, such as humid tropical environments. A correction can readily be made to the OCMA data. OCMA makes no effort to predict the effects of wind or temperature inversions. Both have a considerable part to play in the wide variation in environmental noise levels perceived round a plant during a year. The standard deviation about a mean is likely to be around 4 to 5 dB.

The problems in the accurate prediction of in-plant screening are very considerable. There is still a desparate lack of design data. Reference [17.16] shows that plant screening is a variable quantity depending on direction of propagation and the position of the noise source within a plant. This is the only known published reference and applies to one specific plant. The greater the amount of noise control required on a plant, the more important is a knowledge of in-plant screening attenuation because of the increased cost per decibel of noise control. Legislative pressures to reduce plant noise are increasing, so the requirement for a more accurate in-plant attenuation prediction procedure should be increasing. An overdesign of one plant to the extent of one decibel's worth of unnecessary noise control would more than pay for a substantial investigation of in-plant attentuation.

Surprisingly little information exists on the attenuation which could be expected from storage tanks. This is unlikely to be useful at large distances but could be useful for achieving boundary fence limits which are demanded by some authorities.

17.6 COST BENEFIT ANALYSIS

An essential part of noise control design of plant is to achieve a minimum-cost solution. There can be several possible ways of controlling the noise from individual items of equipment. As an aid to cost benefit analysis, equipment bid requests should demand noise levels and costs of various stages of noise control on each piece of equipment. To make use of this information a computer model of the plant noise is essential. This model should be able to take as input sound power levels and position coordinates of each equipment item, an attenuation and screening effect to be attributed to each item and to calculate sound pressure levels at any position coordinates around the plant. A sophisticated model will also be able to predict in-plant noise levels. Means must be available in the program to make rapid substitutions of equipment and assess their effect on received sound pressure level. Many calculations will be necessary in the search for the most cost-effective solution.

Several companies and other organizations working in the noise control field have suitable computer programs available.

17.7 NOISE DURING CONSTRUCTION

A recent innovation has been the need to control noise during the construction phase of a plant. This poses a range of new problems for contractors. In Britain, the Control of Pollution Act 1974 gives local authorities the power to regulate the noise of construction processes. The *Code of Practice* on construction noise [17.21] gives guidance on calculation procedures for construction noise. In the case of the erection of a large plant with many contractors involved over a lengthy period and a large area of ground construction, noise prediction and control would become a major exercise.

17.8 CONCLUSION

Legislation has become a considerable stimulus to the design of low-noise industrial plants. The stimulus has so far been greater in certain countries of continental Europe than in Britain. Design procedures to predict plant noise have been developed to a sufficient degree for them to be a useful tool for the plant designer. Owing to the lack of data in some areas, design accuracy is not great, but a number of features of present methods tend to err on the side of overdesign, which is safe if expensive. Striving for fine precision of design is not realistic if the vagaries of atmospheric propagation are considered, although an attempt at accuracy may be necessary to fulfil legislative requirements. Comparatively small sums of money need to be invested by the industry to fill in some of the gaps in available knowledge.

REFERENCES

[17.1]   Concawe (1981) *Review of national legislation official guidelines on work area noise,* Concawe report No. 3/81.

[17.2]   Seebold, J. (1981) *Noise Control Engineering,* **17**, No. 1. Sounding board.

[17.3]   Ihde, W. H. (1975) *Noise Control Engineering,* **5**, No. 3. Tuning stubs to silence large air-handling systems.

[17.4]   British Electrical and Allied Manufacturers Association. Publication
         No. 225. (1969).
[17.5]   N.E.M.A. Publication MG1 – 12.49. (1970).
[17.6]   Bolt, R. H., Beranek, L. L. & Newman, R. B. (1952). Vol. 1, Physical
         acoustics, Wright Air Development Centre, Tech. Report 52-204.
         Handbook of acoustic noise control.
[17.7]   Dransfield, P., Stecki, J. S. & Wilkins, P. (1974) *Noise Shock and
         Vibration Conference,* Monash University, Melbourne, Australia.
         Noise in hydraulic control systems.
[17.8]   Small, D. J, & Read, R. B. V. (1975), *Conference on Vibration and
         Noise in Pump Fan and Compressor Installations,* I.Mech.E., South-
         ampton, Paper No. C/14/75. The build-up of broad band high fre-
         quency sound due to reflections in pipes.
[17.9]   Erskine, J. B. & Brunt, J. (1975) *Conference on Vibrations and Noise
         in Pump, Fan and Compressor Installations,* I.Mech.E. Southampton,
         Paper No. C/104/75. Prediction and control of noise in fan instal-
         lations.
[17.10] Concawe (1978). *Method for determining the sound power levels of
         air-cooled heat exchangers.* Concawe report no. 5/78.
[17.11] Beranak, L. L. (1971) *Noise and vibration control.* McGraw–Hill.
[17.12] Anderson, R. E. (1973). 85th ASA Meeting, Paper M10. *Blowdown
         noise – analysis and control.*
[17.13] Smith, T., Rae, J. M. & Lawson, P. (1980) *Applied Acoustics,* **13,** No.
         5. Pipe lagging – an effective method of noise control?
[17.14] Oil Companies Materials Association, London (1972). NWG1, 2 and 3,
         *Procedural specification for the limitation of noise in plant and equip-
         ment for use in the petroleum industry.*
[17.15] Sutton, P. (1980) *J. Sound Vib.,* **8(1),** 33-34. Design of a noise
         specification for process plant.
[17.16] Middleton, A. H. & Seebold, J. G. (1972), *Internoise '72 Washington
         D.C.* Propagation of machine generated sound within and around a
         process plant.
[17.17] Concawe (1981) *The propagation of noise from petroleum and petro-
         chemical complexes to neighbouring communities.* Report no. 4/81.
[17.18] Sutton, P. (1975) *9th World Petroleum Congress,* Process plant noise
         evaluation and control.
[17.19] Concawe (1976) *Determination of sound power levels of industrial
         equipment, particularly oil industry plant.* Report no. 2/76.
[17.20] Piercy, J. E., Embleton, T. F. W. & Sutherland, L. C. (1977). Review
         of noise propagation in the atmosphere. *J. Acoust. Soc. Amer.,*
         61, No. 6.
[17.21] BS 5228: 1975 *Code of practice for noise control on construction
         and demolition sites.*

CHAPTER 18

Road vehicle noise

T. Priede

Institute of Sound and Vibration Research, University of Southampton

18.1 INTRODUCTION

Various noise surveys show conclusively that road traffic is at the present time the predominant source of annoyance; no other single noise is of comparable importance. Such a finding is not surprising because of the large number of automotive vehicles in comparison with other machines. The total horsepower which is 'built in' in automotive vehicles exceeds 20 times the horsepower of all other prime movers combined (aircraft, ships, power stations, etc.).

Having such a wide use (not only by industry, but by private persons), road transport is generally very cost conscious. Economy is therefore one of the prime factors which has so far dictated the development of vehicle design and operational methods. For these reasons, in the commercial field, the more efficient diesel engine has replaced the petrol engine with a reduction of fuel consumption by a factor of two. Diesel engines operate at considerably higher peak combustion pressures and higher rates of pressure rise and thus result in greater noise and vibration. The higher noise and vibration, however, is in no way detrimental to the life expectancy of a diesel engine; its life in general exceeds that of a petrol engine. Improvements in economy are also obtained by reducing the vehicle weight for the same load-carrying capacity, and, also, engines are made smaller and lighter by running at higher speeds to produce the same or greater power.

For all these reasons, the prime mover — the engine — and the vehicle were gradually becoming noisier; therefore it has been made essential to introduce noise legislation in many countries.

Apart from noise, there are now numerous legislations in road transport which have been introduced to improve man's standard of safety and comfort; the laws now cover many aspects of construction and operation of the vehicle. These laws have already had a marked success in dictating basic principles of car and commercial vehicle design. For example, legislation covering exhaust emission has led to great advances and improvements in the combustion system design of a petrol engine. There are also indications that in the near future radical changes in engine and vehicle design will result from noise legislation.

18.2 SUBJECTIVE RATING OF AUTOMOTIVE VEHICLE NOISE

A comprehensive study of the relation between subjective rating of noise emitted by motor vehicles and the objective measurements with a sound-level meter has been made by Mills & Robinson [18.1]. These tests were carried out using 'live' automotive vehicle noise in the open air on one of the test tracks at the proving ground of the Motor Industry Research Association.

The subjects were asked to rate the noises which were presented to them according to a six-point rating scale by verbal description, as shown in Table 18.1.

Table 18.1

A	B	C	D	E	F
0	2	4	6	8	10
–	Quiet	Acceptable	Noisy	Excessively Noisy	–

No descriptions were ascribed to the first and last categories, which the subjects were instructed to regard as extremes. For convenience in expressing results, the verbal categories of the rating scale were first expressed numerically, so that quiet became 2, acceptable became 4, and so on.

The results are shown in Fig. 18.1 for cars, commercial diesel vehicles and motor cycles. In the diagram the average subjective rating for each vehicle is plotted against the recorded sound level in decibels (dB(A)). The scatter is generally small and, to a reasonable approximation, a straight-line relation exists between the subjective rating and the measured dB(A) level. The most significant point on the rating scale is the numerical value 5 which corresponds to the demarcation line between 'acceptable' and 'noisy'.

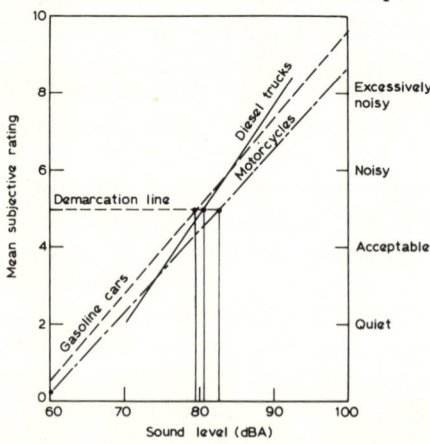

Fig. 18.1 – Relation between diesel truck, gasoline car and motorcycle sound level dB(A) and subjective ratings.

It can be concluded from Fig. 18.1 that a level close to 80 dB(A) fairly represents the demarcation line between 'acceptable' and 'noisy' for most vehicles.

Only a small number of commercial vehicles are able to comply with the 80 dB(A) criterion and over the past ten years a large number of commercial vehicles have now reached the level of 'excessively noisy'; that is, 92 dB(A).

18.3 CHARACTERISTICS OF VEHICLE NOISE

The mechanism of radiation of noise to outside from a vehicle is basically different from the generation of noise inside the vehicle. None of the noise-producing systems of the vehicles are fully enclosed; if anything, they are only partially screened. Thus the noise emitted depends on the relative levels, characteristics, and the interaction of the directly radiated noises from these systems. For current production vehicles the principal noise source is the power unit and its auxiliaries. Other important generators are the transmission system, tyres, and braking system.

Vehicles can be classified from the emitted noise point of view as heavy commercials, light commercials, public service vehicles, small cars, large cars, and high-performance cars. Figs. 18.2 to 18.5 show the spectra of noise obtained from tests under the conditions of the ISO drive past test [18.2].

The ISO test involves accelerating the vehicle at maximum rate in a low gear over a distance of 20 m from three-quarters of the engine maximum power speed past a microphone positioned 10 m from the start of the acceleration and 7.5 m from the centre line of the vehicle. The use of a low gear (2nd or 3rd depending on the vehicle) ensures the maximum engine speed is reached at or shortly after passing the microphone.

This test procedure is the basis of vehicle construction regulations in several countries, and for this reason most investigations of vehicle noise are limited to this test. The drive-past test is frequently supplemented by a stationary test at maximum governed engine speed. It should be noted that this test is only justifiable when the total vehicle noise is engine controlled.

The spectra of a number of heavy commercial vehicles are shown in Fig. 18.2. Most of these are powered by four-stroke cycle diesel engines of cubic capacity from 4 to 16 litres with power output from 80 to 300 bhp. The spectra from most of the vehicles are almost constant at about 80 dB up to 3000 Hz. A broad peak in the frequency range from 800 to 3000 Hz constitutes the characteristic 'diesel knock' emitted by the engine surfaces.

Above 3000 Hz the noise decreases by about 6 dB per octave. The scatter in the low-frequency noise between various vehicles is attributed to variations of the exhaust and inlet silencing systems used.

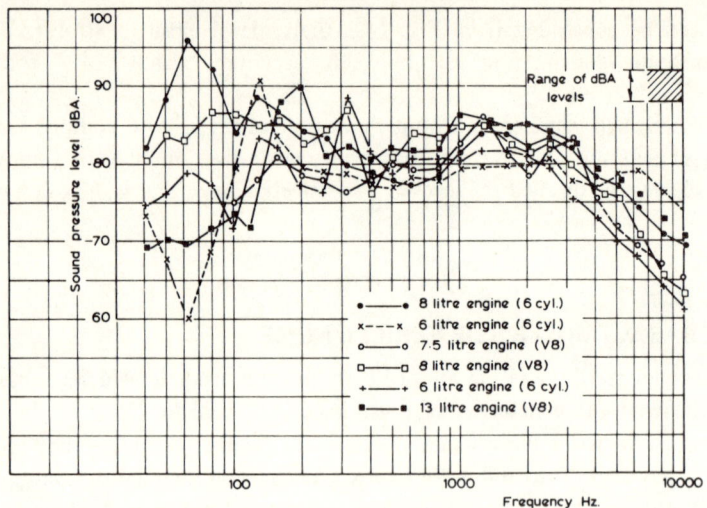

Fig. 18.2 – Spectra of noise emitted by heavy commercial vehicles.

The spectra of light commercial vehicles (less than 3.5 ton gross vehicle weight or with less than 12 seats) are shown in Fig. 18.3. These vehicles are fitted with petrol and smaller capacity diesel engines. As can be seen, the spread of the noise in this class of vehicle is considerably larger, being of the order of 15 to 20 dB. It is generally found that reduction of noise obtained by body shielding is appreciable in this type of vehicle, which largely accounts for greater differences measured.

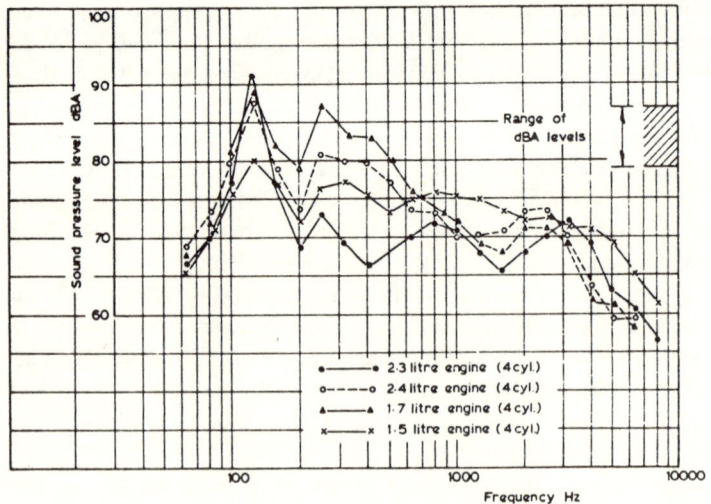

Fig. 18.3 – Spectra of noise emitted by light commercial vehicles fitted with gasoline engines.

The low-frequency noise which is comparable with heavy goods vehicles arises not only from the exhaust and inlet, but also from the noise radiated by the body structure. High-frequency noise from 800 to 3000 Hz is 5 to 8 dB lower than from heavy commercial vehicles.

The spectra from small and large cars fitted with petrol engines are shown in Figs. 18.4 and 18.5. The spectra are reasonably flat, the average spectrum level of the small cars being above 70 dB and for larger cars being somewhat higher. The high-frequency content of the spectra is in no way less significant than that of the diesel engined vehicles.

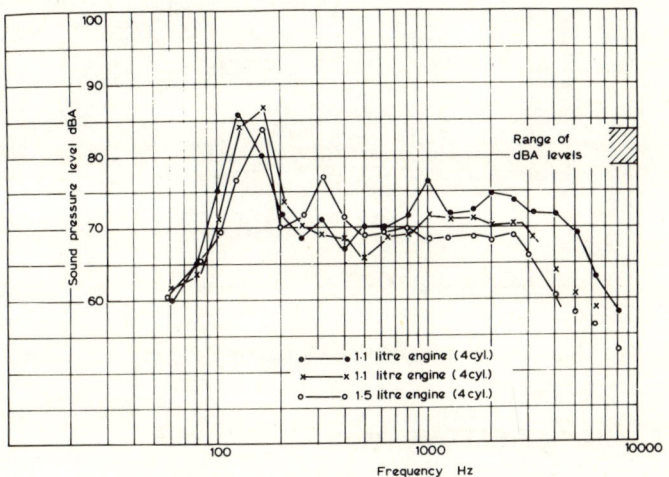

Fig. 18.4 – Spectra of noise emitted by small cars.

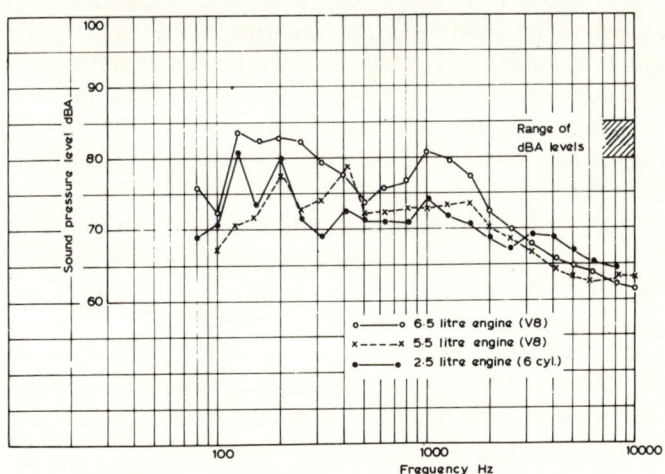

Fig. 18.5 – Spectra of noise emitted by large cars.

The high-performance and sports cars shown in Fig. 18.6 produce very high levels of low-frequency noise (100–200 Hz) which is intentionally allowed by inadequate silencing of the exhaust. The high-frequency noise, however, is almost the same as from other types of car.

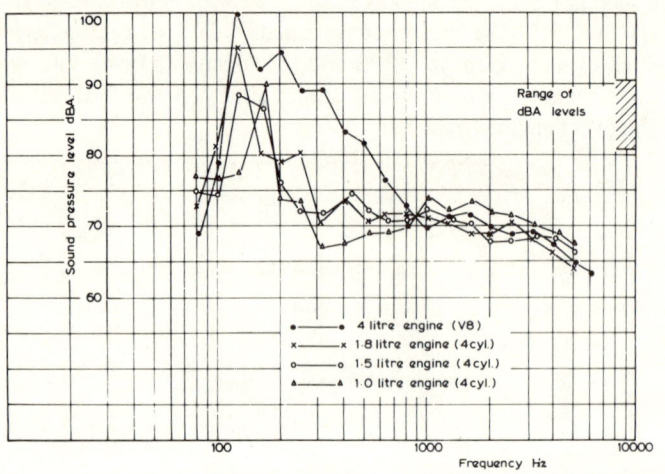

Fig. 18.6 – Spectra of noise emitted by high-performance cars.

18.4 VEHICLE NOISE LEGISLATION

During the 1950's there were great advances made in automotive diesel engine development. It was found that automotive diesel engines could be satisfactorily run at higher speeds and higher specific loads, thus considerably improving the specific power output. Although Daimler Benz successfully introduced a high-speed diesel engine in a car before the war, it was not until the 1950's that a widespread development of the high-speed diesel engine for vans and light industrial applications also took place. The noise of the automotive truck diesel engines produced before 1950, running at moderate speeds, was around 98 to 100 dB(A) measured at a distance of 1 metre. By 1960 the engines were running much faster with often large bore-to-stroke ratios and the noise levels reached 103 to 107 dB(A). The result was an appreciable increase of vehicle noise. Similar trends could be noted also in petrol car engine developments. It is for these reasons that in the mid-sixties the first noise legislations were introduced which since then have remained virtually the same. The present European Economic Community noise legislation for road vehicles is illustrated in Fig. 18.7 together with subjective assessments of vehicle noise [18.1] already discussed in section 18.2. The legislation is usually based on practically observed data; trucks have been found noisier than cars, and therefore they have been permitted higher levels of noise.

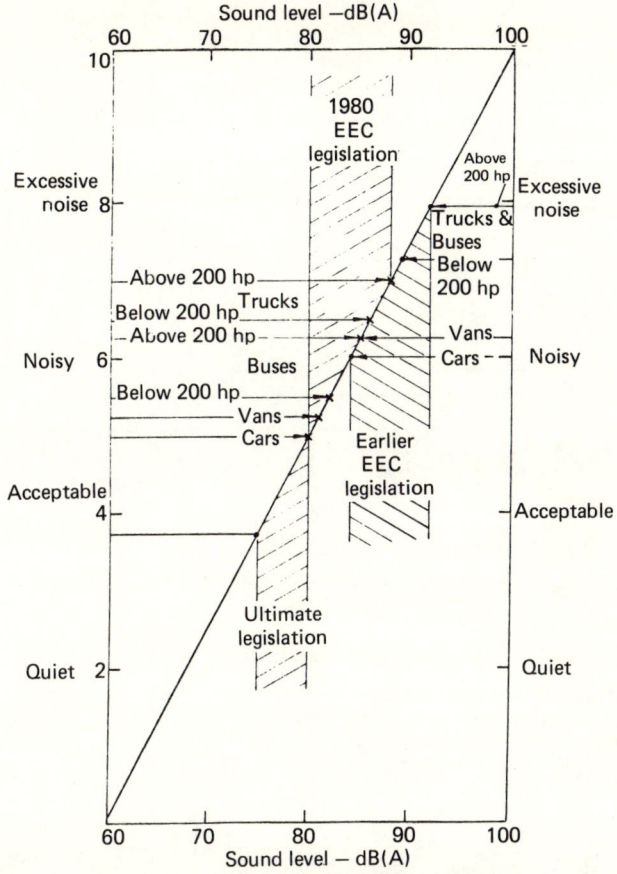

Fig. 18.7 – Legislation and subjective assessments.

The detailed investigations on the origins of road vehicle noise [18.3] have shown that noise radiated by the engine surfaces in various vehicles is the main contributor. Since there are significant differences in the legislated levels the question is whether the individual vehicle in a particular group is quieter because of a quieter engine, or whether it is due to basic principles of vehicle design.

Fig. 18.8 shows the test-bed overall noise at 1 metre distance from the side of the engine, for a large number of engines running at rated conditions (maximum speed and load) and classified in the various vehicle categories. The data presented are from the engines measured at ISVR and AVL Laboratories. The mean values of noise are as follows:

Group 1 – Truck diesel engines above 200 hp   102.4 dB(A)
Group 2 – Truck diesel engines below 200 hp   101.8 dB(A)
Group 3 – Van high-speed diesel engines   101 dB(A)
Group 4 – Car petrol engines   99 dB(A)

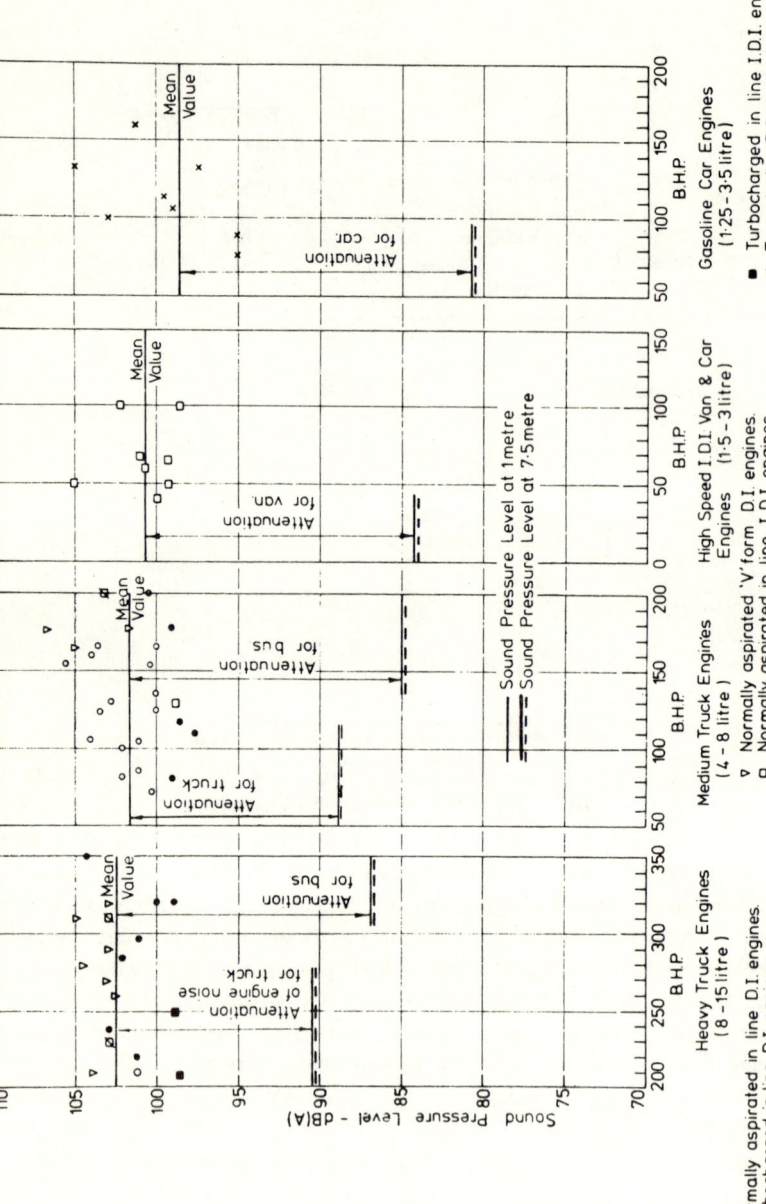

Fig. 18.8 – Attenuation of engine noise by vehicle structure and distance.

As can be seen, the mean levels between various engine groups only differ by 1.4 dB(A). Also the mean noise level of the petrol engines is only 2 dB(A) quieter than the diesel engines.

It may be concluded that differences of the noise vehicle groups are mainly due to principles of vehicle design. The relevant features of various vehicles are illustrated schematically in Fig. 18.9.

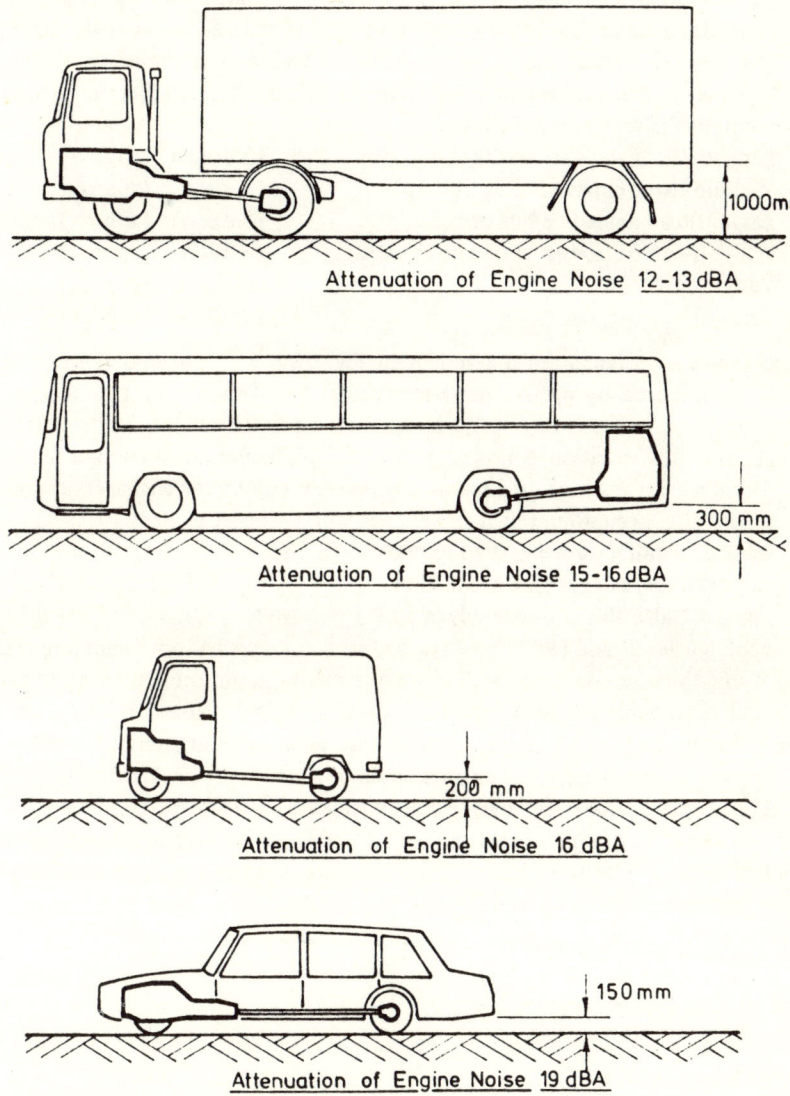

Attenuation of Engine Noise 12-13 dBA

Attenuation of Engine Noise 15-16 dBA

Attenuation of Engine Noise 16 dBA

Attenuation of Engine Noise 19 dBA

Fig. 18.9 – Typical vehicle layouts.

As can be seen, trucks have a high ground clearance, the engine is more exposed, and the vehicle also provides little attenuation of noise. In buses, vans, and cars, the vehicle body provides considerable shielding of engine noise and the ground clearance is considerably smaller.

Investigations on the ISVR open-air pad have shown that attenuation with distance from 1 metre to 7.5 metres also depends on engine size. For large engines attenuation of noise is about 12 dB(A); for medium size engines 13 dB(A); while for smal high-speed diesel and petrol engines about 14 to 15 dB(A).

The general attenuation of engine noise measured as the noise on the test bed at 1 metre distance less the noise from the vehicle, with no specific acoustic treatment, at 7.5 metres is as follows:

Large truck engines above 200 hp	12 dB(A)	(approx)
Medium truck engines below 200 hp	13 dB(A)	(approx)
Large truck engine in a bus or coach	15 dB(A)	(approx)
Medium size truck engine in a bus or coach	16 dB(A)	(approx)
Vans	16 dB(A)	(approx)
Cars	19 dB(A)	(approx)

These values of attenuation are shown in Fig. 18.8 and correspond to values which are imposed by present noise legislation. Fig. 18.8 shows that there is a considerable variation in the noise levels of the various engines in the various groups, and this variation is not only due to a particular engine design feature but also to a great extent to the manufacturer's choice of rated engine conditions.

It has also been found that in vehicles where noisier engines are installed, the vehicle manufacturer has to incorporate in the vehicle design some noise reducing features, such as partial shielding.

Fig. 18.7 also shows the details of EEC Directive No. 77/212/EEC regarding the legislated levels for 1980. New trucks, vans and cars had to obtain a further 3 to 4 dB(A) reduction of noise. This requirement did not present vans and cars with any formidable problem, but trucks using pre-1980 engines required significant shielding techniques. Buses required the greatest reductions (7–8 dB(A)) but this was achieved since bus engines could be more easily enclosed.

The ultimate legislation envisaged is expected to be from 80 dB(A) for trucks down to about 75 dB(A) for cars and vans. To achieve an elegant solution to this problem it can be easily estimated from the data given in Fig. 18.8 that future engine noise should not exceed 93 dB(A) at its rated conditions. This figure can be considered as a target at which present engine designers must aim.

18.5 SOURCES OF VEHICLE NOISE

The origins of vehicle noise are due to sources which fall into two distinct categories,

(a) those related to engine speed, and

(b) those related to road speed.

Fig. 18.10 illustrates the principal sources of noise for a diesel vehicle tractor unit. Those related to engine speed (that is, independent of road speed), are engine, intake, and exhaust noise and cooling fan noise. In this category also is included the contribution of that part of the gearbox which rotates at engine speed and various engine accessories such as air compressors, hydraulic pumps, and electrical generators.

Sources related to road speed include part of the transmission which is not changed by the engagement of different gears, the rolling noise produced by tyres, and aerodynamically generated noise.

As can be seen from Fig. 18.10, the forward control cab design (cab over engine) provides very little shielding to the various noise sources, and they therefore can radiate the noise directly; for example, air intake noise, exhaust outlet noise and also the engine noise radiates via the many openings through the wheel arches, along the road surface, the front and the rear. Passenger cars, buses and vans provide considerably greater shielding and are thus appreciably quieter.

The vehicle structure (frame, cab, or car body), although not in itself a source of noise, is set into vibration by both the engine- and road-induced noise and vibration and thus becomes like a sounding board which emits the noise from both the engine-related and road-related sources.

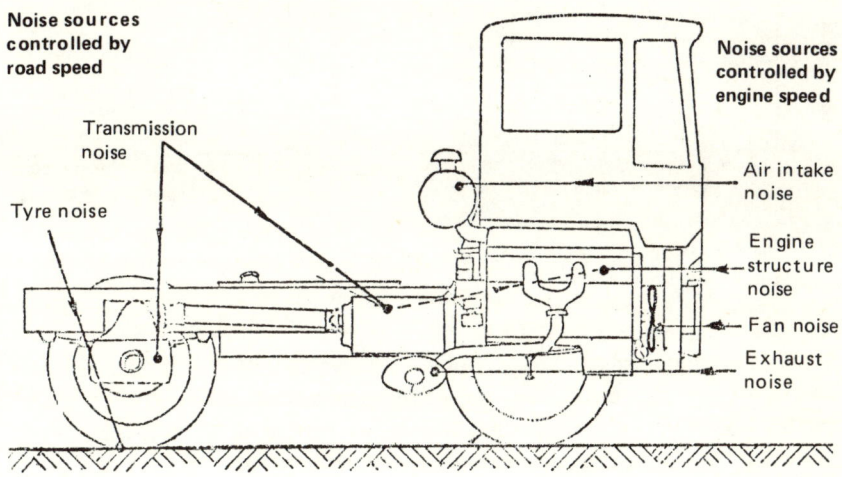

Fig. 18.10 – Vehicle noise sources.

The general effect of vehicle speed and engine speed on noise is illustrated in Fig. 18.11 which shows the noise produced by a fully laden 6 litre diesel engine truck when driven at various constant speeds in various gears past a microphone situated 7.5 metres from the centreline of the truck. The results obtained cover a road speed range of 3-50 mph (5-80 km/hr) and an engine speed range of 1000 to 3000 rev/min. Also shown is the rolling noise obtained by coasting past the microphone with engine stopped and the clutch disengaged.

The test results indicate that, although there is a considerable difference in road speeds between various gears, there is little difference in noise levels. If these figures are plotted against the speed as shown in Fig. 18.11b they reduce to one line. This suggests that the noise of the truck is primarily controlled by the power unit which therefore controls the rate of increase of noise. In this instance the rate of increase of noise of the power unit is 11 dB(A) per doubling of speed, or 37 dB(A) per tenfold increase of speed. Rolling noise, which has a more or less similar rate of increase, will become of equal level to that of the power unit at about 68 m.p.h. (109 km/hr).

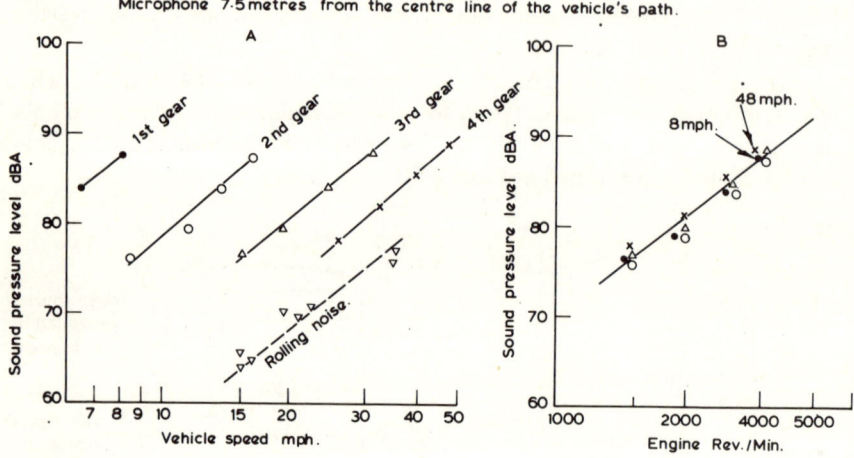

Fig. 18.11a and b – Fully laden constant speed drive-past noise levels of a diesel truck.

Results obtained from a petrol car, shown in Fig. 18.12, illustrate very similar trends. As can be seen, the noise again is to a major extent controlled by the power unit although other sources of noise such as road transmission contribute to a much greater extent. Noise levels plotted versus engine speed (Fig. 18.12b) do not reduce to a single line, but to a distinct band of about some 6 dB(A) width. The rate of increase of noise with engine and vehicle speed is considerably greater, i.e. 15 dB(A) per doubling or 50 dB(A) per tenfold increase of speed.

The different rates of increase of noise with speed of the two types of vehicles, as will be discussed, are due to basic differences of the engines used.

Microphone 7·5 metres from the centre line of the vehicle's paths.

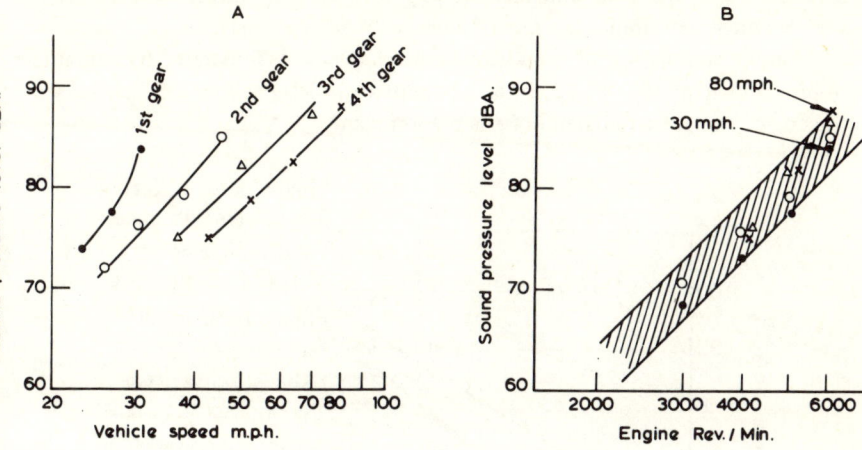

Fig. 18.12a and b – Constant speed drive-past noise levels of a family saloon car.

As shown in Figs. 18.11 and 18.12 the vehicle noise is predominantly determined by only one operating parameter: speed. If the basic parameters of each of the individual sources of noise of the vehicle are considered it can be deduced that speed is again the predominant variable.

The basic relationships are as follows:

	SPEED FACTOR	DESIGN FACTOR
AIR INTAKE	$I \sim N^{3 \text{ to } 4.5}$ rev/min	Valve overlap, valve diameter, cam profile
EXHAUST	$I \sim N^{2 \text{ to } 4.5}$ rev/min	Valve diameter and cam profile. Pressure in the cylinder at the instant of exhaust valve opening
COOLING FAN	$I \sim N^5$ rev/min $\times D^7$	(where D is fan diameter)
ENGINE	$I \sim N^{2 \text{ to } 5}$ rev/min $\times B^5$ (exponent is function of combustion system and mechanical design of engine)	(where B is cylinder diameter)
GEARS	$I \sim N^2$	Tooth profile and tooth overlap
TYRES	$I \sim V^{2.5 \text{ to } 4.5} \times W^3$ miles/hour (exponent is function of tyre and road design)	(where W is tyre width)

The noise for all sources increases by 20 to 50 dB per tenfold increase of speed. It is generally the predominance of any one of the sources of noise of the vehicle which determines the rate of increase of vehicle noise.

Simple summation of the various sources is valid as illustrated by actual test results shown in Fig. 18.13, from a commercial diesel vehicle. As can be seen, the engine structure radiated noise is predominant.

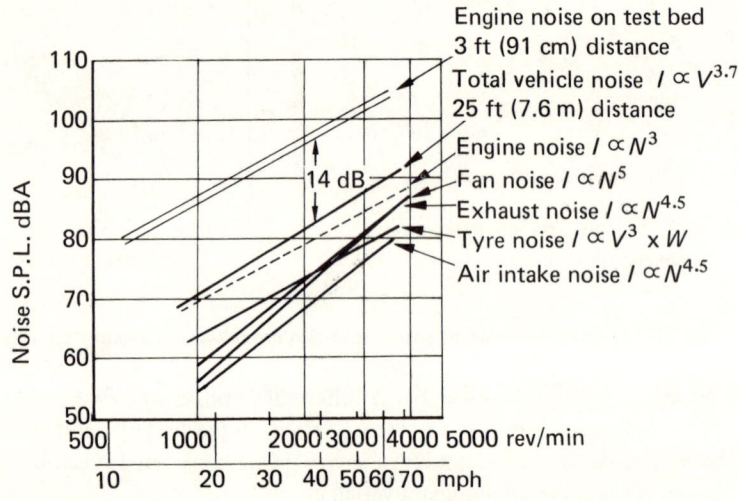

Fig. 18.13 – Summation of various sources.

The overall noise data of various groups of engines shown in Fig. 18.8, are summarized in a graph in Fig. 18.14 which shows the general range of noise levels produced by each group over its operating speed range. It will be seen that in each group there is a significant spread in noise levels between engines of different manufacture by as much as 10 dB(A). This spread is not purely due to size alone but also its details of design.

The heavy truck engines (200 to 350 hp) are only slightly noisier than the small and medium truck engines, if considered on the basis of the same running speed (for example 2000 rev/min). At the rated speed the differences of noise are very small.

The high-speed diesel engines are by about 10 dB(A) quieter over most of the comparable speed range but, owing to higher rated speeds, they ultimately attain similar levels of noise. The same applies to petrol engines which are yet quieter by a further 10 dB(A) at comparable speeds (for example 2000 rev/min).

The significant factor in comparing the engine as a source of vehicle noise is the range of noise level over its useful operating range, i.e. the potential noise range from 1000 rev/min to rated speed.

The average values of the noise ranges are as follows
Group 1   200-350 hp − 1000 to 2600 rev/min − 10 dB(A) range
Group 2   80-200 hp   − 1000 to 3000 rev/min − 15 dB(A) range
Group 3   30- 80 hp   − 1000 to 4500 rev/min − 20 dB(A) range
Group 4   40- 50 hp   − 1000 to 6000 rev/min − 30 dB(A) range

These data explain the main reason why commercial vehicles in traffic, particularly on urban roads, generally produce much higher levels of noise than light commercial vehicles and cars, and under fast highway conditions produce noise levels which are very nearly comparable.

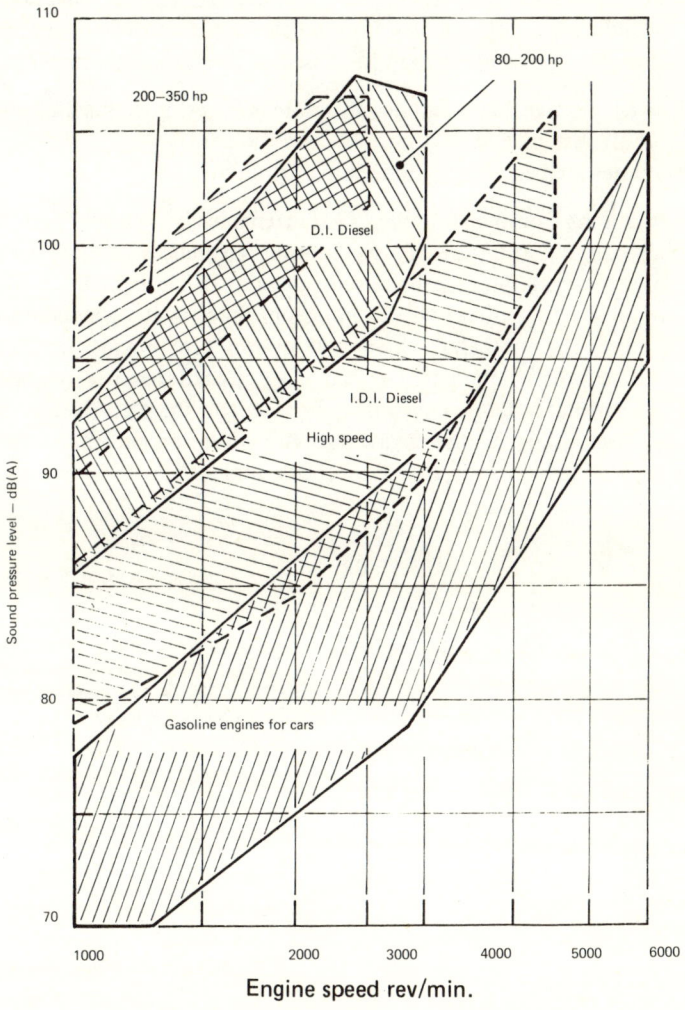

Engine speed rev/min.

Fig. 18.14 − Range of noise levels of various engine types.

This chapter only considers in detail the two predominant sources of noise of the vehicle, namely engine and tyre noise. These sources present the vehicle manufacturer with the most difficult problems. Exhaust, air intake, cooling fan, and gear noise are reasonably understood, and adequate control has been achieved.

18.6 GENERAL CONSIDERATIONS OF THE MECHANISM OF GENERATION OF ENGINE NOISE

Fig. 18.15a shows a cross-section of an automotive diesel engine. By disregarding the timing mechanism (the valve gear) and accessories (e.g. injection system, water and lubricating oil pumps, etc.) which do not contribute appreciably to noise, the engine can be considered to consist of two basic structural elements.

(A) The internal load-carrying structure, i.e. piston-connecting rod–crank-shaft system, and

(B) Outer load carrying cylinder block structure.

The internal load-carrying structure (A) is mechanically separated from the main outer load-carrying structure (B) by running clearances. A simple equivalent system of the engine thus can be developed as shown in Fig. 18.15b illustrating the principal exciting forces which are responsible for the generation of vibration and noise. The outer elastic load-carrying structure provides location for the piston and crankshaft represented as two masses joined together by a spring.

In the engine system as shown in Fig. 18.15b there are two major forces which are responsible for the engine structure vibration and the emitted noise, namely unidirectional forces, P, and reversible forces, F.

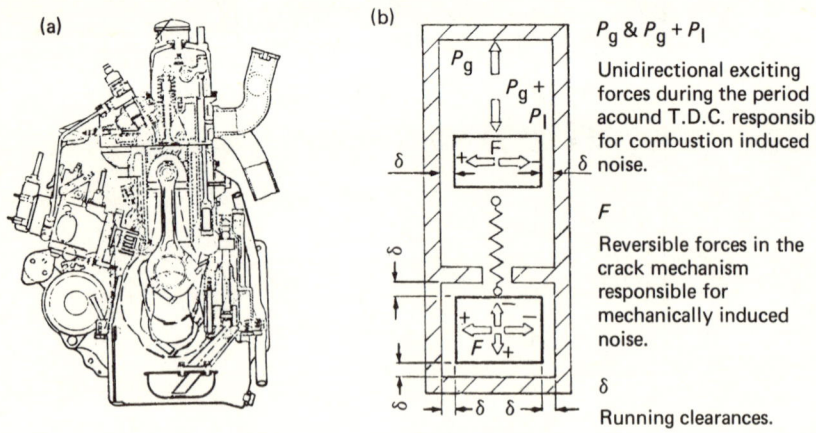

P_g & $P_g + P_l$

Unidirectional exciting forces during the period acound T.D.C. responsible for combustion induced noise.

F

Reversible forces in the crack mechanism responsible for mechanically induced noise.

δ

Running clearances.

Fig. 18.15 – Engine structure and its equivalent system.

18.6.1 Combustion-induced noise – unidirectional force excitation

Unidirectional forces, Fig. 18.16, (P) are only important in the vicinity of TDC on the compression stroke and are produced from compression and subsequent pressure rise resulting from combustion. The clearances in the vertical direction for the equivalent mass of piston, connecting rod, and crankshaft are taken up by these forces, and a linear vibratory system results. Since during this period the force does not change its direction any appreciable vibration can only be produced if there is a rapid change in the magnitude of the force. As shown by the typical diagrams (Fig. 18.16) the rapid change in the magnitude is produced by the onset of combustion in the engine cylinder and thus can be defined as combustion-induced noise. As can be seen the gas force P_g excites the top part of the engine structure (i.e. cylinder head) while the lower part of the structure is excited by the combined gas force and inertia force, $P_g + P_I$.

Investigations have shown [18.3] that in this period engine structure excitation is similar to a simple linear spring mass system, and its vibration can be represented by the equation.

$$M\ddot{x} + C\dot{x} + Kx = P(t)$$

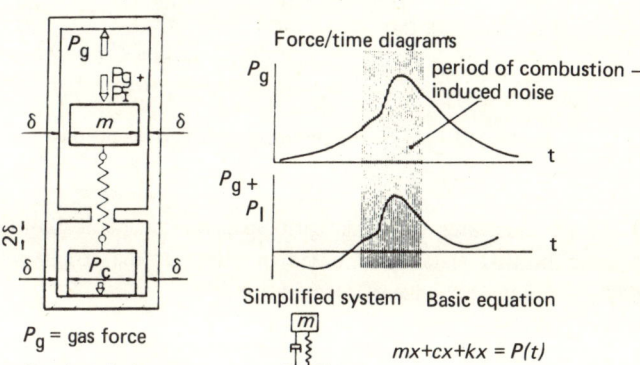

Fig. 18.16 – Combustion-induced noise (unidirectional force excitation).

18.6.2 Mechanically-induced noise – reversible force excitation

Reversible forces (F) which change direction are produced by the engine crank mechanism and associated inertia forces as illustrated in Fig. 18.17. Although these forces change in magnitude, the rate of change is too low to induce any appreciable vibration amplitudes in the comparatively stiff (high natural frequency) engine structure. These forces, however, accelerate the various elements of the internal load-carrying structure across the clearances and thus cause impact which effectively induce engine structure vibration.

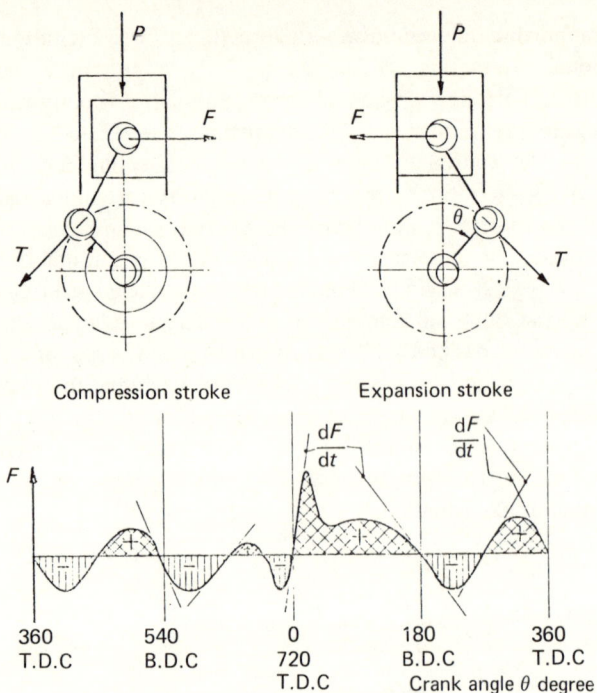

Fig. 18.17 — Reversible forces acting on the piston.

Fig. 18.18 illustrates the basic principles of mechanically-induced noise. The equation in this instance is based on the fact that transverse motion is governed by the rate of change of acceleration.

$$M\dddot{x} = \mathrm{d}F/\mathrm{d}t = \text{const} .$$

$\mathrm{d}F/\mathrm{d}t$ is approximately constant, while the actual component moves across the clearance. For rotational parts the equation is similar except it is presented in terms of angular acceleration (θ) and the moment of inertia (J)

$$J\dddot{\theta} = \mathrm{d}T/\mathrm{d}t = \text{const} .$$

The result of the acceleration of the moving components across the clearances is to impart on contact, as shown in Fig. 18.18, kinetic energy to the structure in the form of an instantaneous step load which depends on the time taken for the component to move across the clearances. Once imparted, the continuing application of the changing force produces further excitation of the structure. Since there is lubricating oil in the clearances, the movement of components in this manner produces impulsive hydraulic loads [18.4].

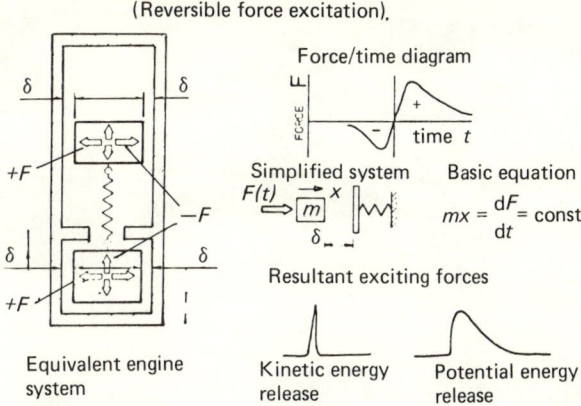

(Reversible force excitation).

Force/time diagram

Simplified system   Basic equation

$$mx = \frac{dF}{dt} = \text{const}$$

Resultant exciting forces

Equivalent engine system

Kinetic energy release

Potential energy release

Fig. 18.18 – Basic principles of mechanically induced noise.

Mechanically-induced noise can again be studied on the non-running engine by simulating the reversible forces with hydraulic force generators [18.4]. In this way the important parameters which control mechanically-induced noise can be readily assessed.

18.7 RELATION BETWEEN NOISE, ENGINE DESIGN, AND OPERATING PARAMETERS

Despite the numerous exciting forces which almost simultaneously excite the engine structure there is some justification to look at the problem in a simpler way. Since the gas force resulting from combustion tends to be the predominant force in most of the engines, the relationship between the gas force character-istics and emitted noise can be used to establish a basic model to identify the effects of fundamental engine design and operating parameters [18.5].

The three basic parameters of an engine are

(a) speed; (b) size; (c) load.

18.7.1 Engine speed

Fig. 18.19 shows a basic model which determines the speed/noise relationship.

The engine structure characteristics can be defined by use of electrodynamic vibration generators, and the broad response readily established as shown by the solid envelope line. It will be seen that when the engine structure is subjected to a constant magnitude sinusoidal force it exhibits maximum response in the high-frequency range from 800–2000 Hz.

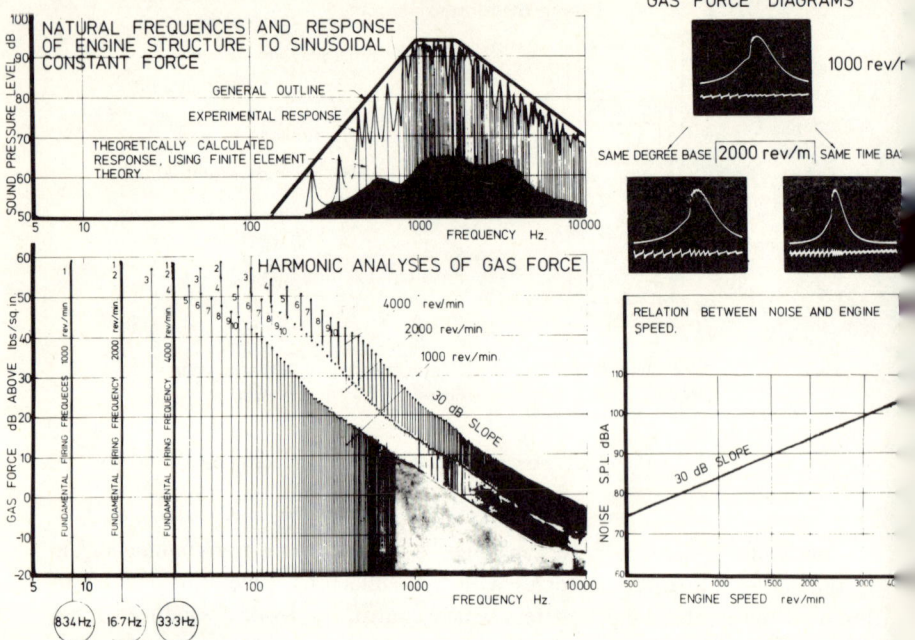

Fig. 18.19 – Relation between the gas force, its characteristics and rate of increase of noise of an engine.

Electronic analyses show in some detail the existence of numerous natural frequencies at which the structure can vibrate. It also indicates that it is reasonbly heavily damped and thus any better resolution of various modes of vibration by any instrument is impossible.

The existence of the high density of various modes has been confirmed by finite element computer calculations of a simple model of a crankcase which confirms that in a very narrow frequency range of one-third octave there can be at least some 20 natural frequencies of the crankcase walls.

The gas force which again in itself is very complex, can be subjected to frequency analyses to quantify its exciting propensities. Analysis of the gas force shows that in the low-frequency range the magnitude of the harmonics is a maximum, gradually decreasing with increasing frequency at higher orders. Comparing this force spectrum with the response of the structure one can see that only the high order harmonics (frequency range 800–2000 Hz) are responsible for the predominant noise of the engine.

If the engine speed is doubled, the pressure diagram, on the degree basis as shown in the inset of Fig. 18.19, remains the same shape and thus the amplitude of all the harmonics should remain the same, i.e. the same force spectrum. Since the actual event occurs in one half the time (see inset of the pressure diagrams at 2000 rev/min) it means that the actual force spectrum is shifted sideways by a factor of two, namely from 8.34 to 16.7 Hz in this example. One can see that the engine structure (frequencies remain the same) is now excited with lower order harmonics which have higher amplitudes. Since the general slope of the force spectrum is about 30 dB/decade an increase of excitation by 9 dB will be obtained. With further speed increases the same pattern is followed.

It can be concluded that the characteristics of force determine the rate of increase of noise with engine speed which in this instance, for a naturally aspirated diesel engine, is 30 dB per tenfold increase of speed.

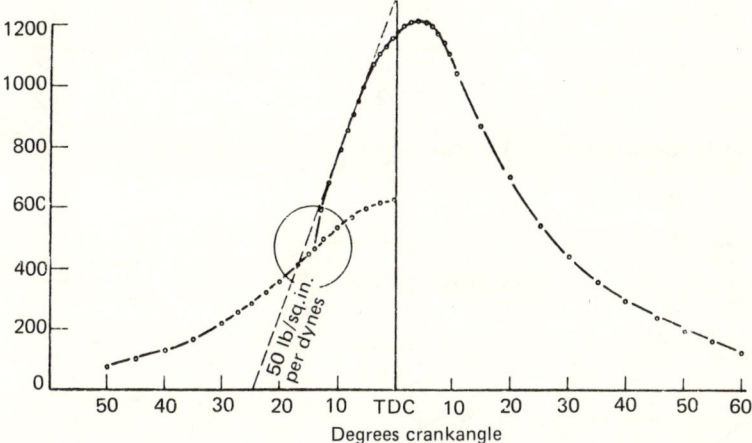

(a) Direct injection engine with toroidal chamber — four hole injector

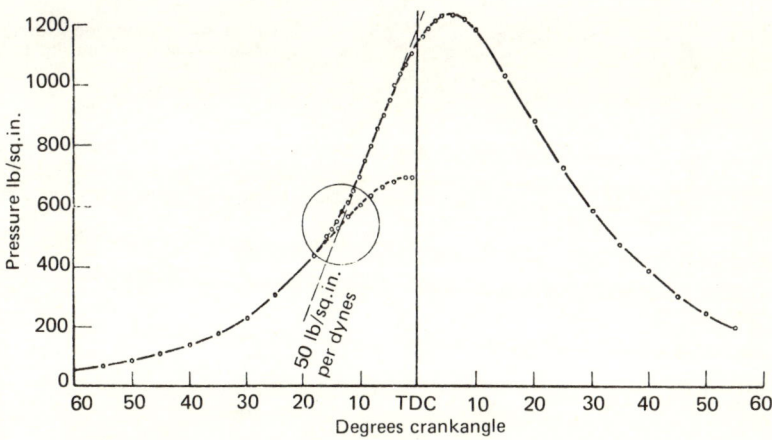

(b) M.A.N. combustion system — two hole injector — fuel sprayed on the wall of combustion chamber

Fig. 18.20 — continued next page.

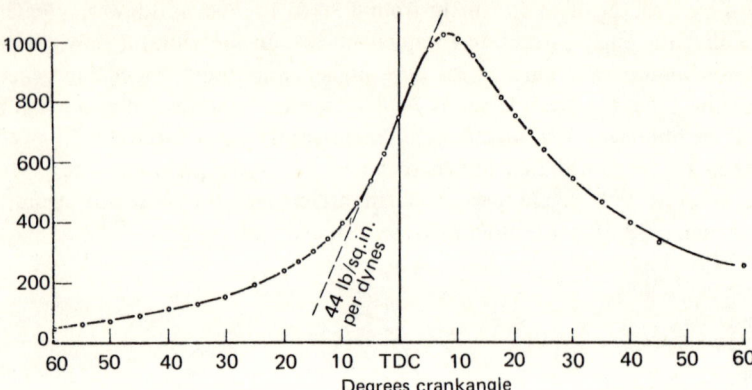

(c) High performance petrol engine

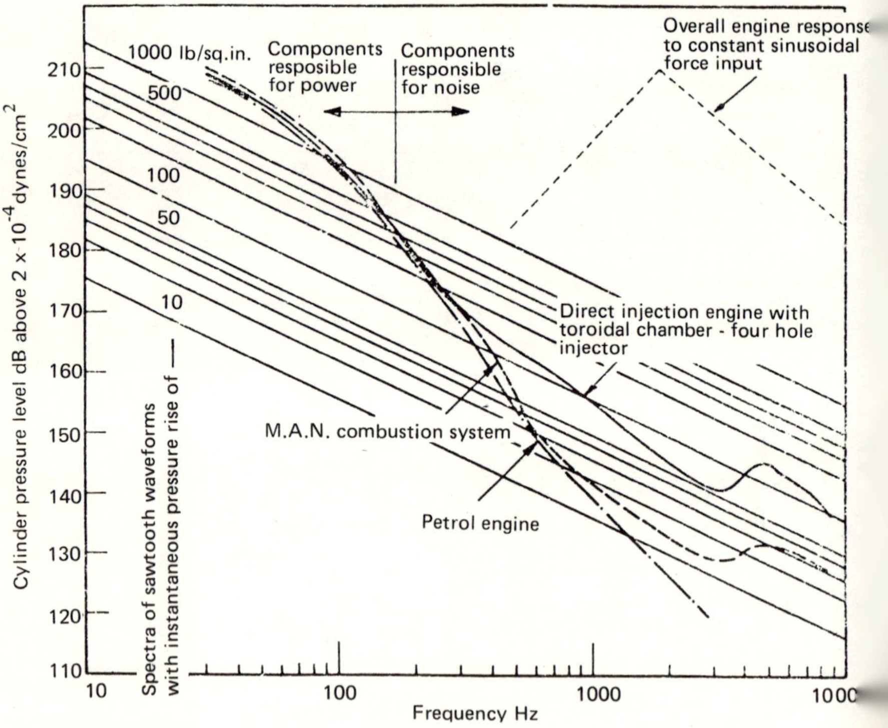

Fig. 18.20 (continued from previous page) – Relation between combustion system cylinder pressure development and its exciting propensities.

18.7.2 Effect of combustion system on noise

Fig. 18.20 illustrates that the amplitude of the high-frequency harmonics in the gas force are produced at the onset of combustion in the diesel engine; that is, the explosion. It requires only slight modification of the diagram in this region (indicated by a circle) to reduce the noise by as much as 20 dB. This part of the diagram has no effect on power developed by the engine. Only a few low frequency harmonics are required to develop the necessary mechanical power.

Fig. 18.21 illustrates the relationships obtained with different engines. It will be seen that the combustion system determines the rate of increase of noise with engine speed. The exponent for a petrol engine is 5, while for the most viscious diesel engine about 2.5. It can be seen that at low speeds there is a possibility of reducing noise by as much as 25 dB(A) by smoothing the development of the gas force. As speeds are increased the lines tend to converge, clearly indicating that in high-speed engines the form of the gas force is irrelevant.

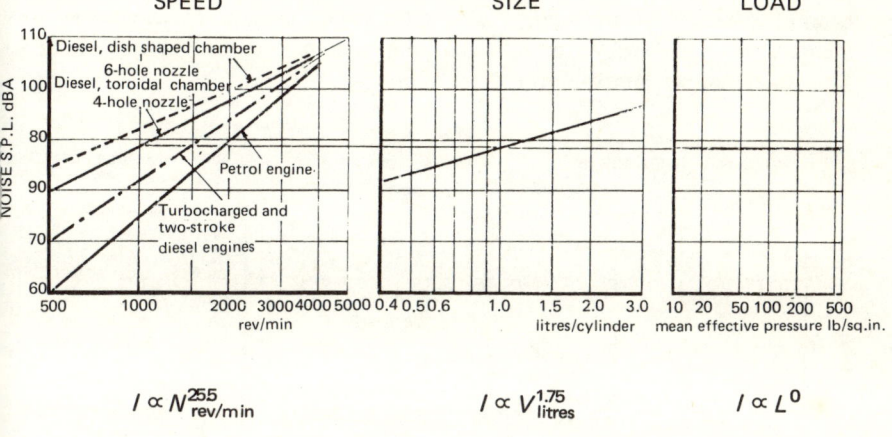

Fig. 18.21 – Effect of engine parameters on emitted noise.

18.7.3 Engine size

Measurements carried out on a large number of engines show that increase of noise with engine size is considerably less. An increase of size to ten times gives an increase of noise of 17.5 dB(A). The detailed investigations now indicate that vibration levels of the engine surfaces are about the same irrespective of their size, thus the increase of noise with size is simply due to a larger radiating surface area.

18.7.4 Engine load

Engine load has no effect on noise, which is in agreement with the findings that noise is simply due to the initial ignition of the fuel. This occurs at the same intensity whether the engine is running at no load at all or full load. It can be concluded that:

(1) The form of the exciting gas force determines the rate of increase of noise with engine speed.

(2) At high engine speeds the form of the gas force has a less significant effect on noise.

(3) Engine noise is independent of the horsepower produced.

18.7.5 Summing up effects of engine design and operating parameters

In the investigation of the variations in noise level produced by engines of similar cylinder capacity it has been possible to find relevant examples which have provided further evidence on the parameters which control engine noise.

Fig. 18.22 shows the noise spectra of three different engines which have been designed by the same design team in the same company. The engines are of identical bore and stroke except that they are built as four-cylinder in-line six-cylinder in-line, and V8 form.

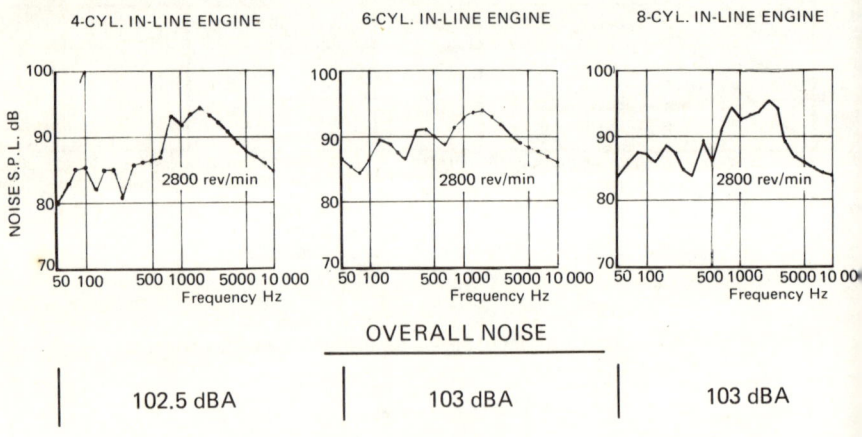

Fig. 18.22 – Noise section of 3 engines of same size but different design.

There are significant differences in the spectra because the natural frequencies of each of the structures are markedly different. The general overall levels of the noise spectra, however, are all within 1 dB.

We can therefore conclude that one can add to the engine any number of cylinders and thus increase the power of the engine without increasing the noise. In this case the power increased by a factor of two.

Another example is shown in Fig. 18.23 for two engines with the same cylinder capacity and the same number of cylinders. Both engines are of V8 form. The only difference between the two engines is the stroke-to-bore ratio. The fundamental differences of the two engines are illustrated in somewhat exaggerated form [18.6].

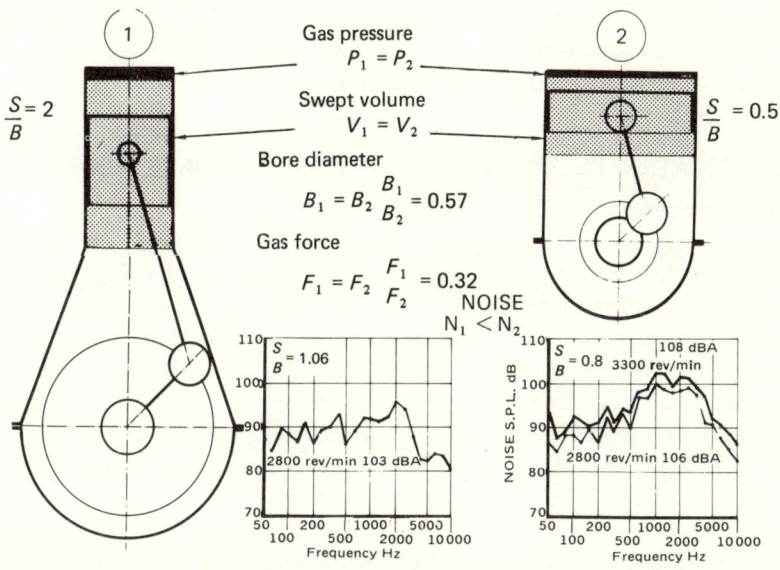

Fig. 18.23 – Effect of bore and stroke ratio on engine noise.

The oversquare engine is very attractive to the vehicle manufacturer because of its reduced engine bulk volume and weight for the same power output. Furthermore it can run considerably faster for the same limiting piston speed. The result, as can be seen, is a marked increase of noise of the oversquare engine despite its smaller size. The increase of noise is from 103 to 108 dB(A). These observations led to the conclusion that the basic parameters which determine the noise of an engine are only its speed and cylinder diameter, and the following relation can be derived.

$$I \sim N^n \times B^5 .$$

By taking into account the level of gas force and engine structure characteristics the engine noise intensity becomes

$$I \sim C_f C_s (N^n \times B^5)$$

where C_f defines the level of the gas force
        C_s defines the structure characteristic
        B is bore diameter
        N is engine speed
        n is the combustion index .

Based on these, and other investigations, the ISVR formula for predicting Direct Injection (DI) engine noise was derived [18.5] which gave noise predictions for automotive engines over wide size and type ranges. The formula was also applied to Indirect Injection (IDI) and turbocharged engines but results were less satisfactory. The reasons for this are considered in the following sections.

18.8 RELATION BETWEEN NOISE AND ENGINE COMBUSTION SYSTEM

To establish the relevant differences in noise characteristics of engines of different combustion systems a more detailed study has been made of the results of all the different engines tested at ISVR laboratories. On each engine gas force diagrams have been taken, and cylinder pressure, noise, and vibration analyses have been carried out. The overall noise in dB(A) at full load conditions of some 44 different engines is summarized in Fig. 18.24. The noise of all the engines show straight-line relationships with speed, except for some small high-speed IDI and petrol engines which show two slopes, a low rate of increase in the low-speed range and a high rate in the high-speed range. If the results are compared on the basis of constant speed the increase of noise with engine size is apparent. At the rated speeds, all engines (except the opposed piston two-strokes) reach about the same level of noise within a band of some 10 dB(A).

An attempt is also made to classify the engines in various groups according to combustion system and fundamental design principles. In each of these groups listed below, there are about seven engines:

(a)   Turbocharged in-line DI
(b)   Naturally aspirated in-line DI
(c)   Naturally aspirated vee-form DI
(d)   Two-stroke DI
(e)   Naturally aspirated IDI
(f)   Spark ignition petrol .

To find some correlation with a simple combustion model the overall noise levels are plotted against engine bore diameter in Fig. 18.25 at a constant speed of 2000 rev/min which represents a realistic speed for all the engines in this wide range. It can be seen that the engines do fall into specific groups:

(a)   All normally aspirated DI engines fit within a 3 dB band of slope (bore) [18.5]. It is clear that there are no differences between the

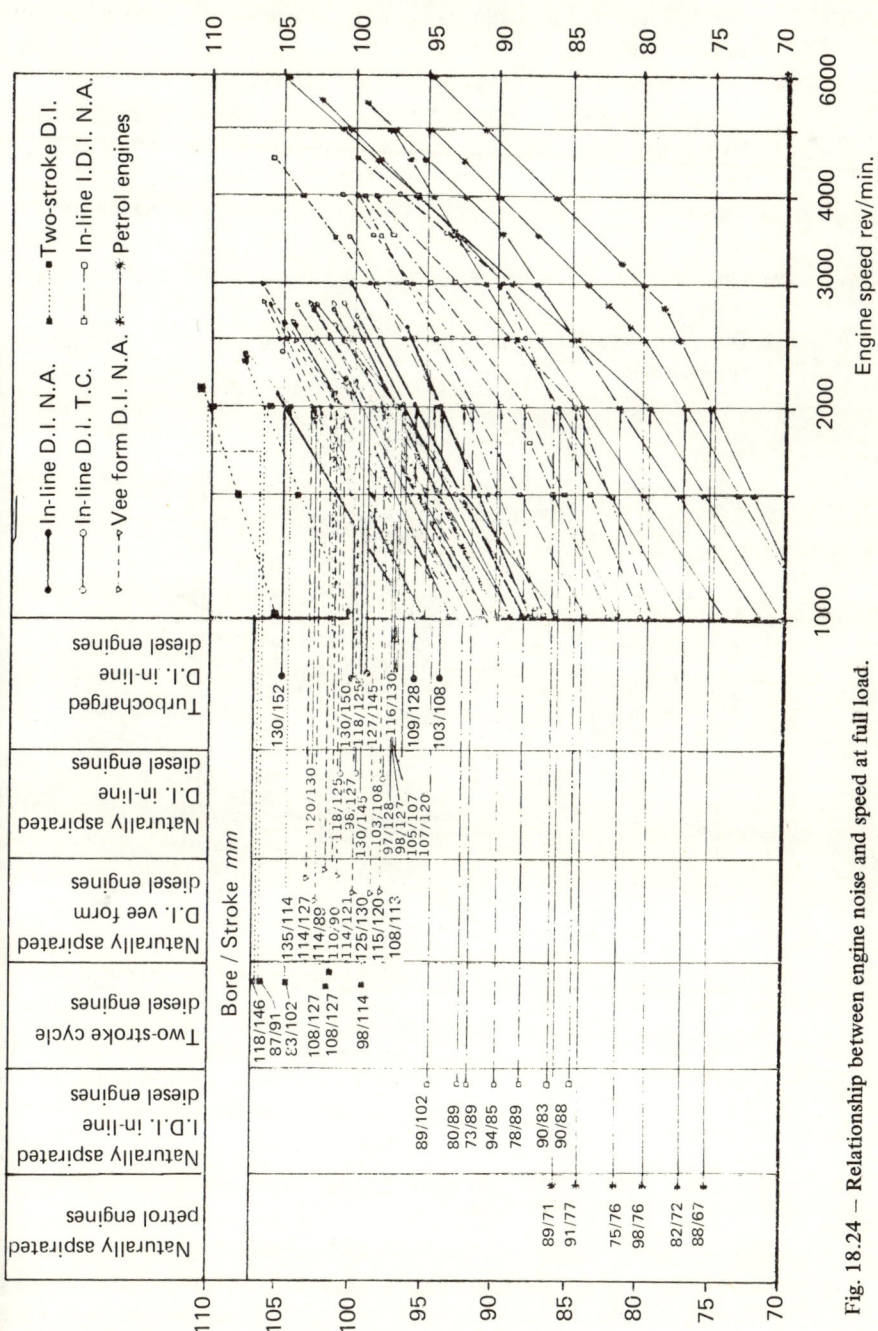

Fig. 18.24 — Relationship between engine noise and speed at full load.

overall noise of vee form and in-line engines. Some of the IDI engines also fall within this same band, but these generally have abrupt or 'advanced' pressure diagrams.

(b) The turbocharged engines occupy a band just below.

(c) The remaining IDI engines fall within a band some 8 dB(A) below the DI engines. These engines generally have smooth or 'retarded' type pressure diagrams.

(d) Two-stroke cycle engines fall within a band some 4 dB(A) higher than the DI engines.

(e) Opposed piston two-stroke cycle engines fall in a band 12 dB(A) higher.

(f) Petrol engines show considerable scatter but are about 15 dB(A) below the DI engines.

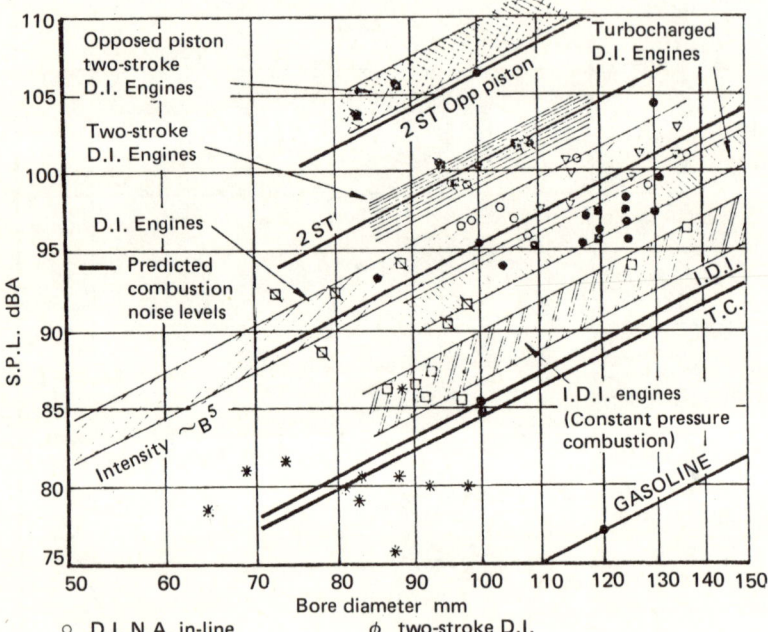

Fig. 18.25 — Relation between measured overall noise and bore size of engines in the various groups at 2000 rev/min.

Many of the differences between these various groups can be explained by the salient features of cylinder pressure development. Typical pressure diagrams are shown in Fig. 18.25 for a normally aspirated DI engine, a turbocharged engine, both 'advanced' and 'retarded' type IDI engines, and a petrol engine. The diagram of the turbocharged engine is extremely smooth but with a high peak pressure of 100-135 bar. The DI engine pressure diagram is abrupt but the peak pressure is considerably lower at 65-80 bar. The advanced IDI diagram is similar to that of the DI, while the 'retarded' diagram has a flat peak or smooth double hump peaking at 65-75 bar. The petrol engine diagram is smooth and the peak pressure very low at 35-50 bar. The different forms of the cylinder pressure development can be fully described by their spectral analyses. The rate of increase of noise of the engine with speed also depends on the cylinder pressure form as cylinder pressure levels increase with increasing speed according to the slope of the spectrum [18.7].

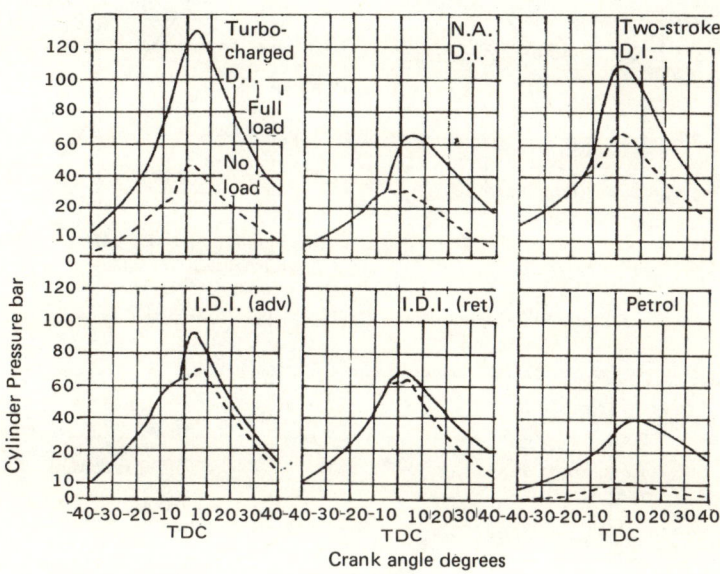

Fig. 18.26 – Typical cylinder pressure diagrams of various combustion systems.

18.8.1 Relation between combustion-induced noise and the overall noise of the engines with various combustion systems

To compare various combustion systems it is necessary to reduce them to a standard form which is independent of speed (simplified normalized spectra). It is the levels of the harmonic components between 800 and 3000 Hz (the predominant range of engine structure natural frequencies) that, to a first approximation, determine the importance of the cylinder pressure spectrum. Fig. 18.27 shows that the cylinder pressure spectrum can be approximated to a straight line representing a best fit over this critical frequency range. Fig. 18.27 shows this simplified spectrum normalized for engine speed which defines the cylinder pressure level as a function of both frequency and speed. Fig. 18.26 also shows measured cyclinder pressure levels for a DI engine taken at various speeds plotted on a base of normalized frequency. As can be seen, all spectra are coincident in the important frequency range and can be approximated by a single straight line. Such spectra have been derived for each engine tested and the results, grouped according to their various combustion systems, are summarized in Fig. 18.28.

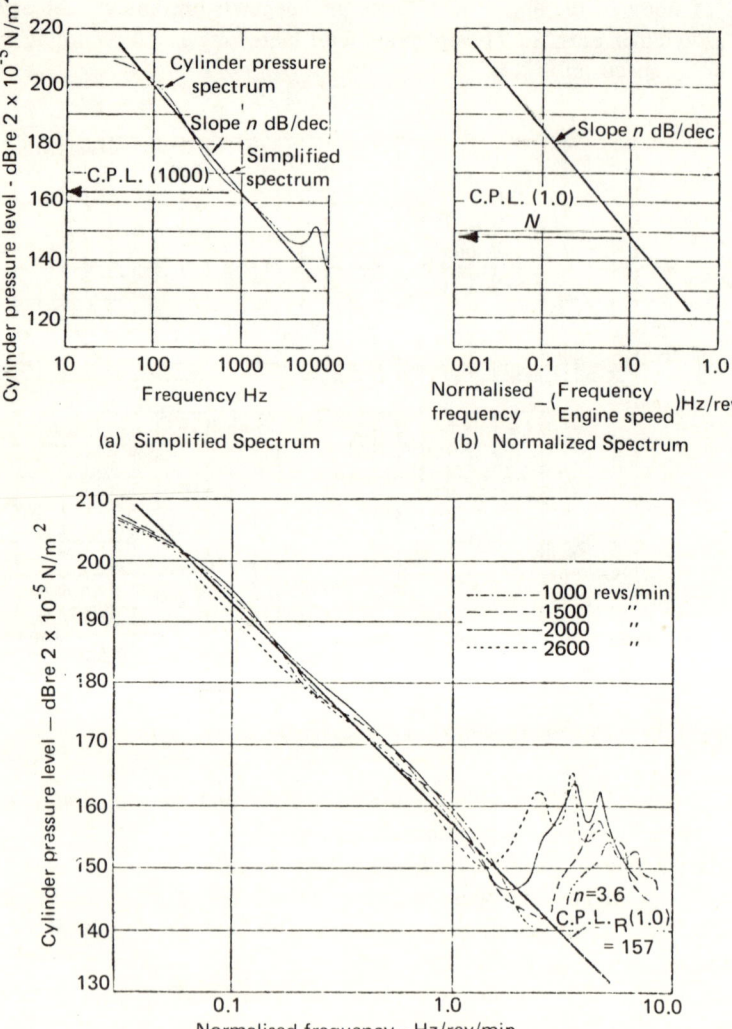

(a) Simplified Spectrum   (b) Normalized Spectrum

Fig. 18.27 – Normalized and simplified cylinder pressure spectra.

As previously stated, noise from the normally aspirated DI engine is mainly combustion-controlled, therefore the noise of engines of other combustion systems should be judged on the basis of the DI engine results. Comparing Fig. 18.28, combustion noise, and Fig. 18.25, overall engine noise, it can be seen that for DI two-stroke and IDI engines there is reasonable agreement between overall noise levels and cylinder pressure levels.

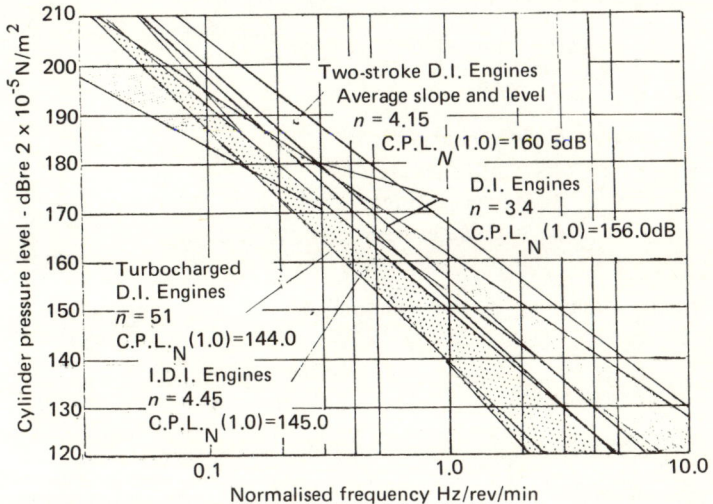

Fig. 18.28 – Full load normalized cylinder pressure spectra grouped according to the class of combustion.

A combustion model, developed at the ISVR [18.8] enables the prediction of combustion induced noise of the engine based on fundamental principles of thermodyanmics and structural attenuation. Using this model, overall noise of all engines can be predicted. Fig. 18.25 illustrates the relative combustion-induced noise levels of engines in the groups shown in Fig. 18.25 against engine size. The model clearly shows that the combustion-induced noise of turbo-charged engines should be considerably lower then DI engines. As the overall noise of the turbocharged engine is relatively higher than would be expected from combustion alone, it suggests relatively high levels of mechanical exciting forces in turbocharged engines.

18.9 RELATION BETWEEN MECHANICALLY-INDUCED NOISE AND THE OVERALL NOISE OF THE ENGINES WITH VARIOUS COMBUSTION SYSTEMS

Prediction of the mechanically-induced noise is more complex because the excitation takes place in many different parts of the engine, such as piston slap, bearing impacts at big end and main crankshaft bearings, etc. The effects produced by the oil film are also at present difficult to predict. It can be assumed, however, that the main parameter which determines the intensity of the mech-anically-induced noise is the rate of change of the reversible force, that is, dF/dt.

In the case of the piston slap [18.8] it can be shown that intensity of noise, assuming only the kinetic energy release on impact, is

$$I \sim \left[\frac{dF}{dt} \delta^4 M.K^3 \right]^{1/3} .B$$

where δ = clearance
M = piston mass
K = liner stiffness
B = cylinder bore

The various parameters in the equation are related to bore in the following manner:

$$\delta \sim B$$

$$M \sim B$$

$$K \sim B .$$

Thus the intensity of noise is approximately given by

$$I \sim dF/dt . B^4$$

Since for a given speed dF/dt is proportional to engine bore, the intensity of piston slap noise will be proportional to the (bore)^5.

That is $I \sim B^5 .$

These considerations explain why overall noise of the engines where combustion induced noise is low, i.e. turbocharged and IDI engines, the predominant mech-anically-induced noise also increases with (bore)^5.

The values of dF/dt for samples from each engine group are plotted against engine speed in Fig. 18.29. It is apparent that dF/dt for each group has distinct values. Turbocharged engines, mainly on account of very high values of peak pressure, show rates of change of reversible forces higher by a factor of two than on naturally aspirated DI engines. IDI engines with smooth pressure dia-grams and petrol engines have values lower than DI engines by a factor of two.

Since the combustion-induced noise of both IDI and turbocharged engines is of similar level, the higher mechanical noise (higher value of dF/dt) must be the cause of the higher overall noise level of the turbocharged engine.

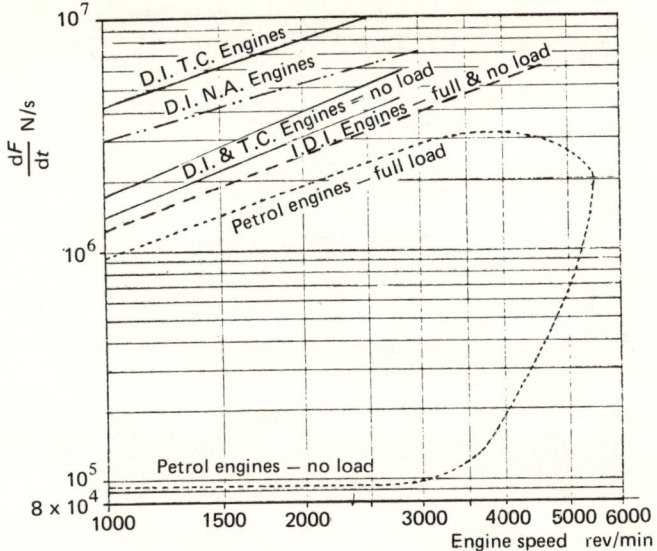

Fig. 18.29 – Rate of change of side force at T.D.C. as a function of speed for various engines.

18.10 LOW-NOISE ENGINE DESIGN

From the point of view of noise legislation it is the maximum level of noise which is important; that is, at its rated conditions of load and speed.

Fig. 18.30 illustrates the relation between this noise and the power developed by the engine for all the engines presented in Fig. 18.24. Most of the points lie in the band between 100 to 105 dB(A). To a first approximation, therefore, the noise of all engines, regardless of size, is independent of horsepower. In Fig. 18.30 also a band of noise levels is shown which is considered necessary to meet future noise legislation for road vehicles, agricultural machinery, and construction equipment.

There is considerable latitude to select engine design parameters for low noise performance; that is, operating cycle, combustion system, and bore size. There are, however, good economic reasons – production costs, gross vehicle weight limitations, and specific fuel consumption requirements – which tend to reduce this latitude in practice and to dictate the type of engine design needed for a specific duty. Thus choice is limited.

Another factor influencing engine noise is the design of the internal parts themselves. That the natural frequencies of the crank system and bearing supports influence the noise radiated by the outer engine structure is clear, but serious research must be carried out in this area before any positive design recommendations can be made.

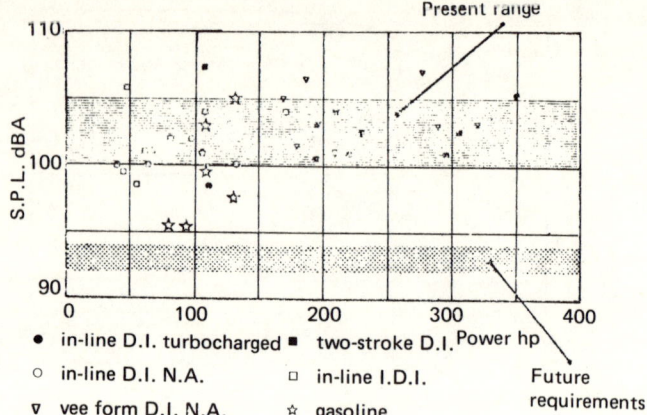

Fig. 18.30 – Relation between engine noise and power.

Some optimism must be derived from the fact that the noise of engines in a specific group covers a certain range – at least 3 dB(A), in most cases more, between the highest and the lowest levels. The factors influencing these variations can be summarized as follows:

(a) *Pistons* Piston slap noise can vary widely, either from chance or design. Prediction procedures are now well advanced, and optimum pin offsets can be readily calculated [18.9].

(b) *Timing systems* Controlled experiments have shown that chain drives are quieter then gears by 2–4 dB(A) [18.10].

(c) *Gear location* Timing systems at the rear of the engine are preferable, where not only are they shielded by the transmission but are located near an antinode of the crankshaft. 2.5 dB(A) has resulted from this one modification.

(d) *Engine layout* The petrol engine has a smaller frontal area than the diesel because the camshaft and distribution drives are usually located on the same side. Diesel engines which have pump and camshaft on the same side have some advantage.

(e) *Non-stressed covers* Experiments have shown that on most engines a 3–6 dB(A) noise reduction would be possible if all cover noise could be completely eliminated [18.11]. Selective use of damping, isolation, and stiffness is the criterion of quiet cover design.

(f) *Engine structure* Minor design differences in casting details can show small advantages. This is particularly true where attention is paid to the parts of the casting to which covers are attached.

It is stressed that where a major noise reduction is required, the optimum design details described above are selected together with a radical structure redesign; that is, every possible detail must be attended to.

An essential prerequisite to the design of diesel engines for low noise is to know how much of the total noise is radiated by individual parts of the engine. The largest noise producers can then be examined, first ensuring that a maximum noise reduction is achieved for the minimum effort.

A simple covering technique using lead sheet lined with sound-absorbing material enables the measurement of the noise radiating characteristics by any individual surface or component of the engine to be evaluated. All the outer surfaces of the engine are first shielded with 1.5 mm thick lead sheet lined with 25 mm thick fibreglass wool. To assess the contribution from a particular surface the lead shielding from that surface is then removed and the resultant noise spectrum compared with both the spectra of the normal and fully shielded engine.

The results of such covering tests (noise balance) for three engines of about 10 litres capacity in terms of percentage acoustic power radiated by the various components are shown in Fig. 18.31. The precise proportion of acoustic power radiated by the same components on different engines is seen to vary greatly. Also shown in Fig. 18.31 is the cumulative noise reduction which would result from the total elimination of each source. To achieve substantial noise reductions a considerable proportion of the total engine must be treated in each case.

As an addition to the lead covering techniques the average surface vibration technique can be used. It is particularly useful where interference to surface vibration (mass loading of thin covers by lead shielding) or lower frequencies are of interest as well as a check on the lead shielding used. The fundamental relationship between the direct radiated acoustic power of a surface S_{rad} of area average mean square velocity $<u^2(f)>$ over the surface is

$$W_{rad(f)} = \rho c \, S_{rad} \, \sigma_{rad}(f) \, <u^2>(f)$$

ρ = density of air

c = velocity of sound in air

$\sigma_{rad}(f)$ = radiation ratio at f Hz.

The sound pressure measured on a spherical surface S_{trav} around the engine will be

$$p(f) = (\rho c)^2 <u^2> (S_{rad/S_{trav}}) \cdot \sigma_{rad}(f) \ .$$

Thus a knowledge of the surface area of the radiating component, the surface area of microphone transverse or position, the area average mean square vibration velocity of the surface, and the radiation ratio of the surface enables the sound power or pressure level contribution of the surface to be obtained as a

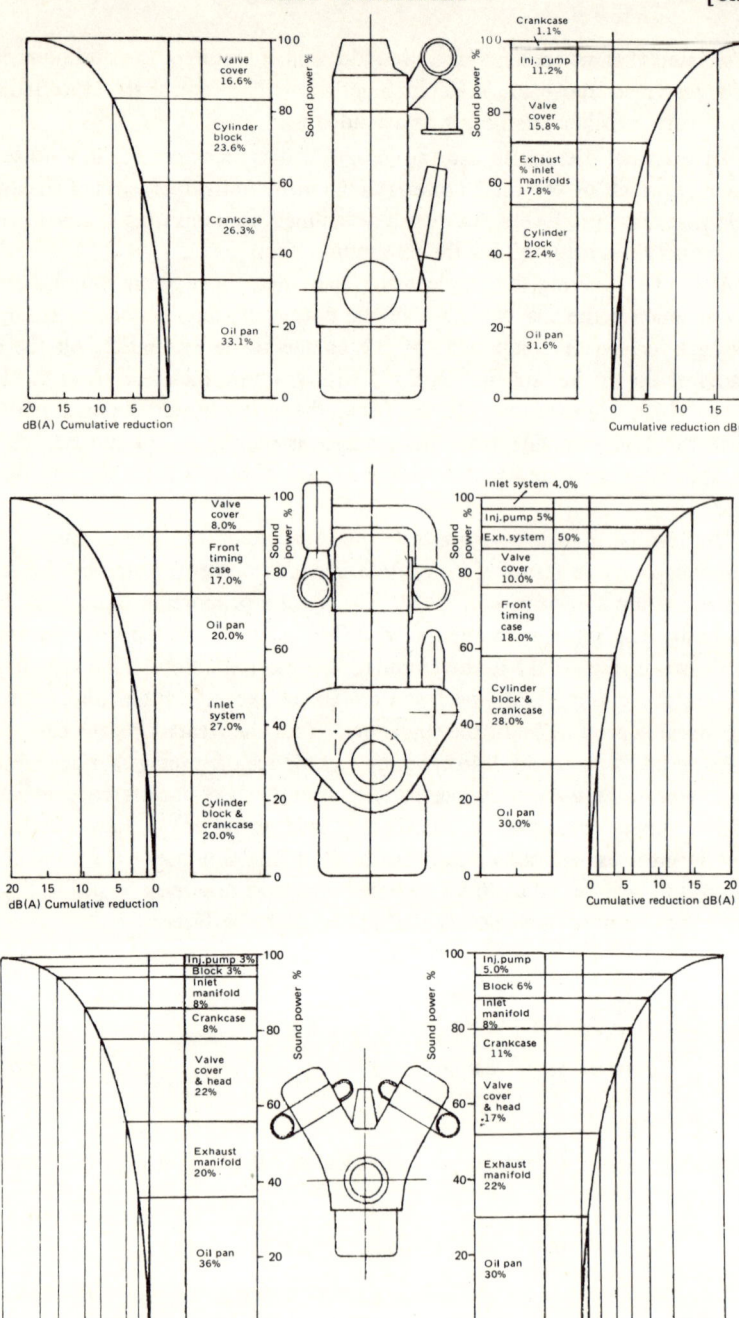

Fig. 18.31 – Sources of noise radiation for three ten-litre engines.

function of frequency or overall level. In this way a noise balance equivalent to and complementing the lead shielding technique can be made. Details of the method are given in references [18.12] and [18.13]. At present the radiation ratio values for the main load-carrying surfaces are well understood, but the situation for small, lightweight covers is not so well documented.

The actual block structure of an engine can be represented by a three-dimensional network of essential plates and rods which by finite element analysis technique enables the overall distortion pattern to be determined when the structure is loaded by typical combustion forces. Modifications can then be made to the various elements of this representation until a more desirable structural deformation is achieved.

At present this exercise can only be carried out in an existing engine structure as certain relevant testing is necessary in order to establish the base-line parameters of the model. These tests are:

(a) *Modal analysis:* Because of damping at sliding surfaces in a running engine, mode shapes are difficult to ascertain, and therefore a non-running procedure is adopted. A bare engine block is suitably suspended and excited electrodynamically. Vibration acceleration measurements are made over a grid on the internal and external surfaces of the block, and essential information of vibration amplitude and phase derived.

(b) *Static deformation of the engine block.* High pressure oil is pumped into the cylinder of a suitably suspended block, and the resultant distortion of the block measured.

To interpret the dynamic relevance of these results this static deflection pattern is regarded as a vibration phenomenon at zero frequency. In practice the static deflected shape is composed essentially of the first few block mode patterns determined from the modal excitation.

The modifications determined by this analytical method usually take the form of altering the load paths to the structure, and thus it may only be necessary to modify certain parts of the engine such as bulkhead and crankcase ribbing layout, and block wall thickness and curvature.

Over a number of years some ten different experimental low-noise engine structures have been built embracing all the engine groups described in this chapter, namely five IDI engines, two normally aspirated DI engines, one vee-form engine, one opposed piston two-stroke engine, and two high-output turbo-charged DI engines. Many different principles of design have been explored.

(1) Skeleton frame, both cast and fabricated, clad with damped materials.

(2) Increased stiffness for the same weight using magnesium alloy.

(3) Reduction of area of cast outer surfaces and replacing with large non-stiff covers.

(4) Increased stiffness using bedplate, ladder frame, or split crankcase principles.

(5) Combinations of increased damping and stiffness.

(6) Partial enclosure of cast surfaces.

Sketches illustrating some of these design principles are shown in Fig. 18.32.

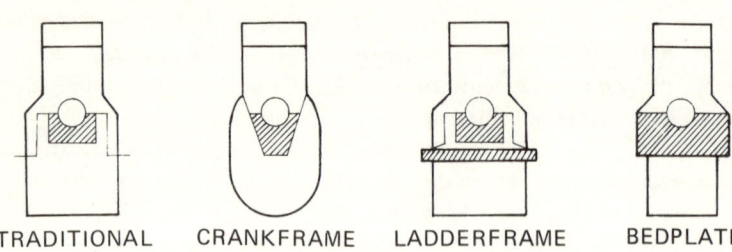

TRADITIONAL   CRANKFRAME   LADDERFRAME   BEDPLATE

Fig. 18.32 – Principles of engine structure design.

The levels measured from these experimental engines are compared with those obtained from orthodox designs in Fig. 18.33. Reductions range from 7 to 11 dB(A) at engine speeds of 2000 rev/min.

General conclusions are hard to draw, but based on economic, production, and durability considerations a universal structure design with the following features is suggested.

(1) An engine with walls of normal thickness cast as flat as practicable with raised flanges at the periphery (top deck, sump flange, and block side edges) to which damped shields can be attached (by screwing, gluing or riveting).

(2) Rear gear location with heavy gear cover to which all ancillary equipment can be attached (fuel pump, compressor, power take-off etc).

(3) Crankcase split at crankshaft axis; the lower portion comprising stiff integral bearing bedplate providing a rigid support to which the side shields and oil sump can be attached.

(4) Damped or isolated rocker cover and sump.

Although this combination of design principles can be applied to all engines it is particularly suited to those of larger size where small increased costs and weight are more acceptable. For the smaller engines, particularly on the IDI range, completely new thinking is required. There is experimental evidence that lighter weight structures of reduced stiffness could well offer a solution, and consequently studies are being made at the ISVR on these lines to produce a lightweight, low-noise passenger car diesel.

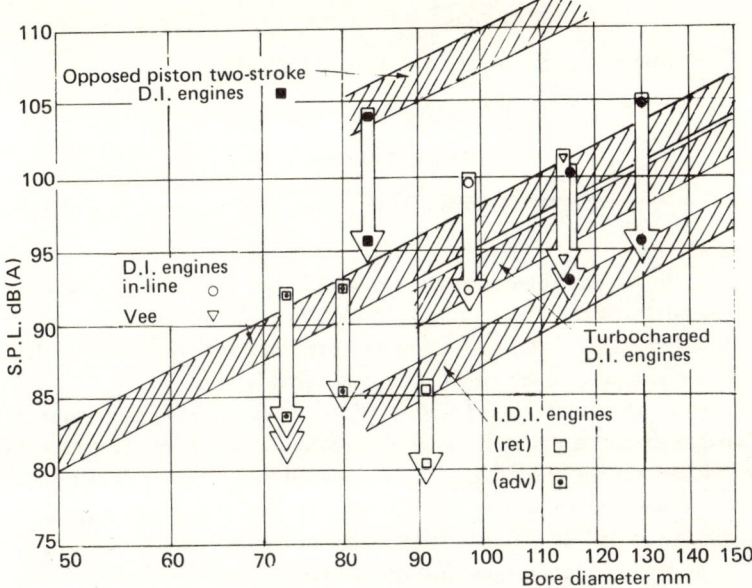

Fig. 18.33 – Overall noise reduction by structure redesign. Engine speed 2000 rev/min.

18.11 THE ORIGINS OF TYRE NOISE

Tyre noise is the second most important component of total vehicle noise which becomes predominant only at high vehicle speeds, namely above 50 km/hour for cars and above 60 km/hour for trucks. Thus by the introduction of quieter engines the overall reduction of vehicle noise will be achieved only at low and moderate vehicle speeds unless the tyre noise is significantly reduced.

Based on the comprehensive studies carried out at the TRRL [18.14] the following general conclusions can be drawn from the various tyre and road surface design combinations. The quietest combination is a smooth tyre on a smooth concrete surface which at a vehicle speed of 100 km/hour gives a 78 dB(A) pass-by level of noise measured at 7.5 metres distance. The noisiest tyre/road surface combination is a five-rib cross-ply tyre on a regular transverse grooved concrete surface which at 100 km/hour gives a pass-by level of 90 dB(A). Thus the difference between the noisiest tyre/road surface and the quietest tyre/road surface combination is about 12 dB(A).

Of all the factors investigated the speed has the greatest influence; a change in speed from 30 to 100 km/hour causes a 20 to 30 dB(A) change of noise.

Results by various investigators show that the following tyre noise-speed relationship is valid:

$$L_A = m \log_{10} V + C$$

where L_A =   sound pressure level (dB(A))

         m =   slope of regression line

         V =   vehicle speed km/hour

         C =   intercept of regression line, i.e. level of noise at km/ hour.

Table 18.2 shows the regression constants obtained by the TRRL which have been calculated for the various tyre/road surface combinations. These are based on at least ten measurements of noise and speed, and the correlation coefficient obtained are in each case very high, averaging about 0.99.

As can be seen, from Figs. 18.34 to 18.37, the presence of rain-water on the road surface increases the level of noise by 5 to 10 dB(A), the highest increase being at lower vehicle speeds. Thus in wet conditions the rate of increase of noise with speed is reduced, (m is reduced) but the value of intercept, c, is considerably increased. If the surface water is removed by providing effective drainage paths in the road surface, this source of noise is more or less completely eliminated. Surfaces that provide such drainage paths are TRRL pervious and the transverse grooved concrete surfaces.

The values for the slope (m) range from 29 to 43, and the values of intercept (C) range from -4 to $+26$. The highest values of m occur when there are low values of C, and the highest values of C occur when there are low values of m. A high value of m shows a strong speed dependence and is associated with regular features in the tyre or road surface.

Figs. 18.34, 18.35, 18.36 show rolling noise of the smooth rib, and traction tyres on the smooth concrete, coarse quartzite, and motorway surfaces. Fig. 18.37 shows the rolling noise of smooth tyres on three road surfaces.

Underwood [18.15] has summarized the effects of numerous parameters which control tyre noise as illustrated in Fig. 18.38.

Vehicle speed is the major parameter which controls the noise. The second, in order of importance, is the effect of the road surface parameters, while the tyre design parameters have negligible effect on noise produced.

Table 18.2

Relationship between speed and peak noise levels for tyres on the TRRL test road surfaces (Values of parameters)

Tyre construction		Radial ply					Cross-ply			
Regression constant	Tyre	\<--- Road surface† ---\>								
		SC	Q	M	RN	RG	SC	Q	M	RN
m	Smooth	35.1	29.6	35.7	35.9	42.8	30.4	27.5	34.9	33.1
	Ribbed 3-rib	36.1	33.8	36.3	37.1	–	37.5	33.0	36.2	–
	5-rib	38.0	31.7	35.7	–	–	42.8	39.0	36.9	–
	Traction	35.5	36.6	40.2	–	–	–	–	–	–
	Mixed rib/block	37.3	31.9	34.5	40.5	–	–	–	–	–
	rib/traction	32.8	34.9	38.0	–	–	–	–	–	–
c	Smooth	7.8	19.8	11.1	14.9	2.2	17.2	25.5	15.2	21.4
	Ribbed 3-rib	8.6	12.7	10.7	10.1	–	6.4	14.6	12.7	–
	5-rib	4.3	15.7	11.5	–	–	-3.8	3.1	8.4	–
	Traction	10.0	9.4	3.8	–	–	–	–	–	–
	Mixed rib/block	5.9	15.7	12.4	4.2	–	–	–	–	–
	rib/traction	14.9	10.8	6.7	–	–	–	–	–	–

†SC – Smooth concrete surface
Q – Coarse quartzite surface
M – Motorway surface
RN – Random transverse concrete grooving
RG – Regular transverse concrete grooving

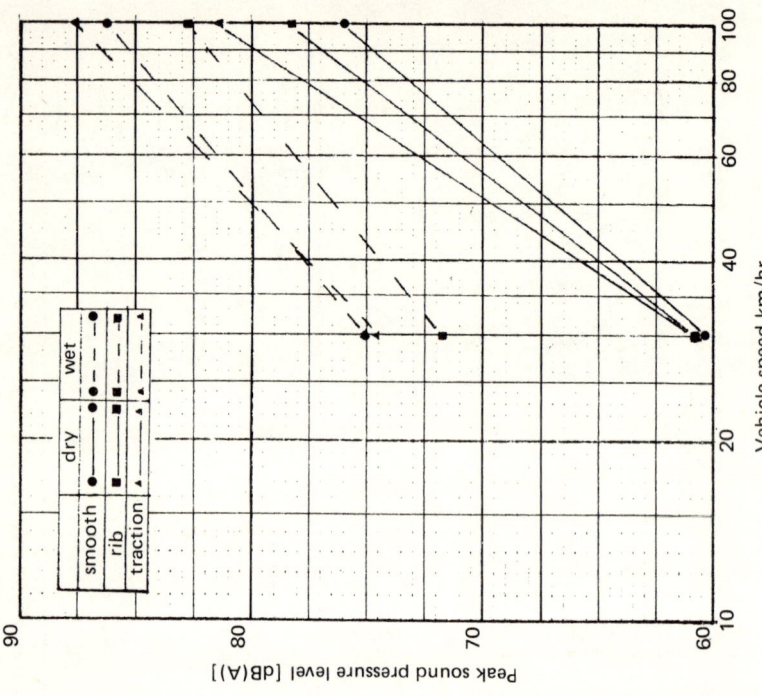

Fig. 18.35 — Rolling noise of smooth, rib, and traction tyres on coarse quartzite surface. (See Table 18.2).

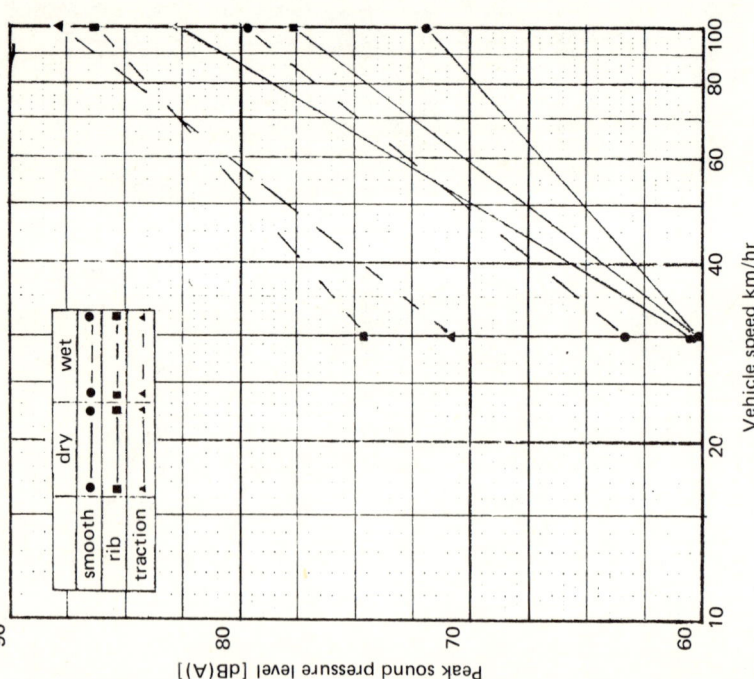

Fig. 18.34 — Rolling noise of smooth, rib and traction tyres on smooth concrete. (See Table 18.2).

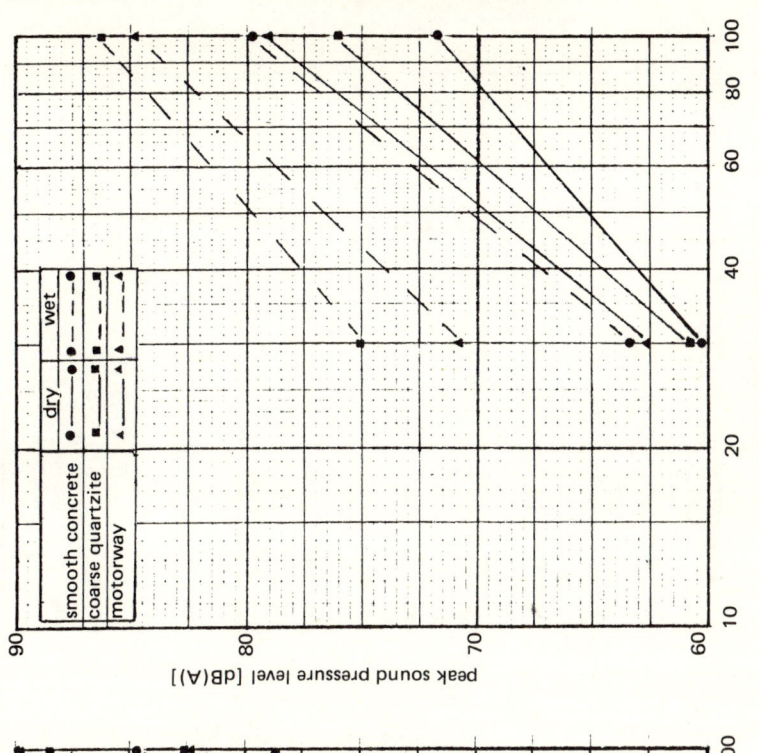

Fig. 18.37 – Rolling noise of smooth tyres on various road surfaces. (See Table 18.2).

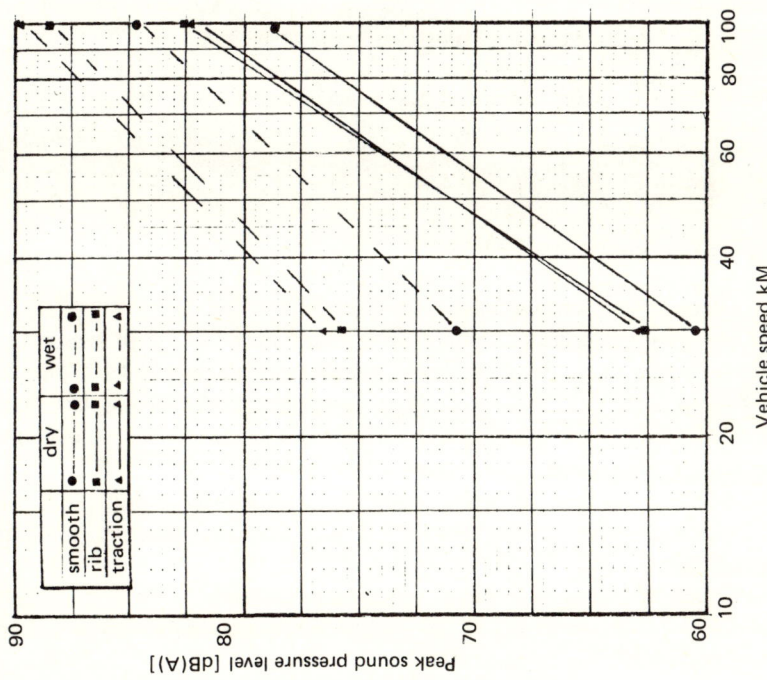

Fig. 18.36 – Rolling noise of smooth, rib, and traction tyres on motorway surface. (See Table 18.2).

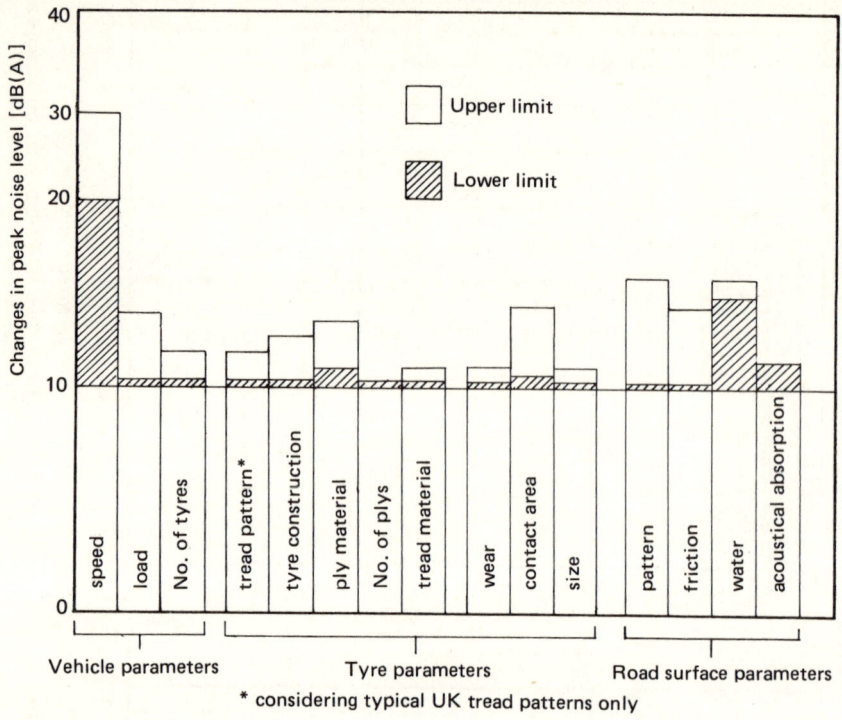

Fig. 18.38 – Estimated changes in peak sound pressure levels with changing
parameters.

18.11.1 Characteristics of tyre noise

Fig. 18.39 shows the typical one-third octave band noise spectra of tyre noise
as received by the trackside microphone at different vehicle speeds. The level
increases throughout the entire spectrum as the vehicle speed increases, but does
not show any peaks shifting in frequency. The maximum levels occur in the 630
to 1000 Hz frequency bands.

The effect of the road surface indicates that a motorway surface produces
greater noise than coarse quartzite which in turn is quieter than the smooth
concrete surface.

The shapes of the spectra are not in any way changed by the road surface
characteristics. The major effect of the road surface occurs in the two predominant frequency bands of 630 and 1000 Hz.

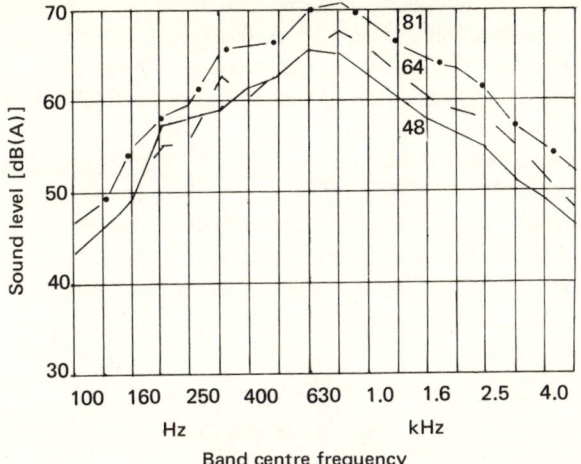

Fig. 18.39a – 1/3 octave spectra of pass-bys on motorway surface at 48, 64, and 81 km/h.

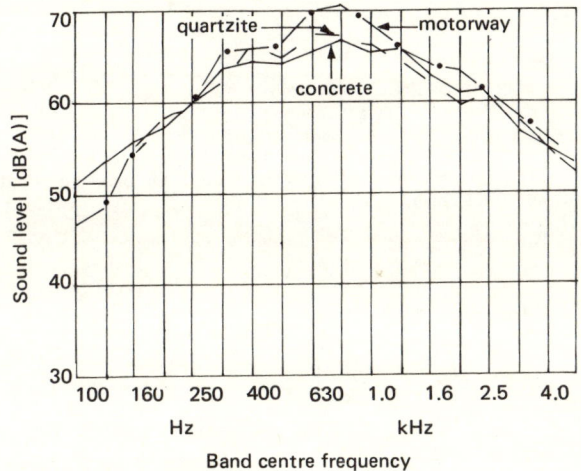

Fig. 18.39b – 1/3 octave spectra of pass-bys on smooth concrete; coarse quartzite amd motorway surfaces at 81 km/h.

18.11.2 Mechanism of generation of tyre noise

There are many controversial opinions concerning the generation of tyre noise. The following main views about the sources and generation are expressed in published literature and include tyre structure vibration, the pumping of air in

Recent studies by Underwood [18.15] who carried out detailed studies of tyre vibration by accelerometers bonded in the tyre tread, tyre shoulder, and tyre sidewall, show that the highest vibration levels in the high-frequency range are those of the tyre shoulder, while the tread and sidewall vibrations are maximum in the low frequency range below 400 Hz.

The one-third octave band spectra of vibration of these tyre elements are illustrated in Fig. 18.40. It is of interest to note that in the predominant frequency bands of 630 and 1000 Hz the sidewall, tread, and shoulder vibrations are of about equal magnitude. Maximum vibration levels are near the contact patch and are some 15 to 20 dB greater than on the top of the tyre.

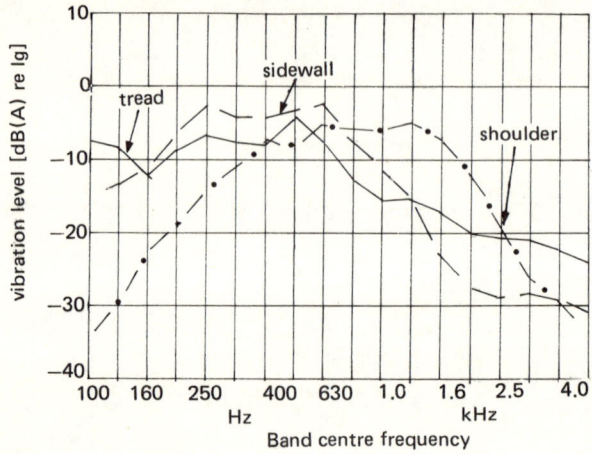

Fig. 18.40a – 1/3 octave spectra of tread. shoulder, and sidewall vibration on motorway surface at 64 km/h.

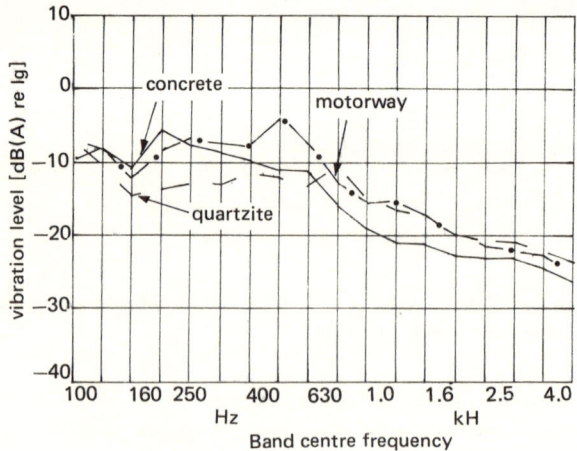

Fig. 18.40b – 1/3 octave spectra of tread vibration on concrete, quartzite, and motorway surfaces at 64 km/h.

It is possible, by using a simple model of oscillating spheres, to calculate the noise radiated by a vibrating tyre, and it has been found that the calculated levels agree with reasonable accuracy to those measured as shown in Fig. 18.41.

It can therefore be suggested that noise resulting from tyre vibrations is a substantial component of tyre noise.

Shielding experiments show that the principal tyre noise source is located close to tyre/road surface contact patch, which is consistent with tyre vibration being the dominant source, since the greatest tyre vibration occurs in this region.

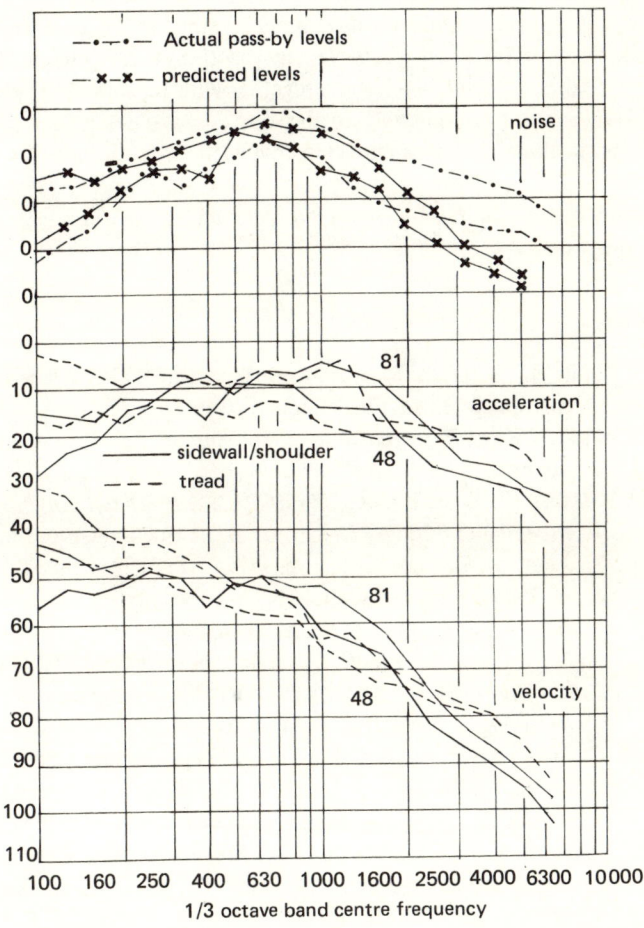

Fig. 18.41 – Comparison of 1/3 octave spectra at 48 and 81 km/h on coarse quartzite surface.

The analysis of the effects of rubber friction on a rolling tyre show that longitudinal slip occurs as the tyre leaves the contact patch, and the amount of slip being determined by the adhesive component of the frictional force acting between the tyre and the road surface. Longitudinal slip can lead to tangential excitation of the tyre which could cause it to vibrate and thereby generate the noise.

18.12 CONCLUSIONS

There are now a number of quiet demonstration vehicles built after close co-operation between various research establishments and companies, and which have been sponsored mainly by government grants. It has been shown that 80 dB(A) levels of noise can be achieved in practice, if the vehicle design incorporates partial or complete enclosures together with considerably quieter engines. So far it has not been possible to demonstrate a quiet vehicle by low-noise engine design alone, but it is expected that in the near future this goal will be achieved.

In the writer's opinion a quieter vehicle will in no way result in increased cost of the product.

REFERENCES

[18.1]   Mills, C. H. G. & Robinson, D. W. (1966) *The subjective rating of motor vehicle noise,* Final Report, HMSO.

[18.2]   (1970) *Road Research Laboratory Report,* LR 357. A review of road traffic noise.

[18.3]   Waters, P. E. & Priede, T. (1972) *S.A.E. Diesel Engine Noise Conference,* Paper 720636. Origins of diesel truck noise and its control.

[18.4]   Haddad, S. D, Pullen, H. L. & Priede, T. (1974) *FISITA Conference Paris,* Paper A-2-4. Relation between combustion and mechanically induced noise in automotive diesel engines.

[18.5]   Anderton, D., Grover, E. C., Lalor, N. & Priede, T. (1970). *ASME Paper* 70-WA/DGP-3. Origins of reciprocating engine noise – its characteristics, prediction and control.

[18.6]   Chan, C. M. P. & Anderton, D. (1974) *ISVR Report* No. 74/2. The effect of engine bore on engine noise, surface vibration and combustion for a six cylinder engine.

[18.7]   Austen, A. E. W. & Priede, T. (1958) *Inst. Mech. Engrs. Symposium on engine noise suppression,* London. Origins of diesel engine noise.

[18.8]   Anderton, D., Grover, E. C., Lalor, N. & Priede, T. (1977) *Inst. Mech. Engrs. Conference on Land Transport Engines – Economics versus Environment,* Paper C14/77. The automotive diesel engine – its combustion, noise and design.

[18.9]   Munro, R. & Parker, A. (1975) *S.A.E. Paper* 750800. Transverse movement analysis and its influence on diesel piston design.

[18.10] Priede, T. & Grover, E. C. (1970) *Inst. Mech. Engrs. Symposium – Acoustics as a Diagnostic Tool.* Application of acoustic diagnostics to internal combustion engines and associated machinery.

[18.11] Priede, T., Austen, A. E. W. & Grover, E. C. (1964) *Proc. Inst. Mech. Engrs.*, **179,** (Pt. 2A), No. 4. Effect of engine structure noise of diesel engines.

[18.12] Chan, C. M. P. & Anderton, D. (1974) *Noise Control Engineering* Winter 1974, Vol. 2, No. 1. Correlation between engine block surface vibration and radiated noise of inline diesel engines.

[18.13] Chan, C. M. P. & Anderton, D. (1975) *1st European Congress on Acoustics,* Paris, September 1975. The surface vibration and radiated noise of machine structures.

[18.14] Underwood, M. C. P. (1973) *TRRL Report* LR 601. A preliminary investigation into lorry tyre noise.

[18.15] Underwood, M. C. P. (1980). PhD thesis, ISVR, University of Southampton. The origins of tyre noise.

CHAPTER 19

Fan noise

S. Glegg

Institute of Sound and Vibration Research, University of Southampton

19.1 INTRODUCTION

This chapter will consider the aerodynamic noise produced by rotating blades in fans, rotors, and compressors. This covers a wide range of applications from modern aero-engines to low-speed ventilation fans in offices, and consequently many different types of flow interactions must be considered. The chapter therefore will take a general approach and concentrate on the principles of the noise generation process.

Most of the work on fan noise has been associated with the aero-engine industry where the noise problem has obviously been most severe. Therefore the majority of the theoretically based prediction schemes have been developed for aircraft noise applications. These include open rotors, propellors, isolated ducted rotors, and ducted rotors with inlet and outlet guide vanes (stators). A typical frequency spectrum from fans of this type shows (Fig. 19.1) that the noise includes significant tones at blade passing frequencies and lower level tones at shaft rotation frequencies which are superimposed on a broadband spectrum. While these tones appear to dominate the spectrum, they only exist over a narrow frequency band, and so in many applications the overall sound level is dominated by the broadband noise.

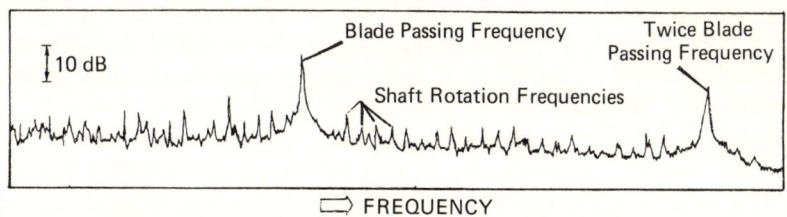

Fig. 19.1 – Spectrum from a 27-bladed rotor. After Ref. [19.51].

The noise produced by fans or rotors is usually caused by the fluctuating loads on the fan blades, which results in a source of dipole order. It was shown by Lighthill [19.1] that a source of this type will scale on the sixth power of the flow speed. Therefore as a generalization which has been shown to be correct in most applications, the overall noise output from a fan or a blower will be proportional to the sixth power of the blade tip speed. This gives a very strong speed dependence and is the reason why the most effective way to reduce fan noise is to reduce the blade tip speed.

When the blade tip speed approaches the speed of sound, transonic flow becomes important and an additional set of source mechanisms is excited. These result in a very high level source being generated by each blade, and a Mach wave can be set up which propagates non-linearly away from the source. This chapter will not include a discussion of transonic rotor noise or 'buzz saw' noise as it is sometimes referred to in specific applications, but the interested reader can find more information on this topic in references [19.49] and [19.50].

19.2 SOURCE MECHANISMS

In general the noise from rotating blades is described by Lighthill's acoustic analogy [19.1] and its extension to flows which include surfaces moving with arbitrary speed [19.2]. The radiation of noise can thus be described in terms of monopole, dipole, and quadrupole noise sources which each have their own characteristic radiation properties, and the classification of fan noise sources into these different types is fundamental to the prediction of the radiated noise.

The simplest type of monopole source is a pulsating sphere which radiates sound due to a time varying volume displacement acting on the surrounding fluid. However, fan blades do not pulsate in this sense, but owing to their rotation they move with accelerated motion which causes an effective time varying injection and extraction of fluid. This is referred to as 'thickness noise' and although the mechanism is apparently that of a monopole, which would give omnidirectional radiation, the simultaneous injection and extraction of fluid results in a directional characteristic which is more akin to that of a dipole, producing a strong pulse when the acceleration of the blade in the direction of the observer is a maximum. Therefore thickness noise peaks in the plane of the rotor and is reduced to zero on the axis of the rotor. In addition to this radiation effect, thickness noise is directly proportional to the volume of the blade and is strongly dependent on the blade tip speed which determines the blade acceleration. However, because of its directional characteristics, thickness noise is only significant on open rotors, and unless the blade tip Mach number is greater than about 0.5 then it is not significant compared with other types of noise mechanism.

The most important type of source on most fans and rotors is of dipole order, and is generated by the fluctuating pressure distribution on the surface of the blade. It was shown by Lighthill [19.1] that a fluctuating point force behaves like a dipole, giving a source strength which is proportional to the magnitude of the force resolved in the direction of the observer. When the blade chord is acoustically compact the surface pressure distribution may be modelled as an effective line force distribution which may be resolved into lift and drag components. These different components radiate most effectively at different angles to the rotor axis, the lift force fluctuations dominating on the rotor axis, and the drag force fluctuations dominating in the plane of the rotor. It should be noted that force fluctuations relative to the observer are the result of not only unsteady aerodynamic effects, but also blade rotation. Steady loadings on the blades can result in radiated sound because the blades are rotating and so have a phase speed relative to the observer. The efficiency of this type of source mechanism depends strongly on this phase speed, and so rotor or fan noise increases rapidly with the tip speed of the fan. The effects of rotation are also important for unsteady force noise radiation, and this point will be discussed in more detail in section 19.3.

In addition to these two classical fan noise mechanisms there are also a number of other mechanisms which result from nonlinear interactions between the blades and the fluid and also between features of the blade wakes. These sources are listed by Morfey [19.3] as:

(i) Fluctuating velocity field interactions resulting in quadrupole type sources. These include sound generation by turbulent wakes, the interaction of the potential flow fields of fixed and moving blade rows, and the interaction of turbulence with a potential flow field. Since these sources are of quadrupole order and the corresponding blade forces give rise to a dipole term, they are unimportant at low Mach numbers.

(ii) Entropy wakes or hot spots interacting with the velocity field of a rotor. Such interactions will occur if there is non-uniform combustion, or in fans if there is a non-uniform heat source upstream.

(iii) Interactions between the pressure field of one blade row and the pressure field, thickness, or loading of another. At low Mach numbers these sources are insignificant compared with the unsteady blade forces.

19.3 ROTATING SOURCES

One of the most important features of the acoustic field from fans and rotors is the directionality which results from the circular motion of the source. In most of the fan noise prediction schemes currently in use, the noise sources on the blade are modelled by a single point source at 80% of the blade span. This approximation is justified by the velocity dependence of the types of source involved, which causes a strong peak in source strength close to the tip of the blade. However, the principle reason for this approximation is to reduce the complexity of the computations required for a complete span-wise integration, but to retain the principal features of the radiated field which result from the

circular motion of the source. In this section, therefore, we will only consider
a ring of sources with a rotating pattern representing the blades, accepting these
as a suitable model for the complete rotor.

19.3.1 Rotational noise

In this section we will consider the periodic components of rotor noise which
cause the blade passing harmonics in the noise spectrum. The original analysis
of rotational noise was presented by Gutin [19.4] who considered the noise
generated by the steady loads on a propeller. However, the more general case of
periodically varying loads was considered by Lowson & Ollerhead [19.5], and a
detailed discussion of rotating sources is given by Barry & Moore [19.6].

The most important sources of rotor noise are the steady and fluctuating
blade loads. Therefore to evaluate the acoustic field we must first consider the
sound pressure, $p(x_i,t)$, generated by a stationary fluctuating point force:

$$p(x_i,t) = -\frac{\partial}{\partial x_i} \frac{F_i(t-r/c_0)}{4\pi r} \tag{19.1}$$

where $x_i = (x_1,x_2,x_3)$ represents the observer location and $F_i(t)$ the fluctuating
force components in each direction. In the case of a rotor blade the loads may
be resolved into their lift and drag components, which for convenience, are
assumed to act at a point displaced by a distance a from the rotor axis. (See
Fig. 19.2).

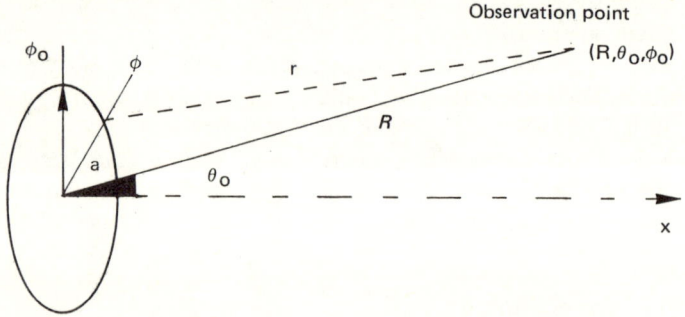

Fig. 19.2 – Coordinates of rotor blade and observer.

The acoustic field generated by these two forces may be expressed in
spherical coordinates as

$$p(R,\theta_0,\phi_0,t) = \frac{\partial}{\partial x} \frac{L(t-r/c_0)}{4\pi r} - \frac{1}{a}\frac{\partial}{\partial \phi_0} \frac{D(t-r/c_0)}{4\pi r} \tag{19.2}$$

Where the lift force $L(t)$ acts parallel to the rotor axis, and the drag force
$D(t)$ acts on the plane of the rotor.

To explain the effect of blade rotation the rotor is modelled by a continuous distribution of sources on a ring of radius a. The source strength on this ring is zero everywhere apart from at the instantaneous locations of each blade, where it is equivalent to the instantaneous blade loading. Since the rotor has an angular velocity Ω, the source strength on the stationary ring will vary with time. At a fixed point ϕ, the source strength variation will consist of a series of impulses which occur every time a blade passes that point.

Thus the contribution to the acoustic field from the source at ϕ from a rotor with B blades will be

$$dp = \frac{\partial}{\partial x} \frac{L(\phi)}{4\pi r} \sum_{m=-\infty}^{\infty} \delta(\phi + \frac{2\pi m}{B} - \Omega(t - r/c_0))$$

$$- \frac{1}{a} \frac{\partial}{\partial \phi_0} \frac{D(\phi)}{4\pi r} \sum_{m=-\infty}^{\infty} \delta(\phi + \frac{2\pi m}{B} - \Omega(t - r/c_0))$$

(19.3)

where $\delta(\phi)$ represents a Dirac delta function.

This series of pulses is periodic as a function of time or azimuth angle ϕ, and its frequency components may be evaluated using the Fourier series expansion

$$\sum_{m=-\infty}^{\infty} \delta(\phi + \frac{2\pi m}{B}) = \frac{B}{2\pi} \sum_{m=-\infty}^{\infty} e^{imB\phi}$$

(19.4)

so that equation (19.3) becomes

$$dp = \frac{B}{8\pi^2} \sum_{m=-\infty}^{\infty} e^{-imB\Omega t} \left[L(\phi) \frac{\partial}{\partial x} - \frac{D(\phi)}{a} \frac{\partial}{\partial \phi_0} \right] \frac{\exp(imB(\phi + (\Omega r/c_0))}{r}$$

(19.5)

To evaluate the acoustic field generated by the complete ring source, equation (19.5) must be integrated over ϕ, and the variation of lift and drag as a function of position accounted for. In the simple case when the steady forces dominate, $L(\phi)$ and $D(\phi)$ may be considered constant; however, in practice this only applies to highly loaded rotors operating in clean aerodynamic inflows. In most cases an inflow distortion of some type causes the loading on the rotor to vary periodically with position, and this can be incorporated into the theory by expanding $L(\phi)$ and $D(\phi)$ in a series of rotor loading harmonics. It is assumed that both lift and drag vary in the same way to a given inflow distortion and may be expanded as

$$L(\phi) = L_0 \sum_{n=-\infty}^{\infty} A_n e^{in\phi}, \quad D(\phi) = D_0 \sum_{n=-\infty}^{\infty} A_n e^{in\phi}.$$

Therefore the acoustic field from the complete ring source is given by

$$
p(R,\theta_0,\phi_0,t) = \frac{B}{8\pi^2} \sum_{m=-\infty}^{\infty} e^{-imB\Omega t} \sum_{n=-\infty}^{\infty} A_n \left[L_0 \frac{\partial}{\partial x} - \frac{D_0}{a} \frac{\partial}{\partial \phi_0} \right] \times
$$

$$
\times \int_{-\pi}^{\pi} \frac{e^{i(mB+n)\phi + imB(\Omega r/c_0)}}{r} \, d\phi \tag{19.6}
$$

To evaluate the acoustic far field the propagation distance r may be approximated by

$$
r \cong R - a \cos(\phi - \phi_0) \sin \theta_0
$$

in the phase term, and in the amplitude term of $1/r$ this approximation reduces to $1/r \cong 1/R$. The integral over ϕ may then be written in terms of Bessel's integram formula as

$$
\frac{1}{R} e^{imB(\Omega R/c_0) + i(mB+n)\phi_0} \int_{-\pi}^{\pi} e^{i(mB+n)(\phi-\phi_0) - imB(\Omega a/c_0)\cos(\phi-\phi_0)\sin\theta_0} \, d(\phi-\phi_0)
$$

$$
= \frac{1}{R} e^{imB(\Omega R/c_0) + i(mB+n)\phi_0} \, 2\pi \, i^{(mB+n)} J_{mB+n}(mB(\Omega a/c_0)\sin\theta_0)
$$

where $J_m(\zeta)$ is a Bessel function of the first kind. The differentiations in equation (19.6) may be easily evaluated since $\partial/\partial x = \cos\theta_0 \, \partial/\partial R$, and by incorporating the source Mach No $M_s = \Omega a/c_0$ the acoustic field is defined as

$$
p(R,\theta_0,\phi_0,t) = \frac{B}{4\pi R a} \sum_{m=-\infty}^{\infty} \sum_{n=-\infty}^{\infty} A_n \, e^{-imB\Omega(t-R/c_0) + i(mB+n)\phi_0} \times
$$

$$
\tag{19.7}
$$

$$
\times \; [mBM_s L_0 \cos\theta_0 - (mB+n)D_0] \, i^{mB+n+1} J_{mB+n} (mBM_s \sin\theta_0) \; .
$$

This shows that the far field may be considered as a set of modes each with its own directionality which is defined by the Bessel function term.

To interpret this result we note that the Bessel functions tend to zero when their arguments $mBM_s\sin\theta_0$ are very much less than their orders $(mB+n)$. Therefore the far field radiation is dominated by the terms in the series for which

$$
M_s\sin\theta_0 > \left| \frac{mB+n}{mB} \right| \; .
$$

If we consider equation (19.6) it can be shown that each term in the series represents a pattern which is rotating with a phase speed of

$$
\Omega_p = \frac{mB\Omega}{mB+n}
$$

and so efficient radiation occurs only when

$$M_p \sin\theta_o = M_s \left| \frac{mB}{mB+n} \right| \sin\theta_o > 1 \qquad (19.8)$$

where M_p is the pattern Mach No. $\Omega_p a/c_o$.

Therefore only patterns which have supersonic phase speeds (i.e. $M_p = \Omega_p a/c_o$ > 1) will radiate efficiently to the acoustic far field. This demonstrates the importance of inflow distortions in the calculation of the acoustic far field since for a subsonic rotor the condition (19.8) will not be satisfied unless $-n$ is of the same order of magnitude as mB. Although the azimuthal variation of the modulation function may only be very small so that $A_{-mB} \ll A_o$, it is apparent from these results that the small variations are most significant to the radiated acoustic field. This highlights the major problem with rotor noise prediction since these small-scale disturbances are the most difficult to predict accurately.

However, we can consider two limiting cases: first when the rotor is highly loaded and has a clean inflow, only the steady loading terms $(n = 0)$ need be considered, and the amplitude of each blade passing harmonic in the noise spectrum will be

$$\left| P_{mB}(R, \theta_o, \phi_o) \right| = \frac{mB^2\Omega}{4\pi R c_o} \left(L_o \cos\theta_o - \frac{D_o}{M_s} \right) J_{mB}(mBM_s \sin\theta_o) .$$

(When a signal of this type is subjected to frequency analysis the r.m.s. value of the signal at this frequency will be $1/\sqrt{2}$ times the sum of the contributions from both the $mB\Omega$ component and the $-mB\Omega$ component. Therefore the amplitude should be increased by a factor of $\sqrt{2}$). At low speeds when $M_s \sin\theta_o \ll 1$ the Bessel function may be approximated by its asymptotic value, and the spectral harmonics will decay rapidly as a function of mB according to

$$\sqrt{\frac{mB}{2\pi}} \left[\frac{eM_s \sin\theta_o}{2} \right]^{mB}$$

Alternatively if the blade encounters a discrete velocity disturbance which increases the lift and drag on the blade to the extent that it dominates the force fluctuations over the rotor plane, the signal generated will be very impulsive and the acoustic field will be similar to that of a point source at the position of the disturbance. This impulsive signal will have a lot of energy in the higher harmonics, and their level will be determined by the lift and drag forces generated by the interaction.

19.3.2 Ducted rotors

When a rotor is in a duct the propagation of sound from the rotor sources is determined by the boundary conditions which the duct imposes on the sound field. Sound propagation along cylindrical ducts has been considered in detail in

Chapter 12 of this book, where it was shown that the acoustic field can be considered as a set of modal patterns which either propagate or decay along the duct depending on whether they are cut-on or cut-off. The modes, which describe the sound field, can be considered as spinning as they pass along the duct, and those with a supersonic phase speed at the wall are cut-on. Those which have a subsonic phase speed at the wall decay very rapidly and, providing the duct is at least one diameter long, will be insignificant to the overall radiated sound. Radiation from the open end of the duct is therefore determined by a set of supersonically spinning modes, which generate an acoustic field in the same way as the open rotor (assuming the boundary conditions at the duct exit do not have an appreciable effect and only the forward arc is considered). Therefore the acoustic far fields from a ducted rotor and an open rotor are both determined by the supersonically spinning modes generated at source.

The duct may be used to reduce fan or rotor noise by using acoustic liners on the duct walls. These can be used to attenuate specific dominant modes provided that the correct liner impedance characteristics are chosen (Vaidya & Dean [19.7]).

19.3.3 Interacting blade rows

Many ducted fan designs include a set of stationary inlet or outlet guide vanes which interact with the flow over the rotors. This interaction process can be described by the modulation of the source strength by a stationary pattern as discussed in section 19.3.1. The spatial modulation of the source strength in the fixed frame of reference results in modes with an angular phase speed $mB\Omega/(mB+n)$. If there are V guide vanes, then n may only take on the values $n = kV$. Therefore the angular phase speeds which are generated will be

$$\Omega_p = (mB/(mB+kV))\,\Omega\ .$$

By the careful choice of B and V the pattern Mach numbers $M_p = \Omega_p a/c_0$ which are generated at a given operating condition will all be subsonic ($M_p < 1$) so that only inefficient sound propagation takes place.

19.3.4 Modulation tones

Equation (19.7) in 19.3.1 describes the sound field generated by a B-bladed rotor, and in its derivation it was assumed that each blade was identical, giving spectral harmonics at the frequencies, $B\Omega$. However, in practice each blade may not be identical or equally spaced, and so the summation (19.4) no longer eliminates terms which are not multiples of B. When this occurs, tones are generated at multiples of the shaft rotation frequency. (See Fig. 19.1).

It was also assumed that the modulation of the rotor sources was not a function of time. However, in practice the inflow may include swirl which results in a modulation function which rotates at a speed Ω_0 and may be described by $L(\phi - \Omega_0 t)$. Expanding this in a Fourier series gives

$$L(\phi - \Omega_0 t) = \sum_{n=-\infty}^{\infty} L_0 A_n \exp(in\phi - in\Omega_0 t) \tag{19.9}$$

which when included in (19.6) demonstrates that additional tones are generated at the frequencies

$$mB\Omega + n\Omega_0 . \tag{19.10}$$

These additional frequencies are not multiples of the shaft rotation frequency and are sometimes referred to as 'modulation tones'.

Finally, if the inflow to the rotor varies with time independently of ϕ then a band of frequencies about each tone will be generated. This may be described by the lift function $L(\phi, t)$ which is related to the frequency domain by

$$L(\phi,t) = \int_{-\infty}^{\infty} L_0 \sum_{n=-\infty}^{\infty} A_n(\omega) \exp(i\omega t - in\phi) \, d\omega/2\pi . \tag{19.11}$$

When this is included in (19.5) it shows how the frequencies $(mB\Omega + \omega)$ are excited. If $A_n(\omega)$ only exists over the range $-\omega_c < \omega < \omega_c$ then the line spectra at $m\Omega$ will be broadened to cover the range

$$m\Omega - \omega_c < \omega < m\Omega + \omega_c . \tag{19.12}$$

19.3.5 Broadband noise sources in circular motion

In the preceding sections noise sources which generate spectral harmonics due to their rotational motion have been described. However, in many applications broadband rotor or fan noise is often more important than the discrete tones, and so in this section the effects of circular motion on broadband rotor noise will be considered.

When a discrete tone such as a blade passing harmonic is generated by a source in circular motion, the single frequency contribution results from the coherence of the source in space and time. In other words the single frequency is generated because the output from the source is periodic. However, when a broadband source moves in circular motion the spatial and temporal coherence is reduced, and the interference effects which take place for the single frequency source no longer occur. It is therefore necessary to adopt a completely different approach to broadband noise sources in circular motion, evaluating the acoustic power spectrum in the far field.

The analysis of a point source with a continuous frequency spectrum in circular motion is given by Ffowcs-Williams & Hawkings [19.8]. They considered a dipole with a source spectrum $G_T(\omega)$ rotating at an angular velocity Ω in a circle of radius a, and obtained the far field power spectrum as:

$$G_p(\omega) = \left(\frac{\omega\cos\theta_0}{4\pi c_0 R}\right)^2 \sum_{m=-\infty}^{\infty} G_T(|\omega - m\Omega|) J_m^2\left(\frac{\omega a \sin\theta_0}{c_0}\right) . \tag{19.13}$$

This result demonstrates how the effects of rotation weights different source frequencies with a different directionality. In fact this is a general result which applies to both broadband and discrete tone generation provided that the correct form of the source spectrum is used. However, it does assume that the source is statistically stationary and that the bandwidth of the spectral analysis is less than the rotational frequency Ω, so that each discrete tone in the spectrum is resolved.

However, this result is difficult to evaluate and has been simplified by Morfey & Tanna [19.9] in the high frequency limit where $\omega \gg \Omega$. The most important conclusion from their approximation was that when $M_s \sin \theta \ll 1$ the effects of rotation could be ignored and the measured power spectrum would take the form of a stationary point source.

$$G_p(\omega) = \left(\frac{\omega \cos\theta_o}{4\pi c_o R}\right)^2 G_T(\omega) \ . \tag{19.14}$$

However, at higher rotational speeds the slope of the source spectrum becomes important and additional terms must be included in this approximation.

19.3.6 The effects of axial flow

In a ducted axial fan the flow through the duct can affect the propagation of sound both upstream and downstream of the fan. This problem has been considered by Morfey [19.3], and it shown that the power spectrum of the radiated noise should include a correction factor of

$$1 \pm \tfrac{3}{4} M_x \cos^2\beta \text{ where } \beta = \tan^{-1}\frac{M_s}{M_x}$$

where M_x is the flow Mach number and the $\pm$ sign refers to the sound radiated in the $\pm x$ direction. This demonstrates how more noise is radiated in the downstream direction, but this will only be important for high speed axial flows.

19.3.7 Conclusion

In this section the effect of source rotation on the radiated field has been considered, giving results for the directionality and frequency dependence which will be observed from a source of known strength. However, to complete the prediction of fan noise it is also necessary to predict the actual source strength, and this will be considered in the next section.

19.4 FLUCTUATING BLADE LOADS

In section 19.2 the various different source mechanisms in fans were discussed, and it was concluded that the most important of these was caused by the fluctuating pressure distribution on the blade surface. In order to predict the noise generated by the fan it is therefore necessary to understand the mechanisms which cause the fluctuating surface pressures, and in this section some of the methods which have been used to evaluate this mechanism will be described.

19.4.1 Inlet turbulence

When a blade encounters a turbulent inflow the effective angle of attack of the blade will change and the loading will fluctuate, creating an unsteady source mechanism. Therefore in order to predict the radiated noise from blade/turbulence interaction it is necessary to know:

(1) Details of the inlet turbulence
(2) The response of the blade to a non-uniform inflow
(3) The acoustic efficiency of fluctuating blade loads.

Of these three factors the most difficult to predict is the inlet turbulence to the fan. Each design of fan will operate in a different inflow environment, and there is no general method which can be used to predict accurately the details of the turbulent inflow to an arbitrary fan. There are, however, detailed models of the turbulence generated by a grid, and for experimental purposes these have been used successfully to check various noise prediction schemes.

Most of the prediction schemes [19.10, 19.11, 19.12] which have been developed to estimate the noise from blade-turbulence interaction follow the same approach. The inflow turbulence is considered as a frozen pattern which is convected past the rotor at a constant speed U. The rotor blades then encounter an unsteady upwash $v(x, y, t)$ (where x is a streamwise coordinate and y a spanwise coordinate), which may be described in terms of the wavenumber spectrum of the frozen pattern.

$$v(x, y, t) = \iiint_{-\infty}^{\infty} w(k_x, k_y, k_z) \exp(ik_x(x - Ut) + ik_y y) \, dk_x dk_y dk_z \quad .(19.15)$$

The response of the blade to a gust of the type $v_g = v_0 \exp(ik_x(x - Ut) + ik_y y)$ is then calculated, giving the fluctuating lift per unit span for a single wavenumber component as

$$L(y, t) = \pi \rho U b v_0 g(k_x b, k_y b) \exp(ik_y y - ik_x Ut) \qquad (19.16)$$

where $g(k_x b, k_y b)$ is the lift response function of the blade due to the sinusoidal upwash v_g, and b is the blade chord. (The response to a streamwise gust has also been considered [19.13, 19.14] but will not be included here.) The fluctuating lift from all wave-number components is then the integral of (19.16) over all wave-numbers, and at the frequency ω only the $k_x = \omega/U$ component contributes.

A considerable research effort has gone into evaluating the correct form for the response function: the result for the two-dimensional problem ($k_y = 0$) was

given by Sears [19.15], and subsequently useful approximations have been made
to Sears' function by Liepmann [19.16] who demonstrated that for $\omega b/2U \gg 1$

$$\left| g\left(\frac{\omega b}{U}, 0\right) \right|^2 \cong (\pi \omega b/U)^{-1} . \tag{19.17}$$

The effects of spanwise gusts $(k_y \neq 0)$ has been considered by Filotas [19.17],
Mugridge [19.18], Graham [19.19], and Amiet [19.12], and have been reviewed
by Morfey [19.20] who notes that the two-dimensional response (19.16) is
useful for noise prediction when the acoustic wavelength λ lies in the range
$b < \lambda < d$ where b is the blade chord and d is the blade span. An approximate
guide to the three-dimensional response is given by

$$\left| g\left(\frac{\omega b}{U}, k_y b\right) \right|^2 \cong \left[2\pi \sqrt{\left(\frac{\omega b}{2U}\right)^2 + \left(\frac{k_y b}{2}\right)^2} \right]^{-1}$$

where $k_y b < 1$; but when the blade span d is less than the acoustic wavelength
the total integrated lift is required which is given by Filotas [19.21].

There are also general expressions which may be used to model the inflow
turbulence [19.22, 19.12 and 19.24]. However, in practice each type of fan will
operate in a different environement, and so usually a better model is obtained
from empirical data (see section 19.5).

19.4.2 Laminar boundary layer/vortex noise

It has been realized for many years that isolated aerofoils in a steady flow can
produce discrete tones (e.g. Tyler [19.25], Gongwer [19.26]); however, it is
only more recently that the mechanism which produces these tones has been explained. An experimental investigation by Paterson et al. [19.27] established
that tones produced by aerofoils in a flow occurred only when the boundary
layer over either surface of the aerofoil was laminar at the trailing edge. Their
results for a NACA 0012 aerofoil in an open jet with low turbulence levels
demonstrated the range of Reynolds numbers and angles of attack where the
tone could be found. In addition to this it was also demonstrated that the
frequency of the tone varied as $U^{3/2}$ (where U is the flow velocity) in general
but had a ladder type structure with a frequency dependence of $U^{0.8}$ on each
step (see Fig. 19.3). As a result of this experimental investigation Tam [19.28]
suggested that the mechanism for the tone generation was a feedback loop which
exists betweem the aerofoil wake and the boundary layer on the surface. Since
the laminar boundary layer will result in unstable Tollmein–Schlichting waves
[19.29] in the wake, Tam argued that the noise produced in the wake would
then feed back to create new disturbances at the trailing edge. This model
explains the ladder type of structure for the frequency variation with velocity,
since maximum amplification will occur when the distance between the peak
wake displacement and the excitation point is an integer number of acoustic

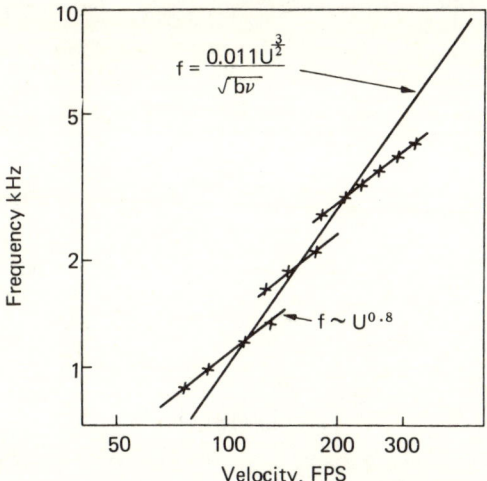

Fig. 19.3 – Effect of velocity on laminar boundary layer tone frequencies (After Ref. [19.27]).

wavelengths. Therefore a dominant discrete tone is produced at a frequency where fluctuations in the wake are phase locked to the acoustic field.

Subsequently it has been demonstrated [19.30] that the general variation of tone frequency with velocity is given by

$$f = 0.092 \left(\frac{\omega' \delta^*}{U} \right) \frac{U^{3/2}}{(b\nu)^{\frac{1}{2}}}$$

where b is the aerofoil chord, ν the kinematic viscosity, and δ^* the boundary layer thickness. The value of $(\omega'\delta^*/U)$, which is a non-dimensional frequency describing the instabilities in the wake, can be calculated from boundary layer theory [19.29] and typically varies between 0.17 and 0.1 in the transitional Reynolds Number range 1×10^5 to 3×10^6. A good approximation to fit all the data has been found as $\omega'\delta^*/U = 0.12$ [19.27].

Although this discussion has been limited to the consideration of an isolated aerofoil in parallel flow, the significance of laminar boundary layer noise in low-speed axial fans and rotors has been demonstrated by several investigations, for instance [19.31, 19.32, 19.33]. The difference between the aerofoil in parallel flow and the rotating blade application is that in the latter, broadband noise rather than a tone is generated over the range of frequencies which are excited by each section of the blade between the hub and the tip. This is illustrated by Fig. 19.4, taken from [19.23], which shows the measured results from a rotor with and without boundary layer trips on the pressure surface of the blades. When the boundary layer has been tripped it no longer remains laminar at the

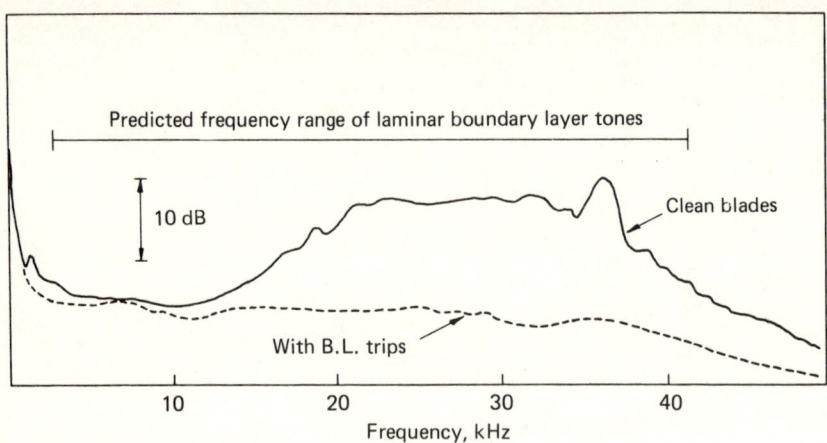

Fig. 19.4 – Spectrum of rotor noise in open jet tunnel with and without boundary layer trips (M_{tip} = 0.47, Tunnel speed = 9.1 m/sec.) (After ref. [19.23]).

trailing edge, and so the mechanism for laminar boundary layer noise is eliminated. Consequently the level of broadband noise at high frequencies is reduced by $\sim$ 15 dB over the frequency range

$$\frac{0.011}{(b\nu)^{\frac{1}{2}}} (U_{root})^{3/2} < f < \frac{0.011}{(b\nu)^{\frac{1}{2}}} (U_{tip})^{3/2}$$

where laminar boundary layer noise is significant.

These results are based on boundary layer calculations for flat plates, and it has been pointed out by Longhouse [19.31] that significant differences can be expected for cambered aerofoils. Perhaps the most important difference is that on cambered aerofoils laminar boundary layer noise is only observed [19.31, 19.33] at negative angles of incidence as the result of a laminar boundary layer on the upper surface.

The prediction of the noise level generated by unsteady laminar boundary layers does not appear to have been given much attention, and the reason for this is probably because it can be eliminated simply by the introduction of boundary layer trips. However, Aravaumden et al. [19.34] have developed a velocity scaling law of $U^{5.8}$, and this agrees with the measurements of Sharland [19.35] and Clark [19.33]. On the contrary Longhouse [19.31] in a detailed set of fan noise measurements found a variation of level with velocity which did not follow any simple scaling law; this was attributed to the involved feedback mechanism taking place, and further investigations are required before the prediction of laminar boundary layer noise levels can be obtained.

19.4.3 Turbulent boundary layer noise

Turbulent boundary layer noise represents the base level of broadband noise which is generated by a rotor operating at high Reynolds numbers (where transition precedes separation). When the incident turbulence to the blade is reduced sufficiently, the pressure fluctuations on the blade surface are dominated by those which result from the turbulent boundary layer and its wake.

The noise generated by turbulent boundary layers has been studied by Mugridge [19.36] for blades which have acoustically compact chords. In this situation the integrated fluctuating pressure over the upper and lower surfaces of the blade must be evaluated, and estimates made of the correlation length scales involved. As a result of his experimental work Mugridge estimated the fluctuating force on the blade caused by the turbulent boundary layer to have a power spectrum.

$$G_T(\omega) = \frac{U_s^5}{\omega^2} l^2 b C_D \times 6.4 \times 10^{-4} \ . \tag{19.18}$$

This result is valid over the frequency range $0.02 < \frac{\omega b}{U} \cdot C_D < 0.6$, and the spanwise correlation length scale of these fluctuations is approximated by $l = 2U_s/\omega$.

In his estimates of the radiated sound, Mugridge only considers the total radiated sound power which eliminates directionality effects and so does not require any estimate of the direction in which this force acts. This simplifies the sound radiation problem and will be discussed in more detail in section 19.5.3.

19.4.4 Secondary flow noise

In addition to the fluctuating load mechanisms described above, there are additional mechanisms which have been identified as potentially important sources of high-frequency broadband noise. One of these is the three-dimensional flow at the tips of ducted rotors. The flow around the tip of a ducted rotor contains at least two types of unsteady vortex flow; the first is the generation of a vortex at the blade tip adjacent to the suction surface due to leakage through the tip clearance gap, and the second is a blade passage vortex which is generated by the deflection of the flow streamlines within the annulus wall boundary layer. These flows are illustrated in Fig. 19.5 from which it can be seen that they oppose each other, and by reducing one independently the total unsteady load may be increased. The design parameters which control these flows are the tip clearance, the duct boundary layer thickness, and the blade loading. The radiated noise is principally at high frequencies owing to the small scales of turbulence which these flows generate.

This noise source has been studied by Mugridge & Morfey [19.37], and they conclude that the noise produced by secondary flow follows very closely the behaviour of the mean tip loss coefficient. They consider the optimum tip

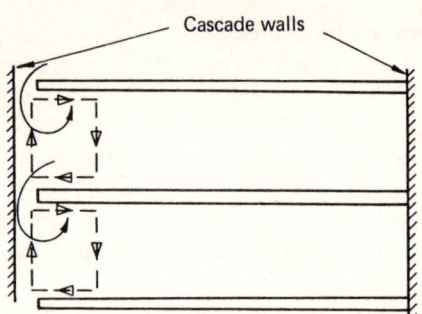

Fig. 19.5 – Leakage and secondary flow in a cascade with non uniform inlet flow
—— Leakage flow. ---------- Secondary message flow; (After ref. [19.37]).

clearance is about 5% of the blade chord, and note that removal of the duct boundary layer may actually increase the radiated noise for a given tip clearance because of the increase in the leakage loss effect. However, there is some experimental work by Moore [19.38] which has shown limited noise reductions using duct wall boundary layer removal, but the relative importance of other sources and the operational conditions of the fan were shown to be very important to the success of this technique.

19.5 FAN NOISE PREDICTION

In sections 19.2, 19.3, and 19.4 we have considered the sources of fan noise, the propagation effects which take place, and various mechanisms which are responsible for the generation of noise. In this section these features will be combined into a set of formulae which can be used to predict the noise levels which are generated by fans and rotors. First the prediction of discrete tone noise will be considered, and then some of the methods for broadband noise will be described.

There is an important distinction between discrete tone noise and broadband noise which is often masked when too wide a bandwidth is used in analysing the frequency spectrum of the radiated noise. In order to make the correct distinction between these two categories the frequency bandwidth must be less than the shaft rotation frequency, and this is usually not the case when 1/3 octave spectra are used. In the prediction schemes given below this distinction is implicitly assumed, and in order to convert to the equivalent wide bandwidth results, the tones and broadband noise must be integrated over the bandwidth which is to be used.

19.5.1 Thickness noise

Rotor thickness noise can be predicted using the method given by Hawkings & Lowson [19.39]. The blade is modelled by a thin aerofoil with a thickness dsitribution $h(\phi)$, and the directional properties of each harmonic are obtained using the approach given in section 19.3.1. The prediction formula for the blade harmonics in the acoustic far field spectrum takes the form

$$|P_{mB}(\bar{x})| = \frac{\rho(mB\Omega)^2}{2\sqrt{(2.\pi R)}} \,|\int_S h(\phi)J_{mB}(mBM_s(r)\sin\theta_o)\, e^{imB\phi}dS| \qquad (19.19)$$

where S is the planform area of one blade.

This source mechanism is absolutely deterministic and requires no empiricisms or estimation of unknown factors. Its accuracy has been demonstrated by an experimental study [19.40] which has shown its relative importance to other sources.

To reduce blade thickness noise the tips of the blade should be thinned since this is the most efficient noise-producing part of the blade. The noise reductions which can be obtained using tapered blades which have a 'thick' root and a 'thin' tip are given in [19.40].

19.5.2 Rotor force noise

Several methods have been developed to predict rotor force noise using the principles described in sections 19.3 and 19.4. In this section prediction formulae will be developed for the acoustic far field from open rotors and axial flow fans. It will be assumed that the far field from both these types is essentially the same, and that propagation from a ducted rotor is not significantly affected by the presence of the duct. This assumption will lead to a 3-6 dB correction which allows for the radiation from the face of the duct, and in principle should only be applied to ducts which are short compared to the acoustic wavelength.

The noise generated from fluctuating forces on the rotor blades is predicted using equation (19.7), and estimates are made of the fluctuating lift and drag components. However, it must be assumed that the blade loads can be lumped at a single spanwise station on the blade. This is usually taken to be 80% of the blade span, and so M_s in (19.7) will be equal to $0.8M_{tip}$.

Lowson and Ollerhead [19.5] present a prediction scheme using equation (19.7) in which they assume $D_o/L_o = 0.1$ and that BL_o is the total thrust generated by the rotor. The mean square value of each rotor harmonic in (19.7) is then evaluated assuming that the modulation components A_n are uncorrelated so that cross terms in the squared summation are eliminated. The fall-off of modulation components is assumed to follow $A_n^2 \sim |n|^{-2.5}$. Therefore, assuming that $J_{mB+n}(\quad) \ll J_{mB-n}(\quad)$ gives

$$|P_{mB}(\bar{x})| = \frac{mB(BL_o)\,\Omega}{2\sqrt{(2\pi Rc_o)}} \left[\sum_{n=0}^{\infty} \, (\cos\theta_o - \frac{0.1(mB+n)}{M_s.mB})^2 \frac{J_{mB-n}^2(mBM_s\sin\theta_o)}{n^{2.5}} \right]^{\frac{1}{2}}$$

This series converges very rapidly when $n > (1 + M_s \sin\theta_o)mB$.

The noise generated by fluctuating lift components which result from the interaction of blades with the wakes of upstream rotors or stators has been discussed by Morfey [19.3], and formulae are given for the in-duct acoustic intensity. These results demonstrate that the intensity falls off with rotor-stator separation x as

$$I_W \propto (1 + \exp(-0.11x/C_D b))^2 \ .$$

Therefore a rapid decay of wake interaction noise is expected as the separation is increased up to a blade chord length; thereafter increasing the separation will have little effect.

Similar results are also given [19.3] for the interaction of an upstream stage with the potential field of a downstream stage. Thus interaction decays as

$$I_P \propto \exp\left(\frac{-2mBx}{a}\right) = \exp(-4\pi m\sigma x/b) \ ,$$

where σ is the solidity of the upstream stage. Therefore for high solidity rotors, or separations which are the order of a chord length, this interaction is not important. More detail of the prediction of rotor stator interaction noise may be obtained from [19.3].

19.5.3 Broadband noise prediction

The prediction of broadband noise from fans is usually given in terms of the total radiated sound power, and the directional features given in section 19.3.5 are not included. This follows the analysis of Sharland [19.36] who considered the total sound power radiated by a line force distribution, with a spectrum $G_T(\omega)$, along a blade with a span a and spanwise correlation length scale l. The total radiated power spectrum from a B-bladed rotor is then

$$\frac{dW}{d\omega} = \frac{\omega^2 G_T(\omega) a \, lB}{12\pi \rho c_0^3} \tag{19.20}$$

where it is assumed that $l \ll a$.

This expression is valid providing the blade chord b and the spanwise correlation length scale l are both much shorter than the acoustic wavelength.

Estimates of the force fluctuations due to the blade boundary layer were given in section 19.4.2, and so the radiated sound power from both upper and lower surfaces of the blades is estimated as

$$\frac{dW}{d\omega} = \frac{1}{3} \frac{\rho c_0^3}{\omega} B.b.a.M_s^6 C_D \times 10^{-4} \ . \tag{19.21}$$

The broadband noise due to incident turbulence may be estimated using the approach given in section 19.4.1. Using equations (19.20), (19.16), and (19.17) we obtain

$$\frac{dW}{d\omega} = \frac{1}{12} \frac{\omega b}{U} \, a \, l \, B \, \rho c_o M_s^4 G_v(\omega) \tag{19.22}$$

where $G_v(\omega)$ is the power spectrum of the upwash velocity fluctuations. Formulations of this type have been used successfully to estimate broadband noise from rotors provided that the inlet turbulence can be correctly estimated [19.10], [19.12], [19.23], [19.24].

In addition Morfey [19.41] has considered the case where the turbulence is generated by the wake of an upstream blade row. The experimental results from a number of multistage high-subsonic axial compressors and transonic fans were analysed, and it was found that the noise spectra collapsed on the non-dimensional frequency $\omega l/c_o$ where l is obtained from the upstream blade drag coefficient as $l = bC_D$. A collapse of the total radiated sound power from an upstream rotor and downstream stator was obtained using:

$$W = \rho c_o^3 B . a . b . M_s^5 C_D^2 \sigma_R \sin^4\beta \quad (M_s < 0.6) \ .$$

where σ_R is the rotor solidity and β is the inflow angle, and all parameters are based on the upstream rotor.

In addition to these formulae there are also a number of empirical techniques for the estimation of broadband noise from industrial fans and compressors which are based on the fan design parameters. These will not be given here but may be found in references [19.42-19.45].

19.6 NOISE CONTROL

The range of ideas for reduction of fan noise at source, particularly within the industry, is legion. Fans have been run with porous surfaces to try to reduce boundary layer noise and with trailing edge suction or blowing to reduce trailing wakes and hence modify rotor wake/outlet guide vane interaction noise. Constant reaction fans have been experimentally operated with many combinations of blade and vane numbers, to reduce interaction tones. It is fair to say that, whilst many such tests have demonstrated that modest changes in noise character can be engineered, the overall noise benefit is usually outweighed by performance, weight, cost, or reliability penalties, which is why aircraft are rarely to be found with such features to their engines. Nevertheless, experimental fans have demonstrated improvements of over 10 dB for certain fan tones by such means as tilted or leaned vanes [19.46, 19.47]. Most of the progress which has been made in the aeronautical field has therefore come from deliberate changes in basic design point and degree of reaction. This trend continues with the demonstrator programmes which have recently been worked on in this country and the USA. Fan relative tip Mach number have been made as low as practicable. Avoidance of superficial flow obviates the buzz-saw noise at take-off thrust, although the advanced turbo-fans currently in-service do not entirely avoid this source. The lower relative speed reduces all the aerodynamic interaction tone and broadband noise mechanisms by at least 9 dB per halving and possibly much more. Choice of free vortex instead of constant reaction fan flow enables the design to dispense with inlet guide vanes with their additional attendant noise excitations for the rotor operating either within their potential flow fields or their wakes and secondary flows. Rotor-outlet guide vane tones have been minimized by choosing the maximum axial spacing subject to adequate strength, acceptable performance, etc., whilst the outlet guide vane (OGV) numbers have been selected to avoid fundamental tone propagation in the smooth inflow found away from the ground running situation. In the latter condition, some distortion noise at the fundamental passing tone frequency is almost impossible to avoid because of the limited contraction in the inflow which can be designed into a flight nacelle.

On the industrial and ventilation fan front, although economic factors inevitably rule out sophisticated measures for noise reduction at source (although reference [19.48] gives several ideas which are worthy of pursuit), at least installations which entail poor aerodynamic inflow can be avoided. All too often, fans are installed much too close to duct bends or corners in process plants (the exception is the centrifugal where the intake may be aligned with the supply duct axis whilst the exhaust duct, which is generally rectangular in section, can be arranged to deliver either upwards or to the side at a right angle to the inflow). It is beneficial to arrange for the flow to contract into the fan inlet, via a well designed and generously radiused flare if the fan is to take in atmospheric air. This minimizes turbulent inflow interactions and flow distortion harmonics. Fans with drive shafts or belt drives across the ducting close to the fan, sometimes partly enclosed in bulky fairings, should be avoided. Electric motor drives within the fan hub, or well designed axial drive shafts well away from inlet duct bends, are preferable.

It is common to separate fan ducts from general trunking for vibration isolation purposes. If flexible rubberized canvas gaiters are used, these bow inwards due to the lower static pressure inside the duct, and can result in a separated wall flow into the fan. Keep such features as far upstream as possible, or use a well-fitting sleeved or sliding joint and keep the sealing gaiter outside of this. It is also as well to remember that even downstream of the fan rotor, severe blockages of flow instabilities will cause extra interaction tone noise. Flow dampers should never be placed at the end of diffusers, nor should they be used close to inlets if they are of the bulky butterfly type and liable to be used in a part-closed setting. Vane dampers are clearly better because of their generally much smaller scale wake structures which will decay more readily before reaching the fan rotor.

It is also vital to choose the right fan for a given task. Temptations are always around to economize and buy the smallest possible machine, perhaps because space in a plant room is at a premium too. Then comes the day when more flow capacity is needed because of the need to increase the output of a process line. The fan speed is increased, or the working point changes so that the fan operates closer to the stall condition, with increased blade drag forces. The consequence is all too often a marked and embarrasing increase in noise. For minimal noise, always choose the largest and slowest rotating machine possible within all the other genuine constraints to the contrary, noting that this will inevitably *not* be the design or working point offering minimum operational costs. Some fan types have settable blade pitch (angle of attack). These can be adjusted within bounds to ensure that optimum noise for a given flow or pressure rise is achieved, but note that unequal pitch settings will give rise to extra multiple tone component noise.

A final note is that by no means all of the noise from some industrial and ventilating fan installations is aerodynamic in origin. Vibration radiation noise from casings, from non-isolated supporting floors, and noise from drive motors, gears, belts, and so forth should be separately considered.

ACKNOWLEDGEMENTS

The portion of this chapter concerning Noise Control was adapted from previous notes written by M. E. House.

REFERENCES

[19.1]   Lighthill, M. J. (1952) *Proc. Royal Soc. (London)*, **A211**, 564–587. On sound generated aerodynamically. I. General Theory.

[19.2]   Ffowcs-Williams, J. E. & Hawkings, D. L. (1969). *Phil Trans. Royal Soc. (London)*, **A264**, 321–342.

[19.3]   Morfey, C. L. (1972) *J. Sound Vib.*, **22**, 445–466. The acoustics of axial flow machines.

[19.4]   Gutin, L. (1936) *Zhurnal teknicheskoi fizik*, **6**, 899–904. (In Russian) Translated as 1948 NACA TM 1195: On the sound of a rotating airscrew.

[19.5]   Lowson, M. V. & Ollerhead, J. B. (1969) *J. Sound Vib.*, **9**, 197–222. A theoretical study of helicopter rotor noise.

[19.6]   Barry, B. & Moore, C. J. (1971) *J. Sound Vib.*, **17**, 207–220. Subsonic fan noise.

[19.7]   Vaidya, P. G. & Dean, P. D. (1977) *AIAA 77-1279*. The state of the art of duct acoustics.

[19.8]   Ffowcs-Williams, J. E. & Hawkings, D. L. (1969) *J Sound Vib.*, **10**, 10–21. Theory relating to the noise of rotating machinery.

[19.9] Morfey, C. L. & Tanna, H. K. (1971) *J. Sound Vib.,* 15, 325-351. Sound radiation from a point force in circular motion.

[19.10] Sevik, M. (1970) *NASA SO-304,* Part II. Sound radiation from a subsonic rotor subjected to turbulence.

[19.11] Mani, R. (1971) *J. Sound Vib.,* 17(2), 251-260. Noise due to interaction of inlet turbulence with isolated stators and rotors.

[19.12] Amiet, R. K. (1975) *J. Sound Vib.,* 14(4), 407-420. Acoustic radiation from an aerofoil in a turbulent stream.

[19.13] Horlock, J. H. (1968) *Basic Engineering: Trans. Amer. Soc. Mech. Eng. (Series D),* 90, 494-500. Fluctuating lift forces on aerofoils moving through transverse and chordwise gusts.

[19.14] Morfey, C. L. (1970) *Basic Engineering: Trans. Amer. Soc. Mech. Eng (Series D),* 92.

[19.15] Sears, W. R. (1941) *J. Aeronaut. Sciences,* 8, 104-8. Some aspects of non-stationary airfoil theory and its practical application.

[19.16] Liepmann, H. W. (1952) *J. Aeronaut. Sciences,* 19, 793-800, and 822. On the application of statistical concepts to the buffeting problem.

[19.17] Filotas, L. T. (1969). University of Toronto *UTIAS REPORT No. 139.* Theory of airfoil response in a gusty atmosphere. Part I – Aerodynamic transfer function.

[19.18] Mugridge, B. D. (1971) *Aeronautical Quarterly,* 22, 301-310. Gust loading on a thin aerofoil.

[19.19] Graham, J. M. R. (1971). *Aeronautical Quarterly,* 21, 182-198. Lifting surface theory for the problem of an arbitrarily yawed sinusoidal gust incident on a thin aerofoil in incompressible flow.

[19.20] Morfey, C. L. (1973) *J. Sound Vib.,* 28(3), 587. Rotating blades and aerodynamic sound.

[19.21] Filotas, L. T. (1971) *Journal of Aircraft,* 8, 395-400. Approximate transfer functions for large aspect ratio wings in turbulent flow.

[19.22] Hinze, J. O. (1959) *Turbulence,* New York: McGraw-Hill Book Company Ltd.

[19.23] Paterson, R. W. & Amiet, R. K. (1979). *NASA CR-3213.* Noise of a model helicopter rotor due to ingestion of turbulence.

[19.24] Aravamundan, K. S. & Harris, W. L. (1979) *J. Acoust. Soc. Amer.,* 66(2), 522. Low frequency broadband noise generated by a model rotor.

[19.25] Tyler, E. (1928) *Phil. Magazine,* 5, 449-463. Vortices behind aerofoil sections and rotating cylinders.

[19.26] Gongwer, C. A. (1952) *J. Applied Mech.,* 19(4), 432-438. A study of vanes singing in water.

[19.27] Paterson, R. W., Vogt, G. V., Fink, M. R. & Munch, C. L. (1973) *Journal of Aircraft,* 10(5), 296-302. Vortex noise of isolated airfoils.

[19.28]   Tam, C. K. W. (1974). *J. Acous. Soc. Amer.*, **55(6)**, 1173-1177.
Discrete tones of isolated airfoils.

[19.29]   Schlichting, H. (1968). New York: McGraw-Hill Book Company
Limited. Boundary Layer Theory. Sixth Edition.

[19.30]   Fink, M. R. (1975) *Journal of Aircraft*, **12(2)**, 118-120. Prediction
of airfoil tone frequencies.

[19.31]   Longhouse, R. E. (1977) *J. Sound Vib.*, **53(1)**, 25-46. Vortex shedding noise of low tip speed axial flow fans.

[19.32]   Grosche, F. R. & Stiewitt, H. (1977) *AIAA Journal*, **16(12)**, 1255.
Investigation of rotor noise source mechanisms with forward speed.

[19.33]   Clark, L. T. (1971) *J. Engin. Power, Series A93*, 366-376. The radiation of sound from an airfoil immersed in a laminar flow.

[19.34]   Aravamundan, K. S., Lee, A. & Harris, W. L. (1978) *J. Sound Vib.*,
57(4), 555-570. A simplified Mach number scaling law for helicopter
rotor noise.

[19.35]   Sharland, I. J. (1964) *J. Sound Vib.*, **1(3)**, 302-322. Sources of noise
in axial flow fans.

[19.36]   Mugridge, B. D. (1971) *J. Sound Vib.*, **16(4)**, 593-614. Acoustic
radiation from aerofoils with turbulent boundary layers.

[19.37]   Mudgridge, B. D. & Morfey, C. L. (1972). *J. Acous. Soc. Amer.*,
51(5), Part I. 1411. Sources of noise axial flow fans.

[19.38]   Moore, C. J. (1972) *J. Acoust. Soc. Amer.*, **55(1)**, 1471. In duct
investigation of subsonic fan rotor alone noise.

[19.39]   Hawkings, D. L. & Lowson, M. V. (1974) *J. Sound Vib.*, **36(11)**,
1-20. Theory of open supersonic rotor noise.

[19.40]   Glegg, S. & Wills, C. R. (1979) *Proc. Acous.*, **20.GS**. High speed rotor
thickness noise.

[19.41]   Morfey, C. L. & Wills, C. R. (1979) *Proc. Inst. Acoust.* **20.GS**. High
speed rotor thickness noise.

[19.42]   Grott, G. C., Schreiner, J. R. & Bullock, C. E. Centrifugal sound
power level prediction. *ASHRE Journal*, October 1967.

[19.43]   Graham, J. B. (1972). *Sound and Vibration, May 1972.* How to
estimate fan noise.

[19.44]   Beranek, L. L. (1960). McGraw-Hill. Noise reduction.

[19.45]   Erskine, J. B. & Brunt, J. (1975). *I.Mech.E. Conference on Vibration
and Noise in pump, fan and compressor installations*, Southampton.
Prediction and control of noise in fan installations.

[19.46]   Benzakein, M. J. (1972). *J. Acoust. Soc. Amer.*, **51(5)**, 1427. Research
on fan noise generation.

[19.47]   Nemec, J. (1967) *J. Sound Vib.*, **6(2)**, 230. Noise of axial fans and
compressors: study of its radiation and reduction.

[19.48]   Yeow, K. W. (1966) *I.S.V.R. Memorandum No. 143.* Centrifugal fan
noise research.

[19.49]   Hanson, D. B. and Fink, M. R. (1979) *J. Sound Vib.*, **62(1)**, 19-38.
          The importance of quadrupole sources in prediction of high speed
          propeller noise.
[19.50]   Morfey, C. L. & Fisher, M. J. (1970) *J. Roy. Aeronaut. Soc.*, **74**,
          579-585. Shock wave radiation from a supersonic ducted rotor.
[19.51]   Mather, J. S. B., Savidge, J. & Fisher, M. J. (1971) *J. Sound Vib.*,
          16, 407. New observations on tone generation in fans.

CHAPTER 20

Linear elastic fracture mechanics

A. Jefferson

Department of Aeronautics and Astronautics, University of Southampton

20.1 INTRODUCTION

20.1.1 Material selection

In general, material selection for the manufacture of a particular component is based upon consideration of certain material properties. These may be mechanical, electrical and/or chemical. Normally the material chosen will be that which gives optimum performance, in the particular service environment, at minimum weight and/or cost. For structural items the mechanical properties likely to be considered include elastic modulus, static strength, fatigue strength, and time dependent properties (e.g. creep, stress corrosion). In the past it has been the intention, by the determination of the properties, to demonstrate that the combination of loads and environments encountered during service are such that the probability of a failure during the service life is acceptably small. This approach, in general, assumes that the material is free from significant flaws or is operating at stress levels sufficiently low to preclude the propagation of flaws.

20.1.2 Damage tolerant design

Increasingly in structural design there is a move towards designing to meet the requirements for Damage Tolerance. Regulating bodies now require to be satisfied that the effects on structural integrity of possible accidental damage, or the presence of crack-like defects, have been taken into consideration. The designer therefore has to admit that damage in the form of crack-like defects may be present in various structural items. This may be the result of accidental damage or of fatigue crack initiation and propagation. The presence of such defects usually manifests itself by a drastic reduction in the static strength, which, if not accounted for, can result in catastrophic failure.

20.1.3 Linear elastic fracture mechanics (LEFM)

In order to demonstrate compliance with the Damage Tolerance Requirements, additional material properties are required by the designer to account for the behaviour of defects. The properties of interest include the rate of propagation of defects and the residual strength in the presence of such defects. It is in this context that LEFM is employed. LEFM can be used to determine the behaviour of structures containing flaws or cracks under a variety of loading conditions. Residual static strength and rate of crack growth, under cyclic loading and stress corrosion cracking environments, can be determined. Reference [20.1] contains a valuable compilation of fracture and crack growth data for high strength alloys.

In this chapter relevant background information on the principles of fracture mechanics as an engineering discipline is given. Its application to the determination of residual strength and fatigue crack propagation is also described. In the damage tolerance approach to design, inspection intervals and methods play an important part in ensuring the continued safe operation of a structural part in service. However, this aspect has not been covered here.

20.2 ELASTIC THEORY

20.2.1 Stress intensity factor (SIF)

The classical theory of elasticity yields a solution, for the stress some small distance r ahead of the crack tip, of the form

$$\sigma = \frac{K}{\sqrt{2\pi r}} + \text{other terms} . \tag{20.1}$$

As $r \to 0$ the stress field is dominated by the $1/\sqrt{r}$ singularity. The coefficient K of this singular term is called the STRESS INTENSITY FACTOR. Stress fields surrounding the tips of simple cracks have the same distribution, but differ in their intensity according to the variation of K which is a function of the geometry and loading. It is assumed that the stress field local to the crack tip controls crack behaviour and since K has a significant influence on these stresses, the characterization of these stresses by K yields a most useful parameter. That is, for those materials for which LEFM is applicable, the assumption is made that the behaviour will be the same every time the stress intensity factor K has the same value. Thus fracture occurs when K achieves a critical value K_c, or a certain increment of crack growth will occur for a given range of SIF, ΔK. [20.2, 20.3].

20.2.2 Effects of geometry

In general the stress intensity factor is expressed in the form:

$$K = \alpha \sigma \sqrt{(\pi a)} \tag{20.2}$$

where α is a factor relating geometry and loading and σ is a reference stress typically the stress normal to the crack of length $2a$. For the case of a through-the-thickness crack of total length $2a$ in an infinite sheet subjected to a uniaxial stress σ normal to the crack, α takes the value of unity and $K = \sigma\sqrt{(\pi a)}$. Other crack geometries and loadings are therefore related to this classic case through the geometric factor α. It must therefore account for such things as the loading conditions and the proximity of boundaries, stiffeners, cut-outs, orientation, and crack shape. Many solutions have been obtained for a range of practical and mathematically tractable configurations and are given in the literature [20.4, 20.5, 20.6].

20.2.3 Mode of cracking

In practice, cracks usually propagate perpendicular to the maximum principal stress, fracture of the material arising as a result of direct opening of the crack (Mode I). Other modes of crack tip deformation are possible (Modes II and III) (see Fig. 20.1). However, because of its importance the vast majority of work is done in terms of Mode I opening.

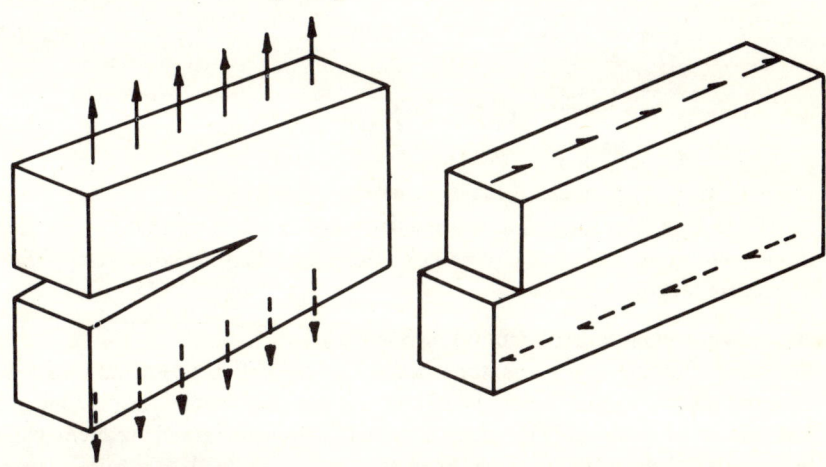

Mode I — Opening Mode   Mode II — Sliding Mode

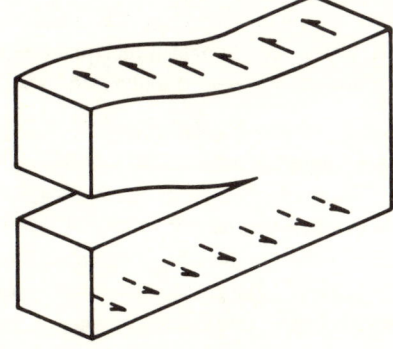

Mode III — Tearing Mode

Fig. 20.1 — Basic modes of crack surface displacements.

20.3 EFFECTS OF PLASTICITY

20.3.1 Size of the plastic zone

The application of LEFM to most practical materials leads to the violation of certain assumptions in the elasticity theory. At the tip of the crack LEFM predicts an infinite stress, which in practical materials will not be realized but will give way to plastic flow and finite stress level. However, it is argued that the relation still holds even in the presence of plasticity provided that the plastic zone size is small compared to the size of the near crack tip elastic field. Further, the use of K has proved to be a successful engineering parameter for correlating toughness and crack growth properties of a wide range of materials. An approximate measure of the plastic zone is a notional radius about the crack tip of:

$$r_p = \frac{1}{2\pi} \left[\frac{K}{\sigma_y} \right]^2 \quad \text{PLANE STRESS} \tag{20.3}$$

$$r_p = \frac{1}{6\pi} \left[\frac{K}{\sigma_y} \right]^2 \quad \text{PLANE STRAIN} \tag{20.4}$$

where σ_y is the yield stress of the material.

These estimates show that the smaller plastic zone is associated with plane strain conditions, and hence these conditions are likely to be more applicable in fracture toughness testing.

20.3.2 Valid plane strain fracture toughness tests

For valid fracture toughness tests according to LEFM a generally accepted criterion has been laid down [20.3] giving the permissible plastic zone size. The specimen thickness (t), crack length (a), and uncracked ligament length $(w-a)$ for plane strain fracture toughness tests should all be greater than 2.5 $\left[\frac{K_{IC}}{\sigma_y} \right]^2$. A typical test specimen for determining fracture toughness is shown in Fig. 20.2; further information on plane strain fracture toughness testing can be found in reference [20.7]. The value of K_c, determined under plane strain conditions, represents a lower limit on toughness for a material denoted by K_{IC}.

20.3.3 Stress corrosion cracking (SCC)

Lower values of toughness may be found in materials susceptible to stress corrosion cracking (SCC) denoted by K_{ISCC}. For such materials, e.g. high-strength steel with yield strengths over 1300 MN/m^2, the best test specimen to use is a self-loaded crack arrest type since this will not then involve tying up an expensive loading machine for a long period (see Fig. 20.3). The specimen is loaded along the crack line, placed in the corrosive environment, and measurements of crack extension are noted with time. As the crack extends over a period the stress intensity factor is reducing. Once a stable crack length has been achieved the corresponding minimum stress intensity K_{ISCC} can be calculated.

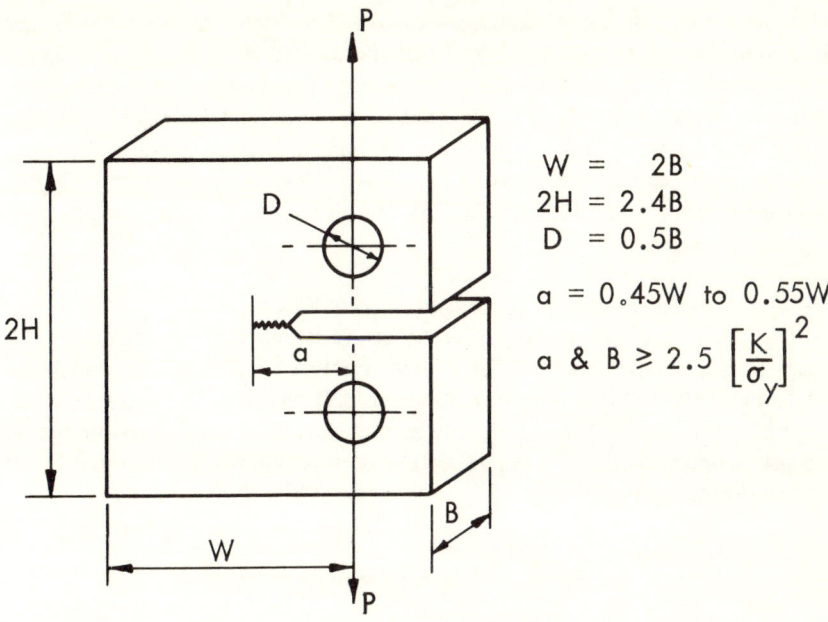

$$W = 2B$$
$$2H = 2.4B$$
$$D = 0.5B$$
$$a = 0.45W \text{ to } 0.55W$$
$$a \, \& \, B \geq 2.5 \left[\frac{K}{\sigma_y}\right]^2$$

Fig. 20.2 – Compact tension specimen for plane strain fracture toughness testing.

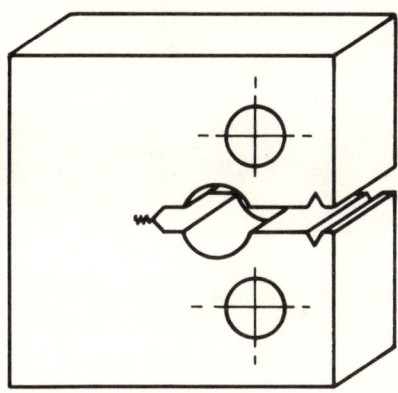

Fig.20.3 – Wedge open loaded specimen for determination of K_{ISCC}.

20.4 FATIGUE CRACK PROPAGATION

20.4.1 Stress intensity factor range ΔK

Taking the SIF K to be the most appropriate parameter available for describing the stress situation at the crack tip, it is reasonable to expect that under cyclic loading conditions the magnitude of the stress intensity factor should correlate with the amount of fatigue damage occurring. The results of numerous fatigue tests over the years have also shown that the applied stress range has a significant influence on the amount of fatigue damage which arises. In the same way, tests have shown that the SIF range ΔK is a useful parameter for correlating fatigue crack growth data. It can also be argued that under fatigue loading conditions the applied stress levels are usually of such a low magnitude that essentially elastic conditions prevail and the difficulties associated with plastic zone size are avoided.

20.4.2 Relationship between ΔK and crack growth rate
Fatigue crack propagation data in terms of crack length (a) versus number of cycles (N) have been analysed for a wide variety of materials and reduced to the form where the rate of propagation (da/dN) has been correlated with the range of stress intensity factor (ΔK). In this way the general shape of the relationship between (da/dN) and (ΔK) has become known and is typically of the form shown in Fig. 20.4.

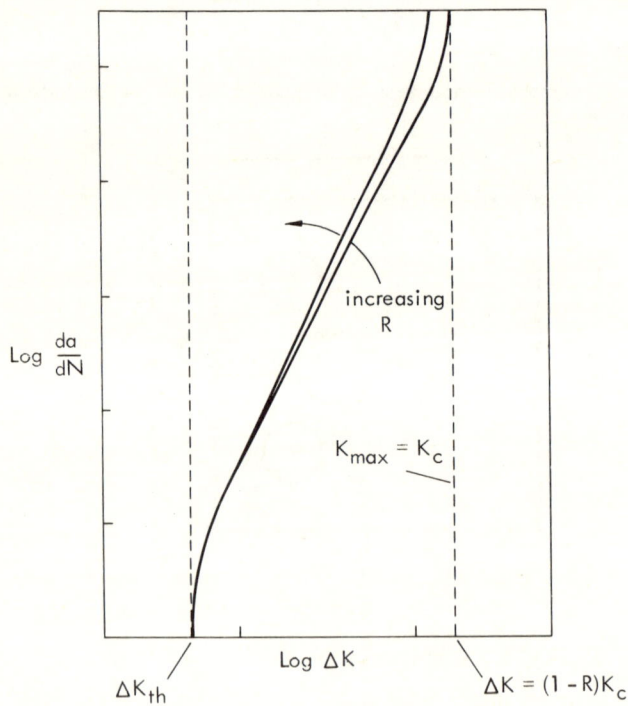

Fig. 20.4 – Fatigue crack growth data in terms of $da/dN \sim \Delta K$.

The crack growth rate is observed to approach two asymptotes:

(a) tending to zero as the threshold ΔK_{th} is approached.
(b) tending to very high growth rates as the maximum stress intensity factor in the cycle approaches the fracture toughness value K_c of the material.

Paris [20.8] was among the first to propose a relationship between crack growth rate (da/dN) and the stress intensity factor range ΔK ($= K_{MAX} - K_{MIN}$) of the cycle. He proposed an equation of the form:

$$da/dN = C \, \Delta K^m \tag{20.5}$$

where m and C are material constants. Experimental data showed that, for a large range of materials, m is of the order 4, but may vary between 2 and 8.

Clearly the equation proposed by Paris does not satisfy either of the two asymptote conditions, and where this form of equation is used its range of validity should be stated together with the stress ratio ($R = \sigma_{MIN}/\sigma_{MAX}$).

An empirical relation put forward by Forman [20.9] which takes account of stress ratio and meets the asymptote condition (b) is:

$$\frac{da}{dN} = \frac{C\Delta K^m}{(1-R)\,K_c - \Delta K} \; . \tag{20.6}$$

This equation has shown good correlation with data obtained from high-toughness alloys except in the region of high values of K. Neither is the correlation as good for low toughness alloys.

Pearson [20.10] modified the above equation and obtained good correlation on high- and low-toughness materials for crack growth under plane strain conditions with an equation of the form:

$$\frac{da}{dN} = \frac{C\Delta K^m}{|(1-R)\,K_c - \Delta K|^{\frac{1}{2}}} \; . \tag{20.7}$$

This equation, like the Forman equation, meets one of the asymptote conditions and also accounts for stress ratio effects.

Of the equations suggested in the literature which meet both asymptote requirements the one proposed by Nicholson [20.11] is perhaps one of the simplest:

$$\frac{da}{dN} = C \left[\frac{\Delta K - K_{th}}{K_c - K_{max}} \right]^m \tag{20.8}$$

Comparison of this equation with experimental data would appear to show a dependence of the threshold ΔK_{th} on the stress ratio R, particularly for tests on steels.

All the above empirical equations were developed from test data obtained under constant amplitude loading and were derived in general from a large amount of data condensed to the form $\Delta K - \mathrm{d}a/\mathrm{d}N$. The test evidence confirms that ΔK, the range of stress intensity factor, is a useful parameter for correlating fatigue crack growth data.

20.4.3 Crack growth prediction under constant amplitude loading
For components subjected to constant amplitude loading at a given mean stress and with appropriate materials data $(\mathrm{d}a/\mathrm{d}N - \Delta K)$ it becomes possible to predict crack growth from calculated values of ΔK corresponding to the loading and assumed defect sizes. In practice the calculation is performed numerically, the number of cycles N required to propagate a crack from an initial size a_0 to some final length a_f being:

$$N = \int_{a_0}^{a_f} \frac{\mathrm{d}a}{\mathrm{d}a/\mathrm{d}N} = \int_{a_0}^{a_f} \frac{\mathrm{d}a}{f(\Delta K)} \, . \tag{20.9}$$

where $f(\Delta K)$ is the functional relationship between $\mathrm{d}a/\mathrm{d}N$ and ΔK.

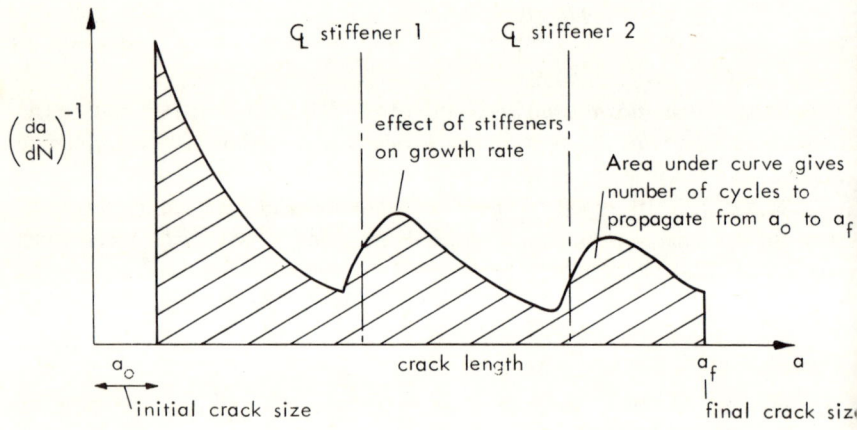

Fig. 20.5 – Computation of crack propagation life-N for a stiffened panel.

20.4.4 Crack closure
Crack closure is the term used to describe the coming together of the upper and lower fracture surfaces of a fatigue crack before complete unloading of the specimen. It was first observed by Elber [20.12] and was attributed to

the plastic field left in the wake of the propagating crack. Crack closure stress has been measured for different applied mean stresses on an aluminium alloy 2024-T3, and it was found that the proportion of the stress range (ϕ) during which the crack was open was given by

$$\phi = 0.5 + 0.4R \ . \tag{20.10}$$

Therefore if it is assumed that the applied stress which causes fatigue damage is that portion of the stress cycle during which the crack is open, then the effective range of stress intensity would be $\phi(\Delta K)$. Making this modification to the Paris equation gives:

$$da/dN = C \,|\,(0.5 + 0.4R)\,\Delta K\,|^m \tag{20.11}$$

which was found to correlate with data on 2024-T3 for various mean stresses.

Although it is not easy to measure accurately the stress level at which crack closure occurs, there is abundant evidence to indicate that it occurs in several materials under tensile load. It is often manifested in crack propagation tests where the initial crack growth rate out of the initial saw-cut is frequently more rapid than subsequent growth following the establishment of the plastic zone region.

20.4.5 Sequence effects
Whilst constant amplitude cyclic loading is an important loading action, which arises in many general engineering situations, it is however a special category of fatigue loading, which is more generally of a purely random nature. Practically all loading histories encountered in service are complex sequences of cycles with varying amplitude and mean values. Tests under such variable amplitude fatigue loading have been carried out by various investigators, and a number of systematic trends have been observed.

20.4.5.1 *Application of a single overload*
The initial application of a high tensile overload prior to fatigue cycling under constant amplitude loading has been found to give a retardation in crack growth rate. An initial compressive overload, however, can lead to an accelerated crack growth rate. These effects have been attributed to the setting up of residual stress in the crack tip region. The concept of the crack closure model also gives a useful illustration of how crack growth acceleration and retardation can arise under the conditions of overload. The overload influencing the level of stress at which crack closure is likely to occur.

20.4.5.2 *Applications of a periodic overload*
Periodic application of a high tensile overload has been found to be extremely effective in increasing the life of a fatigue specimen by retardation of the rate of crack growth. Fig. 20.6 illustrates how the occasional overload leads to crack

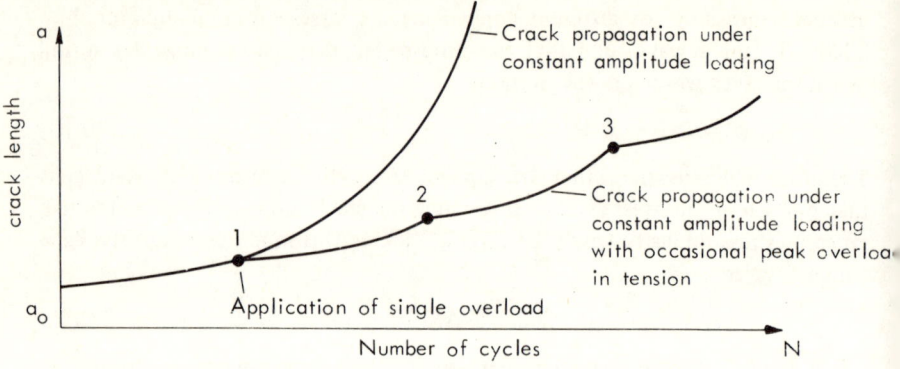

Fig. 20.6 – Sketch showing the influence of occasional peak overloads (tension) on the rate of crack growth during constant amplitude fatigue cycling.

growth retardation and an increase in the overall life. Acceleration of crack growth has also been observed in cyclic tests when occasional peak overloads in compression have been applied. It therefore becomes important when representing service loads on a test programme to ensure that the test loading is as representative as possible.

20.4.6 Crack growth retardation models
A number of models have been proposed to treat retardation in a quantitative fashion. The models are again empirical with little physical basis, and generally contain one or more constants to be derived from variable amplitude crack growth experiments. The two best known models are those of Wheeler [20.13] and Willenborg [20.14].

20.4.6.1 *The Wheeler Model*
In this model the retarded crack growth rate is given by:

$$\frac{\mathrm{d}a}{\mathrm{d}N}_{\text{RETARDED}} = C_p \, f(\Delta K) \tag{20.12}$$

where C_p is the retardation parameter and $f(\Delta K)$ is the appropriate crack growth function. The retardation parameter C_p is bounded by zero and unity and is given by:

$$C_p = \left[\frac{r_{pi}}{a_p - a_i} \right]^n \tag{20.13}$$

where r_{pi} is the extent of the current plastic zone at the ith cycle of the lower stress level and ($a_p - a_i$) is the distance from the crack tip to the elastic/plastic interface caused by the peak overload (see Fig 20.7). The exponent n is a shaping factor.

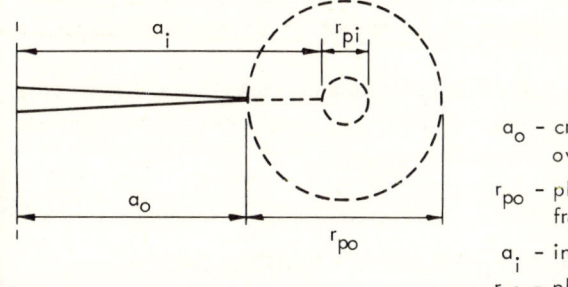

a_o – crack length at time overload applied

r_{po} – plastic zone size resulting from application of overload

a_i – instantaneous crack length

r_{pi} – plastic zone size associated with a_i and reduced level of loading

Fig. 20.7 – Definition of terms used in crack growth retardation models.

There is retardation of the crack growth rate as long as the current plastic zone size due to the more frequent lower levels of stress is contained within a previously generated plastic zone. The assumption is that the retardation process occurs because the crack must thrust its way through the initial large plastic zone produced by the peak overload. The retardation is proportional to the amount by which the retarded crack has extended through the initial plastic zone. When the boundary of the plastic zone r_{pi} coincides with the boundary of the peak plastic zone r_o, the retardation disappears and the non-retarded crack propagation given by the usual crack growth function takes over.

20.4.6.2 *The Willenborg Model*
This model makes use of an effective stress intensity factor such that

$$K_{\text{MAX EFFECTIVE}} = K_{\text{MAX } i} - K_{\text{MAX } o} \sqrt{\left[\frac{a_p - a_i}{r_{po}}\right]} - K_{\text{MAX } i} \tag{20.14}$$

$$K_{\text{MIN EFFECTIVE}} = K_{\text{MIN } i} - K_{\text{MAX } o} \sqrt{\left[\frac{a_p - a_i}{r_{po}}\right]} - K_{\text{MAX } i} \tag{20.15}$$

The maximum and minimum stress intensity factors are reduced by the same amount, and consequently there is no change in ΔK following an overload. The only change to occur is a reduction in the cycle ratio R with a consequential effect on growth rate, This is true as long as $K_{MIN \ EFFECTIVE}$ is greater than zero (i.e. non-negative). Should $K_{MIN \ EFFECTIVE}$ become negative it is set at zero, so that the $R = 0$ and $\Delta K = K_{MAX \ EFFECTIVE}$.

Neither of the two retardation models will account for the effect of occasional negative overloads, producing an acceleration on subsequent cycling at the more frequent stress level.

The models imply a cycle by cycle integration of crack growth. The Wheeler model has been found to give reasonably accurate predictions of crack propagation under a type of variable amplitude loading.

20.5 ANALYSIS OF VARIABLE AMPLITUDE FATIGUE LOADING

In order to achieve reasonable predictions of crack growth rates for components subjected to fatigue loading the anticipated stress spectra need to be determined. This may be determined by measurement or derived from anticipated loading during service. This random stress-time history has to be reduced, using one of a number of counting methods, to a series of stress cycles of various amplitudes and frequencies of occurrence. Using simple one-parameter counting methods, where a count is made of the number of occurrences of a particular event, e.g. peak, a range, or a crossing of a given stress level, all information about the sequence of individual stress variations is lost. Consequently any subsequent analysis in terms of crack propagation rates neglects any effects sequence may have on the crack growth. Two-parameter counting methods can be used to preserve some information regarding the mean stress as well as the alternating stress level.

Crack propagation behaviour may be calculated on a cycle by cycle basis, ignoring interaction effects between individual cycles, or applying one of the crack growth retardation models where appropriate. However, this would be very laborious, and a number of simplifications are possible. One approach is to assess the most damaging features of a typical duty cycle by comparing crack growth rates for each level of stress cycle at a given crack size. It is often possible to identify a particular feature of the stress spectra which contributes most damage, either as a result of the magnitude or the frequency of occurrence. The whole of the stress spectra may then be equated to an equivalent number of cycles of the most damaging type.

20.6 RANDOM LOADING

For a purely random loading the crack will grow under the influence of a randomly varying stress field $\sigma(t)$. The stress intensity factor at the crack tip will vary in the same random way. If the stress variation is described in terms of its rms value σ_{rms}, then the variation of the stress intensity factor can also be defined in rms terms, i.e.

$$K_{rms} = \alpha\,\sigma_{rms}\,\sqrt{(\pi a)}\ .\qquad\qquad(20.16)$$

Attempts have been made to relate crack growth rates da/dN_r to K_{rms} where N_r is equivalent endurance in cycles, taken as half the number of zero crossings. In the same way that constant amplitude loading is seen to be influenced by the stress ratio it might be reasonable to expect an influence of stress ratio in the relation between da/dN_r and K_{rms}. However, this cannot be defined in the same way for random loading, though it may be possible to use a ratio σ_m/σ_{rms} as a measure of the relative severity. There is only limited evidence to confirm the applicability of this approach.

REFERENCES

[20.1] *Damage Tolerant Design Handbook* (1975). Metals and Ceramics Information Centre, Battelle, Columbus Laboratories. MCIC-HB-01. A Compilation of Fracture and Crack Growth Data for High Strength Alloys.

[20.2] Paris, P. C. & Sih, G. C. (1965). American Society for Testing and Materials (ASTM) Special Technical Publication (STP) 381, 30–83. *Stress analysis of cracks – in fracture tougness testing and its applications.*

[20.3] Brown, W. F. & Srawley, J. G. (1966). STP 410, ASTM. *Plane strain crack toughness testing of high strength metallic materials.*

[20.4] Rooke, D. P. & Cartwright, D. J. (1976) *Compendium of stress intensity factors.* HMSO, London.

[20.5] Tada, H., Paris, P. & Irwin, G. (1973) *The stress analysis of cracks handbook.* Del Research Corporation, Hellertown, Pennsylvania.

[20.6] Sih, G. C. (1973) *Handbook of stress intensity factors.* Le High University, Bethlehem, Pa.

[20.7] BS 5447 (1977) *Methods of test for plane strain fracture toughness* (K_{IC}) of Metallic Materials. BSI.

[20.8] *Proceedings of 10th Sagamore Conference* (1964). Fatigue – An interdisciplinary approach. Syracuse University Press.

[20.9] Forman, R. G., Kearney, V. E. & Engle, R. M. (1967) *J. Basic Engineering*, **89**, 459. Numerical analysis of crack propagation in cyclic-loaded structures.

[20.10] Pearson, S. (1972) *Engineering fracture mechanics*, **4**, 9–24. The effect of mean stress on fatigue crack propagation in half inch thick specimens of aluminium alloys of high and low fracture toughness.

[20.11] Nicholson, C. E. (1973). Influence of Mean Stress and Environment on Crack Growth. *Proc. BSC Conference on mechanics and mechanics of crack growth.* Churchill College, Cambridge (April 1973).

[20.12] Elber, W. (1970) *Engineering Fracture Mechanics*, **2**, 37. Fatigue Crack Closure Under Cyclic Tension.

[20.13] Wheeler, O. E. (1972) *J. Basic Eng. Trans. ASME*, **181**. Spectrum loading and Crack Growth.

[20.14] Willenborg, J. D., Engle, R. M. & Wood, H. A. (1971). AFFDL-TM-FBR-71-1, Air Force Flight Dynamics Laboratory. A Crack Growth Retardation Model using an Effective Stress Concept.

CHAPTER 21

Sound-absorbent duct design

R. J. Kershaw and **M. E. House**

Institute of Sound and Vibration Research, University of Southampton

21.1 INTRODUCTION

The control of noise emitted by a machine or process can readily be achieved in many cases by means of enclosure or wrapping. However, such methods are usually inappropriate where a fluid entering or leaving the system transports a significant proportion of the total sound energy emitted. It becomes necessary to reduce the level of sound energy propagating in the flow ducts. In this chapter we shall be concerned with the general principles of the dissipation of sound energy by means of sound-absorbent duct linings, and those factors which must be considered in their design.

In Chapter 12 the basic theory of sound propagation in ducts was presented, and it was made clear that the problem of the determination of sound attenuation in ducts involves consideration of the complete system of source, duct, and termination.

Harmonic wave propagation in a duct with hard (i.e. perfectly reflecting) walls is possible only in certain spatial modes determined by the geometry of the duct. Each mode is associated with a unique frequency, the so-called *cut-off frequency*, below which it does not propagate but attenuates exponentially with distance. Above the cut-off frequency a mode may propagate without attenuation. The cut-off frequency increases with increasing mode order, the zero-order or plane-wave mode having a zero cut-off frequency. Thus at any given frequency there is only a finite number of modes available for the transmission of sound energy without loss, but there is always at least one mode, the zero-order mode, available.

The sound source determines the initial modal distribution of sound energy, while the termination and any discontinuities in the duct will reflect sound energy and cause some development of standing waves in the duct and may cause the output of the source to be modified.

If sound energy transmission is to be reduced then some modification to a hard-walled duct system is necessary. There are two basic means of achieving a reduction; by (a) reflecting sound waves to cause destructive interference, and (b) dissipating sound energy in the duct. Practical systems will usually have elements of both types of reduction but are characteristically separable into two categories. *Reactive* systems are those operating principally by reflection of sound energy. They are typically resonant devices having little or no absorption, and are usually capable of tuning only to relatively narrow frequency bands. Wide application is found, therefore, in reciprocating machine exhausts having a predominantly tonal nature. However, they may be relatively efficiently designed for low-frequency operation and cannot be ignored in other systems. *Resistive* (or *absorptive*) systems operate principally by sound energy dissipation, achieved by the conversion of sound energy into heat by viscous friction in porous layers at the duct walls. They may have a wide frequency range of operation, and they find a broad range of industrial applications. We will be confining our remarks to this type of system in this chapter.

21.2 SOUND PROPAGATION IN LINED DUCTS

21.2.1 General considerations

Whether the absorptive duct system is created as a purpose-built unit or as a modification of an existing element, the same principles apply to its design, and no distinction will be made at this stage.

Replacing a hard-walled duct by one having sound-absorbent lining of the walls affects all aspects of sound wave propagation. The pattern of duct modes and the phenomenon of cut-off are changed. Perfect plane waves are no longer possible, and all modes are subject to attenuation with distance at all frequencies, which means that sound energy cannot be transmitted without loss. The sharp cut-off effect becomes blurred, and it is more appropriate to identify a cut-off frequency region where there is typically a rapid transition in modal behaviour from substantially attenuating to substantially propagating character as frequency is increased. Fortunately, the position of the cut-off region can usually be pre-dicted well by the cut-off frequency of a hard-walled duct with the same airway dimensions.

A second path for energy transmission becomes available through the lining material, which may lead to an effective short-circuit of the attenuating path through the main airway, and steps may have to be taken to prevent this occurring.

The attenuation behaviour of a lined duct is intimately dependent upon the modal energy distribution, which emphasizes the need to consider both source and termination characteristics when assessing the required performance. In addition, the possible effects on the source of introducing the attenuating duct element must be considered at the design stage. Thus, the performance of a sound-absorbing duct will be different if, for example, it is transferred from the inlet of an axial fan to the inlet of an identically-sized centrifugal fan, because each will have in general different distributions of sound energy between the modes propagating along the duct.

However, in many instances little, if any, precise information will be available about the noise source and, in order to proceed with a design, an educated guess must be made at an appropriate modal energy distribution, or some standard initial condition (such as a plane incoming wave) must be imposed at the inlet to the absorbing duct. Furthermore, it is quite usual to ignore, at least in the first instance, the effects of standing waves set up at the duct exit and entrance. Such an approach is justified in part through the tacit assumption that the proportion of energy reflected is not too large and will be attenuated fairly rapidly if a reasonable amount of lining dissipation is achieved. *At low frequencies and in short ducts this may not be a valid assumption.* The result of this lack of information and the assumptions necessary to make progress in design in practical circumstances is that a duct will often be designed in partial isolation from source and termination conditions, which are typically allowed for as additional gross corrections to the predicted attenuation, along with those corrections relating to the effects of discontinuities such as bends in the system.

21.2.2   The major parameters controlling attenuation

At sufficiently low frequencies only the zero-order mode of a lined duct will be effective in transporting sound energy, as higher order modes will be below their cut-off frequencies and will have an essentially evanescent character. At such frequencies there is an oft-quoted formula for attenuation in rectangular ducts lined on all four sides with rigid fibrous material, viz,

$$\Delta dB_f = 1.05 \frac{L.P}{A} (\alpha_f)^{1.4} \tag{21.1}$$

where ΔdB_f is the attentuation at frequency f in a duct of length L whose airway perimeter is P, airway cross-sectional area A, and α_f is the absorption coefficient of the lining at frequency f.

This formula was derived empirically (Sabine [21.1]) and is appropriate to ducts carrying low-speed flows. Obviously, care must be taken when attempting to apply this relation to other types of lining constructions and duct shapes. The formula does serve to illustrate the significance of duct length and airway dimensions. Duct attenuation increases as length, L, and lined perimeter, P, increase and decreases as the cross-sectional area, A, increases. This broad relationship is more or less valid for all modes and frequencies. Formula (21.1) is misleading, however, because the attenuation is shown to depend on the absorption coefficient of the lining, and on no other acoustic property. In a reverberant field the sound pressure does depend solely upon the absorption coefficient, but in a lined duct there is no truly reverberant field, and, correspondingly, the attenuation is not governed by the lining absorption coefficient but by the acoustic impedance of the lining. This distinction is an important one: the mechanisms involved in these two types of sound absorption are different.

Returning to the attenuation formula, for a duct of given length, attenuation is directly proportional to the ratio of airway lined perimeter, P, to airway cross-sectional area, A. Replacing a duct having a single airway by one having a number of smaller airways, formed by lined splitters, having the same total cross-sectional area, will advantageously increase the perimeter/area ratio, P/A, while little affecting the flow through the duct. This property is qualitatively correct for lined ducts at all frequencies, which will be borne out by later sections of this chapter.

The property of a duct liner which governs its behaviour when harmonic waves are propagating in the duct is its specific normal acoustic impedance, Z, which is, in general, a complex quantity usually expressed as

$$Z = R + iX \tag{21.2}$$

where the real part, R, is termed the *specific acoustic resistance* and the imaginary part, X, is termed the *specific acoustic reactance*. The unit of specific acoustic impedance is the Rayl $(\mathrm{Nm^{-3}\,s})$.

The specific acoustic resistance of a thin layer of porous material is closely approximated by its specific flow resistance (being the ratio of pressure difference across the material to the face velocity of the flow through it). For thicker materials the relationship is far more complex, but the resistivity (specific flow resistance per unit thickness) is a significant parameter. The specific acoustic reactance represents the inertia and stiffness properties of the lining, and depends on both the enclosed fluid and the micro-structure of the matrix of porous material. The specific acoustic reactance is closely dependent upon the liner thickness.

In broad terms, the resistive component of impedance governs the level of attenuation achievable and the frequency range over which attenuation is significant. As a rule of thumb, the normalized specific resistance $R/\bar{\rho}c$ should be in the range of 0.5 to 2 for a good level of attenuation over a reasonably wide frequency range. Correspondingly, the reactive component controls the frequency at which maximum attenuation is achieved, the so-called *tuned frequency* of the liner. Typically, the thicker the liner the lower the tuned frequency, although often 'tuning' is not sharply defined. Possibly the most important feature of the role of acoustic impedance in determining attenuation is that the detailed physical nature of the liner is unimportant; that is, provided lining materials exhibit the same surface impedance properties they will produce the same attenuation all other factors being equal.

Other parameters of major importance for duct propagation are the flow speed and temperature and their variations within the duct. Both parameters affect not only the effective acoustic impedance of a lining material but also the behaviour of propagating sound waves. The details of these parameters' influence are complicated and will be discussed more fully below.

21.2.3 Optimization

A natural question to ask is whether it is possible to determine the 'best' lining to choose to give maximum possible attenuation throughout the frequency range of interest. In general, the answer must be that it is not possible, on both theoretical and practical grounds.

There is a principle noted by Cremer [21.2] which allows an optimal choice of acoustic impedance in the case of a 2-D duct lined on one side only, in the absence of flow, for the zero- and first-order mode pair only. This result has been extended by Tester [21.3] to any pair of successive modes, circular ducts, and to the case of mean flow in the duct.

Essentially, Cremer's Principle identifies the least attenuated mode and chooses the lining impedance to maximize this attenuation. Usually, at low frequencies the zero-order mode is the least attenuated, and it happens that its maximum attenuation occurs where it becomes indistinguishable from the first-order mode [21.4]. This enables the optimum impedance Z_{opt} to be found, which for a rectangular duct is given by

$$\bar{Z}_{opt}/\bar{\rho}c = (0.929 - i0.744)\,\ell_y/\lambda \tag{21.3}$$

where ℓ_y is the airway width and λ the freespace wavelength, and for a circular duct

$$\bar{Z}_{opt}/\bar{\rho}c = (1.76 - i0.76)R'/\lambda \tag{21.4}$$

where R' is the airway radius [21.4].

Equation (21.3) applies equally well to symmetrically lined ducts for the first two *symmetric* modes where ℓ_y is the airway half-width.

The variation of the optimal attenuation with frequency (or wavelength) is shown in Fig. 21.1 in terms of decibels of attenuation per duct width, duct half-width, or radius (h), for asymmetrically-lined rectangular, symmetrically-lined rectangular or circular ducts, respectively. An approximate formula is

$$\Delta dB/h = \frac{3.3}{(h/\lambda)} \qquad h/\lambda > 0.5 \tag{21.5}$$

The actual limit as $h/\lambda \to 0$ is $\Delta dB/h = 18.3$ dB (rectangular) and $\Delta dB/h = 26.1$ dB (circular).

Fig. 21.1 clearly shows the advantage of choosing the duct airway measure h to be as small as possible at any given frequency. This emphasizes that the use of lined splitters to reduce airway width will increase attenuation, provided that the liner is chosen optimally for the new configuration.

Tester further demonstrated that a similar optimization principle applies to higher-order successive mode pairs, but also that it is not possible to optimize for more than two modes at a time [21.3]. This is a major limitation of Cremer's concept of optimum impedance, which cannot be extended to overall attenuation over a wide range of frequencies, since this depends on modal energy distribution at source. A second limitation is the extreme sensitivity of attenuation to small changes of impedance in the region of optimum impedance. This would make it extremely difficult to realize the correct impedance value in practice. Finally a most severe limitation is the required frequency variation of the optimum impedance (equations (21.3) and (21.4)), since both real and imaginary parts are directly proportional to frequency. Most duct liner reactances show the opposite trend with frequency or oscillate increasingly rapidly as frequency increases.

However, this optimization principle does serve as a design guide. For example, it supplies an upper limit on achievable attenuation in any particular mode, which may in addition assist in mode identification, often a problem in the analysis of lined duct attenuation behaviour. Moreover the characteristics of modal behaviour near optimum are typical of those seen in the region of maximum attenuation even off optimum, when higher-order modes may be present.

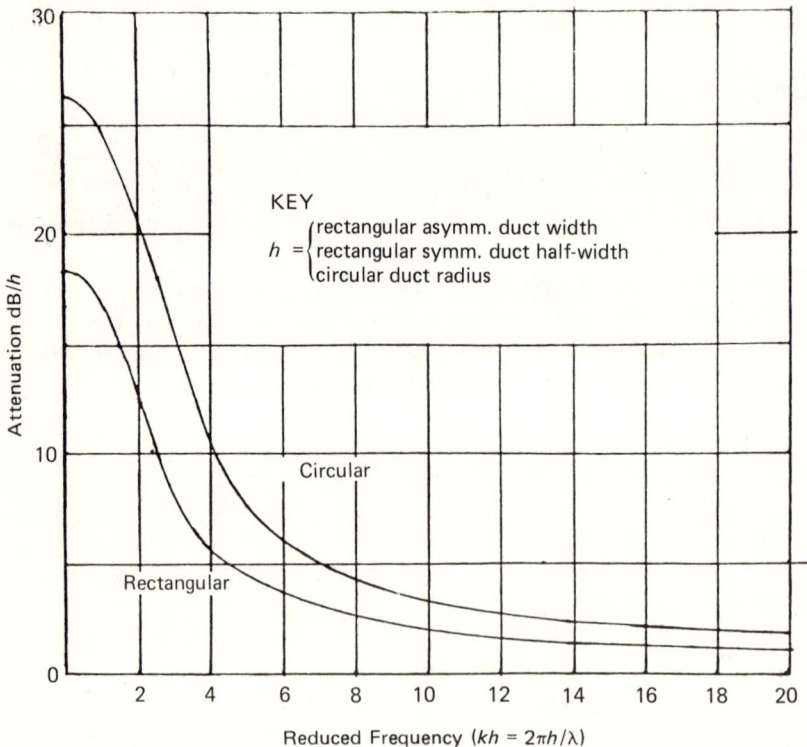

Fig. 21.1 – The variation of optimal attenuation per length h, with h and frequency. [Key in diagram]

21.3 THE INFLUENCE OF THE MAJOR DESIGN PARAMETERS ON ATTENUATION

21.3.1 Typical liner constructions

Within the category of interest here there are two main types of liner: 'bulk' and 'laminar' liners. A 'bulk' lining consists of a thick layer of porous material (e.g. mineral wool) backed by a solid wall. A 'laminar' lining is composed of a relatively thin layer of porous material (which acts almost wholly as an acoustic resistance) suspended over a cavity (acting almost wholly as an acoustic reactance). Of course, there are many instances where a combination of the two types is desirable or necessary, or where several layers of similar types are used to obtain the required performance over a wide frequency range. Historically, bulk fibrous liners were evolved for use in ventilation systems, whereas laminar constructions have been widely exploited in modern aircraft turbo-fan ducts, owing to their potential for high mechanical integrity, and resistance to corrosive substances and erosion. Usually a honeycomb backing in laminar liners both provides structural strength and prevents lining sound propagation. Bulk liners usually need a surface screen (of high porosity) to retain the absorbent material and to resist erosion. Furthermore, it may sometimes be necessary to enclose the absorbing material in bags of fine-gauge plastic film in corrosive, dirty, or otherwise difficult environments.

It is not proposed to discuss in detail the prediction of liner acoustic properties, since this information is readily available, for example in standard noise control texts (e.g. [21.5]), and would require an unjustifiably long discussion in this chapter.

21.3.2 The dependence of attenuation on duct and lining parameters

In this section the major parameters associated with a laminar liner construction will be varied and the resulting attenuation spectra displayed. The characteristic features displayed are, however, sufficiently typical of many lined ducts to be useful in illustrating general features of attenuation behaviour.

Up to quite high frequencies the liner construction allows a particularly simple approximate expression for its specific acoustic impedance in terms of the specific flow resistance, R, of the porous surface layer and the depth, d, of the cavity, viz.,

$$Z = R + i\bar{\rho}c \cot kd \qquad (21.6)$$

where $k = \omega/c$ is the free space wavenumber as before, $\bar{\rho}$ is the mean density of the fluid, c the speed of sound in the fluid. The quantity $\bar{\rho}c$ is termed the *characteristic impedance* of the fluid ($\bar{\rho}c \simeq 410$ Rayls for dry air normal conditions). Commonly, Z and R are referred to this characteristic impedance to give normalized non-dimensional quantities $\bar{Z}$ and $\bar{R}$. In equation (21.6) the convention of a time variation in the form $\exp(-i\omega t)$ has been assumed. For many thin porous materials R may be regarded as approximately independent of frequency.

Figs 21.2 and 21.3 show the effects of varying liner specific acoustic impedance. The normalized resistance $\bar{R}$ has little effect upon the frequency of maximum attenuation, but governs the maximum attenuation achieved and the bandwidth of the attenuation (Fig. 21.2). As $\bar{R}$ increases, the attenuation maximum decreases, but the bandwidth, particularly at low frequencies, markedly increases. It is for this reason that proprietary silencers are designed with highly resistive material, thereby achieving a modest attenuation over a broad frequency range, making the system widely applicable. However, a lower resistance giving a sharper attenuation peak may well be desirable in specific cases. For example, if there is a pronounced tonal component in the source spectrum, as with screw-compressor whine, it would be natural to design for a narrow bandwidth, high peak attenuation, spectrum centred on the worst-offending whine component.

In contrast, changing the specific acoustic reactance of the liner, by changing the cavity depth, d, has little effect on the peak attenuation and attenuation bandwidth, but strongly influences the tuned frequency of the liner (Fig. 21.3). In fact, the greater the depth, the lower the tuned frequency. The significance of this property is that it becomes increasingly unpractical to design for a high level of attenuation as the frequency decreases owing to the large physical dimensions of the cavity required.

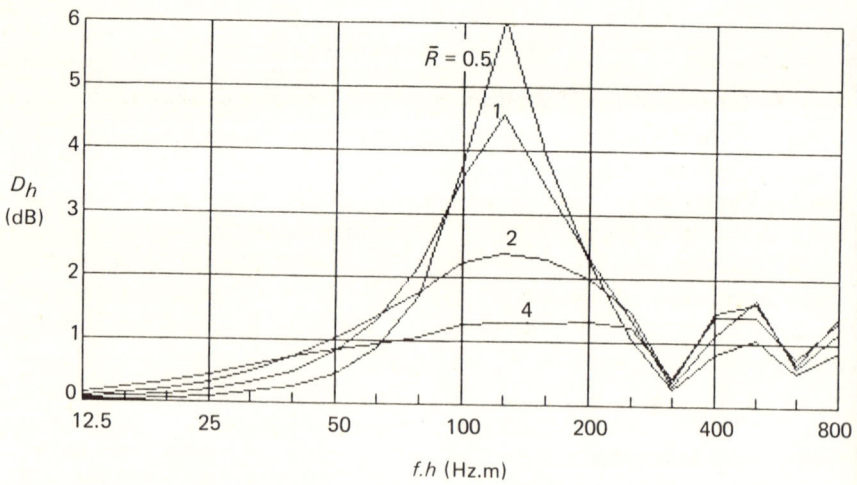

Fig. 21.2 – Effect of varying liner acoustic resistance.
(h = 2 m, d = 0.1 m, l = 3 m)

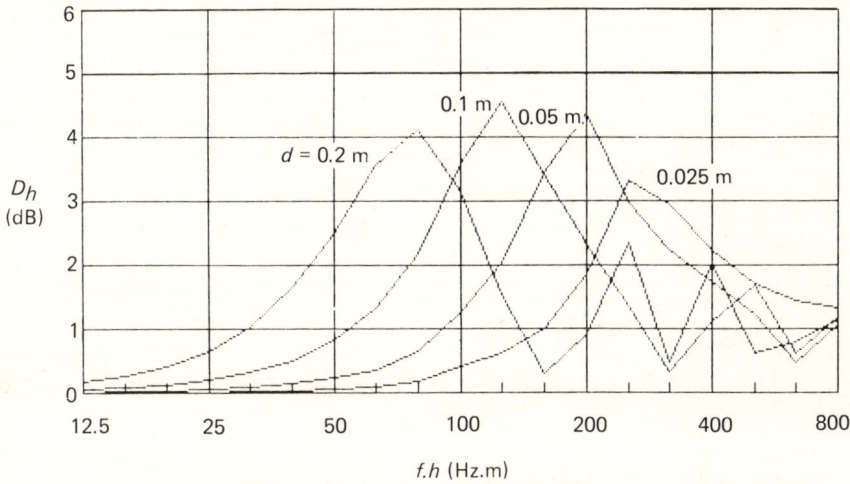

Fig. 21.3 – Effect of varying liner depth on attenuation per duct half-width, D_h.
(h = 0.2 m, l = 3 m, $\bar{R}$ = 1)

The influence of the duct dimensions of airway width and lined length are shown in Figs 21.4 and 21.5. As the duct width increases there are two distinct attenuation changes: a reduction in tuned frequency and a decrease in maximum attenuation (Fig. 21.4). This supports the design principles elicited from the optimization results, and further emphasizes that the tuning cannot be designed in isolation from the duct. As expected, attenuation increases with increasing duct length. However, there is a trend towards a lower tuned frequency as length increases, which is not such an obvious effect. This is caused by the changing influence of the different duct modes contributing to sound propagation. Higher-order modes typically have higher rates of attenuation than lower-order modes. Thus for short duct lengths it is to be expected that attenuation will not be large until a sufficient number of higher-order modes are able to propagate; that is, at higher frequencies. As duct length increases, the more attenuation is available in lower-order modes, while higher-order modes quickly reach the stage where effectively all their sound energy is absorbed. Increasing length will have no further significant effect on the higher-order mode contribution to attenuation, and so the tuned frequency of the duct appears to shift towards lower frequencies. It should be noted that when duct length is of the same order as airway width, then entrance and termination effects will become dominant, and the above considerations may not apply.

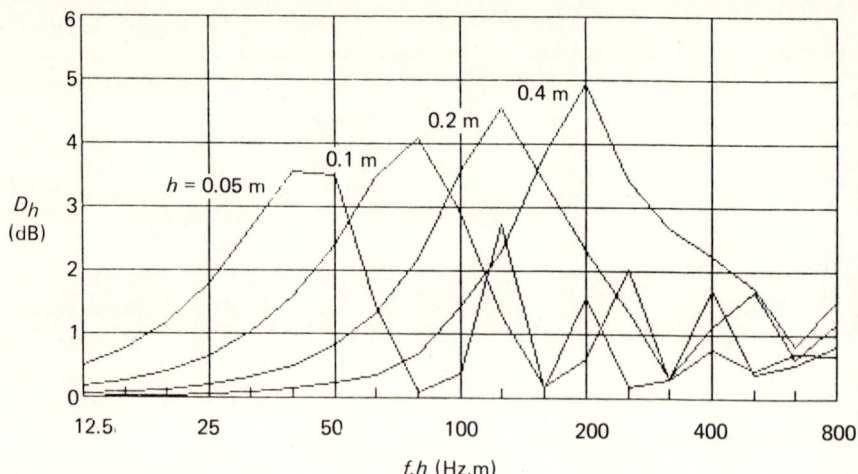

Fig. 21.4 – Effect of varying airway width (2 h) on attenuation per duct half-width D_h. (d = 0.1 m, l = 3 m, $\bar{R}$ = 1)
(Note that both ordinate and abscissa depend on h)

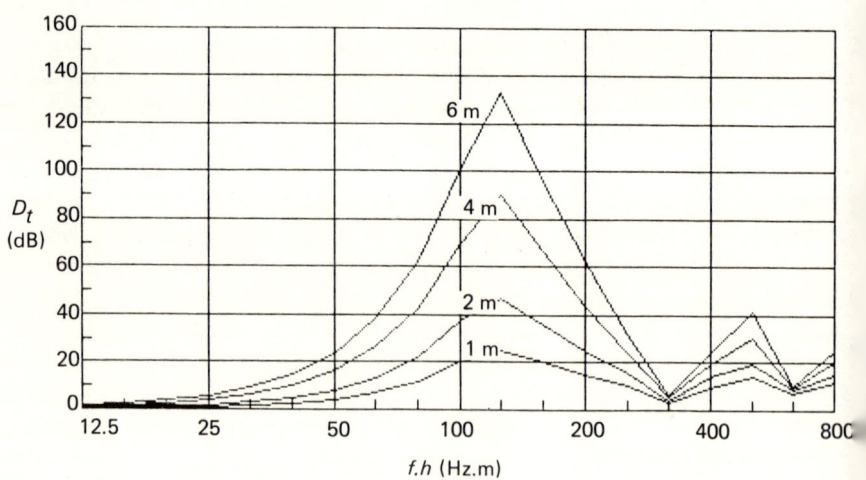

Fig. 21.5 – Effect of varying duct length on total attenuation.
(h = 0.2 m, d = 0.2 m, $\bar{R}$ = 1)

21.3.3   The effect of flow

Up to this point the effects of flow have not been considered explicitly. In many instances duct flow speeds are sufficiently low that flow effects may be neglected in the first approximation: notionally if the Mach number, M, of the flow ($M =$ flow speed/speed of sound) is less than about 0.1 then flow will not be very important.

The principal effect of a mean flow on acoustic propagation in a duct is to increase or decrease the phase velocity of sound waves, thereby increasing or decreasing the effective wavelength in the duct at a given frequency. This is a *convective* effect resulting from the nature of sound waves which occur as perturbations of the ambient fluid motion.

In addition there is a *refractive* effect due to the viscous boundary layers which exist at the duct walls in the presence of flow. There is a gradient of flow speed across the boundary layer as the speed of the main axial flow matches to the zero flow speed at the wall. The phase velocity of sound waves therefore changes in the direction of the wall. When the direction of sound propagation is the same as that of the mean flow (denoted by a *positive* Mach number) the refraction is towards the wall; conversely, when the direction of sound propagation is opposite to that of the flow (*negative* Mach number) the refraction is away from the wall. This refraction away from or towards the wall results in a corresponding decrease or increase in attenuation which is counter to the effect of convection. In general, the convective effect is greater than the refractive effect, which, however, does become of increasing importance at higher frequencies.

Another consequence of duct mean flow is that most absorptive lining materials change their acoustic properties with flow passing over the surface.

Tester extended the concept of optimum wall impedance to rectangular ducts with uniform flow. An asymptotic approximation valid for modes well above their cut-off frequencies, when the axial wavenumber $k_x \to k/(1 + M)$ is

$$Z_{opt}\bigg|_M = \frac{1}{(1 + M)^2} Z_{opt}\bigg|_0 \tag{21.7}$$

The frequency f_{opt} at which maximum attenuation occurs for a given lining acoustic impedance is shifted by the introduction of flow, and when the approximate expression (21.7) for optimum impedance is valid, we have approximately

$$f_{opt}\bigg|_M = (1 + M) f_{opt}\bigg|_0$$

Experimental support exists for this relation; see for example Tester [21.3] and Lowson [21.6].

For very small flow Mach numbers at the wall surfaces, as is the case in many (but no means all) industrial applications, the change of impedance may be quite small, and if so, the dynamic insertion loss may be approximated at low frequency in terms of static insertion loss by factoring by $1/(1 + M)$. That is to say, attenuation is decreased in the direction of flow and increased against the direction of flow. This is amply illustrated by Fig. 21.6 which clearly shows the changes in tuning frequency and attenuation for uniform flow in a 2-D rectangular duct.

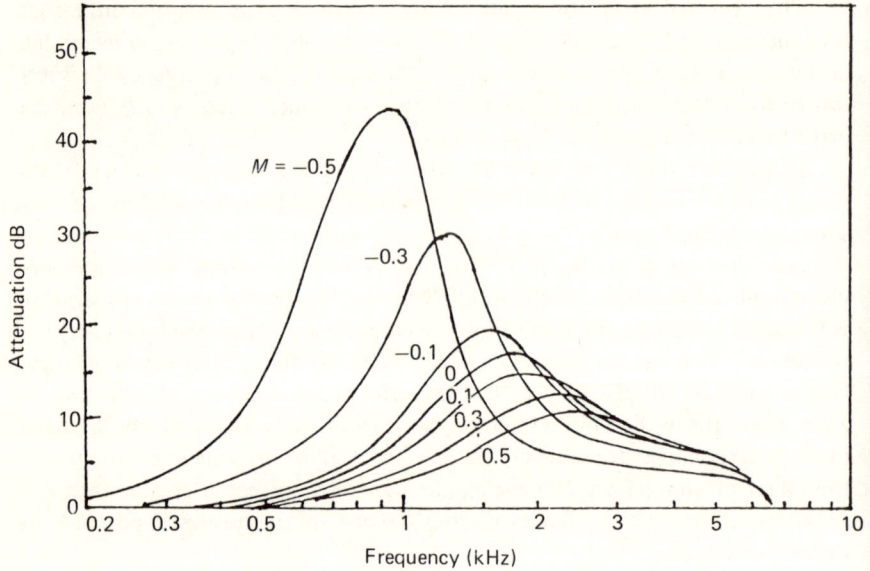

Fig. 21.6 – Effect of mean flow on total attenuation.
(h = 0.19 m, d = 0.025 m, l = 0.4 m, $\bar{R}$ =ʻ1.5) (After Ko [21.21])

21.3.4 The effect of temperature and temperature gradients

The speed of sound is proportional to the square root of the absolute temperature, and so increases with temperature, while the wavelength also increases. The density of fluid is inversely proportional to temperature, and so the characteristic impedance varies inversely as the square root of temperature. These altered values of sound speed, gas density, and characteristic impedance must be employed in the duct analysis, together with the correct values for lining specific acoustic impedance at the prevailing temperature, which may be rather more difficult to determine.

Furthermore, if the duct walls are cooler or hotter than the main body of fluid in the duct, then transverse temperature gradients will exist. These gradients will cause a refraction of sound waves owing to the changing speed of sound across the gradient. Broadly speaking, walls hotter than the mean flow will produce reduced attenuation, and cooler walls increased attenuation, and this effect is independent of flow direction, unlike flow gradient effects.

21.3.5 The effect of wave propagation in the liner

As mentioned previously it is possible for sound energy to propagate along the lining of a duct, unless specific action is taken to prevent this occurrence, for example by compartmenting the liner (bulk liners) or using honeycomb backing (laminar liners).

When a liner does not allow wave propagation along the duct it is said to be *point-reacting*, and its performance is governed only by the normal acoustic impedance at its surface. When a liner does allow wave propagation along the duct, then its properties are dependent also upon the nature of the wave motion in the duct, and it said to be *bulk-reacting*. Lining propagation is likely to become important when the duct is narrow or when the attenuation is high, and it can lead to a reduction in attenuation or a different characteristic behaviour from that produced by the equivalent point-reacting liner.

The problem of the interaction of the liner-borne sound waves with air-borne sound waves in the duct is a complex one, and it is not within the scope of this chapter to consider it in any detail. Scott [21.7] first analysed the effect of isotopic bulk lining materials on wave propagation in a two-dimensional duct without flow. His theory has been verified by several investigators, e.g. Bokor [21.8], while Kurze & Ver [21.9] extended his analysis to anisotropic lining materials. Tack & Lambert [21.10] and Nayfeh *et al.* [21.11] have produced theoretical analyses of the problem of flow in ducts with bulk-reacting lining materials, and it is demonstrated [21.12] that the effects are most important in higher modes and in the lowest mode at low frequencies. It is further suggested in reference [21.12] that an optimum liner could possibly be produced by a combination of bulk-reacting and point-reacting liner materials, although no details are given.

21.3.6 Effects of high incident sound level on linear impedance

A problem that arises with the use of laminar absorbers is the change in impedance, mainly the resistive part, which occurs at high incident sound pressure levels. Research at the ISVR and elsewhere has demonstrated that the influence of sound pressure levels is simply one of increased particle velocity through the liner material. Thus the effect may be studied using steady flow resistance measurements (i.e. measuring the pressure drop across the material for a range of air volume velocities through the material), the ratio $\Delta p/V$ being the flow resistance required. It is found that felted materials, such as sintered layers of fine metal fibres or gauzes, exhibit a weaker dependence of resistance on flow velocity and hence sound pressure levels. This is because the resistance comes mainly from the capilliary friction of the air motion within the micropores, and this is velocity independent.

At higher volume velocities, however, some turbulent flow shedding develops as the air leaves the fibres, and this gives rise to a drag force which depends on V^2. Thus $\Delta p/V$ becomes directly proportional to V.

Perforated sheet laminar absorbent liners generally have very little viscous resistance unless the holes are very small. However, the turbulent loss can be quite large at velocities equivalent to a pressure of 110 dB, continuing to increase proportionally to V as sound pressure level becomes even higher. Since the sound pressure levels inside the ducts close to many high-speed industrial machines (such as power turbines and screw compressors and also in valves operating with a large pressure drop) are usually as high as 160 dB and often approach 190 dB, a usable value of resistance can be obtained from a perforated sheet with as much as 10% open area and 1.5 mm diameter holes. It now becomes obvious that for optimum design application we must know the 'in-duct' sound levels fairly accurately. In terms of most engineering problems, changes in mass reactance of liners with high sound pressure level can be neglected.

Surface flow also affects the apparent resistance of liner materials, both perforate and compressed fibre types. This is owing to the surface boundary layer pressure fluctuations, which act in an almost random manner, introducing their own velocity fluctuations through the lining. This effect is found not to be additive to the sound pressure level effect. In other words, for low sound pressure levels and low surface tangential flows a change in either will have a substantial effect on resistance, but if either sound pressure level or flow is high, then an increase in the other will not have a very significant effect on resistance.

It is generally accepted that the impedance of bulk linings is very much less affected by surface flow, but where a large open area perforated sheet or a coarse wire screen retaining surface is incorporated, there can be moderate changes of both resistance and surface generated noise. Provided such retaining sheets are of sufficient open area, normally over 25%, they have negligible additional resistance in the presence of airflow.

21.4 PRACTICAL DESIGN CONSIDERATIONS

21.4.1 Choice of duct geometry

Selection of a particular overall geometry for a suppressor is very much a matter for individual engineering cases. Usually, however, the cross-section shape is dictated by the nature of the ducting adjacent to the machine in question, *or* for simplicity of construction a transition to rectangular ductwork is made. Frequently the suppressor may need to be built to pressure vessel standards, and here the cylindrical type has obvious merits. For rectangular ducts, as mentioned earlier, splitters usually have to be specified to reduce the airway between lining surfaces to a value giving adequate attenuation. The lining thickness has to be applied each side of the splitter (although not absolutely always), and this effectively doubles the overall lining dimension, small additional thicknesses being necessary for retaining sheets, and sometimes for a hard supporting spine or septum. The overall splitter thickness is thus largely determined by the frequency at which maximum absorption is required, and we have seen that the necessary choice of airway gap to go with this is a matter for trade-off between overall silencer length and overall silencer breadth transverse to the splitters. Where flow losses are not so critical, the smaller airway gap will generally give more attenuation improvement than an equivalent increase in silencer length, in terms of the total use of absorbent materials. Similar arguments apply to cylindrical ducts with a central absorbent core and possibly ring splitters, although this type of design, whilst aerodynamically clean, will be costly to produce. When we have fixed airway gap, splitter thickness, and silencer length, we are free to choose width (along the splitters) and the number of parallel airways (number of splitters plus unity) in order to suit the requirements of flow pressure drop and entertainable overall silencer size. Consideration, as above, of the sound level in duct, airflow past the splitters, and the nature of the fluid and other engineering circumstances, then dictate the choice of the actual splitter absorbent elements and the covering.

21.4.2 Practical problems with resistive suppressors

There are a number of practical points which should be borne in mind when designing or ordering attenuators.

(i) Settling of liner fill materials

The use of certain bulk fibre mat fills at relatively low packing density will, in time, lead to settling or shaking down of the material, so that non-uniformity of linear impedance will arise. This is prevented by incorporating bulkheads at intervals and/or by stitching the fibres into a quilting using some material such as scrim cloth in order not to affect the absorption properties. Provided the resulting resistance remains acceptable, fibres can also be bonded into rigid or semi-rigid slabs, either to be mounted on a back sheet or retained behind a face sheet of large open area.

(ii) Temperature and fire retardation

Some materials (the foam rubbers and polyurethanes in particular), are unsuitable above about 140°C temperature and are unable to satisfy full Class I fire retardance requirements, although many types of foam can now be treated with a retardant coating which will offer reduced tendency to ignite at least for the initial service life of the material. The problem usually lies with accumulation of combustible deposits over many months or years of use, and this can quite obviously be more critical than the initial treatment. Some makes of foam, indeed, combust more vigorously when treated, once the fire has taken hold. Further details are available in standard texts, e.g. [21.5].

(iii) Wicking of fluids and accumulation of deposits

All microporous linings will wick-up water or fuels, and only the perforated sheet lining is really immune from this problem. The same considerations apply to other deposits such as atmospheric grime and dust. Consideration will have to be given to possible damage from icing if water can be soaked up and the silencer is exposed to low temperatures either from the atmosphere or, for example, a process flow expansion. Loose fibre fills and perforates are generally tolerant of such conditions, but bonded slabs may well become damaged. It is often possible to enclose the absorbent fill in very fine gauge plastic film bags such as 'Melinex'. Of course there will be some deterioration in absorption, particularly at high frequency, but in some circumstances the low-frequency performance might be improved since the surface film acts as a panel absorber to a certain extent.

(*iv*) *Erosion by airflow*

Loose fill materials will tend to detach from the splitters, especially in surface flows of order 2.5 ms^{-1} and above. It is normal practice for retention covers of various types to be used to prevent this, or else some suitable non-erodable form of absorbent is used, such as bonded wool slabs or perforate liners. Metal or GRP based laminar liners are largely immune from airflow erosive problems.

(*v*) *Chemical and other interactions*

Certain applications involve corrosive elements within the gas flows through the attenuator. Obviously stainless steel or, wherever appropriate, plastic construction should be used. For the absorbent material, either a resistant material must be employed or the problem avoided by the use of a protective thin face skin or the perforate lining concept.

21.5 SELF NOISE

Self noise will occur in attenuators wherever the flow through the duct leads to substantial velocities passing by the absorbent surfaces. The subject has been relatively little researched, although all the larger product companies have specific data to offer for their own particular designs, which have usually been measured as part of their determination experimentally of dynamic insertion loss, the source of sound being switched off for this purpose.

There is some experimental evidence that flow disturbances caused by blockages, such as splitters or spoilers, at low speeds, generate noise proportional to the 4th power of duct flow speed past the obstruction up to the cut-off frequency of the first higher-order mode, and at higher frequencies proportional to the 6th power of duct flow speed [21.13]. At higher flow speeds, above about Mach 0.3, at the obstruction, these scaling laws change to 6th and 8th power of flow speed respectively [21.13, 21.14, 21.15, 21.16]. The level of noise produced is also dependent on the Strouhal number for the flow at the blockage, and decreases as Strouhal number increases at approximately 20 dB per decade. Thus sound power generated increases as flow speed increases and decreases with increasing frequency or blockage diameter.

Noise generated by turbulent boundary layer flow over perforated metal facings may be considerable, particularly if the associated vortex shedding becomes coupled to liner cavity resonances or to duct modes near to their cut-off frequency [21.13, 21.17]. The sound generated is dependent on a Strouhal number based on hole diameter. High Strouhal numbers are associated with higher levels of noise. Typically, resonant reinforcement will not occur with high flow speeds, small holes, and large duct widths [21.13].

21.6 PRESSURE LOSSES

The absorbently lined walls and splitters impose losses in total pressure head for the fluid flowing through the duct, over and above a plain duct. The losses arise from the same factors that cause the flow noise discussed in section 21.5. There are two main components, surface skin friction losses from all treated walls and splitters, and blockage losses due to the splitter frontal area. Both types of loss increase with the surface velocity squared. Blockage or profile drag losses are related substantially to noise and trailing edge (or evasé) design and to the splitter thickness/airway gap ratio, and therefore do not greatly depend on silencer length. The surface boundary layer losses, however, are naturally only influenced by the surface texture and the treated length for practical airway widths.

Again, the most useful data come from manufacturers, who measure the losses as part of the standard determination of acoustic performance on their flow duct facilities.

REFERENCES

[21.1]   Sabine, M. J. (1940) *J. Acoust. Soc. Amer.,* **12**, 53. The absorption of noise in ventilating ducts.

[21.2]   Cremer, L. (1953) *Acustica,* **3**, 249-263. Theorie der Luftschall-Dampfung im Rechtkanal mit schluckender Wand und das sich dabei ergebende hochste Dampfungsmass.

[21.3]   Tester, B. J. (1972) PhD Thesis, University of Southampton, Sound attenuation in lined ducts containing subsonic mean flow.

[21.4]   Tester, B. J. (1973) *J. Sound Vib.,* **27**, 477-513. The optimization of modal sound attenuation in ducts, in the absence of mean flow.

[21.5]   Beranek, L. L. (Ed.) (1971) *Noise and vibration control,* McGraw-Hill.

[21.6]   Lowson, M. V. (1975) AGARD Lecture Series No. 77, *Aircraft noise generation, emission and reduction,* pp. 7-1 to 7-34. Duct acoustics and mufflers.

[21.7]   Scott, R. A. (1946) *Proc. Phys. Soc. Lond.,* **58**, 358-368. The propagation of sound between walls of porous material.

[21.8]   Bokor, A. (1969) *J. Sound Vib.,* **10**, (3), 390-403. Attenuation of sound in lined ducts.

[21.9]   Kurtze, U. J. & Ver, I. L. (1972) *J. Sound Vib.,* **24**, (2), 177-287. Sound attenuation in ducts lined with non-isotropic material.

[21.10] Tack, D. M. & Lambert, R. F. (1965) *J. Acoust. Soc. Amer.*, **38**, (4), 655-666. Influence of shear flow on sound attenuation in lined ducts.

[21.11] Nayfeh, A. H., Sun, J. & Telionis, D. P. (1974) *AIAA Journal*, **12**, No. 6 (June), 838-843. Effect of bulk-reacting liners on wave propagation in dicts.

[21.12] Nelson, P. A. (1981) PhD Thesis, University of Southampton. Aerodynamic sound production in low speed flow ducts.

[21.13] Gordon, C. G. (1968) *J. Acoust. Soc. Amer.*, **43**, 1041-1048. Spoiler-generated flow noise. I: The experiment.

[21.14] Gordon, C. G. (1969). *J. Acoust. Soc. Amer.*, **45**, 214-223. Spoiler-generated flow noise. II: Results.

[21.15] Szewczyk, V. M. (1974) MSc. Dissertation, ISVR, University of Southampton. An investigation into tailpipe noise in jets.

[21.16] Tsui, C. Y. & Flandro, G. A. (1977) *J. Sound Vib.*, **50**, 315-331. Self-induced sound generation by flow over perforated duct liners.

[21.17] Bauer, A. B. & Chapkis, R. L. (1977) *Journal of Aircraft*, **14**, 157-160. Noise generated by boundary layer interaction with perforated acoustic liners.

Further Reading

Reference [21.6], above, contains a fairly comprehensive review of theoretical and experimental work in the propagation of sound in lined ducts both with and without flow, up to about 1974. The leading reference below contains a good review of theoretical aspects up to 1973.

Attenuation in Lined Ducts

[21.18] Nayfeh, A. M. & Kaiser, J. E. (1973) *AIAA Paper* No. 73-1153, OASI/AII Aeronautical Meeting, Montreal, Canada, Oct. 29-30. The acoustics of aircraft engine duct systems.

[21.19] Snow, D. J. & Lowson, M. V. (1972) *J. Sound Vib.*, **25**, 3, 465-477. Attenuation of spiral modes in a circular and annular lined duct.

[21.20] Rice, E. J. (1968) *Proc. of AFOSR-UTIAS Symposium*, Toronto 20-21 May 1968. Attenuation of sound in soft-walled circular ducts. (Also appears in NASA TMX-52442).

[21.21] Ko, S. H. (1971) *J. Acoust. Soc. Amer.*, **50**, (No. 6, Pt. 1) 1418-1432. Sound attenuation in lined rectangular ducts with flow and its application to the reduction of aircraft engine noise.

[21.22] Ko, S. H. (1972) *J. Sound Vib.*, **22**, 193-210. Sound attenuation in acoustically lined circular ducts in the presence of uniform flow and shear flow.

[21.23] McCormick, M. A. (1975) *J. Sound Vib.*, **39**, 1, 35-41. The attenuation of sound in lined rectangular ducts containing uniform flow.

[21.24] Tester, B. J. (1973) *J. Sound Vib.*, **28**, No. 2 151–203. The propagation and attenuation of sound in lined ducts containing uniform or 'plug' flow.

Acoustic Properties of Materials
[21.25] Delany, M. E. & Bazley, E. N. (1970) *Applied Acoustics,* **3**, 105–116. Acoustical properties of fibrous absorbent materials. (Alternatively (1969) *NPL Aero Report* AC37 March 1969. Acoustical characteristics of fibrous absorbent materials).
[21.26] Christie, D. R. A. (1976) *J. Sound Vib.*, **46**, 3, 355. Measurement of the acoustic properties of a sound-absorbing material at high temperatures.
[21.27] Zwikker, C. & Kosten, C. W. (1949) *Sound-absorbing materials.* Elsevier Publishing Co. Amsterdam.
[21.28] Melling, T. H. (1973) *J. Sound Vib.*, **29**, 1–65. The acoustic impedance of perforates at medium and high sound pressure levels.

Non-linear Propagation
[21.29] Nayfeh, A. H. & Tsai, M. S. (1974) *J. Acoust. Soc. Amer.,* **55** (6), 1166–1172. Non-linear acoustic propagation in two-dimensional ducts.
[21.30] Kurtze, U. J. & Allen, C. M. (1971) *J. Acoust. Soc. Amer.,* **49** (5), Pt. 2, 1943–1947. Influence of flow and high sound level on the attenuation in a lined duct.

Effects of Temperature on Propagation
[21.31] Nayfeh, A. H. & Sun, J. (1974) *J. Sound Vib.*, **34** (4), 505–517. Effects of transverse velocity and temperature gradients on sound attenuation in two-dimensional ducts.
[21.32] Kapur, A., Cummings, A. & Munger, P. (1972) *J. Sound Vib.*, **25**, 129–138. Sound propagation in a combustion can with axial temperature and density gradients.

CHAPTER 22

Noise from industrial machines

E. J. Richards

Institute of Sound and Vibration Research, University of Southampton

A. FUNDAMENTAL RELATIONSHIPS

22A.1 INTRODUCTION

While much thought had been given in the past to the subjective effects on the populace living nearby the noise of aircraft, road vehicles, trains, hovercraft, boats, and construction equipment and environmental rules have been drawn up in most countries, the effects on hearing of the noise inside factories has only recently been studied sufficiently to allow the establishment of codes of safe practice. It has been established for example, that some half a million workers in the United Kingdom are in a work environment which is liable to cause serious deafness during their lifetimes, and the corresponding figure in the United States is well over 3 millions.

Of course, individual workers vary in their susceptibility to deafness and to the way the noise exposure is received, while the degree of deafness which leads to a significant hearing impairment must be specified in quoting such figures. In the United Kingdom, the most practical measure of noise exposure is considered to be $L_{A\text{eq}}$, the average daily dose of A-weighted noise energy received each day [22.1] and charts have been drawn up of the expected hearing loss in decibels (the average shift in hearing threshold for the three frequencies 0.5, 1, and 2 kHz) expected to occur for various fractions of the workers who have been subjected to 85, 90, or 95 dB(A) continuous exposure or its equivalent for various numbers of years of employment. As it is generally accepted that real and practical impairment does not develop until a hearing loss of 25 decibels has occurred, it may be seen from Fig. 22.1 that such a level is reached (because of industrial noise, and not including effects of age) during a working career of 40 years by ten per cent of the workers if the exposure is kept to an $L_{A\text{eq}}$ (i.e. an equivalent continuous A-weighted level averaged over an eight hour day) of 90 dB(A), and the same degree of impairment occurs in 22 years if the level is raised to 95 dB(A). The interpretation of such curves as these is of course the subject of endless arguments, many persons feeling that a 10 per cent statistical population risk is too high and that an 85 dB(A) level of $L_{A\text{eq}}$ should be called for. In the

USA more emphasis is placed upon the peak noises, while different laws, based upon a 5 decibel exchange rate between sound pressure level and duration, are used to obtain equivalent exposure.

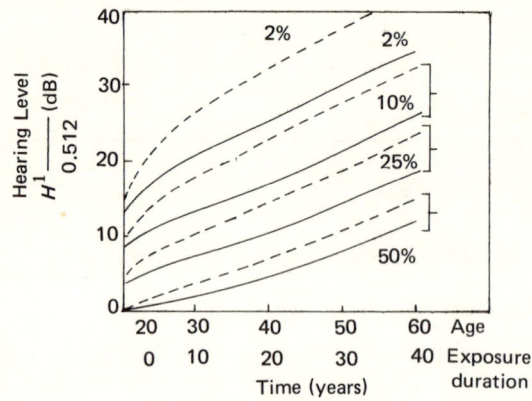

Fig. 22.1 – Hearing impairment levels occurring to various percentages of the working population suffering 90 dB(A) exposure (full lines) and 95 dB(A) exposure (dotted lines).

While the total noise energy exposure laws (based upon the above or some modified form of them) which are now being introduced into the United Kingdom and already exist in the USA are acceptable to the lawyer and the audiologist, we have only just begun to consider how they affect the machine design engineer, and his ability to add such limitations of the total noise energy reaching every operator's ears during the single workday to all the other design requirements. A factory noise environment arises from many machines and not just the one the operator is using at the time. He changes his work role and position from time to time, and leaves his workplace occasionally. Not only that, many of the worst noises are of an impulsive nature, the power needed by the machine to do the work being built up over tenths or even whole seconds, the noise emitted in turn being related to the fracture or impact process and lasting milliseconds at most. There can be as many as thirty decibels difference between the peak noise levels registered at each blow compared with the average noise energy per operation.

Table 22.1 shows such differences obtained by Brüel [22.2] as a result of using different kinds of meters designed to record peaks and averages respectively. It may be seen clearly that if engineers are to include noise as a design parameter on machines involving impacts occurring over just a few milliseconds, and if adequate methods of standard testing are to be evolved, they are going to have to understand the noise creation processes far more fundamentally than they do at present.

Table 22.1

Sound source	Fast dB(A)	Imp dB(A)	Imp hold dB(A) 5×	Peak hold dB(A) 5×	Δ
Sinusoidal pure tone 1000 Hz	94	94	94	97	3
Beat music from a gramophone	90	91	93	97	4
Modern music from a gramophone	102	103	103	105	2
Electric guitar from a gramophone	85	86	86	91	5
Motorway traffic 15 m distance	80	80	81	89	8
Motorway traffic 50 m distance	68	68	68	76	8
Train 70 km/h rail noise 10 m distance	95	96	98	106	8
Train 70 km/h rail noise 18 m distance	85	87	87	94	7
Noise in aircraft type PA 23, cruising speed	90	91	91	100	9
Noise in aircraft type Falco F8, cruising speed	97	98	98	109	11
Noise in aircraft type K23, cruising speed	102	102	103	112	9
Noise in car type Fiat 500, 60 km/h	78	79	79	93	14
Noise in car type Volvo 142, 80 km/h	75	75	76	86	10
Lawn mower 10 HK 1 m distance	97	99	99	116	17
Typewriter IBM (head position)	80	84	83	102	19
Electric shaver, 2.5 m distance	92	92	92	107	15
75 HK diesel motor in electricity generating plant	100	101	101	113	12
Pneumatic nailing machine 3 m distance	112	114	113	128	15
Pneumatic nailing machine near operator's hand	116	120	120	148	28
Industrial ventilator 5 kH 1 m	82	83	83	93	10
Air compressor room	92	92	92	104	12
Large machine shop	81	82	82	98	16
Turner shop	79	80	81	100	19
Automatic turner shop	79	80	80	99	19
40 tons punch press, near operator's hand	93	98	97	121	24
Small automatic punch press	100	103	103	118	15
Numerically driven high-speed drill	100	102	103	112	9
Small high-speed drill	98	101	101	109	8
Ventilator with filter	82	83	83	94	11
Machine-driven saw, near operator's head	102	102	104	113	9
Vacuum cleaner type Hoover, 1.2 m distance	81	81	81	93	12
Bottles striking each other	85	88	90	105	15
Bottling machine in brewery	98	99	101	122	21
Toy pistol (cap)	105	108	108	140	32
Pistol 9 mm, 5 m distance from side	113	114	116	146	30
Shotgun 5 m distance from side	108	110	111	143	32
Saloon rifle 1 m distance from side	107	110	110	139	29

The situation is illustrated rather well in some analysis of recordings made by the Institute of Sound and Vibration Research at Southampton [22.3] of the noise of a typical punch press, and of a pneumatic hammer used successively to cut concrete blocks and a steel ship structure. In order to do the analysis the sound pressure level is taken from the tape recordings at $20\,\mu s$ intervals and recorded (after A-weighting) in a digital data analysis unit. It is then summated with time and plotted as a true summation of A-weighted sound energy during a single operation of the machine; alternatively 10 or 20 ms samples can be analysed to provide frequency related spectra of the noise energy during the consecutive stages of a machine process. Fig. 22.2a shows the growth with time of the total A-weighted energy emission from the pneumatic hammer when it is operating on a concrete structure. In the first instance the noise radiated from the concrete workpiece is small, and Fig. 22.2a illustrates the impulsive nature of the noise from the hammer, dominated by the air exhaust system and lasting only a few milliseconds. Since the instantaneous peak noise level would equate to the maximum slope in the curve, and the $L_{A\text{eq}}$ level to the average slope during any one operating cycle, this method of plotting provides considerable insight into both the design and measurement problems to be dealt with.

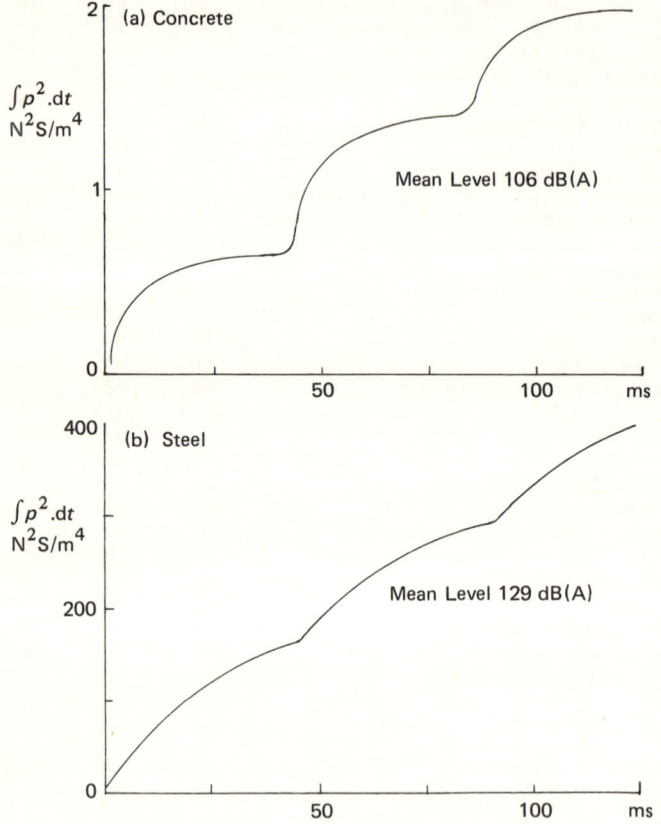

Fig. 22.2 – Pneumatic hammer on different workpieces.

When in turn the same hammer with the same cutting tool operates on steel (Fig. 22.2b) (as it does in shipyard work) the average $L_{A\text{eq}}$ level is increased from 106 dB(A) to 129 dB(A), an increase of 23 dB; the noise energy has therefore increased 200 times, and a worker would only be able to operate the same machine in such an environment for one two-hundredth of the time before developing any specified degree of deafness. Clearly, in this case, the hammer is acting as a mechanical exciter of the large undamped metal sheet, and the dominating noise is that from the workpiece vibrating almost at constant amplitude. The details of the hammer design are unimportant in such a case, feasible changes in the impulsive inputs causing very little changes in noise output. This example raises the question of the design requirements for the hammer. It would be foolish to require standard tests unrelated to the extreme operation of the machine. Yet the specification of some standard tests must be called for which relates chiefly to the design of the hammer itself, rather than the radiation from the material being worked upon.

A similar but more complicated situation arises on a punch press (Fig. 22.3). The noise energy from each operation comprises a sequence of noises each of which has to be looked at separately. From Fig. 22.3, we can deduce that the A-weighted noise energy per cycle is shared over three sources, the clutch engagement, the blanking and piercing process, and the blow-off noise from the valve which reverses the clutch machinery mechanism. The time-integrated noise energy emitted from each of these actions may be seen to be about the same, though the peak noise level is very large as the piercing occurs. Thus, emphasis on reducing the peak need not do much to the total emitted noise per machine cycle, and a study of peaks may be misleading in the diagnosis of the design changes that might be needed. Since the height of the ordinate is the actual A-weighted sound energy at that frequency and moment of time (and not its logarithm) the relevant radiated noise energy is given by the volume under the 'carpet', and can be thought of in this way.

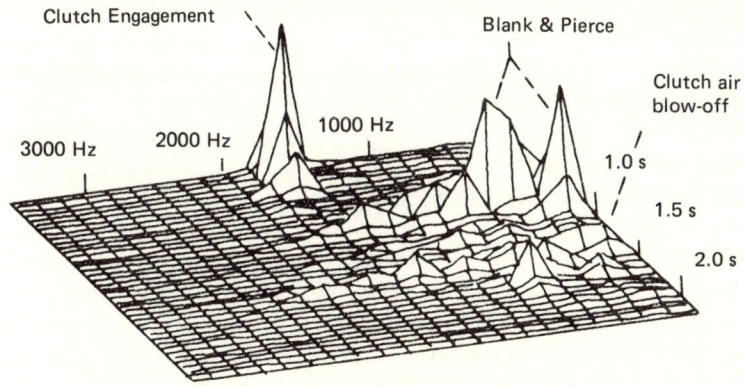

Fig. 22.3 – 450 ton manually cycled power press.

It may be deduced that the clutch noise arises from a vibrational resonance of some part of that mechanism at about 1.7 kHz but that the weighted energy emitted is less than the quite complicated blanking or piercing process which seems to excite a number of parts of the machine at a frequency below 1 kHz. The noise from the blow-off valve is similarly described and suggests that the noise is not wholly jet-based but provides some discrete frequency whistle component.

The above examples have been given, not to present a technique of analysis which is always useful, but to illustrate the need for design engineers to be able to identify the sources of noise in impact machines with the same preciseness which is now used to trace high stress points, areas of excessive vibration or of insufficient stiffness in machinery elements. Only if this can be done can we expect to be able to deal with noise control requirements with the kind of preciseness which will be required if we are to give noise limitations its proper place in the design process generally. How much or how little do we know about machinery noise sources?

All noise arises from the creation of pressure perturbations in the air which in turn arise from the movement of solid surfaces either pulsating (to create fluctuating volumes) or cutting through the air (to create fluctuating forces). The exception to these is the pressure perturbations on the surrounding air created by regions within itself of turbulent air or gas which are in unsteady motions until a new state of equilibrium is attained.

Examples of the first would be the noise from a transformer, that from a large panel in vibration, that from the ringing of a large anvil when struck, or from a violin box. These noises are categorized as pure sources, and can be treated analytically by the mathematical methods outlined in other chapters.

Examples of the second mechanism, that is, of a thin surface cutting through the air, displacing no mass, but exerting a fluctuating force on the surrounding air, are the noise of a fan, aircraft propeller, or any moving solid whose opposite sides move together in such a way that no 'volume' change occurs. Such sources of noise can be categorized as 'dipoles' and can be treated as such mathematically as described elsewhere in this book.

An example of noise being created by turbulent air in the process of achieving equilibrium is that from a jet engine, from an industrial blow-off valve, or from an industrial cooling jet.

In practice, the noise of industrial machines can be composed of any one or any combination of these sources; the structural pulsations can be continuous, or transient, and as often as not, many areas of the structure are vibrating together. Thus it is seldom possible to establish theoretical methods of prediction on practical machines, the categorization being useful as a general guide to the effects of design changes rather than to theoretical prediction.

It is important to realize also that while the ear recognizes pressure changes, the noise radiated from a machine is related to the average value of the product of the instantaneous pressure and the 'in phase' velocity perturbation; that is, of the energy radiated from the machine. Close to a machine or a fan surface or a vibrating panel, the pressure and the velocity can be out of phase, so that though the pressure fluctuations can be large, and heard as such if an operator is present, these are very local and are not propagated as sound waves.

In general this so called 'near field' is limited to a distance from the source of no more than half a wavelength of the sound, and microphones measuring the mean squared pressures are unlikely to be in error if held more than this distance from the radiating surface. It does bring out the two points, however: first that the amplitude of a panel vibration is not necessarily directly related to the noise from it. A 'radiation efficiency' of the panel has to be established. Secondly, we must always be wary of situations where an operator is not so much in an acoustic so much as a 'reactive' aerodynamic environment. Two examples of the latter come to mind; an operator standing very close to a jet, and an operator in direct line with an air efflux from a drop hammer. These are unusual, however, and in general the significance of near field pressure fluctuations are more relevant to measurement techniques and panel radiation efficiencies than they are to operational interpretations.

22A.2 MECHANICAL NOISE SOURCES

Apart from noise radiating jets or fans, industrial machines are characterized by the use of mechanical energy to reshape some piece of material (workpiece) – for example, pressing; to take parts out of it – for example, machining; or to reconstitute it in some other form – for example, die casting. In most instances, the energy level needed for the process is such that it has to be built up over a period of hundreds of milliseconds, and is then used in the process over a period of hundreds of microseconds; that is, the process is impulsive. Such processes are far more common than is generally realized, sawing, planing, stamping, forging, and rivetting all being impulsive processes with varying levels of pulse and repetition rates. The firing of reciprocating engines, pneumatic drills, textile operations and so on are also examples of such mechanisms. In all such systems, noise sources arise from various factors in the process. These may typically be listed as follows:

(1) the energy container (e.g. hammer, anvil, punch, saw tooth, etc), decelerates or accelerates inelastically on impact, giving rise to a pressure perturbation which radiates as a single pulse.
(2) the workpiece changes shape and causes a pressure perturbation. This can be due to volume change, or to the excess energy in the workpiece causing semi-continuous noise radiating vibration.
(3) the initial or final energy container may vibrate elastically and radiate sound (ringing of the hammer or anvil).

Table 22.2

Classification of machine or process	Noise level dB(A) (PERA) Ref. [22.16]	Jet emission	Fan	Source classification					
				1 Hammer deceleration	2 Workpiece distortion or vibration	3 Anvil or case ringing	4 Suppprting structure	5 Air ejection	6 Blow-off valves
Air cleaning bench for machine castings	109	√√	√						√
Blower for heating of blanks	104	√							√√√
Chain rivetting machine	111			√		√			√√
Chipping hammer	118				√√		√√		√
Chisel, pneumatic	116						√√		√
Dicing machine	100		√√						
Diesel generator set	110	√	√			√			
Dust collector plane	98	√			√√	√√	√		
Feed slide for tins	100	√√						√	
Forge plants	107	√√		√		√√	√√		√√
Hammer (pneumatic)	122			√√√		√√√	√		√√√
Handling and transport of billets	98			√	√	√√	√√√		√√
Mill-alumina	109			√		√√	√√√		√√
Mill-coal	107			√		√√	√√√		√√
Air line for cleaning down	102	√	√		√				√√
Ball mill for powderizing materials	106						√√		
Concrete pipe making machine	97								
Crusher plane in quarry	110				√	√	√		√√
Die casting machine	97		√		√	√		√√	
Drop hammer (peak)	111			√		√√√	√√√√		
Fan (for tank cooling)	97								√
Fettling tool	104						√√		√

Table 22.2 – *contd.*

Source classification

Classification of machine or process	Noise level dB(A) (PERA) Ref. [22.16]	Jet emission	Fan	1 Hammer deceleration	2 Workpiece distortion or vibration	3 Anvil or case ringing	4 Supporting structure	5 Air ejection	6 Blow-off valves
Grit blasting machine	104	✓							
Hammer (for plate levelling) (peak)	110			✓		✓✓✓	✓		
Lathe (wheel)	108				✓✓	✓	✓	✓	
Linishing machine	104			✓	✓✓		✓✓✓	✓✓	
Mill (flint)	101		✓						
Mixing plant (plastics) (peak)	110								✓
Mould machine in foundary	101	✓							
Nut runner (air operated)	106					✓			
Perforating machine	101			✓	✓	✓	✓	✓	
Power hammer	108			✓	✓	✓	✓	✓✓	✓✓✓
Press – cropping	100			✓	✓	✓	✓	✓	
Press – laminations	98			✓	✓	✓	✓		
Wet riddle	106		✓	✓	✓	✓	✓		
Woodworking circular saw	104				✓	✓	✓		
Woodworking double end tenoner	100					✓	✓		✓✓
Nailing machines	95	✓✓							
Peening gun	97								
Plasma arc profiling machine	110			✓		✓	✓		
Saw bench – plasterboard	106			✓		✓	✓		
Saw – friction	118			✓	✓	✓	✓		
Saw – for batches of tubes	105			✓			✓	✓	
Saw – slotting	112			✓	✓		✓	✓✓	
Woodworking planer	103			✓	✓	✓	✓	✓	

(4) some of the energy may be transferred to the supporting structure or to the floor which then vibrates and radiates sound.

(5) the deceleration process may be accompanied by air ejection (or in the case of gears, oil ejection) which causes a pressure perturbation, which may be aerodynamic (reactive) or acoustic (radiative).

(6) the mechanical parts may have to be put in reverse at this stage by reversal of pneumatic jacks etc. This is usually accompanied by valve blow-off noise.

Table 22.2 lists an array of noisy machines and identifies the noise sources according to the above classification. It is clear that each noise mechanism dominates on some machine or other, and that in order to understand and predict machinery noise, the correct recognition of the dominating mechanism is essential. The rest of the chapter is therefore devoted to examining each mechanism.

As acousticians, we tend to separate such noise creation processes into two categories. The first is 'acceleration' or 'deceleration' noise arising from the quick change of velocity of various parts of the machine or workpiece during the period of impact. As this forced geometric movement is essentially related to the way in which the machine energy is transferred to the workpiece and is fundamental to the process (for example, the rate of deceleration of a drop hammer as it strikes and manipulates its workpiece), this noise is inherent in the process and is developed only during the 'contact' time taken to dissipate it into the workpiece.

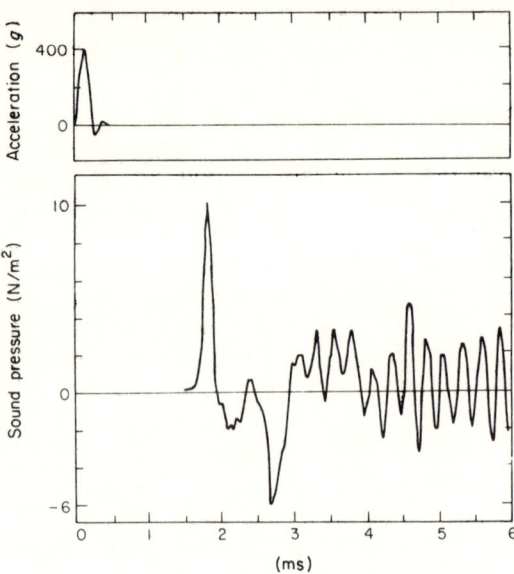

Fig. 22.4 – Typical acceleration and sound pressure signals
from colliding cylinders.

Typically, it is this mechanism which determines the amplitude of the first peak shown in Fig. 22.4, even though it may or may not contribute greatly to the total noise energy. We therefore need to understand it in cases where peak noise is our dominating interest.

The second mechanism is 'ringing noise'. During the work process, some energy will go into vibration of the tool, vibration of the workpiece or the surrounding structure. The radiated noise resulting from this must depend upon the vibration level and the internal damping available. It will continue until all the vibrational energy is dissipated, and the fraction radiated as sound will depend upon the ratio of acoustic to structural damping, and the rate of propagation into other parts of the machine and into the floor.

These two categories of noise are in a way analogous to the process of forced and natural vibration and produce a signal of the kind shown in Fig. 22.4. Often the acceleration noise pattern overlaps that due to ringing, but the advantage of considering them separately remains valid. Acceleration noise does not depend upon damping or isolation; ringing noise does.

22A.3 ACCELERATION NOISE

It has been shown [22.4] that if a sphere of volume $\frac{4}{3}\pi a^3$ and mass m accelerates from rest to a speed v_0 sufficiently slowly not to incur compressibility of the air, an amount of energy equal to $\frac{1}{2}(m + \frac{m'}{2})v_0^2$ has to be supplied to sustain the motion of the solid where m' is the 'virtual' mass of the air displaced by the solid; that is, enough energy to provide the kinetic energy of the solid plus that of the mass of air equal to half that displaced by the body. This latter is called the 'virtual mass'. If the motion is brought back to rest, this energy is regained from the solid body and the air around it. If, however, the motion of the sphere occurs instantaneously (i.e. with infinite acceleration) an additional amount of energy equal to that arising from the virtual mass has to be provided to overcome the compressibility of the air arising from the rapidity of the motion, and this is not necessarily restored if the restoration to a stationary situation occurs. This additional energy must be propagated as acoustic waves away from each part of the surface and will be heard as a true soundwave. Thus we can quantify the maximum possible sound energy radiated from a suddenly decelerating or accelerating body as equal to that of $\frac{1}{2}\rho_0 v_0^2 \times$ the $\frac{1}{2}$ volume of the body. If the acceleration is not infinite and lasts a time t_0 the degree to which it can restore the energy of compression back into the body will be a function of how far the sound pressure arising from the acceleration can travel in this time relative to the size of the sphere. Thus the actual noise energy radiated is a function of $\frac{ct_0}{\pi a}$ where a is the radius of the sphere. It is therefore to be expected that in the most elementary event of a sphere accelerating from one velocity to another rapidly we will have some noise energy (E_{acc}) radiated, and this will be a function of $\frac{1}{2}\rho_0(\text{vol})\,v_0^2$ and $\frac{ct_0}{\pi a}$; for other shapes we define non-dimensional measures of this acceleration noise as $\mu_{acc} = E_{acc} \div \frac{1}{2}\rho_0(\text{vol})\,v_0^2$ and the size parameter δ as $\frac{ct_0}{2(\text{vol})^{1/3}}$ where c is the speed of sound in air.

Similarly, if we must design for acoustic pressure peaks, it can be shown that the instantaneous peak acoustic pressure p_0 can be examined uniquely in the form of a variation of $p_0 r/\rho_0 c a v_0$ against $\frac{ct_0}{\pi a}$ or intensity $\frac{p_0^2 r^2}{\rho_0 c}$ against $\rho_0 c\, a^2 v_0^2 f(\frac{ct_0}{\pi a})$; that is, the effective radiation efficiency σ_{rad} will be a function f of $\frac{ct_0}{\pi a}$. The first step in the prediction of machinery noise must be the establishment of broadly based laws for the variations of μ_{acc} and $\frac{p_0^2 r^2}{\rho_0 c}$ for impact situations in as simple terms as possible.

22A.4 PREDICTING ACCELERATION NOISE

While the maximum acceleration noise output possible from say a drop hammer impact can be equated to $\frac{1}{2}\rho_0(\text{vol})v_0^2$ (since we have two bodies to consider, the hammer and anvil or workpiece), in reality we cannot consider the impact to occur in zero time, as work has to be done on the workpiece, and even in the case of die to die clashing, the contact time is finite. We have therefore looked at real cases at the ISVR using the elementary rig shown in Fig. 22.5 and have both calculated and measured the total unweighted acoustic energy and first peak pressure p_0 emanating from two spheres, and from cylindrical and conical hammers impacting one with the other. The impact time t_0 has been varied independently by inserting small (too small to radiate) metal or plastic pellets between the hammers and measuring the impact times electronically. In this way we consider that we can relate the actual radiated noise surrounding each hammer as a function of $ct_0/V^{1/3}$ where V is the hammer volume, and its cube root gives us the typical dimension needed for our purpose. It is only possible at present to calculate the noise energy for simple shapes, but since acoustic output is the gross result of the radiation from all the surfaces of a body, large differences are not to be expected from other than extreme changes of shape.

Fig. 22.5 – ISVR rig used to investigate the influence of impact
time on noise levels.

22A.5 PREDICTION OF AMPLITUDE OF FIRST PRESSURE PEAK

When a sphere of radius a accelerates from zero to a velocity v_0, the instantaneous acoustic far field pressure $p(r,\theta,t')$ only includes the azimuth angle θ in a form $p(r,\theta,t') = P_0(r,t')\cos\theta$, where t' is the retarded time $t - (r/c)$. This is not true if we consider the interaction of the acoustic pressure field of two spheres impacting slowly since acoustic cross-coupling terms appear. Nevertheless these are only significant if the impact is slow ($\delta = ct_0/V^{1/3} \gg 1$), and in this case the impact noise peaks are small anyway; it is acceptable to ignore such coupling. When we do so, an expression for $P_0(r,t')$ can be written [22.5] for $t' \leqslant t_0$:

$$P_o = \frac{\rho_0 a c\, v_0}{2\sqrt{(4\delta^4 + 1)}}\frac{1}{r} \times$$

$$\left\{(2\delta^2 - 1)\cos\frac{\pi t'}{t_0} + 2\delta \sin\frac{\pi t'}{t_0} + \left[(1 - 2\delta^2)\cos\frac{ct'}{a} - (2\delta^2 + 1)\sin\frac{ct'}{a}\right]e^{-ct'/a}\right\}$$

where P_0 is the pressure along the axis of impact, and if we write

$$\tau = \frac{t'}{t_0}, \quad \gamma = \sin^{-1}\frac{2\delta}{\sqrt{(4\delta^4 + 1)}}, \quad \chi = \sin^{-1}\frac{2\delta^2 + 1}{\sqrt{2}\sqrt{(4\delta^4 + 1)}}$$

the expression for the pressure peak P_0 can be expressed more simply as

$$P_0 = \frac{\rho_0 c a\, v_0}{2\sqrt{(4\delta^4 + 1)}}\frac{1}{r}\left[\cos(\pi\tau - \gamma) + \sqrt{2}\,e^{-\delta\pi\tau}\cos(\pi\delta\tau + \chi)\right]$$

and for any value of δ, the pressure P_0 can be calculated for values of $\tau \leqslant 1$ and the peak pressure obtained.

An expression similar to that above for the far field pressure at any point (r, θ, t) after $t = t_0$, can be written in the form

$$\frac{P_0 r}{\rho_0 c a\, v_0} = \frac{1}{2\sqrt{(4\delta^4 + 1)}}\left\{e^{-\pi\delta(\tau - 1)}\cos\left[\pi\delta((\tau - 1) + \chi)\right] + e^{-\pi\delta\tau}\cos(\pi\delta\tau + \chi\right.$$

$$\text{and} \quad \tau = \frac{t'}{t_0} \geqslant 1.$$

These expressions have been evaluated and the peak values of P_0 obtained. Fig. 22.6 shows the agreement obtained between·this theory and the sphere experiments at the ISVR and by others [22.5]. In themselves the expressions quoted are of limited value since they apply only to spheres, but a little careful thought leads to the formulation of a more general expression. The terms within $\{\}$ are all periodic or exponential, and at no stage can their numerical values exceed unity. They represent in detail the way in which the pressures adjust themselves over the sphere, and different adjustments would be obtained for different shapes of bodies. The term $1/\sqrt{(4\delta^4 + 1)}$ on the other hand is a size factor which is based upon acoustic path distances, and would apply for other shapes if we were to replace πa by one half a peripheral length; that is, by $2V^{1/3}$ where V is the body volume. A maximum value of P_0 can therefore be deduced for all solid shapes in the form

$$\frac{P_0 r}{\rho_0 c a\, v_0} \sim \frac{P_0 r}{\rho_0 c\, v_0(\frac{3}{4}\pi)^{1/3}\,\text{vol}^{1/3}} = \frac{1 + \sqrt{2}}{2\sqrt{4\delta^4 + 1}} \quad \text{where} \quad \delta = \frac{c t_0}{2\,\text{vol}^{1/3}}.$$

In this expression, we should more logically have referred the pressure peak P_0 to the square root of the surface area of the body; however, this is not generally as well known to designers as the volume or mass of the body, and the error is in any case small.

The most convenient form of the above expression is:

$$\frac{P_o^2}{P_{ref}^2} r^2 = \frac{(\rho_o c)^2}{P_{ref}^2} \times 2.24 \times (\text{Vol})^{2/3} \times v_o^2 \times \left[\left(\frac{ct_o}{\text{Vol}^{1/3}} \right)^4 + 4 \right]^{-1},$$

and if $P_{ref} = 2 \times 10^{-5}$ Pascal, the Sound Pressure Level, defined as

$$10 \log_{10} P_o^2/P_{ref}^2$$

can be obtained from the expression:

$$SPL + 20 \log r - 6.67 \log (\text{Vol}) - 20 \log v_o =$$

$$= 149 - 10 \log \left[\left(\frac{ct_o}{V^{1/3}} \right)^4 + 4 \right]$$

where v_o is in metres/second, volume in cubic metres, and the distance r in metres. We can either take the constant to be the theoretical figure of 149 or, as our expression for P_o is a generalized maximum, we can take an empirical constant based upon experiment.

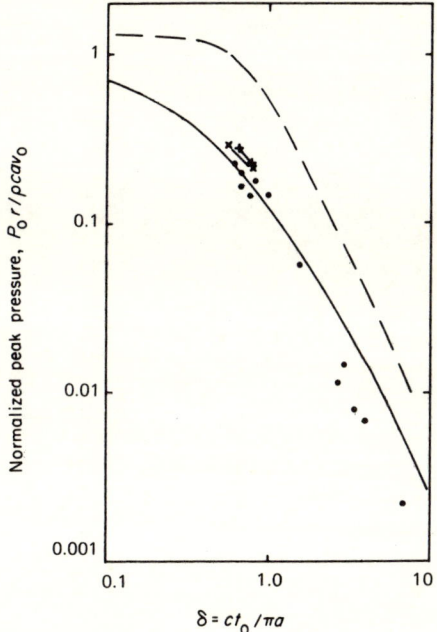

Fig. 22.6 – Normalized peak pressure vs. δ. $- - \dfrac{1.207}{\sqrt{4\delta^2 + 1}}$, $-\!-$ theory, ● 10 cm spheres, x 5 cm spheres, + 5 cm spheres (Koss & Alfredson).

One further simplification can be envisaged: the establishment of two regimes; that for which $ct_0/V^{1/3}$ is large; that is, the impact is soft; and that for which $ct_0/V^{1/3}$ is small compared with 4. Thus we have two very simple expressions for the size of the first peaks:

$$SPL + 20 \log r = 143 + 20 \log v_0 + 6.67 \log (\text{Vol}) \text{ for } ct_0 \ll 2V^{1/3}$$

$$= 143 + 20 \log v_0 + 6.67 \log (\text{Vol}) - 40 \log \frac{ct_0}{(\text{Vol})} 1/3 \text{ for } ct_0 \gg 2 \text{Vol}^1$$

22A.6 EXPERIMENTAL VERIFICATION OF PEAK SOUND PRESSURE LEVELS

The peak sound pressure levels arising from impacting bodies are highly sensitive to direction, and agreement between experiment and theory would not be expected in directions well away from those of the maxima. Often, the operator positions are in such regions, and the peaks arise from reflections and from ancillary structural vibrations. Even so, some preknowledge of the likely levels of peak noise from a machine impact can be useful, particularly in indicating whether such a process is inherently dangerous.

A typical variation of the polar distribution of peak pressures when two cylindrical hammers (Fig. 22.5) impact upon each other is shown in Fig. 22.7. It may be seen that the peaks are easily recognizable in the direction of impact, and that at no other angle is the level any higher.

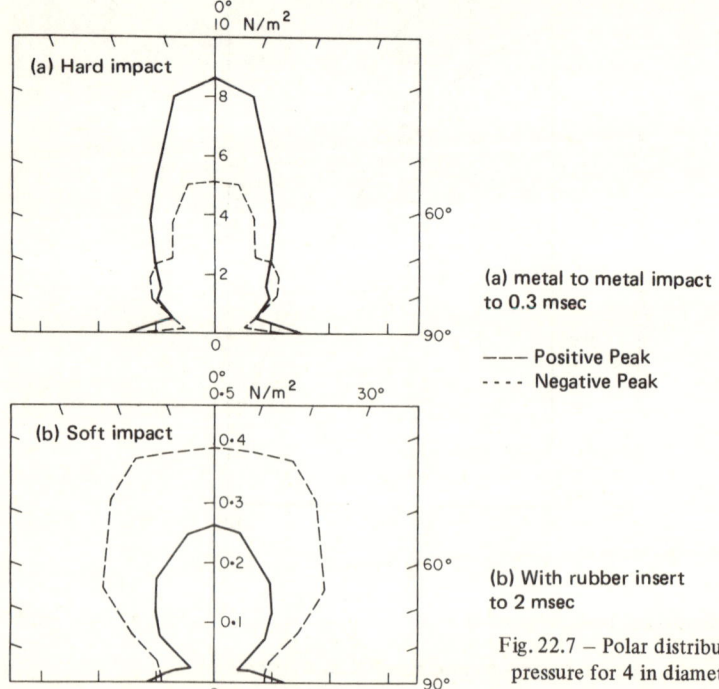

(a) metal to metal impact to 0.3 msec

––– Positive Peak
- - - - Negative Peak

(b) With rubber insert to 2 msec

Fig. 22.7 – Polar distribution of peak pressure for 4 in diameter cylinders.

In order to obtain validification of the generalized curves put forward in the previous section, the peak noise and the time related signals has been measured around spherical, cylindrical, and conical hammers, using the rig shown in Fig. 22.5, and both the peaks, and the total radiated noise energies from acceleration and ringing have been calculated.

A range of impact velocities v_0 were used, and the results normalized to provide experimental measures of both $P_{0\,\text{max}}\, r \div \rho_0\, c a v_0$ and $\mu_{\text{acc}} = \dfrac{E_{\text{acc}}}{\frac{1}{2}\rho_0(\text{Vol})v_0^2}$ for a range of non-dimensional impact times. As our interest lies in the prediction of noise for soft as well as hard blows involving respectively workpiece forming and steel to steel clashing, the range of δ examined has been extended by the insertion of small metal and rubber work pellets designed not to sound themselves but to give a wide range of values of δ. Since under these circumstances the collisions are not fully elastic, the acceleration form is not truly Hertizian. The relative impact velocities and impact times were measured electronically, and the measurements confined to the quarter circle to the rear of the direction of impact, the total noise energy being then calculated on the basis of symmetry.

The agreement between theory and experiment for spheres was indicated in Fig. 22.6. More significant, though, is the agreement between the generalized theoretical 'maxima' formula and the experimental first peak sound pressure levels obtained from cylinders and cones impacting in both hard and soft modes. These are shown in Fig. 22.8 as plots of $(SPL + 20 \log r - 20 \log v_0 - 6.67 \log \text{Vol})$ against δ, now defined as $c t_0 \div V^{1/3}$, and are compared with the theoretical 'maximum' curve and with the two asymptotic expressions obtained for $\delta \ll 1$ and $\delta \gg 4$. It may be seen that the curves are well validated by the points, representing not only impacts of spheres, but also of cylinders and, in two cases, cones hitting base-to-base and apex-to-apex, respectively. The only instance where the experimental points are in gross disagreement with this master curve is that from two cones striking base-to-base. Since the acceleration noise radiated from the apex of a cone is a somewhat indeterminable quantity, this disagreement should be taken as a general warning against excessive expectations of agreement for extreme shapes, rather than as proof of inapplicability of such a general formula for normal shapes.

The actual time histories associated with these two cone impacts is illustrated in Fig. 22.9 which shows the peak pressures and the summated noise energy with time for both the point-to-point and base-to-base impacts. The maximum acoustic pressures are in fact not very different in amplitude but are of different sign and arise in both cases from the pressure generated at the base. The total integrated noise energy differs by a factor of two, but the main difference is that of the ringing noise energy associated with the base-to-base impact.

Some industrial validation of the generalized curves of Fig. 22.8 have been sought from records made of the noise radiated from an industrial drop hammer situated at the Drop Forge Research Association in Sheffield. Readings could

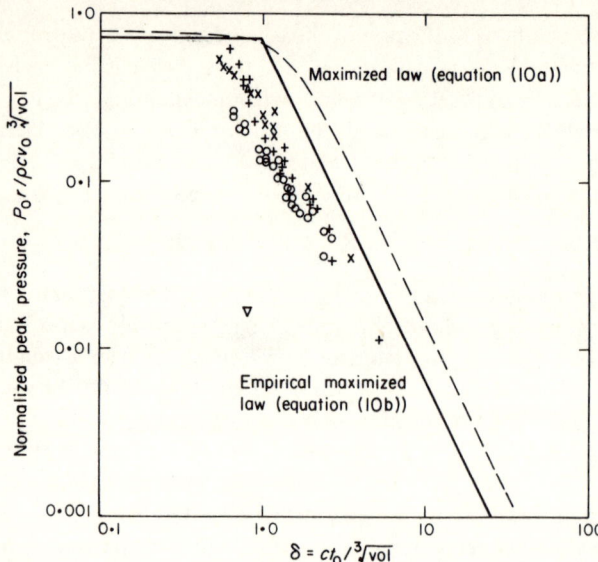

Fig. 22.8 – Normalized peak pressure vs. δ for cylinders and cones. $\bullet$ 3" dia. cylinders; + 4" dia. cylinders; $\times$ 6" dia. cylinders; ∇ cones point to point; $\triangle$ cones base to base; – – – – Maximised theory.

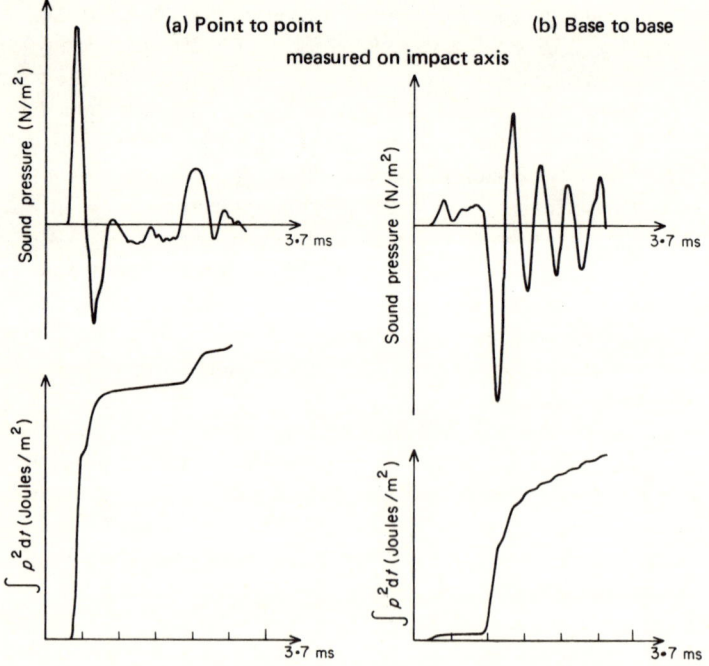

Fig. 22.9 – Typical sound pressure and energy integration signals for colliding cones.

not be taken at the expected position of the peak sound pressure level, but were taken at an angle of 30 degrees to the line of impact. At this angle, the values of the peak amplitudes would be expected to be 0.87 of the largest values. In practice, they represent more nearly the peaks at the operator's position in this particular installation. Fig. 22.10 shows the values obtained of $\dfrac{p_{0\,\text{max}} \cdot r}{\rho_0\, c\, v_0\, (\text{vol})^{1/3}}$ for a number of drop heights, and for both soft and hard blows plotted against δ. It may be seen that even though the tup (hammer) was to some extent conical, the actual and predicted peak levels are in broad agreement.

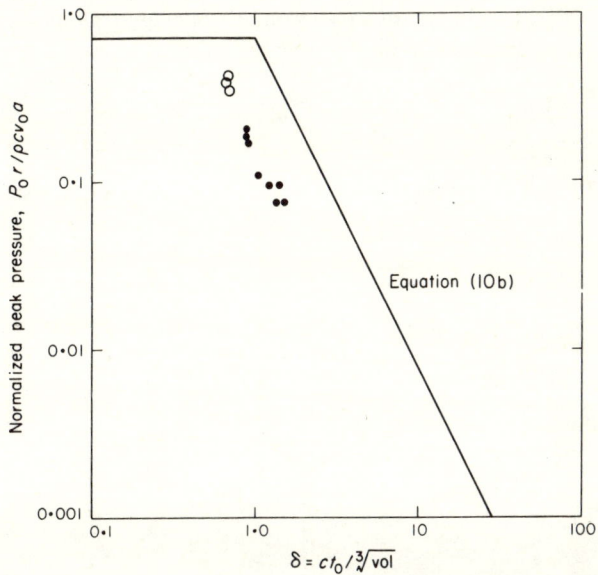

Fig. 22.10 – Normalized peak pressure vs. δ for small friction drop stamp. o Die to die blows; • Forging blows; – – – Maximised theory.

22A.7 THE PRACTICAL SIGNIFICANCE OF FIG. 22.8

The noise regulations in factories, while calling for a limiting value of the total emission of A-weighted noise energy per working day, also specify limitations regarding the peak noise pressures to be tolerated. In the United Kingdom, a maximum peak noise level of 150 dB(A) is permitted (135 dB(A) (Fast response)) while in the United States one suggestion permits a maximum of exposure to a level of 140 dB(A) with a complicated reduction for daily peaks over one hundred. It is interesting to examine in broad terms the implication of these laws in terms of machinery design.

The above formulae refer to real time peak pressure perturbations with no allowance for the effects of A-weightings; the A-weighting is a function of frequency, and the spectral density obtained by transforming a single pulse is very much a function of the pulse shape. Koss & Alfredson [22.6] give a frequency $f = 76.1/a$ Hz for the peak amplitude for metal-to-metal clashing spheres (a is the radius in metres). Thus a body would need to have a typical dimension of less than 10 centimetres not to be affected to some extent by the A-weighting procedure, and a still smaller size for soft blows involving much larger contact times. The relation $f = 76.1/a$ is the same as $f = 1/2\,t_0$ and may well have been derived in this way. Using this relationship we can express the peak sound pressure likely to arise from a steel-to-steel impact process by rewriting our earlier equation in the form:

$$SPL_{max} = 10 \log \frac{P_0^2}{P_{ref}^2}$$

for $\delta < 1$,

$$SPL\ (dB(A)) + 20 \log r = 117 + 20 \log v_0 + 6.67 \log (\text{Mass}) - A$$

and

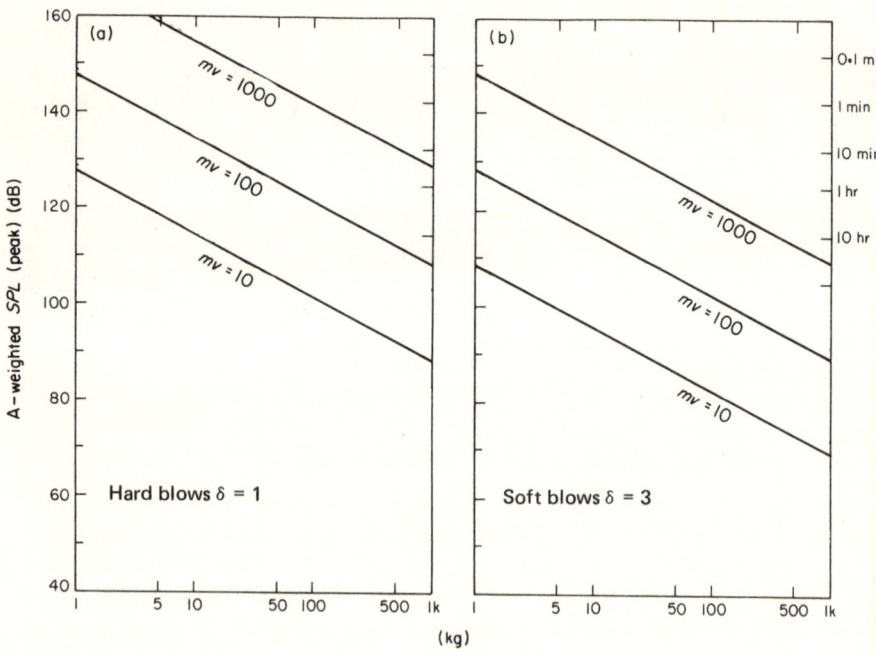

Fig. 22.11 – Predicted A-weighted SPL(peak) at 3 m distance for various impulses and permitted daily exposure time at 1 pulse per second under OSHA proposals [22.34].

for $\delta > 1$,
$$SPL \, (\text{dB(A)}) + 20 \log r = 117 + 20 \log v_0 + 6.67 \log (\text{Mass})$$
$$- 40 \log \delta - A$$

where A is a variable weighting (in dB) which depends on the frequency $1/(2 t_0)$. Here we have replaced the volume by the mass of the impactee, impactor, or workpiece, and have of necessity to relate the A-weighting to the actual impact time rather than to the basic parameter δ. For $\delta < 1$ the impact time only enters the expression, as an A-weighting correction.

Figs 22.11a and 22.11b illustrate the way in which this expression can be useful. The peak A-weighted predicted SPL at a distance of 3 metres is plotted for a series of mass and impulse (mV_0) conditions for hard and soft blows corresponding to $\delta = 1.0$ and 3.0, respectively. For each of these values $f = 1/2 t_0$ has been evaluated for each value of the impacting mass, and the relevant A-weighting correction added. The daily permitted work periods in the USA are also given to correspond to these SPLs.

22A.8 TOTAL ACCELERATION NOISE ENERGY AS A FUNCTION OF IMPACT DURATION

While the peak noise level likely to be radiated in any impact process is important, especially for processes with high peaks which are repeated relatively infrequently, the form of noise legislation in the United Kingdom and elsewhere is firmly linked with the concept of time integrated noise energy arising from the acceleration or deceleration of impacting bodies, and a generalized understanding of how this quantity varies with design parameters is important.

For very short impact durations, the same considerations apply as for continuous high-frequency vibration, the 'radiation efficiency' is unity, and each sphere can be considered independently of the other. Thus the radiated acceleration noise energy from two impacting spheres can be taken simply as twice Longhorn's [22.4] expression for a single sphere.

As the contact duration is increased, the efficiency of radiation of each sphere separately falls because each part of the sphere operates against a modified ambient pressure, and in addition the radiation efficiency of a pair of spheres is modified because the one sphere acts in a pressure field modified by the motion of the other. Thus if the pressures from spheres 1 and 2 at any far point from both spheres are designated p_1 and p_2, the resultant total noise energy radiated through a control surface S is:

$$E_{\text{acc}} = \int_{t'} \int_s \frac{(p_1 + p_2)^2}{\rho_0 c} \, dS \, dt'.$$

E_{acc} consists of three terms

$$E_1 = \int_{t'=0}^{\infty} \int_S p_1^2 / p_0 \, dS \, dt';$$

$$E_2 = \int_{t'=0}^{\infty} \int_S \frac{p_2^2}{\rho_0 c} \, dS \, dt'$$

and

$$E_{12} = \int_{t'=0}^{\infty} \int_S \frac{2p_1 p_2}{\rho_0 c} \, dS \, dt'.$$

E_1 and E_2 will be equal, by symmetry, even though the time variations will be reflected images of each other. The term E_{12} will, however, be dependent on the interaction of the two pressure fields and their build-up with time.

We have calculated E_{12} for various values for S for equal spheres [22.5] and have concluded that only when acceleration noise is low does the acoustic cross-coupling become important. Under such circumstances, the two spheres can be treated as a longitudinal acoustic quadrupole, with very poor radiative efficiencies.

For short non-dimensional contact durations (metal-to-metal clashing) given by $S = ct_0/V^{1/3} < 2$, it is reasonable to predict acceleration noise energy radiation as twice that from a single body decelerating at the relevant rate. This expression has been computed for a range of values of S, and $\dfrac{2E_1}{\rho_0(\text{vol}) v_0^2}$ is given in Fig. 22.12. This expression is the value obtained for μ_{acc} if acoustic coupling terms E_{12} are ignored.

In order to validate the generality of this curve the acceleration noise radiation was measured around the half periphery of the spheres, cylinders, and cones referred to earlier, and the experimental values of μ_{acc} plotted against δ $\left(\delta = \dfrac{ct_0}{V^{1/3}}\right)$. It may be seen (Fig. 22.13) that broad agreement with the theoretical results can be claimed, and that now we can include with much greater confidence such extremes of shapes as cones, impacting nose-to-nose or base-to-base.

Once again, the value of Fig. 22.13 lies in the degree to which we can generalize the result to practical cases, and to do this, it is convenient to simplify the nature of the non-dimensional curve as much as possible. If the noise energy radiated from each body separately is required, a reasonable expression that we can deduce from Fig. 22.13 is:

$$\frac{\mu_{\text{acc}}}{2} = \frac{E_1}{\rho_0(\text{vol})^{1/3} v_0} = 0.7 \text{ for } \delta = \frac{ct_0}{\text{vol}^{1/3}} < 1$$

$$= 0.7\delta^{-4} \text{ for } \delta > 1,$$

and for cold metal-to-metal clashing, δ can be assumed to be < 1.0.

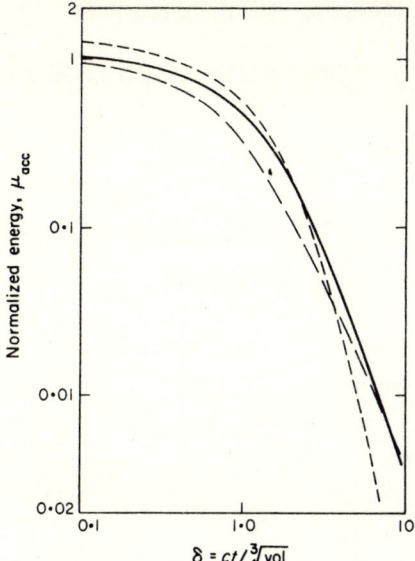

Fig. 22.12 – Normalized energy vs. δ. -----, $E_{acc}/½\rho_0$ (vol)v_0^2 (with cross coupling terms); – – –, energy from Longhorn's results: ———, $2E_1/½\rho_0$ (vol)v_0^2 (without cross coupling terms).

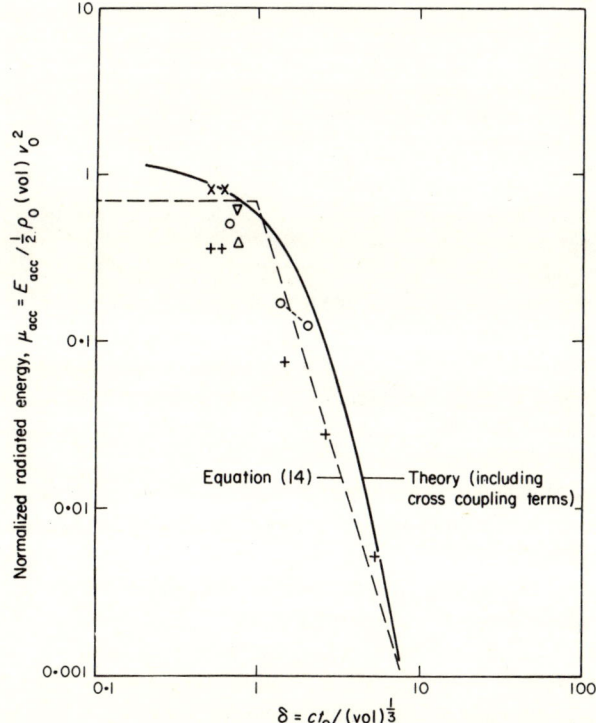

Fig. 22.13 – Normalized sound energy μ_{acc} vs. δ. ○, 75 mm diameter cyclinders; +, 100 mm diameter cylinders; ×, 150 mm diameter cylinders; ●, 100 mm diameter spheres; Δ, cones point to point, ∇, cones base to base.

22A.9 ACCELERATION NOISE FROM AN INDUSTRIAL DROP FORGE

In order to check the validity of this master curve, we have identified and measured the acceleration noise energy radiated from a drop hammer at the Drop Forge Research Association at Sheffield; the normalized results shown in Fig. 22.14 were obtained [22.7].

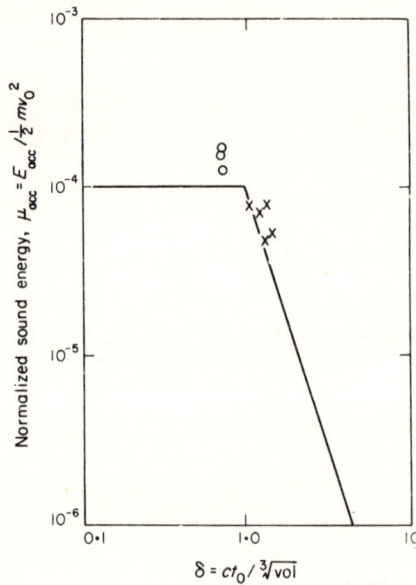

Fig. 22.14 – Normalized sound energy $\mu_{acc} = E_{acc}/\frac{1}{2} m v_0^2$ vs. δ for friction drop stamp. $\circ$, Die to die blows, $\times$, forging blows.

It may be seen that a gratifying agreement with our predictive curve occurs. Each point involves a series of results from a number of drop heights and involves identifying from an oscilloscope trace the acceleration noise component and integrating it roughly around the machine. A limited variation of values of the contact time parameter δ is provided, though not enough to confirm in industrial practice the fall-off in the master curve.

Even so this agreement leads us to consider with some confidence the significance of such a noise mechanism in drop forge and other practices. Two deductions can be made, one from the constancy of values of the ordinate for short contact times, the second arising from the sharp fall away above a certain value of the non-dimensional contact time 'δ'.

As the need to carry out a specified amount of distortion by use of a drop hammer implies a constancy of kinetic energy used, we can from the constancy of the value of μ_{acc} for low values of δ calculate the maximum possible A-weighted or unweighted L_{eq} receivable per day from drop hammers of various sizes and impulses. Making some reasonable assumptions regarding the relationship between the total energy emitted and that passing the operator's position, and assuming that a hammer takes three times as long to rise as it does to fall, the maximum values of L_{eq} and L_{Aeq} for a day's operation have been plotted in Fig. 22.15 for a range of hammer weights and specific impulses. The interesting point emerges that provided liability to deafness is couched in terms of A-weighted L_{eq} levels, it is difficult to see how a real drop hammer can make more 'acceleration' noise than a daily immission L_{Aeq} level of 90 dB(A).

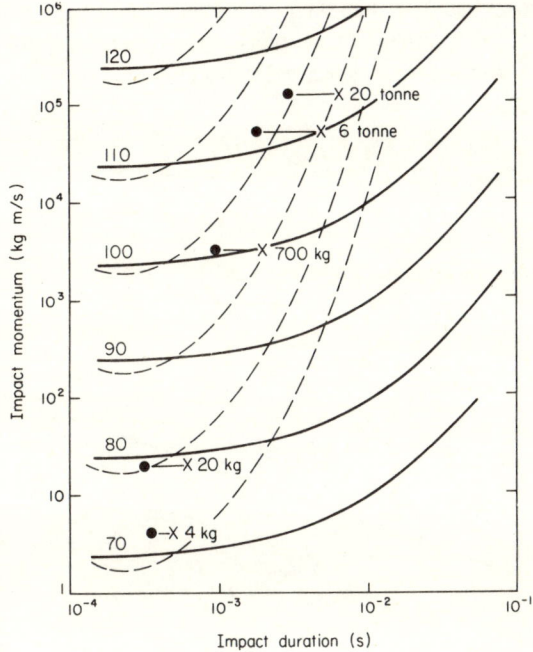

Fig. 22.15 – Impact momentum vs. contact duration for various impacts.
Maximum L_{eq} from gravity process – – – dB(A) —— dB lin.

This is not a ridiculous conclusion; the total noise radiated in these tests with soft blows from the drop forge was some 10 decibels higher than the 'acceleration' noise level owing to the noise radiation arising from the subsequent vibration of the machine. The point which must be made is that the inescapable level of noise which is inherent in the drop hammer process is not such as to cause deafness – it is the ringing noise which is the cause of the problem, and this is not an inescapable adjunct to the process.

The second conclusion from the master curve of Fig. 22.13 arises effectively from the existence of a value of a critical non-dimensional contact time, above which the acceleration noise falls away rapidly. Taking this value of $ct_o/V^{1/3}$

as 3, a separation line of the kind shown in Fig. 22.16 can be plotted, and the liability of a machine mechanism being on the quiet or noisy side of this curve can be established. In the graph it may be seen that it is only the solid die blows which are conducive of high radiated acceleration noise energies, and that soft blows on hot billets are unlikely to be worrying. Indeed other unintended clashes arising from backlash and wear have very much shorter impact times than do the primary processes in many machines, and may dominate the situation.

Such criteria are of course applicable to other machines than drop forges, and we can use the time factors to examine the noise-making potential of such things as backlash, bad fits, handling techniques, tumbling, poor gears, and impacts generally, provided we relate the contact time (or more correctly the acceleration time) to the correct surface dimension.

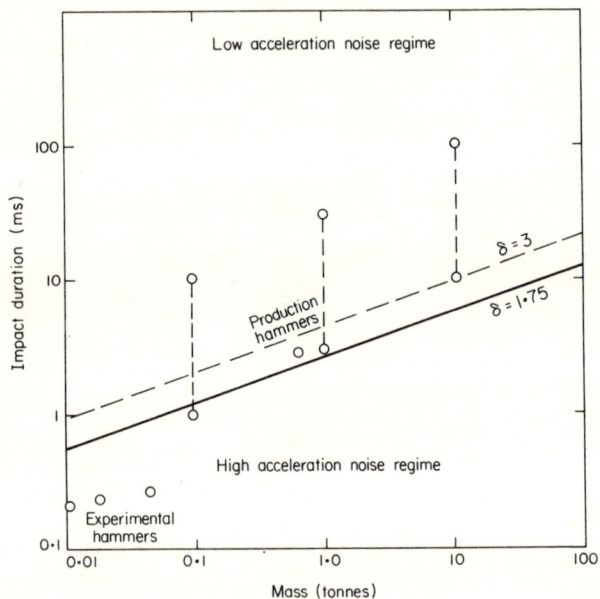

Fig. 22.16 – Impact duration vs. mass.

22A.10 RINGING NOISE

In our sphere and hammer experiments, it became clear very quickly that while 'acceleration' noise could be quantified and studied on bodies in which no one dimension exceeded any other by more than a factor of two, and that in such cases, the acceleration noise was of the same order as the subsequent ringing noise, this ceased to be so the moment an element of flexure appeared in the oscillations. Table 22.3 illustrates this; the total radiated noise energy arising from the natural vibration of the hammers was obtained for typical frequencies of longitudinal and flexural modes of oscillation, by measuring the initial ringing amplitude integrated around the body acoustically and recording its duration from acceleration measurements. It may be seen that while the relative amplitudes of ringing noise energy to the acceleration noise energy was of the same order as far as longitudinal modes were concerned, the ringing energy levels associated with flexural frequencies were as many as 23 dB higher than the acceleration noise, even for the least elongated of the cylindrical hammers examined.

Table 22.3
Ratio of solid body acceleration noise (E_{acc}) and ringing noise (E_{vib})

Impacting bodies	Ringing mode	Frequency (kHz)	E_{vib}/E_{acc}
10.2 cm (4 in) diameter steel spheres	Fundamental	26	−9 dB
15.2 cm (6 in) diameter by 30.5 cm (12 in) long steel cylinders	Fundamental longitudinal	7.9	−4.5 dB
	Fundamental flexural	4.9	+10.9 dB
10.2 cm (4 in) diameter by 30.5 cm (12 in) long steel cylinders	Fundamental longitudinal	8.44	−1.1 dB
	Fundamental flexural	4.02	+25 dB
7.6 cm (3 in) diameter by 30.5 cm (12 in) long steel cylinders	Fundamental longitudinal	8.49	+1.5 dB
	Fundamental flexural	3.2	+20.9 dB

Spectral measurements of the radiated noise suggested that while the oscillating modes of the sphere were essentially longitudinal and that dumb-bell modes, when they occurred, were of very high frequency and carried little energy, the modes which dominated the vibrations of cylinders were flexural, even though we had taken care to excite them longitudinally. The weak coupling at our support system was sufficient to initiate the flexural motions, and because of the slow progressive speed of such flexural waves, energy could be amassed into these modes without leading to the rebound process needed to complete the contact time. Our calculations have now shown that the lateral accelerations recorded are in keeping with the level of ringing noise energy measured by integration around the hammers. The conclusion must be drawn that flexural oscillations set up in supporting frames, mechanical links, and plate-like structures drain energy away from the manipulative processes and may well be the predominating source of noise in machines, even though these may play little part in the actual manipulative process itself. Such flexural motions may incidentally not even be related to the process, but arise from sharp impulsive forces occurring anywhere in the circuit because of backlash.

22A.11 THE CONTROL OF RINGING NOISE

There are three steps to be taken to minimize ringing noise: (a) reducing the excitation to as nearly a uniform load as possible, and to discourage all forms of impulsive action when not required for the process, (b) to add structural damping matched to the nature of the likely oscillations and particularly near sources of flexural vibrations, and (c) to reduce the radiation efficiency of structural members liable to vibrate flexurally by greater knowledge of the relationship between vibration, frequency, and noise.

The first two methods have been subject to much study already and it is difficult to discuss them further without reference to specific machine configurations. The third item, radiation efficiency, however, remains something of a mystery to most engineers and is well worth examining further, especially as a good and comprehensive knowledge of the radiation efficiencies of engineering components will tend to transfer the mystery from one of acoustics to one of specific components' vibration, a distinct advantage in the design process. The radiation efficiency of a block, rod, or plate is the ratio of actual radiation of sound energy (in unit time) compared with that from a surface of the same area and the same mean square normal surface velocity (averaged over both time and space) radiating perfectly. Such a surface is an infinitely large flat plate radiating into its half-space and moving normally to itself with a mean square speed of $\overline{v^2}$. In such a case the acoustically radiated energy into the air per unit surface area is given by $\rho_0 c \, \overline{v^2}$. Consequently we define the radiation efficiency σ_{rad} for any body as being equal to

$$\sigma_{rad} = \frac{E_{rad}}{\rho_0 c \, S \langle \overline{v^2} \rangle}$$

where S is the surface area of the body, $\langle \overline{v^2} \rangle$ is the time averaged squared normal velocity at any point ($\overline{}$) averaged over the whole surface ($\langle \, \rangle$), ρ_0 is the air density, and c the speed of sound in it.

It is usual to express the radiated sound energy in logarithmic form so that

$$10 \log E_{rad} = 10 \log \langle \overline{v^2} \rangle + 10 \log S + 10 \log \sigma_{rad} + \text{Constant.}$$

As it is seldom that σ_{rad} is greater than unity (though it can be) we normally plot $10 \log \sigma_{rad}$, or, if the subjective A-weighted noise is required we sometimes plot $10 \log (A \sigma_{rad})$, where A is the subjective weighting corresponding to any particular frequency.

22A.12 CALCULATIONS OF RADIATION EFFICIENCY OF SOLID BODIES IN RINGING [22.8]

Many calculations have been made of this quantity σ_{rad}, but it has seldom been presented in a form which is easily usable by engineers. For block-like solid bodies which pulsate or oscillate rigidly as solid bodies with a known circular frequency $\omega = 2\pi f$ the radiation efficiency can be reasonably assumed to approximate to that for a sphere of the same volume, and this depends not on the frequency but on the wave-number $k(= \omega/c)$ times the radius or any other typical dimension; that is, on the value of $ka = 2\pi a/\lambda$ (where λ is the wavelength of the sound). Thus ka represents the number of waves in a distance equal to the circumference or peripheral length of the sphere. For large ka, that is for high-frequency radiation with many acoustic waves encompassing a body, σ_{rad} becomes unity and such a body (e.g. a diesel engine) becomes a good radiator. On the other hand a small loudspeaker radiates poorly at low frequencies where the number of wavelengths across a typical speaker diameter is small. This is illustrated in Fig. 22.17a plotted for bodies of typical sizes (defined by the cube roots of their volumes). It may be seen that small solids whose typical dimensions are below 10 cm are poor radiators when in vibration below 800 Hz, but that

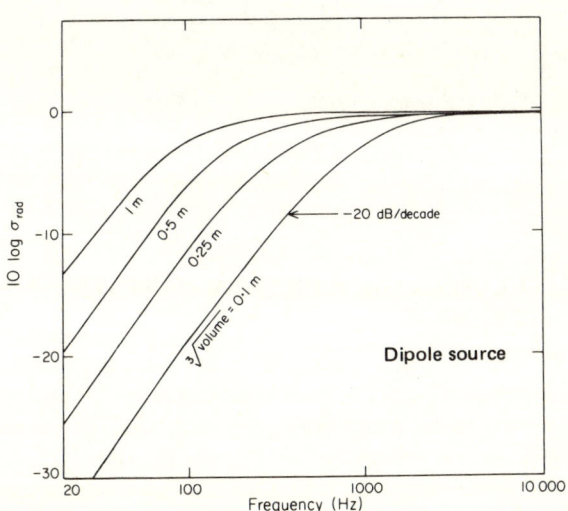

Fig. 22.17a − The variation of $10 \log \sigma$ with frequency for vibrating bodies.

large bodies are efficient as low as 100 Hz. These charts can usefully be used as engineers' guides; if A-weighting is needed on transferring vibration levels to those of radiated noise, the corresponding chart for $10 \log A \sigma_{rad}$ shown in Fig. 22.17b can be used. Since the A-weighting is related to frequency and not to the acoustic parameter ka, it is necessary to plot different curves for different sizes of (in the case shown) vibrating spheres. It is also worth noting that the growth of inefficiency of radiation and the subjective (A scale) with reducing frequencies are additive and that A-weighted slopes are larger than unweighted ones. It follows that even on large bodies the A-weighted radiation efficiency is quite poor at frequencies below 500 Hz.

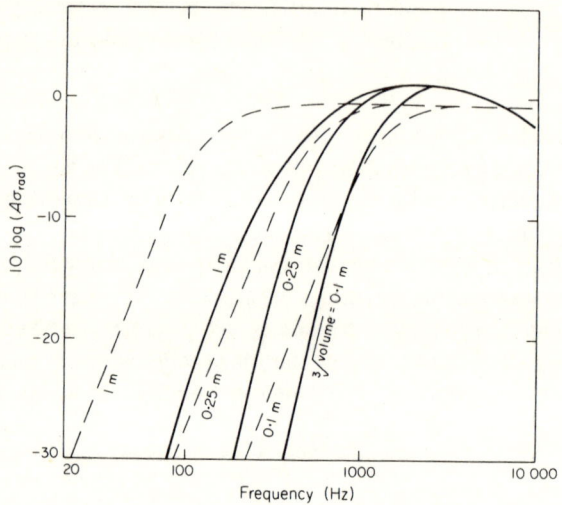

Fig. 22.17b – A-weighted radiation efficiency of vibrating bodies.

22A.13 CALCULATING THE RADIATION OF PLATES AND RODS IN FLEXURE

The use of the above engineer's charts will allow the radiated noise energy to be obtained from values of $\langle v^2 \rangle$ for solid bodies, and the frequency of oscillation is known. However, as explained earlier, we must look for our sources of noise to panels and to beams rather than to solid bodies. Here we have to recognize that the ability of the surface to radiate acoustically depends upon the way the acoustic air pressures develop along the beam or plate as well as around it. Ideally, it can be shown [22.8] that on an infinite plate vibrating flexurally, no sound is radiated acoustically below what is known as the coincidence frequency, that is, that frequency where the bending or tension surface wave travels along the plate at the same speed as does an acoustic wave in the air adjacent to the plate. This cut-off frequency and the ratio of the actual radiating frequency to it is a crucial parameter. Below this critical frequency some radiation does occur on finite plates because the energy transfer processes which ensure that no net energy is radiated from infinite plates below coincidence does not occur near the edges, and in practice 'edge' and 'corner' radiation does occur. This situation is indicated in Fig. 22.18 and makes it clearly impossible to be precise in specifying the amount of radiation from finite plates on which the detailed matching of preferred natural edge or corner modes to the excitation can make differences of as much as 10-15 dB. Even so, we find in practice that excitation is not usually confined to narrow frequency bands, and the engineers' charts [22.8] of radiation efficiency shown in Fig. 22.19 a,b and c can usefully be used to translate vibration levels to those of acoustic output. These cover the range of sizes, thickness, and frequency normally found in engineering machinery problems.

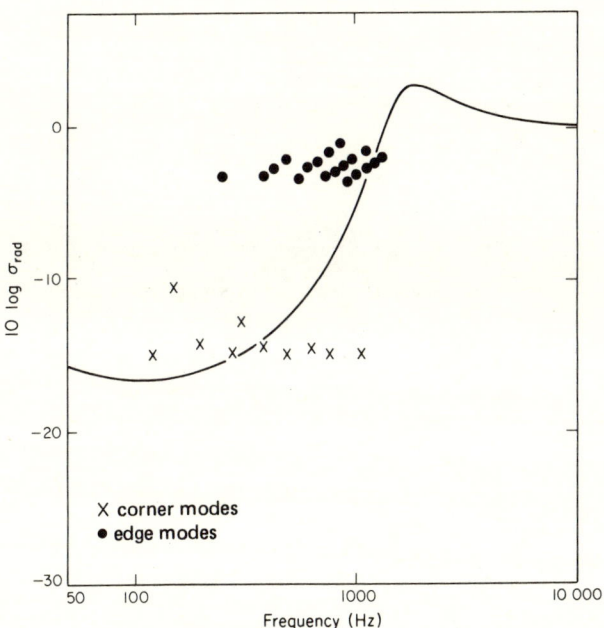

Fig. 22.18 – Radiation efficiency of a 1 metre square plate (of ¼ in thickness) when it is excited at one of its natural frequencies. (These values are calculated by using the formulae given in NASA Report CR.160 [22.35]).

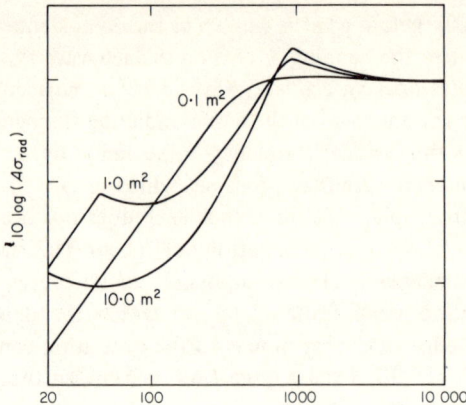

Fig. 22.19a $-\frac{1}{2}$ in thick steel or aluminium plates, simply supported along its edges.

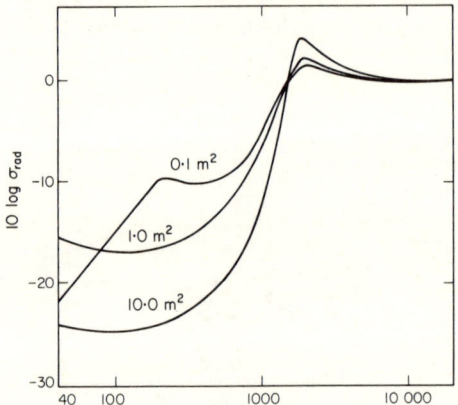

Fig. 22.19b $-\frac{1}{4}$ in thick steel plates, simply supported along its edges.

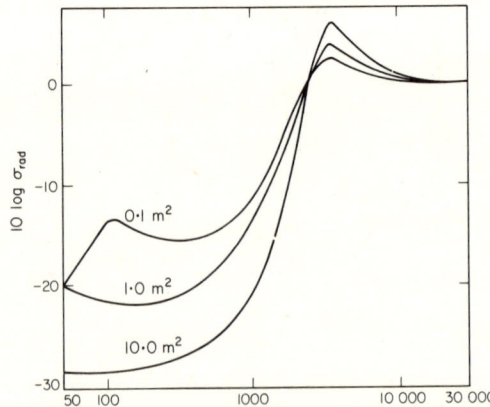

Fig. 22.19c $-\frac{1}{8}$ in thick steel plates, simply supported along its edges.

Although these have been calculated for square panels, we do not need to recognize this limitation as is shown in Fig. 22.20 which compares the results [22.8] for finite square and rectangular plates. It may be seen that the differences are insignificant in practical terms and can be ignored, particularly as these charts are more likely to be used for broad diagnostic rather than predictive purposes. This is still more true if we bear in mind the need to add an A-weighting to the noise energy in each frequency band. Fig. 22.21 shows the variation of $10 \log A \sigma_{rad}$ with frequency for panels of three typical surface areas and with a thickness of a quarter of an inch, a not untypical machinery situation. The

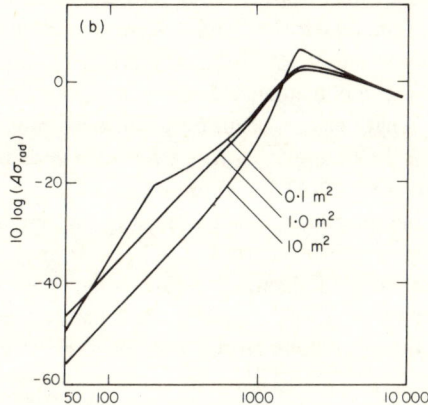

Fig. 22.20 – Radiation efficiencies of $\frac{1}{8}$ in thick square plates and rectangular plates of equal areas. Both types are simply supported along the edges.

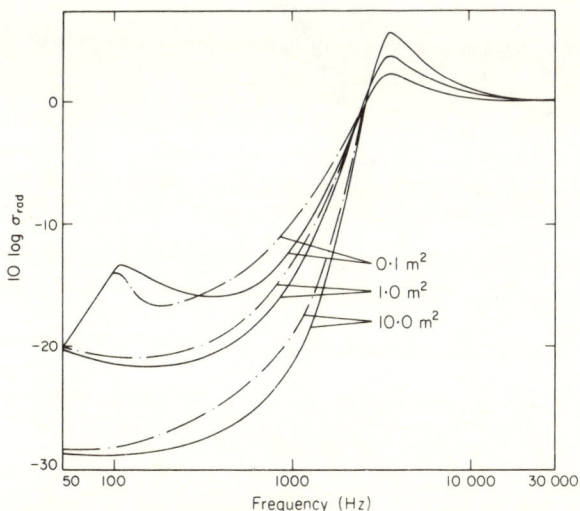

Fig. 22.21 – A-weighted efficiency of $\frac{1}{4}$ in thick plates.
– – – rectangular plates, —— square plates.

coupling of the A-weighting factor with the fall in σ_{rad} below the coincidence frequency (approx. 2 kHz) for this particular thickness of plate suggests that we can well ignore any vibration below 1 kHz unless the amplitude of vibration is very high.

The most difficult but probably most useful area of radiation efficiency calculation is that for rods, beams, tubes, etc. as they tend to be numerous and are designed intuitively to give the right stiffness rather than for their vibration qualities. Acoustically they are difficult to study since they combine acoustical flows around their circumference (ka), with flows along their lengths (if flexing), and are relatively short (not infinite). They have varying cross-sectional shapes, from circular and rectangular to I, Γ and L shapes, and vibrate in both planes. Fortunately it has been found [22.8] that the effect of finiteness is confined to conditions well below the coincidence frequency of the flexural waves in rods, and that these finite length effects occur only when the rods are radiating poorly anyway. The radiation loss factor η_{rad} (per cycle) for rods of any length can be taken to be given by Fig. 22.22 for finite circular beams, and an engineering approximation for the resultant efficiencies σ_{rad} for circular rods of various cross-sectional diameters are given in Fig. 22.23. It may be seen that to all intents and purposes, rods of diameter greater than 8 cm radiate perfectly at all frequencies above 1 kHz, though equally the curves explain the inefficiencies of radiation of such things as violin strings and telegraph wires.

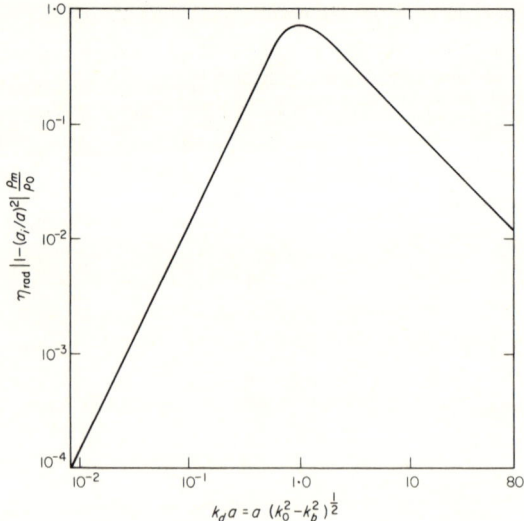

Fig. 22.22 – Radiation loss factor for long circular cylinders in flexural motion.
k = air wave number, k_b = surface wave number.

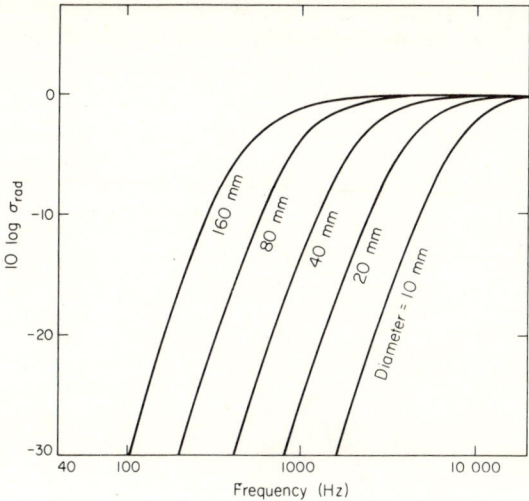

Fig. 22.23 – The radiation efficiencies of long circular beams in flexural vibration.
$\sigma_{\text{rad}} = 2ka/\pi(k_d a)^2 |\bar{H}_1(k_d a)|^2$.

22A.14 THE RADIATION EFFICIENCY OF RECTANGULAR AND ELLIPTIC BEAMS

It is often inconvenient or wasteful to use circular members in machinery design, and the radiation properties of flat beams of differing breadth-to-depth ratios are of equal if not greater design interest. Approximate methods have been evolved [22.8] to determine their acoustic loss factors, but a greater insight into the systematic variation can best be obtained by examining the radiation efficiencies of a family of elliptic cylinders whose eccentricities are such as to allow them to represent a series of beams which on the one hand approximate to flat plates of finite width, or are circular or elliptic with their major axis along the direction of the vibration. The method of attack [22.9] remains the same although the Helmholtz wave equation is written in elliptical coordinates. The solution obtained for the radiation loss factor (η_{rad}) involves Mathieu Functions which can be expressed in summation of product of Bessel functions multiplied by suitable coefficients. Once again the radiation efficiency cannot be described explicitly in terms of a single variable. Fig. 22.24 shows a graph of

$$\eta_{\text{rad}} \frac{\rho_m}{\rho_o} \left(1 - \frac{a_i b_i}{ab}\right),$$

the radiation loss factor plotted against $k_d a$ or $k_d b$, the relevant axis being that at right angles to the direction of oscillation. Here a_i and b_i are the inner semi-axes while a and b are outer semi-axes of the elliptical beam. From this figure the radiation efficiency has been obtained for solid elliptical cylinders with different breadth-to-depth ratios. The engineering chart for $10 \log \sigma_{\text{rad}}$ against frequency is shown in Fig. 22.25 for a series of eccentricities of rod with one particular major axis: any extrapolation to hollow pipes can be made in accordance with Fig. 22.24.

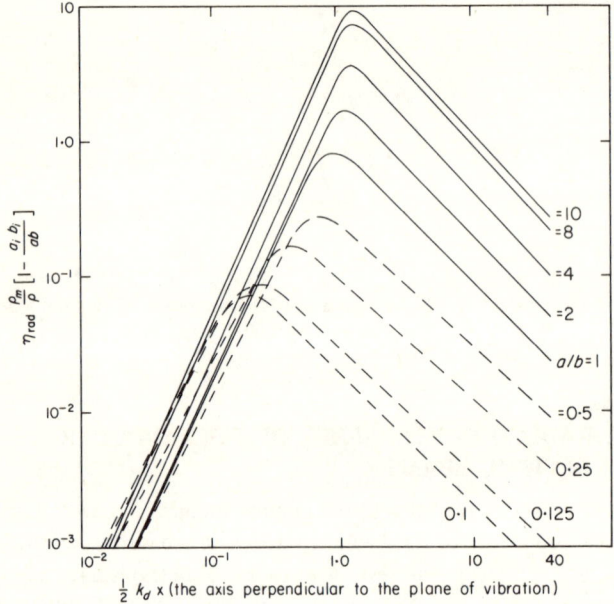

Fig. 22.24 – Acoustic loss factor η_{rad} for elliptical cylinders.

It may be seen that as the flatness of the rods is increased, the radiation characteristics approach more and more those of large flat plates, a rapid increase occurring above coincidence. Thus flat plates with open edges can, to some extent, be treated as simply supported plates but with the coincidence frequency raised in proportion to the equivalent cylinders circumscribing their edges.

As elliptic rods are not often used as such, but have been examined here as a means of studying the acoustic properties of rods arbitrary cross-sectional shapes, it is important that we generalize Fig. 22.24 in such a way as to allow its use for rods and pipes generally. To do so, we must replot it in terms of generalized parameters related to acoustic path lengths and effective radiation areas.

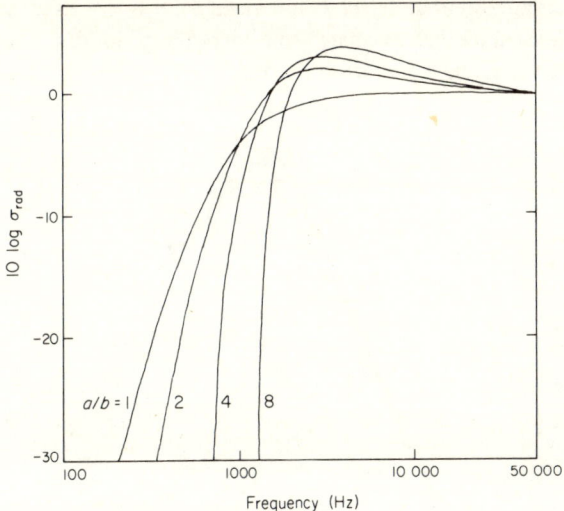

Fig. 22.25 – The radiation efficiencies of elliptical cylinders of different aspect
ratios but of constant major axis of 80 mm.

As η_{rad} is defined as $W_{rad}/\omega E_s$ where E_s is the sum of the kinetic and strain
energy of vibration at any moment which is equal to $M\langle\overline{v^2}\rangle$ where M is the mass
of rod per unit length, it is clearly going to vary with the degree to which the
major surface faces the direction of vibration. A quantity which is more likely to
be invariant with the eccentricity of the ellipse is therefore $\eta_{rad}\cdot(b/a)$. Fig. 22.26
shows the much more universally applicable measure of the variation of the
radiation loss factor, with wave-number, by plotting

$$\eta_{rad}\,\frac{\rho_m}{\rho_o}\cdot\left(1-\frac{a_i b_i}{ab}\right)\frac{b}{a}$$

against $k_d \times$ circumference and a recommended master curve suitable for use
with arbitrary cross-sectional shapes is given in Fig. 22.27. Here a/b can be taken
as the ratio of the frontal areas of the rod in, and at right angles to, the direction
of oscillation.

22A.14.1 Rectangular beams

Johnson & Barr [22.10] have calculated the approximate radiation loss factors
for a series of rectangular bars and have plotted η_{rad} for a series of width-to-depth
ratios. Without access to their detailed calculations it has not been possible to
obtain great accuracy from their figures in non-dimensionalizing their curves
to match the variables of Fig. 22.26; however, the agreement between their
points and the master curve of Fig. 22.27 shows that this latter figure can be
used for rectangular rods and pipes with no great loss of accuracy, provided that

the relevant circumferential lengths and frontal areas are used, and provided that the radiation does not emanate from local flange modes. Some validation of this procedure is given in the next section.

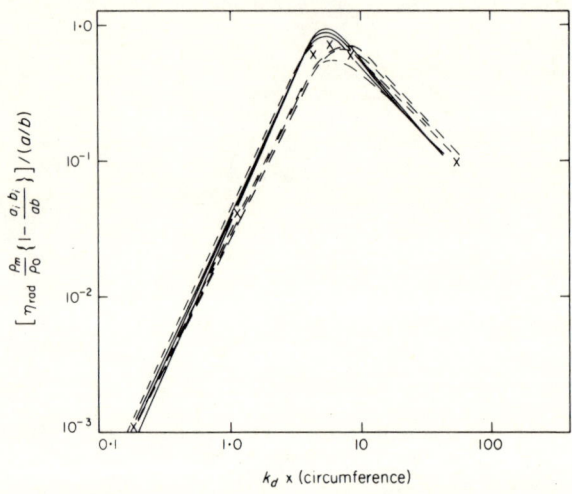

Fig. 22.26 – Normalized loss factor for elliptical beams of various aspect ratios
—— a/b = 10, 4, and 2, – – – circle, — — a/b = 0.1, 0.25 and 0.5, x points for
master curve.

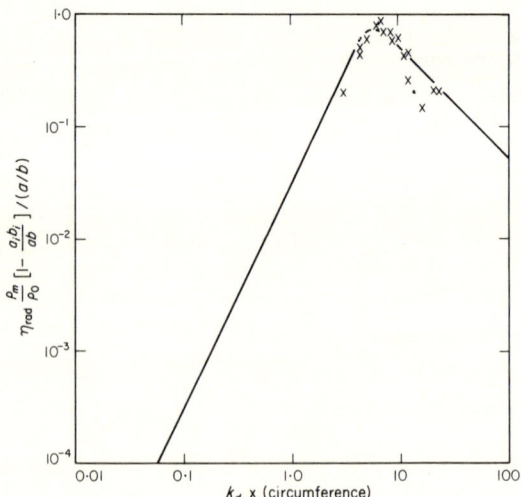

Fig. 22.27 – Master curve for elliptic and rectangular beams x points for
rectangular beams of different aspect ratio.

22A.15 VALIDIFICATION EXPERIMENTS OF RINGING NOISE PREDICTIONS

The value of the above expressions and the corresponding engineers' curves for radiation efficiency rests on the extent to which they can be validated for real cases, extending from elementary components of shapes close to those investigated analytically to assemblies of parts and to whole machines. This section is concerned with verification experiments on practical structural members which are like those which have been treated theoretically but which suffer the usual experimental deficiencies of finiteness, slight shape differences, and of liability to local distortions not treated in the theory. We have therefore examined experimentally and theoretically the very short cylindrical hammers referred to in [22.8], to see whether the observed ringing noise can be accounted for adequately by the observed flexural motions of short cylinders; we have also studied the vibration and radiated noise from long circular cylinders twenty diameters in length both straight and including a right angle bend excited continuously in a reverberant chamber (a configuration which is not untypical of constructional practice), that from two rectangular plates of different length-to-breadth ratios not bound by rigid baffles, and the expected and observed radiation from square I· beams vibrating in both planes at a series of natural frequencies, and at the frequencies corresponding to local vibrations of the webs and flanges.

22A.15.1 The ringing noise of short circular cylinders

The ringing noise energy from the short circular cylinders described in [22.5] was calculated from experimental observations of the initial acoustic levels coupled with measurements of the rate of decay of the levels of vibration of the cylinders at each identifiable longitudinal and flexural frequency, as the rig could not be installed in an anechoic chamber. Our tentative conclusion that acceleration noise energy was exceeded significantly at frequencies corresponding to flexural modes of oscillation must now be justified by the production of evidence that such flexural oscillations with measured mean square velocity amplitudes would be expected to give the recorded sound outputs. Such agreement would not be expected to be close on very short cylinders where end effects will overshadow that calculated from flexural movements.

Table 22.4 shows the calculated and measured acoustic outputs for the three cylinders, the decay time being the measured vibrational decay time in both cases. It must be pointed out that the small amount of structural damping involved was determined by the exact fit of the spring support system and was found to be different from that obtained in the ringing experiments referred to in [22.5]. Nevertheless, the comparison is a fair method of validating the values of σ_{rad} obtained for short cylinders. In each experiment it was found that σ_{rad} was effectively unity, and that the best method of indicating predictability lay in indicating in decibel form the differences in measured and calculated L_{eq} for the fundamental flexural ranging of the cylinders. It may be seen in Table 22.4 that for the two thinner cylinders agreement is complete within 1-2 dB, but that for the 15 cm (6 in) cylinder, the radiated noise exceeds that which we have calculated from $\langle \overline{v^2} \rangle$ by ten decibels. This has to be attributed to cylinder end effects, the cylinders not being mounted on their flexural axes, and to experimental difficulties in a condition where flexural movements were very small and the total ringing noise energy was very little greater than the acceleration noise (in this repeat experiment). Our conclusion that noise arising from impacts often arises from flexural motions of flexing structures is nevertheless upheld by the evidence of the thinner cylinders.

Table 22.4
Radiated energies for impact velocity of 0.7 m/s

	E_r (calculated) from $\sigma_{\mathrm{rad}}, k, \langle v^2 \rangle$ (1)	E_r (measured) from $P_o, k, A(\theta)$ (2)	$10 \log \dfrac{(1)}{(2)}$
75 mm dia. cylinders	6.1×10^{-3}	7.6×10^{-3}	-1.0 dB
100 mm dia. cylinders	1.7×10^{-3}	1.1×10^{-3}	$+1.8$ dB
150 mm dia. cylinders	8.9×10^{-6}	1.9×10^{-4}	-13 dB

22A.15.2 Finite circular rods in bending
Circular rods are handled in factories by rolling, impact between each other being a constant source of high level noise. They constitute frameworks of many textile machines and gantry frames, and because of their high stiffness in every direction, circular, rectangular, or square box structures are very popular as machinery structures generally, either as solid members or in a hollow form. The acoustic radiation from such members depends only on the external geometry once the flexural patterns have been prescribed, and these are a function of the cross-sectional radius of gyration. Thus in validating the expressions we have outlined earlier for solid circular, elliptic, or rectangular rods, this may be assumed to provide validation for hollow rods as well, provided that local vibrations of individual webs do not dominate.

Three sets of solid circular cylinders have been vibrated at their various resonant frequencies in a reverberant room [22.8]. The choice of resonant or preferred frequencies arose from the need to develop a large enough amplitude of vibration with limited excitation sources, and the need to establish the exact nature of each mode. Figs 22.28a, b show comparisons of the measured and predicted values of σ_{rad} with frequency for two straight solid circular steel cylindrical bars of length-to-diameter ratios 40 and 20 respectively. Also compared with the same theory are the experimental values of $10 \log \sigma_{\mathrm{rad}}$ for rods of similar lengths containing right angled bends in their centres.

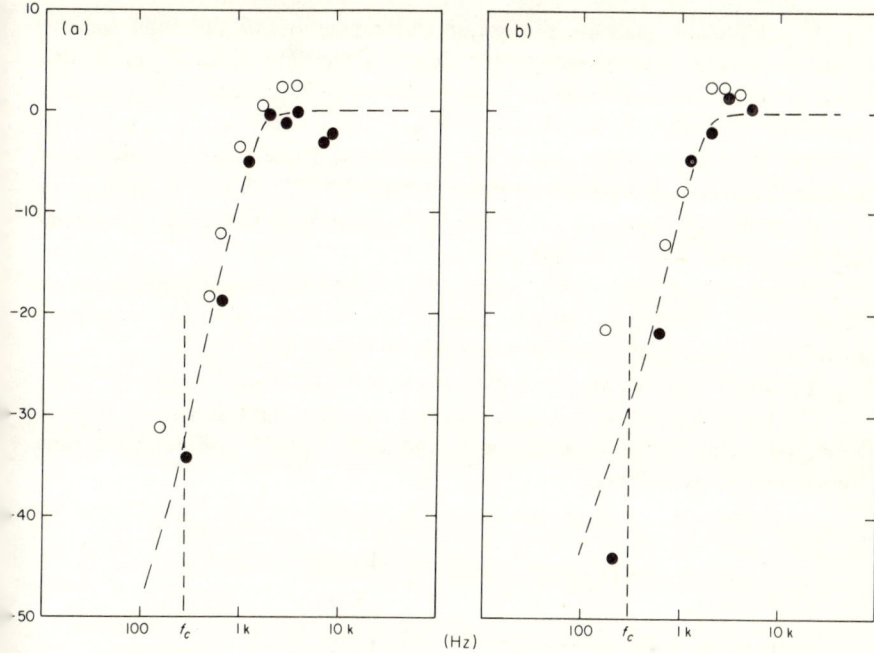

Fig. 22.28 – a, measured radiation efficiency vs. frequency for a 2 m long × 51 mm diameter steel bar; b, measured radiation efficiency for a 1 m long × 51 mm diameter steel bar; • straight bar; ○ with 90° bend; – – – theory for a straight bar.

Four or five points of agreement need to be sought between experiment and theory if such curves are to be used to predict the radiated noise once the mean square vibration velocities are known. Well above coincidence, which frequency must be predicted accurately, experimental values of zero for $10 \log \sigma_{rad}$ are required; the rapid fall-away with decreasing frequency of vibration should occur not from the flexural coincidence frequency, but from a value corresponding to $k_d a = 1$, and the rate of fall should correspond to that predicted from the theory. Effects due to the finiteness of the length would not be expected to be significant in these experiments, but lastly modifications to σ_{rad} arising from local acoustic couplings near the bends must be seen to be small.

Above the coincidence frequency, and for a radiation efficiency range of 1000:1, quite satisfactory agreement may be seen concerning all these four points. Below the flexural coincidence frequency agreement is less satisfactory, but at this frequency the accuracy is poor, both because of the low acoustic output and uncertainties in technique in the reverberation chamber; in any case finite rod effects are to be expected at this low level of radiation efficiency as explained earlier. It can be said therefore that these experiments justify the use of the engineers' charts under all but low noise radiation conditions.

If component analysis as a method of prediction of the total acoustic radiation is to be generally acceptable, details of the configuration and acoustic cross-coupling and corner effects must be shown to be unimportant, at least in conditions of high radiation. Further measurements were therefore carried out with the same bars when bent in half at right angles. These results are also shown in Figs 22.28 a, b. It may be seen that agreement with theory is again good, (within 4 dB) for both straight and bent bars above coincidence, but greater differences occur below coincidence.

The acoustic cross-coupling close to the corner must depend upon the exact nature of the vibrational mode. In these experiments the rods were suspended at the bend, and two types of vibrational mode were found:

(a) antisymmetric, where the bend remains at right angles

(b) symmetric, where the bend does *not* remain at right angles.

Vibrational amplitudes, however, were the same in each half for all natural frequencies. (Figs 22.29 a,b).

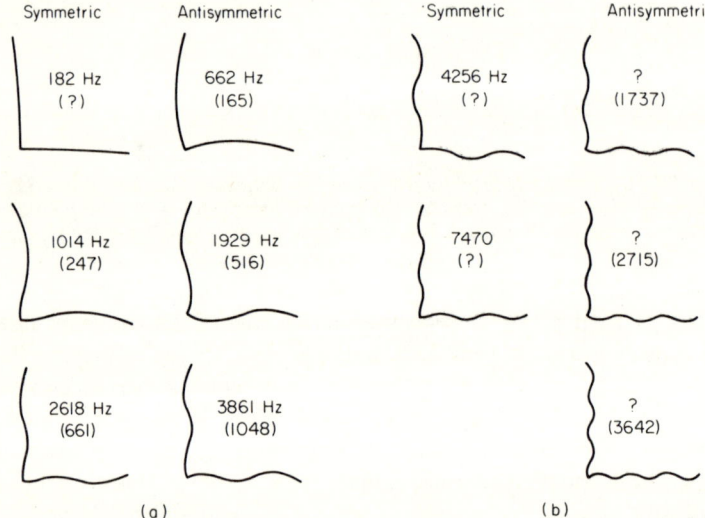

Fig. 22.29 – a, measured mode shapes for vibrating bent cylinders 1 m long (frequencies for corresponding mode shapes of 2 m cylinders are bracketed; b, measured mode shapes for vibrating bent cylinders.

Examination of Figs. 22.28 a,b shows the measured radiation efficiencies for the bent rods to be somewhat higher than those for the straight ones above f_c, but that there is no consistent difference between the two modal types described above. The errors involved in such a broad approach are of the same order as these differences. It may be concluded therefore that to the same order of accuracy as is feasible in the analysis of any practical configuration, straight rod computations can be used, particularly in high radiation regimes, even when bends occur in their geometry.

22A.15.3   Rectangular plates

As part of our validation programme, two rectangular plates, one of length/ breadth ratio 2, the other of ratio 6, have been tested in a freely hanging condition, and the measured radiation efficiencies are shown in Figs 22.30 and 22.31. Agreement with the theoretically estimated values for simply supported baffled plates would be expected to be reasonably good above coincidence where each element radiates independently, but agreement below coincidence would be expected to be progressively poor because the freedom for cancellation to occur around the edges would suppress corner and edge mode radiation. Thus the general characteristics below the critical frequency would be expected to resemble more those of a flat rod than those of a simply supported plate. This tendency may be observed in Figs 22.30 and 22.31, the simply supported baffled plate theory overestimating the radiation efficiency below the critical frequency by over 10 dB in some instances. Even so, flat plate theory predicts the critical frequency reasonably well for free plates, with the maximum radiation occurring at a somewhat higher frequency (say $f = 1.5\,f_c$) as anticipated by the flat rod theory illustrated in Fig. 22.25.

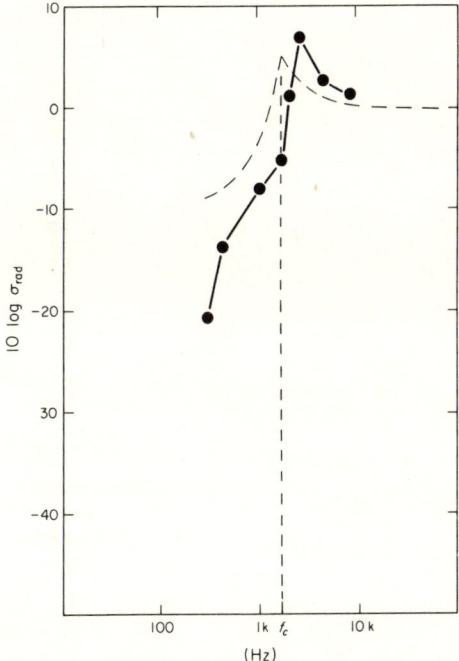

Fig. 22.30 – Measured radiation efficiency of a freely suspended rectangular plate, 0.609 m × 0.305 m. ● measured values, – – – theoretical curve for simply supported baffled plate.

The two lowest excitation frequencies indicated in Fig. 22.31 do undoubtedly represent rod rather than plate modes, with no distortion across the plate, so that the experimental measurements can be taken to indicate the acceptability of the flat rod theory in principle even though the theoretical calculations have not been extended to such an extreme case.

No experiments were carried out on plates with simply supported or clamped edges in a large baffle, as it was felt that such experiments have been done and reported elsewhere. Unbaffled plates appear in many industrial processes, including vibration from workpieces, lathe frames, and from stillages and metal containers generally; it was to study these cases that our present plate experiments were designed.

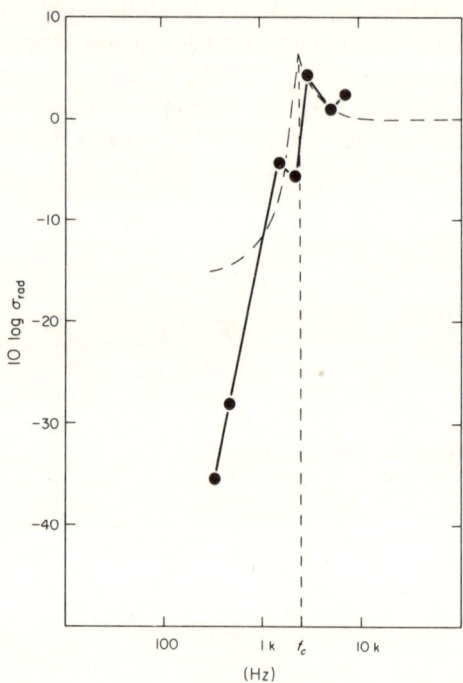

Fig. 22.31 – Measured radiation efficiency of a freely suspended rectangular plate 1.37 m × 0.22 m, 4.75 mm thick (long strip plate). ● measured value, – – – theoretical curve for simply supported baffled plate.

22A.15.4   The radiation efficiency of I beams

No analytical method exists by which the radiation efficiency of complex industrial beam shapes, such as I and Γ beams, can be calculated, and the value of Fig. 22.27 must depend upon the degree to which it can be used for such a prediction, irrespective of the analytical niceties involved.

Several simple approximations can be made based upon different physical concepts of the nature of the radiation sources. The first is to assume that the flow in the recesses between the two flanges is entirely reactive, both flanges moving in phase with the same velocity at all times. Thus, provided the progressive flexural wave speed and the same enveloping circumferential lengths are obtained, the same radiation efficiency should be obtainable as from a completely enclosed circular or square bar. The provision of the same coincidence velocity can be obtained by choosing a different volume density or different hollowness to give the same sectional radius of gyration, while also providing the same peripheral length.

If it is suspected that radiation occurs as a result of individual flange motions, unrelated to each other, the approximations can be related to the radiation from elliptic cylinders of high eccentricities.

We have therefore tested an I beam of depth/width ratio 1, exciting it flexurally in both planes, and have compared the experimentally determined values of σ_{rad} with those obtained from Fig. 22.27 for the following models having the same flexural wavespeed and modeshapes, (Fig. 22.32).

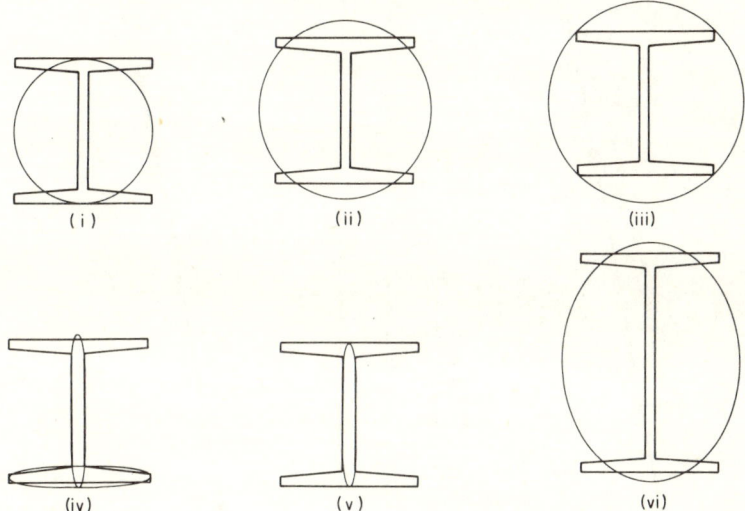

Fig. 22.32 – Models for the prediction of radiation efficiency of I-beams.

(i) a circular cylinder of the same diameter as the beam width and depth.

(ii) a circular cylinder having the same periphery as a square of the same beam depth and width.

(iii) a cylinder just circumscribing the beam.

(iv) an elliptical beam having the same periphery and a/b ratio as a single flange or web.

(v) an elliptical beam inscribed within the flange or web.

Figs 22.33 a, b show a comparison between experimental values for an I beam vibrating in its direction of greatest strength, and theoretical values for the models described above. Agreement is reasonable, the best fit being for the elliptical models, although an error of 10 dB can arise at low frequencies. A much higher radiation efficiency than would be expected was obtained at the frequency corresponding to $k_d \ell = 1$, and this is difficult to explain. Values of σ_{rad} greater than unity are usual on baffled plates, but these rapidly fall to unity with increasing frequency. Unfortunately confirmation at higher frequencies could not be obtained, as local distortion modes of the flanges began to appear which cannot be expected to be predictable without detailed knowledge of their nature. Fig. 22.34 shows one typical structural mode of each type as measured experimentally and the variation of σ_{rad} in this frequency range. Further examination of local modes will be carried out and this discrepancy studied further.

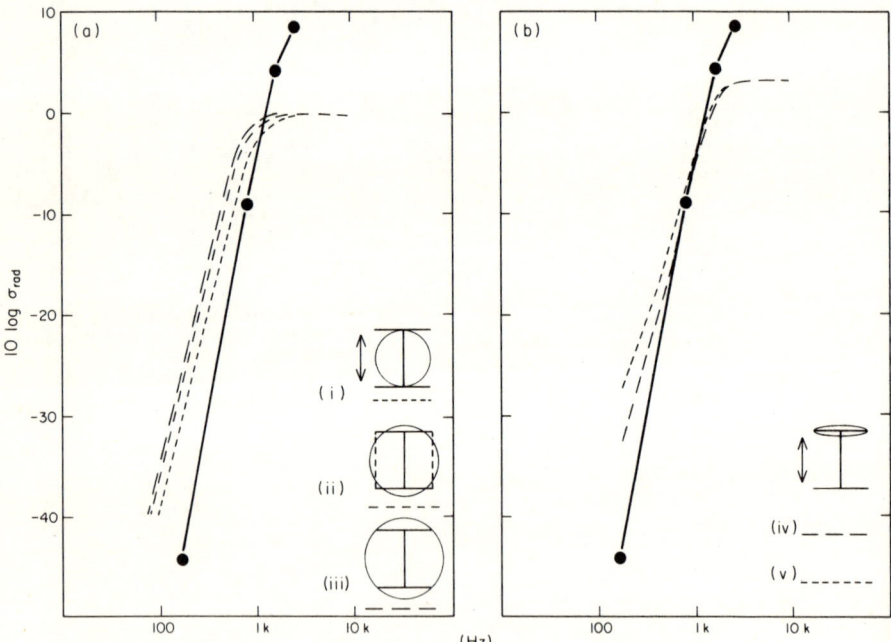

Fig. 22.33 – Measured radiation efficiency of an I-beam compared to (a) theoretical values for three cylindrical modes (stiffest direction), (b) theoretical values for 2 elliptical models (stiffest direction).

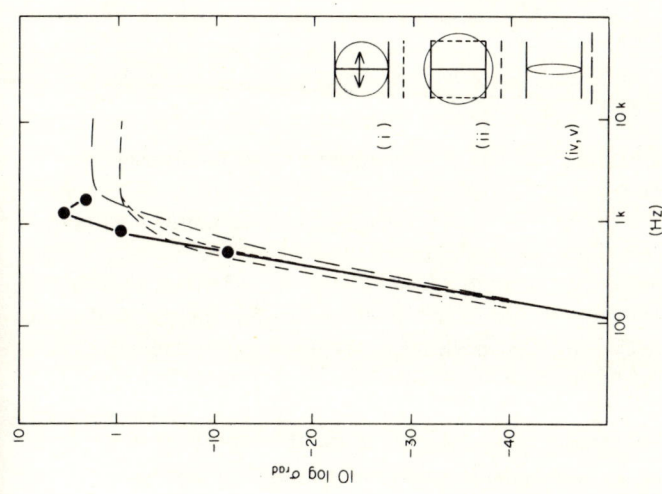

Fig. 22.34 – Typical structural modes for an I-beam with measured radiation efficiencies for each type.

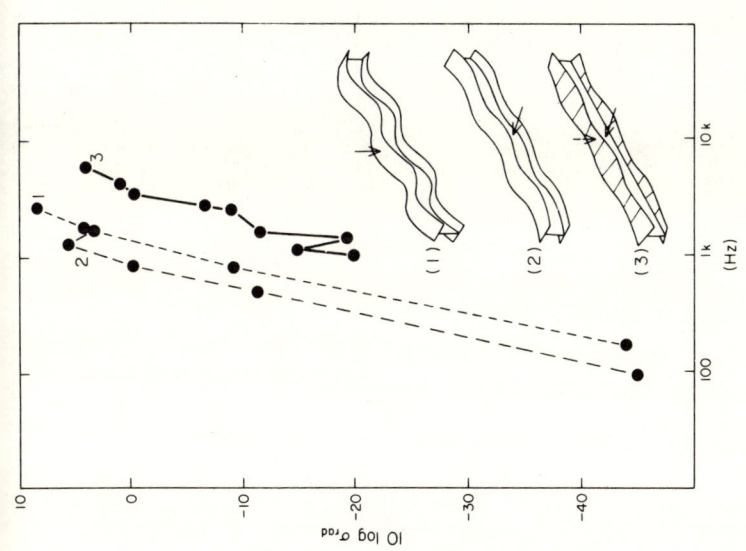

Fig. 22.35 – Measured radiation efficiency of an I-beam compared to theoretical values for 3 models (least stiff direction).

Fig. 22.35 shows a comparison between theoretical models and experiment when the I beam is vibrating in its weaker plane. Agreement here is much more satisfactory both in respect of frequency variation, and in absolute terms. Such a good agreement is to be expected in that the flanges are acting as baffles and that the representation is closer to that of a rectangular rod than is true in the comparison shown in Fig. 22.33 where assumptions had to be made regarding the radiation properties of the air gaps.

22A.15.5 The radiation efficiency of a rectangular frame of square box type members

As part of another research programme concerned with the noise radiation from fabricated steel containers, or stillages, as they are called, rectangular base frames consisting of welded box girders were made up and tested for their basic radiative properties. As such a configuration represents an industrial combination of square box beams with right angled corners, the same reverberation room technique was used to compare their radiation efficiencies with that predicted from straight rod theory. Fig. 22.36 shows this comparison for a 0.7 m square frame of 38 mm square box section steel, with the theoretical axes for cylinders of various diameters and the same flexural wavespeed and modeshapes.

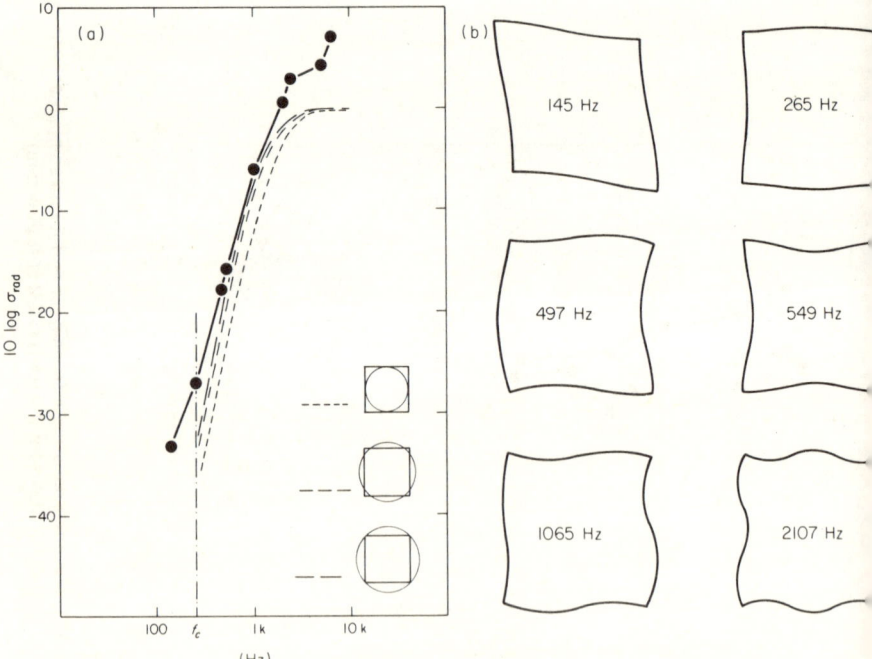

Fig. 22.36 – a, measured radiation efficiency of a 0.7 m square steel frame of 48 mm square box section compared with theoretical curves for 3 cylindrical models; b, measured mode shapes of vibrating square frame (48 mm square hollow section).

While the shape of the variation with frequency is predictable and qualitative agreement is good, there remains the frequency shift and the excessive maximum radiation to explain. Further studies are continuing, even though the initial purpose of the calculations, to show that frame noise in stillages was small, is still acceptable.

22A.16   ENERGY ACCOUNTANCY AND THE PREDICTION OF L_{eq} IN ANY FREQUENCY RANGE

Having studied acceleration noise, and the radiation efficiency of vibrating components, it is now possible, by choosing the relevant values of the A-weighted radiation efficiency, to relate directly the instantaneous noise output at any frequency from any machine or component of a machine to its vibration level averaged over a short period and over the component structure. To obtain an $L_{A\,eq}$ contribution of noise energy per event, the vibration levels must be recorded against time and the total time integrated vibration level at any frequency related to $L_{A\,eq}$ at that frequency. This can be done, and some examples will be referred to later. However, more can be gained in terms of diagnostic techniques if the contribution to $L_{A\,eq}$ per event in an impact machine can be related, at any frequency, to the initial impulsive vibrational energy or force input and to the damping and bulkiness characteristics of the components, or of the machine as a whole, expecially if these can all be separated out in the analysis and treated individually.

We therefore complete this part of the chapter by examining in broad terms what happens to that function of the input energy W_{input} escaping from the work area into the machine, into the workpiece, or into any particular component which is being studied as a result of a single impact in the machine. We do not consider the energy escape into the floor, the floor being considered either to be isolated or as an additional component of the machine with its own radiative and damping properties.

22A.16.1   The energy accountancy equation

Impact machines function by building up energy slowly and distributing it rapidly into the workpiece or throughout the machine. Thus, in a drop hammer, the energy is contained in the hammer (tup) and is quickly transferred into the workpiece or into the machine structure. For soft blows, ninety-five per cent of the energy is absorbed in deforming the billet. For the final coining blows, ninetyfive per cent of the energy is distributed into the structure as vibrational energy and into the ground. On a punch press, the structure is loaded relatively slowly, and a rapid redistribution of strain energy follows fracture.

The noise radiated from a drop hammer has a typical signature of the kind shown in Fig. 22.37. The first double pulse is related to the forced deceleration arising from the tup coming to rest, the 'ringing' noise arising for the vibration which follows from that part of the stored vibrational energy which is dissipated acoustically as the structure vibrates. As the latter must depend upon how quickly the internal damping of the structure and the transmission into the floor can dissipate the vibrational energy of the machine, the first 'acceleration' noise and the subsequent 'ringing noise' must be treated as separate contributors to L_{eq} or L_{Aeq} even though they are related in terms of vibration characteristics.

As the 'quasi-steady' 'ringing' noise dominates in the determination of L_{eq} (though acceleration noise determines peaks) let us deal with the parameters involved in the energy balance accountancy associated with ringing.

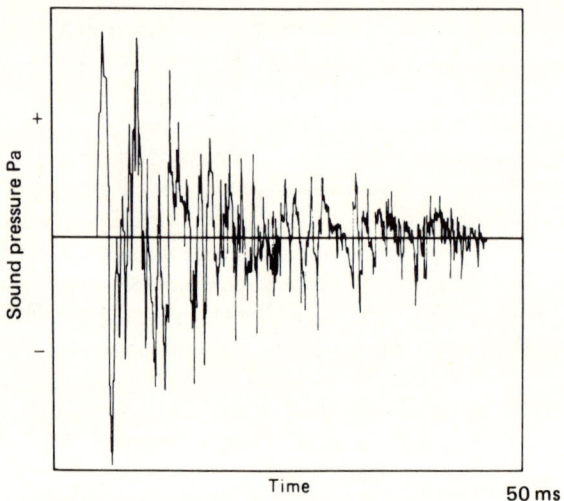

Fig. 22.37 – Drop hammer noise signature.

22A.16.2 The energy balance equation in terms of the vibrational escape energy

As a large plane plate vibrating with a mean square normal surface velocity $\overline{v_n^2}$ radiates per second a noise power W_{rad} which is equal to $\rho_0 c S \, \overline{v_n^2}$ (where ρ_0 is the density of the surrounding air, c is the speed of sound in air, and S is the plate radiating surface area) we define a radiation efficiency 'σ_{rad}' for any other body such that, for the body, the radiated sound power W_{rad} is $\sigma_{rad} \, \rho_0 c S \, \langle \overline{v_n^2} \rangle$ where ‾ represents an average over many cycles of the square of the surface normal velocity and $\langle \rangle$ represents an average over the surface. If the averaging time chosen is δt then $\sigma_{rad} \, \rho_0 c S \, \langle \overline{v_n^2} \rangle . \delta t$ is the power radiated in this time. The radiation efficiency σ_{rad} becomes unity when the vibration is so fast that each part of the body can be considered as a separate and independent radiator. As energy can be put back into the body in the so-called 'near field', which extends for a distance of about half a wavelength from any point of emission, σ_{rad} will depend upon the ratio of the circumferential length of a body to the wavelength; or in the case of a plate, to the ratio of the speed of a flexural wave along the surface to the speed of sound in the adjacent air (see sections 22A.12–22A.15).

The fraction of internal energy in the structure dissipated into heat per radian of vibration is called the damping factor η_s and tends to be constant with frequency. The rate of dissipation of vibrational energy per second is therefore $2\pi f.\eta_s \times E_{vib}$ where E_{vib} is the instantaneous internal vibrational energy at any moment. For a body of mass M, this energy is shared at any moment between elastic and kinetic energy so that for plate-like and rod-like structures (which comprise most machine parts) it is usual to describe the average vibrational energy in a structural component as $M \langle \overline{v_n^2} \rangle$ or $\rho_s S \langle \overline{v_n^2} \rangle$ where S is the radiating surface area, and ρ_s is the surface density ($\rho_m d$) where d is the average 'thickness' of the component.

For any single impact and energy escape E_{escape} we therefore have equations which describe the total energy radiated as sound (E_{rad}), that absorbed in the structure (E_{struct}), and the balance between these quantities and the energy loss to the ground and machine work, viz:

$$E_{rad} = \sigma_{rad} \rho_0 c S \int_0^\infty \langle \overline{v_n^2} \rangle \, dt$$

$$E_{struct} = \rho_m d \eta_s S \, 2\pi f \int_0^\infty \langle \overline{v_n^2} \rangle \, dt$$

and

$$E_{escape} = E_{input} - E_{work} - E_{ground} = E_{rad} + E_{struct}$$

$$= [\sigma_{rad} \rho_0 c S + \rho_m S \eta_s d \, 2\pi f] \int_0^\infty \langle \overline{v_n^2} \rangle \, dt \qquad (22.1)$$

and hence

$$\frac{E_{rad}}{E_{escape}} = \frac{\sigma_{rad}}{\left(\sigma_{rad} + \dfrac{\rho_m}{\rho_0 c} 2\pi f \, d\eta_s\right)} = \frac{\sigma_{rad}}{\sigma_{rad} + 1.23 \, d\eta_s f} \qquad (22.2)$$

for steel structures, d being measured in centimetres.

If there is no structural damping in the machine or components, η_s is zero and $E_{rad} = E_{escape}$; that is, the radiated sound energy is independent of the radiation efficiency of the surfaces. Under such circumstances, E_{rad} can only be reduced by isolating such components from the source of vibrational energy, that is, by reducing E_{escape}. Examples which come to mind are gearbox casings, flywheels, some guardrails, hammer anvils, machines mounted on girders, and to a greater extent than is generally realized, concrete floors themselves. In such cases, even small amounts of damping can be beneficial if the radiation efficiency is low; that is, if $\sigma_{rad} \ll 1$.

In most practical fabricated machinery structures, however, it is seldom that a figure of η_s less than 0.01 occurs, with figures as high as 0.05 being measured in punch presses. Under these circumstances ($\eta_s = 0.01$) and for typical structural thickness of 1 cm, the value of the expression $1.23 d.\eta_s.f$ will be 1.23 at 100 Hz, 6.15 at 500 Hz, and 24.6 at 2 kHz. As σ_{rad} is likely to be significantly less than unity below 500 Hz, no great loss in accuracy will occur and a far easier understanding of the general laws of noise emission will be obtained if we ignore σ_{rad} in the denominator and write in terms of power per event

$$\frac{E_{rad}}{E_{escape}} = \frac{\sigma_{rad}}{1.23\,d.\,\eta_s.f.} \quad \text{or in logarithmic A-weighted dB terms,}$$

$$L_{Aeq} \text{ per event at any frequency } f = 10 \log E_{escape}\,(f) + 10 \log (A\sigma_{rad})$$
$$- 10 \log f - 10 \log \eta_s - 10 \log d$$
$$+ \text{Constant} \qquad (22.3)$$

If E_{escape} is specified in terms of a percentage band width, so will E_{rad}.

This method of writing the energy balance equation has the advantage that the contributions arising from changes, respectively, in the vibrational escape energy per event, E_{escape}, the time history of this input (the pulse shape), or the spectral content of the vibration, the acoustically frequency sensitive terms, the damping factor, and the bulkiness factor or average thickness (internal energy is related to the volume of material, the radiated sound to the surface area) can be added linearly and studied separately. Thus we have:

$$L_{Aeq}(f) = 10 \log E_{escape}\,(\text{total}) + 10\log(s.c.)$$

↑	↑	↑
'Deafness' contribution per event	Total escape energy into structure	Spectral content at any frequency

$$+ 10[\log A\sigma_{rad} - \log f] - 10\log \eta_s - 10\log d + \text{Constant}$$

↑	↑	↑
A-weighted and frequency sensitive joint radiation/ damping efficiency	Damping level	Machine bulkiness or solidity

(22.4)

For a given component structure geometry, these terms are independent of each other, and the effects of machinery change can be examined term by term and summed linearly. Thus the first term on the right-hand side represents the total energy escaping into the structure per impact; the second term depends upon the impulse shape of the vibration escape energy and of the structural response, and can be described in the form of a spectral distribution curve; the third term includes the radiation efficiency of any component of the structure, allowance being made for the sensitivity of the ear and for the shorter ringing period of the higher frequency vibrations; the fourth term represents the structural damping (per radian of oscillation) associated with the component and its interface with other parts of the machine; the fifth term represents the ratio of the radiating surface area to its volume or mass, i.e., the average thickness of webs, flanges, plates, etc.

This formulation of the energy dissipation mechanisms has most relevance to machinery noise diagnostics rather than to analysis; it tells us at once that apart from the use of enclosures the art of noise reduction on a machine of previously prescribed geometry and therefore of unalterable radiation efficiencies of each member lies (a) in the reduction in the magnitude of the excitation and (b) in the mismatching of its frequency spectrum with that of the modified radiation efficiency of those components which carry this vibration. Fig. 22.38 shows some typical curves of the 'modified' radiation efficiency [$10 \log A\sigma_{rad} - 10 \log f$] against frequency for solid bodies, plates, and bars. It may be seen that because structural damping increases with frequency so that acoustic radiation from a single impact is less prolonged, these curves reach peaks at frequencies depending upon their geometries and then fall off. As the total $L_{A\text{eq}}$ contribution to deafness exposure summates the weighted noise energy in each frequency band linearly, whereas the logarithmic additions imply the multiplication of energy levels in the formula at any one frequency, the advantage of keeping separate the frequencies of the peaks in each term is self evident (Fig. 22.39). Thus, in theory at least, there might be an advantage on some structures whose peaks in modified radiation efficiencies occur at low frequencies in hardening impacts to raise the frequency of excitation. For most machines, however, this frequency is dictated by the contact time of cold metal-to-metal impacts and cannot easily be raised. In any case transient forces excite lower modes in addition to those related to the impact time alone. In general, therefore, the art of noise reduction reduces to that of reducing vibration intensity, particularly at the higher frequencies where the 'modified' radiation efficiency is at its peak, of softening or lengthening the impacts to lower the preferred frequencies of vibration, of adding damping, particularly at low frequencies, of devising structures with lower 'modified' radiation efficiencies at the excitation frequencies, and of increasing the bulkiness factor, defined here as the average thickness of the relevant machinery members. In many instances such a 'bulkiness' factor will be difficult to define, but can be introduced into the analysis if it is remembered that vibrational energy in a machine is a function of its volume, and the sound radiated is a function of its surface area. Bulkiness must therefore in all cases be a measure of the ratio of the volume to surface area or to its stiffness.

As indicated in Table 22.5, contributions to $L_{A\text{eq}}$ at any frequency of any of these parameters can vary greatly, and a careful accountancy of noise energy change can prevent abortive changes in design. The above $L_{A\text{eq}}$ formulation has a further advantage which must not be overlooked: the help it gives to the formulation of experimental techniques of noise control. Thus, for example, it is possible to proceed upon a noise control programme based upon reducing the

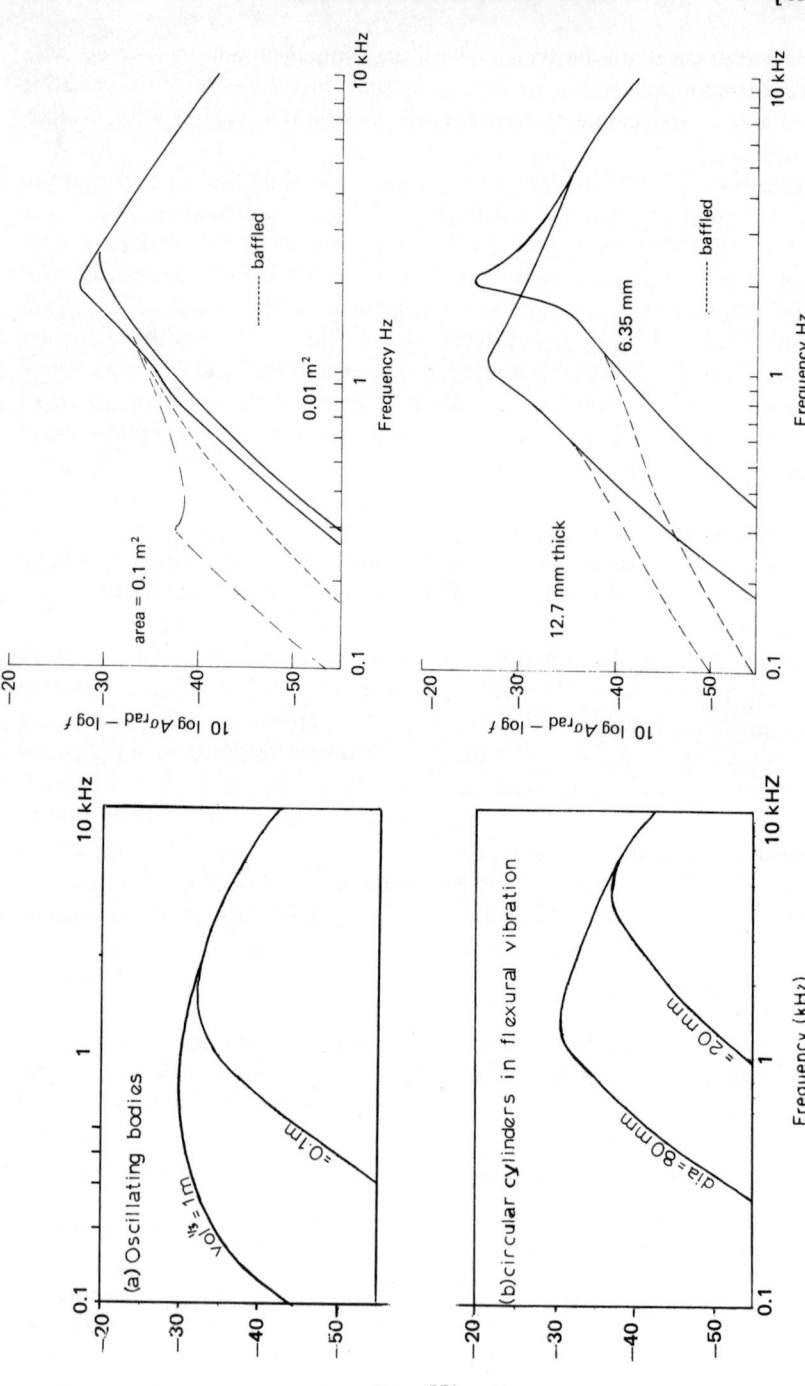

Fig. 22.38 – The variation of 'modified' radiation efficiency of

escape energy input and/or its frequency distribution without concerning oneself with the exact location of the noise sources. It can simply be taken that provided the noise is related to the energy impulse being studied, and not to a further nonlinear impulse arising from backlash in the bearings, there will be a simple relationship between E_{escape} and $L_{A eq}$.

Table 22.5

	Sum logarithmically at any freq. arithmetically to obtain L_{eq} (total)
$L_{A eq}(f) = \begin{bmatrix} 10 \log E_{escape} \text{ (total)} \\ + 10 \log \text{(spectral distribution)} \\ \text{or } 10 \log f'^2 + 10 \log \\ \text{(transfer function)} \end{bmatrix}$	Any value depending upon vibration level or f'^2 level, $(= W_{eq}(f))$
$+\ 10 \left[\log_{rad} A - \log f\right]$	-10 to -70, typically -30
$-\ 10 \log \eta_s$	$+30$ to $+10$
$-\ 10 \log d \text{ (cm)}$	0 to -10
$+\ \text{Constant}$	

$$\text{Total } L_{eq}(f) = W_{eq}(f) \quad 0 \text{ to } -60$$

If *high* damping and *low* modified radiation efficiency then $L_{eq}(f) \sim W_{eq}(f) - 60..$
If *low* damping and *high* modified radiation efficiency then $L_{eq}(f) \sim W_{eq}(f)$.

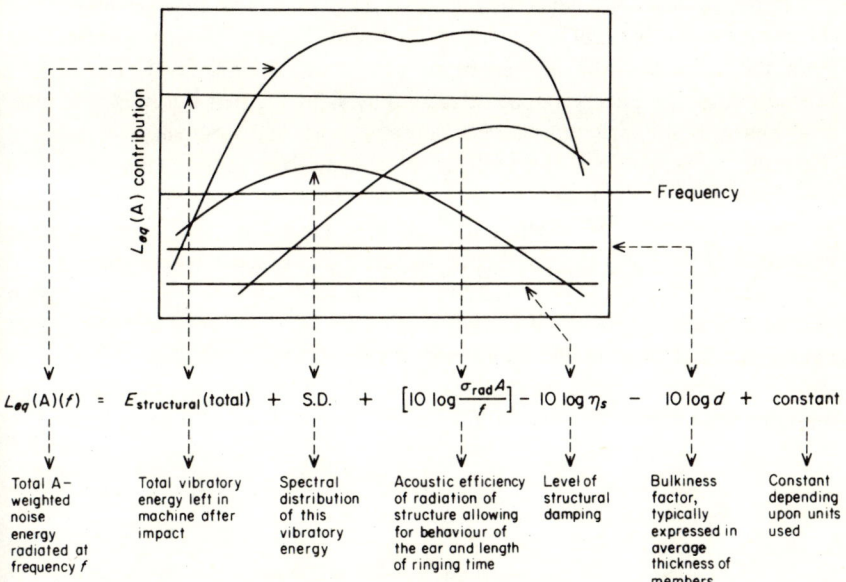

Fig. 22.39 – The make-up of machinery noise at any particular frequency (f).

22A.16.3   The energy balance equation in terms of applied force derivatives

While equation (22.4) is convenient as a diagnostic tool particularly as E_{escape} cannot be related to the initial input energy if deformation energy is required to shape or fracture the workpiece, nevertheless, it is sensible to try to recast the equation in terms of applied forces or their time derivatives, if they can be measured during and at the end of the impact time.

Such time-dependent forces or the time derivatives of such forces can be transformed into the frequency plane, and at each frequency they will be related by the multiplication of the one by the frequency raised to relevant power. If we take the forces as our relevant quantity, it will be associated with a spectral distribution curve which falls sharply from left to right (see Fig. 22.40), with an excitation level at the frequency of maximum modified radiation efficiency (Fig. 22.38) which may be 30 dB lower than the static (zero frequency) value. It is therefore more sensible in acoustical investigations to work in terms of a time derivative of the force which, while still retaining a physically comprehensible significance, will provide a flatter frequency distribution curve which is easier to relate diagnostically in the context of equation (22.4). Such a variable is the rate of change of force $f'(t)$, or for less impulsive motions some still higher derivative of the force. Physically this is the same as saying that no noise arises from the application of a time independent force to a machine, but that there will be if a time dependent force is applied whose frequency matches, to some extent, the frequency of peak radiation efficiency adjusted to provide for the effects of damping and the behaviour of the ear.

If the applied force derivative occurs over a very short time at one place in the machine, the vibrational energy built up in the machine can be calculated from the force and the movement of the point of application. The rate of dissipation of this energy can be calculated by assuming that it is predominantly absorbed into the structure as heat, and the noise radiated can be calculated from the movements of the surfaces of the machine during this time. The assumptions involved in such a procedure are the same as that of ignoring σ_{rad} in the denominator of equation (22.2), that the relation between the point excitation of the force and the average surface excitation can be ascertained.

This is done in Appendix 22.1, and the following relevant relationship obtained which relates the L_{eq} contribution at any frequency and percentage bandwidth, not only to the excitation of the structure but also to the point structural response characteristics $H_c(f_o)$ to the structural damping factor, the modified radiation efficiency, and the bulkiness or stiffness of the machine:

$$L_{eq}(A, f_o \Delta_f) = 10 \log \left[|F'_c(f_o)| \right]^2 + 10 \log \left\{ \text{Re} \left| \frac{H_c(f_o)}{j} \right| \right\}$$

$$+ 10 \log \left[\frac{A \sigma_{rad}}{f_o} \right] - 10 \log \eta_s - 10 \log d + c'. \tag{22.5}$$

The first term now consists of a logarithmic measure of the square of the rate of change of applied force when transformed into the frequency plane. Depending upon the frequency (f_0) being considered, $10 \log \left[|F'(f_0)| \right]^2$ will be determined by the magnitude of the impact and its shape. Much of the value of the formulation must therefore depend upon the degree to which the variation in magnitude and frequency distributiion can be predicted.

Meier-Dornberg [22.11] gives a simple method of obtaining the Fourier transform $G(f)$ of any transient time function $g(t)$ whose characteristics consist of a pulse time to a maximum value $g_{max}(t)$ and a maximum time derivative $g'_{max}(t)$ (Fig. 22.40a). The expression for $G(f)$ can be represented (Fig. 22.40b) by three straight lines hinged at frequencies f_1 and f_2 where $f_1 = \dfrac{g(t)_{max}}{\text{area} \times 2\pi}$ and $f_2 = \dfrac{g'(t)_{max}}{2\pi g(t)_{max}}$. Since the area under the transient time function can be approximated to $g(t)_{max} \times \dfrac{t_1}{2}$, f_1 can be taken to be $\dfrac{1}{\pi t_1}$ where t_1 is the pulse time and

$$f_2 = \frac{g'(t)_{max}}{2\pi g(t)_{max}} = \frac{1}{\pi t_2},$$

where t_2 is a notional time over which the large value of the derivative $g'(t)$ is recorded. The time t_1 of the impact force differs in different types of machine, and as often as not is sufficiently long to make f_1 low compared with the frequency for peak modified radiation efficiency. The same is not true for time t_2, however. In a diesel engine, the quick firing occurs over only a few microseconds, as does the collapse rate of cavitation bubbles, fracture times in punch presses, the impact of bearings and small solid bodies. Thus f_2 is often recorded in the kilohertz frequencies.

Since in the case of a time derivative of a force function, $|F'(f)|^2$ is the square of the Fourier transform of the derivative of $g(t)$, a representation of the variation of $|F'(f)|^2$ with frequency (f) can be obtained by tilting $10 \log G(f)$ through a slope of 20 dB per decade, making the variation between f_1 and f_2 flat (Fig. 22.40c). The justification for using the form $|F'(f)|^2$ lies in the expectation that this flat position of the impulse shaping term overlaps the frequencies for high radiation efficiencies.

In Table 22.6 some typical impact times and corresponding frequencies associated with the main pulses and with the sharp deviations from smoothness for various kinds of operation are given. The interesting point to notice is that there are very few impact processes in which the whole work pulse is important and that, providing we work in terms of the rates of change of forces relating to the microstructure of the operation, we can usually take the pulse spectrum $|F'(f)|^2$ as flat, up to and exceeding the frequencies corresponding to peak modified radiation efficiencies. Beyond f_2, the correction to be applied to account for pulse shape will reduce at 20 dB per decade.

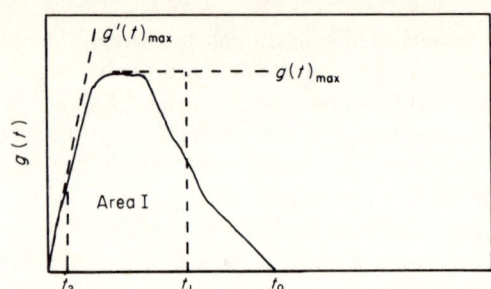

Fig. 22.40a – Impulse shape $g(t)$ as a function of time. t_0 is impact time, $t_1 = \dfrac{2I}{g(t)_{max}}$ and $t_2 =$ notional time over which $g'(t)$ is large.

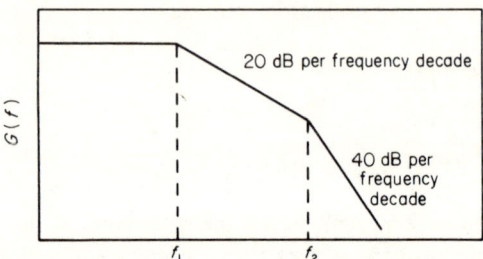

Fig. 22.40b – Impulse function $G(f)$ as a function of f_1 and f_2. f_1 can be equated approximately to $\dfrac{1}{\pi t_1}$, f_2 approximately to $\dfrac{1}{\pi t_2}$.

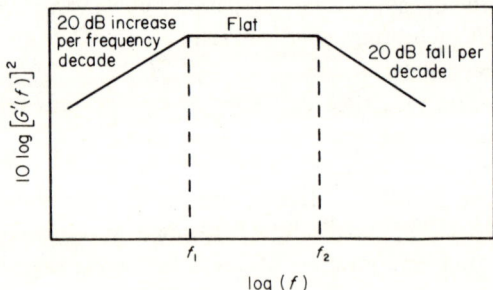

Fig. 22.40c – Variation of $10 \log |F_c'(f)|^2$ with $\log f$.

Evensen [22.12] has found that on punch presses, it is reasonable to assume that the motion is made up of a series of sharp impacts each of which are sufficiently impulsive to justify an assumption that $F''(f_0)$ is flat with frequency f_0; that is, that $|F'(f_0)|^2$ will be falling with frequency at 20 dB per decade of increase in frequency, in the relevant frequency range. Under these circumstances it is possible to relate $F''(f_0)$ to the rate of change of force in the time plane, each sharp and independent change in force being treated as separate events.

This is described later in one of the case studies comprising this programme of work of the group.

Table 22.6
Impact times and corresponding frequencies for various machine operations

Operation	t_1 (seconds)	f_1	t_2 (seconds)	f_2
Impacting 4″ steel balls	2×10^{-4}	1.5 Hz	10^{-4}	3 kHz
Drop hammer (large)	2×10^{-3}	150 Hz	2×10^{-4}	1.5 kHz
Punch press	1.5×10^{-2}	25 Hz	5×10^{-4}	600 Hz
Petrol engine combustion		150 Hz		300 Hz
Diesel engine combustion	1.7×10^{-1}	2 Hz	10^{-5}	30 Hz
Diesel engine piston slap	1.7×10^{-1}	2 Hz	2×10^{-4}	1.5 kHz
Hydraulic pump (constant displacement)	3×10^{-3}	100 Hz	2×10^{-4}	1.5 kHz
Backlash (no lubrication)	2×10^{-4}	1.5 Hz	10^{-4}	3 kHz
Backlash (packed grease)	2×10^{-2}	15 Hz	10^{-3}	300 Hz

Any lowering of f_2 by softening (increasing t_2) the hard 'knock' components of the impact, will modify the noise contribution at any frequency f_0, and the lowering of f_2 towards f_1 will obviously raise parts of the curve near the low frequency end in order to provide the same total area under the curve. It is to be expected, therefore, that general softening of impacts can and does increase the lower frequency content of $L_{A\text{eq}}(f)$ and can lead sometimes to an increase in the total emitted noise dose. This is referred to in some of the case studies which follow (see Fig. 22.40).

The term $10 \log \text{Re} \left[\dfrac{H_c(f_o)}{j} \right]$ represents solely the structural response at the point C of the application of the force, and is not a function of the pulse shape but of the machine geometry. It will be the same for continuous excitation as for impulsive excitation, and it can be measured most easily by such a continuous discrete frequency excitation or be calculated if the machine modes are known.

The response of the structure can be expressed as the sum of all the contributions from each mode of vibration of the structure. If m_i is the modal mass and K_i the modal stiffness of the structure when vibrating in the ith mode, η_s being the structural loss factor, then,

$$\text{Real} \left[\frac{H_c(f)}{j} \right] = \sum_{i=1}^{\infty} \frac{-\eta_s \omega_1^2 / m_i}{(\omega_i^2 - \omega^2)^2 + \eta_s^2 \omega_i^4}$$

where $\omega_i^2 = K_i / m_i$; i^{th} mode resonance frequency.

For values of ω less than the first modal frequency ($\omega < \omega_1$)

$$\text{Real}\left[\frac{H_c(f)}{j}\right] = \frac{-\eta_s/M}{\omega_1^2(1 + \eta_s^2)}; \quad m_1 = M \text{ the mass of the structure,}$$

and when $\omega = \omega_1$

$$\text{Real}\left[\frac{H_c(f)}{j}\right] = \frac{1}{\eta_s\omega_1^2 M}.$$

Therefore, the response curve will take the form of a constant up to an arbitrary frequency which is less than the first modal frequency. A sudden increase in the response between this arbitrary frequency and the first resonance, which will be about 20 dB per decade in slope depending on damping. Above the first resonance mode, frequency averaging between successive resonances, the average value for Real $[H_c(f)/j]$ will have a slope of between -10 dB and -15 dB per decade, depending on the structure and type of excitation. For $\omega > \omega_1$, the average value of Real $[H_c(f)/j]$ will be independent of structural loss factor (η_s) and mass of structure.

Cremer *et al.* [22.13] shows that the ratio of the spatial averaged velocity squared, to the applied point force, above the first modal frequency, can be expressed as

$$\frac{\langle \overline{v^2} \rangle}{F^2} = \frac{1}{m^2} \frac{\pi}{2\eta_s\omega} \frac{\Delta N}{\Delta\omega}.$$

($\Delta N/\Delta\omega$) is the modal density, which is constant in certain instances (beam in longitudinal vibration, plate in flexure, and for a thin-walled tube above the circumferential resonance frequency), but may vary with $\omega^{-\frac{1}{2}}$ in others (beam in bending, ring excited radially, etc.).

It can be shown from power considerations that there is a direct relation between Real $[H_c(f)/j]$ and $\langle \overline{v^2} \rangle/F^2$ above the first modal frequency. The ratio between the two being the mass of the structure and the loss factor. Thus above the first harmonic, both Real $[H_c(f)/j]$ and $\langle \overline{v^2} \rangle/F^2$ will have the same fall-off rate, that is, between -10 dB and -15 dB per frequency decade. The final equation of L_{eq} can therefore be summated as shown in Fig. 22.41.

In such calculations of the operation of a punch press, we have found that there are at least ten modes occurring below 1 kHz and that the modal density is quite high in the frequency range where the modified radiation efficiency is at its peak. Broad energy methods are therefore applicable, if applied with caution, particularly on highly damped machinery or if broad diagnostic results are being sought on an existing machine in which it is only the excitation which is being changed. Goyder & White [22.14] have examined the way mobility of one part of the structure is related to that in other parts containing isolators with various characteristics. As one of the noise control methods available to us in machinery

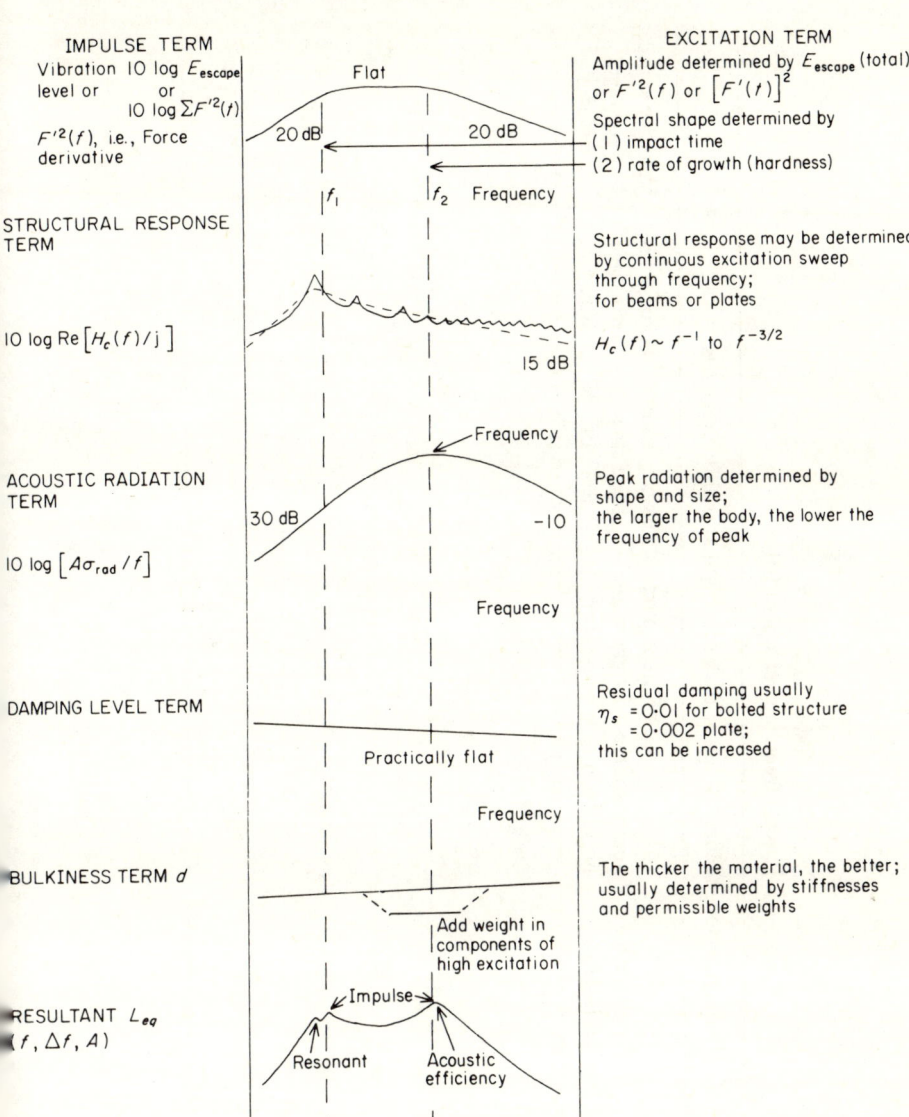

IMPULSE TERM
Vibration 10 log E_{escape}
level or   or
         10 log $\Sigma F'^2(f)$
$F'^2(f)$, i.e., Force
derivative

EXCITATION TERM
Amplitude determined by E_{escape} (total)
or $F'^2(f)$ or $\left[F'(f)\right]^2$
Spectral shape determined by
(1) impact time
(2) rate of growth (hardness)

Flat

20 dB   20 dB

f_1 f_2 Frequency

STRUCTURAL RESPONSE
TERM

10 log Re $\left[H_c(f)/j\right]$

Structural response may be determined
by continuous excitation sweep
through frequency;
for beams or plates

$H_c(f) \sim f^{-1}$ to $f^{-3/2}$

15 dB

Frequency

ACOUSTIC RADIATION
TERM

10 log $\left[A\sigma_{rad}/f\right]$

Peak radiation determined by
shape and size;
the larger the body, the lower the
frequency of peak

30 dB   -10

Frequency

DAMPING LEVEL TERM

Residual damping usually
η_s =0·01 for bolted structure
      =0·002 plate;
this can be increased

Practically flat

Frequency

BULKINESS TERM d

The thicker the material, the better;
usually determined by stiffnesses
and permissible weights

Add weight in
components of
high excitation

RESULTANT L_{eq}
$(f, \Delta f, A)$

Impulse

Resonant   Acoustic
             efficiency

Fig. 22.41 – Summation of contributions to L_{eq}.

design is that of isolating strong acoustic radiators from their source of excitation, this work on power flow through structures is relevant in the context of the value of $[H(f)]$ and its variation with frequency.

Goyder & White [22.14] have estimated the power–frequency characteristics in various simple beam and plate structures incorporating single and two-stage

Table 22.7
Power – frequency characteristics

Foundation type	Frequency dependence of power flow spectrum at high frequencies					
	Force source no isolator	Velocity source no isolator	Force source single stage isolator	Velocity source single stage isolator	Force source two stage isolator	Velocity source two stage isolator
Beam in flex vibration due to force excitation	$\dfrac{1}{\omega^{\frac12}}$	$\omega^{\frac12}$	$\dfrac{1}{\omega^{4\frac12}}$	$\dfrac{1}{\omega^{2\frac12}}$	$\dfrac{1}{\omega^{8\frac12}}$	$\dfrac{1}{\omega^{6\frac12}}$
Beam in flex vibration due to moment excitation	$\omega^{\frac12}$	$\dfrac{1}{\omega^{\frac12}}$	$\dfrac{1}{\omega^{3\frac12}}$	$\dfrac{1}{\omega^{1\frac12}}$	$\dfrac{1}{\omega^{7\frac12}}$	$\dfrac{1}{\omega^{5\frac12}}$
Plate in flex vibration due to force excitation	constant	constant	$\dfrac{1}{\omega^{4}}$	$\dfrac{1}{\omega^{2}}$	$\dfrac{1}{\omega^{8}}$	$\dfrac{1}{\omega^{6}}$
Plate in flex vibration due to moment excitation	ω^{3}	$\sim\dfrac{1}{\omega^{3}}$	$\sim\dfrac{1}{\omega^{3}}$	$\sim\dfrac{1}{\omega^{3}}$	$\sim\dfrac{1}{\omega^{7}}$	$\sim\dfrac{1}{\omega^{5}}$

isolators, and these are shown in Table 22.7. It can be deduced that if no isolator system is present the variability in $H(f)$ with frequency is weak, if present at all, but that once spring isolators occur in the structure, the variability is extensive, the power flow across such a region being proportional to the inverse of the frequency to a high order. Such effects are to be noticed in the keying systems of drop hammers, in punch-die sets, and in the flexibility of linear walls in diesel engines.

B. APPLICATIONS

22B.1 INTRODUCTION

In part A of this chapter we developed the formulae needed to understand the parameters underlying the practical study of machine noise at source. We developed expressions for the peak noise levels, and for the L_{eq} contribution arising from the ringing of the structure subsequent to the impact. Part B now deals with practical applications of these ideas, and with the prediction of noise once the vibrational or force derivative levels are known.

In doing this we must separate our studies into two categories: (a) those in which the vibratory amplitudes of each component are known throughout and following the impact, and (b) those in which only the initial vibratory level or the value of the force derivative is known during the time of impact. In the first case, we need only multiply the value of $\langle \overline{v^2} \rangle$ throughout the decay period by the surface area and value of $A\,\sigma_{rad}$ at the frequency being examined to obtain the predicted noise energy being emitted. In the second case, if we only know the input energy or the initial impulse we have to bring in the expected level of structural response, damping and the bulkiness of the machine in accordance with equations (22.4) and (22.5) in Part A.

22B.2 CATEGORY (a) PREDICTING NOISE RADIATION FROM MEASUREMENT OF SURFACE VELOCITY

In Part A and References [22.5] and [22.8] we showed how acceleration noise peaks and L_{eq} depended upon the kinetic energy at impact and on the non-dimensional contact time $ct_0/(\text{vol})^{\frac{1}{3}}$. We also indicated that the ensuing ringing noise could be related to the mean square of the surface velocity of any component averaged over time and surface and proceeded to calculate the relationship between the noise radiated for any shape of component. Needless to say, one cannot do this for very complicated shapes, but acoustically most shapes can be represented as solids, plates, or rods, especially when they are radiating with a high efficiency. This efficiency (σ_{rad}) is described in terms of the actual noise radiation compared with a large flat plate radiating normal to itself with the same value of $\overline{v^2}$. Thus we have:

$$W_{rad} = \sigma_{rad} \times \rho_0 cS \langle \overline{v^2} \rangle.$$

Knowledge of σ_{rad} can be very valuable in cases where we know $\langle \overline{v^2} \rangle$, and this technique of noise prediction from vibration measurement is coming into

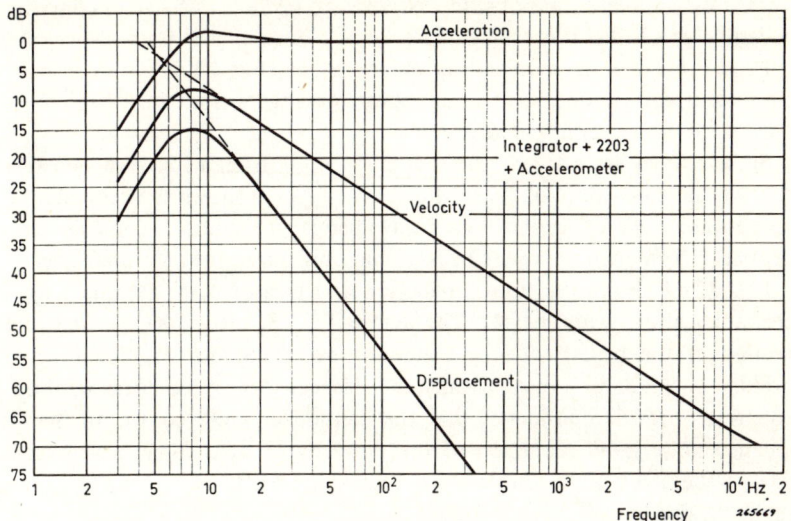

Fig. 22.42 – Frequency response curves with the Precision Sound Level Meter type 2203 and an accelerometer having capacitance 1000 pF.

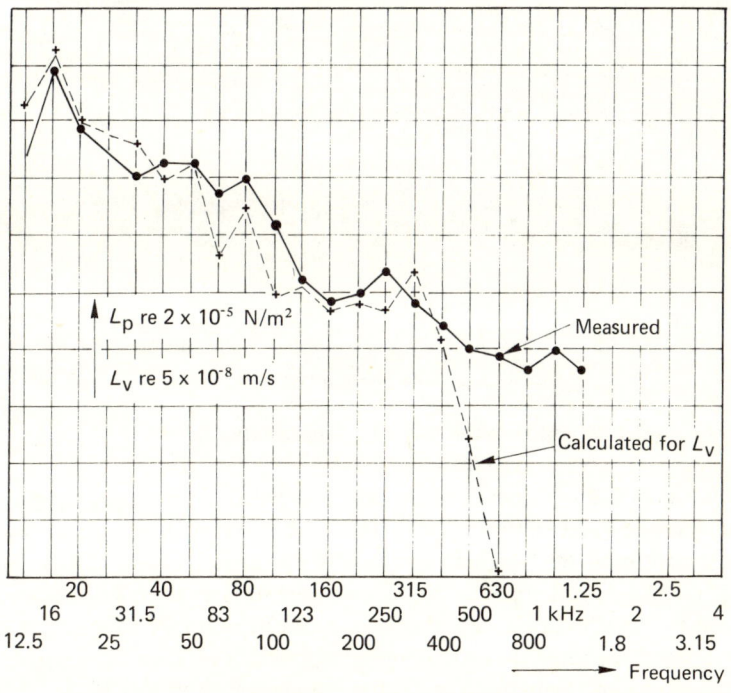

Fig. 22.43 – Calculated and measured noise energy from boiler wall.

increased use in many areas of study. Needless to say, it is not always convenient to measure $\langle \overline{v^2} \rangle$, but as this can be related at any frequency to both displacement on the one hand or to acceleration on the other (Fig. 22.42), predictions of noise output can be obtained (even though directivity and the noise levels at operators' positions cannot be determined).

Six examples are offered: the first is the noise from a high-pressure steam boiler whose walls were vibrating. Fig. 22.43 [22.17] shows the estimated and measured noise energy in one-third octave bands and indicates good agreement except at high frequencies where background noise raises the measured levels.

Fig. 22.44 shows the noise prediction and measurement for the noise of a steam injection water heater. The author [22.18] attributes the high noise measured at the high-frequency region as being due to additional sources providing background noise. Clearly, in this case, the prediction method is adequate to provide confidence that the method of noise control of the low-frequency noise is by reduction of the unsteady internal forces acting on the inside of the walls of the heater, but that above 1 kHz other sources of noise must be tackled.

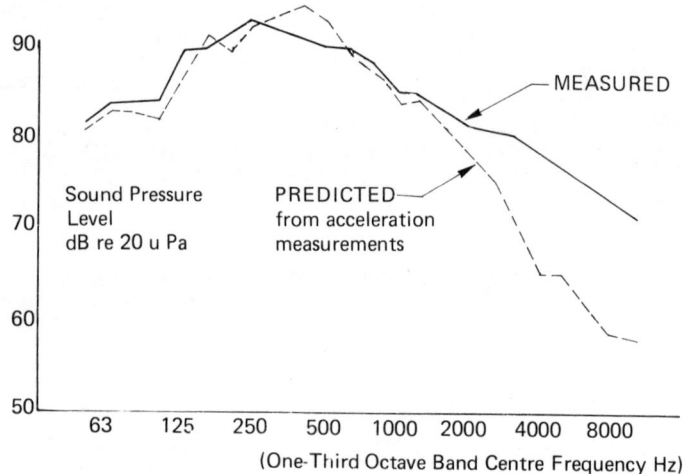

Fig. 22.44 – Sound levels close to a blancher (after J. F. Wilby [22.18]).

Our third example concerns the noise radiated from various parts of a sewing machine [22.19] (Fig. 22.45). Here the vibration levels of the various panels constituting the machine were measured directly, (Fig. 22.46) and by using (Fig. 22.47) a microphone very close to the surface (the near pressure field is related to the vibration levels, though not directly). It may be seen that using either method indicates that the dominating noise arises from the bedcover of the machine rather than from any other component. This can be reduced by isolating it from the source of excitation (the motor) or by reducing its ability to radiate by replacing it by a perforated plate (Fig. 22.48). Here the excitation frequency is well below the coincidence frequency, and its ability to radiate is significantly reduced by cancellation of pressure perturbations through the holes. Above coincidence each element of the surface acts independently, so that perforation does not help a lot. Furthermore, the under surface of the motor cover then becomes a dominant source. By a careful combination of using modified covers and reduced excitation from changed belts and gears, a total reduction of some 9 dB (Fig. 22.49) was achieved.

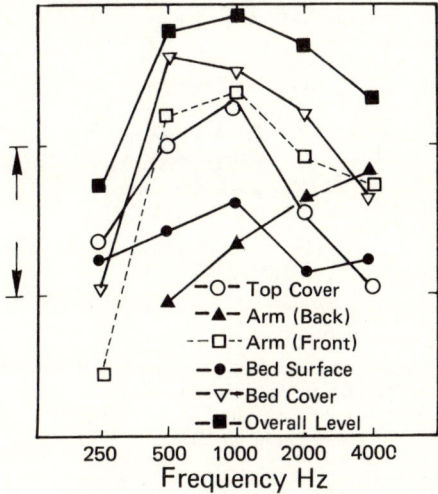

Fig. 22.45 – Surface vibrations are used in this figure to estimate the sound radiated by various surface elements. The ranking is generally the same as indicated in Fig. 22.47, with the bed cover dominant. Lower level sources rank differently, however. Vibration estimates are probably more reliable than are near-field pressure scans in ranking the weaker radiators, but are more cumbersome to make because of the greater variability in vibration amplitudes.

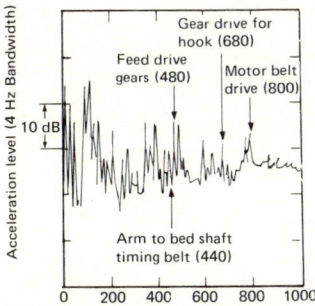

Fig. 22.46 – This constant bandwidth acceleration spectrum (taken at a location on the work surface) shows that certain tooth frequencies of the mechanisms are prominent in the vibration. Other peaks, some identifiable, may be due to other sources or major structural resonances (the peak at 100 Hz is known to be a structural resonance).

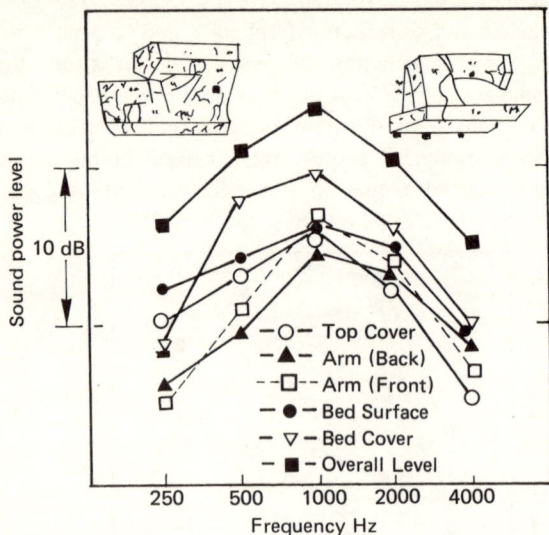

Fig. 22.47 – The relative contributions of various machine surfaces to the overall radiated sound power in octave bands is shown here, as determined by using near field pressure scans. Contours of equal pressure level in the 500 Hz octave band are shown in the inset. The dominance of the bed cover as a radiator is evident form this data.

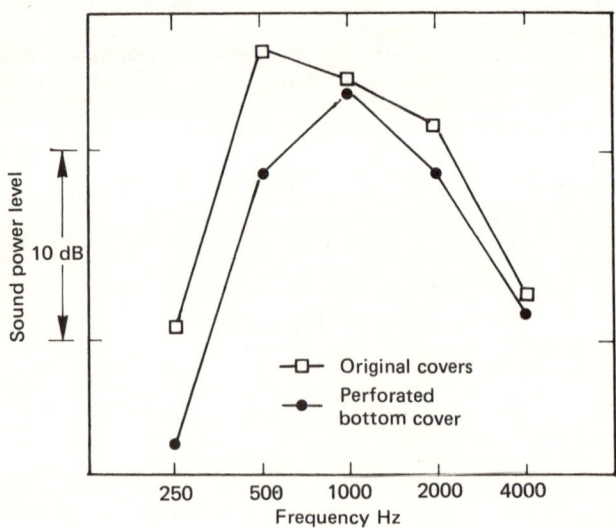

Fig. 22.48

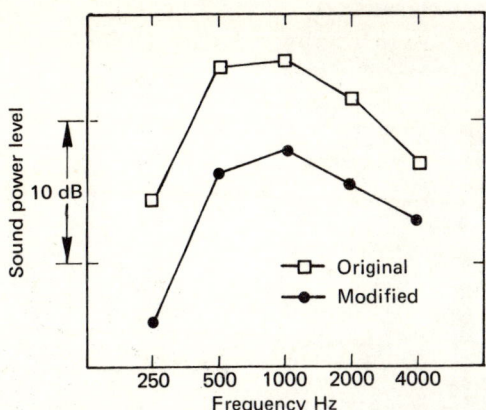

Fig. 22.49 – Comparison of octave band radiated power levels for original and modified machines. The reduced radiation is due to a number of modifications dealing with gears and belts, casting structure, and covers. Any single change in the machine produces limited benefits in sound reduction.

A fourth example deals with the noise from a lathe at the ISVR. Fig. 22.50 shows the sound pressure level arising at the operator's head position from various parts of a lathe based upon predictions of radiation efficiency and

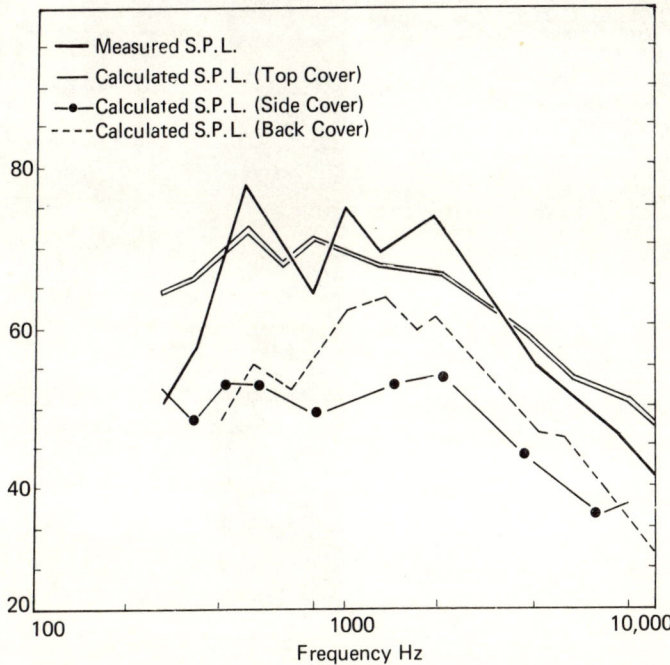

Fig. 22.50 – Measured and predicted sound pressure levels for a Lathe.

Fig. 22.51a – Friction drop hammer at
Drop Forge Research Association
Laboratories (Scheffield).

Fig. 22.51b – One-third scale model dro
hammer in ISVR Laboratorie
(Southampton

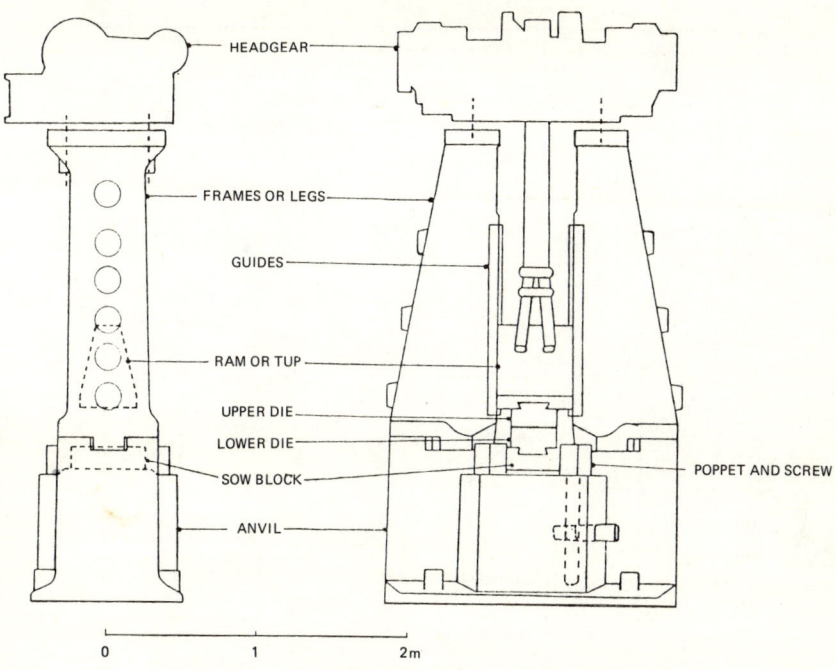

HEADGEAR

FRAMES OR LEGS

GUIDES

RAM OR TUP

UPPER DIE

LOWER DIE

SOW BLOCK

ANVIL

POPPET AND SCREW

0   1   2m

Fig. 22.51c – Components of a typical friction drop hammer.

measured vibration levels. These are compared with the measured figures. It may be seen that the dominating source of noise is the lathe skirt, and that if this were isolated from the various excitation sources significant reductions could be achieved.

A topic which has involved us deeply at the ISVR [22.20] and in which we have used the radiation efficiency method of diagnosis is our fifth example and involves the measurement of both vibration and noise on the Massey drop hammer at the DFRA and on a one-third scale model of it in Southampton. Both of these are shown in Fig. 22.51a,b, while a diagram illustrating the various components is given in Fig. 22.51c.

It is always difficult to measure noise and acceleration on impulsive machinery in regular use, as reverberation occurs in the factory space, accelerometers tend to strip from their fastenings, and factory inspectors are loath to permit any great amount of tampering with machinery. We are therefore grateful for being able to measure noise and vibration full-scale at the DFRA and to be able to build a one-third scale model at the ISVR. Fig. 22.52 shows the comparison of the total radiated noise energy in octave bands as measured by microphones and summated on the full-scale drop hammer, and also by vibration measurements, allocating to each component a radiation efficiency taken from the most relevant figure of [22.5, 22.8]. It may be seen that agreement is reasonable, especially in view of the reverberant noise reflected from the walls, floor,

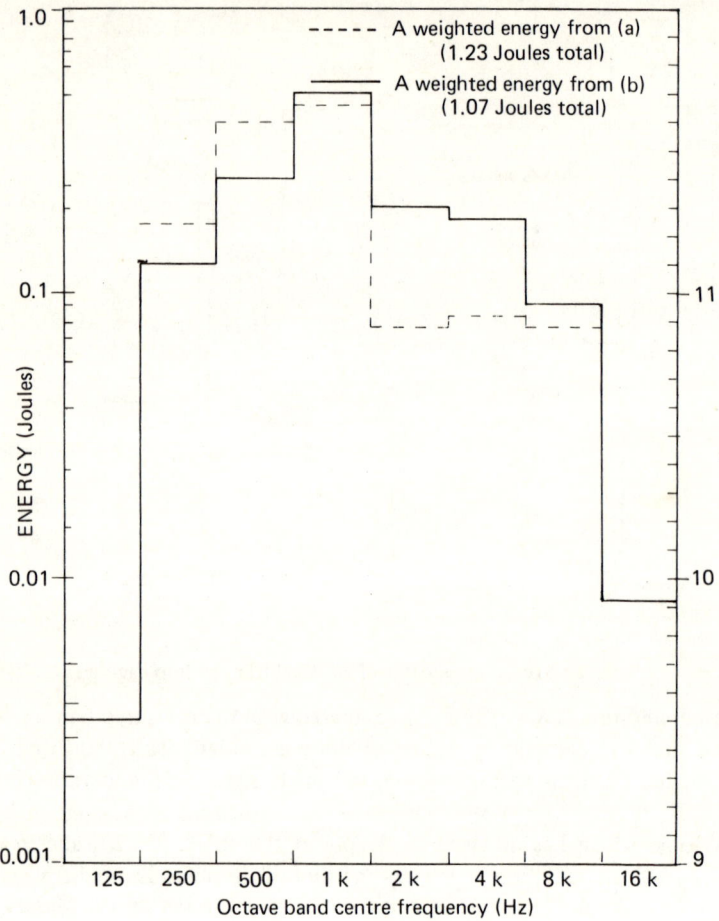

Fig. 22.52 – Comparison of total radiated noise per blow from full scale drop hammer, using (a) component $\langle \overline{v^2} \rangle$ and (b) integrated sound measurements.

and roof. An extremely interesting and useful method of presentation is given in Table 22.8, which lists the noise calculated to be radiated from each component in each one-third octave of frequency and summates the estimated radiated sound energy in milliJoules per blow. Several things are apparent:

(a) agreement between sound measurement using microphones, and calculated from vibration using radiation efficiency is gratifyingly close, the difference of 0.6 dB being more fortuitous than can be expected generally;

(b) the agreement obtained suggests that we can look at Table 22.8 in some detail for information on how to reduce the noise. All items in which the estimated noise radiated in a one-third octave band is larger than

Table 22.8 — Radiated energy for full size drop hammer components (estimated from $\overline{v^2}$ measurements). Energies (mJ) are A-weighted and weighted for radiation efficiency.

Frequency	Top of concrete base	Anvil	Columns	Sowblock and lower die	Poppet heads	Tup and upper die	Total
250	3.67	1.62	2.69	0.42	1.44	55.44 †	65.28
315	2.49	1.20	6.01	0.37	1.37	54.73 †	66.17
400	4.64	1.00	6.11	2.65	1.45	56.92 †	72.77
500	5.70	2.86	5.58	4.00	2.61	66.68 †	87.43
630	13.75 †	21.24 †	16.64 †	4.81	30.24 †	88.12 †	174.58
800	5.55	17.22 †	18.56 †	3.52	40.40 †	85.13 †	170.38
1000	2.31	23.90 †	76.15 †	2.34	3.74	43.91 †	152.35
1250	1.74	4.30	18.55 †	4.29	4.28	21.86 †	54.75
1600	0.63	1.47	13.67 †	3.81	2.94	4.30	25.89
2000	0.54	3.28	13.25 †	3.34	0.72	1.61	22.69
2500	0.37	8.75	10.25 †	3.49	1.12	4.58	28.56
3150	0.39	2.67	3.94	1.96	0.73	12.05 †	20.74
4000	0.52	3.52	5.19	1.27	0.96	15.87 †	27.33
5000	0.65	4.43	6.54	1.60	1.20	20.00 †	34.42
6300	0.73	4.97	7.34 †	1.79	1.35	22.41 †	38.39
8000	0.56	3.77	5.57	1.36	1.08	17.03 †	29.37
Total	44.23	105.85	215.85	40.01	94.91	570.58	1071.52 = 120.3 dB (ref = 10^{-12} J)

Radiated energy measured by microphone = 120.9 dB (ref. 10^{-12} J).

one-hundredth of the total have been marked with a †. In this way, non-dominating sources can be identified for further examination as the dominating sources are eradicated;

(c) the dominating sources consist of the tup (or hammer), the columns, and the anvil. As there is little likelihood of the hammer vibrating at low frequency, the low-frequency noise is most likely to be acceleration noise analysed into its third octave components, and will need special consideration, as hammer deceleration is inherent in the forging process;

(d) the column noise can be, and has subsequently been, eliminated by fitting resilient underlayers to the guide columns and to the base of the uprights;

(e) the poppet head vibration is a feature of this design of hammer and is not always present. It arises from the fact that locating systems involve large stresses, elastic systems, and friction damping, and the vibration is very much a function of the microslip and damping in the screw threads. This noise has been eliminated in the model;

(f) even if all the marked sources were eliminated, the total noise radiated would still be high in decibel terms.

As part of our investigations, a one-third dynamic model was built and investigated; after relevant scaling was applied, the agreement is excellent between full scale and model and is shown in Fig. 22.53.

Noise control measures have been taken on this model to reduce all noises not directly related to the forging process, and the resulting noise output in milliJoules is illustrated in Table 22.9. Again, each component radiating more than one-hundredth of the original total (on the model) has been marked for further investigation. It may be noted that the agreement between the noise calculated from vibration measurements, and that from microphone measurements is closer than ever. As this one-third scale model was located in a semi-anechoic chamber, this is a very gratifying agreement, and probably relevant in this case.

The dominating noise now stems clearly from two sources, acceleration noise from the tup or hammer, and from the anvil. As the lowest natural frequency of the tup is above 6 kHz, it is clear that the tup noise arises from its transient deceleration as a solid inelastic body, and the methods to be used to eliminate it must be related to details of the deceleration pulse and not to ringing. That this is true can be seen in Fig. 22.54, showing the spectrum of the tup acceleration noise with and without the impact period included. It may be seen that if the impact period is excluded, the vibrational energy averaged during the ringing period consists only of high-frequency resonances.

Further work carried out on the tup by cladding it with a steel cover which was separated by a resilient layer of damped material shows (Fig. 22.55) that the cladding itself gives out low-frequency noise, and that the net result of such cladding is a total noise reduction of 1.5 dB at most. It is obvious that cladding

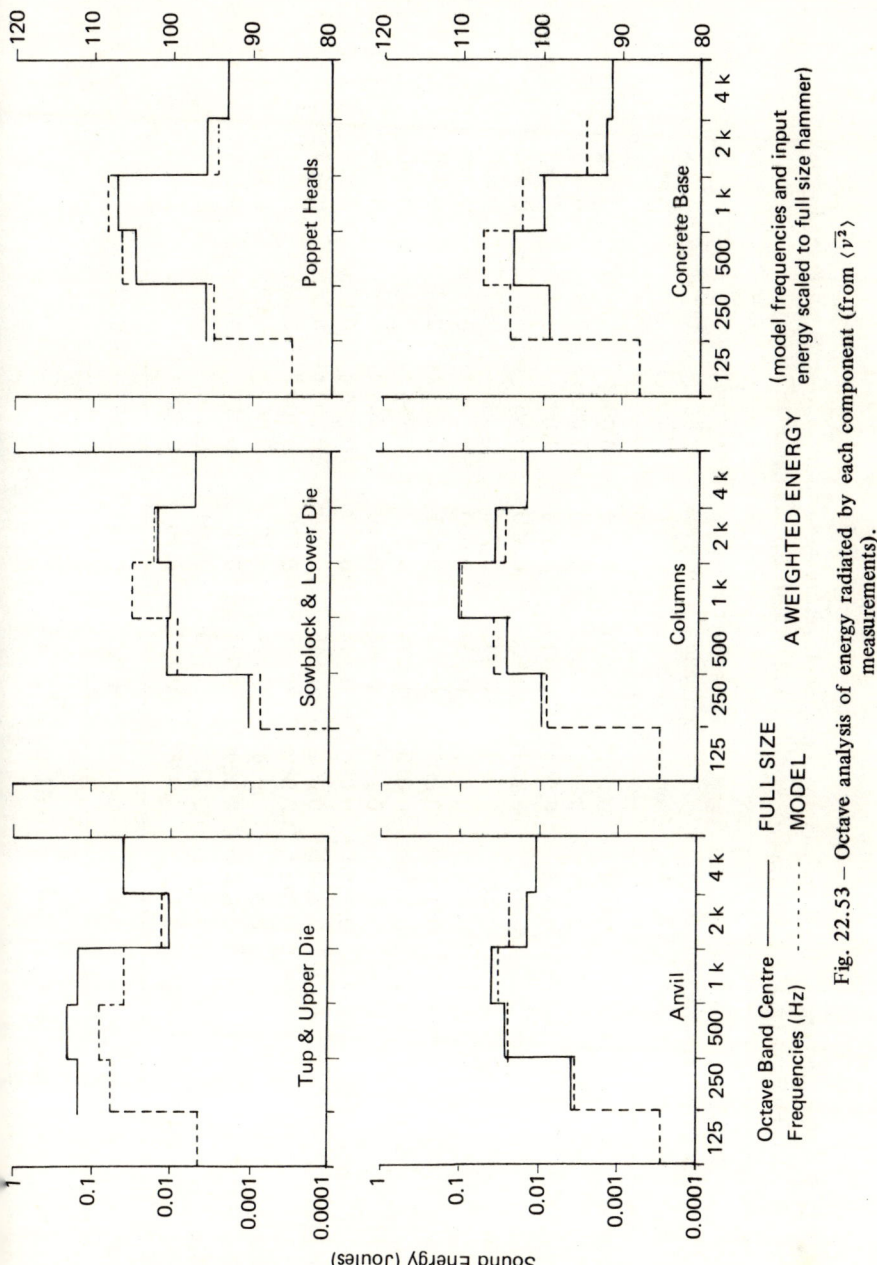

Fig. 22.53 – Octave analysis of energy radiated by each component (from $\langle \overline{v^2} \rangle$ measurements).

Table 22.9 – Radiated energy for drop hammer components (estimated from $\overline{v^2}$ measurements) ($\frac{1}{3}$rd scale). Energies (mJ) are $\frac{1}{3}$rd scale A-weighted and weighted for radiation efficiency.

Frequency	Small concrete block	Large concrete block	Anvil	Columns	Sowblock and lower die	Poppet heads	Tup	Upper die	Total
250	—	—	0.006	0.001	0.000	0.000	0.001	0.000	0.008
315	—	—	0.005	0.002	0.000	0.000	0.003	0.000	0.010
400	—	—	0.006	0.004	0.000	0.000	0.001	0.005	0.016
500	—	—	0.010	0.003	0.000	0.000	0.063	0.000	0.076
630	—	—	0.046	0.019	0.002	0.000	0.243	0.000	0.310
800	—	—	0.026	0.011	0.007	0.000	0.198 †	0.001	0.243
1000	—	—	0.028	0.004	0.011	0.001	0.231 †	0.001	0.276
1250	—	—	0.028	0.010	0.034	0.003	0.264 †	0.001	0.340
1600	—	—	0.128	0.134	0.065	0.010	0.265 †	0.002	0.604
2000	—	—	0.755 †	0.056	0.134	0.107	0.256 †	0.005	1.313
2500	—	—	0.200 †	0.084	0.148	0.019	0.207 †	0.006	0.664
3150	—	—	0.342 †	0.012	0.153	0.020	0.102	0.005	0.634
4000	—	—	0.067	0.024	0.129	0.007	0.044	0.045	0.312
5000	—	—	0.128 †	0.009	0.091	0.004	0.015	0.010	0.257
6300	—	—	0.353 †	0.013	0.120	0.004	0.016	0.019	0.525
8000	—	—	0.725 †	0.028	0.127	0.007	0.553 †	0.042	1.482
Total	—	—	2.851	0.415	1.021	0.182	2.46	0.142	7.07

= 98.5 dB
(ref = 10^{-12} J)

Radiated energy measured by microphone = 6.8 mJ = 98.3 dB.

introduces low-frequency ringing in the process of smoothing out the sharp deceleration pulse (see Fig. 22.56).

The solution to the ultimate problem of the drop hammer must lie therefore in the direction of establishing whether the forging process depends crucially upon the sharpness of final forging blows. If methods (such as hydraulic damping of the sharp decelerations) can be accepted, the noise of the tup can be eliminated.

The sixth and last example [22.21] is that of predicting the peak noise from bottle impacts. As peaks stem from the acceleration during the impact, and not from the subsequent vibration (dealt with later), it is interesting to observe the agreement or otherwise of the peak noise radiated from bottles via impacts of varying softness. Fig. 22.57 shows how the levels fall neatly between those predicted for spheres and the maximized values for other shapes, and shows clearly the fall-off in peak noise with $ct_0(\text{vol})^{1/3}$. As the subsequent vibrational level tends to commence at the level of peak noise, the advantage to be obtained from softening impacts in general is apparent (Fig. 22.58).

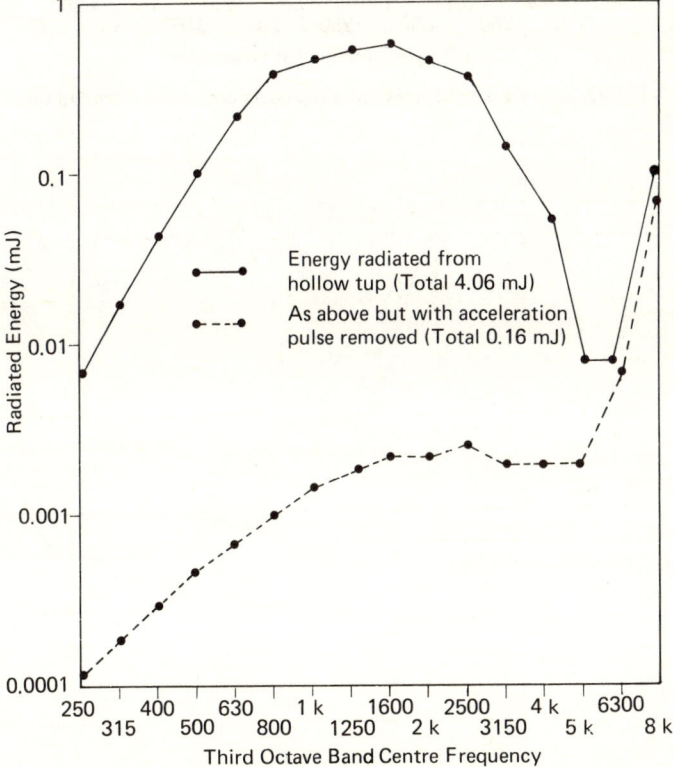

Fig. 22.54 – Effect of removing acceleration pulse from energy calculations on tup.

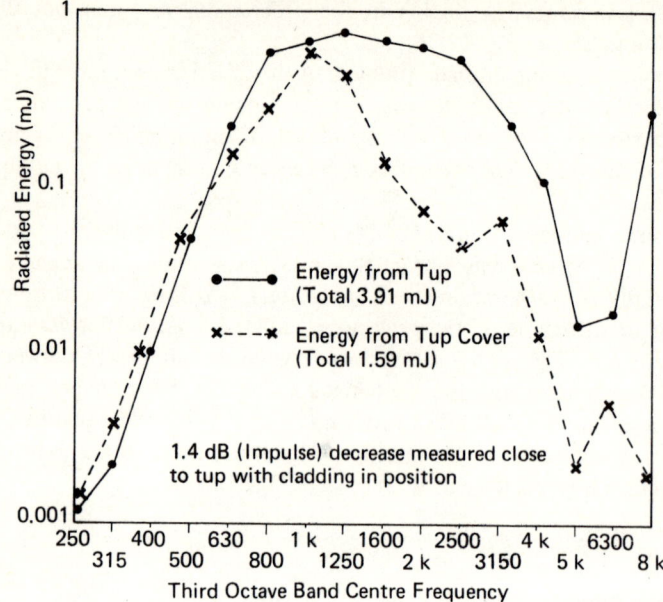

Fig. 22.55 – Radiated energy from tup cover compared to unclad tup.

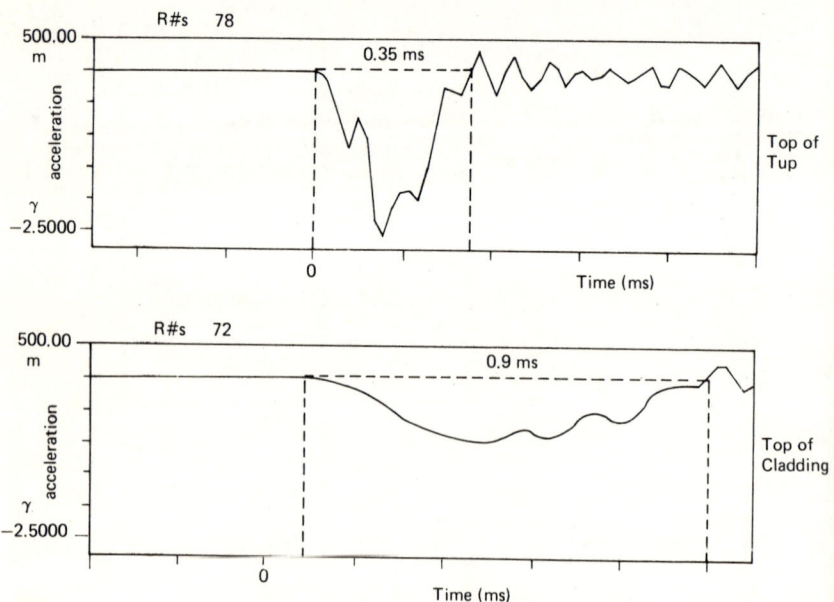

Fig. 22.56 – Acceleration pulse on tup cladding.

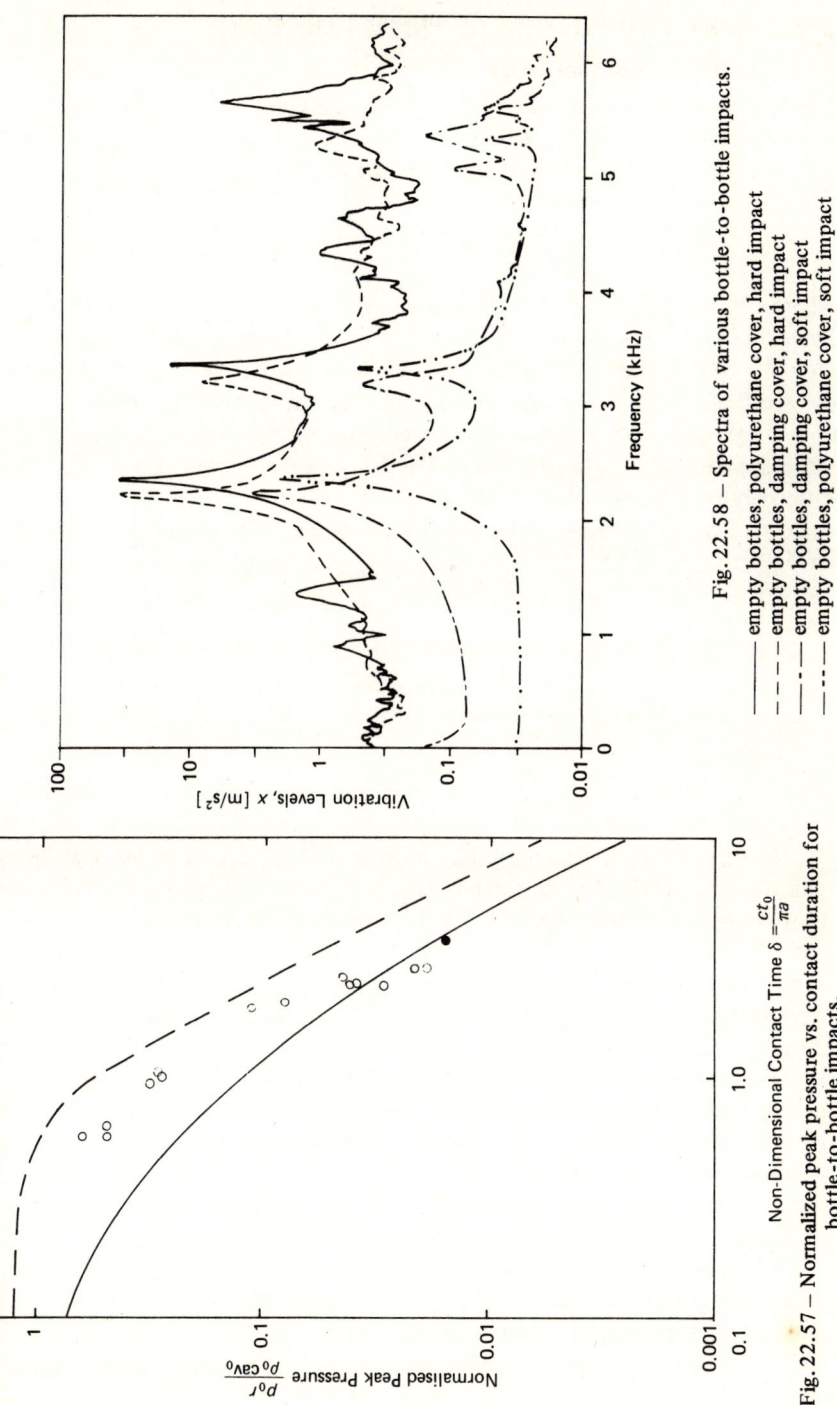

Fig. 22.58 — Spectra of various bottle-to-bottle impacts.

——— empty bottles, polyurethane cover, hard impact
– – – empty bottles, damping cover, hard impact
–··– empty bottles, damping cover, soft impact
–···– empty bottles, polyurethane cover, soft impact

Vibration Levels, x [m/s²]

Frequency (kHz)

Non-Dimensional Contact Time $\delta = \dfrac{ct_0}{\pi a}$

Normalised Peak Pressure $\dfrac{p_0 r}{p_0 c a v_0}$

Fig. 22.57 — Normalized peak pressure vs. contact duration for bottle-to-bottle impacts.

22B.3 SOME CASE STUDIES OF PULSE TAILORING

22B.3.1 Punch press tool design

Evensen in an associated study [22.22] has examined theoretically the noise radiated from a punch press and concludes that the value of L_{eq} per event averaged over a time T can be expressed in the form:

$$L_{eq}(T) = E_1 + 10 \log \tilde{F''^2}$$

where $(\tilde{F''^2}) = |F''(f, \tilde{r})|^2$ and is constant with respect to frequency, $F''(f, \tilde{r})$ being the Fourier transform of $f''(t, \tilde{r})$. Thus $F'(f_0)$ falls linearly with frequency and $[F'(f_0)]^2$ falls at 20 dB per decade, and he is dealing with frequencies above f_2 (see Table 22.2). Here $f(t)$ is used to represent force, as well as frequency.

In this case E_1 is a transfer term depending both upon the response of the structure to such a force derivative, the damping factor, and also the acoustic efficiency of the particular radiating surface. Clearly E_1 could not easily be calculated, and the value of the above formula lies in its ability to predict the change in L_{eq} with $(\tilde{F''^2})$. Evensen goes on to show that δ, the second derivative of $f(t)$ $(= f''(t))$, is sufficiently impulsive for $|F''(f)|^2$ to be represented by a series of the form

$$|\hat{F}''(f)|^2 = \sum_{n=0}^{N} b_n^2$$

where b_n is the amplitude of the changes in peak force derivatives in the time domain.

In the case of the punch press, Evensen [22.22] shows in his analysis of pulse shapes $f(t)$ and their derivatives that the second derivatives of the force history are sufficiently impulsive for the equivalent sound level averaged over a time T to be related to the independent changes b_n of the force derivative of $f(t, r)$. Thus it would be expected that, provided the geometry of a machine and its workpiece is unaltered, the L_{eq} value would be related to the square of the maximum rate of change of force applied by the tool in fracturing the material.

Burrows and Herbert [22.23] [22.24] have carried out experiments on the noise radiated from a punch press, varying the thickness and the material being cut, the percentage clearance between the punch and die, the punch stroke position at the time of fracture, and the eccentricity of the clearance between the punch and die. Other workers [22.25, 22.26, 22.27] have experimented with other methods of progressive cutting aimed at decreasing noise.

Burrows and Herbert have recorded the variation of the punch force with time, and values of $f'^2_{max}(t)$ have been evaluated for a wide range of punch arrangements; the values of the total L_{Aeq} radiated per event have also been measured experimentally and a unique curve of L_{Aeq} plotted against

L'_f (= 10 log Σ $\dfrac{f'^2_{max}}{f^2_o}$, f_o being a reference force derivative of 1 MNs^{-1}.

Fig. 22.59 shows this relationship, the best fit of all resultant points giving a slope of 1.02 rather than that of unity indicated in equation (22.5). More significant is the small standard deviation of ±1.5 dB indicated in these results.

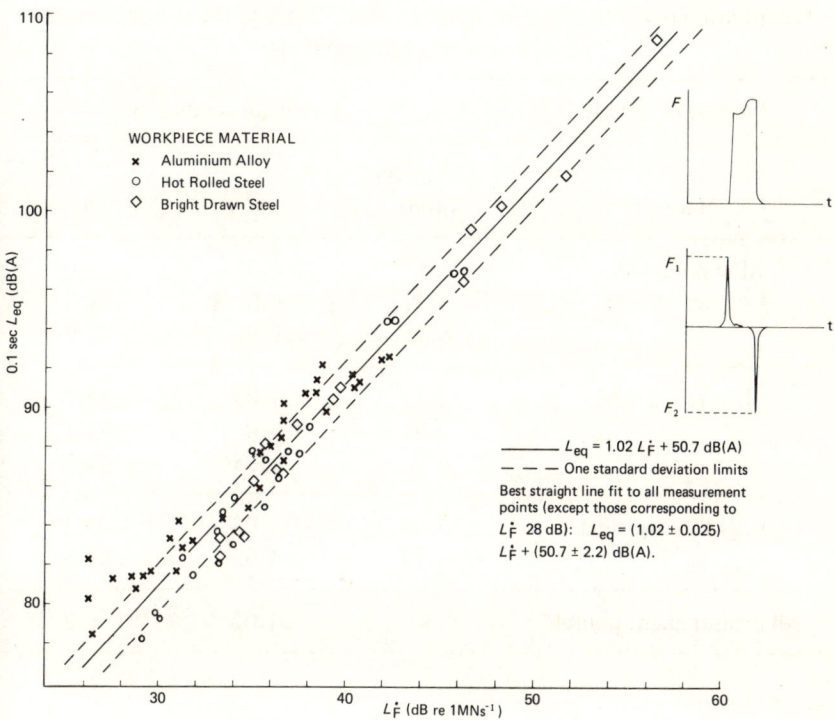

Fig. 22.59 – Relationship between change in force derivatives (L_F) and punching noise (L_{eq}) for all measurement conditions.

It should be emphasized that the term L'_f includes the peak force derivatives at both penetration of the metal and at fracture, since both impacts contribute independently to L_{Aeq}. If only the peak value of f'^2 at fracture is measured, agreement is still good for high-noise–high-fracture impact conditions (Fig. 22.60), but the fracture impact noise includes other noise sources if the fracture impact force derivative has been reduced to a level below that arising from the initial penetration or from some other secondary impacts arising from backlash in the bearings. Table 22.10 shows the values of the constants obtained to give the best fit between the noise and the force derivative function with the punch cutting a number of different thicknesses of different metals. It may be seen that the only case where the constant a differs significantly from unity is that for thin cuts of soft aluminium. In this case, the noise arising from fracture is small compared with that from secondary unrecorded backlash

Table 22.10

Relationship between punching noise L_{eq} and change in force derivatives L_f' for different workpiece materials.

Workpiece details		Fitted constants	
Material	Thickness (mm)	a	b
Alluminium alloy	1.58	0.67	61.9
	3.22	0.81	58.3
	6.28	0.95	52.9
Hot rolled steel	2.09	1.09	47.2
	3.01	1.07	48.1
	6.20	1.08	48.7
Bright drawn steel	3.19	1.03	49.2
	6.32	1.05	49.6
All measurement points*	–	1.02	50.7

$$L_{eq} = aL_f' + b$$

where constants a and b are as given above. Data obtained from measurements on the effect of punch-die clearance on noise levels, except * which also includes data from tests on tool height setting and the use of an eccentric die.

impacts, and the value of the constant is unrelated to the fracture process as such. Fig. 22.61 shows the values of the experimental points obtained with varying percentage clearance before and after the backlash in the press bearings had been eradicated. Before renovation of the press the total noise output appears to be independent of percentage clearance and die setting, largely because the dominating noise for all but high tool clearances arises from secondary backlash impacts. Thus, values of $L_{A eq}$ greater than that given by the expression $1.02 L_f' + 50.7$ dB(A) can be generated if secondary impacts dominate.

It may be concluded that impulse shaping presents a very strong mechanism for noise reduction in punch presses, if such presses are operating at high load,

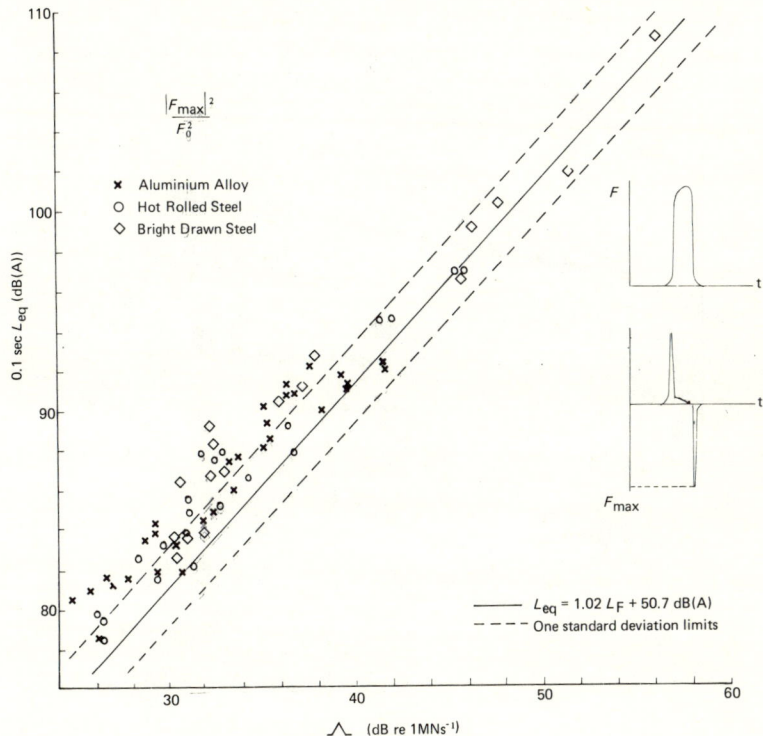

Fig. 22.60 – Relationship between maximum change in force derivative (A_F) and punching noise (L_{eq}) for all measurement conditions.

and that successful piercing can be achieved even with punch arrangements which give extreme L_{Aeq} reductions of as much as 30 dB(A). A more realistic anticipation is 10 dB(A) reduction by tool design, an amount which is generally sufficient to remove the punch process from the category of noise-dangerous mechanisms in a factory environment in which air emission noise and that from handling or ancillary machine clatter also contribute to deafness.

This example can be taken as an application of equation (22.5) in Part A of this chapter, but with the second force derivative used in the frequency plane rather than the first, i.e., we replace $|F'(f)|^2$ by $|F''(f)|^2/f^2$. As we are working in the frequency range $f > f_0$, this implies that $|F''(f)|^2$ will be independent of frequency and equal to $\Sigma f_{max}'^2(t) (= \Sigma b_n^2)$. In such a case equation (22.5) will need to be altered to account for the $-20 \log f$ term introduced into the equation (i.e., E_1 will be different).

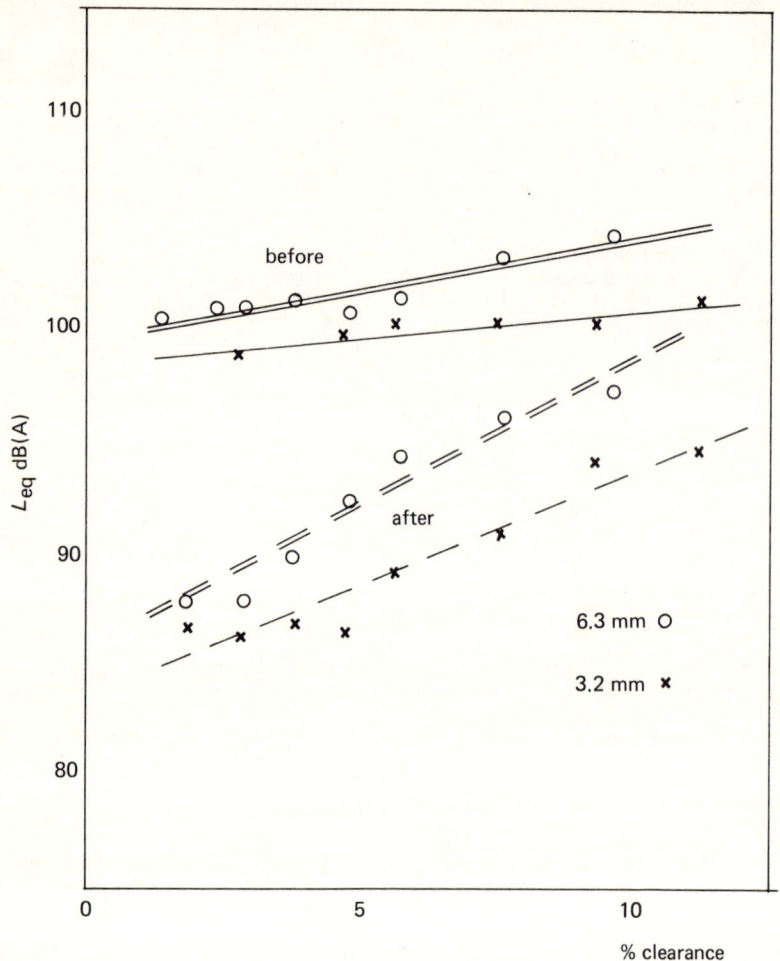

Fig. 22.61 – The variation of $L_{A\,\text{eq}}$ with percentage clearance before and after elimination of backlash in bearings.

22B.3.2 A test case for equation (22.5)

Before we proceed further with case studies, therefore, it is interesting to investigate the degree to which equation (22.5) can be used in its entirety, not only to permit the prediction of changes in $f'(t)_{\text{max}}$ or $F'(f)$ but also those associated with changes of damping factor $10 \log \eta_s$, structural response $\text{Re}\left[\dfrac{H(f)}{j}\right]$, modified radiation efficiency $10 \log\left[\dfrac{A\sigma_{\text{rad}}}{f}\right]$, etc. Damping can be increased, structural response can be modified by impacting through a heavy mass possibly attached via a spring, and the value of $|F'(f)|^2$ can be reduced by softening the impact and extending the impact time.

This has been done by impacting a plate, a plate with attached mass, and a plate with an attached mass coupled to the plate via a spring, and the damping factor η_s has been measured for two damping conditions, with a calibrated hammer which measures acceleration. Thus $f(t)$ and $F(f)$ and $F'(f)$ can all be measured. In the simple plate configuration, experimental values of η_s with frequency are shown in Fig. 22.62 a, the calculated values of $10 \log A\sigma_{rad}/f$ for the plate in Fig. 22.62 b, and the theoretical structural response term

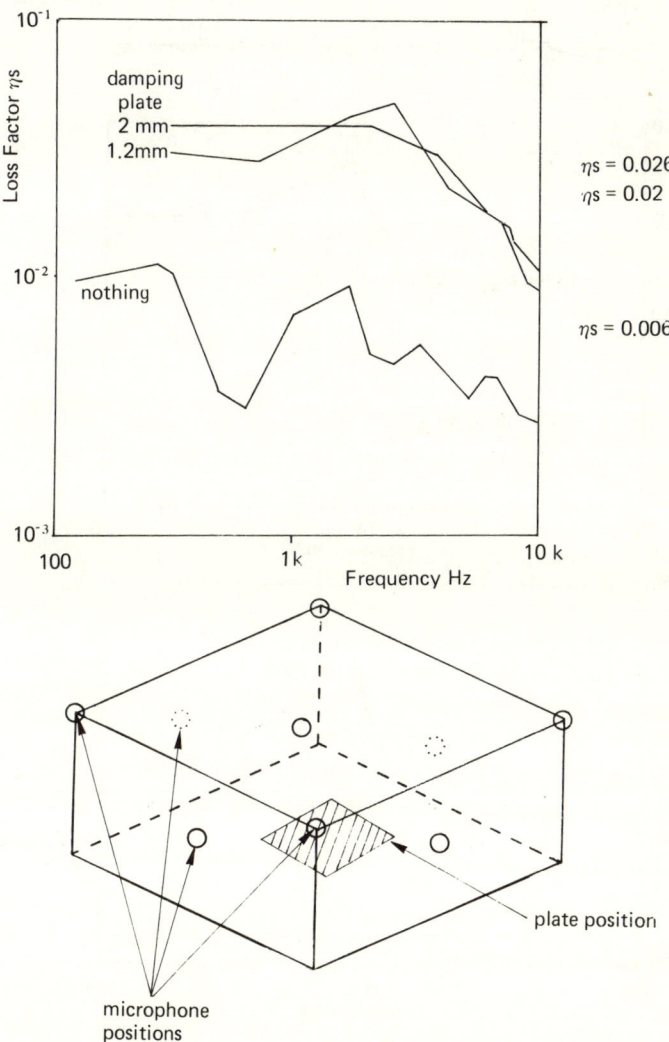

Fig. 22.62 a – Experimental values of η_s with frequency for simple plate.

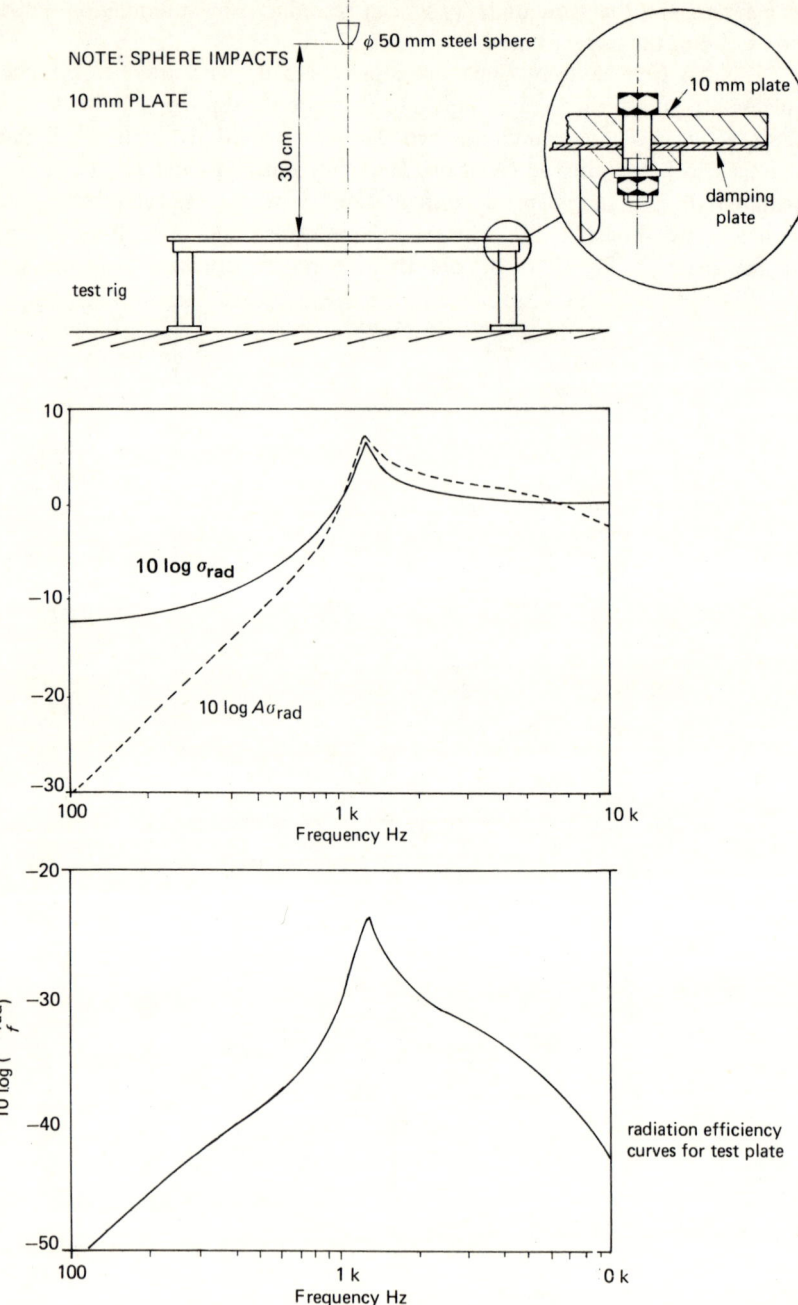

Fig. 22.62 b – Radiation efficiency curves for test plate.

$10 \log \operatorname{Re} \left[\dfrac{H(f)}{j} \right]$ at the point of impact is compared with the experimental response in Fig. 22.62 c. The force pulse shapes associated with hard and soft impacts, respectively, are recorded both as experimental and approximate theoretical values in the frequency domain in Fig. 22.62d. Similar figures have been obtained for the other conditions of attached mass and mass plus spring.

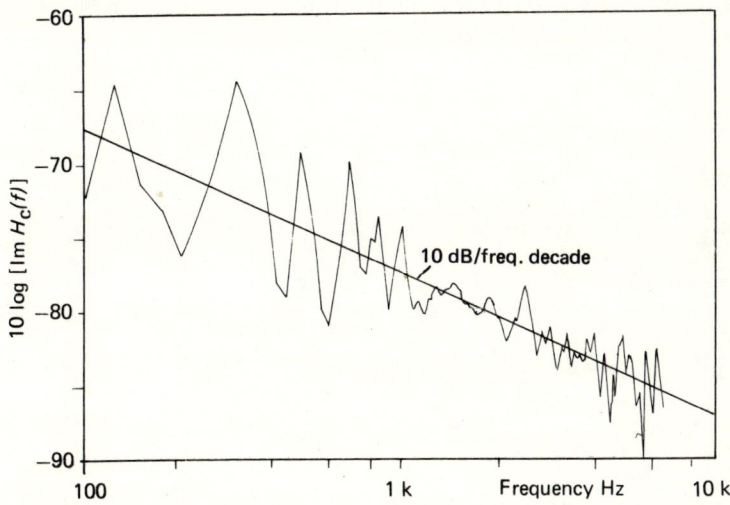

Fig. 22.62 c – Comparison of theoretical and measured response for simple plate.

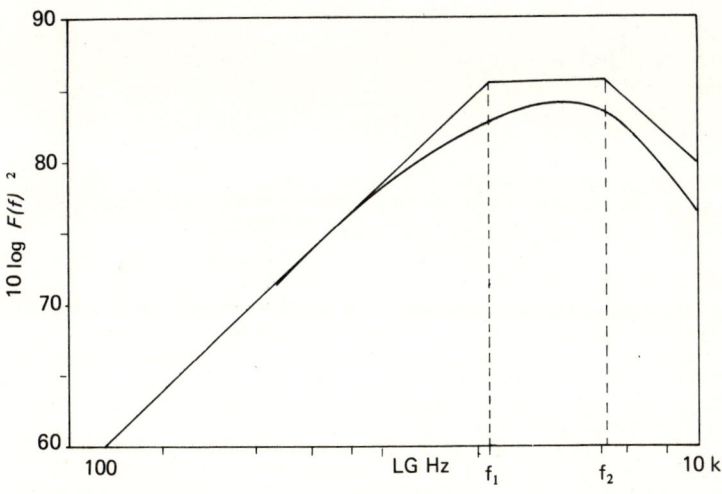

Fig. 22.62 d – Comparison of experimental and theoretical pulse shapes for hard and soft impacts.

The interesting comparisons of predicted noise are shown in Fig. 22.63, which shows the progressive changes in $L_{eq}(f)$ spectra, both predicted and measured acoustically, as we move from direct impact of the plate above to impacts via a heavy mass and via a resiliently separated heavy mass. It may be seen that agreement is good and that large reductions of noise can be predicted quite accurately. The effect of a threefold increase in structural damping, also predicts (Fig. 22.64) correctly a 5 dB fall in noise throughout the frequency range.

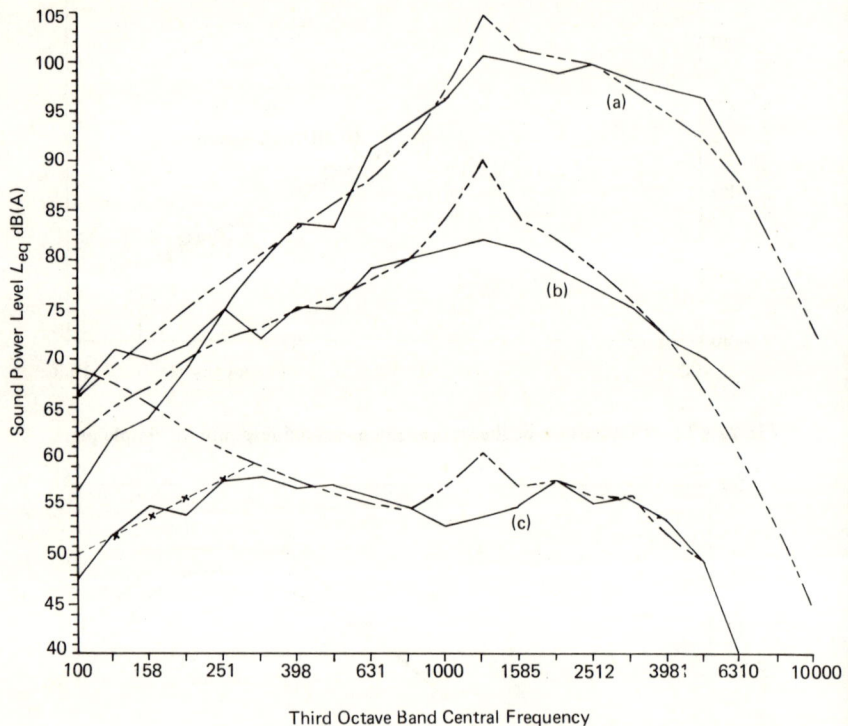

Third Octave Band Central Frequency

Fig. 22.63 – Radiated noise energy for equal momentum change, for different excitation of plate. Showing the high attenuation due to change in structure response. ————— Measured. ——‧‧—— Estimated. (a) Plate directly excited. (b) Excitation through blocking mass. (c) Excitation through blocking mass coupled to plate via isolators.

With this confirmation of the diagnostic value of equation (22.5) we now proceed to consider generally the practical noise reductions made possible by changes in damping, resilience, and radiation efficiency.

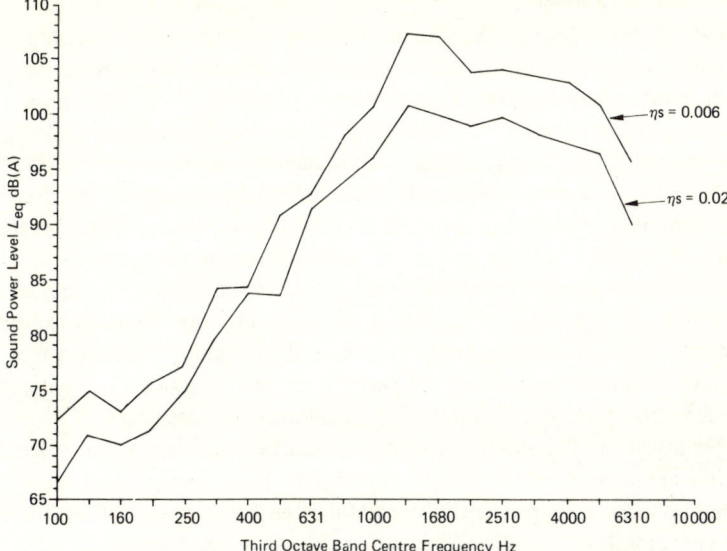

Fig. 22.64 – Metal-to-metal impact for different plate damping.

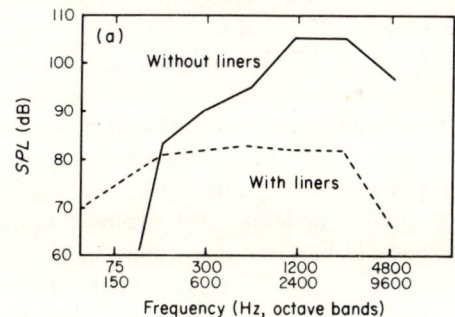

Fig. 22.64 a – Noise reduction from metal castings tumbling drum through using rubber liner.

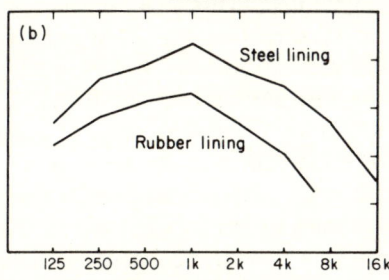

Fig. 22.64 b – Noise reduction obtained by addition of rubber lining in cement manufacture.

22B.3.3 Handling noise

While metal fracture dominates noise mechanisms in highly loaded punch presses, or croppers, there exists also a wide variety of handling or forming mechanisms in which the vibrational energy input arises from hard metal-to-metal impacts, or from the impacts of glass or rock with metal. Examples of metal-to-metal impacts include those of metal tubes or billets being handled across a factory area, drop hammer mechanisms, tumblers, textile and braided wire-making shuttle arresters, swaging hammers, pile drivers, and a variety of dockyard hammering processes. Hard impacts between glass and hard metal conveyor belts, between metal balls and rock fragments in cement making tumblers, between manufactured forgings and metal containers (stillages), and between the vibratory excitation mechanisms and the rotating metal drums used in making concrete tubes, are just a few further examples of the way vibrational energy and noise can be introduced into processes without full cognisance by designers of the noise principles involved. The noise radiated as a result of backlash in plain bearings is another application which will be referred to later. In each and every case, noise can be reduced significantly by careful attention to any or all of the terms in the expression for $L_{A\text{eq}}$ given in equations (22.4) or (22.5).

The above experiments on a ball hitting a plate explain the noise reductions achieved by the addition of rubber liners inside rotating tumbler drums used for deburring metal stampings [22.28] (Fig. 22.64 a) and for crushing the rock fragments in cement making [22.29]. In each case the noise is measured as a continuous sound pressure level, but since the rotational speed and the cage geometries inside the drums have been unaltered by the introduction of the softer liners, these readings reflect the L_{eq} changes (per impact) quite accurately.

If such drums contained no intrinsic structural damping, the reductions would have been explicable in terms of the increase in damping due to the rubber. It is seldom possible, however, to assemble fabricated structures without some considerable amount of damping (say $\eta_s = 0.005-0.010$) so that the addition of the small amount of damping in the rubber is not in itself a worthwhile palliative. The noise reductions shown in Figs 22.64 a, b must be attributed to change of impulse shape rather than to material damping.

This detuning of the maxima (see Fig. 22.62 d) in the impulse spectral content from that of the peak in modified radiation efficiency so that their energy contributions are added linearly over the frequency bands rather than multiplied together at coinciding frequencies must be accepted as one of the strongest mechanisms for noise control available to us. In general, it leads to the general conclusion that great reductions are possible if every attempt is made to soften impacts, whether these be the primary ones associated with the operation of the machine or whether they be secondary and unintended ones arising from backlash, roughness, conveyor belt clatter, or something else. This applies also to gear design, diesel engine frame excitation, hydraulic pumps, saws, and planers, and to a series of other machinery mechanisms which at first sight appear to create continuous sounds. Each impact can be treated separately and more easily understood if the repetition rate is not allowed to become a primary factor in the analysis.

22B.3.4 Noise arising from backlash in mechanisms

Unintended secondary impacts are often the sources of the dominant noise in mechanisms in which the severe basic impact has been designed out of the system and wear has led to excessive backlash in the bearings.

Fig. 22.61 shows the noise emanating from the Southampton punch press before and after overhaul. Prior to overhaul, the reduction of noise to be expected from a reduction of percentage clearance in the punch-die set is not evident, as the dominant noise stems from the high-frequency structural vibration arising from the bearing backlash. Once these have been eliminated (Fig. 22.61) the noise reduction obtained by reducing percentage clearance agrees with that to be expected from the reduction of F'^2_{max} in the punch.

Noise arising from backlash can in turn be predicted from equations (22.4) and (22.5) if the vibratory velocity or F' level arising from these impacts are measured. Herbert & McWhannel [22.30] indicate the kinds of acceleration histories obtainable (Fig. 22.65) for the bearing accelerations with small and excessive clearances and without and with heavy grease lubrication and rubber bushes. It may be seen that the combined effects of no lubrication and a tenfold increase in bearing clearance is to increase the acceleration level some tenfold;

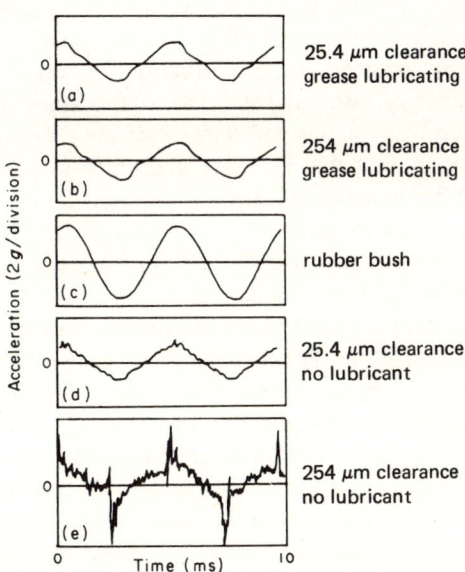

Fig. 22.65 – Vibration signatures from the push rod lever mechanism.

the noise increase arising from such wear (shown [23.30] for a full-scale mechanism in Fig. 22.66) is in keeping with the pattern shown in Fig. 22.65. The noise increase is greatest at high frequencies corresponding to the short metal-to-metal impact times rather than at the low frequencies associated with the smooth repetition cycle of the initial mechanism when the radiation efficiency of the surrounding structure is poor.

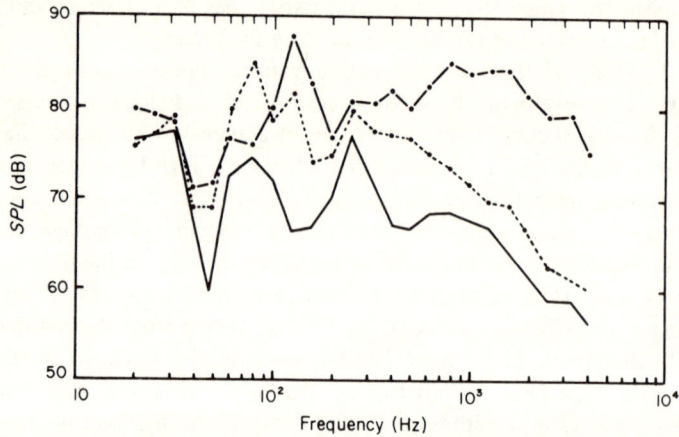

Fig. 22.66 – Noise levels radiated from a full-size mechanism.

– · – · – worn mechanism   – – – – new mechanism   ——— background noise

In terms of equation (22.4), we can say that the noise due to backlash is not that associated with the cyclic time, nor of the broad pulse shape of the backlash (that is of the frequency f_1), but rather of the very short impact times t_2 and high frequencies f_2, associated with the infrastructure of the impact arising from roughness, local flat points, and the lack of any cushioning grease lubrication at the points of impact; that is, of the magnitude of $[F'(f)]^2$ at the frequencies of high modified radiation efficiencies (see Table 22.11).

This factor allows us to draw attention to the fact, seldom referred to in bearing and backlash analysis, that the noise increase due to wear is a function of two variables, the kinetic energy transferred from the housing and the link to the nearby radiating structure, and the non-dimensional impact *time* over which the deceleration of the link occurs. In other words, the magnitude *and* the detailed shape of the impulse must be studied. This is illustrated in Fig. 22.65 where it may be seen that if a bearing is packed with grease, a large clearance can be tolerated without any increase in acceleration levels, because the rate of transfer of load and the kinetic energy developed by the link in 'free flight' is minimized and the sharpness of the pulse is not increased. This is in keeping with the well-known fact that backlash noise can be quite small while lubricating oil is cold and thick. The relevant impact times t_1 and t_2 and the corresponding frequencies associated with $[F'(f_0)]^2$ are shown in Table 22.11.

Table 22.11

Impact times and corresponding frequencies for various machine operations

Operation	t_1 (seconds)	f_1	t_2 (seconds)	f_2
Impacting 10.2 cm (4 in) steel balls	2×10^{-4}	1.5 kHz	10^{-4}	3 kHz
Drop hammer (large)	2×10^{-3}	150 Hz	2×10^{-4}	1.5 kHz
Punch press	1.5×10^{-2}	25 Hz	5×10^{-4}	600 Hz
Petrol engine combustion		150 Hz		300 Hz
Diesel engine combustion	1.7×10^{-1}	2 Hz	10^{-5}	30 kHz
Diesel engine piston slap	1.7×10^{-1}	2 Hz	2×10^{-4}	1.5 kHz
Hydraulic pump (constant displacement)	3×10^{-3}	100 Hz	2×10^{-4}	1.5 kHz
Backlash (no lubrication)	2×10^{-4}	1.5 Hz	10^{-4}	3 kHz
Backlash (packed grease)	2×10^{-2}	15 Hz	10^{-3}	300 Hz

Earles & Wu [22.31] have put forward an expression for the impulse arising from a joint with clearance, in terms of the loading conditions and the geometry in a system without backlash. If R is the minimum absolute value of bearing reaction force (in Newtons) and $\dot{\gamma}$ is the time rate of change of the direction of R (in radians per second), Earles' and Wu's experiments suggest that if $\dot{\gamma}/R > 1$, contact will be lost during the cycle, and that the severity of the impact will increase as the product of two terms, $\log_e \dot{\gamma}/R$, and $K(2.8^r - 1)$ where r is equal to $1000 \times \dfrac{\text{bearing clearance}}{\text{bearing diameter}}$ and K has a value which depends upon the lubrication. Thus:

$$K = 0.61 \text{ (unlubricated)}$$
$$K = 0.48 \text{ (boundary lubricated)}$$
$$K = 0.16 \text{ (film lubricated)}.$$

As the noise energy induced will be proportional to f'^2_{max} or to (acceleration)^2 for a given frequency, the decrease of noise level may be expected to vary as $20 \log K$, which between the extreme cases of $K = 0.16$ and $K = 0.61$ gives a noise reduction of 12 dB. This will of course be accompanied by a lengthening of the impact duration and a reduction of the peak excitation frequency; the exact amount of reduction to be expected at a given frequency must however, depend upon whether the reduction in acceleration and therefore of $f'_{max}(t)$ will be accompanied by a longer time of acceleration and an unchanged total energy transfer. Intuitively, it seems reasonable to assume that the presence of thin oil will not alter the kinetic energy developed by the link while floating across the contacts, and that the main purpose of the oil will be to increase the contact time at the end of its travel; $\langle v^2 \rangle$ in the neighbourhood of the bearing surface will be unaltered, and the L_{eq} reduction will in practice only be that associated with the reduction of the modified acoustic efficiency term arising from the general lowering of the excitation frequency associated with increasing the impact time.

Reduction in the general level of the total impulse can be achieved only if the oil can carry momentum across the gap throughout the period of travel of the pin, and this requires the use of heavy grease or a highly pressurized hydrostatic bearing. Under these circumstances, the bearing has a greater stiffness (proportional to the cube of the film thickness) and is free of the radial slackness associated with other lubrication configurations.

22B.3.5 Stillage design

In many factory situations, the dominating noise arises from the transfer of heavy metal objects into bins or stillages. These objects are often hot and are thrown with considerable energy into the steel stillages which must be robust, mobile, cheap, and capable of accepting red-hot materials, They are often therefore made of relatively thick steel whose radiation efficiencies are high at all but very low frequencies.

Several methods of noise reduction are available to us, all indicated in the terms of equation (22.4). The total initial energy of impact can be reduced by some prior soft impacts with a series of deflectors, or springs. The softness of the final impact can be increased by making the wall and floor surfaces of the bin more resilient (reducing frequency); structural damping can be added; the walls can be perforated, thus reducing surface area and also reducing radiation efficiency below the critical speed; while the bulkiness factor can be increased by thickening the walls, specially those carrying the highest vibration densities.

Experiments have been carried out at the ISVR [22.32] to establish the scope of the reductions which are possible, and the worthwhileness of such measures, bearing in mind that once the stillages are partially filled, the dominating noise comes from the impact of the heavy objects upon each other rather than on the walls and floor of the containers.

Only robust installations have been examined, as such bins or stillages are subject to extremely rough use in emptying, and are often moved from one factory area to another where the duties involved are very different. Thus wire gauze or expanded metal bins, wooden bins, and some of those consisting of poorly bonded double metals have been excluded from our investigations. Our experiments have included the use of two thicknesses of steel (to examine the relative noise reduction from increased bulkiness factors and increase from reduced resilience and increased radiation efficiencies), of a perforated metal (to examine the gain from reduced radiation efficiency and increased resilience without change of damping), of point-welded double sheets and sound-deadened double sheets containing a visco-elastic damped interlayer (to examine the effects of fluid and visco-elastic damping respectively), and finally a woven system of relatively broad metal strips so arranged as to provide increases in resilience and friction damping and reductions in radiation efficiency and surface area.

Experiments have included elementary strikes by a steel sphere [22.32] (a sphere was used to minimize the ringing noise in the workpieces themselves), similar experiments with complete bins using 'tilt and drop' as well as workpiece impact tests, and in some cases complete factory-based lift tests have been included [22.33].

Table 22.12 – Panel impact results

| Panel type | Peak SPL | | $L_{eq}(0.2\,s)$ | |
	dB	db(A)	dB	dB(A)
1	118	116	102	102
2	113	111	99	96
3	110	106	96	86
4	105	103	84	83
5	103	102	74	73
6	103	101	88	84
7	103	103	84	84

Panel Construction

Panel	Material
1	10 gauge mild steel sheet (3.25 mm)
2	14 gauge mild steel sheet (2.02 mm)
3	2.7 mm Sound Deadened Steel – ambient grade (18 g steel/damped interlayer/18 g steel).
4	Overlaid 16 gauge × 20 mm wide mild steel strips (two layers at right angles, each strip butted to the next).
5	Woven 16 gauge × 20 mm wide mild steel strips (20 mm × 20 mm 'holes' in weave).
6	Three 22 gauge mild steel sheets spot-welded together at 100 mm centres.
7	10 gauge perforated mild steel sheet – Associated Perforators and Weavers type 567 (¼″ holes, 40% 'open' area).

A summary of the peak noise and the L_{eq} reductions obtained in the controlled impact experiments (shown in Fig. 22.61) are given in Table 22.12. As the coincidence frequency falls with increasing plate thickness and as the slope of the impulse will increase directly with the plate stiffness (or inversely with the resilience), thickening the plate must result in a reduction in coincidence frequency and a greater radiation of the high-frequency vibration energy which will itself have been raised by the increased stiffness. Thus the small improvement associated with an increase in the bulk factor (10 log 1.5) must be expected to be overshadowed by the hardening of the blow and the increase in efficiency of radiation. This may be seen in Fig. 22.67a,b and c which show respectively a comparison of the growth of radiated noise power with time, the noise energy content in any given 40 Hz bandwidth, and a cumulative summation of the noise energy with increasing frequency.

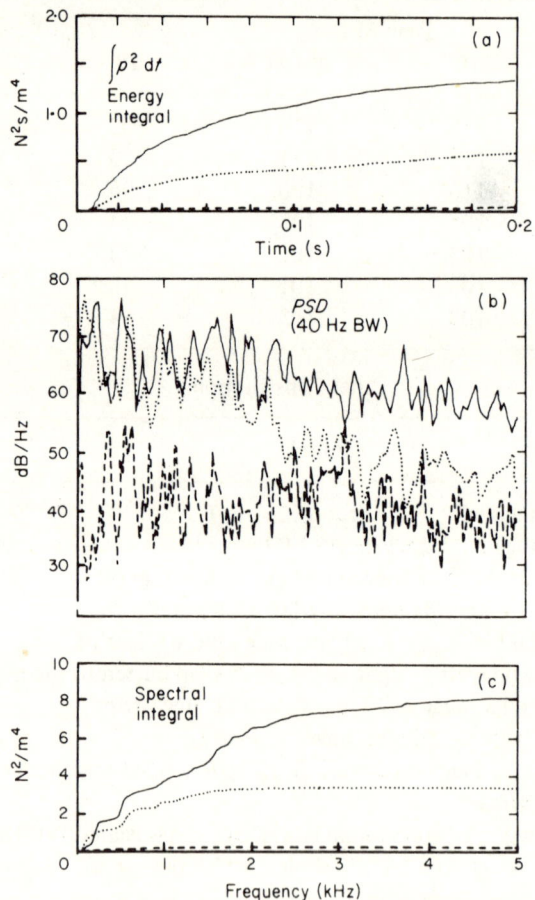

Fig. 22.67 – a, The variation of noise energy output with time; b, with frequency; and c, cumulative with frequency.

———— panel 1 (thick) ········ panel 2 (thin) – – – – panel 7 (perforated, thick)

It may be seen that there is a little effect of reduced thickness on damping (seen as the rate with time of the noise energy build up during ringing), but that the use of thin material (and an increased resilience) leads to a concentration of energy in the low-frequency range, and the elimination of noise energy at those frequencies in which high radiation efficiencies would be expected.

This also explains why reducing the material thickness gives greater reduction in L_{Aeq} than if the energy is summated linearly with frequency, even though the changes in peak sound pressure levels are unaltered by changing the acoustic weighting. This characteristic is noticeable to a much greater extent with material in which damping has been deliberately added; for example the sound-deadened steel (SDS) double plates tested in our experiments. It may be seen in Table 22.12 that whereas the L_{Aeq} is reduced very significantly by virtue of the increased damping and increased resilience of the double (but thinner) steel sheets incorporating a visco-elastic interlayer, the same reduction is not obtained if the summation over the different frequencies is carried out linearly. Equally the peak sound pressure level is not decreased greatly by the use of an A-weighting filter.

This same phenomenon was found to occur in factory tests carried out on orthodox and sound-deadened steel (SDS) stillages, Table 22.13 shows the

Table 22.13

The measurement stiffness, estimated surface density and calculated coincidence frequency for various panel types

Panel	$B(N.m^2)$	$\rho_s(Kg/m^2)$	$f_c(Hz)$
10 g SS	700.27	25.513	3512
10 g PS	217.55	15.308	4880
SDS	289.76	19.625	4780
Woven	44.11	13.345	10120
3 layers	128.19	16.721	6645

Table 22.14

The total A-weighted and linearly weighted noise energy radiated from a standard and sandwich type stillage during drop, and tilt tests (empty). Energy input to microphone per impact (J/m^2)

Stillage	Drop test		Tip test	
	A	Linear	A	Linear
Standard	9.5×10^{-3}	2.4×10^{-2}	1×10^{-2}	2×10^{-2}
SDS	2.4×10^{-3}	3×10^{-2}	4.1×10^{-3}	4.1×10^{-2}

A-weighted and linear noise energy outputs arising from dropping workpieces into standard and SDS type stillages, and from dropping the stillages while empty from a tilted position. It may be seen that whereas the L_{Aeq} is reduced by changing from single steel sheet to the SDS configuration, the total noise energy emitted, summed linearly with frequency, is actually increased by using the sound-deadened material. This again is in keeping with equation (22.4); the reduction in L_{eq} from the increased damping term being nullified by the increased values of $[F'(f_o)]^2$ at low frequencies as f_2 is decreased.

In view of the differences in the nature of the noise laws in different countries, some of which emphasise peaks, and the general view now being expressed that a flatter frequency weighting curve would correspond more with hearing statistics, it is worthwhile dwelling on this anomaly for a moment. Fig. 22.68 shows the damping factor η_s we have measured for our array of panels

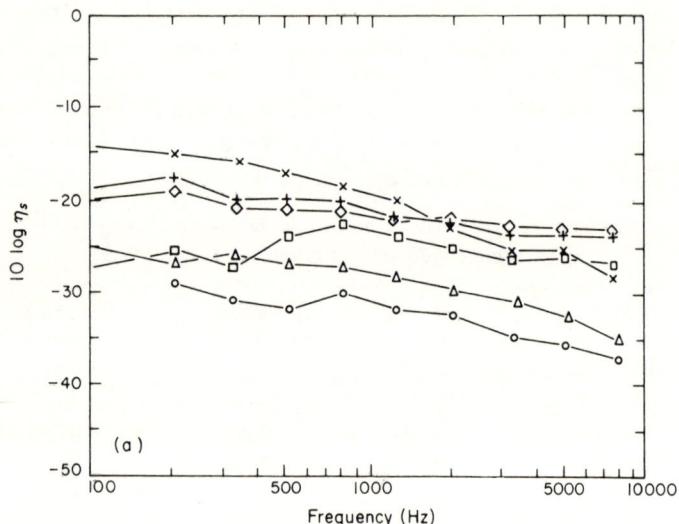

Fig. 22.68 – Damping factors for various panel configurations.

x—x 3 layers, +—+ SDS, ◊—◊ strips spot welded, □—□ woven, △—△ 10 g S sheet, ○—○ 10 g perforated sheet.

[22.32] when mounted in their stillage frames, and when vibrated both by one third octave white noise. The accuracies are not great, and variations from octave to octave are less important than the broad generalization we can make (that η_s is about 0.01 for damped panels), whether of SDS, spotwelded or woven construction, and is about ten times as great as the values occurring for the single skin construction, whether perforated or not. Since an η_s of 0.01 (per radian) implies an absorption of about a sixteenth of the vibrational energy per cycle, the high-frequency vibrational energy associated with sharp jumps in the pulse shapes will be rapidly absorbed, and the continued ringing which dominates the L_{eq} contribution will arise from the poor damping associated with the low frequencies. This of course is incorporated into the modified radiation frequency terms shown earlier and included in equations (22.4) and (22.5). Thus the effect of increased damping below 500 Hz is minimal on L_{eq} whereas the peak noise is related directly to the initial pulse shape and is not related to the damping level.

It may be seen from Table 22.12 that a very significant reduction in L_{eq} both linear and A-weighted is obtainable by the use of one-quarter inch perforations which leave a 60% closed area. As the structural damping level is unaltered from that of a single sheet (based upon actual surface area) this reduction must be explained by the change in pulse shape arising from the increased resilience, and by the increase in the coincidence frequency and the elimination of corner and edge modes below this coincidence frequency.

Table 22.13 indicates the changes in the measured stiffness and the resultant calculated coincidence frequency for all the panels investigated [22.31]. It may be seen that the decreased stiffness implies an increased coincidence frequency from 3500 to 4900 Hz.

Probably of more importance is the increased difficulty for a perforated panel to establish any degree of radiation below this frequency owing to the cancellation of any pressure difference through the holes. It may be seen in Part A that this inefficiency arising from the elimination of edge and corner modes below coincidence can amount to a very significant reduction in radiative power at the likely peak excitation frequencies corresponding to heavy metal-to-metal impacts. Thus the use of heavily perforated conveyor systems for billet carriage can be advantageous.

While the results shown for the perforated sheet are correct for our laboratory conditions, the same success has not been achieved in the field, probably because the exact amount of structural damping in the structure is small and depends very much on the way the sheet is welded to the frame. A construction which contains damping throughout its structure is therefore preferable, though our laboratory experiments on perforated forms imply that this may not be essential.

Following this finding, an attempt was made to optimise the L_{eq} reduction obtainable when an open mesh construction and increased resilience is coupled with an increased damping level. Fig. 22.69 shows a stillage made by a woven strip system wherein good strength is coupled with high frictional damping, a high resilience, an open mesh, and a very high coincidence frequency (Table 22.13). It may be seen in Table 22.12 that an $L_{A eq}$ of almost 30 dB reduction compared with a 14 gauge single sheet panel and a peak sound pressure level reduction of some 15 dB is achievable. Similarly, a combination of three thin sheets spotwelded to give some degree of fluid damping in the air interlayers reduces the L_{eq} by a very appreciable amount (Table 22.12).

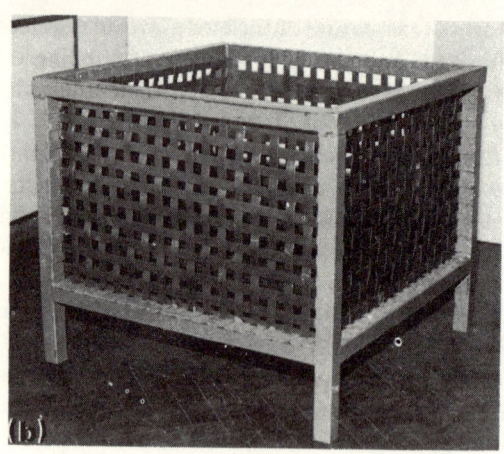

Fig. 22.69 – Woven stillage.

If we draw together all these results we can argue that additional damping to *very* poorly damped installations can provide 10 dB reduction, and impulse shaping (increasing resilience) by the use of a thinner skin, or by using a soft woven metal fabric will lead to up to a further 10 dB reduction in L_{eq}. All these reductions can be additive provided we bear in mind the noise reduction and frequency mismatching mechanisms that are involved and incorporated in the terms of equation (22.4), that is, that damping effectiveness depends upon the damping level being low initially, that impulse shaping does not necessarily eliminate vibration but moves it away from the high radiative frequency regimes, and that much of the reductions in radiation efficiency depends not only upon raising the coincidence frequency but also reducing its value below the critical frequency by eliminating corner and edge radiations.

22B.3.6 Bottle clatter

It is well known that bottle clatter, an early-morning environmental noise nuisance, as well as a bottling factory hearing hazard, can be eliminated by replacing the glass by paper or plastic materials, or by carrying the glass bottles in plastic lined containers. What is not known is the relative importance of structured damping and resilience in obtaining this solution. Does the noise reduction arise from the adding of even a minimal amount of damping to a bottle or is the mechanism one of softening impacts and the lowering of the excitation frequency with its subsequent noise reduction? Table 22.15 shows [22.33] the measured damping coefficients, the impact times, the change in peak noise levels, and the change in short-time L_{Aeq} measured when two bottles were struck glass-to-glass, and through relatively thin thermal and visco-elastic resilient layers respectively, both of which modified the impact shape considerably.

Table 22.15(a) – Bottle impacts – effect of change of hardness of impact (with constant damping)

	t_o change m/sec	ΔL_{peak} dB	ΔL_{Aeq} (0.08) dB	Change of η_s	Frequency shift kHz
Use of soft damped impact point	0.45–1.50	13	16	7.0–7.3×10^{-3}	3.3–1.0
Use of soft damped cover	0.50–1.60	16	18	15–14×10^{-3}	3.0–0.94
Use of soft polyuethane impact point	0.45–1.90	21.5	27	6–6.2×10^{-3}	3.3–0.79
Use of soft polyurethane cover	0.45–1.80	24	25	6.4–6.0×10^{-3}	3.0–0.79

Table 22.15(b) – Bottle impacts – effect of damping (hardness constant)

	η_s change	t_o change	Frequency change	ΔL_{peak} dB	ΔL_{Aeq} dB
Hard impact-adding damped cover	7.0–15.0×10^{-3}	$0.45 \times 0.50 \times 10^{-3}$	3300–3000	0	2
Soft impact-adding damped cover	7.3–14.0×10^{-3}	1.5–1.60×10^{-3}	1000–940	3	2
Hard impact-adding polyurethane cover	6.0–6.4×10^{-3}	0.45–5.0×10^{-3}	3300–3000	-2	1
Soft impact-adding polyurethane cover	6.2–6.0×10^{-3}	1.9–1.9×10^{-3}	790–790	0.5	-1

It may be seen that doubling the damping while keeping the impact time constant provides a small reduction of 2 dB in L_{Aeq} (compared with 3 dB prediction from equation (22.4), while no consistent reduction in peak level is recorded. On the other hand, a change in the resilience (or hardness) of the impact (while keeping damping constant) makes a difference of the order of 20 dB both in peak level and L_{Aeq}.

Fig. 22.57 showed the variation of the peak pressure perturbation assoçiated with deceleration during impacts of different durations. It may be seen that the points all lie between the theoretical curve for spheres and the maximized values for any other shape (and which were derived in Part A). The variations of spectral content with damping and with the softening of impacts, for whatever reason, provided $F'^2(f)$ is reduced provides a uniform reduction throughout the frequency range. Thus the reductions both in peak noise and L_{Aeq} (Fig. 22.58) obtained in the experiments can be traced to the change in the effective level of excitation (the excited preferred frequencies were reduced from about 3000 Hz to 900 Hz), and the reduced ability of the bottles to radiate efficiently both in the transient 'deceleration' and the 'ringing' modes of oscillation.

The reductions in noise are effectively as great when a soft low damping cover is added, as when the cover is of specially prepared damped visco-elastic material, the mechanism of noise reduction being that of changing impact time, not of energy absorption. Thus any conveyor system or bottle carrying system needs to rely on the need for softness rather than elaborate structural damping. This will probably be true also of any other similar application. It will be the softening at the points of impact that matter.

22B.4　NOISE FROM CONVEYOR SYSTEMS

The previous section illustrates the principles and parameters involved in impact noise prediction. There are many instances in which these parameters cannot be evaluated in detail, but where the application of the general principles of noise control, enunciated earlier, can pay very considerable dividends if they are incorporated into the early design thinking. Probably the greatest of these areas of design lie in the field of conveyor design, not just those carrying bottles and tins, but also those involved in carrying heavy steel tubes and billets across factory floors. A not-unusual method of transfer of metal billets across a floor is illustrated in Fig. 22.70, each slip process involving the basic characteristics which typify the problem; that it is next to impossible to add damping to the workpiece, and that in many instances the products themselves are used to transmit the motive power to propel each other along the conveyor system. Bends and expanded collection areas are commonplace as a means of ensuring a continuous and full flow of bottles into the filling machines, even though such techniques ensure periodic accelerating and high-energy impacts between bottles. Thus, some degree of impact is inherent in the design of conveyor systems for bottles, tins, tubes, and the like. The sources of noise are a product of these impacts and the subsequent undamped ringing of the bottles, tins, and tubes while they dissipate the vibrational energy implanted into them.

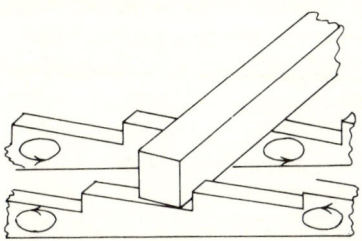

Fig. 22.70 – Method of handling large ingots across factory floor.

The practical possibilities of noise control must depend upon the application, but, apart from the obvious one of total or partial enclosure, must involve one or more of the following principles:

(1) Reduce the impacts to a minimum by careful study of the minimum energy needed for propulsion of the workpieces, and use other methods of propulsion of the workpieces through the conveyor system rather than through the workpieces themselves.

(2) Lengthen the impact times by using resilient materials for the mechanisms of halting and propagation.

(3) Prevent ringing of the conveyor system itself by using damped and resiliant materials and reducing its radiation efficiency.

(4) Discard collection systems which are required to dissipate considerable energy by successive impacts of the workpieces, for example, by letting tubes fall excessive distances into collection slings or by relying on 'can-to-can' transfer of energy to turn corners in conveyor systems.

(5) Add damping or resilience to or between workpieces if at all possible; for example, by covering bottles with resilient labels (as used in USA on Coca-Cola bottles, or by hanging damping strips in the paths of bottle lines.

There is little point in dwelling on the details involved in applying such principles to practical cases, as no two installations will be similar. There are some broad conclusions to come to, however, as a result of studying such applications.

The first conclusion is that 90% of ancillary impacts are unnecessary in factories, and alternative solutions can be found once the noise source has been identified. For example, mechanical designers have too easily assumed that wear will be excessive in installations other than those giving metal-to-metal contacts. The replacement of one or other of the impacting surfaces by a relatively hard non-metalic material has too often been turned down on the basis of an assumed short life. Rubber or plastic pads designed to be replaceable, can often eliminate excessive impact noise in chutes and conveyor corners where directional changes are achieved by impact mechanisms.

Similarly, vibration is often used as a smoothing mechanism where stickage tends to occur, without the realization that the elimination of stickage in itself encourages multiple impacts between cans, balls, bottles, and that the major part of the vibrational energy introduced to overcome local stickage finishes up as noise. Magnetic conveyor belts designed to hold the base of tins against the vibratory or unsteady forces inherent in a mechanical conveyor system can eliminate much of the clatter.

REFERENCES

[22.1] *Code of practice for reducing the exposure of employed persons to noise*. (1972) London, HMSO.

[22.2] Brüel, P. V. (1976) *Proc. Inter-Noise 76*, 111–116, Do we measure damaging noise correctly?

[22.3] Richards, E. J. & Herbert, A. G. (1977) *Proc. Inter-Noise 77*, Zurich, A new method of analysing impact noise energy.

[22.4] Longhorn, A. L. (1952) *Quarterly Journal of Mechanics and Applied Mechanics*, **5**, 64–81. The unsteady motion of a sphere in a compressible inviscid fluid.

[22.5] Richards, E. J., Wescott, M. E. & Jeyapalan, R. K. (1979) *Journal of Sound and Vibration*, **62**, 547–575. On the prediction of impact noise, Part I: Acceleration noise.

[22.6] Koss, L. L. & Alfredson, R. J. (1973) *Journal of Sound and Vibration*, **27**, 59–75. Transient sound radiated by spheres undergoing elastic collision.

[22.7] Richards, E. J. (1978) *Proc. Inter-Noise 78*, San Francisco. Impact noise from industrial machinery: some general laws regarding its magnitude.

[22.8] Richards, E. J., Wescott, M. E. & Jeyapalan, R. K. (1979) *Journal of Sound and Vibration*, **65**, 3, 419–451. On the prediction of impact noise, Part II: Ringing noise.

[22.9] Jeyapalan, R. K. & Doak, P. E. (1980) *Journal of Sound and Vibration*, **72**, (3). Sound energy calculation of transient sound sources by the radiation method.

[22.10] Johnston, R. A. & Barr, A. D. S. (1969) *J. of Mec. Eng. Science*, **11**, 117–127. Acoustic and internal damping in uniform beams.

[22.11] Meier-Dornberg, K. E. (1969) *Die Beschreibung von Stossvergängen durch ihre Zeitsfunctionen*. Fourier und Schockspectrum. VDI Verichte 135.

[22.12] Evensen, H. A. (1980) *Journal of Sound and Vibration*, **68**, 3. A fundamental relationship between force wave form and the sound radiated from a power press during blanking and piercing.

[22.13] Cremer, L., Heckl, M. & Ungar, E. E. (1973) *Structure borne sound.* New York, Springer Verlag.

[22.14] Goyder, H. G. D. & White, R. G. (1980) *J. Sound Vib.* **68**, No. 1, p. 59–118. Vibrational power flow from machines into built-up structures.

[22.15] Priede, T. (1978) *Course notes – Industrial and machinery noise control,* ISVR, Transient Sources of Noise in Machinery.

[22.16] Healiss, K. (1975) Problems of noise in the production engineering industry. *Proc. of Indust. Noise Control Conf.,* London. 7/8 th May.

[22.17] CONCAWE (1977) Report No. 8/77 *Measurements of vibrations complementary to sound measurements.*

[22.18] Wilby, J. F. (1975) *90th meeting of Acoustical Society of America.* Paper 49, Noise of steam injection water heaters.

[22.19] Lyons, R. (1978) *Proc. Inter-Noise 78,* San Francisco. Noise reduction by design – an alternative to machinery noise control.

[22.20] Richards, E. J. & Carr, I. On the prediction of impact noise. Part V: The noise from drop hammers. Accepted for publication in *Journal of Sound and Vibration.*

[22.21] Richards, E. J. & Lenzi, A. On the prediction of impact noise. Part VI: Bottle impacts. Accepted for publication in *Journal of Sound and Vibration.*

[22.22] Evensen, H. A. (1980) *Journal of Sound and Vibration,* 68/3. A fundamental relationship between force wave form and the sound radiated from a power press during blanking and piercing.

[22.23] Burrows, J. M. (1979) Southampton University, MSc Thesis. The influence of tooling parameters on punch press noises.

[22.24] Herbert, A. G. & Burrows, J. M. (1979) *Proc. Inter-Noise 79,* Warsaw. Noise in blanking and piercing.

[22.25] Stewart, N. D., Bailey, J. & Daggerhart, A. (1974) *Experimental investigation of noise control of a 60-ton power press.* Centre for Acoustical Studies.

[22.26] Stewart, N. D. Bailey, J. & Daggerhart, A. (1975) *Noise and Vibration Control Engineering.* Study of the parameters influencing punch press noise.

[22.27] Koss, L. L. & Alfredson, R. J. (1974) *Journal of Sound and Vibration,* **34**, 11–33. Identification of transient sound sources in a punch press.

[22.28] American Industrial Hygiene Association (1966) *Industrial noise manual* 2nd Edition.

[22.29] Koolshaw, G. L. (1970) IEEE *Transactions on industry and general applications,* **IGA-6**, No. 5, Reducing machinery noise in cement plants.

[22.30] Herbert, R. G. & McWhannel, D. C. (1976) ASME Paper 76-DET-36. *Shape and frequency composition of pulses from an impact pair.*

[22.31] Earles, S. W. E. & Wu, C. L. S. (1976) *Inst. Mech. Eng. Tribology*

Convention. A clearance magnitude relationship for plain bearings in oscillating systems.

[22.32] Lenzi, A. (1978) Southampton University, MSc Thesis. Institute of Sound and Vibration Research, Study of parameters affecting component/stillage impacts.

[22.33] Croker, M. (1978) *DFRA Contract*. Results from impact testing seven different types of panel construction for a 'quiet' stillage.

[22.34] Occupational Safety and Health Administration (OSHA) of the US Department of Labor (1974). US Federal Register 1974, Vol. 39, 207. *Occupational noise exposure: proposed requirements and procedures.*

[22.35] Smith. P. W. & Lyon, R. H. (1965) *NASA Contract Report CR 160*. Sound and structural vibration.

APPENDIX 22.I

A FORMULA FOR THE NOISE ENERGY RADIATED FROM A SIMPLE STRUCTURE DUE TO AN IMPULSE

Consider a structure which has many natural modes, e.g. a beam. It is excited by a transient normal force $f_c(t)$ at point C. The type of structure we consider is one in which the only velocities of movement are normal to the surfaces. The response velocity $v_c(t)$ at the point C can be written

$$v_c(t) = \int_{-\infty}^{\infty} h_c(\tau) . f_c'(t - \tau) \, d\tau$$

where $h_c(\tau)$ is the impulse response function relating to the force input and displacement output at point C. i.e.

$$x_c(t) = \int_{-\infty}^{\infty} h_c(\tau) f_c(t - \tau) \, d\tau.$$

In the frequency domain

$$V_c(f) = H_c(f) F_c'(f)$$

where $V_c(f)$, $H_c(f)$ and $F_c'(f)$ are the Fourier transforms of $v_c(t)$, $h_c(t)$ and $f_c'(t)$.

The energy supplied to the structure during the period T of the impulse is

$$E = \int_0^T f_c(t) . v_c(t) \, dt = \int_{-\infty}^{\infty} f_c(t) . v_c(t) \, dt$$

since $f_c(t)$ is zero except between $t = 0$ and $t = T$, and this can be written in the frequency domain as

$$E = 2 \int_0^{\infty} \text{Real}[F_c^*(f) . V_c(f)] \, df.$$

This gives the total energy absorbed by the structure during the impulse provided there is no plastic deformation at the point of application. The energy at frequency f_0 and in a frequency band Δf will then be

$$E(f_0) = 2 \int_{f_0 - \frac{\Delta f}{2}}^{f_0 + \frac{\Delta f}{2}} \text{Real}[F_c^*(f) V_c(f)] \, df = 2\Delta f . \text{Real}[F_c^*(f_0) . V_c(f_0)]$$

and as

$$V_c(f) = H_c(f).F'_c(f)$$

$$E(f_o) = 2\Delta f. \text{Real}[F_c^*(f_o).F'_c(f_o).H_c(f_o)]$$

and as we can replace

$$F_c^*(f_o) \text{ by } \frac{F_c^{*'}(f_o)}{j2\pi f_o}$$

$$E(f_o) = 2\Delta f. \text{Real} \; |F'_c(f_o)|^2 \frac{H_c(f_o)}{j2\pi f_o}$$

Now this energy will be dissipated predominantly as heat into the structure with a small amount going into radiated sound. $E(f_o)$ can therefore be equated largely to the vibrational energy dissipated in the structure between $t = 0$ and $t = \infty$. That is, at frequency f_o

$$E(f_o) = 2\pi f_o \eta_s . \rho_s S . \int_0^\infty \langle v_n^2(f_o, \Delta f, t)\rangle \, \mathrm{d}t$$

$$= 2\pi f_o . \eta_s . \rho_s S \int_{f_o - \frac{\Delta f}{2}}^{f_o + \frac{\Delta f}{2}} \langle |V_n(f)|^2\rangle \, \mathrm{d}f$$

using Parseval's theorem.

$$E(f_o) = 2\pi f_o \eta_s \rho_s S \langle |V_n(f_o)|^2\rangle \Delta f.$$

Equating the two expressions for $E(f_o)$ we obtain the space averaged velocity squared term:

$$\langle |V_n(f_o)|^2\rangle = \frac{1}{2\pi f_o \eta_s \rho_s S . \Delta f} 2\Delta f \; \text{Real} \left[|F'_c(f_o)|^2 \frac{H_c(f_o)}{j2\pi f_o} \right]$$

$$= \frac{1}{\pi f_o \eta_s \rho_s S} \text{Real} \left[|F'_c(f_o)|^2 \frac{H_c(f_o)}{j2\pi f_o} \right].$$

The A-weighted radiated sound is obtained simply by multiplying the surface vibration level by the A-weighted radiation efficiency $(A\sigma_{\text{rad}})$ at frequency f_o and by the characteristic impedance $\rho_o c$. That is,

$$E_{\text{rad}}(f_o) = A\sigma_{\text{rad}} \rho_o c S \int_0^\infty \langle v_n^2(f_o, \Delta f, t)\rangle \, \mathrm{d}t$$

$$= 2\rho_0 c S A \sigma_{\text{rad}} \int\limits_{f_0 - \frac{\Delta f}{2}}^{f_0 + \frac{\Delta f}{2}} \langle |V_n(f)|^2 \rangle \, df$$

$$= \Delta f \frac{2\rho_0 c S A \sigma_{\text{rad}}}{\pi f_0 \eta_s \rho_s S} \text{Real} \left[|F_c'(f_0)|^2 \frac{H_c(f_0)}{j 2 \pi f_0} \right].$$

That is,

$$E_{\text{rad}}(f_0) = \frac{A \sigma_{\text{rad}}}{f_0 \eta_s} \left(\frac{\Delta f}{f_0} \frac{\rho_0 c}{\pi^2 \rho_m} \right) \frac{1}{d} \text{Real} \left[|F_c'(f_0)|^2 \frac{H_c(f_0)}{j} \right]$$

and $L_{\text{eq}}(A, f_0, \Delta f)$

$$= 10 \log \text{Real} \left[|F_c'(f_0)|^2 \frac{H_c(f_0)}{j} \right] + 10[\log(A \sigma_{\text{rad}}) - \log(f_0)]$$

$$- 10 \log \eta_s - 10 \log d + C' \tag{4c}$$

or $L_{\text{eq}}(A, f_0, \Delta f)$

$$= 10 \log |F'(f_0)|^2 + 10 \log \text{Real} \left[\frac{H_c(f_0)}{j} \right] + 10 \log \frac{A \sigma_{\text{rad}}}{f_0}$$

$$- 10 \log \eta_s - 10 \log d + C', \tag{22.5}$$

where $C' = 10 \log \left[\frac{\Delta f}{f_0} \frac{\rho_0 c}{\pi^2 \rho_m} \right]$ will depend upon the bandwidth being examined.

This expression is similar in form to equation (22.4) except that instead of it containing terms in total vibrational energy and its distribution over the frequency spectrum, it contains two terms, an impulse shaping term $|F'(f_0)|^2$ and a structural response term $- \text{Real} \left[\frac{H_c(f_0)}{j} \right]$ both of which may be functions of frequency f_0.

Depending upon the impulsiveness of $f_c(t)$, $|F'(f_0)|^2$ may fall with frequency from left to right at any rate. The success of the method of analysis and prediction contained in this depends upon the degree to which this is known. Thus, as referred to in one of the case histories, Evensen [22.12] shows that the fracture process in a punch press permits $|F''(f_0)|$ to be assumed constant over all frequencies, while Priede [22.15] shows that the force history associated with the firing of a diesel engine combustion system suggests that $10 \log |F(f_0)|^2$ falls at a rate of 30 dB per decade of frequency so that $10 \log |F'(f_0)|^2$ falls at a rate of 10 dB per decade.

The term $10 \log \text{Real} \left[\frac{H_c(f_0)}{j} \right]$ relates solely to the dynamic compliance at the point of application C, and is unrelated to the impulse shape; indeed the expression can be used to apply to a continuous excitation at frequency f_0 provided that $10 \log |F'(f)|^2$ is evaluated correctly with the correct repetition rate.

It will be noticed that the terms are not all non-dimensional and that the constant C must depend upon the units used. The advantage of this form of equation is that it interpretes L_{eq} in terms of the time rates of change of force, and an impulse unity correction, both of which can be conceived physically by the engineering designer.

CHAPTER 23

Application of signal processing techniques to machinery health monitoring

R. M. Stewart

Stewart-Hughes Ltd., Chilworth Centre for Advanced Technology, Southampton

23.1 INTRODUCTION

The ease with which machinery problems such as gear, bearing, and rotor failure may be detected and diagnosed by vibration analysis, and so by implication through signal processing, has been found to depend crucially on the balance of five factors, namely

(i) rotational speed of the component, denoted S,
(ii) the distance from monitoring transducer to component, A,
(iii) the background noise level, N,
(iv) the degree of load sharing within the component under fault conditions, L,
(v) the dynamic interaction between the component and its support or drive arrangement, D.

For rolling-contact bearings the problem is essentially three-dimensional in S, A and N; for gears, it is S, A and D, and rotors S, L and D. The most difficult problem of all is generally reckoned to be the axial compressor blade (the detection of fatigue cracking in particular) which is five-dimensional in S, A, N, L, and D. However, to a greater or lesser extent the techniques of signal processing may be used to alleviate the problems incurred by the introduction of each new dimension. This survey of current state-of-the-art technology deals in particular with the treatment of gears, rolling-contact bearings, and large rotors. Detailed examples of the detection and diagnosis of epicyclic gearbox, aero-engine mainshaft bearing, and large alternator rotor faults are given.

23.2 THE PRIMARY GOAL OF DIAGNOSTIC SIGNAL PROCESSING

Within industry at large the ability to detect and possibly diagnose faults in machinery has many uses, and one of the main ones is to permit the 'on-condition' operation or maintenance of complex machinery systems such as helicopters, compressor sets and electricity generators. By this is meant the policy to operate without interruption until such times as the monitoring system indicates the presence of a fault. In practice this does not usually mean the total abolition of scheduled or 'hard-time' maintenance, rather an ability to extend these times significantly and at a faster rate of increase. Also, the economics of the monitoring process can be very finely balanced between the cost of installation and the gains to be made from increased utilization of plant, especially if the latter is already reliable. The propensity of the monitoring system to produce false alarms can therefore be as significant as its inability to detect faults. Finally, the 'quality' of diagnostic information must be such that the resulting maintenance actions can be well planned. In many cases, therefore, the role and accuracy of the diagnostic process can turn out to be extremely important, affecting in turn such things as basic operating/maintenance policy, decisions on whether or not to 'spare' key machines and size of the maintenance group. What therefore should we be aiming for in this respect?

The simple answers are ease of operation, low cost, and high accuracy/ definition. These are, however, the features of a mature well-engineered system, and at the monitoring system design stage, which is what this chapter is mainly about, the aims must be phrased in terms that have direct meaning to the technologies of vibration, performance, debris, etc. involved. For instance, the set appropriate to vibration analysis might be as follows:

(i) the production of signatures or techniques able to isolate one component from another,
(ii) the determination of sensitivities with respect to common faults,
(iii) the recommended back-up or support technique for situations where either better overall sensitivity or corroboration is required.

For a number of reasons it is the first that is usually most important. The ability to inspect a bearing or gear, say, within an operating machine as though it were running on a simple test stand leads to the more efficient build-up of experience (as one bearing or gear then looks much like any other), the more rapid determination of sensitivity and a stronger case for the development of possible alternatives.

This ability to (at least in concept) 'remove' the component from the machine is therefore the primary goal of signal processing's application to the problems of machinery health monitoring.

23.3 THE INSPECTION OF GEARS

The gear is almost unique amongst mechanical components in that its vital parts, i.e. the teeth, are required to function once, twice, or perhaps thrice per revolution, and then only for a relatively short time. Gear faults may, however, be considered to fall into three categories, namely

(i) individual tooth faults which manifest themselves as damage to individual teeth,

(ii) 'whole' gear faults such as misalignment which affect the signature produced by the entire gear,

(iii) externally-induced faults which impose high dynamic loads on the gears, often caused by too rapid a start-up or running too near to some torsional natural frequency.

Some faults will obviously be more important than others. For example, a gear may run misaligned for some considerable time without tooth damage, but a high-speed epicyclic gear that loses a tooth may run only for fractions of a second. From the point of view of vibration analysis the problem of diagnosis is made even more difficult by the perverse fact that it is generally the less serious faults which produce the greatest vibrations.

Analysis of a gearbox's vibration signal generally starts with an assessment of its amplitude modulations. Generally speaking, these may be of two kinds – (i) synchronous with gear rotation, and (ii) asynchronous with rotation. For the vast majority of gears individual tooth faults emerge most clearly as the result of a synchronous analysis. For stationary gears, as opposed to marine drives, so do most 'whole' gear faults, and in particular misalignment.

The obvious signal processing technique to use for synchronous analysis is *signal averaging,* and the method of application is shown in Fig. 23.1. Two signals must be taken from the gearbox, namely casing vibration and rotational speed. A signal average is then derived for each gear in the system, which in turn involves the electronic manipulation of the available tacho signal to produce the necessary once-per-rev synchronization pulse. The accuracy of this manipulation is obviously crucial, with phase errors of less than 1 part in 1000 being looked for.

The signatures that stem from signal averaging of the casing vibration tend to be of two general types. First, there are those that may be called 'regular' (see Fig. 23.1b) and produced in the main by fixed-axis and low-speed epicyclic gears. The predominant component within them is gear meshing frequency (the average is performed across the frequency band from 0 Hz to greater than 1.5 X meshing frequency). Whole gear and external faults tend to modulate this frequency in a major way, whereas individual tooth faults may or may not. In Figs. 23.1c, d, and e are shown the modulations produced by various common faults. The trick is therefore to have a suite of pattern recognition techniques for detection of the various modulations, and the more mutually exclusive these are the better. The other type of average has a much more random character and analysis of these will be discussed in sections 23.3.1 and 23.3.2. The randomness is most often the result of low dynamic isolation of the gear, as opposed to the 'regular' averages which stem from pure forced vibration.

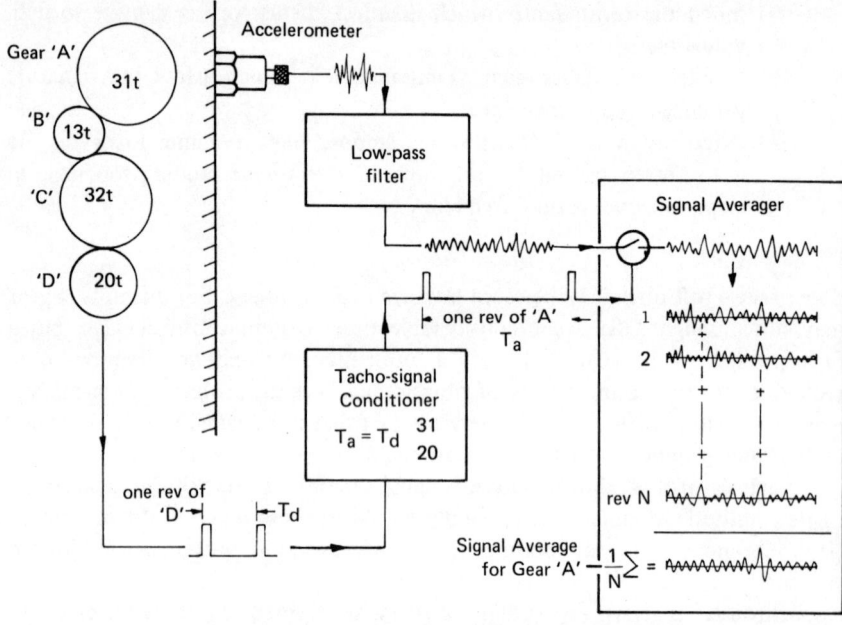

(a) Signal averaging applied to gearboxes

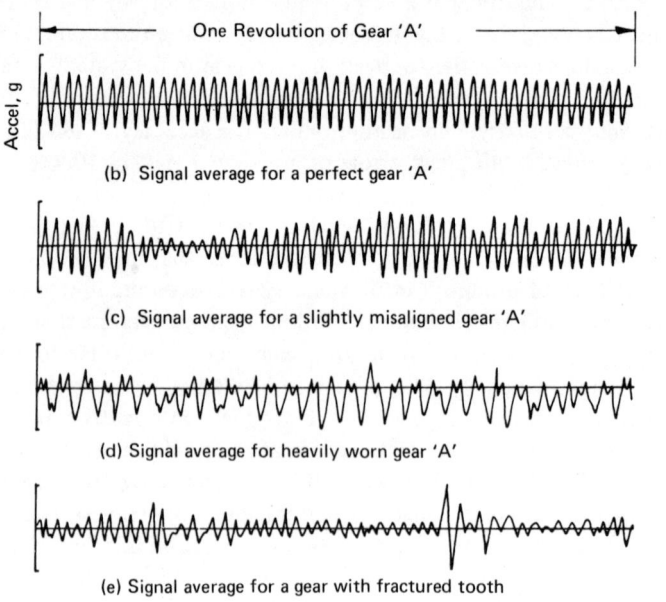

(b) Signal average for a perfect gear 'A'

(c) Signal average for a slightly misaligned gear 'A'

(d) Signal average for heavily worn gear 'A'

(e) Signal average for a gear with fractured tooth

Fig. 23.1 – Signal averaging applied to gearboxes.

The precise nature of the fault signature depends on the dynamics of the gear train, hence the introduction of the dimension D in section 23.1. For individual tooth faults two effects can occur, namely (i) modulation of the meshing frequency waveform and its harmonics and (b) excitation of sensitive natural frequencies, usually torsional. Faults involving either severe loss of metal or weakening of the tooth through fatigue cracking may cause both to happen, whereas for minor surface damage involving loss of metal (as opposed to displacement produced by chipping, say) only the latter may occur, and often to an extent hidden by the powerful elements within the average at meshing frequency and its harmonics.

Two distinct forms of amplitude modulation must therefore be looked for:

(i) overall modulation which in effect can be seen in the envelope of the raw average,

(ii) internal modulation which has to be sought out through methods referred to as bootstrap reconstruction [23.1] or adaptive decomposition. By definition it is not necessary for internal modulations to be observable in the envelope of the raw average.

The signatures of Fig. 23.1c and 23.1e are typical examples of overall modulation. The depths of both modulations depend on gear train dynamics, and the reasons for this are instructive in the sense of pointing out the limitations of vibration analysis and signal processing. First, the slow, once-per-rev modulation produced by misalignment is in fact a secondary effect resulting from the fact that misalignment generally acts to remove backlash from the mesh and so turn the normally weak eccentricity sources of vibration into very powerful ones. This of course stems from the well known ability of involute gearing to accommodate changes in centre distance without ill effects, provided that sufficient backlash exists. Secondly, the pulse produced by the cracked or lost tooth arises from a displacement form of excitation, and this in turn depends on the inertias of the gears and stiffnesses of the shafts. Therefore, in vibration analysis little or nothing is absolute.

23.3.1 The analysis of overall modulations

The modulations produced by whole gear faults are relatively simple to deal with. One merely envelopes the signal average, low-pass filters the envelope function to remove tooth frequency components, and match filters the end result with the appropriate modulation pattern. The pattern is a function of the gearbox's geometry in the way shown below:

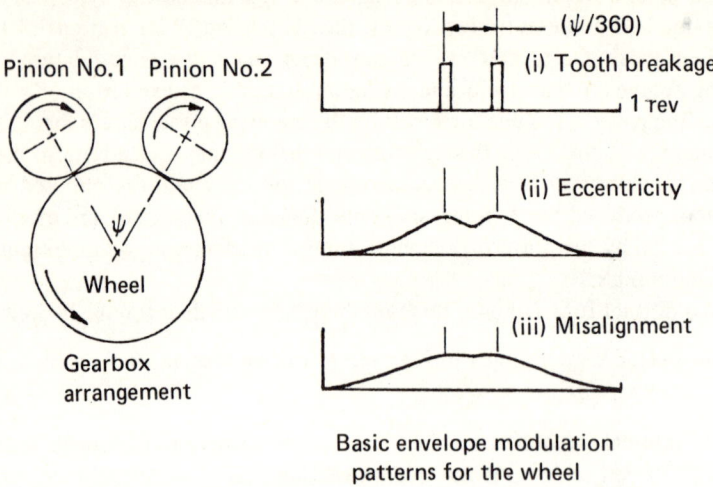

Basic envelope modulation
patterns for the wheel

It is generally thought sufficient to analyse for modulations up to a shaft order frequency of twice the number of meshes.

The modulations produced by individual tooth faults are treated in much the same way, but with the envelope filtering being set to eliminate the low frequencies at the expense of the higher tooth spacing ones. The exception is the epicyclic gearbox, for there the process of matched filtering used above is replaced by straightforward correlation.

Figs. 23.2 and 23.3 show an example of the technology applied to a high-speed epicyclic gearbox. The actual unit is shown in Fig. 23.2. The monitoring positions are marked $P_1 \ldots P_4$ and the fault was chipping of two teeth on the sun gear. From a vibration analysis point of view this is an extremely difficult problem to treat, for the following reasons. First, the sun gear has no bearings of its own on which to run and merely takes up an equilibrium position between the three planet gears it meshes with. There is therefore no direct path which the vibration signal can take from the site of generation to the monitoring position. Also, the path which it eventually must take — sun gear to planet gear to planet carrier to casing — is a time varying function due to rotation of the planet carrier. Secondly, in order to bring about an even load distribution between the planet gears, the structures of epicyclic gearboxes are generally designed to be highly compliant, and this has the effect of greatly influencing the mesh-generated vibration. Whereas for rigid, fixed-axis gears the signal average is generally a regular function (see Fig. 23.1), for epicyclics that operate at any reasonable speed the signal averages tend to be much more random. Here again we are having to work under the influence of the dimension D.

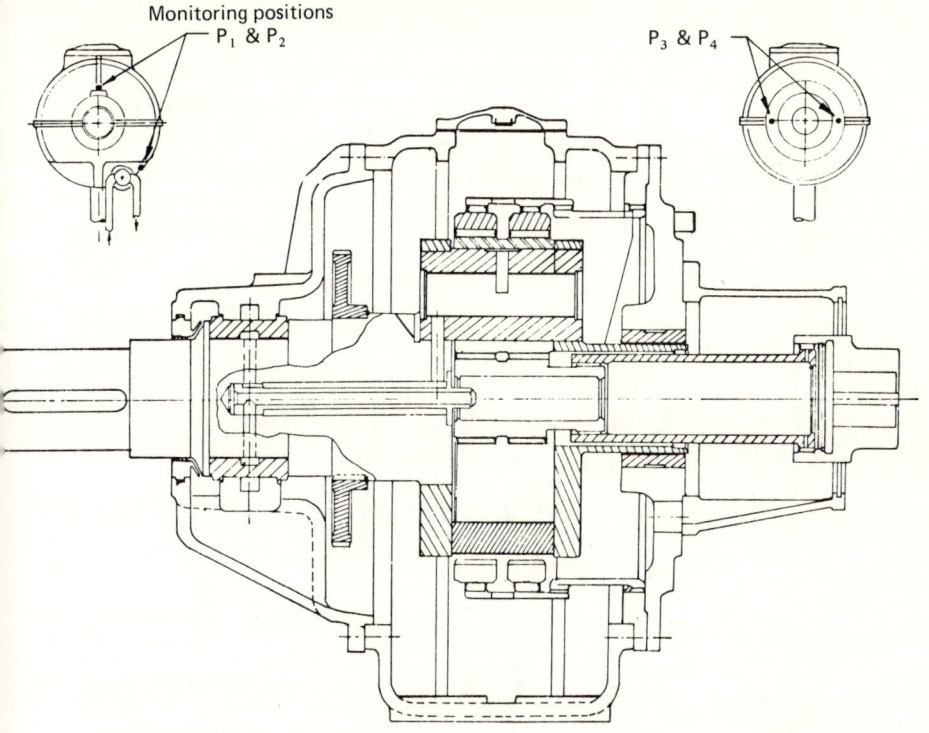

Fig. 23.2 – The epicyclic gearbox with floating sun gear.

The signal average for the sun gear is shown in Fig. 23.3a. The gear in question had 19 teeth, and there is little or no evidence of this frequency component in the average. The search for tooth damage then proceeds as follows:

(1) Enveloping of the average using a process of full-wave rectification and low-pass filtration, the cut-off frequency of the latter having been set to not less than 1.5 × meshing frequency (see Fig. 23.3b).

(2) Removal of the envelope's d.c. component and calculation of its standard deviation.

(3) Clipping of the a.c.-coupled envelope to preserve components above a threshold level of one standard deviation (see Fig. 23.3c). This particular threshold level is not immutable, and it is used because experience has shown it to be the most satisfactory. To clip below this seems to allow too much signal through to the next stage of correlation, and to clip above it does not allow enough through. Much the same form of argument has been applied to certain pitch detection algorithms used in speech processing.

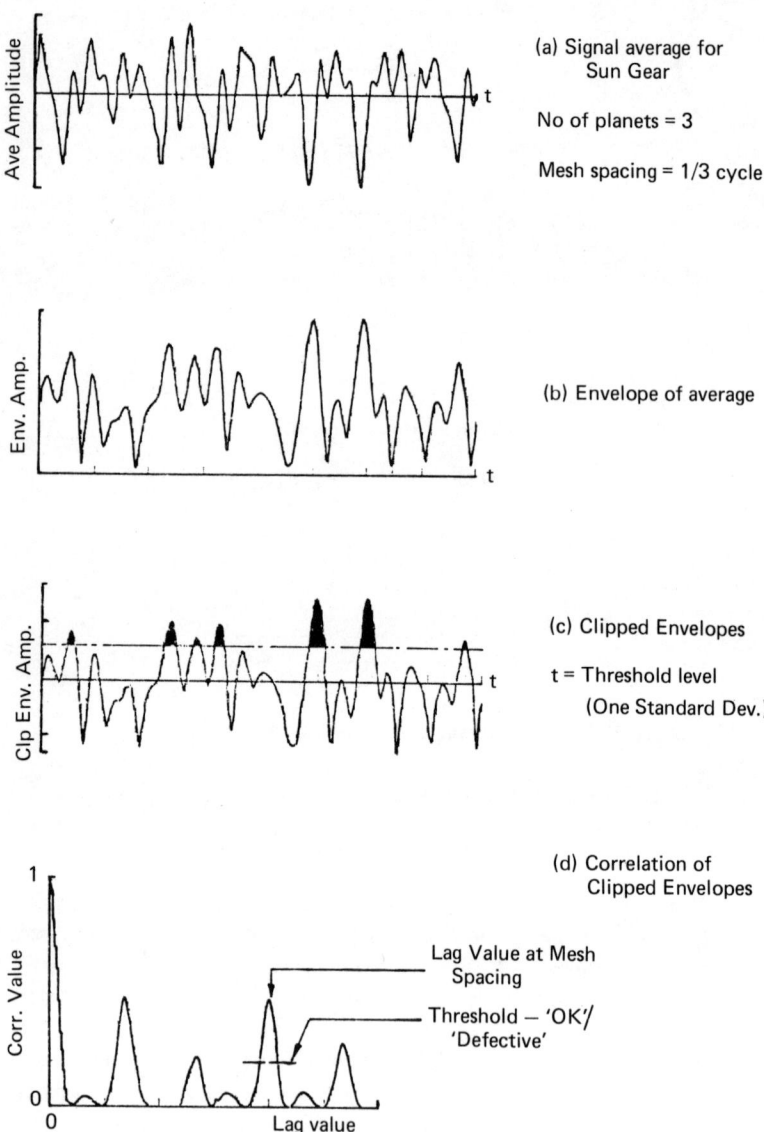

Fig. 23.3 – The detection of tooth chipping on the sun wheel of an epicyclic gearbox.

(4) Correlation of the clipped envelope to ascertain whether or not any of the peaks occur at interval of mesh spacing. The logic behind this is obviously that tooth damage will excite the system only when it passes through the mesh (three times per rev in the case shown), whilst under under normal circumstances the gear in good condition will produce an envelope with Gaussian statistics. Experience has shown the latter to be true, and also that the correlation level at the mesh point of 0.2 represents a suitable division between 'good' and 'bad'. The correlogram of Fig. 23.3d is a relatively simple signature to interpret. The high peak at the lag value corresponding to mesh spacing is in fact quite obvious from the clipped envelope itself, and the two minor peaks beside it reflect the existence of two damaged teeth rather than one. In practice, it is the robustness of the general method which is important because to be economic all such work must be performed automatically by a computer.

Hence, in analysing the overall modulations of the signal average we have brought to bear a number of quite simple signal processing techniques. In each case the objective has been to reduce a complex signature to a simple measure of some geometrically interpretable pattern. Experience has shown that the nearer such patterns are to the physical picture of gears the simpler and more accurate will be the subsequent diagnosis.

23.3.2 The analysis of internal modulations

It has just been shown how quite minor tooth damage could be diagnosed using a combination of signal averaging, enveloping, statistical clipping, and correlation. On reflection, the power of the method is obviously based on the combination of signal averaging and correlation. The former gave the all-important 'cross-section' of the component, and the latter told us that the significant peaks were spaced at the crucial distance apart. Sometimes, however, averages such as shown in Fig. 23.3 are produced by single-mesh gears, and we then have to search for signs of damage without the aid of correlation. Two techniques may then be used to recover the situation, namely bootstrap reconstruction [23.1] if the average is 'regular' and/or adaptive decomposition if it is random.

The technique of adaptive decomposition is carried out as follows:

(1) The energy spectrum of the average is computed and the ranking of shaft order numbers from maximum to minimum energy worked out.

(2) Starting with the most powerful shaft order, the average is complex heterodyned at that particular frequency and the envelope (i.e. modulus) of the result worked out.

(3) The fourth moment of the envelope is computed and added to a total. Also, if the fourth moment (kurtosis) exceeds a certain specified level a count parameter (known as the Event Ratio) is incremented by 1.

(4) The frequency component is then knocked out of the average, the standard deviation of the averaged increased to compensate for the reduction in energy, and the process repeated for the next highest component. Decomposition carries on until either a certain number of components or a given fraction of the original energy has been 'consumed' or analysed. The process is obviously a long-winded one.

Some synthetic results of the decomposition process at work are shown in Fig. 23.4 and Table 23.1. Fig. 23.4a shows the average (in fact a real one), and Fig. 23.4b the synthetic pulse. The superimposed energy spectra of the two appear in Fig. 23.4c, and it is clear that the energy of the pulse exceeds that of the signal average only in a very small band of frequencies around 1 kHz. Figs. 23.4d, e, and f then show the composition waveforms and overlayed spectra for the same average and pulses to twice, four and eight times the amplitude respectively. Only in Fig. 23.4f is the pulse evident as an overall modulation. The statistical results of adaptive decomposition are given in Table 23.1 below for an analysis involving 50 loops around the circuit described above.

Table 23.1
Results of an adaptive decomposition

Fig. Nos. (Fig. 23.4)	Normal signal energy $V{**}2$	Pulse energy V	Pulse signal $V{**}2$	Pulse/Normal signal dB $V{**}2$	Event ratio
a, b, c	1.0	0.5	0.0033	−24.8	3 in 50
d	1.0	1.0	0.0135	−18.8	13 in 50
e	1.0	2.0	0.054	−12.7	19 in 50
f	1.0	4.0	0.216	−6.6	27 in 50

The important parameter is the Event Ratio, which for the worst case of Gaussianly distributed average should give a count of 0 or 1. Thus the process of adaptive decomposition has achieved detection of the basic pulse (Fig. 23.4b) at the computed signal to noise ratio − 24 dB. Obviously, as the pulse size increases so too should the Event Ratio.

Signal processing of this kind must often be resorted to if the objective is to direct *localized* pitting/spalling damage at an early stage. Other techniques handle the distributed case.

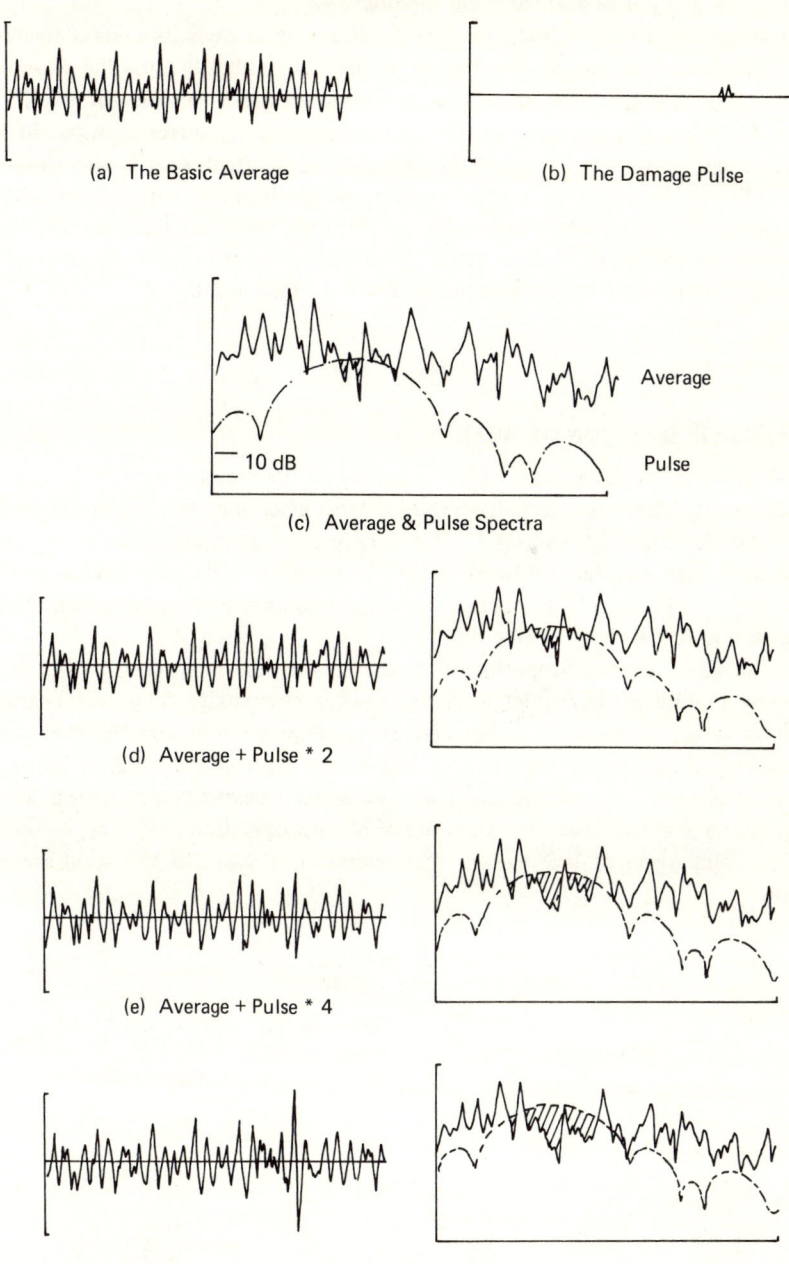

(a) The Basic Average

(b) The Damage Pulse

Average

10 dB

Pulse

(c) Average & Pulse Spectra

(d) Average + Pulse * 2

(e) Average + Pulse * 4

(f) Average + Pulse * 8

Fig. 23.4 – Adaptive decomposition.

23.3.3 The analysis of asynchronous modulations

In comparison to synchronous ones, modulations of an asynchronous character are much harder to analyse. They occur, in the main, either because the gearbox structure is flexing or the drive system is transmitting a variable torque. The worst offenders in this respect are the main propulsion gearboxes of ships which twist owing to hull flexure in roll and wind up as the result of propeller torque fluctuations. The problem is not, however, so much one of catering for those asynchronous motions but rather the effect that they can have on the all-important synchronous ones. It can in practice be very difficult to obtain stable signal averages from a marine gearbox monitored whilst at sea.

23.4 THE INSPECTION OF BEARINGS

Of all components associated with monitoring of the rotating machine it is the rolling-contact bearing which has received most attention. The reasons for this are basically threefold, namely (i) it is simply the commonest component, (ii) because it generally fails through fatigue, it possesses a definite catalogue life with high scatter, and (iii) because of its high tolerance to abuse it is then often abused, and so fails more frequently.

The signal processing problems associated with the monitoring of bearings increase rapidly as the parent machine grows in complexity. Thus, the bearing such as might be found at either end of an electric motor can be monitored using very simple equipment, in contrast to the aeroengine mainshaft bearing, which might require some very sophisticated signal processing equipment indeed. Why this is so stems from the mechanics of bearing operation.

A comparison of bearing and gear operation is shown in the small sketch below:

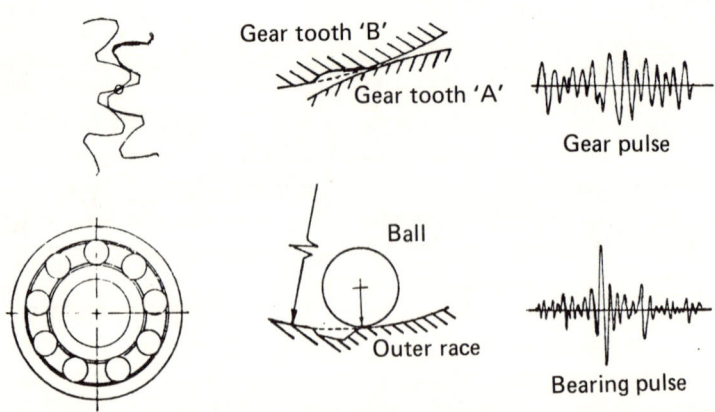

Gear tooth 'B'

Gear tooth 'A'

Gear pulse

Ball

Outer race

Bearing pulse

The factor which determines the vibration produced by the good component, either bearing or gear, is the stiffness of displacement modulation produced during the process of 'hand-over'. Hand-over for the gear occurs when the responsibility for transmitting torque passes from one tooth to another, and it tends to take the form largely of a displacement modulation primarily because of limitations imposed during the manufacturing process and the subsequent introduction of what are called 'transmission errors'.

The rolling contact bearing on the other hand modulates in both stiffness (referred to as the variable compliance source) and displacement (due to manufacturing errors referred to as 'waviness', or, in extreme cases, 'lobing'). However, the hand-over vibration produced by a good bearing is orders of magnitude less than that of a good gear, and also less dominated by narrowband tones at the fundamental hand-over frequency and its harmonics (i.e. the spectrum is much whiter). The signature produced by a good bearing can then rarely be seen through the vibration produced by the supported gear, or, for that matter turbine, compressor, or pump.

The introduction of a spalling fault into the bearing does, however, bring about a radical change in the picture. Because of the very low surface conformality of the bearing in comparison to the gear (see sketch above), the loss of a piece or metal from the bearing rolling-element or raceway creates the perfect conditions for the generation of high crest factor spikes. A similar loss of metal from the equivalent gear (i.e. the gear that might be supported by the bearing) would generate almost no detectable change in vibration simply because both surfaces happen to have a large radius. It generally takes the loss of a large part of a tooth for the gear pair to generate spikes similar in size to those of the bearing.

From the signal processing point of view there are, however, two other vital factors to be recognized. First, the signals produced by the bearing are asynchronous, hence techniques such as signal averaging cannot be used. Secondly, the signal produced by early damage (at least spalling damage) are generally highly coherent, but become entirely random as the damage spreads around the bearing. Attached to this second phenomenon is the fact that the frequency band associated with the high crest factor spikes changes with the progression of damage. Experimental results gathered in the frequency range 0 to 50 kHz indicate, as a general trend, the first signs of damage as highly coherent impulsiveness in the range 2–5 kHz. However, as the damage spreads the impulsiveness is seen to translate towards higher frequencies and become less coherent. Finally, with high percentage raceway or rolling-element damage the signal is impulsive only at frequencies in excess of 20 kHz, and is also totally incoherent. Overall energy levels increase at all frequencies (though some time after the onset of impulsiveness) and typically by factors in excess of 20 from undamaged to heavily damaged.

23.4.1 Detection of bearing damage in simple arrangements

Here we must first of all mention the rotational speed qualification of operation between 100 and 10 000 rev/min. In essence, therefore, if the monitoring transducer can be placed close to the bearing a number of very simple methods can be used. Simplest of all is just to trend analyse the level of vibration in the frequency band 100 to 10 kHz and watch for the accelerating rise in this level. This is not however the quickest way to detect incipient failure, and it does involve the manual task of logging and trending results.

One step further on are the techniques that set out to detect the impulsive signals so characteristic of bearing track damage. There are now some half dozen of these available, and their operating ranges extend from 2 kHz to 100 kHz. One such technique, developed by the researchers at the Institute of Sound and Vibration Research, involves use of the signal's normalized fourth moment, otherwise known as its kurtosis value. The principal of operation is shown in Fig. 23.5, and in one of its many possible implementations involves the use of three multiplying devices. The simplest way to visualize the process is to think of the first two as acting to produce a signal normalized to an energy level of approximately 0.11 volts-squared. This section of the circuit has therefore been called the Normalization Block. So, should the signal be 'white' in statistical terms (see top signal trace), only the odd spike will rise above the 1 V threshold, and in passage through the third squarer the vast bulk of the signal will be 'squelched' to much less than 1 V. The impulsive signal will, on the other hand, have many spikes that pass through the 1 V level, and so far less of the normalized signal will be squelched by the final squarer; a good deal of it will in fact be amplified. The kurtosis values can then be directly related to the integrated output of the third squarer by the mathematical formula:

$$\text{Kurtosis}, K = \frac{\int\limits_{\infty}^{\infty} (x - \bar{x})^4 \, p(x) \, \mathrm{d}x}{\sigma^4} \quad \begin{array}{l} = 3.0 \text{ for no impulsiveness} \\ > 3.0 \text{ for impulsiveness} \end{array}$$

where
x = signal level
$\bar{x}$ = signal mean
$p(x)$ = probability density distribution of x
σ = signal standard deviation

or by the more practical time average estimator:

$$K = \frac{\dfrac{1}{T} \int\limits_0^T (x(t) - \bar{x}_n)^4 \, \mathrm{d}t}{\sigma^4}$$

which is actually what the analogue circuitry shown in Fig. 23.5 implements.

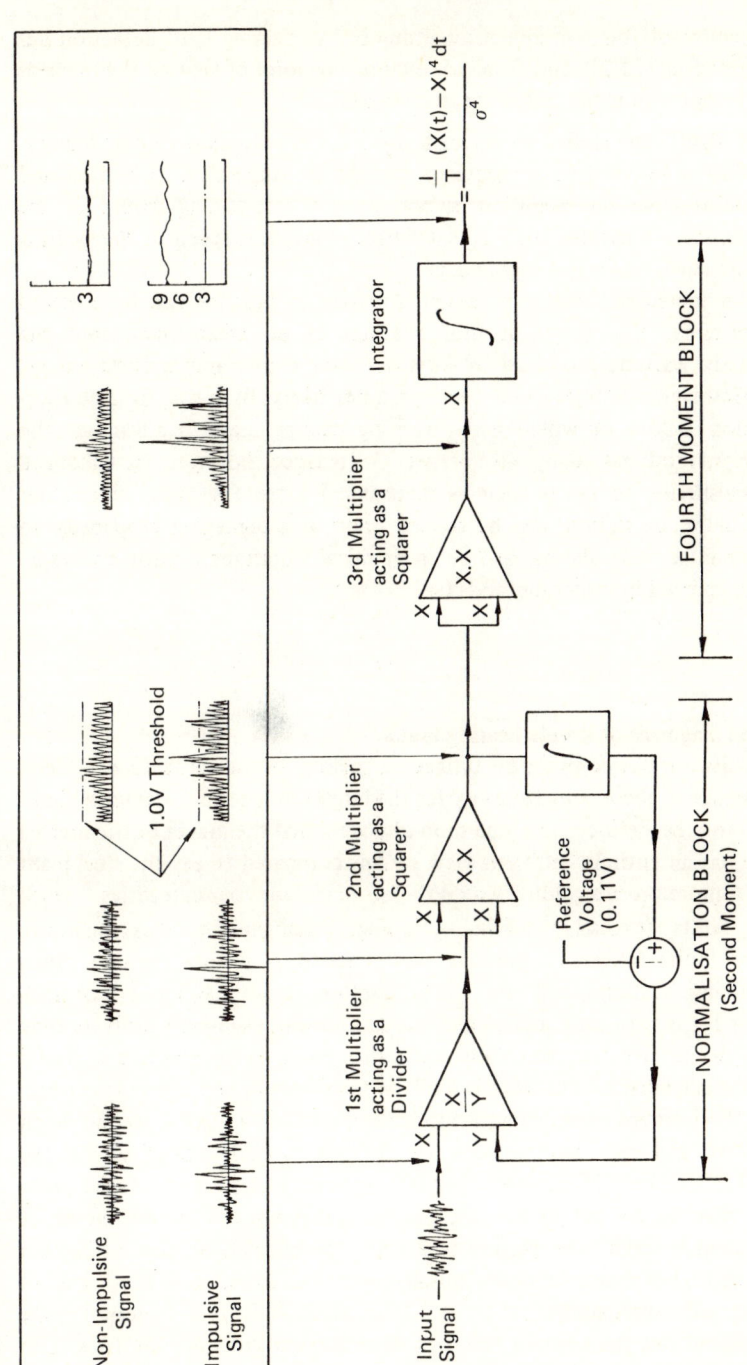

Fig. 23.5 – Measurement of impulsiveness using the kurtosis parameter.

The application of this non-dimensional number to bearing fault detection has been described in [23.2], and from an engineering point of view of the kurtosis functions brings with it the following advantages:

    (i) A significant ability to detect bearing faults regardless of past history. This is referred to for obvious reasons as 'single-shot' detection and stems from the non-dimensional, pattern recognizing nature of the number. Variables such as installation, load, and speed of the bearing then assume far less significance.

    (ii) An increased sensitivity to faults, which in practice can be a mixed blessing. For run-of-the-mill bearings of no great importance the early warning produced by kurtosis value is probably a disadvantage. However, for high-speed bearings, ones likely to suffer from lubrication failure or where early warning can be used to advantage, the K-method has many advantages. On balance, however, its enhanced sensitivity to faults such as mechanical looseness, rubs, lubrication failure, cavitation, can be looked upon as a damaging propensity to produce false alarms, and so in practical situations it must always be supported by other methods (see below).

23.4.2 The diagnosis of simple bearing faults

Bearing faults are not confined to fatigue or spalling of the contacting surfaces, though damage to these constitutes by far the largest proportion. The term 'fault' must therefore be defined, and one soon discovers that the maker of the inertial gyroscope has an entirely different view of this compared to say the steel plant operator. A more precise grading would therefore include four categories — track geometry faults (excessive waviness, lobing, misalignment, cross-location), lubrication faults (excessive greasing, low viscosity), surface damage faults (electric arcing, spalling), and cage faults (fracture, instability, loss of control). The proper list is very long indeed and the monitoring engineer's problem may well be to decide which of the many signal processing techniques to use. By far the simplest category to deal with is track damage (mentioned earlier in connection with the kurtosis method) with lubrication perhaps a distant second. Both the geometry and cage fault groups are difficult to deal with in all but the simplest (e.g. quietest) of situations. In fact, geometry and cage faults are probably best checked for out of the machine, using a low-speed dynamometer, of which a number exist. The picture is further complicated by the interaction between faults. For example, heavy misalignment will often cause heating up of the bearing and subsequent creation of a lubrication fault. It is therefore easily possible to find one fault as the result of detecting another, and the trick is to find the primary one before irreversible damage occurs.

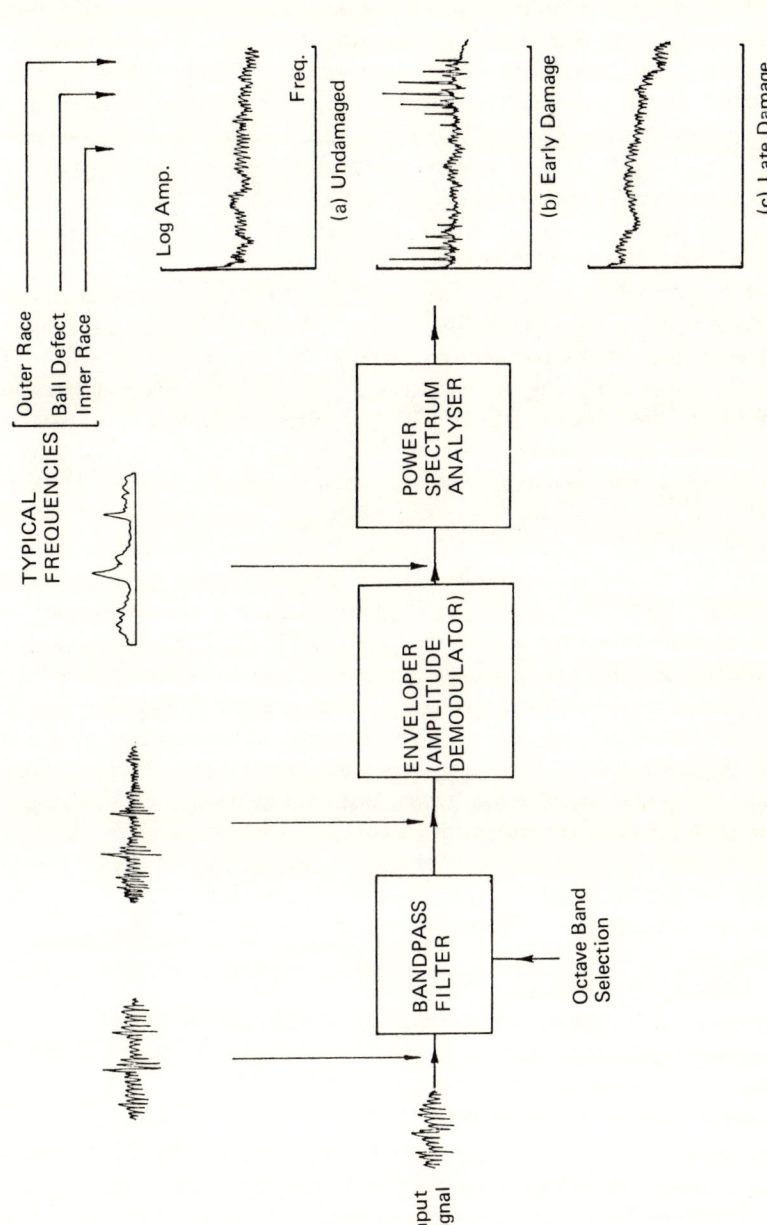

Fig. 23.6 — The envelope power spectrum.

For simple bearing arrangements (see [23.3] for further details) it is probably true to say that vibration analysis, and hence, by implication, signal processing, can cope with detection of at least 90% of all terminal faults. However,
as the dimensional complexity of the arrangement increases, this percentage
tends to fall dramatically. Background noise, very low or very high speed, and
poor access are the main enemies. Techniques such as overall level measurement
and kurtosis obviously then cannot be relied upon, and some recovery of the
position is necessary. Experience indicates that the best way to go about this is
through joint application of oil supply debris analysis and the vibration-based
techniques of enveloping and power spectrum analysis.

The enveloping/spectrum approach is laid out in Fig. 23.6. In common with
the kurtosis method, the signal is first of all octave filtered to remove the major
influences of the casing's transmission function and the obscurative effects of
gear meshing tones etc. The filtered signal is then amplitude demodulated in
some way and the power spectrum of this produced. 'Information' carried as
modulations of high-frequency resonances can in this way be transferred from
those high frequencies where spectral resolution is generally poor to the base-
band where it is much better. In the vast majority of cases the spectrum of the
envelope function is a far simpler thing to interpret than its equivalent for the
raw signal (compare Fig. 23.7a with 23.7b). However, returning to the point
made earlier about the signals becoming more random as damage spreads, it
must be recognized that the pattern so clearly seen in Fig. 23.6b holds only for
early damage, and that the envelope spectrum returns to its flat shape (albeit
at much higher amplitude) with extensive damage (see Fig. 23.6c), even though
the filtered raw signal may still retain its impulsiveness (as measured by either
kurtosis or crest factor) at very high frequency. Therefore, in practice, envelope
spectrum analysis, so useful for diagnosis, is seldom used on its own and never
without some means of measuring the power and impulsiveness of the incoming
raw signal. One sensible procedure for use on bearing installations of low to
middling complexity is therefore to combine simple r.m.s. level and kurtosis
measutement for detection with envelope spectrum analysis for diagnosis.

23.4.3 Detection and diagnosis under difficult conditions
By the end of 1981 the kurtosis/envelope/PSD technology or its several other
equivalents represented the limit of practical technology and equipment existed
to implement it at various cost levels from $3000 to $20 000. The technology
does, however, have its weak points, notably in relation to installations with
poor access and/or a high level of background noise, typified by the mainshaft
bearings of many small, high-speed aeroengine gas turbines. In such cases the
level of background noise can easily be too high for the kurtosis method (or for
that matter any other of the simple techniques, e.g. shock pulse) to work,
thereby forcing far greater reliance on enveloping and spectrum analysis. It is
therefore unfortunate that the envelope-spectrum method very quickly runs

into trouble on two counts, namely

(i) the rising level of the white noise floor (see Fig. 23.6a), sometimes caused by non-linear operation of the enveloper circuitry, but more often through the presence of intense white-noise sources, e.g. turbine or combustion chamber, in the immediate vicinity of the bearing;

(ii) the entry into the envelope power spectrum of corrupting or obscuring non-bearing-related tones, caused almost invariably by rotor systems in general and gears/turbines in particular.

Both have the effect of reducing the bearing's signal-to-noise ratio, and to a limited extent the problems can again be overcome by application of signal processing techniques. One of these involves computation of a function known as the 'metacepstrum,' or more often in practice its autocorrelation equivalent. The basic reason for its use lies in the need to detect the presence of very weak harmonic families in the envelope power spectrum.

The calculation procedure is shown diagramatically in Fig. 23.7. Starting with the envelope power spectrum of the octave of bandpass filtered vibration signal, the low-frequency fluctuations of the spectrum are first of all removed by a combination of median filtering and subtraction (see Figs. 23.7b and 23.7c). This is roughly equivalent to the logarithmic operation of cepstrum analysis in that a compression of dyanmic range is being effected. In practice, it is a much better way of compressing the signal as the result is totally devoid of low-frequency ripple, an artefact which can cause considerable trouble with standard cepstral analysis. The final stage in this part of the process is a clipping operation which removes the vast bulk of the new white noise base, see Fig. 23.7d. Clip levels of around 3 dB are generally applied, so making it possible to detect very weak bearing tones.

The next important operation is removal of the corrupting rotor tones. This is normally a simple operation as these occur at multiples of rotor rotational speed (i.e. as shaft orders).

Finally, the flattened, clipped, and combed spectrum is autocorrelated to reveal the relative strengths of tone families residing within it (see Fig. 23.7e). Autocorrelation is used because the tones are best treated as delta function pulses. The standard cepstral techniques use the Fourier transform for this purpose, but experience has shown it not to give such good results.

Both the processed envelope spectrum (Fig. 23.7d) and its autocorrelation (Fig. 23.7e) are then used for the purpose of identifying bearing faults. To do this the analysis computer must be fed details of the bearing's basic geometry, its likely range of contact angles (if appropriate) and slip percentages, for somehow it must know where to look in the spectrum for the tone patterns symptomatic of damage. When faulty the rolling contact bearing has the potential to generate some half dozen or so unique tonal patterns, so these must each be looked for in turn.

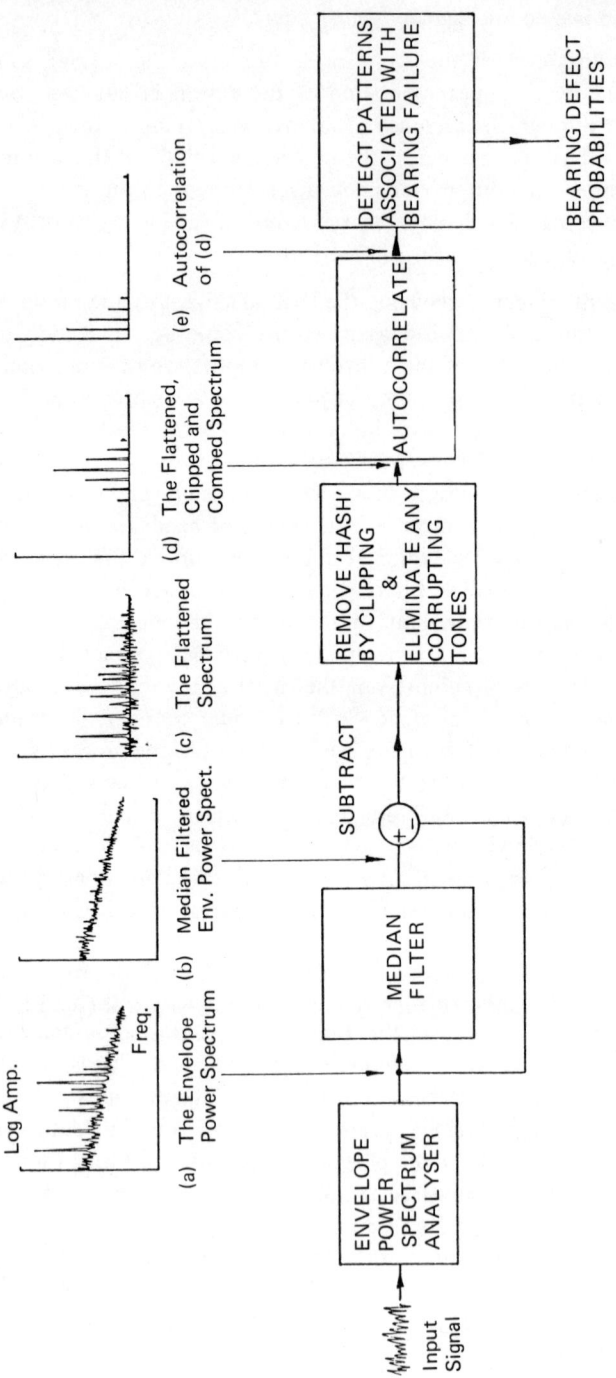

Fig. 23.7 — The analysis of envelope spectra produced by complex machinery.

23.5 THE INSPECTION OF ROTORS

It is invariably the case that the large industrial rotor has to be considered as part of a larger system which includes the bearings, casing, foundations, and sometimes the process fluid. Therefore, in a diagnostic sense, whereas we may have felt fairly confident about treating the gear or the rolling-contact bearing as an isolated unit, the rotor is an entirely different matter. Furthermore, the very largest rotors, such as those to be found in turbines and alternators, are mechanisms in their own right.

The archetypal rotor problem is the so-called 'rough runner' or rotor which either persistently or suddenly begins to generate either more or different vibrations. Questions of misalignment, thermal bending, coupling failure, blade damage, and even rotor transverse cracking may then arise, and in a not insignificant number of cases an intractable problem results. In part this comes about because most rotor faults manifest themselves as changes in the once, twice, thrice, and fourth orders of rotor rotation. Gears and bearings, it will be remembered, spread their defect-related vibrations across a much wider bandwidth. Rotor diagnostics are therefore hindered by ambiguities resulting from this, and recourse is sometimes therefore made to observations of run-up and run-down characteristics of the rotor in the hope that some 'identifier' will emerge.

The signal processing technique most appropriate to rotors is the power spectrum, the reason being that for both steady and transient analysis it pulls out the information on rotor order vibrations better than any other. Non-synchronous vibrations can also of course be seen through it. Techniques such as signal averaging have their specific uses, e.g. balancing, but not the same overall degree of usefulness.

The normal power spectrum has, however, no ability to distinguish between amplitude and frequency modulating effects within the signal, so basically the result that one ends up with is a function of the transducer used, its orientation with respect to the rotor, and the impedance of the casing. Thus a radially mounted accelerometer will sense mainly amplitude modulations, and it requires special processing equipment (see section 23.6) for the frequency modulating components of much lower level to be extracted.

The published literature on rotor vibrations is truly enormous, yet only a small percentage of it deals with fault diagnosis. This is due partly to our general lack of theoretical models concerning defective rotors (treatment of the non-defective case is difficult enough) and partly to the dynamic interaction problem mentioned earlier. It is therefore unfortunate that the pattern based methods outlined earlier for gears and bearings find little scope within the rotor field, which essentially remains one of parameter identification.

One interesting example of parameter identification applied to rotors is the detection of transverse cracking. It is easy to appreciate the fact that a fault of this type will introduce flexural asymmetry and nonlinearity of stiffness and damping into the rotor. The questions then are (i) how to detect these, and (ii) estimate the extent and location of the crack.

A number of simplified analyses of the rotor cracking problem have been carried out, and in several of these the approach has been to linearize the essentially nonlinear equations/matrices which emerge from rigorous formulation of the problem. Linearization is generally possible because of the small changes in stiffnesses involved (thpically 10-15%), the high damping attributable to the bearings, and the forcing mechanism which acts to open and close the crack [23.4]. The results which emerge then predict that the major response to a crack will be observed at the third order and all odd harmonics above it [23.5]. It would have been difficult to predict such an effect without recourse to modelling of the rotor and close inspection of the model's parameters.

In addition, such analyses can predict the interacting effects of rotor eccentricity, gravity, imbalance, and the crack. For instance, at rotational speeds above the main critical speed (i.e. the operating speed of most large rotors) the eccentricity dominates and the shaft takes up a steady rotating deflected shape, so keeping the crack either open or closed. From a fault detection point of view this is an unfortunate result, for confusion may then arise between simple bending/balance problems and cracking. To separate the two, the machine must be taken off line and run down through its main critical speed in order to 'exercise' the crack, and in so doing produce the third -order response. In practice, the situation may be much more complicated, yet the principle still holds.

Different types of rotor do of course demand different diagnostics. The helicopter rotor is, for example, in a class of its own owing to blade lead/lagging etc. Also, the high-speed pump is worth mentioning because of its very quick run-down times and the scrambling of flexural resonance data that they produce. The bulk of rotor problems are, however, fairly trivial and in the end corrected by such things as re-balancing and realignment.

23.6 SIGNAL PROCESSING EQUIPMENT

Wherever large concentrations of machinery exist, such as in power stations, petrochemical plant, and ships, large numbers of the components such as those dealt with in the preceding sections will be found. To bring the best detection/ diagnostic power to bear on these therefore requires a signal processing and data management system of some considerable sophistication. One such system is the so-called Mechanical Systems Diagnostic Analyser (MSDA), designed and developed for use by the Navy on ships at sea, but now also finding its way into much land-based plant. A picture of it is shown in Fig. 23.8.

In essence the unit consists of five highly integrated sub-systems, namely (i) an executive microcomputer, (ii) mass storage device, (iii) digital array processor, (iv) analogue signal processor, and (v) communications/display/

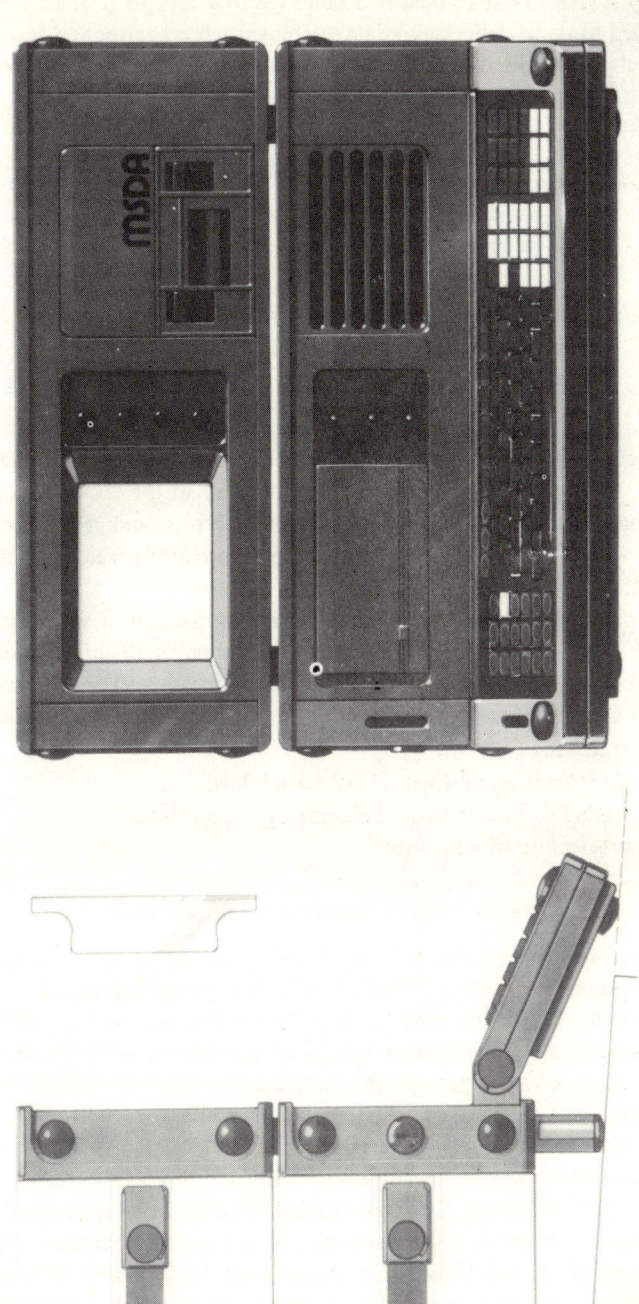

Fig. 23.8 – The S.H.L. Mechanical Systems Diagnostic Analyser (MSDA).

hard-copy system. Programmed in a language that accepts primarily mechanical (as opposed to signal processing) data on things such as bearing and gear geometry, the typical process plant or marine unit will monitor plant in the following way:

Main machinery – Continuous monitoring of perhaps 32 dynamic and 128 static sensors at a 2-3 s scan rate.

Auxiliary/spared machinery – Periodic monitoring using hand-held data logger/vibration/corrosion monitoring equipments that can send single-valued results (as opposed to signatures) back to MSDA via modems etc.

Remote off-site machinery – The attachment to remote or off-site machinery of permanent monitoring systems which can transmit low-frequency vibration, temperature, seal cavity pressure, fire warning data, etc. back to MSDA via low-power f.m. or telemetry links.

The key element in the system is of course the diagnostic and data management software. MSDA, for example, can implement all of the techniques so far discussed together with others for performance, debris, and trend analysis. The software would, however, be useless without supporting hardware, and in this respect MSDA is unique.

The principle behind its operation is implementation of something referred to as the Test Specification, a file of data containing all information relevant to the testing of a component. The test specification is therefore analogous to biological DNA in that it controls an entire replication process, this being made up of the following elements:

    (i) identification of the machine, speed, load, etc.,
    (ii) setting up and control of the analogue processor,
    (iii) production of a signature,
    (iv) interpretation of the signature,
    (v) trend of results + search for corroborative evidence.

Shown in Fig. 23.9 is the section of the specification which deals with the analogue processing system. This particular sub-system consists of a number of I/O processing devices or modules (e.g. filters, heterodyner) grouped together around a matrix switch. Parallel strings of modules can be put together under software control to implement the operations needed as part of envelope or 'zoom' spectrum analysis, etc. The particular route shown in Fig. 23.9 is the one used for envelope spectrum analysis. Hence filters 1 and 2 (modules FL1 and FL2) combine to give a bandpass function and this is followed by amplitude demodulation (ENV) and anti-alias filtering (FL3 and FL4) prior to digitization (CH1). Also, concurrent with processing of the analogue vibration signal there is processing of the machine's tacho signal (modules TAC and 2XN) so that precise, machine-controlled digitization of data takes place (module SAM produces the clock pulse for analogue-to-digital converter module CH1).

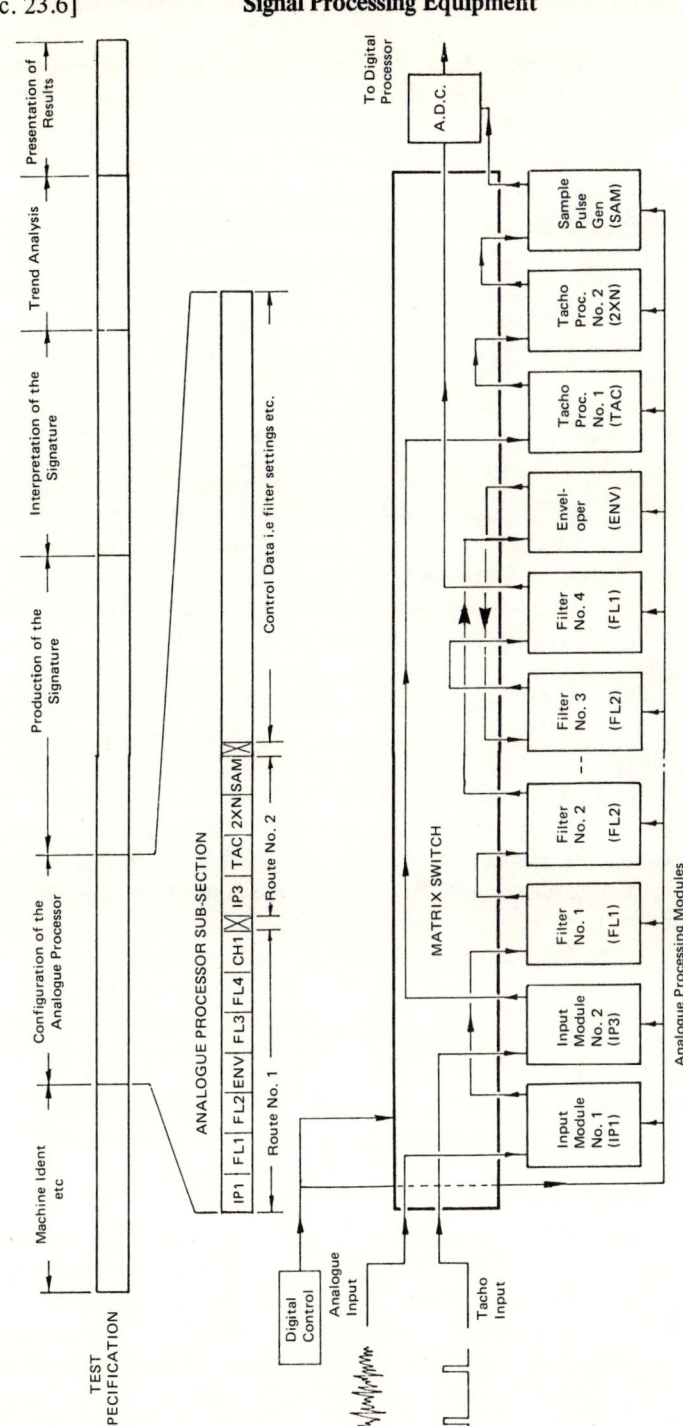

Fig. 23.9 — Configuration of MSDA's analogue signal processor.

Having set up the correct route, the executive microcomputer would then proceed to adjust all signal levels within the analogue processor by way of digitally-controlled gain amplifiers sited on the output lines of all modules. Internal gain control is therefore taken care of automatically, an essential feature if signal-to-noise ratios are to be maintained.

The analogue signal processing module kit for MSDA is extremely comprehensive on account of the diversity of monitoring techniques and the need for high-speed processing. Despite the swing towards digital methods, analogue processing of vibration data to filter, heterodyne, amplitude/frequency demodulate and tacho signal multiply/divide is still either mandatory or highly desirable in many situations.

REFERENCES

[23.1]   Stewart, R. M. (1977) *Application of Time Series Analysis,* ISVR, University of Southampton, Sept. 1977. Some useful data analysis techniques for gearbox diagnostics.

[23.2]   Dyer, D. & Stewart, R. M. (1978) *ASME J. Mech. Des.,* **100,** 229–235. Detection of rolling-element bearing damage by statistical vibration analysis.

[23.3]   Stewart, R. M. (1979) *2nd National Conference on Condition Monitoring,* London. The development of monitoring technology for rotating machinery.

[23.4]   Mayes, I. W. & Davies, W. G. R. (1980) *Proc. 2nd International Conference on Vibration in Rotating Machinery,* Cambridge. A method of calculating the vibrational behaviour of coupled rotating shafts containing a transverse crack. Published by the Institution of Mechanical Engineers, London.

[23.5]   Henry, T. A. & Okah-Avae, B. E. (1976) *Proc. of Conference on Vibrations in Rotating Machinery,* Cambridge. Vibrations in cracked shafts. Published by the Institution of Mechanical Engineers, London.

CHAPTER 24

Measurement and diagnosis of machinery noise

P. D. Wheeler

Institute of Sound and Vibration Research, University of Southampton

24.1 INTRODUCTION

This chapter discusses current and forthcoming standards for the measurement of noise from machinery and quantifies some of the errors involved in free-field, semi-reverberant, and reverberant measurements. The range of techniques available for the location or diagnosis of noise sources in machinery is also reviewed.

UK Standards have recently been revised, with the publication of BS 4196 Part 0-6, replacing the 1967 *Guide to the selection of methods of measuring noise emitted by machinery*. The new British Standard closely follows ISO 3740-6, as shown in Table 24.1.

A number of other, more specific, British Standards have been published relating to areas such as *Measuring noise from machine tools*, BS 4813:1972 and *Cartridge operated fixing tools (recoil noise)*, BS 4078:1966. Various trade association and industry standards also exist. References [24.1] and [24.2] list relevant European and US standards.

24.2 MEASUREMENT METHODS

Measurement methods may be categorized (BS 4196 Part 0) as precision, engineering, or survey, with a further subdivision in respect of environment into anechoic, reverberant, or semi-reverberant conditions. The characteristics of the noise source, i.e. whether broadband, narrowband, or discrete frequency and whether steady or fluctuating, are considered in a final classification subdivision.

Of these various test combinations, the two precision methods for sound power determination, namely, hemispherical free-field radiation and diffuse field (for steady broadband sources only), are well-established and need only a brief mention here.

Under hemispherical free-field radiation, the sound power level may be determined by sound pressure level measurements made over an enveloping surface, which is conventionally a concentric hemispherical or parallelopiped surface. Then

$$L_W = \overline{L_P} + 10 \log_{10} S \qquad (24.1)$$

Table 24.1 – ISO standards

International Standard No.	Classification of method	Test environment	Volume of source	Character of noise	Sound power levels obtainable	Optional information available
3741	Precision	Reverberation room meeting specified requirements		Steady, broad-band	In one-third octave or octave bands	A-weighted sound power level
3742			Preferably less than 1% of test room volume	Steady, discrete-frequency or narrow-band		
3743	Engineering	Special test room		Steady broad-band narrow-band, discrete-frequency	A-weighted and in octave bands	Other weighted sound power levels
3744	Engineering	Outdoors or in large room	No restriction: limited only by available test environment	Any	A-weighted and in one-third octave or octave bands	Directivity information and sound pressure levels as a function of time; other weighted sound power levels
3745	Precision	Anechoic or semi-anechoic room	Preferably less than 0.5% of test room volume	Any		other weighted sound power levels
3746	Survey	No special test environment	No restrictions: limited only by available test environment	Steady, broad-band, narrow-band discrete frequency	A-weighted	Sound pressure levels as a function of time; other weighted sound power levels

where L_W is the machine sound power level

$\overline{L_P}$ is the mean sound pressure level, measured over the enveloping surface S.

The errors involved in this type of measurement are discussed later. In a reverberant room of suitable dimensions, giving a diffuse sound field, the sound power level of the machine is given by:

$$L_W = \overline{L_P} - 10 \log_{10} T + 10 \log_{10} V - 14 \qquad (24.2)$$

where $\overline{L_P}$ is the mean reverberant sound pressure level (usually measured at, at least, six positions)

T is the reverberation time of the test room under the conditions of measurement (seconds)

V is the volume of the test room (m^3) .

The survey method normally takes the form of near-field measurements made at a prescribed distance from the machine (usually 1 metre) and generally does not include the derivation of a sound power level. The engineering method of substitution by a standard power source in a reverberant environment is again an established method and is described in BS 4196.

This leaves us with two areas of interest, namely, diffuse field methods for discrete frequency sources, and engineering out-door or semi-reverberant methods for sound power determination. The former subject has received considerable attention over the past few years, and following research by ISO TC 43/SCI/ Working Group 6, including a 'round robin' test programme, a standard has been issued which describes a qualifying procedure for reverberant rooms. References [24.3] and [24.4] describe this work in depth. The qualifying procedure involves the computation of the standard deviation of reverberant level measurements made at a large number of discrete frequencies, using a calibrated loudspeaker as a source (see BS 4196 Part 2).

An engineering method for free-field radiation over a reflecting plane is the subject of BS 4196 Part 4/ISO 3744 – still in drafting at the time of writing [24.5] . This document describes an enveloping surface method in which measurements are made using precision-grade instrumentation. A qualification procedure for the environment is described in which measurements made over other concentric surfaces are used to judge the validity of the 'free-field' assumption. A correction is applied for small divergences from a truly free-field environment. Section 24.5 of this chapter describes this procedure in greater detail.

24.3 SOURCES OF ERROR IN FREE-FIELD MEASUREMENTS

Hubner [24.6] identifies three sources of error for sound power determinations based upon sound pressure level measurements made under free-field conditions or in the presence of a reflecting ground plane. He defines his three partial errors as:

Δ_1, the near-field error, arising from the fact that the near-field sound pressure p and the particle velocity v may not be in phase;

Δ_2, the finity error, arising from calculating the mean square pressure over the measurement surface from a series of point measurements, rather than by integration over the entire surface;

Δ_3, the actual measurement error, arising from instrument calibration errors, observational error, meteorological conditions, etc.

Δ_3 is not included within the scope of this chapter and is not further discussed here.

24.3.1 Δ_1 The near-field error
Δ_1 arises from the approximation of $p.v$ by $p^2/\rho c$

where p is the sound pressure at a point
             v is the particle velocity at the point
             ρ is the density of air
             c is the velocity of sound in air.

The true sound power P is given by the integration of the product pv over the enveloping surface S, bearing in mind that v is a vector quantity which may not be normal to that surface:

$$P = S \iint \overline{pv}^t \, \mathrm{d}S \tag{24.3}$$

The approximate sound power level P_n is obtained from the measurement of $\overline{p^{2t}}$, thus

$$P_n = \frac{1}{\rho c} S \iint \overline{p^2}^t \, \mathrm{d}S \ . \tag{24.4}$$

Δ_1 is defined as the quotient P/P_n, and may be evaluated for specified source and enveloping surface combinations [24.6].

Fig. 24.1 shows the magnitude of the error Δ_1, expressed in decibels ($10 \log_{10}\Delta_1$) for spherical sources of various orders and a concentric enveloping surface, as a function of measuring distance. It will be seen that for a monoopole source ($n = 0$), Δ_1 is always unity.

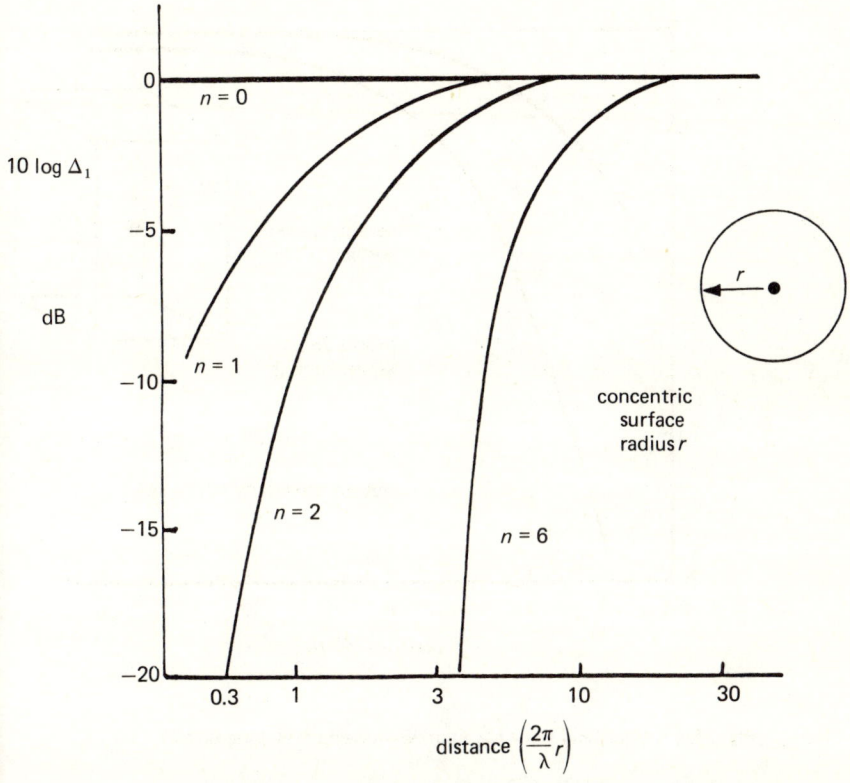

Fig. 24.1 – Near-field error for a spherical enveloping surface.

For a quadrapole source ($n = 2$), the error is less than 5 dB for distances greater than λ/π (i.e. 1 m for 100 Hz). Note also that the error is always of negative sense (as is also the case for the parallelopiped surfaces considered later), and thus the machine sound power level is always overestimated.

Fig. 24.2 shows the errors for a parallelopiped surface enveloping mono-pole, dipole, and quadrapole sources. It will be seen that for a monopole source the error is constant with measurement distance, whereas for higher order sources, the error decreases with distance. For all sources, the asymptotic error depends upon the aspect ratio of the parallelopiped and is least for a cubic surface.

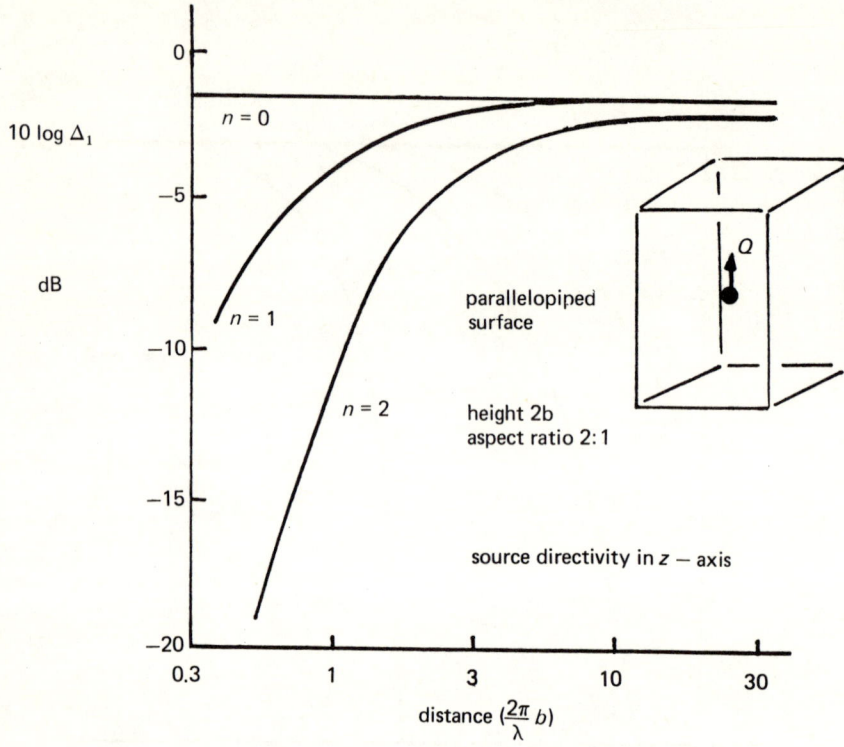

Fig. 24.2 – Near-field error for a parallelopiped enveloping surface.

Nonconcentric measurement surfaces are also considered by Hubner; Fig. 24.3 shows the error for monopole and dipole sources, enveloped by a nonconcentric spherical surface.

Summarizing, the near-field error for a dipole source and spherical measurement surfaces is less than 3 dB for values of the product rf (r measured from the source centre) of greater then 50 ms^{-1}, and less than 1 dB for values greater than 100 ms^{-1}. For a quadrapole source, these values become 120 ms^{-1} and 250 ms^{-1} respectively. Thus for the 63 Hz octave band the measuring distance, for errors of 1 dB or less, for dipoles and quadrapoles, must be greater than 2 m and 5 m respectively.

Since the error decreases as the frequency increases, the use of A-weighted sound power levels will tend to minimize near-field inaccuracies.

The parellelopiped errors encompass two factors, the inequality of v and $p/\rho c$, and the 'geometric error' arising from wavefront propagation non-normal to the measurement surface. This latter error is addressed separately by reference

[24.7]. Since, for propagation at an angle β to the normal, the true sound power is related to $p^2 \cos \beta$, the error, in decibels, will be given by $10 \log_{10} \cos \beta$, and will be less than 1 dB for $\beta < 37°$, and less than 2 dB for $\beta < 51°$. This theory is then developed to consider common situations such as sound radiated diffusely by an opening or wall, and an unlocated point source within a machine surface. For the latter case, the theory above indicates that if the entire machine surface subtends an angle of less than $2 \times 37° = 74°$ at the measurement point, the maximum possible geometric error will be less than 1 dB. It is interesting to note that recent legislation for noise abatement zones [24.8] requires that boundary measurements are made at sufficient points such that the subtended angle for a given boundary segment is 90–130°.

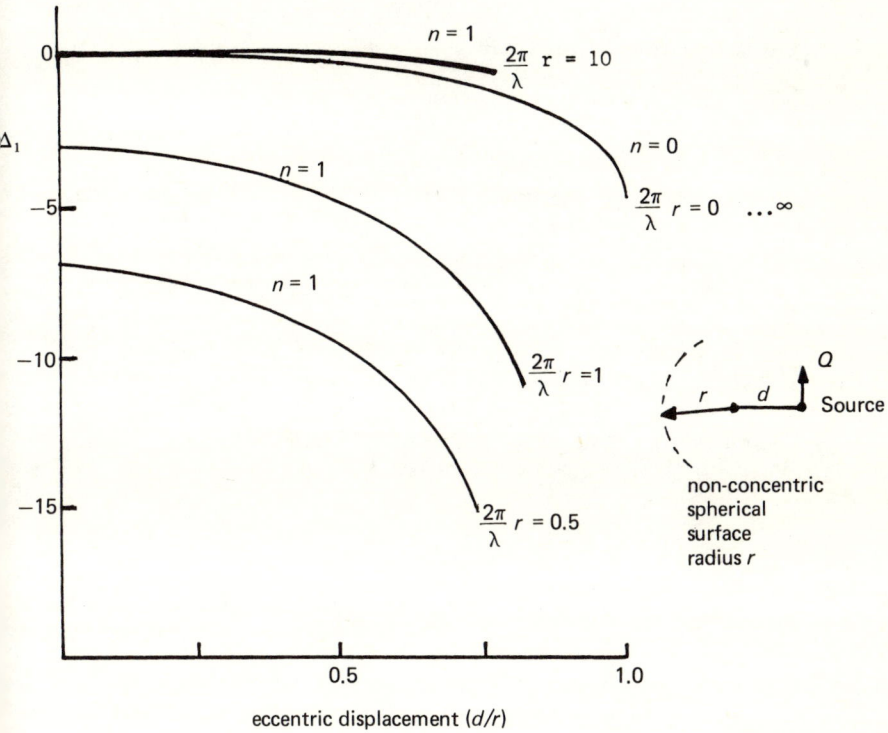

Fig. 24.3 – Near-field error for a nonconcentric spherical enveloping surface.

24.3.2 Δ_2 The finity error
Whereas the true sound power is obtained by integration over the entire surface area, in practice a finite number of measurement points are used, forming an array distributed over that surface. Thus, the approximated sound power P_{na} is given by

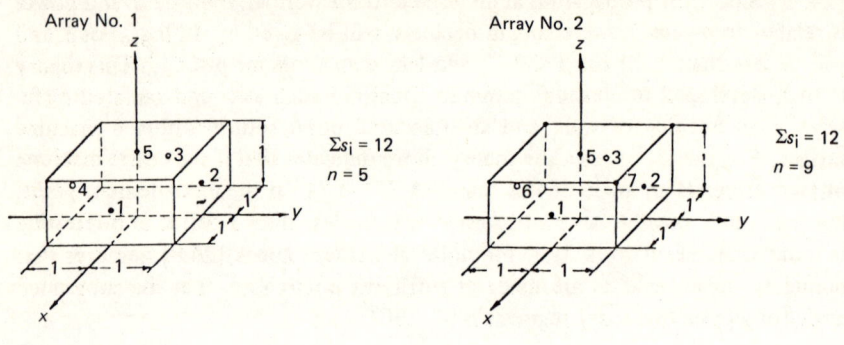

Error for omnidirectional source 2.05 dB

Error for omnidirectional source 0.7 dB

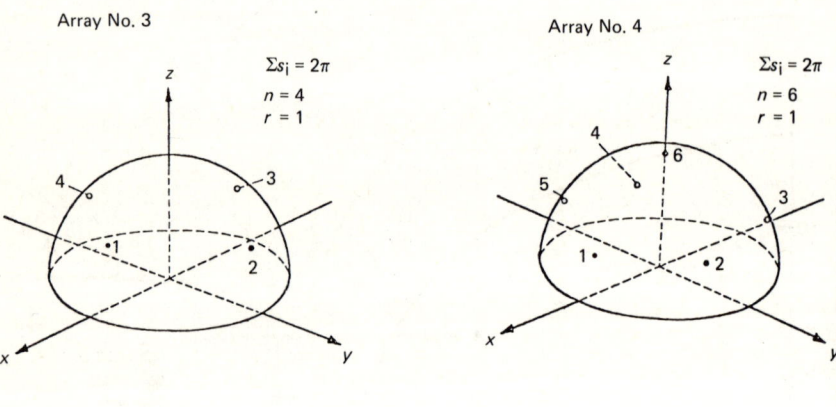

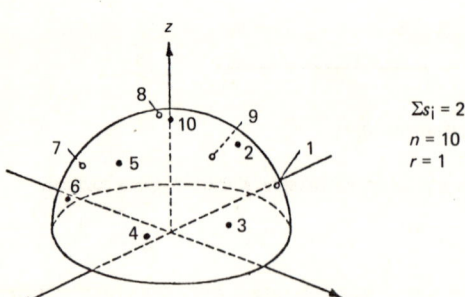

Fig. 24.4 – Various free-field enveloping surface arrays.

$$P_{na} = \frac{1}{\rho c} \; \frac{1}{N} \; \overset{N}{\Sigma} \; \overline{p^2}^t \; S \; .$$ (24.5)

Δ_2 is defined in terms of P_a/P_{na}.

Reference [24.7] tabulates the range of errors found for a number of arrays, for directional sources of order 1-6 whose orientation falls in line with, or between, measurement points. For arrays of 4 or more suitably placed points (Fig. 24.4) the error is always less than 1 dB for dipole and quadrapole sources.

In addition to these orientation errors, arrays 1, 2 (parallelopiped) suffer from systematic 'geometric' errors of the type described earlier arising from non-normal wave propagation. For arrays 1, 2 these systematic errors are -2.05 dB and -0.70 dB respectively.

24.4 SOURCES OF ERROR IN REVERBERANT CONDITIONS

A number of authors have reported apparent systematic differences between the sound power level of well-defined sources as measured in free-field and reverberant conditions. The difference is found to be a reduction in PWL at low frequencies for the reverberant measurement, typically some 4-6 dB in the 63 and 125 Hz bands, reducing to zero difference above 500 Hz. Brüel [24.9] suggested three contributory sources for this discrepancy, which are described in the following subsections.

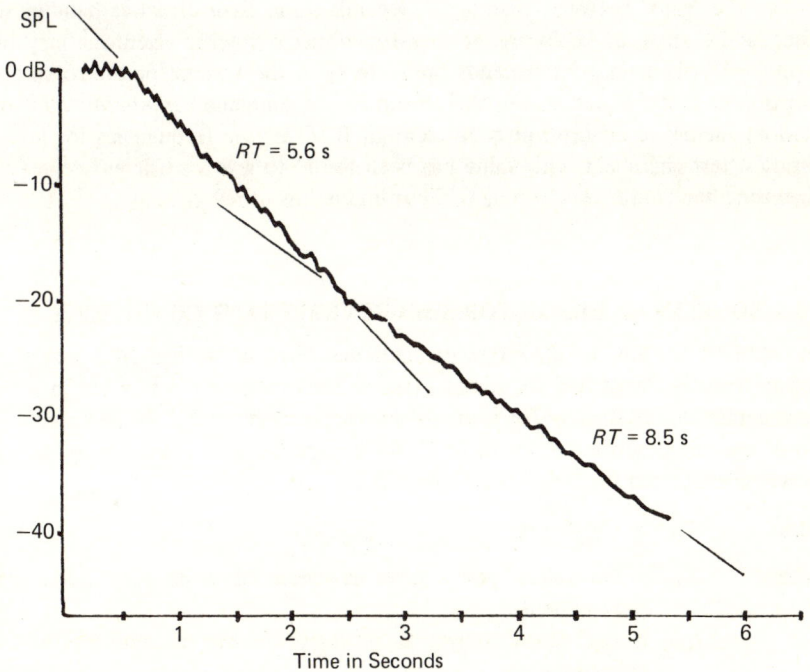

Fig. 24.5 – A typical low-frequency reverberant decay.

24.4.1 Reverberation decay slope

Evidence exists that low-frequency reverberation decays may be concave in practice, with a break-point at $-10/-15$ dB (Fig. 24.5). The ISO method for PWL requires the use of the -5 to -35 dB portion of the decay, whereas the discrepancy is minimized by use of the 0 to -10 dB decay in the determination of reverberation time upon which the PWL calculation depends.

24.4.2 Room impedance

Depending upon the nature of the noise source, its output may be affected by the impedance of the room into which it drives; and, of course, this is very much a function of position at low frequencies where the room dimension is not much greater than the wavelength.

Results are quoted by Brüel for a fan source which exhibited a variability of ± 2 dB in PWL at distances of less than 1.5 m from a wall, for the frequency range 100–200 Hz, as compared to $\pm \frac{1}{2}$ dB at 2000–10 000 Hz. For a loudspeaker source, the loading effects were found to be greater, with ± 3 dB below 200 Hz even for distances of greater than 1.5 m from a wall.

24.4.3 Eigentone spacing

At low frequencies, where eigentones are few and discretely spaced in the frequency domain, not all the source output may be able to support a room mode. The 'gaps' between eigentones depends upon their effective bandwidth (that is the range of input frequencies for which a specific eigentone may be supported) which in turn depends upon the Q of the system, or the room absorption at that frequency. For this reason ISO recommend a *minimum* value of random incidence absorption coefficient of 0.16 at low frequencies for reverberation test chambers. This value has been found to give a sufficient degree of eigentone bandwidth overlapping without undue loss of reverbance.

24.5 SOURCES OF ERROR FOR SEMI-REVERBERANT CONDITIONS

In addition to the three sources of error described in section 24.3 a further source may be identified for sound pressure level measurements made in semi-reverberant conditions, as for some *in situ* measurements. Δ_4, the environment error, may be described in terms of K, the contribution to the measured sound power arising from the reverberant sound field.

Thus $L_W = L_{W_i} - K$ (24.6)

where L_W is the sound power level as measured in an ideal acoustical environment,

L_{W_i} is the sound power as measured *in situ* in semi-reverberant conditions.

If L_{W_i} is measured by the enveloping surface method then

$$L_{W_i} = \overline{L_p} + 10 \log_{10} S \qquad (24.7)$$

where $\overline{L_p}$ is the mean sound pressure level as measured on the enveloping
surface array,

S is the area of the enveloping surface.

K will be a function of the particular enveloping surface chosen, and can be
determined by comparative measurements *in situ* and in ideal conditions. However, such comparative measurements are often not possible, for a given source,
and so it is necessary to calculate or determine K from measurements made of
the *in situ* environment.

Hubner [24.10] shows that for an empty room having its sound absorption
on its boundaries

$$K = 10 \log \left(1 + \frac{4S}{A}\right) \qquad (24.8)$$

where A is the total absorption in the room (m^2).

If sound pressure level measurements are made on concentric enveloping
surfaces at greater or lesser distances from the source (having greater or lesser
areas) then K may be determined from the difference in mean sound pressure
level, $\overline{L_p}$.

ISO [24.5] proposes two measurement surfaces S_2, S_3 of lesser and greater
area respectively than the primary surface S_1. Then, if

$$K_2 = |L_{W_2} - L_{W_1}| \qquad (24.9)$$

and $$K_3 = |L_{W_3} - L_{W_1}| \qquad (24.10)$$

the environmental correction factor K is taken as the lesser of K_2 or K_3.

K_2, K_3 are defined in terms of the difference between the mean surface
sound pressure levels, $\overline{L_{p_1}}, \overline{L_{p_2}}, \overline{L_{p_3}}$.

$$K_2 = 10 \log \frac{S_1/S_2 - 1}{M - 1} \qquad (24.11)$$

where $M = 10^{0.1 \Delta L_2}$

and $\Delta L_2 = \overline{L_{p_2}} - \overline{L_{p_1}}$:

Similarly,

$$K_3 = 10 \log \frac{1 - S_1/S_3}{1 - M} \qquad (24.12)$$

where $M = 10^{0.1 \Delta L_3}$

and $\Delta L_3 = \overline{L_{p_3}} - \overline{L_{p_1}}$.

In reference [24.5], K_2, K_3 are shown graphically, as functions of S_1/S_2, S_3/S_1, ΔL_2, ΔL_3.

Hubner [24.10] quotes relationships for K for other sound field configurations such as high aspect ratio rooms, having absorptive or reflective surfaces and containing scatter-absorption objects. He also quotes results of a series of comparative measurements for K as determined (i) theoretically, (ii) by the ISO method, and (iii) experimentally by comparison with ideal testroom conditions. For the range of situations encountered he finds that K_{exp} is always greater than K_{ISO} (by never more than 3 dB), thus when using the ISO method, the room correction will be underestimated and sound power levels will err on the high side.

24.6 NOISE SOURCE IDENTIFICATION IN PRACTICE

A range of available techniques, both simple and sophisticated, for identifying noise sources in machinery are discussed in the following sections of this chapter. Both physical location in space, and location by association with machine mechanisms, are described, and source level estimation is also included. The usefulness of each technique in current practice is discussed, and references to more comprehensive descriptions are given.

24.6.1 Listening to noise sources

Obvious as it may seem, much useful information may be derived from listening to the noise emitted by a machine. The unoccluded ears are excellent at recognizing, and discriminating between, noise signatures, and are moderately good at directional location in a free-field. By contrast, the ear is, of course, rather poor at judging level. At high noise levels, the wearing of ear defenders may prevent the reliable directional location of sources.

It is a relatively simple matter to modify a sound level meter in order to provide headphone listening, and this technique is frequently used in practice. When tape recording noise data, it also ensures successful recordings are made. The omission of this simple task can lead to useless recordings, and its importance cannot be stressed too highly.

24.6.2 Directional microphones

Since the microphone of the sound level meter is usually onmidirectional in nature, directional information is lost by this technique. The use of a directional microphone is normally recommended in these circumstances, although our experience with a range of proprietary directional microphones suggests that they are not always of benefit. One such microphone is illustrated in Fig. 24.6. Inspection of its directivity characteristics (Fig. 24.7) indicates the reason for such comments — although it provides good front-to-back discrimination, its major lobe is broad at middle frequencies.

Purpose-built microphone arrays provide more suitable directivity characteristics, as discussed in section 24.6.12

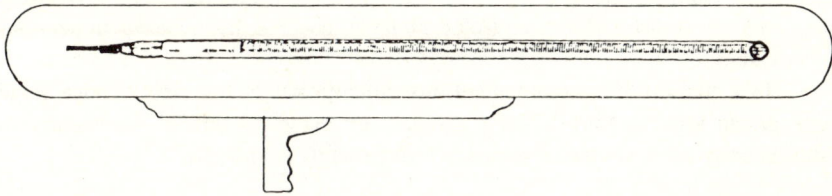

Fig. 24.6 – A directional 'gun' microphone of the type used by outside broadcast
reporter.

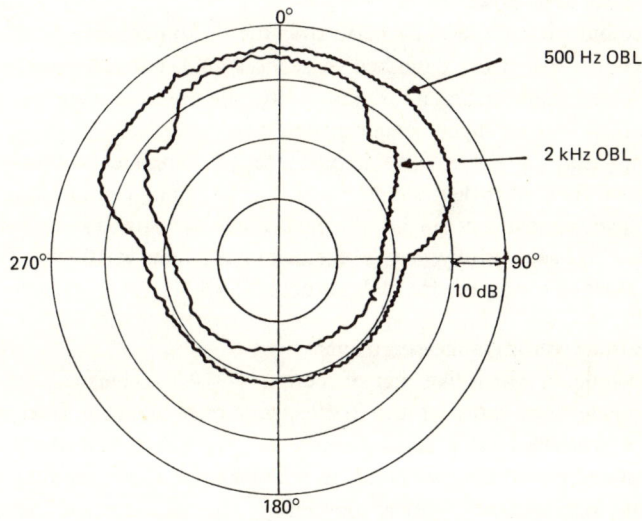

Fig. 24.7 – Polar response of the microphone shown in Fig. 24.6.

24.6.3 Source modification – variation of speed or load

The technique of varying the speed or loading of machinery, associated with
narrowband real-time frequency analysis, is well established. In practice, attempts
to use narrowband analysis for diagnosis (via frequency spotting) often yield
a surfeit of information from which it may be difficult to extract the wanted
information. Two cyclic mechanisms, at slightly differing frequencies, may
appear to 'beat' together showing a characteristic sideband structure.

24.6.4 Source modification – partial hooding

In this method, all the potential noise sources in the machine are fitted with close-fitting enclosures or hoods. These are removed one by one, noting the changes in noise emission which occur. The technique works best for high-frequency sources, above, say, 500 Hz, since these may be more effectively enclosed by a soft, vibration isolation layer covered by a heavy impervious skin material.

High stability of measurement instrumentation is important, since small changes in level in time-differing samples are being considered. Consideration should be given to statistical accuracy requirements in analysis.

24.6.5 Partial power level estimation

Sound pressure level measurements made in close proximity to a surface noise source, and averaged over the surface area involved, can be used to estimate the contribution of that noise source to the total radiated sound power.

The estimated power level, $PWL_i = L_i + 10 \log S_i$ where L_i is the time-average sound pressure level averaged over the ith source, area $S_i\ m^2$. This procedure ignores the errors described in section 24.3, but reference to that text will show that such errors are minimized by the use of A-weighted levels, and are acceptably low for higher frequency sources.

Comparison of the relative importance of component noise sources on the basis of their near-field sound pressure levels alone can be misleading, and partial power estimation can be of considerable use in noise control planning. The author has successfully used a shielded microphone at 100 mm from wood-working machinery sources for such purposes [24.11].

24.6.6 Surface vibration measurements

In this technique, which like that of section 24.6.4 is extensively used in automotive engine noise work at the ISVR, the surface vibration level of machine surfaces is measured using an accelerometer. The assumption is made that the radiated sound power is proportional to the mean square surface velocity.

Below the surface's critical frequency, the radiation efficiency of the surface has to be estimated; above the critical frequency this may be taken as unity in the expression below

$$PWL = L_s - 20 \log f + 10 \log S + 10 \log \sigma_{rad} + 150\ dB$$

where PWL is the sound power level radiated by surface

L_s is the mean square acceleration level over the surface (dB re 1 g r.m.s.)

S is surface area (m^2)

f is frequency (Hz)

σ_{rad} is radiation efficiency.

Reference [24.12] describes the use of the technique in engine noise research at the ISVR.

24.6.7 Time history event analysis
When considering the noise generated by, for example, a punch press or teletype, a time history analysis is often used. If linked to some timing signal derived from a discrete event in the machine's operating cycle, the technique becomes quite powerful.

Slow-speed replay of tape recordings can assist in identifying discrete events by ear.

24.6.8 Sound intensity measurement
Recently developed sound intensity meters will allow direct measurement of sound energy flow and direction, thereby indicating the true sound emission of sources, rather than the sound pressure distribution in their vicinity. Reference [24.13] reviews recent developments in this field.

24.6.9 Signal processing techniques
Over the past few years, sophisticated methods of source location using signal processing techniques, notably correlation, have been developed. Much of the work has been concerned with source location in jet engine noise, but the techniques are suitable for more general application. These techniques are briefly reviewed, from a noise control engineers viewpoint, in the following subsections. A more critical review will be found in references [24.14] or [24.15].

24.6.10 Signal processing techniques – causality correlation
This technique allows the relative importance of a number of known sources, in respect of the far-field noise at a particular receiver location, to be assessed. The cross-correlation coefficients of the signals derived from probes placed at the far-field receiver, and at each noise source in turn, are compared. Propagation paths may also be determined, in the presence of discrete reflections, by an examination of the cross-correlation coefficient as a function of time delay. Fig. 24.8 illustrates the technique, which is described in greater depth in [24.16].

In practice, problems may arise from source probe self-noise – unless a remote sensing technique such as the Laser Doppler Velocimeter is used. In addition the magnitude of the cross-correlation coefficients to be measured may not be very great, being dependent on the reciprocal of the square root of the number of locally correlated regions in the total source.

Although the technique is a powerful analytical tool it is normally considered as more appropriate to fundamental research work than practical engineering source location work.

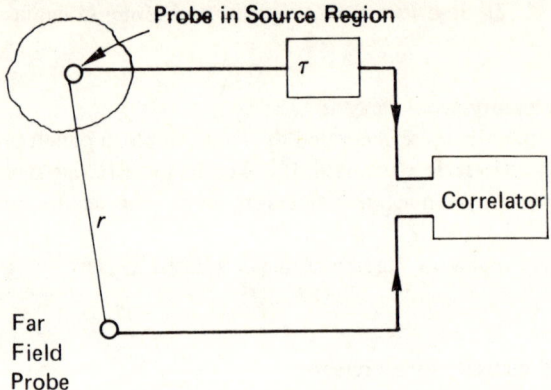

Fig. 24.8 – Causality correlation ($\tau = r$/velocity of sound).

24.6.11 Signal processing techniques – acoustic mirror

This is not, in fact, a signal processing technique, but rather an essentially simple method of creating an acoustic telescope (reflecting). Fig. 24.9 illustrates parabolic and elliptical mirrors, giving line and point foci respectively.

The major limitation is the required size of mirror, being some 50%–75% of the measurement distance for good resolution. It has been estimated that a mirror diameter of the order of 8 metres would be required for aero-engine source location.

Reference [24.17] describes the use of a parabolic mirror.

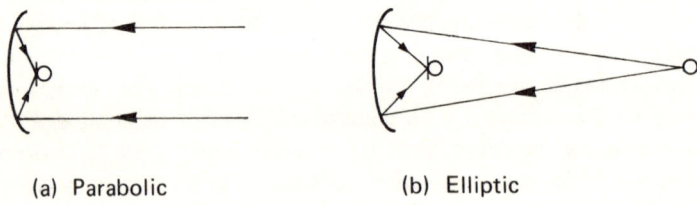

(a) Parabolic   (b) Elliptic

Fig. 24.9 – Acoustic mirrors.

24.6.12 Signal processing techniques – acoustic telescope

An array of microphones can provide an exceptionally directional detector, at least in the plane of the array. The side lobes of its directivity pattern may be minimized by amplitude tapering along the array.

If variable time delays are introduced into each microphone line, and the outputs summed, the array may be effectively 'steered' to swing its major axis in any direction, as shown in Fig. 24.10. The technique has been used in jet noise research [24.18], but its disadvantages are that many instrumentation channels are required, and that the array's optimum dimensions are a function of frequency, and therefore should be varied as analysis proceeds.

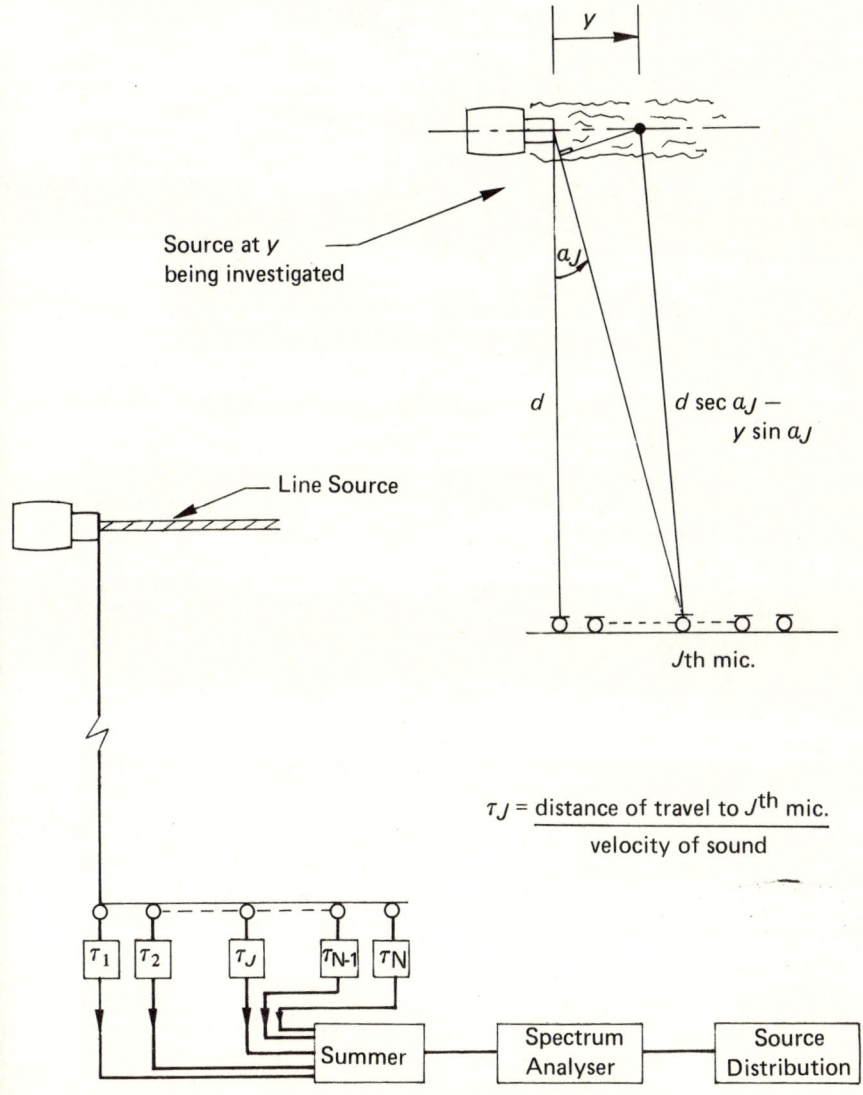

Source at y being investigated

Line Source

a_J

d

d sec $a_J -$
y sin a_J

Jth mic.

τ_J = distance of travel to J^{th} mic. / velocity of sound

τ_1 τ_2 τ_J τ_{N-1} τ_N

Summer

Spectrum Analyser

Source Distribution

Fig. 24.10 − Acoustic telescope.

24.6.13 Signal processing techniques – polar correlation

The ISVR have developed this technique for jet noise research. Fig. 24.11 illustrates the essential features of the method, which requires precise positioning of the microphones on an arc about the source area being investigated.

The cross-power spectral density of signals from two microphones is derived, as indicated in [24.19], which gives a comprehensive description.

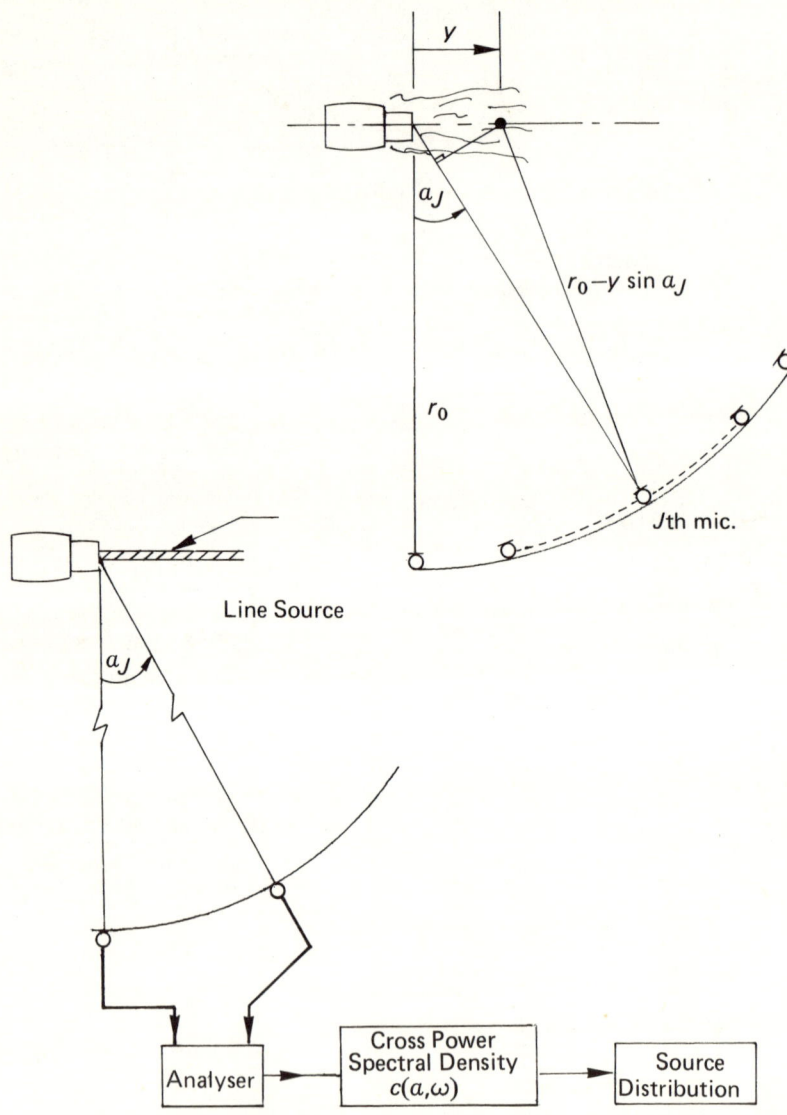

Fig. 24.11 – Polar correlation (after Fisher & Glegg ([24.14]).

REFERENCES

[24.1] Brüel & Kjaer, Denmark (1971) *Standards, formulae and charts*.

[24.2] Lang, W. W. (1973) *J. Acoust. Soc. Am.*, **54**, 4, 960–966. Noise measurement standards for machines *in situ*.

[24.3] Ebbing, C. E. & Maling, G. C. (1973) *J. Acoust. Soc. Am.*, **54**, 4, 935–949. Reverberation room qualification for determination of sound power of sources of discrete-frequency sound.

[24.4] Francois, P. P., Ebbing, C. E. & Maling, G. C. (1973) *Inter-noise proceedings G23Z17*. Results from an international sound power round robin concerning measurements in reverberation rooms.

[24.5] International Standards Organisation (1975) *ISO/DIS 3744*. Determination of sound power levels of noise sources–engineering methods for free-field conditions over a reflecting plane.

[24.6] Hubner, G. (1973) *J. Acoust. Soc. Am.*, **54**, 4, 967–977. Analysis of errors in measuring machine noise under free-field conditions.

[24.7] Concawe (1976) *Report 2/76*. Determination of sound power level of industrial equipment, particularly oil industry plant.

[24.8] HMSO 1976. Statutory Instrument No. 37. *The Control of Noise (Measurement and Registers) Regulations*, 1976.

[24.9] Brüel, P. V. (1978) *Brüel & Kjaer Technical Review*, **3**. The enigma of sound power measurements at low frequencies.

[24.10] Hubner, G. (1973) *Inter-Noise Proceedings G2 3X13*. Qualification procedures for free-field conditions for sound power determination of sound sources.

[24.11] Wheeler, P. D. (1977) *Proc. 9th ICA*. Wood-radiated noise in woodworking machinery.

[24.12] Chan, C. & Anderton, D. (1974) *Noise control engineering*, **2**, 1, 25–29. The correlation between engine block surface vibration and radiation noise of in-line diesel engines.

[24.13] Fahy, F. J. & Elliott, S. J. (1981) *Proc. 1st International conference on acoustic intensity measurements*, CETIM, Senlis, France. Practical considerations in the choice of transducers and signal techniques for sound intensity measurements.

[24.14] Fisher, M. J. & Glegg, S. (1977) *Applications of time series analysis*, Southampton University. Source location.

[24.15] Damms, S. M. (1977) *R.A.E. Tech. Report 77032*. The shielding method for noise source location and a review of alternative methods.

[24.16] Siddon, T. E. (1971) *Proc. 7th ICA*. New correlation method for study of flow noise.

[24.17] Chu, W. T., Laufner, J. & Kao, K. (1972) *International Conference on Noise Control, Washington D.C.* Noise source distributions in subsonic jets.

[24.18] Billingsley, J. & Kinns, R. (1976) *J. Sound Vib.*, **48**, 485-510. The acoustic telescope.

[24.19] Fisher, M. J., Harper-Bourne, M. & Glegg, S. (1976) *J. Sound Vib.*, **51**, 23-54. The polar correlation technique.

CHAPTER 25

Vibration control (I)

D. J. Mead

Department of Aeronautics and Astronautics, University of Southampton

25.1 INTRODUCTION

The control and reduction of unwanted vibration levels must always begin with a study of the source of vibration, and attempts must be made to minimize the magnitude of the source levels. This may involve the reduction of noise output from a jet engine, the balancing of a rotating machine, the smoothing of a turbulent aerodynamic flow, etc. However, there is always a limit to the amount of reduction which is possible, so the structure which is being made to vibrate must be designed to respond to the minimum extent. This is the art of vibration control.

Several different methods of vibration control are available, and each has its optimum sphere of application. They may be summarized as:

Vibration control by: Structural design
Material selection
Localized additions
Artificial damping
Resilient isolation

The first four of these are dealt with in this chapter, and the last in Chapter 26.

The level of structural vibration due to a given source is controlled by the threefold characteristics of mass, stiffness, and damping. Different structural responses under different conditions of excitation will depend on these characteristics in different ways. For example, one response quantity may be highly dependent upon the mass, while another may be quite independent of it. It is therefore necessary to understand the particular problem to be dealt with, and to know which are the important structural characteristics which govern the vibration levels. A few special examples will next be considered to identify the controlling structural parameters.

25.2 FACTORS AFFECTING VIBRATION LEVELS

25.2.1 Single degree of freedom systems

Consider first a structure whose response can be represented by that of a single-degree-of-freedom mass-spring-damper oscillator. Let it be excited at the mass by a *harmonic force* of given amplitude. The equation of motion is:

$$m\ddot{w} + c\dot{w} + kw = F_0\, e^{i\omega t} \qquad (25.1)$$

where the damping is assumed to be viscous. The amplitudes of response, w_0, at different frequencies relative to the natural frequency, $\omega_r\ (=\sqrt{k/m})$ are:

Table 25.1

ω :—	$\ll \omega_r$	ω_r	$\gg \omega_r$
w_0 :—	$\approx \dfrac{F_0}{k}$	$\dfrac{F_0}{\omega_r c}$	$\approx \dfrac{F_0}{\omega^2 m}$

Evidently, at low frequencies the response is controlled by a stiffness k, at high frequencies by the mass m, and at the natural frequency by the damping c.

When the same system is excited by a *random force*, with a power spectral density of $S_F(\omega_r)$ at the system natural frequency and a mean square value of $<F^2>$, the r.m.s. values of the displacement (w_{rms}), velocity ($\dot{w}_{rms}$), and acceleration ($\ddot{w}_{rms}$) are given as follows:

Table 25.2

w_{rms}	$\dot{w}_{rms}$	$\ddot{w}_{rms}$	
$\sqrt{\dfrac{\pi}{2}\dfrac{S_F(\omega_r)}{kc}}$	$\sqrt{\dfrac{\pi}{2}\dfrac{S_F(\omega_r)}{mc}}$	$\dfrac{1}{m}\sqrt{<F^2> + \dfrac{\pi}{2}\dfrac{S_F(\omega_r)(1-\eta^2)k}{c}}$	

In these, η is the system loss factor ($= c/\sqrt{mk}$).

Notice that the random displacement is independent of mass, and the velocity is independent of the stiffness. Both are inversely proportional to the *square root* of the damping, in contrast to the harmonic resonant response which is inversely proportional to the damping. The random acceleration is inversely proportional to the damping if the damping is small, but is independent of it if the damping is large. In both cases, this response is inversely proportional to the mass.

Next consider the *harmonic transmissibility* across the simple system, i.e. the force transmitted to the base-attachment point of the spring and damper $\div F_0$. This is identical to the acceleration of the mass $\div$ base acceleration when the system is excited through its base:

$$\frac{F_{base}}{F_0} = \frac{\ddot{w}_0}{\ddot{w}_{base}} = T(\omega) = \frac{k + i\omega c}{k - \omega^2 m + i\omega c}. \qquad (25.2)$$

The absolute values of this at different frequencies are:

Table 25.3

ω	$\ll \omega_r$	ω_r	$\sqrt{2}\,\omega_r$	$\gg \sqrt{2}\omega_r$
$\lvert T(\omega)\rvert$	1.0	$\sqrt{1+\dfrac{1}{\eta^2}}$	1.0	$\dfrac{c}{\omega m}$

At ω_r, the transmissibility is reduced by increasing the damping (η), and this is true (to a decreasing extent) until $\omega = \sqrt{2}\,\omega_r$. Above this frequency, the transmissibility increases with increasing damping, but decreases with increasing mass and frequency.

25.2.2 Multiple degree of freedom systems

When a single resonating mode contributes most of the vibration response, the parameters which control the vibration level will be the same as in the previous section. If the excitation is random, and the resonant frequencies are well separated, then the total mean square response will (to first order) be the sum of the mean square responses in each mode. The controlling structural parameter will then be the modal (or generalized) stiffness, masses and/or dampings as indicated by Table 25.2.

At high frequencies, the resonant frequencies are no longer well separated, and adjacent resonant peaks tend to merge. A detailed analysis of modal responses is probably impracticable, and spatially averaged responses are more meaningful. Cremer & Heckl [25.1] quote one such example. They consider a uniform plate of total mass M excited by a random force of mean square value F_Δ^2 in a narrow bandwidth $\Delta\omega$, centred on the frequency ω. The average mean square plate velocity (averaged over the whole plate *and* over all possible force locations on the plate) is shown to be given by:

$$\dot{w}_\Delta^2 \approx \frac{F_\Delta^2}{M^2}\,\frac{\pi}{2\eta\omega}\,\frac{\Delta N}{\Delta\omega} \approx F_\Delta^2\,\frac{\pi}{2}\,\frac{1}{Mc}\,\frac{\Delta N}{\Delta\omega}\,. \tag{25.3}$$

In this, ΔN is the number of modes having resonant frequency in the bandwidth $\Delta\omega$. (In the limit, $\Delta N/\Delta\omega \to dN/d\omega$, which is the modal density of the structure. Notice that equation (25.3) yields, in effect, the total mean square response from ΔN modes, vibrating randomly, each one of which yields a mean square velocity as given by Table 25.2). The r.m.s. velocity is inversely proportional to the square root of both mass and damping, and stiffness has no influence. The r.m.s. displacement, on the other hand, does depend on the stiffness ($\propto 1/\sqrt{\text{stiffness}}$) and not on the mass.

At very high frequencies, a damped finite system behaves like a damped infinite system. Some results from Chapter 9 then become applicable. The complex displacement response of an infinite beam at the point of harmonic force loading is:

$$w(o) = -P(1+i)/4Dk^3 \qquad (25.4)$$

where $k^4 = \omega^2 \rho h/D$ and $D = E'(1+i\eta)h^3/12(1-\nu^2)$. Damping of the beam has been introduced into this by the inclusion of the complex term $(1+i\eta)$ with E. η is the material loss factor. If $\eta^2 \ll 1$, the above expression yields:

$$|w(o)| \approx \frac{3P}{16(\omega h)^{3/2}} \left(\frac{1-\nu^2}{E'\rho^3}\right)^{1/4} \frac{1}{(1+\eta^2/8)} . \qquad (25.5)$$

This shows that the beam damping has scarcely any effect upon the response at the point of loading. The beam mass (through ρ) has quite a strong effect, but the stiffness (through E) has a much smaller effect. The response at a point remote from the loading point is, however, quite strongly affected by the damping, since the outward-going propagating wave decays in proportion to $\exp(-k\eta x/4)$, where x is the distance from the loading point.

The response of the same beam to a convected pressure field depends on mass density (ρ), stiffness (E), and damping (η) in a different way still. Important also in this problem is the relative magnitude of the pressure field wavelength and the free flexural wavelength of the beam at the excitation frequency. The latter is a function of the beam mass and stiffness. This problem will not be elaborated here. Sufficient has been shown to emphasize that different vibration problems depend on the system parameters in different ways. It cannot always be said that increasing the damping (or mass, or stiffness) will reduce the vibration level. Each problem must be understood and treated on its own merits.

Further expressions for different system response quantities have been listed in references [25.2] and [25.3]. Many more could be added to these.

25.3 VIBRATION CONTROL BY STRUCTURAL DESIGN

By this we imply that the structure is designed with magnitudes and distributions of mass and stiffness which minimize the vibration generated by the operating environment. Artificial damping is not considered at this point. In the simplest case of a structure subjected to a fixed-frequency harmonic force, the design would ensure that no resonant frequency coincided with the excitation frequency, but the excitation frequency lay in a trough in the response curve. This is the 'detuning' process, being carried out at the design stage.

Fixed frequencies of excitation seldom occur, and the design must allow for variable frequency or random excitation. In this case, resonant frequencies will be encountered, and resonant (or random-resonant) vibrations must be minimized. If the structure is excited by a single-point force, a mode which has a nodal point at the excitation point cannot be made to respond. If the structure could be designed such that several important modes had nodal points at, or close to, the excitation point, low vibration levels would result. High-frequency flexural modes of beams have nodal points close to large attached masses. The addition of a large mass at an excitation point will therefore reduce high-frequency responses, but could have a detrimental effect upon the fundamental mode response unless it detunes the structure at the fundamental frequency.

Considerable attention has been given to the design of stiffened, thin-plate structures to minimize vibration levels due to acoustic excitation. This has been largely in the context of jet-noise excited aerospace structures which have been liable to severe fatigue failure due to the high fluctuating stress levels. Careful attention to detail design is required to avoid stress concentrations which promote fatigue failure, but the techniques for this will not be discussed here. Clarkson [25.4] has further shown that the stiffener design can be adapted to minimize vibration and stress levels. The torsional stiffness of the stiffener should be low, and the flexural stiffness should be high. If both torsional and flexural stiffness are high, the modes of vibration of the stiffened plate occur in narrow frequency bands, and this closely-packed property of the natural frequencies leads to large response levels. If they are more widely spread, the response levels are lower.

From the vibration response point of view, it would be ideal for the flexural stiffness also to be very low − but then the stiffeners would be unable to fulfil their essential requirement of stabilizing the plate under static loads. The high flexural stiffness of the stiffener provides light-weight, efficient stiffening of the whole plate, and this in turn leads to a gross reduction of the plate coincidence frequency. The stiffened plate is then much more responsive to acoustic excitation at a coincidence frequency much below that of the unstiffened plate.

Some computer results based on periodic structure theory reinforce these points. Suppose the plate is stiffened at regular intervals by identical, parallel stiffeners which have infinite flexural stiffness, but finite torsional stiffness κ_r. A random pressure field from a turbulent boundary layer convects across the plate, perpendicular to the stiffener, at a non-dimensional speed of CV. (See Chapter 13 for symbol definitions). The r.m.s. curvature at the centre or edge of a plate bay (whichever is highest) is shown in Fig. 25.1. This is the curvature generated by random vibration in the first propagation band only.

Notice that above $CV = 6$, the lowest r.m.s. curvature is obtained with the lowest torsional stiffness, but with CV less than 4, the reverse is true. If the structure is subjected to a diffuse sound field, with a spectrum of convection velocities from the speed of sound up to infinity, the overall curvature response will be governed by the high CV characteristics so that low torsional stiffness will lead to the lowest curvature and stress.

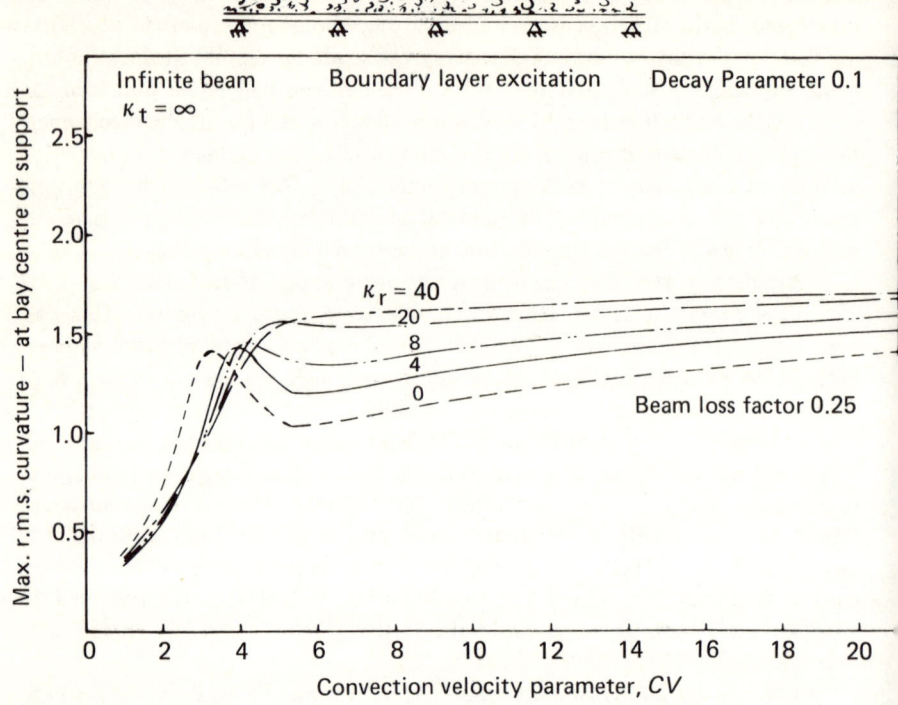

Fig. 25.1 – Effect of torsional stiffness, κ_r, on r.m.s. curvature.

It is important to consider the influence of the stiffener spacing (l) and plate thickness (h) on the plate response. The transverse deflection of the stiffened plate is proportional to the r.m.s. pressure $\times$ (l^4/D) $\times$ function of CV. D is the plate flexural rigidity, $Eh^3/12(1-\nu^2)$. For a given value of CV, the deflection is then proportional to l^4/h^3, so that reducing the spacing and increasing the thickness have great effects. The r.m.s. bending stress generated by the vibration is proportional to the r.m.s. curvature $\times$ plate thickness, and this is proportional to l^2/h^2, for a given CV. The effect of changing l and h is still substantial, though not as great as the effect on the deflection. These responses are influenced also by the plate damping, but this will be dealt with later. The rather obvious conclusion can then be drawn that for minimum response, we require closely spaced stiffeners and a thick plate.

When the plate is reinforced with an orthogonal array of stiffeners with low torsional rigidity and high flexural rigidity, its response to convected random loading depends upon the direction of convection relative to the stiffeners. If the stiffeners subdivide the plate into rectangular (and not square) bays, the response is least when the convection is in the direction of the longer side of the bay. Hence, a stiffened plate which has to resist a random convecting pressure field must be stiffened such that the stiffeners parallel to the convection are closer together than those perpendicular to it. The side-panels of aeroplane fuselage structures satisfy this requirement, when their response to boundary layer pressure fluctuations is considered.

25.4 VIBRATION CONTROL BY MATERIAL SELECTION

This has been discussed comprehensively in reference [25.3]. Suppose the simple system represented by equation (25.1) derives all its damping from internal hysteresis of the structural material. When the excitation is harmonic, such damping is more accurately represented in the complex stiffness form. We then have:

$$m\ddot{w} + k(1 + i\eta)w = Pe^{i\omega t} \qquad (25.6)$$

where η is the system (and material) loss factor, which can be defined by the energy relationship

$$2\pi\eta = \frac{\text{Energy dissipated per harmonic displacement cycle of given amplitude}}{\text{Maximum energy stored in the cycle}} . \qquad (25.7)$$

The generalized stiffness k will be proportional to the Young's modulus, E, of the material, and the generalized mass to the density ρ. Write

$$k = \alpha_s E , \quad m = \alpha_m \rho . \qquad (25.8)$$

The amplitude of vibration at the resonant frequency $\omega_r (= \sqrt{k/m})$ is

$$|w_{\text{res}}| = \frac{P}{k\eta} = \frac{P}{\alpha_s . E\eta} . \qquad (25.9)$$

If there is a choice of material out of which the structural system can be made, then the lowest vibration amplitude under harmonic excitation at resonance will be obtained with the material having the highest value of $E\eta$. The harmonic strains in the system will be proportional to w, and the harmonic stresses to Ew, i.e. to $1/\eta$. Thus, the material with the highest loss factor, η, undergoes the lowest stress amplitudes.

As in an earlier section of this chapter, it can be shown that different response quantities due to different conditions of excitation depend upon E, η, and ρ in different ways. For example, the r.m.s. displacement due to a random exciting force is found to be

$$w_{\text{rms}} = \left\{ \frac{\pi}{2} \frac{S_F(\omega_r)\omega_r}{k^2\eta} \right\}^{\frac{1}{2}} = \left\{ \frac{\pi}{2} S_F(\omega_r) \right\}^{\frac{1}{2}} \frac{1}{\alpha_s^{3/4} \alpha_m^{1/4} (E^{3/2} \rho^{\frac{1}{2}} \eta)^{\frac{1}{2}}} . \qquad (25.10)$$

Thus, the material with the highest value of $E^{3/2}\rho^{\frac{1}{2}}\eta$ gives the lowest r.m.s. displacement. If the r.m.s. *stresses* are to be minimized, the material having the highest value of $(E^{-\frac{1}{2}}\rho^{\frac{1}{2}}\eta)^{\frac{1}{2}}$ is to be preferred — provided its fatigue resistance is adequate.

Quantities such as $E^{3/2}\rho^{\frac{1}{2}}\eta$ may be called 'figures of merit' for particular materials in particular vibratory environments. Expressions for many other figures of merit can be derived [25.3]. However, their usefulness is restricted to those materials and structures in which the greatest part of the total damping stems from internal hysteresis of the material, and not from the structural joints or interfaces. If the joints and interfaces contribute most to the damping, the system damping is not easily related to the material properties and depends heavily on the quality of the joint, the presence of jointing compound, degree of fretting, erosion, and corrosion which has occurred, etc.

At this point, we should quote some typical values of E, ρ, and η for structural materials. Whereas E and ρ for structural metals do not vary over a range of more than 10:1 (at the very most), the values of η can vary over several orders of magnitude. Furthermore, η is often quite heavily dependent upon stress amplitude even when the material appears otherwise to behave elastically. For many materials, the loss factor can be expressed in the simple form

$$\eta = \frac{E.J}{\pi}\,\sigma^{n-2} \tag{25.11}$$

where n normally lies between 2 and 3, but can rise to 30 for some materials. Lazan [25.5] has discussed this in detail, explaining the mechanisms involved and quoting values of J and n. Some typical values of ρ, E and η are quoted in Table 25.4.

Table 25.4

Material	Density ρ kg/m³	Young's Modulus E kN/mm²	Loss factor, η	
			Low stress	High stress
Aluminium	2 700	71	5×10^{-5}	7×10^{-3}
Brass	8 500	104	2×10^{-5}	3×10^{-4}
Cast iron	7 700	105	3×10^{-2}	7×10^{-2}
Lead	11 300	16.5	2×10^{-3}	2×10^{-1}
Steel (stainless)	7 700	195	5×10^{-4}	1×10^{-2}
Steel (Ni–Cr)	7 700	195	2×10^{-4}	2.5×10^{-3}
Concrete (dense poured)	2 300	27	5×10^{-3}	2×10^{-2}
Copper-manganese	7˙280	63	3×10^{-2}	7×10^{-2}
Magnesium alloy	1 740	44	5×10^{-4}	8×10^{-3}

The Young's moduli and loss factors of composite materials (e.g. carbon fibre reinforced plastics) depend upon the relative proportions of the constituent materials. For such a material with the fibres aligned in one direction, and the material being uni-axially stressed in that direction, the composite Young's modulus and loss factor are given by

$$E_c = E_1\alpha_1 + E_2\alpha_2$$

$$\eta_c = \frac{E_1\eta_1\alpha_1 + E_2\eta_2\alpha_2}{E_1\alpha_1 + E_2\alpha_2} .$$

In this E_1, E_2 are the component Young's Moduli and η_1, η_2 are the component loss factors. α_1, α_2 are the fractions of the total volume occupied by each component ($\alpha_1 + \alpha_2 = 1$). Fig. 2 (from reference [25.28]) shows some typical values of E_c and η_c. If component 2 is the resin matrix, $E_2 \ll E_1$, and $E_2\eta_2 \gg E_1\eta_1$ (the damping capacity of the matrix is much greater than that of the highly elastic fibre); then

$$E_c \approx E_1\alpha_1 , \qquad \eta_c \approx \eta_2 E_2\alpha_2/E_1\alpha_1 .$$

The stiffer the composite the lower is the loss factor. Notice that if $\alpha_1 > 20\%$, the loss factors are less than 0.004.

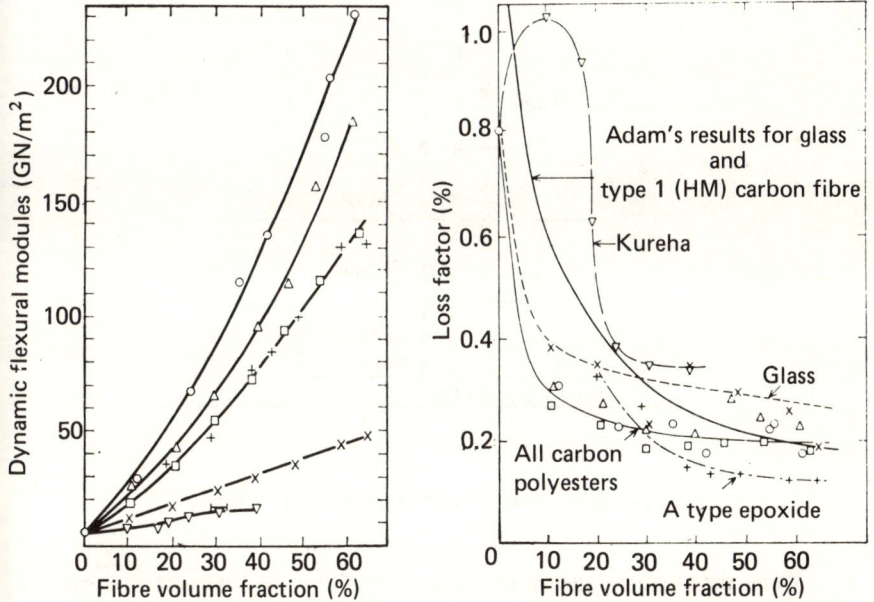

Fig. 25.2 – Modulus vs. volume fraction for all fibre-resin combinations at 25°C. o, HM–S carbon + polyester; Δ HT–S carbon + polyester; □, A carbon + polyester; +, carbon + epoxide; ×, E glass + polyester; ∇, KCF–100 carbon + polyester; –, random E glass + polyester.

25.5 VIBRATION CONTROL BY LOCALIZED ADDITIONS

The attachment of a 'dynamic absorber' to a vibrating system has long been known as an effective means of annulling vibration at a particular troublesome frequency. The absorber (better called a 'neutralizer', see Crede [25.6]) constitutes an 'auxiliary system', added at a particular location, and therefore a special form of a 'localized addition' to the system. The simplest localized addition is the point mass or spring.

Before summarizing and reviewing the different forms of such additions, we shall consider the general effect on the vibration of a system of adding a linear auxiliary system.

Consider the general linear system of Fig. 25.3 (a beam, say) which is excited by a harmonic source $P_s e^{i\omega t}$ at point S. An auxiliary system is to be added at point A with a view to controlling the vibration level at a receiver point R. Let the displacement at A be $w_A e^{i\omega t}$, and at R, $w_R e^{i\omega t}$. The receptances of the system, before the auxiliary system is added, are $\alpha_{SS}, \alpha_{RS}, \alpha_{AS}, \alpha_{RA}, \alpha_{AA}$ where the typical receptance α_{RS} is the complex harmonic displacement at point R due to unit harmonic force at point S. (For the definition of a receptance, see Chapter 4. For an exposition of receptance theory, see Bishop & Johnson [25.7]).

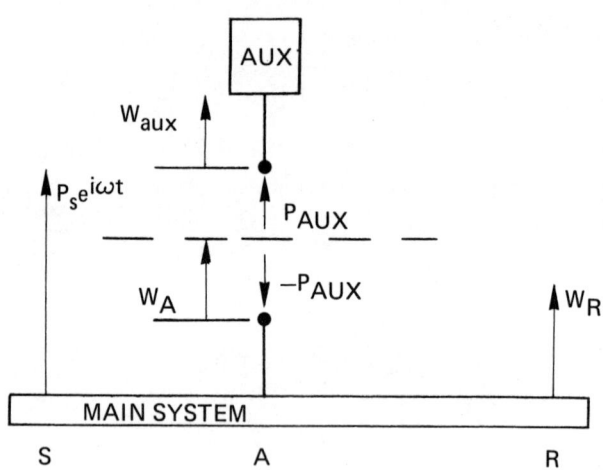

Fig. 25.3 – Diagram of main and auxiliary systems.

When the auxiliary system (receptance α_{aux}) is added at point A, it exerts a harmonic force on A by virtue of the displacement it undergoes. The force is related to the displacement w_{aux} through

$$w_{aux} = \alpha_{aux} P_{aux} .$$

(25.12)

The force exerted back on the main system is $-P_{aux}$, and since the auxiliary and main systems become coupled together,

$$w_{aux} = w_A .$$

Hence $P_A = -P_{aux} = -w_a/\alpha_{aux} .$ (25.13)

The total displacement at A is now due to P_A and P_S. Hence

$$w_A = P_s\alpha_{AS} + P_A\alpha_{AA}$$ (25.14)

or $$w_A = P_S \frac{\alpha_{AS}\alpha_{aux}}{\alpha_{AA} + \alpha_{aux}}$$ (25.15)

and $$P_A = -P_S \frac{\alpha_{AS}}{\alpha_{AA} + \alpha_{aux}} .$$ (25.16)

Likewise $$w_r = P_S\alpha_{RS} + P_A\alpha_{RA} = P_S\left\{\alpha_{RS} - \frac{\alpha_{RA}\alpha_{AS}}{\alpha_{AA} + \alpha_{aux}}\right\} .$$ (25.17)

The displacement at the source (point S) is found from equation (25.17) by replacing suffix 'R' by 'S'. Using the reciprocal relationship $\alpha_{AS} = \alpha_{SA}$, we then obtain

$$w_S = P_S\left\{\alpha_{SS} - \frac{\alpha_{AS}^2}{\alpha_{AA} + \alpha_{aux}}\right\} .$$ (25.18)

The 'auxiliary effectiveness' may be defined as the ratio of the displacement at a point before the auxiliary system is added, to the displacement at the same point after it is added. The displacement before the auxiliary system is added is found from the above expressions with $\alpha_{aux} = \infty$. Thus, with no auxiliary system

$$w_A = P_S \cdot \alpha_{AS} , \qquad w_R = P_S \cdot \alpha_{RS} , \qquad w_S = P_S \cdot \alpha_{SS} .$$

Hence the corresponding auxiliary effectivenesses at the points A, R, and S are

$$E_{AUX,A} = 1 + \frac{\alpha_{AA}}{\alpha_{aux}}$$ (25.19)

$$E_{AUX,R} = \frac{1}{1 - \left\{\dfrac{\alpha_{RA}}{\alpha_{AA} + \alpha_{aux}}\right\}\dfrac{\alpha_{AS}}{\alpha_{RS}}}$$ (25.20)

$$E_{AUX,S} = \frac{1}{1 - \left\{\dfrac{\alpha_{AS}}{\alpha_{AA} + \alpha_{aux}}\right\}\dfrac{\alpha_{AS}}{\alpha_{SS}}} .$$ (25.21)

If the auxiliary system is added directly at the source, $\alpha_{AS} = \alpha_{SS}$, $\alpha_{AA} = \alpha_{SS}$, and equation (25.21) reduces to equation (25.19). These equations will now be used to consider the effects of certain simple auxiliary systems.

(a) The point mass added to an infinite beam

The receptance, α_{AA}, of an infinite uniform beam is $-(1 + i)/4EIk_w^3$, where k_w is the natural wavenumber at the frequency of excitation (see equation (9.18) of Chapter 9). The receptance of a point mass M is $-1/\omega^2 M$. The auxiliary effectiveness at the position of the mass is then

$$E_{AUX,A} = 1 - (1 + i)\frac{M}{m}k_w \qquad (25.22)$$

where m is the mass per unit length of the beam. Now $k_w = (\omega^2 m/EI)^{1/4}$, so at high frequencies,

$$|E_{AUX,A}| \rightarrow \frac{\sqrt{2}\,M\omega^{1/2}}{m^{3/4}\,(EI)^{1/4}} \; . \qquad (25.23)$$

Equation (9.13) of Chapter 9 yields also the cross-receptances α_{AR}, α_{RS} etc. If the points A, R, and S are all widely separated, the displacement at one due to a unit force at another is given approximately by

$$\alpha_{AS} = \frac{-i \exp(-ik(x_s - x_a))}{4EI\,k^3} \qquad (25.24)$$

where x_s and x_a are the x-wise coordinates of points A and B. Similar expressions hold for α_{RS}, α_{AR} etc. These receptances can then be used in equation (25.20) to find the auxiliary effectiveness at point R. Thus we obtain

$$|E_{AUX,R}| = \{1 + \mu^2/(4 + \mu)^2\}^{\frac{1}{2}} \qquad (25.25)$$

where $\mu = Mk_w/m \; .$ (25.26)

This cannot have a greater value than $\sqrt{2}$ at very high frequencies or with very large masses. The reasons for this have already been mentioned in Chapter 9.

(b) The linear, discrete spring added to the infinite beam

The receptance of a simple linear spring is $1/K$, where K is the spring rate. The auxiliary effectiveness is found by the same method as above to be

$$|E_{AUX,R}| = \left\{1 + \frac{\epsilon^2}{(4 - \epsilon)^2}\right\}^{\frac{1}{2}} \qquad (25.27)$$

where $\epsilon = (K\omega^2)k_w/m \; .$ (25.28)

Once again, we see that this effectiveness cannot exceed $\sqrt{2}$ as K gets very large.

(c) The undamped dynamic neutralizer added to a general system

This auxiliary system consists of a simple mass M_a attached through a spring K_a to the system at A. The receptance of this mass-spring system, wheh excited at the base of the spring, is

$$\alpha_{aux} = \frac{1}{K_a} - \frac{1}{\omega^2 M_a} .$$ (25.29)

This goes to zero at the natural frequency $\omega_a = (K_a/M_a)^{1/2}$. The auxiliary effectiveness at its point of attachment, A, is evidently infinite at this frequency (see equation (25.19)), but this is not necessarily so for the other points R and S (equation (25.20) and (25.21)). However, if the system is added at the point S (at the source of excitation), then $E_{AUX,S}$ is infinite, so the point of excitation is stationary. Furthermore, we also find $E_{AUX,R}$ is infinite, so all other points are stationary too.

Hence, when the undamped dynamic neutralizer is added to a system at the point where the system is being harmonically forced, motion *throughout the whole main system* will be prevented at the natural frequency of the neutralizer. The neutralizer mass, however, vibrates with an amplitude given by P_s/K_a. From the practical point of view, sufficient clearance must be allowed around the mass to permit this motion. In effect, the mass and spring are exerting upon the main system a force which is equal and opposite to the exciting force $P_s e^{i\omega t}$.

Consider now the motion at A at any other frequency. The displacement amplitude is given by equation (25.15), which shows clearly that $w_A = 0$ when $\alpha_{aux} = 0$ (at the frequency ω_A). However, w_A approaches infinity wherever $\alpha_{AA} + \alpha_{aux} = 0$, i.e. when $\alpha_{AA} = -\alpha_{aux}$. Fig. 25.4 shows typical curves of α_{AA} and $-\alpha_{aux}$, the intersection of which give the new frequencies at which the combined system will resonate. The addition of the neutralizer produces one more resonance of the system than there was before. Notice from Fig. 25.4 that the resonant frequencies below ω_a are *depressed* by the addition of the neutralizer, whereas those above ω_a are *raised*.

No motion will occur anywhere in the system at ω_a if the undamped neutralizer is attached at the point of excitation. If there is more than one source of excitation a single neutralizer will enforce a node at its point of attachment at its own natural frequency, but motion is possible elsewhere. When the motion to be prevented can be identified as a particular natural mode of vibration, which is excited by several exciting forces of a given troublesome frequency, then a single neutralizer should be attached at an antinode of the mode. If several neutralizers can be added, they should be added at the excitation points to annul (neutralize) the excitation at source.

If the neutralizer is added at a point in the main system where a node exists at a particular frequency, then clearly the neutralizer has no effect on the system at that frequency.

Since the neutralizer adds an additional resonant frequency to the whole system, it must be ensured that the responses at these resonances are minimized, if they are encountered. This is done by adding a damper to the neutralizer, as follows.

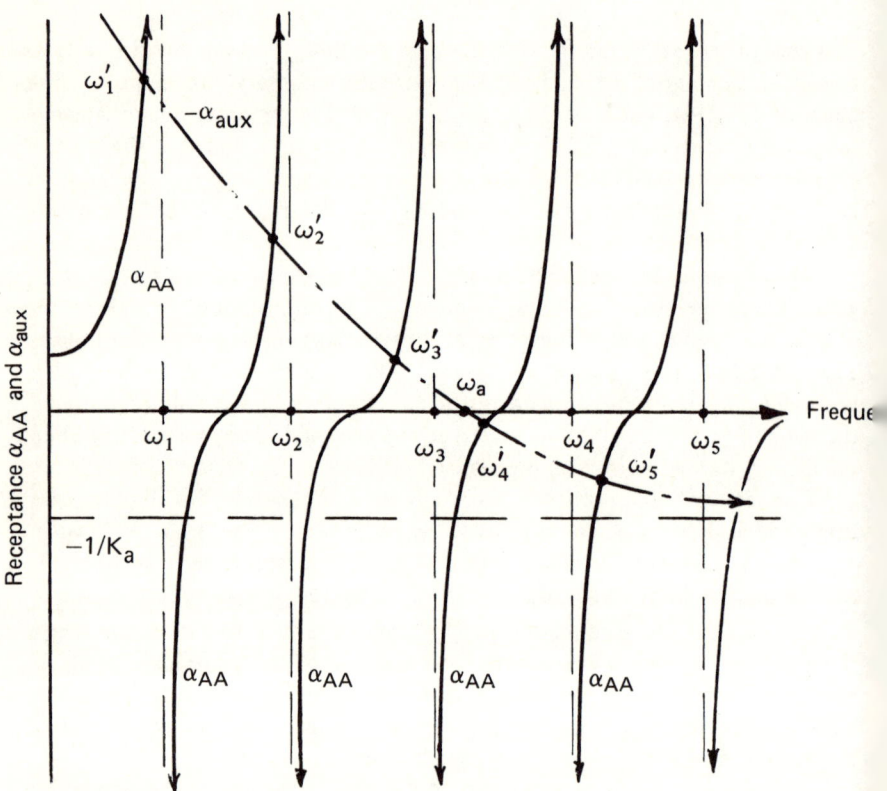

Fig. 25.4 – Typical receptance diagram

(d) The damped, tuned dynamic neutralizer (the 'tuned damper)
Diagrammatically, this is represented by the figure below (Fig. 25.5). The analysis of the system response is identical to that of the previous section, but the receptance α_{aux} must be modified to include the effect of the damping.

We find

$$\alpha_{aux} = \frac{1}{K_a + i\omega C_a} - \frac{1}{\omega^2 M_a}$$

$$= \frac{1}{K_a} \left\{ \frac{1}{1 + 4\Omega^2 \zeta_a^2} - \frac{1}{\Omega^2} - i \frac{2\zeta_a \Omega}{1 + 4\Omega^2 \zeta_a^2} \right\} \qquad (25.30)$$

where $\quad \Omega = \omega/\omega_a, \quad \zeta_a = C_a/2\omega_a M_a$. $\qquad (25.31)$

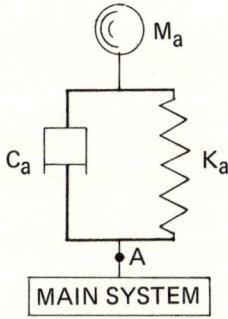

Fig. 25.5 – Diagram of a damped, tuned dynamic neutralizer.

The theory of the tuned damper is well known and is treated in most standard vibration textbooks. For a good summary of the results see Chapter 6 of Ref. [25.8]. There is an optimum value of the damper rate C_a to minimize the resonant responses at one or other of the two resonances on either side of ω_a. The theory is a simple one, assuming that the main system has only one degree of freedom, and can be represented by a simple mass M_m and spring K_m. M_a and K_a are first chosen so that ω_a is equal to the troublesome exciting frequency, or in the middle of the range of exciting frequencies. A convenient optimum value of ζ_a is

$$\zeta_{opt} = \sqrt{\frac{3\mu'}{8(1 + \mu')}} \qquad (25.32)$$

where $\mu' = M_a/M_m$. If the main system is vibrating in a mode whose *generalized* mass is M_m the auxiliary mass M_a would increase this by m_a, say. In this case $\mu' = m_a/M_m$.

Jones & Trapp [25.9] have discussed the use of small tuned dampers on aircraft skin panels subject to random acoustic pressures, on a radar antenna, and on beam-like structures. In these applications, the damping is provided by a viscoelastic damping material which also contributes to the stiffness K_a of the whole auxiliary system. The auxiliary system may consist of a simple metal beam with a damping material attached to it, or of a small rigid mass supported by the viscoelastic material. Optimization of the device follows the general principles set out above.

(e) The untuned damped neutralizer (the 'auxiliary mass damper')
Removing the spring from the mass-spring-damper neutralizer leaves us with the *untuned damped neutralizer*. Its receptance is given by

$$\alpha_{aux} = -\frac{i}{\omega C_a} - \frac{1}{\omega^2 M_a} . \tag{25.33}$$

When this is coupled to a single-degree-of-freedom main system, the optimum value of the damping depends on the auxiliary mass (as above) and is given by

$$C_{opt} = 2m_a M_m \, \omega_m \sqrt{\frac{1}{2(2+\mu')(1+\mu')}} \tag{25.34}$$

where ω_m is the natural frequency of the main system before the auxiliary system is attached. The new coupled system now yields the maximum displacement response to harmonic forcing at the frequency

$$\omega_{ud} = \omega_m \sqrt{\frac{2}{2+\mu'}} . \tag{25.35}$$

The amplitude of displacement is then $\dfrac{P_s}{K_a} \cdot \dfrac{2+\mu'}{\mu'}$

where P_s is the *generalized* force amplitude

and K_a is the *generalized* stiffness.

The auxiliary damper can be optimized in conjunction with one mode only of the general multi-freedom system. When any other mode is excited, the auxiliary system operates at non-optimum conditions. At frequencies well above ω_{ud}, the auxiliary system acts almost like a pure damper, C_a. At frequencies well below ω_{ud} it acts more like a pure mass with a small amount of damping.

25.6 VIBRATION CONTROL BY ARTIFICIAL DAMPING
Every real vibrating structure is subjected to damping actions whereby vibrational energy is lost into heat within the structure or into acoustic/hydrodynamic

energy in the surrounding medium. Artificial damping covers those means of increasing the structural damping by the addition of extra damping materials and mechanisms. It is a useful means of vibration control only if:

(a) the vibration problem is one which is sensitive to the magnitude of the damping (see section 2 of this chapter);
(b) the damping can be increased sufficiently above the initial damping to produce sufficient attenuation.

Before we consider the methods and amounts by which the damping can be increased, we shall consider the mechanisms and magnitudes of the initial damping. These mechanisms will be categorized into (i) structural damping and (ii) acoustic damping.

25.6.1 Initial structural damping

This originates in:

(a) hysteresis of the structural material,
(b) friction at structural joints,
(c) friction with attached non-structural items (e.g. cables, pipes, stowed equipment, etc.),
(d) viscous damping at lubricated sliding surfaces (as in engines and machinery).

It is convenient once again to quantify the damping in terms of the dimensionless loss factor, η_m, corresponding to the given mode of vibration. Then

$$\eta_m = \frac{1}{2\pi} \frac{\Sigma \text{ Energy dissipated per harmonic cycle by each source}}{\text{Maximum energy stored in the structure in the course of one cycle}}.$$

$$(25.36)$$

The contribution, η_s, to this from each source is obvious from this expression, so

$$\eta_m = \sum_{\text{all sources}} \eta_s .$$
$$(25.37)$$

It must be recognized that each contribution, and hence the total, will vary from mode to mode of the structure. Modal loss factors can vary over quite wide ranges. Thin aluminium stiffened-plate structures can have modal loss factors from 0.004 (model structure, tested by Higginson [25.10]) to 0.04 (aeroplane elevator panel, tested by Clarkson & Ford [25.11]). Machinery sliding surfaces can have modal loss factors as high as 0.12.

Material hysteresis does not contribute appreciably to the loss factor unless the structure is of cast iron, copper-manganese, or a ferro-magnetic material. Aluminium has a material loss factor of the order of 0.00005 at small strains, so little energy is lost through hysteresis in the above examples.

Friction at structural joints is often the major contributor to structural damping. A study of the energy dissipated in structural joints is reported in reference [25.12], where it is also shown how the modal loss factors can be calculated if the joint damping characteristics are known. Ungar [25.13] presents an alternative method, and gives a wide-ranging review of the damping of built-up structures. In order to predict the loss factor from joint friction, much detailed information is required concerning the joint dissipation characteristics and the joint loads. It is not yet a practicable proposition to undertake such a prediction, so preliminary estimates of joint friction damping can only be made on the basis of data from similar structures. For aeroplane structures, the ESDU data sheets [25.14] provide such information.

The damping due to attached non-structural items has not yet been quantified, but will obviously vary greatly depending on the rigidity of the attachment, and on the mass, stiffness and damping of the item itself. A welded-steel ship hull has very low structural damping (the joint damping is low), but the attached machinery, pipes, floors, and furnishings add enormously to the loss factors. Likewise, a steel car shell has low damping, but the added seats, carpets, engine, suspension, etc., increase this greatly.

Viscous damping at lubricated sliding surfaces is the major contributor to the damping of many modes of machinery damping. The typical loss factor of 0.12 already quoted for such a machine is many times greater than the modal damping when the sliding component (e.g. crankshaft and pistons) is removed.

25.6.2 Acoustic damping

This originates from the resistive part of the sound pressure generated at the vibrating surface by the surrounding acoustic medium. Chapter 11 gives the basic theory. The corresponding acoustic loss factor is $(1/2\pi) \times$ sound energy radiated per cycle $\div$ maximum energy stored in the structure in the course of the cycle. We shall be concerned only with surfaces vibrating in flexural waves; and for a surface of uniform mass per unit area, the 'radiation loss factor' due to acoustic radiation is found to be

$$\eta_{rad} = \sigma \frac{\rho c}{\omega_n \rho_m h} \tag{25.38}$$

where σ is the radiation efficiency, ρ and ρ_m are the densities of the acoustic medium and the plate respectively, c is the speed of sound in the medium, h is the plate thickness, and ω_n is the natural frequency of the mode. The above expression assumes that sound is being radiated from only one side of the plate. Chapter 11 quotes some radiation efficiencies. Crocker & Price [25.15] have developed some of Maidanik's work [25.16] to yield the following radiation efficiencies for flat plates vibrating in flexural modes:

$$x = (c_p/c)^2 \; ; \quad c_p = \text{flexural wave speed in the plate}$$

$x < 1$	$x = 1$	$x > 1$
	Greatest of	
$\sigma = \dfrac{64c^2}{A\pi^2\omega^2}g_1(x) + \dfrac{Pc}{\pi A\omega}g_2(x)$	$\sigma = \dfrac{Pf_c}{c}$ or $\sqrt{10}$	$\sigma = \dfrac{x}{(x-1)^{\frac{1}{2}}}$

$$g_1(x) = \frac{x^{3/2}\,(1-2x)}{(1-x)^{\frac{1}{2}}} \text{ for } x < \tfrac{1}{2}, \quad \text{ or } = 0 \text{ for } x > \tfrac{1}{2}$$

$$g_2(x) = x\frac{[(1-x)\ln((1+\sqrt{x})/(1-\sqrt{x})) + 2\sqrt{x}]}{(1-x)^{3/2}},$$

P = plate perimeter; A = plate area; f_c = critical (coincidence) frequency of plate.

The radiation loss factors of rectangular plates vibrating in their fundamental modes may be calculated more simply [25.17]. An aluminium plate, vibrating at a low-frequency (acoustic wavelength $\gg$ plate wavelength) has a radiation loss factor which varies approximately linearly with aspect ratio from 0.005 for a square plate to 0.013 for a plate with aspect ratio = 7. (Divide this by 2.9 for a steel plate.)

The radiation loss factors of plates stiffened by attached beam-like members can be considerably greater than those of unstiffened plates. (See Chapters 11 and 13 and reference [25.17]).

25.6.3 Artificial damping through added damping layers

A beam or a plate which vibrates in flexure can be damped by the appropriate addition of a layer of damping material. The methods of adding the layers are discussed in the next two sections. As the whole system vibrates, the layer undergoes cyclic strain and thereby dissipates energy.

The damping material is usually a synthetic rubber (PVA/PVC, filled or unfilled) or (for cheapness and lower efficiency) a bitumen-based compound. The material has a high capacity for dissipating energy, and this is conveniently quantified again by the material loss factor. The complex cyclic stress and strain are related through the complex moduli, thus

Direct stress $\sigma = E'\,(1 + i\eta_d)\epsilon$

Shear stress $\tau = G'\,(1 + i\beta_d)\gamma$

where ϵ and γ are the direct and shear strains respectively. η_d and β_d (the loss factors in direct and shear strain respectively) are usually taken to be equal.

E' and G' are the 'storage moduli', and the maximum energy stored per unit volume in a strain cycle of amplitude ϵ or γ is $U = E'\epsilon^2/2$ or $G'\gamma^2/2$. The corresponding energy dissipated per cycle is

$$\Delta W = \pi E'\eta_d\epsilon^2 \quad \text{or} \quad \pi G'\beta_d\gamma^2 \; . \tag{25.39}$$

For a given strain amplitude, the energy dissipated per cycle is proportional to the *product $E'\eta$ or $G'\beta$.*

As the frequency of the cycle or temperature of operation vary, E' and η_d (and G' and β_d) vary. For a given temperature, the product has a maximum value at a frequency near the middle of the 'transition zone'. Fig. 25.6 shows diagrammatically the variation of E' and η_d separately. Manufacturers of damping materials have optimized the material content to yield maximum values of $E'\eta$ and $G'\beta$ over a broad frequency and temperature range, centred on specified values. Unfortunately, the breadth of the peaks can only be obtained at the expense of the height, so materials with the highest damping capacity can only give this over narrow temperature and frequency bands. Certain silicone rubbers have been found to have better broadband characteristics [25.18], but the quest to improve the situation continues.

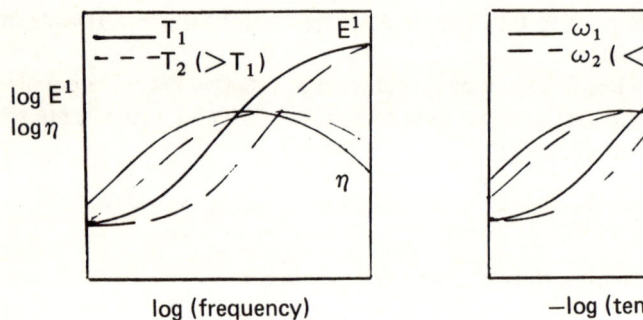

log (frequency)   −log (temperature)

Fig. 25.6 − Dependence of E' and η on frequency and temperature.

At very high temperatures, these synthetic rubbers become useless. However, porcelain enamels have very high damping capacity at very high temperatures (e.g. a loss factor of nearly 2 has been quoted for a temperature of 635°C) and the possibility of using this in turbines has been considered. Jones [25.19] considers the general problem of artificial damping at high temperatures.

When a damping material is added to a vibrating structure, and the amplitude of direct (or shear) strain in the material varies throughout the volume $(\bar{\epsilon} = \bar{\epsilon}_d (x, y, z))$, the total energy dissipated by the damping material is

$$\text{E.D.} = \pi E'\eta_d \int_{vm} \bar{\epsilon}_d^2 (x, y, z) \, \mathrm{d}V \tag{25.40}$$

where the integration extends throughout the whole volume of the damping material. (This assumes that energy is dissipated by virtue of the direct strains in the material. If the shear strain dissipation is important, a similar term involving G', β, γ_d should replace or add to the above.) The loss factor of the mode of vibration is

$$\eta_m = \frac{E'\eta_d \int\limits_{Vm} \overline{\epsilon_d^2}(x, y, z)\mathrm{d}V}{E_s \int\limits_{Vs} \overline{\epsilon_s^2}(x, y, z)\mathrm{d}V + E' \int\limits_{Vm} \overline{\epsilon_d^2}(x, y, z)\mathrm{d}V} \qquad (25.41)$$

where the integral over Vs extends throughout the volume of the whole structure, excluding the damping material.

25.6.4 The damping of flexural vibrations by unconstrained layers

An unconstrained layer is a layer of damping material added to the surface of a beam or plate with its outer surface perfectly free and unconstrained. Such layers can be glued on in the form of plastic tiles, or sprayed on as a wet 'plaster' which subsequently dries and hardens.

As the beam or plate bends, the layer is subjected to direct bending strains which are proportional to the local plate curvature (see Fig. 25.7). The layer dissipates energy principally by virtue of the bending strain cycles it undergoes. For maximum energy dissipation, then, the material must have a high value of $E'\eta_d$.

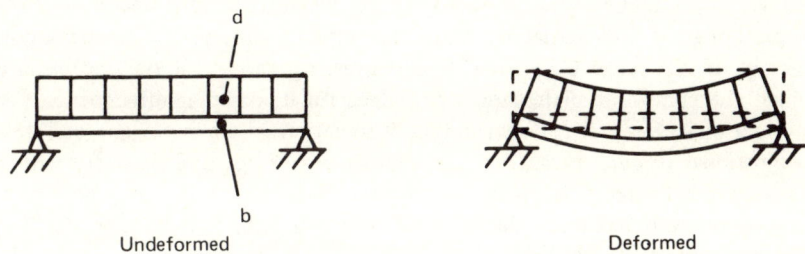

Undeformed   Deformed

Fig. 25.7 – Deformation of the unconstrained layer.

Consider a layer of uniform thickness attached to a beam or plate of uniform cross-section. The loss factor of flexural vibration is given by

$$\eta_m = \eta_d \frac{E'I_d}{E_sI_s + E'I_d} \qquad (25.42)$$

where I_d, I_s are the second moments of area of the cross-sections of damping material and baseplate respectively about the composite neutral surface of bending. This expression may be used for damping layers on just one or both sides of the baseplate.

The stiffness of the plate is also increased by the addition of the damping material. The ratio of the stiffness after and before the addition of the material is

$$s = \frac{E_s I_s + E' I_d}{E I_o} \tag{25.43}$$

where I_o is the second moment of area of the original cross-section about the neutral plane. It should be noticed that η_m and s do *not* depend upon the mode of vibration when the damping material is uniformly applied over the whole surface.

Equations (25.42) and (25.43) apply to uniform beams of any cross-section, as well as to plates. For plates with damping layer thickness h_2 and baseplate thickness h_1, these equations become

$$\eta_m = \eta_d \frac{e\,[12\tau_{21}^2 + \tau_2^2(1+e)^2]}{[1+e]\,[12e\tau_{21}^2 + (1+e)(1+e\tau_2^2)]} \tag{25.44}$$

where $e = E'/E_s$, $\tau_2 = h_2/h_1$, $\tau_{21} = (h_1+h_2)/2h_1$. For thin damping layers $(h_2 \ll h_1)$ this reduces to

$$\eta_m \approx \eta_d\,3\frac{E'}{E_s}\cdot\frac{h_2}{h_1}. \tag{25.45}$$

Reference [25.20] shows how η_m/η_d varies with τ and e. The loss factor η_m reaches a maximum value of about 0.8 η_d when $h_2 \cong 2h_1$, and $E' = E_s/10$ (typical of good commercial materials on aluminium alloy plates). Such materials have η_d in the range 0.35 to 0.70. However, η_m is not the only criterion by which the composite plate should be judged for its damping effectiveness. The product of its flexural stiffness and η_m is sometimes more important, and this is proportional to $s\eta_m$. Figs. 25.8 and 25.9 show how η_m and $s\eta_m$ vary with τ_2, for a particular damping material and plate. It is evident that $s\eta_m$ goes on increasing monotonically as τ_2 increases, so with increasing τ_2 there is an increasing damping benefit [25.2].

The effectiveness of unconstrained layers can be greatly increased by separating the damping layer from the baseplate by a shear-stiff spacer layer. This places the damping layer further away from the composite neutral axis and effectively magnifies the direct strain in the layer for a given plate flexural displacement. It also greatly increases the plate flexural stiffness. This is considered in reference [25.20]. The spacer layer must be stiff, but even then at short wavelengths there is a degradation in damping effectiveness.

When only one mode is to be damped greater efficiency of unconstrained layers can also be obtained by removing the damping material from regions of low bending curvature and adding it in regions of high curvature. This ensures that the damping material is being subjected to the greatest direct strains, and so is dissipating the most energy. Reference [25.21] investigates this in detail. The technique can be used only when the modes of vibration are known, and must not be pushed too far, or degradation of damping performance will result (i.e. do not try to concentrate the damping material over too small an area).

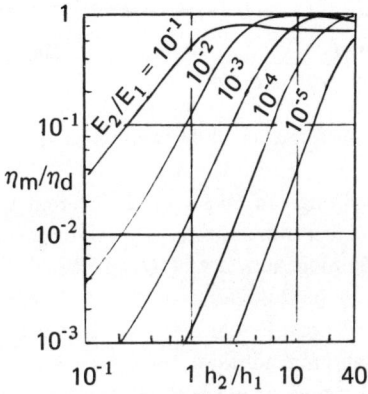

Fig. 25.8 – Variation of η_m with h_2/h_1 and E_2/E_1. (Unconstrained damping layer).

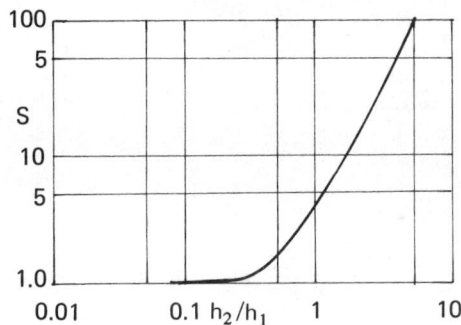

Fig. 25.9 – Variation of s with h_2/h_1 for a given E_2/E_1. (Unconstrained damping layer).

25.6.5 The damping of flexural vibrations by constrained layers

The damping layer, adhering to an outer surface of the beam or plate, now has another stiff layer attached to its outer surface to sandwich the damping layer. When the whole assembly bends, the damping layer is subjected to shear strains, and this cyclic shearing dissipates energy in the layer (see Fig. 25.10).

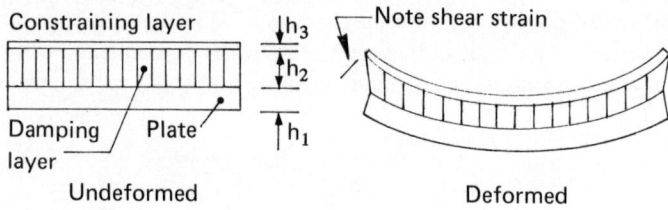

Fig. 25.10 – Deformation of the constrained layer.

Constrained layer damping can take several different forms. The simplest is made from 'damping tape' – a thin aluminium foil stuck to the base plate with a pressure-sensitive high-damping adhesive. (Manufactured by the 3M Co). The foil and damping layer may be 0.002–0.01 in thick. Alternatively, a deep vibrating beam (of I section, say) can be damped by another beam bonded on to its top surface by a high-damping adhesive, or by a bitumen-based material. In another form, two metal plates of equal thickness are bonded together with a high-damping rubber, and are used from the initial design stage for the prevention of vibration. In the first two examples, the damping material and constraining layers can be added after the vibration problem is encountered.

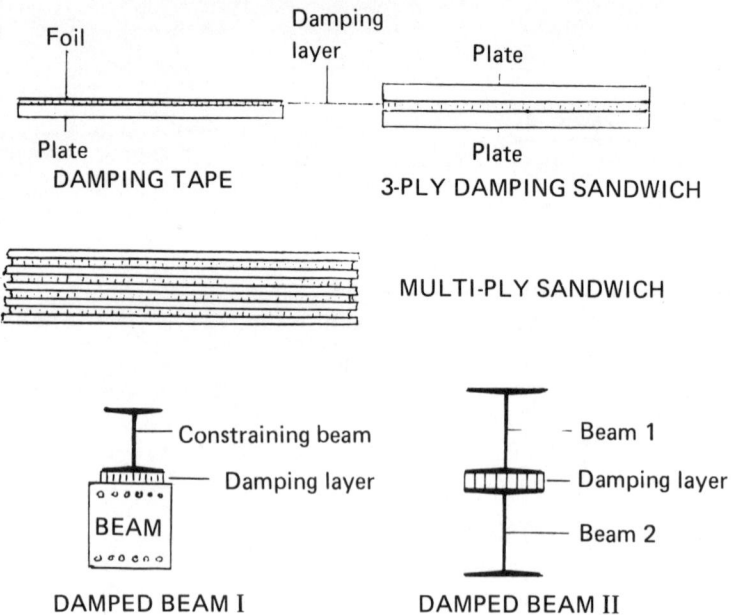

Fig. 25.11 – Some constrained-layer damping arrangements.

The simple constrained layer system (involving three layers in all) can be extended to a multiple layer system, with N damping layers and $N+1$ constraining elastic layers. In all cases, shearing of the damping layer dissipates energy and damps the flexural vibration. Fig. 25.11 shows the sections of these damping arrangements.

In contrast to the unconstrained layer damping, constrained layer damping does depend upon the mode of vibration. At very long wavelengths, there is very little shear strain in the damping layer and very little damping. For other reasons, there is very little damping at very short wavelengths. The damping and flexural stiffness both depend upon the flexural wave number K_B, on the damping layer complex shear modulus and thickness, and on the extensional and flexural stiffness of the constraining layer and baseplate (or beam). There are two important governing non-dimensional parameters.

$$\text{The shear parameter, } g = \frac{G'b}{h_2 K_B^2}\left\{\frac{1}{E_1 A_1} + \frac{1}{E_3 A_3}\right\} \tag{25.46}$$

where h_2 and b are the thickness and width respectively of the damping layer; E_1 and E_3 are the Young's Moduli of the baseplate/beam and constraining layer respectively, and A_1 and A_3 are their cross-sectional areas (N.B. for a uniform plate with a uniform constraining layer, put $b=1, A_1 = h_1, A_3 = h_3 =$ thickness of constraining layer).

$$\text{The Geometric Parameter, } Y = \frac{(E_1 A_1)(E_3 A_3)d^2}{(E_1 A_1 + E_3 A_3)(E_1 I_1' + E_3 I_3')} \tag{25.47}$$

where d is the distance between the centroids of plates/beams 1 and 3; I_1' and I_3' are the second moments of area of plates/beams 1 and 3 about their own neutral axes. Reference [25.22] describes the physical significance of Y.

The effective complex flexural stiffness of the 3-layered configuration can be shown to depend on g and Y as follows:

$$(EI)_{\text{total}}(1 + i\eta_f) = (E_1 I_1' + E_3 I_3')\left\{\frac{1 + g(1 + i\beta)(1 + Y)}{1 + g(1 + i\beta)}\right\}. \tag{25.48}$$

Rationalizing this, one finds the flexural loss factor η_f to be

$$\eta_f = \beta g Y/[1 + g(2 + Y) + g^2(1 + Y)(1 + \beta^2)] \tag{25.49}$$

and the stiffness ratio, s, to be

$$s = \frac{(EI)_{\text{total}}}{(E_1 I_1' + E_3 I_3')} = 1 + \frac{g Y(1 + g(1 + \beta^2))}{1 + 2g + g^2(1 + \beta^2)}. \tag{25.50}$$

The stiffness $\times$ loss factor product, $s\eta_f$, is

$$s\eta_f = \beta g Y/[1 + 2g + g^2(1 + \beta^2)]. \tag{25.50}$$

That η_f, s, and $s\eta_f$ all depend on g shows that they all depend upon the wavenumber K_B, as well as on the shear modulus, G', of the damping layer. Since K_B varies with frequency, so also do η_f and s. The wavenumber, K_B is related to the real part of the flexural stiffness through

$$K_B^4 = \frac{\omega^2 \mu}{(EI)_{total}} \tag{25.52}$$

where μ is the mass per unit length of the beam. This equation leads to a cubic equation for K_B in terms ω^2, Y, g [25.24].

Two special cases lead to some simplification. For damping tape, in which the constraining layer is much thinner than the baseplate, $h_3 \ll h_1$, and $Y \approx 3(E_3/E_1)(h_3/h_1)$. Then

$$\left.\begin{array}{l} \eta_f \approx \beta g Y/[1 + 2g + g^2(1+\beta^2)] \\[2mm] \text{and } s \approx 1 \ . \end{array}\right\} \quad \begin{array}{l} \text{very thin} \\ \text{constraining} \\ \text{layer} \end{array} \tag{25.53}$$

For a sandwich plate with two identical outer layers, $Y = 3(1+h_2/h_1)^2$. Hence $Y = 12$ if $h_2 = h_1$ (all layers of equal thickness); $Y = 27$ if $h_2 = 2h_1$; if $h_2 \ll h_1$, $Y \approx 3$. and

$$\left.\begin{array}{l} \eta_f \approx 3\beta g/[1 + 5g + 4g^2(1+\beta^2)] \\[2mm] s \approx 1 \ \text{(low } g) \\[2mm] \approx 4 \ \text{(high } g) \end{array}\right\} \quad \begin{array}{l} \text{very thin damping} \\ \text{layers; equal face} \\ \text{plates.} \end{array} \tag{25.54}$$

Fig. 25.12a and 25.12b show how η_f and s vary with g and Y for a damping layer which has $\beta = 0.3$. η_f passes through a maximum value as g varies, but generally increases with increasing Y. The optimum value for maximum η_f is given by

$$g_{opt, \eta} = (1 + Y)^{-\frac{1}{2}} (1 + \beta^2)^{-\frac{1}{2}} \ . \tag{25.55}$$

This optimum value is obtained in practice by the appropriate choice of material or of thickness for the damping layer. The corresponding maximum loss factor is

$$\eta_{f, max} = \beta Y/[(2 + Y) + 2(1 + Y)^{\frac{1}{2}} (1 + \beta^2)^{\frac{1}{2}}] \ . \tag{25.56}$$

For the damping tape, with $E_3 = E_1$ and $Y \cong 3h_3/h_1$, this gives

$$\eta_{f, max} = \beta \frac{3}{2} \frac{h_3}{h_1}/[1 + (1 + \beta^2)^{\frac{1}{2}}] \ . \tag{25.57}$$

Now $\beta = 1.0$ is a typical value for damping tape layers; if $h_3 = 0.005$ in. and $h_1 = 0.036$ in., then $\eta_{f, max} = 0.083$ (i.e. 4.2% of critical damping). For the sandwich plate with identical outer layers and a very thin damping layer with $\beta = 1.0$, $\eta_{f, max} = 0.28$. If the damping layer thickness is equal to a face-plate thickness, $\eta_{f, max} = 0.5$.

(a)

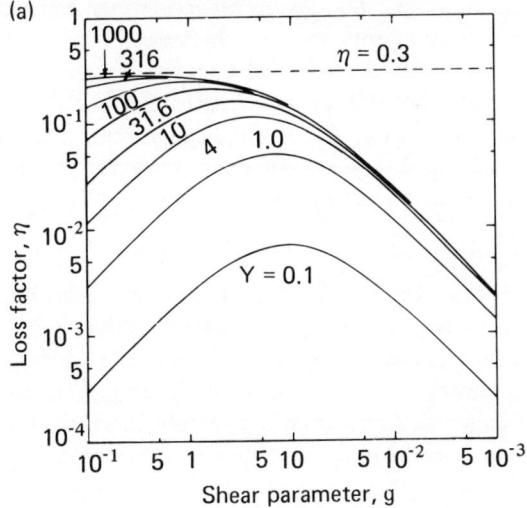

(b)

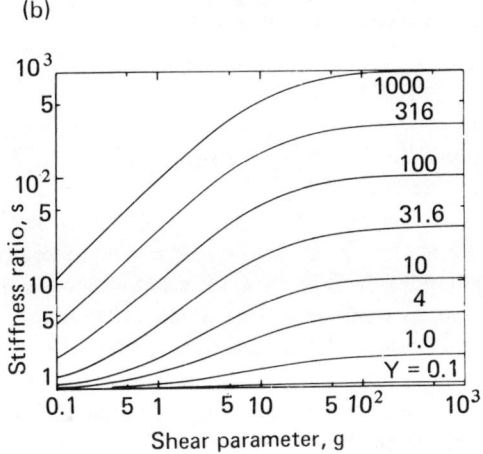

Fig. 25.12 – η and s for simply-supported sandwich beams; $\beta = 0.3$.

If the damping of the plate (beam) is maximized for one particular mode (i.e. for a given K_B and the corresponding g_{opt}) it will not be as great for another mode with a different K_B. Thus, a given beam on simple supports may have optimum damping in its fundamental mode. The same beam, fully-fixed at each end, has a different K_B and non-optimum damping – less than the damping for the simple-supported beam. However, the peak of the η_f vs g curve (i.e. η_f vs $1/K_B^2$ curve) is quite broad, so modes with K_B values close to the optimum value will not have η_f values much below the maximum.

Fig. 25.12b shows that the flexural stiffness of the constrained layer beam/plate increases to an asymptotic value as g increases, i.e. as K_B decreases, and the flexural wave*length* increases. Hence at large wavelengths and low frequencies the stiffness is high, but at small wavelengths and high frequencies it is low. The product stiffness × loss factor ($s\eta_f$, called the 'energy dissipation parameter' in reference [25.23]) has a maximum value at $g = (1 + \beta^2)^{-\frac{1}{2}}$ (independent of Y!). The maximum value is given by

$$(s\eta_f)_{max} = \beta Y/2[1 + (1 + \beta^2)^{\frac{1}{2}}] \ . \tag{25.58}$$

If $\beta = 1$, this is approximately equal to $Y/5$ and increases in proportion to Y. This contrasts with $\eta_{f, max}$, which approaches the value of β as Y gets large. For those vibration problems in which $s\eta$ is the parameter which governs the vibration level, large values of Y (thick damping layers) are required. If η alone is the important parameter, increasing the damping layer thickness does not give an appreciable increase in $\eta_{f, max}$. A three-fold increase in $\eta_{f, max}$ is all that is obtained by increasing h_2/h_1 from 0.1 to 10 (Y from 3.63 to 363!).

The above expressions for η_f, s, etc can be applied directly to constrained layer beams. They can also be used for plates provided the correct value of K_B is used in g. If the plate vibrates with x and y-wise wave numbers K_x and K_y, then

$$K_B^2 = K_x^2 + K_y^2 \ .$$

The effect of fixed and elastically restrained boundaries
This is considered in references [25.23] and [25.25]. Broadly, the maximum loss factors which can be obtained from fully-fixed or elastically supported beams and plates are almost the same as for simply-supported beams/plates, but in the fundamental modes these occur at different values of g. For example, the fully-fixed beam requires a damping layer which is five times as stiff as that for the simply-supported beam, if maximum η_f is to be achieved. If the beam or plate is free at the boundaries, the maximum loss factor for the fundamental mode is obtained with a damping layer which has one-fifth of the stiffness of that for the simply-supported beam.

Stiffeners added along the plate boundaries, or elsewhere across the plate, can have different effects depending on the method of attachment (does the attachment restrain the shearing of the damping layer or not?), or the magnitude of the rotational restraint, etc., etc. Reference [25.23] considers a plate with restraints at the boundaries, and reference [25.25] considers a plate with stiffeners at periodic intervals across it. Some general conclusions as to the effect of the stiffening are drawn.

Multiple-layer sandwich plates
The results so far apply to plates and beams with two elastic layers and one damping layer. Suppose now there are N_E elastic layers and $N_E - 1$ damping

layers. Let the elastic layers all be identical with thickness h_1 and Young's modulus E_1. Likewise, the damping layers are identical with thickness h_2 and complex shear modulus G' $(1 + i\beta)$. Provided that shear deformation of the elastic layers and all longitudinal/rotary inertia is negligible, the following expression applies for s and η_f:

$$s\,(1 + i\eta_f) = 1 + \frac{24(1 + \tau_2)}{(1 + S)\sinh N\alpha - \sinh (N-1)\alpha} \; \frac{1}{N_E} \sum_{r'=1}^{N'} r'\sinh r'\alpha$$

(25.59)

where $\tau_2 = h_2/h_1$, $S = Eh_1 h_2 K_B^2\, G'\,(1 + i\beta)$

$\alpha = \cosh^{-1}(1 + S/2)$

$N = (N_E - 1)/2$, $s = (EI)_{\text{total}}/N_E . EI_1$

$N' = N$, $r' = r$ for N_E odd

$N' = N_E/2$, $r' = r - \frac{1}{2}$ for N_E even .

Calculations for a particular combination and total weight of materials have suggested that the greatest damping and reduction of response occur when many thin elastic and damping layers are used, rather than a few thicker layers. The effect of longitudinal/rotational inertia and shear deformation in the elastic layers is always to degrade the damping performance.

25.6.6 Practical applications of constrained layer damping
The proceedings of a conference on viscoelastic technology [25.26] describe some recent applications of layered damping. These include multiple-layers of tapes applied to the inlet guide-vanes of a jet-engine compressor to reduce torsional vibrations and fatigue failure (Rogers & Parin); unconstrained-layer space damping and constrained layers on printed circuit boards (Medaglia & Stahle); damping tapes applied to the B-1 rear fuselage structure (Tipton); viscoelastic shear damping of skyscrapers, constrained-layer damping on circular-saw blades, 'stave' damping of pipe vibrations (Caldwell). These proceedings also list many useful references.

Large concrete structures have also been successfully damped by constrained damping layers. See Grootenhuis [25.27].

REFERENCES

[25.1]   Cremer, L. Heckl, M. & Ungar, E. E. (1973) *Structure-borne sound.* New York: Springer Verlag, first edition.

[25.2]   Mead, D. J. (1961) *Noise control,* 7, 27–38. Criteria for comparing the effectiveness of damping materials.

682   Vibration Control (I)   [Ch.

[25.3]   Adams, R. D. & Mead, D. J. (1972) *Machinery* **120**, Selection of materials for vibrating structures.

[25.4]   Clarkson, B. L. (1962) *J. Royal Aeronautical Soc.*, **66**, 603–613. The design of structures to resist jet noise fatigue.

[25.5]   Lazan, B. J. (1968) *Damping materials and members in structural mechanics.* Oxford and London: Pergamon, first edition.

[25.6]   Crede, C. (1965) *Shock and vibration concepts in engineering design.* London: Prentice-Hall International.

[25.7]   Bishop, R. E. D. and Johnson, K. L. (1980) *The mechanics of vibration.* Cambridge: Cambridge University Press, second edition.

[25.8]   Harris, C. M. & Crede, C. E. (1976) *Shock and vibration handbook.* New York: McGraw-Hill Book Co., Inc., second edition.

[25.9]   Jones, D. I. G. & Trapp, W. J. (1971) *J. Sound & Vib.*, **17**, 157–186. Influence of additive damping on resonance fatigue of structures.

[25.10] Higginson, R. F. (1964). MSc thesis, University of Southampton. The structural damping of vibration.

[25.11] Clarkson, B. L. and Ford, R. D. (1962) *J. Royal Aeronautical Soc.*, **66**, 31–40. The response of a typical aircraft structure to jet noise.

[25.12] Mead, D. J. (1979) *AGARD Conference Proceedings No. 277.* Damping effects in aerospace structures. Paper 2: Prediction of the structural damping of a vibrating stiffened plate.

[25.13] Ungar, E. E. (1973) *J. Sound & Vib.*, **26**, 141–154. The status of engineering knowledge concerning the damping of built-up structures.

[25.14] ESDU (ENGINEERING SOCIETIES DATA UNIT) (1973). Item No. 73011. *Damping in acoustically excited structures.*

[25.15] Crocker, M. J. & Price, A. J. (1969). *J. Sound & Vib.*, **9**, 469–486. Sound transmission using statistical energy analysis.

[25.16] Maidanik, G. (1962) *J. Acoustical Soc. America*, **34**, 809–826. Response of ribbed panels to reverberant sound fields.

[25.17] ʼMead, D. J. (1965). Chapter 26 in *Acoustical fatigue in aerospace structures.* (Ed. W. J. Trapp and D. M. Forney). New York: Syracuse University Press.

[25.18] Coote, C. T. (1972) *J. Sound & Vib.*, **21**, 133–147. Measurement of the damping properties of silicone-based elastomers over wide temperature ranges.

[25.19] Jones, D. I. G.. (1979) *Shock and vibration digest*, **11**, 13–18. High temperature damping of dynamic systems.

[25.20] Ross, D. Ungar, E. E. & Kerwin, E. M. (1959). Section 3 in *Structural damping.* (Ed. J. E. Ruzicka). New York: Pergamon Press. Damping of plate flexural vibrations by means of visco-elastic laminae.

[25.21] Mead, D. J. & Pearce, T. G. (1964). USAF Report No. ML-TDR-64-51. *The optimum use of unconstrained layer treatments.*

[25.22] Mead, D. J. & Markus, S. (1970) *J. Sound & Vib.*, **12**, 99–112. Loss factors and resonant frequencies of encastre damped sandwich beams.

[25.23] Mead, D. J. (1972) *J. Sound & Vib.*, **24**, 275-295. The damping properties of elastically supported sandwich plates.

[25.24] Mead D. J. & Markus, S (1969) *J. Sound & Vib.*, **10**, 163-175. The forced vibrations of a three-layer, damped sandwich beam with arbitrary boundary conditions.

[25.25] Mead, D. J. (1976) *Trans. American Soc. Mech. Engineers,* **98**, Series B, 75-80. Loss factors and resonant frequencies of periodic damped sandwich plates.

[25.26] Rogers, L. C. (1978). USAF AFFDL-TM-78-78-FBA. *Conference on aerospace polymeric visco-elastic damping technology for the 1980s.*

[25.27] Grootenhuis, P. (1972). *Proc. Symposium on applications of experimental and theoretical structural dynamics,* pp. 17.1-17.26. Damping mechanisms and structures and some applications of the latest techniques.

[25.28] Wright, G. C. (1972) *J. Sound & Vib.*, **21**, 205-212. The dynamic properties of glass and carbon fibre reinforced plastic beams.

CHAPTER 26

Vibration control (II)

R. G. White

Institute of Sound and Vibration Research, University of Southampton

26.1 INTRODUCTION

Machinery-induced vibration is often a cause of concern or annoyance in structures such as buildings, ships, aircraft, etc., and vibration control is therefore a topic of continuing study. In any such problem, the first approach is usually to modify the source and to isolate it from the supporting structure via resilient vibration isolators. The vibration transmission characteristics of the structure between the source and the area of unacceptable vibration levels must then also be examined and appropriate vibration control procedures carried out. It can thus be seen that, from the vibration control point of view, there is continuing interest in vibration isolation and structural vibration transmission.

It would perhaps appear from perusal of vibration textbooks and published literature over the past thirty years that vibration isolation techniques are so well established that continuing study or further research on this topic is unnecessary. This is not the case, as more detailed study will show, because a great proportion of the reported work concerns the use of mass-spring models of the machine and isolator, or equipment and isolator in the complementary problem, with a rigid substructure. Such simplifications which do not allow for deformation of the systems which are coupled by the isolator only have very restricted validity and limited range of application to practical cases. This is particularly apparent in the case of high frequency vibration isolation because in this case, as described in more detail later, resonances of the machine or equipment and structure occur which make estimates of transmissibilities based on simple mass-spring models essentially meaningless. This need to consider the dynamic characteristics of both the source and receiver of vibration has led to the development and application of dynamic structural analysis techniques based on frequency response methods for predicting coupled system performances in a manner analogous to that used by electrical engineers for circuit analysis. This type of approach to structural analysis, for example, has been well reviewed together with description of suitable measurement techniques [26.1, 26.2, 26.3]. The coupled source, isolator, receiver problem may be studied in detail, and such quantities as force or motion transmissibilities evaluated with consideration of the dynamic characteristics of source and receiver. The mobility method and its applications are discussed here; however, there is also a need for a simplified approach to enable transmissibility studies to be made for isolators coupled to structures upon which frequency response measurements have not been taken. The approximate mobility method is therefore also outlined.

In the case of vibration transmission between points on a distributed structure, measurement of transfer functions alone will not yield enough information to enable transmission paths to be identified and suitable remedial action (that is, vibration control procedures) to be carried out. For this reason, attention has turned in recent years to the development of methods for measuring the power flow through structures. This approach to the understanding of transmission mechanisms with the objective of transmission reduction has also been complemented by study of the power flow through isolator systems. The power flow concept can thus be seen to unify vibration control methods by seeking to minimize the power input to the structure and then to minimize the power transmission through the structure.

A review is made in this chapter of the mobility approach to structural analysis and its application to the study of vibration isolation problems, including the use of approximate mobility methods; recently developed power flow measurement techniques are also described.

26.2 THE MOBILITY APPROACH TO STRUCTURAL ANALYSIS AND VIBRATION ISOLATION

26.2.1 Concepts

Mobility is a particular form of the scaled frequency response of a mechanical system which is defined as the ratio of velocity to force. This complex function must generally be measured or estimated by some means. A variety of measurement techniques may be used [26.2] using deterministic or random excitation. For three-dimensional systems, translational velocity may be measured in the three coordinate directions, and rotational responses to a given force may also be measured; this is of particular importance, for example, in the investigation of machinery seatings. Most of the dynamic testing currently carried out is, however, concerned with the measurement of response to an applied force in the same coordinate direction. With reference to Fig. 26.1, mobility having been defined as the ratio of velocity to force $\dfrac{V(i\omega)}{F(i\omega)} = M(i\omega)$, then a driving point mobility is for example $\left[\dfrac{V(i\omega)x}{F(i\omega)x}\right]_{\mathrm{B}}$, and cross mobility is $\left[\dfrac{V(i\omega)y}{F(i\omega)x}\right]_{\mathrm{B}}$, and a transfer mobility is $\left[\dfrac{V(i\omega)xC}{F(i\omega)xB}\right]$ etc.

If coupled system performance is to be estimated, as stated in the introduction, concepts similar to those used in electrical circuit analysis may be used [26.4]. Mechanical elements may be connected in series or in parallel and their forced response evaluated, i.e.:

For series elements

$$M_s \, (i\omega) = \sum_{j=1}^{n} M_j \, (i\omega)$$

For parallel elements

$$\frac{1}{M_p \, (i\omega)} = \sum_{j=1}^{n} \frac{1}{M_j \, (i\omega)}$$

Evaluation of the above expressions for combinations of lumped elements such as springs, masses, and viscous dampers is simple: the vibration isolation problem can therefore be readily examined via the mobility approach, and the general case of a coupled source–isolator–receiver system can be formulated.

26.2.2 The vibration isolation problem in mobility terms

The vibration isolation problem is outlined in Fig. 26.1 and, although a general statement of the problem may be arrived at when the point of interest (C) is remote from the point of attachment of the isolator to the substructure (B), let us consider here, for necessary brevity in the first instance, the case when points B and C coincide. In other words, only the vibration isolation problem will be examined in this section; the transmission of vibrational power from B to C is examined later.

The coupling of the isolator to the substructure at B and system layout may cause forces and moments to act such as to produce deflections along all three coordinate axes and rotations (angular velocities) about these axes. The motion at B will therefore generally be complicated and complete description of all the point and cross mobilities is required if the induced structural responses are to be estimated. The effects of force and moment excitation are both considered later, so for simplicity, and to act as an illustration of the mobility approach, the translation case only will be detailed here initially.

For the case of an isolator which can be considered as massless, that is for frequencies at which inertia effects in the isolator are negligible, the forces at each end of the isolator are equal and opposite and the modified transmissibility T_m is:

$$T_m = \frac{\text{Receiver } |\text{velocity}|}{\text{Source } |\text{velocity}| \text{ before receiver and isolator added}}$$

$$= \frac{M_R}{M_S + M_I + M_R} \tag{26.1}$$

and the isolator effectiveness E is:

$$E = \frac{\text{Receiver } |\text{velocity}| \text{ when connected directly to source}}{\text{Receiver } |\text{velocity}| \text{ when connected through isolator to source}}$$

$$= 1 + \frac{M_I}{M_S + M_R} \tag{26.2}$$

where M_I = the isolator mobility

M_S = the source mobility ⎫ which are, in general, complex functions

M_R = the receiver mobility ⎭

It can be seen from (26.1) and (26.2) above that source and receiver mobilities must be known, measured, or estimated by some means, if an isolator is to be chosen to give a required transmissibility. In particular, E should be as high as possible for good isolation, and expression (26.2) shows very clearly that for good isolation it is required that:

$$M_I \gg M_S + M_R .$$

It can be seen that for good isolation, high isolator mobility is required and, of paramount importance, E will become low if the source or receiver mobilities become large. Similarly, E will be reduced if wave effects occur in the isolator and M_I becomes low. The former point highlights the need for knowledge of source and receiver mobilities.

26.3 APPROXIMATE POINT MOBILITIES OF RECEIVERS (STRUCTURES)

The influence of source and reciever mobilities on the effectiveness of isolator systems has been outlined above. At the present time there is little information concerning the mobilities of sources such as machines and this is a subject of continuing research. However, although some simplifications may be made concerning the nature of sources by assuming that they are of either the force or velocity type, perhaps the most significant problem at present concerns mobility characteristics to be ascribed to receivers (substructures) such as floors, bulkheads, machinery seatings, etc., composed of beams, plates, or beam and plate combinations.

Several approaches are available to the practical structural dynamicist:

(a) to measure the required frequency response data,
(b) to develop a theoretical model using a finite number of normal modes,
(c) to use a single (often fundamental) mode approach.

It is possible to make accurate mobility measurments over a wide frequency range (that is, (a) above) to provide the required data for use in isolation calculations, and if it is not possible to obtain all the required data during the experiment it is possible to construct a theoretical model using a limited amount of experimental data [26.5] in order to predict required functions which have not been measured. This latter technique also offers the great advantage that quantities, such as cross mobilities, which are difficult to measure, may be estimated.

The normal mode approach, (b) above, may be used, but it again usually requires the use of computational facilities and will be of limited validity particularly at high frequencies. The single mode method will obviously be even less accurate.

It can therefore be seen that if a complete set of measured or predicted data is not available, assessment methods will be necessarily approximate. Beause of these problems, and the need for very simple isolation system assessment procedures, which may be applied rapidly to a wide range of structures without recourse to the computer, very simple, approximate formulae have been developed for estimating the point mobilities of common structural elements.

If we consider a typical substructure composed of beam and plate elements, at low frequencies discontinuities cause vibrational waves to be reflected and resonances occur. At higher frequencies, resonant behaviour is not so clearly apparent, either because the structure exhibits a very high modal density and the point mobility does not exhibit clearly defined, well-separated resonances and anti-resonances, or because the vibrational waves are transmitted through the local boundaries and power is radiated or absorbed in remote parts of the structure: an illustration of the effect is given in Fig. 26.2. It has been shown that in these cases a structure of finite dimensions may be approximated to by an equivalent structure of infinite extent with no reflecting devices [26.6]. This approach may be used for the prediction of the point mobilities of practical structures at high frequencies where modal contributions have coalesced. At low frequencies, where the resonances are well separated, this approximation is much less accurate although it is still valuable since it gives an average level of the frequency response magnitude. If broadband sources are being considered, the approximation in the low frequency range is of lesser concern than if discrete frequency sources are of interest because, in the former case, only estimates of mean square values of transmitted quantities are required, and these can be estimated with an acceptable degree of accuracy. In the latter case, the approximate mobility function may still be used but upper limits of mobility due to resonant behaviour must be noted, for reasons given in section 26.2.2 above.

The driving point mobilities of infinite structures are given for force and moment excitation of various structural elements in Table 26.1 (a list of symbols used in Table 26.1 is included in Appendix 26.1) [26.7]. Following the notation of [26.7], when approximate structural mobilities are being considered, mobility is denoted by $\beta(i\omega)$ which is substituted for $M(i\omega)$ used in the exact formulation. The mobility formulae given apply to infinite structures within which no resonant behaviour can occur. The point mobilities of finite structures will approximate to these values at high frequencies above that given in the second column of the second part of Table 26.1 which shows the frequency at which the frequency interval between resonances is less than the half power point bandwidth of a resonance. When a finite structure is being represented by an equivalent infinite structure the largest error in the estimated mobility will occur at a resonance frequency in the low frequency region. The largest peaks in the mobility of finite beams and plates have been calculated; these are included in Table 26.1 together with a list of the moduli of the ratios of the peak point mobility of the finite structure to the point mobility of the infinite structure, which in most cases is inversely proportional to the loss factor.

In the latter calculations for torsional and logitudinal waves the response was calculated at the midpoint of a clamped–clamped beam, whereas for the flexural case a simply-supported beam with excitation at the midpoint was used. In the case of moment excitation the second resonance frequency was used because this type of excitation applied at the midpoint of a beam does not excite the first mode. The finite plate considered was rectangular and simply-supported on all edges with central excitation. Again it was necessary to use the second resonance frequency in the moment excitation case.

26.4 POWER TRANSMISSION THROUGH ISOLATORS AND STRUCTURES

In the case of vibration transmission between points on a distributed structure, for example B–C in Fig. 26.1, measurement of transfer functions alone will not yield enough information to enable transmission paths to be identified and suitable remedial action taken, that is, vibration control procedures to be applied. For this reason, attention has turned in recent years to the investigation of power flow through structures. This approach to the understanding of trans- mission mechanisms, with the objective of transmission reduction, has also been complemented by study of the power flow through isolator systems. The power flow concept can thus be seen to unify vibration control methods by seeking to minimize the power input to the structure and then to minimize the power transmission through the structure. Structural power flow and its measurement are therefore also discussed here.

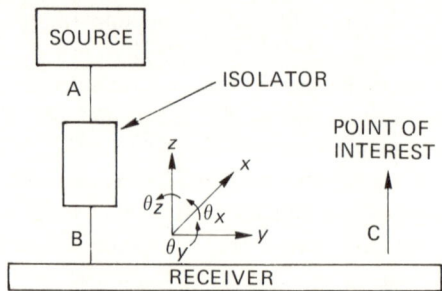

Fig. 26.1 – The vibration isolation and transmission problem.

26.4.1 Power flow through isolators

The objective in applying a vibration control procedure, such as isolation, should be to reduce the power flow from a source to the receiver. The power flow (P) into a structure may be expressed via the point mobility according to the nature of the source in the following way:

For a force (F) source, $P = \frac{1}{2}|F|^2 \, \text{Re}\,\{M\}$ (26.3)

or for a velocity (V) source, $P = \frac{1}{2}|V|^2 \dfrac{\text{Re}\,\{M\}}{|M|^2}$ (26.4)

where $\text{Re}\,\{M\}$ denotes the real part of the mobility of the structure.

In the isolation problem, in its most simple form, the source may be represented by a rigid mass (m) supported on a simple massless spring isolator of stiffness (k) which isolates the source from a receiver of mobility (M). A power flow transmission spectrum (Q_f) for the case of a force source may be employed [26.8] according to:

$$P(\omega) = Q_f(\omega)\,|F(\omega)|^2$$ (26.5)

where $Q_f(\omega)$ is real and only dependent upon the properties of the source, isolator and receiver. If the excitation is in the form of a single harmonic then P in equation (26.5) is the total power flow, or if the excitation is defined via a spectral density then $P(\omega)$ is the power flow spectral density. For the case considered above, the power flow is given by:

$$P(\omega) = \frac{\text{Re}\,\{M\}}{2\,|\,1 - (\omega^2/\omega_0^2) + i\omega m\,M\,|^2}\,|F|^2$$ (26.6)

where $\omega_0^2 = k/m$ is the undamped natural frequency of the system with the flexible foundation clamped. Now, the power flow into a substructure may be estimated by using one of the simple expressions for the receiver mobility given in the preceding section. For example, if a force source, mass-spring system is coupled to a plate then:

$$P(\omega) = \frac{|F|^2}{2m\omega_0}\,\frac{\alpha}{|\,1 - \Omega^2 + i\Omega\alpha\,|^2}$$ (26.7)

where $\alpha = \dfrac{m\omega_0}{\sqrt{8\,B_p\rho h}} = m\omega_0 \times (\text{foundation mobility}, \beta)$

and $\Omega = \dfrac{\omega}{\omega_0}$ where ω = frequency in radians/sec.

Equation (26.7) is an expression for the power flow from a force source through a single-stage isolator system. Similar expression can be derived for velocity sources and the various substructures (see Table 26.1).

Table 26.1

Properties of infinite system (notes: *, torque applied about axis parallel l_2; time dependence of form $e^{i\omega t}$ assumed)

System	Driving point mobility	Power flow into system (P_s) force or torque source	Power flow into system; velocity or angular velocity source
Beam longitudinal wave motion; force excitation	$\dfrac{\dot{\xi}}{F} = \dfrac{1}{2A\sqrt{E\rho}}$	$P_s = \dfrac{|F|^2}{4A\sqrt{E\rho}}$	$P_s = 4|\dot{\xi}|^2 A\sqrt{E\rho}$
Beam torsional wave motion; torque excitation	$\dfrac{\dot{\theta}}{T} = \dfrac{1}{2\sqrt{GQJ}}$	$P_s = \dfrac{|T|^2}{4\sqrt{GQJ}}$	$P_s = 4|\dot{\theta}|^2\sqrt{GQJ}$
Beam flexural wave motion; force excitation	$\dfrac{\dot{\xi}}{F} = \dfrac{(1-i)}{4A\rho}\sqrt{\omega}\left(\dfrac{A\rho}{EI}\right)^{1/4}$	$P_s = \dfrac{|F|^2}{8A\rho}\sqrt{\omega}\left(\dfrac{A\rho}{EI}\right)^{1/4}$	$P_s = |\dot{\xi}|^2 A\rho\sqrt{\omega}\left(\dfrac{EI}{\rho A}\right)^{1/4}$
Beam flexural wave motion; torque excitation	$\dfrac{\dot{\theta}}{T} = \dfrac{(1+i)}{4EI}\sqrt{\omega}\left(\dfrac{EI}{\rho A}\right)^{1/4}$	$P_s = \dfrac{|T|^2}{8EI}\sqrt{\omega}\left(\dfrac{EI}{\rho A}\right)^{1/4}$	$P_s = \dfrac{|\dot{\theta}|EI}{\sqrt{\omega}}\left(\dfrac{\rho A}{EI}\right)^{1/4}$
Plate flexural wave motion; force excitation	$\dfrac{\dot{\xi}}{F} = \dfrac{1}{8\sqrt{B_p\rho h}}$	$P_s = \dfrac{|F|^2}{16\sqrt{B_p\rho h}}$	$P_s = 4|\dot{\xi}|^2\sqrt{B_p\rho h}$
Plate flexural wave motion; torque excitation	$\dfrac{\dot{\theta}}{T} = \dfrac{\omega}{8B_p(1+L)}$ $\times\left[1 - \dfrac{i4}{\pi}\ln ka + \dfrac{i8L}{\pi(1-\nu)}\left(\dfrac{h}{\pi a}\right)^2\right]$	$P_s = \dfrac{\omega|T|^2}{16B_p(1+L)}$	$P_s = \dfrac{4|\dot{\theta}|^2 B_p(1+L)}{\omega\left\{1+\left[\dfrac{4}{\pi}\ln ka - \dfrac{8L}{\pi(1-\nu)}\left(\dfrac{h}{\pi a}\right)^2\right]^2\right\}}$

System	Onset of infinite behaviour	Largest point mobility of finite system	Ratio of finite system maximum to infinite system	Wavenumber (k)	Displacement of structure
Beam longitudinal wave motion; force excitation	$\omega > \dfrac{\pi}{\eta l}\sqrt{\dfrac{E}{\rho}}$	$\beta_l = \dfrac{2}{\pi A \eta \sqrt{E\rho}}$	$\dfrac{|\beta_l|}{|\beta_\infty|} = \dfrac{4}{\pi\eta}$	$k = \omega\sqrt{\dfrac{\rho}{E}}$	$\xi(y) = \dfrac{-iF\,e^{-iky}}{2\omega A\sqrt{E\rho}}$
Beam torsional wave motion; torque excitation	$\omega > \dfrac{\pi}{\eta l}\sqrt{\dfrac{GQ}{J}}$	$\beta_l = \dfrac{2}{\pi\eta\sqrt{GQJ}}$	$\dfrac{|\beta_l|}{|\beta_\infty|} = \dfrac{4}{\pi\eta}$	$k = \omega\sqrt{\dfrac{J}{GQ}}$	$\theta(y) = \dfrac{-iT\,e^{-iky}}{2\omega\sqrt{GQJ}}$
Beam flexural wave motion; force excitation	$\sqrt{\omega} > \dfrac{4\pi}{\eta l}\left(\dfrac{EI}{\rho A}\right)^{1/4}$	$\beta_l = \dfrac{2l}{\pi^2\eta\sqrt{\rho A EI}}$	$\dfrac{|\beta_l|}{|\beta_\infty|} = \dfrac{4\sqrt{2}}{\pi\eta}$	$k = \sqrt{\omega}\left(\dfrac{\rho A}{EI}\right)^{1/4}$	$\xi(y) = \dfrac{-iF}{4EIk^3}[e^{-iky} - i\,e^{-ky}]$
Beam flexural wave motion; torque excitation	$\sqrt{\omega} > \dfrac{4\pi}{\eta l}\left(\dfrac{EI}{\rho A}\right)^{1/4}$	$\beta_l = \dfrac{2}{l\eta\sqrt{\rho A EI}}$	$\dfrac{|\beta_l|}{|\beta_\infty|} = \dfrac{2\sqrt{2}}{\pi\eta}$	$k = \sqrt{\omega}\left(\dfrac{\rho A}{EI}\right)^{1/4}$	$\xi(y) = \dfrac{T}{4EIk^2}[e^{-iky} - e^{-ky}]$
Plate flexural wave motion; force excitation	$\omega > \dfrac{8}{\eta l_1 l_2}\sqrt{\dfrac{B_p}{\rho h}}$	$\beta_l = \dfrac{4l_1 l_2}{\pi^2\eta\sqrt{\rho h B_p}(l_1^2 + l_2^2)}$	$\dfrac{|\beta_l|}{|\beta_\infty|} = \dfrac{32 l_1 l_2}{\pi^2 l(l_1^2 + l_2^2)}$	$k = \sqrt{\omega}\left(\dfrac{\rho h}{B_p}\right)^{1/4}$	Valid in far field only $\xi(r,\phi) = \dfrac{-iF}{8B_p k^2}\sqrt{\dfrac{2}{rk\pi}}\,e^{-i(rk-\pi/4)}$
Plate flexural wave motion; torque excitation	$\omega > \dfrac{8}{\eta l_1 l_2}\sqrt{\dfrac{B_p}{\rho h}}$	$\beta_l = \dfrac{16 l_2}{\eta\sqrt{\rho h B_p}\,l_1(2l_2^2 + l_1^2)}$ *		$k = \sqrt{\omega}\left(\dfrac{\rho h}{B_p}\right)^{1/4}$	Valid in far field only $\xi(r,\phi) = \dfrac{T}{8B_p k}\sqrt{\dfrac{2}{rk\pi}}\,e^{-i(rk-\pi/4)}\sin\phi$

Table reproduced by permission of Academic Press; taken from Ref [26.7].

Clearly, the power flow is strongly dependent on the magnitude and phase of the foundation mobility. In general, straight-line approximations of the type indicated in Fig. 26.2 may often be made to the modulus of a measured mobility spectrum when plotted on a log–log scale. Thus the modulus of the foundation mobility may be represented, to a good approximation, by a law of the form

$$|\beta| = A\omega^s \qquad (26.8)$$

where β is the equivalent 'infinite' foundation mobility and A is a positive constant. Approximation of this form is useful if trends in high frequency power transmission characteristics are to be studied. The exponent s is a real constant which may be estimated experimentally from the slope of a log–log plot of a mobility spectrum or by calculation. A complicated mobility spectrum may be represented adequately by a number of lines, each of the form of equation (26.8). However, when considering power flow it is necessary to know both the modulus and the phase or alternatively the real and imaginary parts of the mobility rather than the modulus alone. By means of Hilbert transforms [26.9] it has been shown by Bode [26.10] that the phase characteristics of a point mobility may be deduced from its modulus spectrum, and application of the method has been deomonstrated [26.11]. A mobility spectrum of the form of equation (26.8) will have a phase given by

$$\phi(\omega) = s\pi/2 \qquad (26.9)$$

If equation (26.8) represents the mobility over only a finite frequency interval and outside of this interval the modulus is different, then the phase will vary significantly from the above value only at the ends of the interval. The point mobility of a foundation which has a straight-line characteristic when plotted on log–log scales may thus be written as

$$|\beta| = A e^{is\pi/2} \omega^s = A\omega^s [\cos(s\pi/2) + i\sin(s\pi/2)] \qquad (26.10)$$

As has been shown above, the power flowing into a structure due to a harmonic force is dependent upon the real part of the mobility, and in this case the formula for the power flow is

$$P = \tfrac{1}{2} \operatorname{Re}\{\beta\} |F|^2 = \tfrac{1}{2} A \omega^s |F|^2 \cos(s\pi/2) , \qquad (26.11)$$

where F is the amplitude of the force applied to the foundation. Since the foundation is passive, no power may flow out of the structure at the driving point and thus P may never be negative. This implies that the phase angle is restricted to lie between $+\pi/2$ and $-\pi/2$ and thus s must lie in the range

$$-1 < s < 1 . \qquad (26.12)$$

When $s = \pm 1$ the mobility corresponds to a stiffness or mass line which represents the extreme cases between which all point mobilities of this form must lie. Thus if a straight-line approximation is made to the mobility modulus plotted on a log–log scale the phase angle is automatically determined and must be in the range $-\pi/2$ to $+\pi/2$.

It should be noted that in its detailed behaviour the slope of a log–log mobility spectrum may be outside the range of relation (26.12) as for example near a resonance. It has been assumed above, however, that an average line has been drawn through resonances so that the overall nature of the mobility is represented.

Substituting the value of β corresponding to a simple power law (equation (26.10)) into equation (26.7) for the power flow enables the power flow transmission spectrum to be written [26.8] for unit force as

$$Q_{\mathrm{f}} = [\gamma\,(\omega/\omega_{\mathrm{o}})^{s}/2m\omega_{\mathrm{o}}\,|\,1 - (\omega^{2}/\omega_{\mathrm{0}}^{2}) + i(\gamma+i\delta)\,(\omega/\omega_{\mathrm{o}})^{s+1}\,|^{2}] \quad,(26.13)$$

where $\gamma = m\omega_{\mathrm{0}}^{s+1}A\,\cos(s\pi/2)$ and $\delta = m\omega_{\mathrm{0}}^{s+1}A\,\sin(s\pi/2)$.

It is not possible to find a method for normalizing this equation in any general manner; the procedure adopted [26.8] has been to normalize in terms of the properties of the isolator and machine so that the effects of various foundation mobilities can be seen. γ and δ are the normalized real and imaginary parts of $\beta(\beta = A\cos(s\pi/2) + iA\,\sin(s\pi/2))$; large values of γ and δ correspond to a very mobile foundation, while small values indicate a rigid foundation.

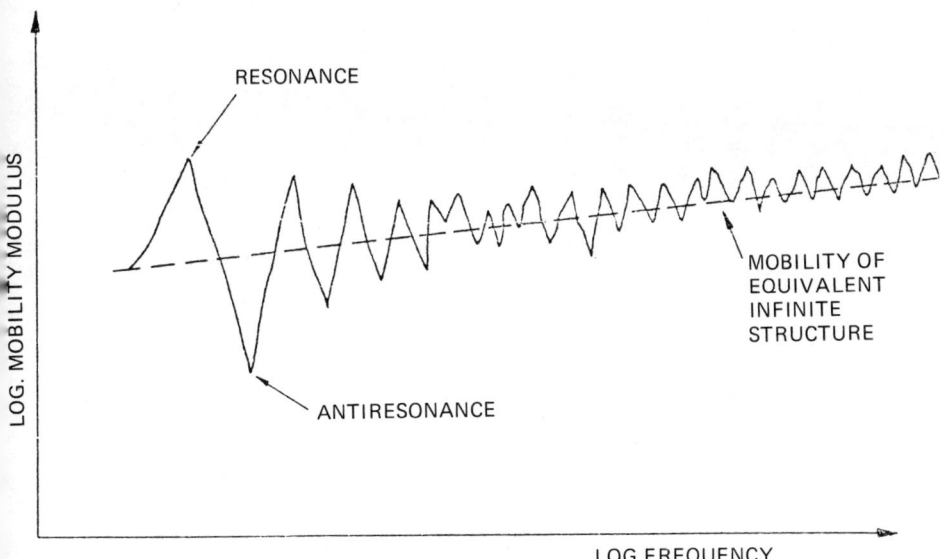

Fig. 26.2 – An illustration of possible high-frequency trends in point mobility data.

Table 26.2
Power flow into foundations from machinery sources on isolators

	Power flow into foundation	Approximate power flow
Force source single stage isolator	$\omega_0 = \sqrt{(K/M)}$; foundation mobility $= [(\gamma+i\delta)/M\omega_0](\omega/\omega_0)^s$ $P \doteq \dfrac{\lvert F\rvert^2}{2M\omega_0}\gamma\left(\dfrac{\omega}{\omega_0}\right)^s \Big/ \left\lVert 1 - \dfrac{\omega^2}{\omega_0^2} + i(\gamma+i\delta)\left(\dfrac{\omega}{\omega_0}\right)^{s+1}\right\rVert^2$	$\omega/\omega_0 \ll 1$ $P = \dfrac{\lvert F\rvert^2}{2M\omega_0}\left(\dfrac{\omega}{\omega_0}\right)^s$
Force source two stage isolator	$\omega_0 = \sqrt{(K_2/M_2)}$; ω_1, ω_2 resonance frequencies with base clamped; foundation mobility $= [(\gamma+i\delta)K_1/M_1M_2\omega_0^3](\omega/\omega_0)^s$ $P = \dfrac{\lvert F\rvert^2 K_1}{2M_1M_2\omega_0^3}\gamma\left(\dfrac{\omega}{\omega_0}\right)^s \Big/ \left\lVert\left(1-\dfrac{\omega^2}{\omega_1^2}\right)\left(1-\dfrac{\omega^2}{\omega_2^2}\right) + i(\gamma+i\delta)\left(\dfrac{\omega}{\omega_0}\right)^{s+3}\left(\dfrac{\omega_1^2+\omega_2^2-\omega_0^2}{\omega^2}-1\right)\right\rVert^2$	$\omega/\omega_1 \ll 1$ $P = \dfrac{\lvert F\rvert^2 K_1\gamma}{2M_1M_2\omega_0^3}\left(\dfrac{\omega}{\omega_0}\right)^s$
Velocity source single stage isolator	foundation mobility $= (\gamma+i\delta)\omega^s/K$ $P = \dfrac{\lvert V\rvert^2 K}{2}\dfrac{\gamma\omega^s}{\lvert i\omega+\omega^s(\gamma+i\delta)\rvert^2}$	$\omega^{2-2s} \ll \gamma^2 + \delta^2$ $P = \dfrac{\lvert V\rvert^2 K\gamma}{2(\gamma^2+\delta^2)\omega^s}$
Velocity source two stage isolator	$\omega_0 = \sqrt{(K_2/M_2)}$; ω_1 resonance frequency with base clamped; foundation mobility $= [\omega_0(\gamma+i\delta)/K_2](\omega/\omega_0)^s$ $P = \dfrac{\lvert V\rvert^2}{2\omega_0 K_2}\left(\dfrac{\omega_0}{\omega_1}\right)^4\left(\dfrac{\omega_1^2}{\omega_0^2}-1\right)\gamma\left(\dfrac{\omega}{\omega_0}\right)^s\dfrac{\omega_0^2}{\omega^2}\Big/ \left\lVert 1-\dfrac{\omega^2}{\omega_1^2}-i(\gamma+i\delta)\left(\dfrac{\omega_0}{\omega}\right)^{1-s}\left(1-\dfrac{\omega^2}{\omega_1^2}-\dfrac{\omega_0^2}{\omega_1^2}\right)\right\rVert^2$	$\omega/\omega_1 \ll 1$ $P = \dfrac{\lvert V\rvert^2\gamma\omega_0^2}{2K_2\omega_0(\gamma^2+\delta^2)(\omega_1^2-\omega_0^2)}\left(\dfrac{\omega_0}{\omega}\right)^s$

	Approximate power flow	Power flow at $\omega = \omega_0$ (peak if $\gamma^2+\delta^2 < 1$)	Power flow at intermediate frequencies $\gamma^2+\delta^2 > 1$	Low-frequency break point $\gamma^2+\delta^2 > 1$	High frequency break point $\gamma^2+\delta^2 > 1$
Force source single stage isolator	$\omega/\omega_0 \gg 1$ $P = \dfrac{|F|^2}{2M\omega_0}\left(\dfrac{\omega_0}{\omega}\right)^{4-s}$	$P = \dfrac{|F|^2}{2M\omega_0}\dfrac{\gamma}{\gamma^2+\delta^2}$	$P = \dfrac{|F|^2}{2M\omega_0}\dfrac{\gamma}{\gamma^2+\delta^2}\left(\dfrac{\omega_0}{\omega}\right)^{2+s}$	$\left(\dfrac{\omega}{\omega_0}\right)^{2s+2} = \dfrac{1}{\gamma^2+\delta^2}$	$\left(\dfrac{\omega}{\omega_0}\right)^{2-2s} = \gamma^2+\delta^2$
Force source two stage isolator	$\omega/\omega_2 \gg 1$ $P = \dfrac{|F|^2 K_1\gamma}{2M_1 M_2 \omega_0^2}\dfrac{(\omega_1^2\omega_2^2)^2}{\omega^4}\left(\dfrac{\omega}{\omega_0}\right)^s$	Power flow at $\omega = \omega_1$ $P = \dfrac{|F|^2 K_1\gamma}{2M_1 M_2 \omega_0^3(\gamma^2+\delta^2)}\left(\dfrac{\omega_0}{\omega_1}\right)^{6+s}\left(\dfrac{\omega_1^2}{\omega_2^2-\omega_0^2}\right)^2$	$P = \dfrac{|F|^2 K_1\gamma}{2M_1 M_2 \omega_0^3(\gamma^2+\delta^2)}\left(\dfrac{\omega_0}{\omega_2}\right)^{6+s}\left(\dfrac{\omega_2^2}{\omega_1^2-\omega_0^2}\right)^2$	Power flow at $\omega = \omega_2$	
Velocity source single stage isolator	$\omega^{2-2s} \gg \gamma^2+\delta^2$ $P = \dfrac{|V|^2 K\gamma}{2\omega^{2-s}}$	$\omega^{2-2s} \geqslant \gamma^2+\delta^2$	Break point between low and high frequency behaviour $\omega^{2-2s} = \gamma^2+\delta^2$		
Velocity source two stage isolator	$\omega/\omega_1 \gg 1$ $P = \dfrac{|V|^2(\omega_1^2-\omega_0^2)\,\gamma}{2K_2\omega_0^3}\left(\dfrac{\omega_0}{\omega}\right)^{6-s}$	Power flow at $\omega = \omega_1$ $P = \dfrac{|V|^2(\omega_1^2-\omega_0^2)\,\gamma}{2K_2\omega_0^3(\gamma^2+\delta^2)}\left(\dfrac{\omega_0}{\omega_1}\right)^s$			

Table reproduced by permission of Academic Press; taken from Ref [26.8].

Table 26.2 gives the essential behaviour of equation (26.13). The power flow transmission spectra may be approximated on a log–log scale by straight lines at high ($\omega/\omega_0 < 1$) frequencies. If $\gamma^2 + \delta^2$ is small there will be a peak at $\omega/\omega_0 = 1$ where the two lines intersect. Alternatively if $\gamma^2 + \delta^2$ is large a third straight line may be drawn, the three lines having break points at frequencies above and below $\omega/\omega_0 = 1$. The criteria for small or large $\gamma^2 + \delta^2$, the height of the peak, the behaviour of the three lines, and the positions of the break points, are given in Table 26.2. The criterion for small or large $\gamma^2 + \delta^2$ is obtained by establishing whether the peak at $\omega/\omega_0 = 1$ is greater or less than the value of the power flow spectrum at the intersection of the two lines for high- and low-frequency dependence. Fig. 26.3 gives a sketch of a power flow transmission spectrum for $s = -\frac{1}{2}$ (corresponding to a beam-like foundation) for small and large values of $\gamma^2 + \delta^2$. Exact spectra are shown in Fig. 26.4 for the same values of s and $\gamma^2 + \delta^2$.

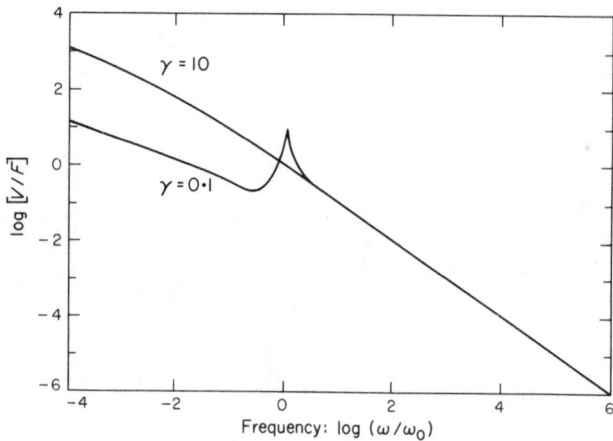

Fig. 26.3 – Power flow transmission spectra for a single-stage isolator with force source and beam-like foundation ($s = -\frac{1}{2}$; $m\omega_0 = 1$): approximate values based on Table 26.2 [26.8].

In general, equations (26.6) and (26.13) show that the overall levels of the power flow transmission spectra are controlled by the foundation mobility, while the spectrum shape is governed by the stiffness and mass of the isolator and machine. Since the foundation mobility always appears in the numerator of the power flow equations it is necessary to choose a small value of A (equation (26.10)) in order to minimize the power flow. When determining the shape of the power flow spectrum there are two alternative extremes. Either $\gamma^2 + \delta^2$ is chosen to be small so that there is a peak in the spectrum or, alternatively, $\gamma^2 + \delta^2$ is large and there is no peak but the levels are increased at high frequencies. This choice will be governed by the nature of the force spectrum in which there may, for example, be specific harmonics which must be avoided.

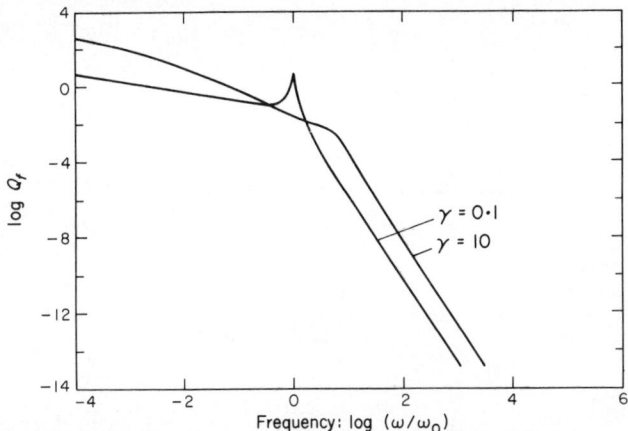

Fig. 26.4 – Power flow transmission spectra for a single-stage isolator with force source and beam-like foundation ($s = -\frac{1}{2}$; $m\omega_0 = 1$) [26.8].

The motion of the machine is important since too large a velocity may be unacceptable. The source velocity is given by

$$V = (iF/m\omega) [(\omega^2/\omega_0^2) - i(\gamma + i\delta) (\omega/\omega_0)^{s+1}]/$$

$$[1 - (\omega^2/\omega_0^2) + i(\gamma + i\delta) (\omega/\omega_0)^{s+1}] \qquad (26.14)$$

The low- and high-frequency dependences are given in Table 26.3 where it may be seen that the velocity may be approximated on a log–log scale by a straight line for all values of $\gamma^2 + \delta^2$ for frequencies greater than the resonance frequency. Examples of velocity spectra given by equation (26.14) for two values of $\gamma^2 + \delta^2$ are given in Fig. 26.5. Velocity and force sources, applied to one- and two-stage isolation systems, are considered in [26.8] and results are given in Tables 26.2 and 26.3.

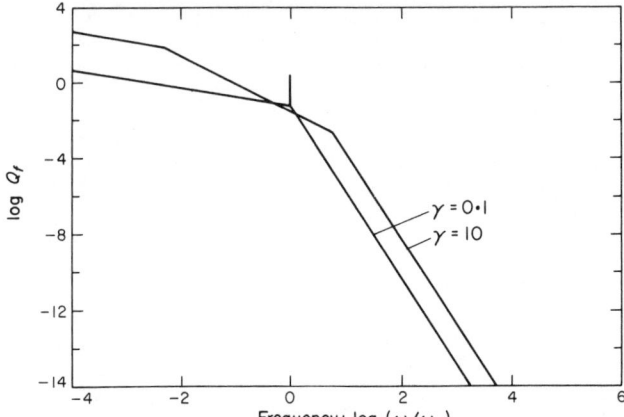

Fig. 26.5 – Velocity of machine on single-stage isolator with beam-like foundation and force source [26.8].

Table 26.3
Velocity of components of isolation systems

Force source single stage isolator — Exact velocity

$\omega_0 = \sqrt{(K/M)}$;   foundation mobility $= [(\gamma+i\delta)/M\omega_0](\omega/\omega_0)^s$

Machine

$$V = iF\left[\frac{\omega^2}{\omega_0^2} - i(\gamma+i\delta)\left(\frac{\omega}{\omega_0}\right)^{s+1}\right] \Big/ M\omega\left[1 - \frac{\omega^2}{\omega_0^2} + i(\gamma+i\delta)\left(\frac{\omega}{\omega_0}\right)^{s+1}\right]$$

Force source two stage isolator — Exact velocity

$\omega_0 = \sqrt{(K_2/M_2)}$; ω_1, ω_2 resonance frequencies with base clamped;   foundation mobility $= [(\gamma+i\delta)/M_1 M_2 \omega_0^3]\, K_1(\omega/\omega_0)^s$

Machine

$$V_1 = \frac{F\omega\left\{i\left(1 - \frac{\omega^2}{\omega_1^2}\right)\left(1 - \frac{\omega^2}{\omega_2^2}\right) + i(\gamma+i\delta)\left(\frac{\omega}{\omega_0}\right)^{3+s}\left(\frac{\omega_1^2}{\omega^2} + \frac{\omega_2^2}{\omega^2} - \frac{\omega_0^2}{\omega^2} - 1\right) + \frac{\omega_1^2\omega_2^2}{\omega^4}\left[(\gamma+i\delta)\left(\frac{\omega}{\omega_0}\right)^{3+s} + \frac{i\omega^4}{\omega_1^2\omega_2^2}\left(\frac{\omega_1^2}{\omega^2} + \frac{\omega_2^2}{\omega^2} - \frac{\omega_0^2}{\omega^2} - 1\right)\right]\right\}}{K_1\left[1 - \frac{\omega^2\omega_0^2}{\omega_1^2\omega_2^2}\left(1 - \frac{\omega^2}{\omega_1^2}\right)\left(1 - \frac{\omega^2}{\omega_2^2}\right) + i(\gamma+i\delta)\left(\frac{\omega}{\omega_0}\right)^{s+3}\left(\frac{\omega_1^2}{\omega^2} + \frac{\omega_2^2}{\omega^2} - \frac{\omega_0^2}{\omega^2} - 1\right)\right]}$$

Isolator mass

$$V_2 = \frac{F\omega_1^2\omega_2^2\left[(\gamma+i\delta)\left(\frac{\omega}{\omega_0}\right)^{s+3} + \frac{i\omega^4}{\omega_1^2\omega_2^2}\right]}{K_2\omega^3\left[\left(1 - \frac{\omega^2}{\omega_1^2}\right)\left(1 - \frac{\omega^2}{\omega_2^2}\right) + i(\gamma+i\delta)\left(\frac{\omega}{\omega_0}\right)^{s+3}\left(\frac{\omega_1^2}{\omega^2} + \frac{\omega_2^2}{\omega^2} - \frac{\omega_0^2}{\omega^2} - 1\right)\right]}$$

Velocity source two stage isolator — Exact velocity

$\omega_0 = \sqrt{(K_2/M_2)}$; ω_1 resonance frequency with base clamped;   foundation mobility $= [\omega_0(\gamma+i\delta)/K_2](\omega/\omega_0)^s$

Isolator mass

$$V_2 = \frac{V_1(\omega_1^2 - \omega_0^2)\left[1 - i(\gamma+i\delta)\left(\frac{\omega_0}{\omega}\right)^{1-s}\right]}{\omega_1^2\left[\left(1 - \frac{\omega^2}{\omega_1^2}\right) + i(\gamma+i\delta)\left(\frac{\omega_0}{\omega}\right)^{1-s}\left(1 - \frac{\omega_0^2}{\omega_1^2} - \frac{\omega^2}{\omega_1^2}\right)\right]}$$

	Approximate velocity	Approximate velocity
Force source single stage isolator	$\omega/\omega_0 \ll 1$	$\omega/\omega_0 \gg 1$
Machine	$V = \dfrac{F(\gamma+i\delta)}{M\omega_0}\left(\dfrac{\omega}{\omega_0}\right)^s$	$V = \dfrac{-iF}{M\omega}$
Force source two stage isolator	$\omega/\omega_1 \ll 1$	$\omega/\omega_2 \gg 1$
Machine	$V_1 = \dfrac{F\omega_1^2\omega_2^2(\gamma+i\delta)}{K_1\omega_0^3}\left(\dfrac{\omega}{\omega_0}\right)^s\left[\dfrac{\omega_1^2}{\omega_0^2}+\dfrac{\omega_2^2}{\omega_0^2}-1-\dfrac{\omega_1^2\omega_2^2}{\omega_0^4}\right]$	$V_1 = \dfrac{-iF\omega_1^2\omega_2^2}{K_1\omega_0^2\omega}$
Isolator mass	$V_2 = \dfrac{F\omega_1^2\omega_2^2(\gamma+i\delta)}{K_2\omega_0^3}\left(\dfrac{\omega}{\omega_0}\right)^s$	$V_2 = \dfrac{iF\omega_1^2\omega_2^2}{K_2\omega^3}$
Velocity source two stage isolator	$\omega/\omega_1 \ll 1$	$\omega/\omega_1 \gg 1$
Isolator mass	$V_2 = V_1$	$V_2 = \dfrac{-V_1(\omega_1^2-\omega_0^2)}{\omega^2}$

Table reproduced by permission of Academic Press; taken from Ref [26.8].

Moment (torque) excitation as well as force excitation of foundations is an important consideration in this type of analysis. This is because the method in which a machine is mounted often results in torques being the mechanism by which the foundation is excited. For example, torques occur when there is rocking of a machine on a horizontal foundation or, alternatively, when a machine is mounted as in Fig. 26.6, where the support acts as a lever, creating flexural wave motion in the vertical member. When a lever of length l is the cause of flexural wave motion, the relationship between the force and velocity at the tip of the lever and the torque and angular velocity at the root of the lever (see Fig. 26.6.) may be written as

$$V/F = \beta_f = l^2 \dot{\theta}/T = l^2 \beta_T \ , \tag{26.15}$$

where β_f and β_T are the point mobilities for a driving force at the end of the lever and for a torque applied to the member at the root of the lever, respectively. For equation (26.15) it is assumed that the angle through which the lever rotates remains small.

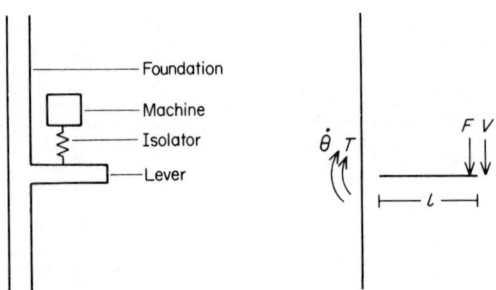

Fig. 26.6 – Method of mounting machinery which results in torque excitation of foundation [26.8].

A comparison between typical beam- and plate-like foundations excited by torques or forces is of interest. The power flow into a foundation is given by

$$P = \tfrac{1}{2} \, \mathrm{Re} \, \{\beta_f\} \, |F|^2 \ \text{or} \ P = \tfrac{1}{2} \, l^2 \, \mathrm{Re} \, \{\beta_T\} \, |F|^2 \tag{26.16}\,(26.17)$$

where in this case F is the force applied either to the foundation or to the end of a lever of length l. Fig. 26.8 gives power flow transmission spectra for an infinite beam and plate of the dimensions shown in Fig. 26.7 and with a lever of length 1.0 m. Clearly in this case a plate-like foundation is inferior to a beam, the best foundation being one in which the force is applied to a beam in a perpendicular direction. From equation (26.17) it may be seen that the length of the lever is important, a long lever being particularly poor.

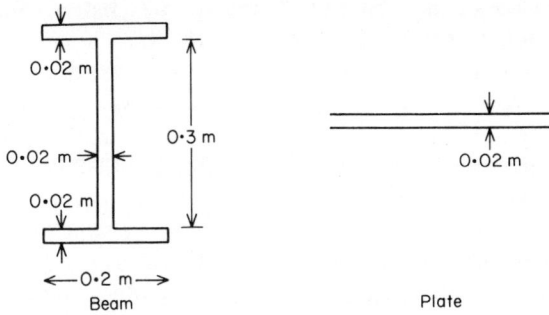

0·02 m

0·02 m →|←

0·02 m

0·3 m

0·02 m

←—0·2 m—→
Beam

Plate

Fig. 26.7 – Dimensions of beam and plate used in Fig. 26.8 [26.8].

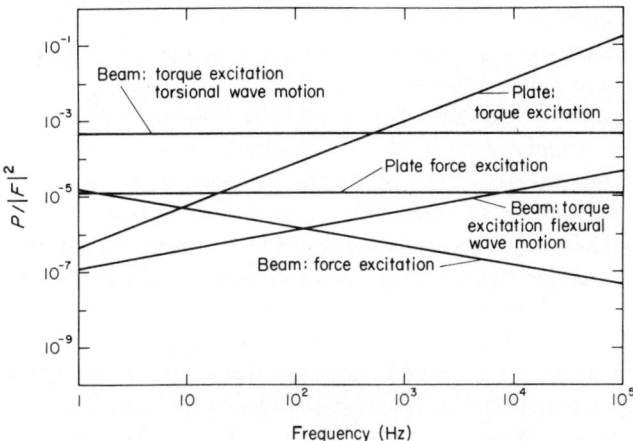

Beam: torque excitation
torsional wave motion

Plate:
torque excitation

Plate force excitation

Beam: torque
excitation flexural
wave motion

Beam: force excitation

$P/|F|^2$

Frequency (Hz)

Fig. 26.8 – Power flow transmission spectra for a beam and plate (dimensions, Fig. 26.7) with a force source applied directly or as a torque by means of a lever of length 1.0 m [26.8].

The introduction of isolators between the machine and foundation always results in a reduction in power flow as long as the excitation spectrum does not include those frequencies which are the resonance frequencies of the system. The use of a two-stage isolator will always produce a significantly greater reduction in power flow than a single-stage isolator; however, the introduction of a second peak in the power flow transmission spectra may result in this form of isolation being appropriate only for high-frequency excitations. If the excitation spectrum does extend over the resonance frequencies, then the relatively large amount of power flow resulting is independent of the foundation characteristics, and the use of an isolator may not be beneficial.

The importance of the nature of the source of excitation within a machine has been repeatedly demonstrated here. Clearly it is important to establish whether excitation by a particular machine has the nature of a force or velocity source and is independent of the isolation system or whether a source mobility should be included when designing isolators. In addition, the spectral distribution of the excitation must also be known before an effective isolation system can be designed which will reduce power flow.

Other factors which affect the power flow into a structure require study. Specifically no attempt has been made in [26.8] to evaluate the result of resonances (wave effects) within isolators or the levers by which machines are supported. Also, owing to the size of most machines, a multipoint isolation and support system is necessary which will produce additional complications when studying power flow.

The above discussion is drawn from references [26.7 and 26.8] to which the reader is directed for further detail.

Studies have been made [26.12] of methods for measuring the power flow through vibration isolators. Several methods have been proposed for measuring the power input to a structure; for example measurement may be made via evaluation of equation (26.3) which necessitates accurate measurement or estimation of the real component of the point mobility of the structure. Experiments have been conducted on an 'infinite' structure composed of a steel beam, 6 mm $\times$ 50 mm $\times$ 6.21 m with its ends embedded in sand boxes designed to be anechoic terminations [26.12]. The beam was used to represent a substructure to which an isolator could be connected. Removal of the sand boxes gave a resonant substructure with free-free boundary conditions.

Fig. 26.9 shows the measured point mobility of the 'infinite' beam in the usual form of presentation, i.e. modulus and phase. The measured and theoretical values agree well at frequencies above about 100 Hz. The real and imaginary components of the mobility of the infinite structure are equal in magnitude, giving rise to a constant phase of $-\pi/4$. The low frequency data were corrupted by noise.

The second set of mobility measurements were made on the beam after the sand boxes had been removed, as in Fig. 26.10. Both sides of this finite beam were covered with damping tape in order to give more representative damping levels. The loss factor, as calculated from the decay times of free vibration, was found to have a frequency independent value of 0.008 between 0 and 1 kHz, the frequency range considered. The same measurement conditions applied as for the 'infinite' beam. The measured real component of mobility is shown in Fig. 26.10. The value of the real component of mobility of the infinite beam lies logarithmically between the peak and trough values. In order to find the average value of the real component of mobility a running integral of Fig. 26.10 was evaluated, which is shown in Fig. 26.11. It can be seen that this integral is close to that calculated for an infinite beam over the same frequency range.

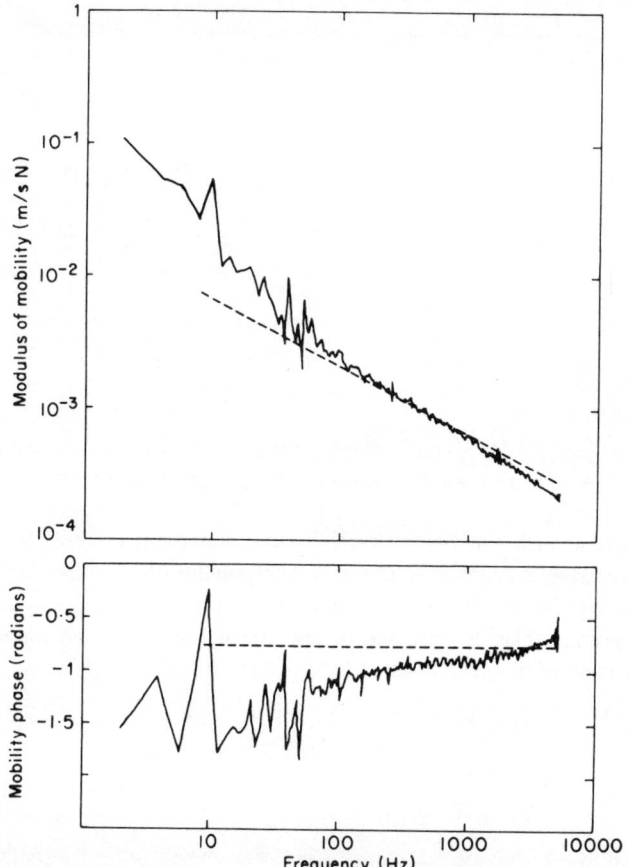

Fig. 26.9 – Measured values of mobility modulus and phase of an experimental 'infinite' beam [26.12].

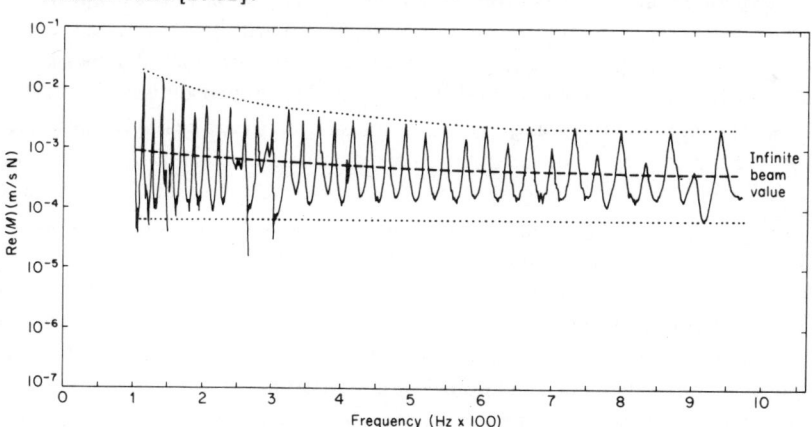

Fig. 26.10 – Real component of finite beam point mobility [26.12].

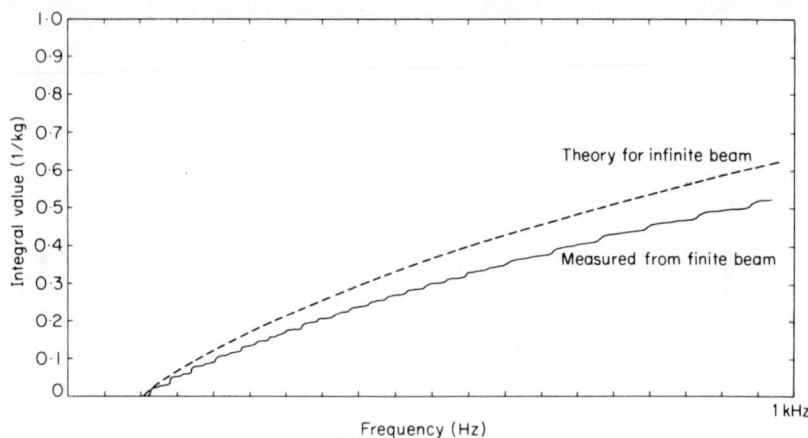

Fig. 26.11 – Running integral across Fig. 26.10 [26.12].

This is consistent with the theoretically proven statement in [26.12] that for frequency averaged calculations, the real component of mobility of a finite beam may be represented by that of an infinite beam. If a structure is driven by a force of random time dependence of spectral density G_{FF}, giving rise to a velocity at that point of spectral density G_{VV}, the power/Hz is given as:

$$P/\text{Hz} = G_{FF} . \text{Re} \{M\} \tag{26.18}$$

$$= \text{Re} \{G_{FV}\} \tag{26.19}$$

$$= G_{VV} . \text{Re} \{Z\} \tag{26.20}$$

where G_{FV} is the cross spectral density between the force and velocity and Z is the point impedance.

Experimental work has been carried out [26.12] using a mass supported on rubber blocks to simulate a source-isolator system which was connected to a beam. The mass was then excited by a random force. Fig. 26.12 shows results for tests with a finite beam as a substructure, and a comparison is made between the power input to the isolator with that transmitted to the beam. It can be seen that, around the region of maximum mean power transmission, only a fraction of the power input to the isolator was transmitted to the beam. This is because in this region when $\omega = \omega_o$, the point mobility at the 'free' end of the spring and mass in series is a minimum. This low value is less than the beam point mobility, and so the spring and mass act as a velocity source on the beam, thereby transmitting most power when the beam impedance is a maximum, i,e, near the beam anti-resonances when Re $\{M\}$ is small. It can also be noted from Fig. 26.12 that the overall power transmitted to the receiver from a broadband source may be adequately predicted on the basis of the infinite structural representation of the receiver.

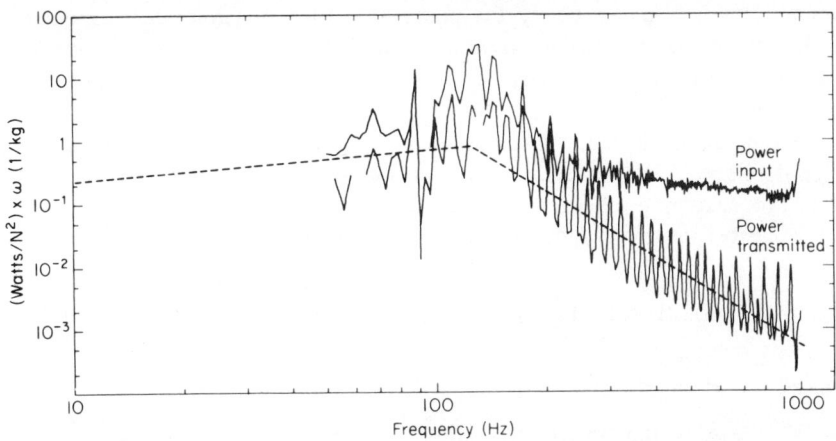

Fig. 26.12 – Power X ω, input to isolator transmitted to the finite beam for
unit force spectral density, O Hz – 1 kHz [26.12].

26.4.2 Power flow through structures

Returning now to Fig. 26.1 it is clear that the preceding sections have been
concerned only with the power flow from the source into the structure at
point B. The further problem is to consider the transmission of power through
the structure from B to C.

If vibration control on the structure is to be attempted it is important
that the mechanisms governing power transmission be identified. All too often
some local change is made to a structure, perhaps the addition of a damping
treatment or stiffening in a chosen local area between B and C, only to find
afterwards that the local vibration level may have changed but the level at
the point of interest was substantially unchanged. Both theoretical and experimental approaches have therefore been made in the study of structural power
flow in order to determine the relative importance of various power transmission
mechanisms.

Theoretical studies carried out so far have been on necessarily simple
models. Power transmission in grillage-type structures has not been examined.
The wave propagation and power flow due to force and moment excitation
applied to the beam have been studied at the driving point and in the far field of
an infinite plate with a simple, line stiffener [26.13]. It has been found that the
motion at the driving point is largely controlled by the beam. If the beam is
excited by a force or moment so that flexural wave motion is induced, then the
power transmitted by these waves will initially be associated with the beam. As
the waves travel away from the source they radiate into the plate so that in the
far field more power is transmitted by the plate than the beam. A strongly direc-

tional wave is also carried by the plate. Power transmission due to moment excitation producing torsional motion of the beam tends to be predominantly due to the beam at high frequencies with the plate being more significant at low frequencies.

Experimental techniques for measuring the power flow in structures have been developed [26.14, 26.15]. Consider the case of a one-dimensional bending wave propagating in a beam along the x direction and it is required to measure the intensity 'power flow per unit width', which is given by:

$$p(x, t) = B \left[\left(\frac{\delta^3 y}{\delta x^3} \right) \frac{\delta y}{\delta t} - \left(\frac{\delta^2 y}{\delta x^2} \right) \frac{\delta^2 y}{\delta t \delta x} \right] \tag{26.21}$$

where B = the flexural rigidity = EI

         E = modulus of elasticity

         I = second moment of area of cross-section

and y = displacement.

It is necessary to determine the time and spatial derivatives of the motion in order to evaluate the term within the brackets in equation (26.21). The time derivatives may be obtained from the signals from transducers mounted on the structure, and the spatial derivatives may be approximated to via a finite difference method. For these purposes a line array of four velocity transducers may be used centred around a point $x = x_0$ with separation Δ between them. If the transducers are denoted by numbers 1 to 4 consecutively, the intensity through the point x_0 may be estimated using the following approximations

$$y \approx \tfrac{1}{2} (y_2 + y_3)$$

$$\frac{\delta y}{\delta x} \approx \frac{1}{\Delta} (y_2 - y_3)$$

$$\frac{\delta^2 y}{\delta x^2} \approx \frac{1}{2\Delta^2} (y_1 - y_2 - y_3 - y_4) \tag{26.22}$$

$$\frac{\delta^3 y}{\delta x^3} \approx \frac{1}{\Delta^3} (y_1 - 3y_2 - 3y_3 - y_4) \ .$$

The instantaneous intensity may then be approximated as:

$$p = \frac{B}{\Delta^3} [\dot{y}_3 \int (\dot{y}_1 - 2\dot{y}_2 + \dot{y}_3) \, \mathrm{d}t - \dot{y}_2 \int (\dot{y}_2 - 2\dot{y}_3 + \dot{y}_4) \, \mathrm{d}t] \ , \tag{26.23}$$

The average intensity of a one-dimensional flexural wave may be determined by use of an expression derived from equation (26.23):

$$<p>_T \approx \frac{B}{\Delta^3} \left[\dot{y}_2 (4y_3 - y_4) - \dot{y}_1 y_3 \right] . \qquad (26.24)$$

It is noteworthy that in this case either velocity or displacement is required from each point thus enabling velocity transducers to be used at points 1 and 2 and displacement transducers at points 3 and 4.

In the same work [26.14] an arrangement of ten transducers was proposed to enable simultaneous measurements to be made of the intensities in two perpendicular directions in plates. The necessary signal processing techniques have been described and an error analysis carried out [26.15]. It has also been shown that the presence of other types of wave motion, e.g. longitudinal, during flexural measurements should not significantly affect measured intensities. The method not only permits directional intensity measurements to be made but yields estimates of the separate components due to shear, bending, and twisting in a plate. The technique has been used to estimate the power flow in a plate; a digital computer was used to perform the necessary data manipulations.

A large steel plate (approximately 1.2 m × 2.4 m) was used for the experiment [26.15]. The plate was supported on cables at one end, and the other end was embedded in sand in order to provide an absorbing termination. Ten accelerometers were used at each of three measurement areas as shown in Fig. 26.13, i.e. the response was measured at thirty stations. The plate was excited during transient testing [26.16] (see also Chapter 27 for discussion of experimental technique) by an electrodynamic exciter, driven by a rapid sweep oscillator via a power amplifier, the exciter being mounted on the vertical axis of symmetry; the applied force and driving point velocity were also measured. Intensity vectors for each of the three measurement areas are plotted in Fig. 26.13; the table in the same figure shows the contributions due to shear, bending, and twisting components of the intensity in two directions for each measurement position. The averaged power flow into the absorbing termination was estimated by assuming symmetrical behaviour across the plate and was found to agree with the averaged input power within 20%. It is hoped that, in the future, simplified, multipoint measurement methods for structural intensity determination will form the basis of an approach to the problem of vibration transmission mechanism identification in built-up structures.

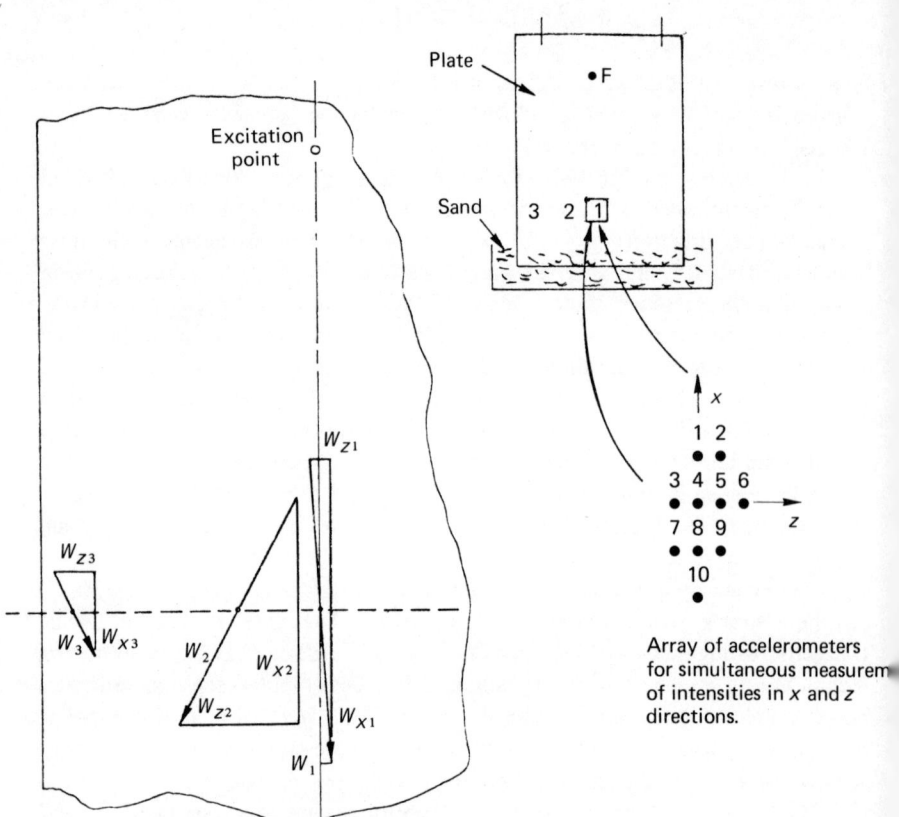

Array of accelerometers
for simultaneous measurem
of intensities in x and z
directions.

Position	Intensity W_x, N$\bar{\text{s}}^1$				Intensity W_s, N$\bar{\text{s}}^1$			
	shear	bending	twisting	total	shear	bending	twisting	total
1	−136.8	−104.0	−15.3	−256.1	10.4	−2.5	7.7	15.6
2	−91.8	−71.5	−28.0	−191.3	−25.5	−53.5	−30.6	−109.6
3	−61.2	−33.1	17.8	−76.5	20.4	35.7	−17.6	38.5

Fig. 26.13 − Distribution of measured intensities in a plate [26.15].

REFERENCES

[26.1] Ewins, D. J. (1975) *J. Soc. Environmental Engineers*, Dec. 3-12. Measurement and application of mechanical impedance data: Part I Introduction and ground rules.

[26.2] Ewins, D. J. (1976) *J. Soc. Environmental Engineers*, March 23-33. Measurement and application of mechanical impedance data: Part II Measurement techniques.

[26.3] Ewins, D. J. (1976) *J. Soc. Environmental Engineers*, June 7-17. Measurement and application of mechanical impedance data: Part III Interpretation and application of measured data.

[26.4] Harris, C. M. & Crede, C. E. (Eds.) (1961) *The shock and vibration handbook*. McGraw-Hill.

[26.5] Goyder, H. G. D. (1980) *J. Sound & Vib.*, **68(2)**, 209-230. Methods and application of structural modelling from measured structural frequency response data.

[26.6] Skudrzyk, E. (1958) *J. Acoust. Soc. Am.*, **30**, 1140-1152. Vibrations of systems with a finite or infinite number of resonances.

[26.7] Goyder, H. G. D. & White, R. G. (1980) *J. Sound & Vib.*, **68(1)**, 59-75. Vibrational power flow from machines into built-up structures: I Introduction and analyses of beam and plate-like foundations.

[26.8] Goyder, H. G. D. & White, R. G. (1980) *J. Sound & Vib.*, **68(1)**, 97-117. Vibrational power flow from machines into built-up structures. III Power flow through isolation systems.

[26.9] Skudrzyk, E. (1971) *The foundations of acoustics*. Wien: Springer-Verlag.

[26.10] Bode, H. W. (1945) *Network analysis and feedback amplifier design*. New York: Von Nostrand.

[26.11] Goyder, H. G. D. (1980) *Proc. Conference on Recent advances in structural dynamics*, ISVR, Southampton University. Some theory and applications of the relationship between the real and imaginary parts of a frequency response function provided by Hilbert transforms.

[26.12] Pinnington, R. J. & White, R. G. (1981) *J. Sound & Vib.*, **75(2)**, 179-197. Power flow through isolators to resonant and non-resonant beams.

[26.13] Goyder, H. G. D. & White, R. G. (1980) *J. Sound & Vib.*, **68(1)**, 77-96. Vibrational power flow from machines into built-up structures: II Wave propagation and power flow in beam-stiffened plates.

[26.14] Pavic, G. (1976) *J. Sound & Vib.*, **49(2)**, 221-230. Measurement of structure borne wave intensity. Part I: formulation of the methods.

[26.15] Pavic, G. (1976) PhD Thesis, ISVR, University of Southampton. Techniques for the determination of vibration transmission mechanisms in structures.

[26.16] Holmes, P. J. & White, R. G. (1972) *J. Sound & Vib.*, **25(2)**, 217–243. Data analysis criteria and instrumentation requirements for the transient measurement of mechanical impedance.

[26.17] Dyer, I. (1960) *J. Acoust. Soc. Am.*, **32**, 10, 1290. The moment impedance of plates.

APPENDIX 26.1

List of symbols for Table 26.1

ξ = Amplitude of harmonic displacement

θ = Amplitude of harmonic angular displacement

A = Cross-sectional area of beam

E = Modulus of Elasticity

ρ = Volume density

GQ = Torsional stiffness

J = Polar moment of inertia per unit length

I = Second moment of area of cross section of beam

B_P = Bending stiffness of a plate $= Eh^3/12(1-\nu^2)$

ν = Poisson's ratio

h = Plate thickness

a = Radius of disc over which moment is applied to plate

L = Parameter from [26.17] which tends to unity for large a/h

l = Length of finite beam

$l_1; l_2$ = Length of sides of finite rectangular plate

η = Loss factor

r, ϕ = Polar coordinates for plate

ω = Frequency in radians per sec.

CHAPTER 27

Vibration testing

R. G. White

Institute of Sound and Vibration Research, University of Southampton

27.1 INTRODUCTION

Theoretical analyses of structures are necessarily approximate because of the inherent limitation in dynamic modelling via representation by a finite number of degrees of freedom. The complexity of built-up structures also often precludes analysis in the time available to the designer. There is therefore a need for vibration testing, either to validate a mathematical model by comparing the measured parameters with calculated values, or to provide data for practical structures. There is also the intermediate requirement for a test in which the effects of deviations by structural modification from a previously proven design are to be studied. The ultimate objective in each case is, of course, the prediction of dynamic behaviour in service. The dynamic response of a structure is characterized by its modes of vibration, associated natural frequencies, and damping. Vibration testing is therefore often concerned with providing estimates of these detailed characteristics, although much experimental work now involves the measurement of scaled frequency response characteristics through broad frequency ranges.

Damping measurement is particularly important as the damping of structures cannot be predicted, although estimates may be made from previously measured data for similar structures. Calculations must always be based on experimentally determined values. The need for the accurate resonance test is therefore apparent, the degree of accuracy being important because unpredicted modes of vibration often appear in practice which might not be detected or could be dismissed as a spurious effect. An example of this type of activity is the flutter test used in the analysis of aircraft structures in which extremely accurate measurements of structural and aerodynamic damping are required.

Although the need for the accurate resonance test can be seen, it is sometimes the case that only approximate estimates of the characteristics of a particular troublesome mode of vibration of a structure are required, perhaps in order that a method of alleviating the problem be deduced. This is the *ad hoc* test in which the stringent requirements of the resonance test are relaxed for simplicity of technique and speed of data acquisition.

A further activity in the field is environmental testing in which the structure is subjected to a prescribed excitation in order that weakness of design or construction may be detected which would cause failure or unsatisfactory performance in service. This is therefore a type of proof test in which, for example, large responses in the expected excitation range may be searched for; however, this type of test, the resonance search, is usually included as one aspect of the wider specifications for the environmental test. In this type of testing, the structure could be a part of a built-up structure such as vehicle, or it could be electronic equipment. The range of systems to be tested is therefore wide, but the experimental procedure is essentially the same: to subject the structure to a prescribed excitation for a given duration and to interpret the response. Worthy of mention also in the classification of tests is the fatigue test which although outside the scope of this chapter, does require a vibration exciter, sometimes in conjunction with a resonant mechanical system, but dynamic analysis as outlined below is not involved. There are also many variations of the basic frequency response test which have been developed for research experiments, for instance using acoustic excitation [27.1].

Most of the vibration testing carried out in practice falls into the category of frequency response measurement. Techniques and analysis procedures for vibration testing are outlined here. Structural frequency response measurement is principally considered for reasons given above, but for completeness some attention is given to environmental testing.

27.2 STRUCTURAL TESTING

Structures, because they are continuous systems, possess an infinite number of degrees of freedom. The assumption of a finite number of degrees of freedom simplifies theoretical analyses, but the problem of analysis by dynamic testing still appears complicated. The problem which confronts the experimenter is to excite the system, measure the response, and present the derived information which should be readily understood. The assumption of even a limited number of degrees of freedom is not helpful in this direction. The analysis of multi-degree of freedom systems is, however, generally simplified by applying single degree of freedom system theory to each resonance, the complete response being represented by superposition. This representation is applicable of course only to linear systems, and the assumption of linearity is made here, as is almost always the case in experimental testing, because the dynamic analysis of non-linear systems is complicated both theoretically and experimentally. However, as structures behave linearly for small deflections, if care is taken to generate only low response levels then nonlinear effects may be ignored [27.2]. Using the single degree of freedom system analogy for response in each mode of vibration, measurements of natural frequency and damping may be made by simple analysis procedures. Some deficiencies in this simplification are apparent, however, because the principal modes of practical structures may be coupled by the damping present, and the closeness of resonances may cause overlapping of responses in several modes.

The latter point, closeness of resonances, is caused by the high modal density of practical structures, particularly if modes of several types associated perhaps with flexural and torsional vibration occur. The effects of close resonances may to some extent be overcome by the use of improved analysis procedures, and response predominantly in one mode of vibration may be elicited by the use of multi-point excitation. The damping mechanisms present in structures are not completely understood, but it is usually assumed in resonance testing, by statement to this effect or more often lack of theoretical argument, that the modes are uncoupled. One of the principal conclusions of a critical study of the theory of resonance testing [27.3], which also reviewed much experimental work, indicates that it is reasonable to assume that modes are not coupled because of the lack of experimental evidence to the contrary. This generally accepted assumption is made here.

The response of a structure may be specified in terms of its mode shapes, natural frequencies, and damping. Natural frequencies and damping may be derived by frequency domain analysis. Mode shape measurement obviously involves spatial analysis, but this is generally achieved by repeating frequency domain analysis at various points on a structure. The frequency response characteristics of the single degree of freedom system will therefore first be examined to indicate how the required data may be derived from the frequency response test.

Frequency response is the relationship of the response to steady-state sinewave excitation. The frequency response function depends on the form of damping assumed. Two simple means of representing the damping of structures are in common usage: these are viscous and hysteretic. The two representations give the dimensionless frequency response function $H(i\omega)$ of a single degree of freedom system as

$$H(i\omega)_{\text{VISCOUS}} = \frac{1}{1 - \dfrac{\omega^2}{\omega_n^2} + i\,2\zeta\,\dfrac{\omega}{\omega_n}}$$

$$H(i\omega)_{\text{HYSTERETIC}} = \frac{1}{1 - \dfrac{\omega^2}{\omega_n^2} + i\eta}$$

$$\text{(27.1)}$$

where ω = frequency
   ω_n = undamped natural frequency
   ζ = viscous damping ratio
   η = loss factor.

If the damping is light, as is often the case in built-up structures, then the difference between the shapes of the response curves in the region of resonance due to the assumption of either mechanism is negligible [27.4]. The hysteretic mechanism has been postulated as being most representative of structural damping, but the viscous model is probably most often used in experimental data reduction. The similarity of the response curves in the region of resonance enables response prediction to be made according to either notation by observing that for light damping, $\eta \equiv 2\zeta$.

27.2.1 Analysis procedures
The frequency response function may be plotted as modulus and phase versus frequency as shown in Fig. 27.1, for a viscously damped system. This leads to the simplest approach in resonance testing known as the 'peak amplitude method'.

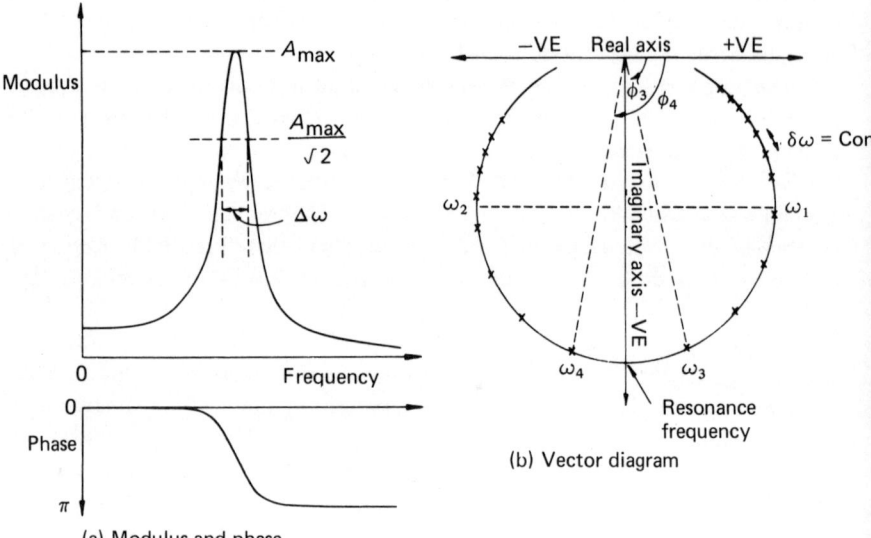

(a) Modulus and phase
(b) Vector diagram

Fig. 27.1 – Frequency response function of a single degree of freedom system.

It is assumed that the resonance frequency, the frequency at which maximum response occurs, is a natural frequency of the system. Maximum displacement response occurs at

$$\omega = \omega_n \sqrt{(1 - 2\zeta^2)} \qquad (27.2)$$

which gives a satisfactory indication if the modes are not close and damping is light. If the damping is heavy the maximum response will not occur at the natural frequency, and if close natural frequencies are present, the responses may coalesce in the modulus versus frequency diagram. Measurement of damping is achieved in this type of presentation by measuring the bandwidth of the curve at the level where the response is maximum response $/\sqrt{2}$ as shown in Fig. 27.1(a). The viscous damping ratio is then given by

$$\varsigma = \frac{\omega_2 - \omega_1}{2\omega_n} = \frac{\Delta\omega}{2\omega_n} \tag{27.3}$$

where $\Delta\omega$ is known as the 'half power point bandwidth'.

The basic shortcoming of the peak amplitude method is that phase information is neglected. Kennedy & Pancu [27.5] recognized that the high rate of change of phase in the region of resonance could be used to facilitate a better form of response presentation. It is accepted that the Kennedy & Pancu method appears to be the best available at present and is most widely used in structural testing, the complex response being plotted as a 'vector diagram'.

In the vector diagram, the modulus and phase information is presented as one curve by plotting the real part of the frequency response function versus the imaginary part. The vector diagram for a lightly damped, single degree of freedom system is approximately circular in shape, and if the response vector is plotted at equal frequency increments, the natural frequency is indicated at the point on the curve where the spacing between the plotted points is a maximum, as shown in Fig. 27.1(b). Damping ratios are measured by fitting the best circle through the plotted points for each resonance, the half power point bandwidth being noted by drawing the diameter normal to the resonance diameter. The damping ratio is then measured according to equation (27.3). The improved method developed by Mead [27.6] may also be used in which the phase difference between two plotted points close to maximum response is measured, whence according to Fig. 27.1(b),

$$\varsigma = \frac{\omega_4 - \omega_3}{\omega_n} \cdot \frac{1}{\phi_4 - \phi_3} \tag{27.4}$$

where ϕ_4 and ϕ_3 are the phase angles at ω_4 and ω_3 respectively. It has been shown by Pendered [27.7] that the Kennedy & Pancu method is the best available for resolving close natural frequencies, it being superior to the peak amplitude and other methods – an example is illustrated in Fig. 27.2. Two close natural frequencies are present, and damping measurement by the peak amplitude method in Fig. 27.2(a) is impossible. The existence of the two resonances is much clearer in Fig. 27.2(b), shown by the arc length maxima at A and B. Damping measurement is possible from the vector diagram by dawing the best circle through each loop of the curve, either manually or by a computer-based curve fitting technique [27.8], and using the phase difference method above. The advantage of using the vector diagram method on structures with close natural frequencies has been demonstrated in [27.4] where experience showed that the use of this method greatly reduces the likelihood of missing a resonance in practical tests.

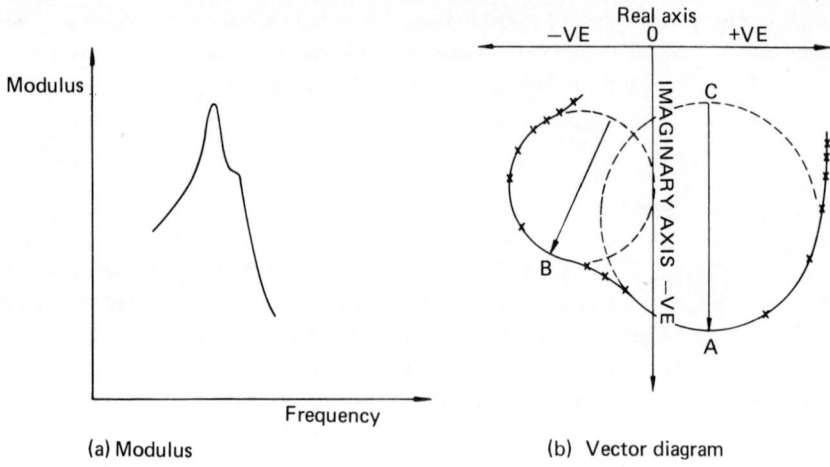

(a) Modulus   (b) Vector diagram

Fig. 27.2 – Frequency response function of a system with close natural frequencies.

Although the above discussion has been concerned with frequency domain analysis, it must be noted that it is possible, perhaps by one of the oldest and most simple forms of analysis, to obtain damping estimates from time data. If the free vibration of a single degree of freedom system is examined, the damping ratio may be measured according to the logarithmic decrement method. For a lightly damped system

$$\zeta = \frac{\log_e \dfrac{x_1}{x_n}}{2\pi n} \tag{27.5}$$

where x_1 and x_n are the amplitudes of free vibration spaced n cycles apart. This simple method does still have its place in vibration testing of single degree of freedom systems and is often used to enable rapid measurements to be made in repeated experiments for parameter studies [27.9, 27.10]. Refined electronic techniques have been evolved for processing free vibration data [27.11] from simple systems. Decrement type analysis for damping measurement may also be carried out on decaying waveforms from multi-degree of freedom systems [27.12]. Fourier transforms of equal data record lengths but with their starting points separated by time t are evaluated. The modulus spectra are calculated and the levels of each resonance in the spectra noted. If f_i is the natural frequency of the mode of interest and the maximum levels of the modulus spectra are $|F(f_i)|_1$ and $|F(f_i)|_2$ for the two successive transforms, then the damping ratio is given by

$$\zeta_i = \frac{\log_e \dfrac{|F(f_i)|_1}{|F(f_i)|_2}}{2\pi f_i t} \qquad (27.6)$$

The method is free from inherent error caused by dynamic range considerations and has been shown to work for a simulated system with ten natural frequencies using dynamic signal ranges as low as 2 dB [27.12]. The ease with which Fourier transforms of time data may be evaluated makes the method attractive, particularly in tests where sustained vibration might cause undesirable effects such as heating or fatigue.

27.2.2 Mode shape measurement

The simple approach to mode shape measurement is to excite the structure at a single point and measure the response at various stations by traversing the response transducer across the structure. However, if a single exciter is used, some resonances might not be excited because the excitation point could be at a node for some modes of vibration. The peak amplitude method of analysis is not completely satisfactory for multi-degree of freedom systems because off-resonant contributions from modes other than that of interest are included in the amplitude diagram. Again the vector diagram may be used to advantage as it offers a method for separating response in the resonant mode from the total measured response. This is illustrated in Fig. 27.2(b). The complete response at resonance in mode A is given by the vector OA, but the diameter CA of the circle represents peak response in that mode alone. The extraneous vector OC generally varies in magnitude from point to point on a structure. A displaced origin for the vector diagram in each mode may thus be constructed by fitting the best circle through each loop of the curve. Good estimates of mode shapes may be obtained by plotting the vector diagram at various stations on the structure. The phase information at resonance is derived from the orientation of the vector diagram on the axes at each response measurement point. This is of particular relevance in resonance tests on multi-bay structures which have groups of close natural frequencies, sometimes having an equal number of half waves in each bay, but with a different phase relationship between the motion in adjacent bays.

The above simple technique is easily applied in practical tests but the contribution in the vector diagram from the mode of interest is often small. Another approach is to use a complicated excitation system composed of many exciters with a relative spatial magnitude and phase distribution of force such that response predominantly in one mode is produced at the resonance frequency of that mode. It must be noted, however, that a given mode need not be excited at its resonance frequency in the multi-excitation technique, for if the force distribution is correct, it should be possible to produce response purely in one mode at any frequency. This is a crucial test of the technique. The ability to excite a pure mode of vibration enables the signal-to-noise ratio to be maximized and a 'circular' vector diagram to be produced at each measurement station. The method has been investigated by several workers. Lewis & Wrisley [27.13] theoretically analysed, designed, and constructed a system using twenty-four exciters and transducers which was successfully applied to excite pure modes of structures, the necessary force magnitude and phase distribution being iteratively achieved by manual operation. A fully automated servo-controlled system has been developed [27.14], but experience has shown that for practical tests a semi-automated system with automatic frequency control and manual control of force distribution is most useful, and such a system has been built [27.15] which used five exciters at frequencies up to 1 kHz. An example of its use is given in [27.16] which shows the improvement in mode shape and vector diagram analysis compared with single point excitation, the tests being carried out on a motor car body.

27.2.3 Scaled force-response relationships
The preceding discussions has been concerned with frequency response, a dimensionless quantity, and this approach suffices to illustrate analysis procedures. In practice, however, a scaled quantity is measured, that is, the relationship between excitation and a particular physical response. Considerable confusion exists, owing to the profusion of terms used to describe the excitation-response relationship, and indeed there are many relationships which could be measured, for instance the strain response to a given acoustic pressure. For three-dimensional systems, displacement, velocity, or acceleration may be measured in the three coordinate directions, also rotational responses to a given force may be measured; this is of a particular importance, for example, in the investigation of machinery seatings. Some current definitions [27.17] used in frequency response testing are given in Table 27.1.

It is important that conventional terminology should be used, as in the case of impedance and mobility, for example, there are direct analogies between electrical and mechanical quantities which permit the use of common concepts in the prediction of coupled system response. Impedance magnitude characteristics of structural elements and of a single degree of freedom system are given in Fig. 27.3. Methods for the assessment and interpretation of measured frequency response data are very well explained in [27.18, 27.19, 27.20].

There are three further definitions describing the positions and directions of the measured force and response which apply to all of the six quantities noted above. For example, with reference to mobility, the point mobility concerns the velocity measured at the same point and in the same direction as the applied force; the transfer mobility uses the velocity at any point other than that at which the force was applied. Finally, the cross-mobility uses the force and velocity at the same point but acting in orthogonal directions, e.g. the horizontal velocity due to a vertical force.

It is apparent from above that measurement of apparent mass or inertance would yield all the other noted forms of force-response relationship via
integration. There are unfortunately always practical considerations concerning
the choice of response transducer and dynamic range or signal-to-noise ratio
limitations which restrict the frequency range of accurate integration.

Table 27.1

Definitions of force–response relationships

$\dfrac{\text{Force}}{\text{Displacement}}$	DYNAMIC STIFFNESS
$\dfrac{\text{Displacement}}{\text{Force}}$	RECEPTANCE
$\dfrac{\text{Force}}{\text{Velocity}}$	IMPEDANCE
$\dfrac{\text{Velocity}}{\text{Force}}$	MOBILITY
$\dfrac{\text{Force}}{\text{Acceleration}}$	APPARENT MASS
$\dfrac{\text{Acceleration}}{\text{Force}}$	INERTANCE

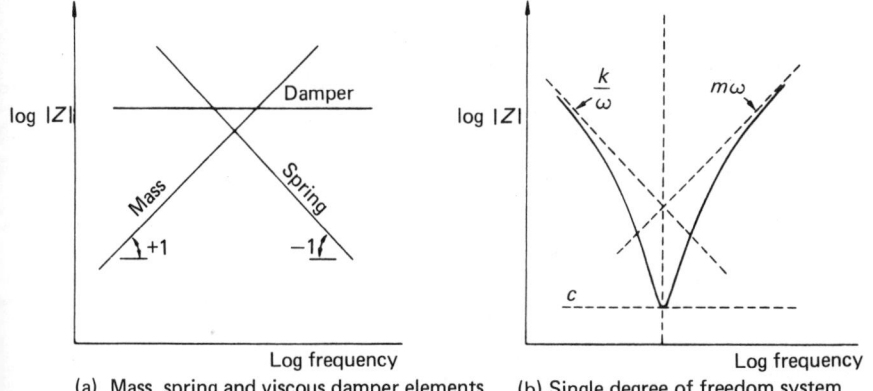

(a) Mass, spring and viscous damper elements   (b) Single degree of freedom system

Fig. 27.3 – Mechanical impedance characteristics.

27.2.4 Test techniques

The frequency response function of a system is the steady-state response to sine-wave excitation. The same function also relates the frequency content of transient excitation to the frequency content of the transient response. It is apparent, therefore, that it is possible to carry out frequency response tests in the steady and transient states. These techniques are discussed here together with the quasi-steady state method.

27.2.4.1 *Steady and Quasi-steady state methods*

In the steady-state, the structure is excited by a sinewave of constant amplitude and frequency; when steady-state response has been achieved, the amplitude and phase of the response with respect to the excitation are noted. This method is tedious and time-consuming if many resonances of a lightly damped system are to be examined, and is not often used. The excitation system and response

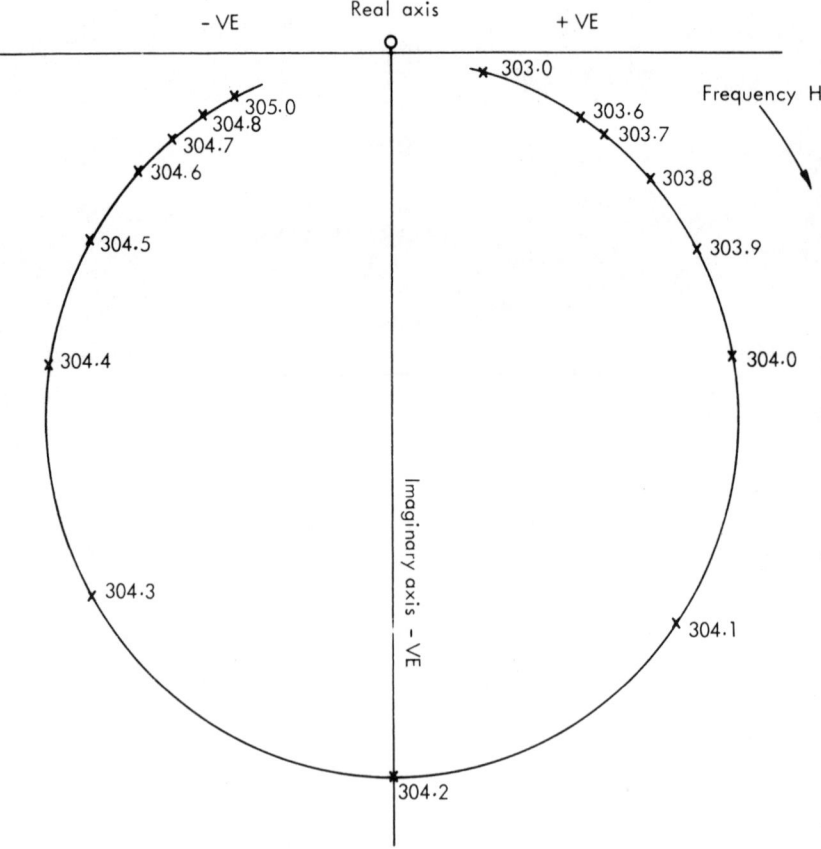

Fig. 27.4 – Vector diagram derived from a steady-state test on a structure.

measurement instrumentation are required to be free from drift over long periods. An example of a vector diagram plotted from the strain response of a structure in a steady-state test is given in Fig. 27.4, where resonance is indicated at 304.2 Hz. The accuracy achieved in the steady-state test using manual control of instrumentation is, of course, dependent upon the analysis procedures and instrumentation used, but also to a large extent on the skill and patience of the operator. The steady-state method has, however, been refined recently by the introduction of computer control [27.21], and incremental frequency response measurements may be made automatically through chosen frequency ranges. There is, however, often insufficient time for steady-state testing, and for this reason the quasi-steady state test is used.

In the quasi-steady state technique, the excitation frequency is slowly varied through the range of interest, and it is assumed that the system response attains steady-state levels. It is possible, however, for quite serious errors to occur in the derived frequency response data because the response may not reach steady-state values if the sweep rate is too high, and the characteristics of the function analyser used in the analysis may cause errors in the plotted response vector. The response of a single degree of freedom system subjected to a frequency varying linearly with time has been examined by Hok [27.22] who showed that the principal effects of varying the excitation frequency are that the response at resonance is less than the steady-state maximum and the frequency at which maximum response occurs is shifted in the direction in which the excitation frequency is changing. If the sweep rate is too high, the measured damping coefficient will also be greater than the true value. These effects are indicated in Fig. 27.5 for a system excited by a frequency which is increasing through the range of interest. If the sweep rate is very high, the beating effect in the high frequency region of Fig. 27.5 occurs because of interaction between the free and forced vibration of the system. For sweeps with frequency decreasing through the range of interest, the effects in the frequency domain are reversed. Hok showed that the apparent bandwidth, that is the apparent damping of a given resonance, of the measured response is related to a single, nondimensional parameter defined as

$$\bar{\alpha} = \frac{\zeta \omega_n}{\sqrt{(2\dot{\omega})}} \qquad (27.7)$$

where $\dot{\omega}$ is the rate of change of excitation frequency with time. The larger the parameter $\bar{\alpha}$, the smaller the errors introduced by the effect of the sweeping excitation. It has been suggested that $\bar{\alpha}$ should be maintained in the range of 1.4 to 4.0 for errors in damping to be approximately 10% and 1% respectively. Some refinements in technique were proposed by Reed [27.23] who developed the 'λ law' sweep in which the percentage change in frequency per cycle is constant. If several modes are present in the frequency range, then the variation of frequency with time may be made such that $\bar{\alpha}$ is always equal to or greater than the chosen critical value. It has been shown that a given frequency range may be swept faster by this function than by logarithmic or linear variation or frequency with time, the predicted accuracy being the same in each case. The inherent averaging time of the resolver used to derive vector diagrams does, however, also cause similar effects to those noted above [27.24].

The errors due to sweep rate can be significant and may only be allowed for if the system damping, or at least the lightest damping present, is known in advance. In practice, an arbitrary sweep rate is often chosen, the test conducted and then repeated at a lower rate to observe effects in the plotted response. A response–dependent, variable sweep rate technique has been established which improves the accuracy of the quasi-steady state method [27.25]. The technique involves sweeping at a variable sweep rate such that the rate is low through resonance bandwidths, and high elsewhere. A servo-controlled system was used in [27.25] to detect resonant response and the sweep rate adjusted accordingly; the results were encouraging, a significant reduction in sweep time being achieved by this method for a given measurement accuracy.

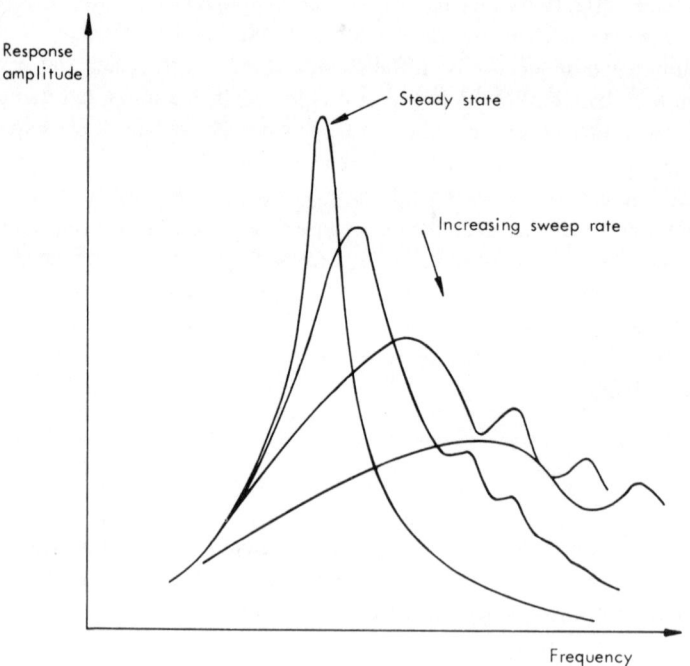

Fig. 27.5 – Effect of sweep rate on the response of a resonant system in a quasi-steady state test (frequency increasing through range of interest).

27.2.4.2 *Transient testing*

Interest in transient testing as a quantified structural test technique arose due to the development of digital computer-based data analysis systems with fast Fourier transform (FFT) routines. Some initial work concerning consideration of pulse-type forcing functions for transient testing [27.26] showed that pulses of simple shape are not completely suitable for excitation in structural testing because of zeros or minima in the modulus spectra which give regions of zero or low response and uncertainty in measured data derived by division of Fourier transforms. It is also extremely difficult, if not impossible, to control simple pulses, either using mechanical impact or electrodynamic force generators in order to achieve desirable energy spectrum characteristics. The linear rapid frequency sweep, which has an essentially flat modulus spectrum and high cut-off rates at the starting and stopping frequencies, has been found to be highly suitable for this purpose, the excitation being applied to the test structure through a conventional power amplifier − electrodynamic exciter − force transducer system. The setting-up time for this equipment is the same as for the other types of tests mentioned above, although hand held exciters have been used to advantage [27.27]; but the total time during which data must be acquired is very short, being typically about 5 seconds maximum. Use of the rapid frequency sweep test technique in the dynamic analysis of structures is discussed in this section and it is demonstrated how scaled frequency response measurements can be obtained form the test data, the method having been used quite widely and become established as a practical procedure.

Now, the frequency response function, $H(i\omega)$ of any linear system may be derived from the response $y(t)$ to any transient excitation $f(t)$ according to

$$H(i\omega) = \frac{Y(i\omega)}{F(i\omega)} \tag{27.8}$$

where $F(i\omega)$ and $Y(i\omega)$ are the Fourier Transforms of $f(t)$ and $y(t)$ respectively; that is, in integral form, if $f(t) = 0$ and $y(t) = 0$ for $t < 0$, (in the practical case if the structure under test is initially at rest and the excitation is applied at $t = 0$),

$$F(i\omega) = \int_0^\infty f(t)e^{-i\omega t} \, dt \text{ and } Y(i\omega) = \int_0^\infty y(t)e^{-i\omega t} \, dt \ . \tag{27.9}$$

Although correlation and cross-spectral density methods [27.28, 27.29, 27.30, 27.31] may be used in the analysis of signals from structural tests, the direct Fourier transform division approach outlined above is the procedure which is most often applied in the measurement of scaled frequency response data. Some limitations and sources of error in structural testing using rapid frequency sweep excitation and this type of analysis are discussed in detail below; a full description of the method, including a simple guide to practical testing, is given in [27.32]. However, before proceeding further, some justification of the very basic assumption of linear dynamic behaviour during transient testing must be made because both the data analysis procedures discussed and the representation of dynamic structural characteristics depend on this condition.

In transient tests, the level of the response in each structural mode cannot be readily controlled by the operator. If, for instance, a wide frequency range is swept, then the level of the response at each resonance is not apparent until the test data have been analysed, and at that stage it may not be possible to repeat the test if it has been found that the response has been forced into the nonlinear region. For these reasons an investigation has been carried out [27.33] into the effects of nonlinearity due to large deflections on the transient measurement of the frequency response characteristics of structures. A perturbation solution and a more accurate computer solution were used to derive the impulse response of a structure. The existence of harmonics of the natural frequency (based on linear theory) in the impulse response and changes in the level and damping ratio of the frequency response function, derived from the Fourier transform of the impulse response, were predicted if the response amplitude should extend into the nonlinear region. It was also shown that the vector diagram could become severely distorted if the structural response becomes nonlinear and that measurements of damping ratio and natural frequency would be made difficult, if not impossible. Steady-state tests on a structure showed that the theoretically predicted effects [27.2] of nonlinearity on $H(i\omega)$ do occur at large deflections; it was noted, however, in transient tests on the same structure, that even at very moderate sweep rates, the response amplitude did not reach a sufficiently high level for nonlinear effects to become apparent. It may therefore be concluded that structures may be considered as linear systems in rapid frequency sweep testing because the response may only become nonlinear if such low sweep rates are used that the response reaches almost steady-state conditions, and this state will never be approached in the transient test.

The frequency sweep, or swept sinewave, is defined by

$$f(t) = F_0 \sin \Phi(t) , \qquad 0 \leqslant t \leqslant T . \tag{27.10}$$

The function is therefore of constant amplitude and time dependent frequency, the instantaneous variation of frequency with time being given by

$$\omega(t) = \frac{d\Phi(t)}{dt} . \tag{27.11}$$

The use of the swept sinewave in structural frequency response measurement has been examined [27.26], and the linear frequency sweep has been shown to be most suitable as a transient forcing function because the modulus spectrum is 'rectangular', for example, see Fig. 27.6. For frequency ranges of approximately one decade, sweep times, T, of the order of one second have been shown to be sufficient to allow enough energy to pass into structures to produce measurable response levels. A linear variation of frequency with time is given by

$$f(t) = F_o \sin (at^2 + bt) \tag{27.12}$$

where $\omega(t) = 2at + b$ and $d\omega(t)/dt = 2a$, with $a = (\omega_2 - \omega_1)/2T$, $b = \omega_1, \omega_1$ and ω_2 being the initial and final frequencies.

To examine the spectral characteristics of the function defined in equation (27.12) it is necessary to evaluate the Fourier transform, defined by

$$F(i\omega) = \int_{-\infty}^{\infty} f(t)e^{-i\omega t} \, dt \ . \tag{27.13}$$

The Fourier transform of the linear frequency sweep is evaluated in [27.34], and the mean modulus spectrum level between the initial and final frequencies is

$$\overline{|F(\omega)|} = F_o \sqrt{\frac{\pi}{4a}} \ . \tag{27.14}$$

The above characteristics are demonstrated in the modulus spectrum of the Fourier transform of a sweep from 20 to 220 Hz in 1 s shown in Fig. 27.6, and the flatness of the spectrum between the peaks at the extremities is apparent. The predicted mean spectrum level from equation (27.14) and the predicted levels [27.34] at frequencies f_1 and f_2 are also shown in the same figure. It can be seen that there is good agreement between the predicted and measured characteristics. Thus, in rapid frequency sweep tests, the excitation level may be predicted, if required.

Equation (27.14) may be rewritten to show that the modulus spectrum level of the linear rapid frequency sweep is

$$\overline{|F(\omega)|} = \frac{F_o}{\sqrt{2}} \sqrt{\frac{\pi}{\dot{\omega}}} \tag{27.15}$$

where $\dot{\omega}$ is the rate of change of frequency with time. Thus, if a given bandwidth is swept, the spectrum level within the bandwidth is controlled by F_o and the time taken to sweep through the bandwidth. This characteristic has been examined further in [27.34] where a technique has been demonstrated for shaping the modulus spectrum of rapid sweeps via use of time-varying sweep rates. However, the linear rapid frequency sweep is generally used in structural frequency response measurements.

(a)　Waveform

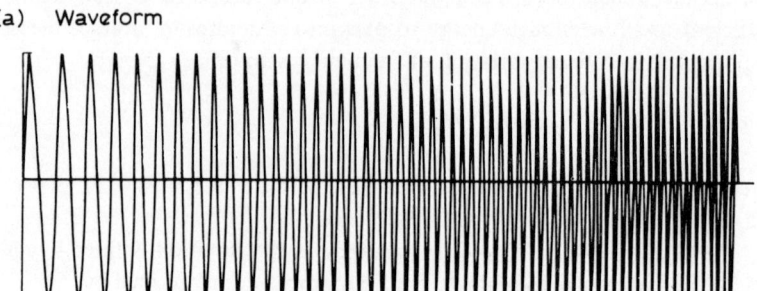

(b) Fourier transform

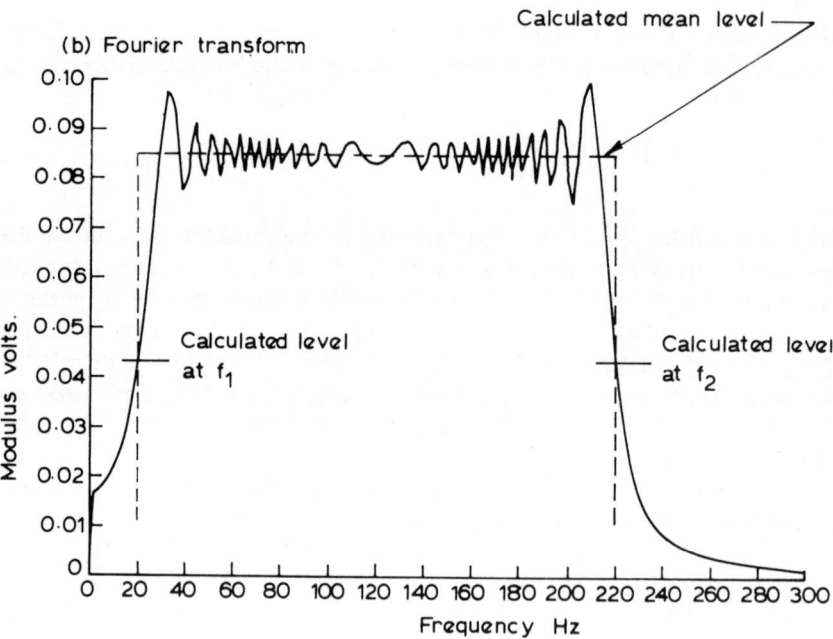

Fig. 27.6 – A rapid frequency sweep and its Fourier transform.

The displacement response of a single degree of freedom system to a linear sweep is obtained by convolving the forcing function $f(t)$ with the impulse response, $h(t)$ using the convolution integral

$$x(t) = \int_0^t f(t-\tau)\, h(\tau)\, d\tau \tag{27.16}$$

where $f(t) = F_0 \sin(at^2 + bt)$,

$$h(t) = \frac{1}{m\omega_d} \cdot \exp(-\zeta\omega_n t) \sin \omega_d t \;,$$

$\omega_n =$ the undamped natural frequency

$$\omega_d = \omega_n \sqrt{(1 - \zeta^2)}$$

$\zeta =$ the viscous damping ratio.

This integral is evaluated in [27.32] which gives a solution in terms of 'error functions'. Using asymptotic values the maximum response amplitude is found to be

$$x_{max} = \frac{F_0}{2m\omega_d} \sqrt{\frac{\pi}{a}} \tag{27.17}$$

where $m =$ the mass.

This is exactly the same as that of a single degree of freedom system excited by an impulse of magnitude.

$$I = \frac{F_0}{2} \sqrt{\frac{\pi}{a}} \tag{27.18}$$

which is the modulus of the Fourier transform of the forcing function (equation 27.14). Thus it can be said that a rapid frequency sweep impulsively excites structural resonances, and the maximum response is independent of the shape of the forcing function and only dependent upon the modulus of the Fourier transform in the region of the resonance frequency.

A typical system for structural testing by the rapid frequency sweep method is given in Fig. 27.7. The swept sinewave is generated by a sweep oscillator which has the facility for setting the initial frequency, final frequency, and sweep time. Instrumentation developed at the ISVR enables sweeps to be made through any part of a 0 to 20 kHz frequency range with a variety of sweep times from 0.25 s to 25 s. The instruments also have a repetitive condition with an adjustable delay up to 5 s between sweeps to enable preset rapid sweeps to be applied periodically to the structure for the purpose of setting amplifier gains, charge amplifier sensitivities, etc., during preliminary tests prior to data acquisition or for time domain averaging of signals, as described later. The interval between sweeps must be chosen such that the structural response has decayed and become immeasurable before the onset of the next sweep. Assume, for the present, however, that the intention is to generate test data due to one sweep using the system of Fig. 27.7.

The objective is to acquire two signals of the form shown in Fig. 27.8, one proportional to force and the other proportional to response, which are un-distorted, have as high a signal-to-noise ratio as possible, and which, when Fourier transformed, enable, via their spectral quotient, to yield estimates of one of the six scaled, complex frequency response quantities given in section 27.2.3. The instrumentation in Fig. 27.7 is essentially conventional with charge amplifiers and voltage amplifiers used to give suitable signals of voltage levels compatible with A to D converter characteristics (or tape-recorder in field trials).

The use of charge amplifiers is generally recommended because of their calibration insensivity to cable length between transducers and their input; this is not the case for voltage preamplifiers, i.e. cathode or emitter followers. Anti-aliasing (low-pass) filters are included as is always the practice in digital data processing techniques. In the rapid frequency sweep test, although the cut-off rate of the modulus spectrum of the sweep is high at the upper frequency limit, there is some spectral content above ω_2; there may also be harmonic distortion in the excitation system.

The block diagram in Fig. 27.7 indicates the sweep oscillator as a separate instrument. Such an oscillator has been developed at the ISVR specifically for transient structural testing, and its controls are arranged in a suitable way for this purpose. A very useful combined instrument has also been built which contains all the required instrumentation for rapid frequency sweep testing, except for the vibrator and force transducer. Specifications of the sweep oscillator and the combined instrument are given in [27.32].

The system shown in Fig. 27.7 would enable apparent mass or inertance to be measured. Single or double integration of the response signal, either in the time or frequency domain, would enable the other four quantities given in section 27.2.3 to be derived. Analogue integration is often carried out via use of integrators combined with the preamplifier, but accurate integration is usually only achieved over a very narrow frequency range − both correct magnitude and phase weighting are required in the 'integration' process. It should perhaps be noted at this stage that the wave shapes of the signals, and hence their *complex* spectra, have to be preserved as accurately as possible in the frequency range of interest in order that reasonable estimates of structural frequency response be obtained. The other approach to the integration problem is always to acquire signals proportional to acceleration into the digital analysis system and perform the integration in the frequency domain by division of the complex acceleration spectrum by $i\omega$ or $-\omega^2$ to obtain velocity or displacement spectra. It is the author's experience that, because digital operations are being carried out, the experimenter tends to forget that many of the problems inherent in analogue integration occur in the equivalent digital weighting. However, integration in the frequency domain is generally superior to analogue conditioning; wider bandwidths of acceptable integration can be achieved, and data manipulation to overcome drift and associated problems is performed more readily than in the time domain.

The preceding discussion shows how, in principle, instrumentation is set up to carry out a rapid frequency sweep test. The intention is to apply a sweeping force to the structure through the required frequency range and to acquire the force and response signals, as much as possible of the data manipulation and conditioning being carried out digitally to produce accurate estimates of frequency response data. The following sections describe some practical considerations and limitations of the method.

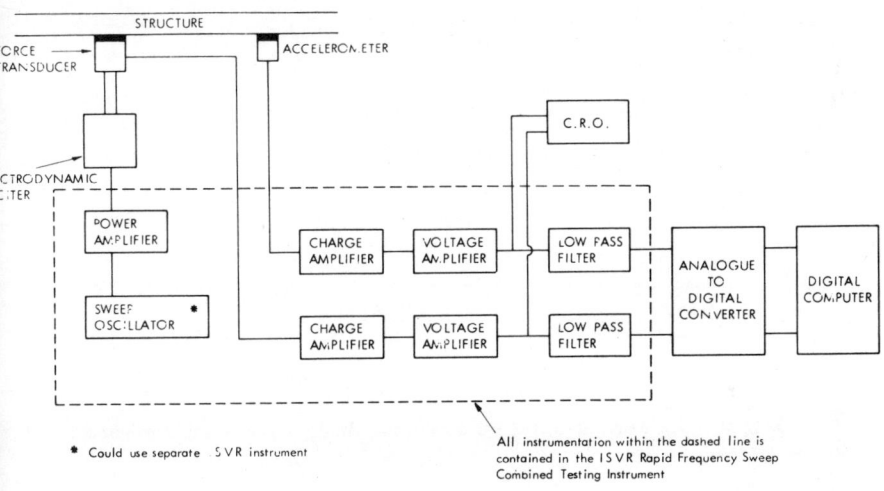

Fig. 27.7 – A system for the measurement of inertance or apparent mass by the rapid frequency sweep method.

The dynamic ranges of transient data derived from the rapid frequency sweep test are not the same in the time and frequency domains, as is the case in steady-state and slow sweep testing. In the transient testing of lightly damped structures the response signal will always be of longer duration than the force signal because of the free vibration components of the response; see Fig. 27.8, for example. It is not possible to take infinitely long data record lengths as required by equation (27.9), and obviously the response record will be truncated in the time domain, and this will influence frequency response characteristics deduced via frequency domain analysis.

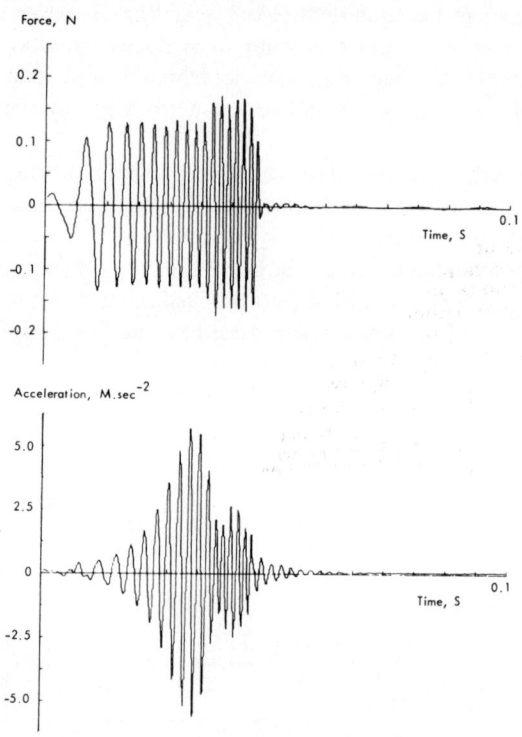

Fig. 27.8 – An example of typical waveforms obtained from a rapid frequency sweep test.

The effects of truncation may be examined by studying the frequency response function $H(i\omega)$ of a single degree of freedom system derived via Fourier transformation of its impulse response, the record length being limited to a finite integration time. The vector diagram derived from the truncated impulse response has a form such that the resonance frequency may be accurately identified as the frequency at which the rate of change of arc length with frequency is a maximum, as in the ideal case, but the damping derived from the truncated time function will be greater than the true value [27.1, 27.26]. The effect is clear in Fig. 27.9 which shows, for example, that if the dynamic range of the measuring, recording, or analysis system is greater than 40 dB then the error in measured damping ratio will be less than 10%. Thus in a practical test the dynamic range of the envelope of the response signal would be observed in order to avoid significant truncation effects.

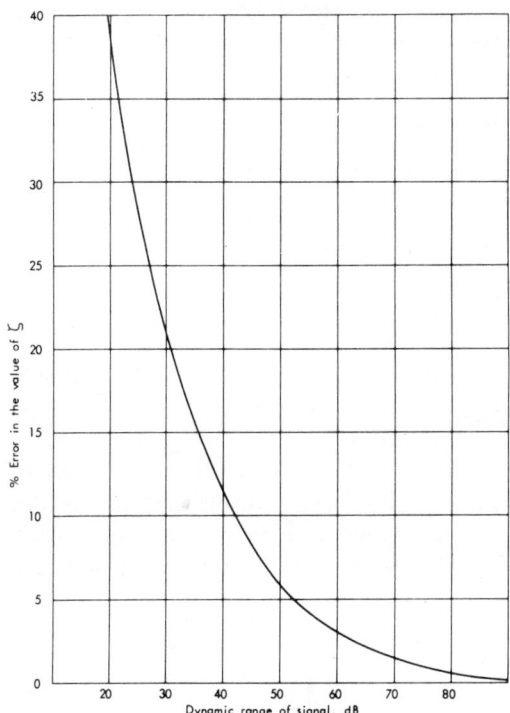

Fig. 27.9 – Percentage error in damping ratio measured from the vector diagram produced from time data with limited dynamic range.

The frequency response plots produced by a digital computer are not continuous but consist of discrete values spaced at equal frequency increments, the magnitude of the increment being termed the 'frequency resolution'. The frequency resolution must be fine enough to give a good representation of rapid changes in impedance level, for example at a lightly damped resonance.

Now, the signal record length controls the frequency resolution, and a technique often employed in transient analysis is to extend the time record by zeros within the computer. This will increase the resolution and cause interpolation of the frequency data. It will not increase the accuracy of the measured data if they are derived from truncated signals, but it is a better technique for increasing resolution in a plot than to acquire extra durations of noisy time data into the computer after the signals have decayed and become immeasurable. Conversely, if a structure is very lightly damped, the duration of its free vibration response could exceed the maximum record length which can be Fourier transformed by the analyser. In some circumstances this problem can be overcome by using multi-part Fourier transforms; the technique is discussed in [27.32].

The results of theoretical studies and complementary experimental work suggest that impedance, for example, can be measured by the rapid sweep technique to within the following limits:

Natural frequency	$\pm 1\%$
General impedance level away from resonance and anti-resonance peaks	$\pm \frac{1}{2}$ dB

Impedance level at resonance and anti-resonance peaks:

for $\zeta > 0.001$	± 3 dB approx
for $\zeta < 0.001$	± 5 dB approx
Phase angle	$\pm 3°$

The natural frequencies of built-up structures are not always well separated, and the problem of resolving close natural frequencies is of paramount importance. The problem is that the frequency response functions of close resonances may coalesce in the data plotted as a result of Fourier transformation of time signals from a test.

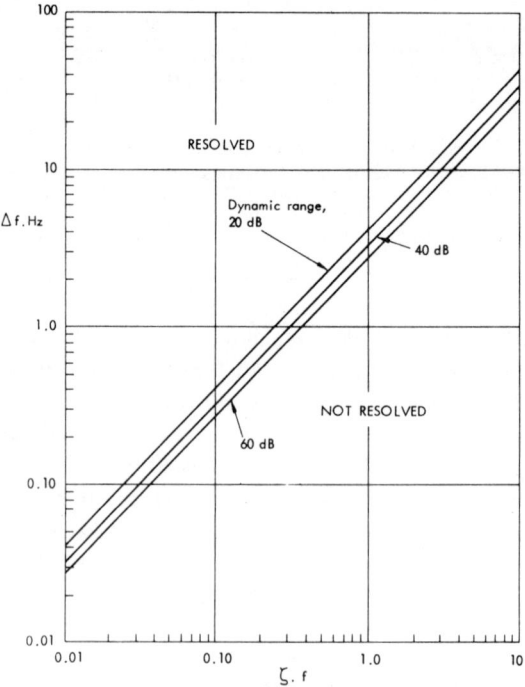

Fig. 27.10 – The resolution limits in transient testing.

A study is reported in [27.35] which considered the resolution problem which depends upon the relative modal contributions in terms of magnitude and phasing, damping, frequency separation, and the dynamic range of the acquired signal. A criterion for the worst case of resolution of two close natural frequencies for a given dynamic range has been established, and the resolution limits are presented in Fig. 27.10, where ζf is associated with the greater of the two decays present in the impulse response.

Point and transfer function measurement (with reference to mobility)
Measurement of point frequency response functions is very often carried out via use of the so-called 'impedance head' which is a combined transducer with acceleration- and force-indicating elements. It is sometimes observed that the point mobility of a heavy structure has a stiffness characteristic imposed upon it at high frequencies. This can have two sources that are related to the impedance head, namely, the local distortion of the material in the contact area on the structure and the elasticity of the force crystal and impedance head body between the structure and the accelerometer.

The contact stiffness, K, is a function of the contact area, and is estimated in [27.36] to be

$$K = \frac{2aE}{(1 - \nu^2)} \qquad (27.19)$$

where E and ν are the modulus of elasticity and the Poisson's ratio respectively, and a is the radius of a circular contact area.

This effect is illustrated in Fig. 27.11 where the point mobility of a suspended mass was examined using impedance heads of different contact areas. It is shown that at low frequencies the mobility of the mass is measured, but a frequency is reached when the mobility of the mass equals that of the local stiffness, causing an anti-resonance. Above this frequency the mobility is that of the local stiffness.

In addition to the distortion of the local structure the elasticity of the impedance head body and force crystal cause an additional stiffness term, as discussed in [27.37]. However, it is to be noted that although these stiffness terms affect the modulus of the measured mobility, the real component of mobility is unaffected, as the mobility of a spring is purely imaginary.

Quite small exciters may be used to obtain point characteristics, assuming linear behaviour to occur. For transfer measurements it would obviously be useful to employ as large a vibrator as possible in order to produce clearly defined response signals with a good signal-to-noise ratio at the remote station. However, there are usually problems associated with transporting and installing large vibrators in field work. Hence, with only moderate forces being generated in many practical situations such as for frequency response studies in buildings, on ships and aircraft, although the force signal will generally be very well defined, the signal-to-noise ratio may not be very high in the response signal, particularly in transfer measurements. The rapid frequency sweep technique involves the use of excitation which produces a deterministic response signal which may be contaminated by extraneous noise components. Frequency domain averaging or correlation techniques could be used for noise rejection, but the transient test may be repeated many times, and hence the method lends itself readily to time domain averaging. Exact repetition of the excitation is highly desirable, and this is achieved by the oscillator. However, phase shift between records will be inevitable because of timebase instability in the recording process or triggering errors in the acquisition and digitization of the analogue data.

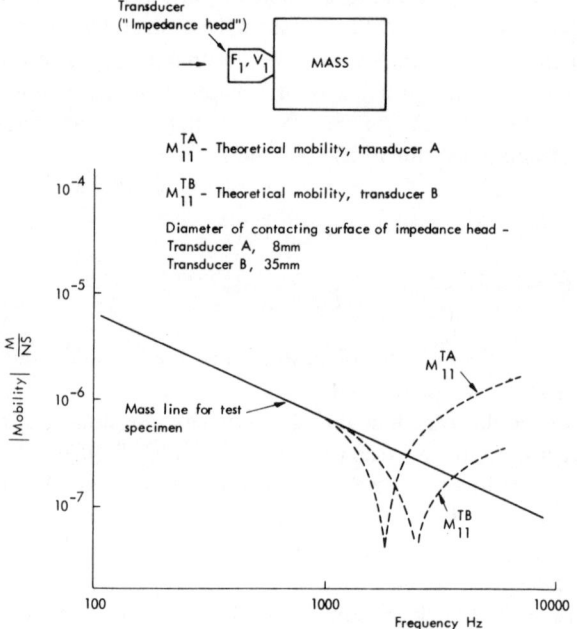

Fig. 27.11 – Effect of transducer contact area on point mobility.

If the signal to be analysed is contaminated by a random noise component of zero mean value, the deterministic (transient) part of the signal can be extracted by ensemble averaging many data records. There must be some reference point, a triggering pulse or the force signal if it has a high signal-to-noise ratio, in each set of data in order to obtain a reasonably correct average. After averaging r data records with a random Gaussian noise component, the signal-to-noise ratio will be improved by $\sqrt{r}$. That is, if the signal-to-noise ratio is indicated by S/N,

$$\left(\frac{S}{N}\right)_r = \sqrt{r}\left(\frac{S}{N}\right)_{\text{original}} .$$ (27.20)

For noise with unknown statistical amplitude distribution, the number of records must be increased, and it is suggested in [27.38] that the number be increased by 3.5 times to maintain the same signal-to-noise ratio as in the above case.

The signal averaging technique applied in point frequency response measurements is illustrated in Fig. 27.12a, b, c [27.38]; the structure tested was a beam. Fig. 27.12a shows the result of excitation by a single linear swept sinewave through the frequency range 10-1000 Hz in 0.25 s, the mobility modulus spectrum being derived by division of Fourier transforms. The most severe case, considered here, is when noise is present as an additional excitation; in the practical case it is then not possible to quantify the noise contamination since the level of the extraneous force is unknown and measurement accuracy cannot be estimated from r.m.s. levels of extraneous components or by other methods such as coherence function measurement.

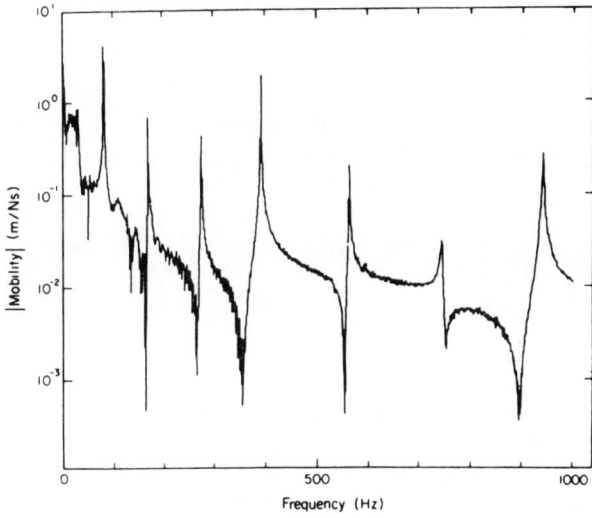

Fig. 27.12a − Measured mobility − no extraneous excitation added − 1 test.

For a signal-to-noise ratio of 0.5, as defined in Fig. 27.12b, the result of a single transient test became completely incomprehensible. Ensemble averaging results of 100 tests vastly improved the clarity of the time histories and the frequency response plot giving a signal-to-noise ratio improvement of 20 dB, corresponding to the expected value from equation 27.20. Resonance frequencies and damping ratios can be determined from the averaged data in Fig. 27.12c. Comparison of the peak spectral levels and half power point bandwidths for the resonance peaks shown in Fig. 27.12a for a single transient test and Fig. 27.12c for the noise contaminated case confirm that the ensemble averaged result provides good quantitative agreement with the reference frequency sweep test result of Fig. 27.12a.

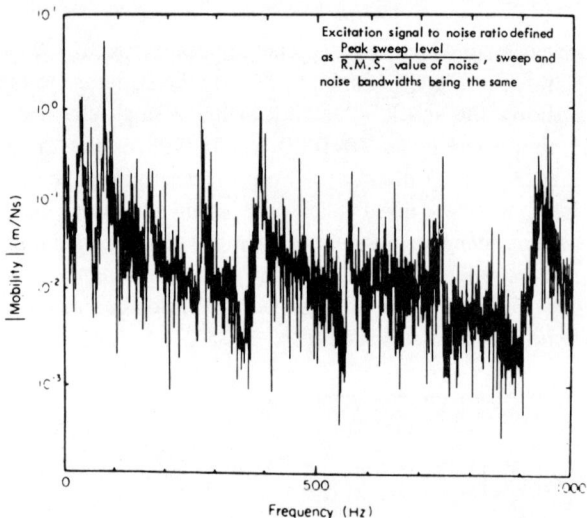

Fig. 27.12b – Measured mobility – excitation S/N ratio = 0.5 – 1 test.

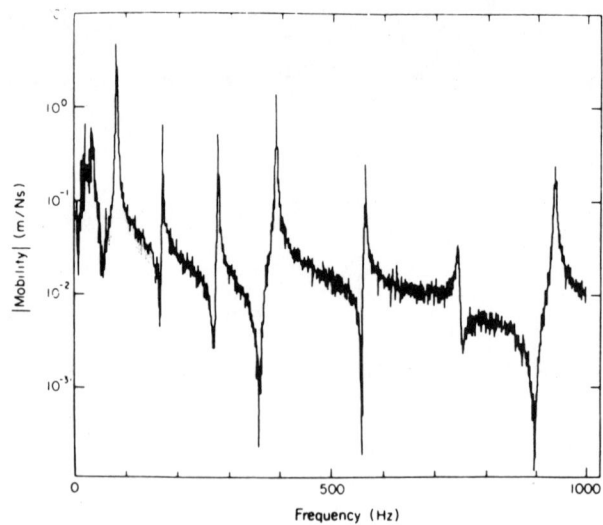

Fig. 27.12c – Measured mobility for the test conditions of Fig. 12b averaged from 100 tests.

It is usual in rapid frequency sweep testing to sweep from low to high frequency, which minimizes record lengths by exciting low-frequency resonances first. If a time domain average is to be formed of many sweeps from low to high frequency it can be seen that small delay errors in data acquisition will cause errors in both the averaged time signal and the derived spectrum, and these will be most evident in the high-frequency region. The error caused by timebase jitter will tend to reduce the amplitude of the ensemble averaged signal. Fig. 27.13 shows how the time history and modulus spectrum amplitudes may vary when ensemble averaging swept sinewaves.

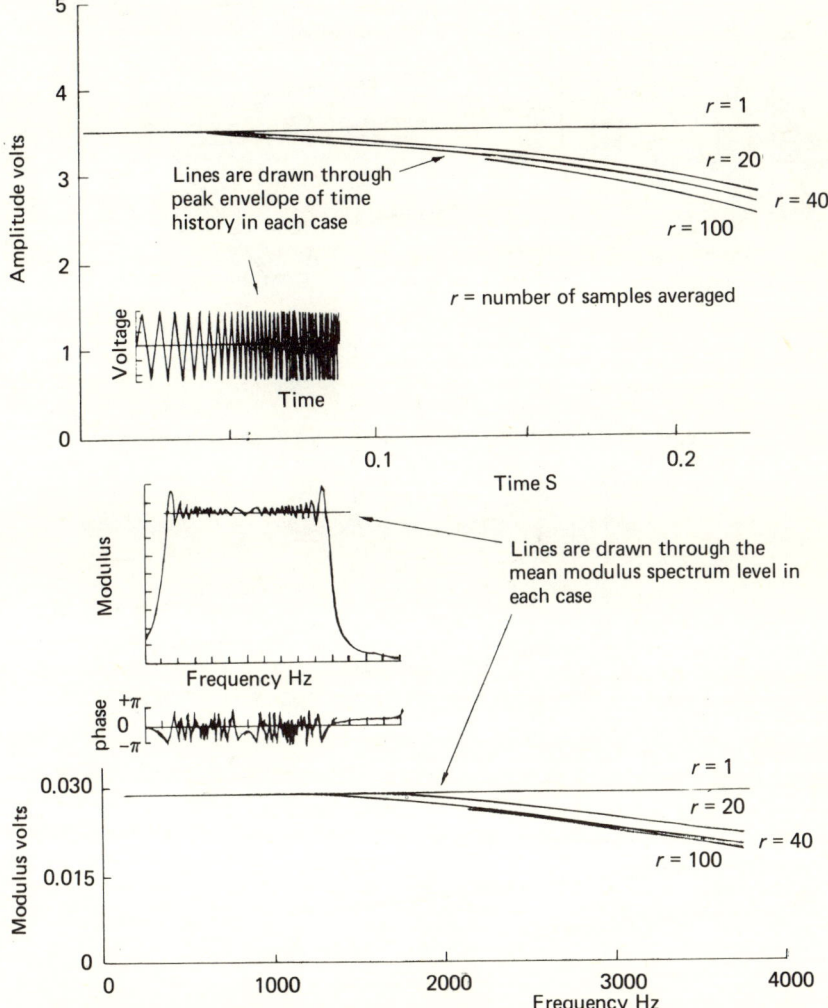

Fig. 27.13 — An illustration of variation in signal and spectrum levels obtained when averaging swept sinewaves.

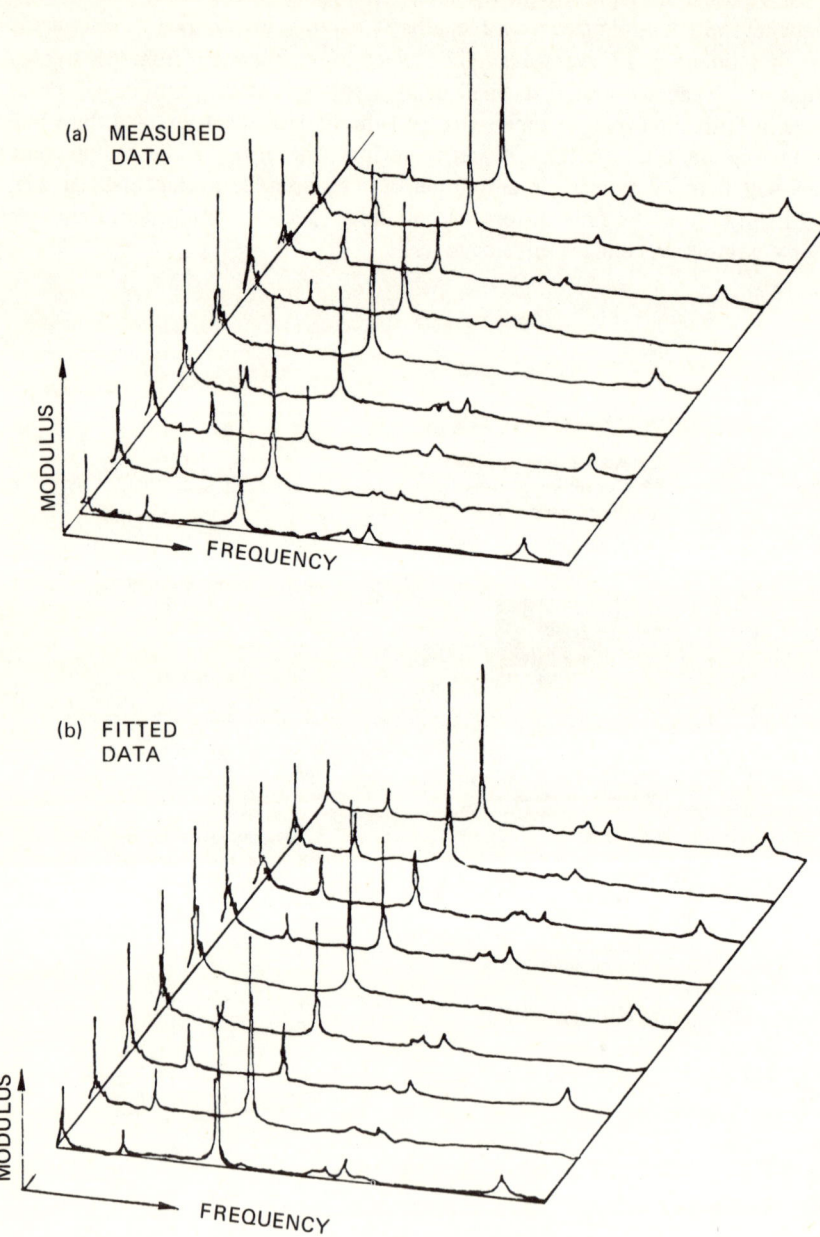

Fig. 27.14 − Measured and fitted frequency response data.

27.3 STRUCTURAL MODELLING

It will by now be apparent that it is possible to obtain a considerable amount of experimental structural frequency response data in a relatively short time, for example by using transient excitation and computer analysis. This can lead to the situation where many graphs of required quantities, e.g. mobility, have been produced and it is difficult to understand the complete dynamic behaviour of the test structure. Conversely, if test programmes are very limited in scope, only a limited amount of experimental data may be acquired. In either case, it may be required to predict the response of a structure to some specified force, i.e. excitation by a machine mounting, at points which have not been examined experimentally. For these reasons, a structural modelling technique has been developed, based on experimental data. The objective is to characterize the structure via conventional dynamic response equations and mode shapes by curve-fitting to experimental data. In this way, an accurate dynamic model of the test structure can be formed and any reasonable response prediction carried out.

Such a study has been carried out [27.39] based on the assumptions that the structure is linear and possesses hysteretic damping. The technique consists of exciting the structure with a rapid frequency sweep, measuring the response at several stations, then processing the data to model the structure. A set of experimental data is shown in Fig. 27.14a and 'curve-fitted' data in Fig. 27.14b. Complex frequency response functions can be estimated from such a model, and experimental verification has shown that such estimates can be made very accurately. Mode shapes can obviously be derived from plots such as in Fig. 27.14 by the inclusion of phase information. Damping information can of course be derived from vector diagrams, and the modelling technique may be used to advantage for interpolation on the plots.

27.4 VIBRATION TRANSMISSION PATH IDENTIFICATION

It is important that the paths by which power is transmitted from the source to the place of interest in a built-up structure may be identified. In a complex structure, there will of course be many types of path involving transmission through air and a variety of structural materials, which involve several types of wave propagation. If vibration transmission paths are to be studied, the frequency response approach described above is not completely adequate. The measurement of transfer functions, such as transfer mobility, does not alone yield enough information for transmission paths to be identified. It is therefore sometimes necessary to also use time domain analysis.

The most simple method would be to transmit a pulse through a system under test and monitor the response at different points in the system. However, the presence of multiple transmission paths and frequency dependent transmission characteristics generally precludes use of the method because the pulse shape becomes significantly modified during transmission. In such situations, cross-correlation may be used to advantage. The autocorrelation function of a rapid frequency sweep is impulse-like, and if a sweep of wide frequency range is used, then this type of forcing function may be used with cross correlation analysis. The cross-correlation function between the force and response will exhibit peaks corresponding to the various time delays in the transmission paths with heights related to the attenuation factors associated with each path. This method of analysis has been examined in depth for various types of transmission path [27.40].

In order to interpret cross-correlograms from practical structures for transmission path identification, the form of correlograms for ideal systems must first be studied. These may then be used to facilitate the analysis of data from multipath systems by combining the idealizations to represent the more complex behaviour of the practical system. In a structure, such as a building, power transmission may take place through both acoustic and structural paths. In a simple model, the acoustic paths may be considered as delay elements and the structure as being composed of resonant elements. Although these representations involve considerable simplification, the approach permits the examination of problems in which frequency response measurements alone are unsatisfactory. Fig. 27.15 shows systems composed of delay and resonant elements. The delay systems have lags τ_d and frequency response functions $D(i\omega)$; the resonant system has a frequency response function $H(i\omega)$. The cross-correlation function for each type of system is shown and peaks indicating the lags can be clearly seen in Fig. 27.15a, b. The cross-correlation function in 27.15c approximates to the impulse response of the system, and the effects of system combination are clear in Fig. 27.15d and Fig. 27.15e. A resolution criterion [27.40] has been derived for delay measurement and path identification in case Fig. 27.15e. The frequency response functions of the various systems are also given in Fig. 27.15, and the effects of delay on phase spectra and combined delays on modulus spectra are clear.

The application of the above technique has been demonstrated by exciting a metal plate by a loudspeaker in an anechoic room; this represented a combined delay and resonant system. The excitation signal was recorded from a microphone placed near the loudspeaker, and the response from an accelerometer on the plate. The plate was excited by rapid frequency sweep excitation from 0 to 700 Hz in 0.25 s and the acoustic path length was 2 m. The cross-correlation function between the excitation and response and the frequency response function $C(i\omega)$ of the combined system are shown in Fig. 27.16. The delay associated with the complete system is evident in the cross-correlation function. The many resonances of the plate are apparent in the modulus diagram, and evidence of the delay due to the acoustic path can be seen in the phase plot. The time and frequency domain methods of data presentation together therefore facilitate system identification.

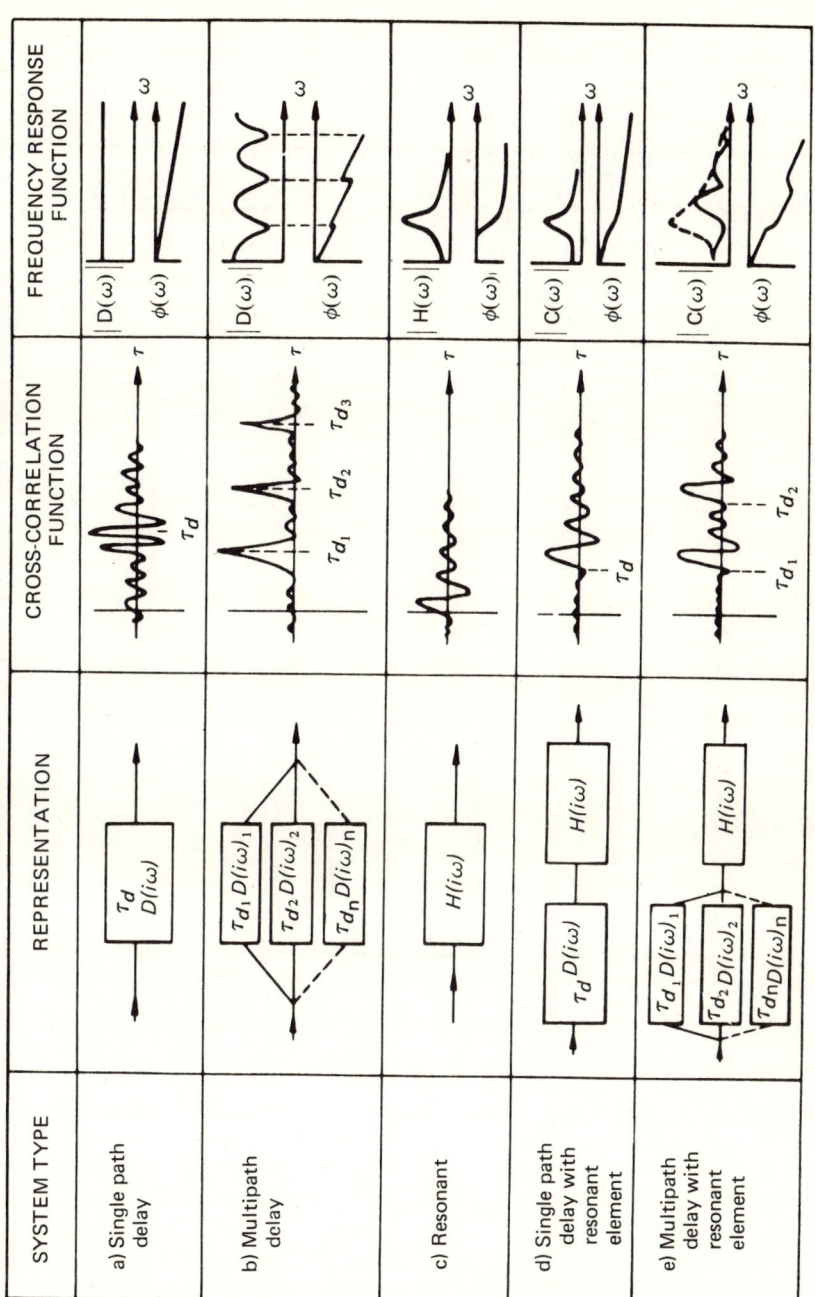

Fig. 27.15 – Lumped element systems.

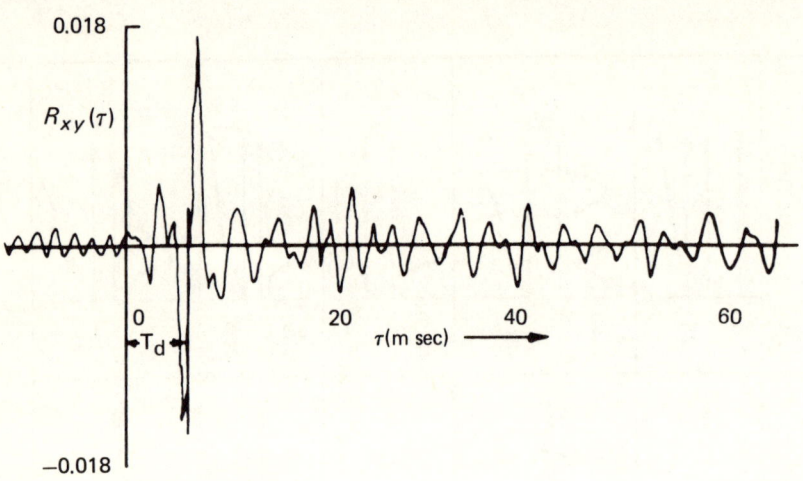

(a) Cross correlation function

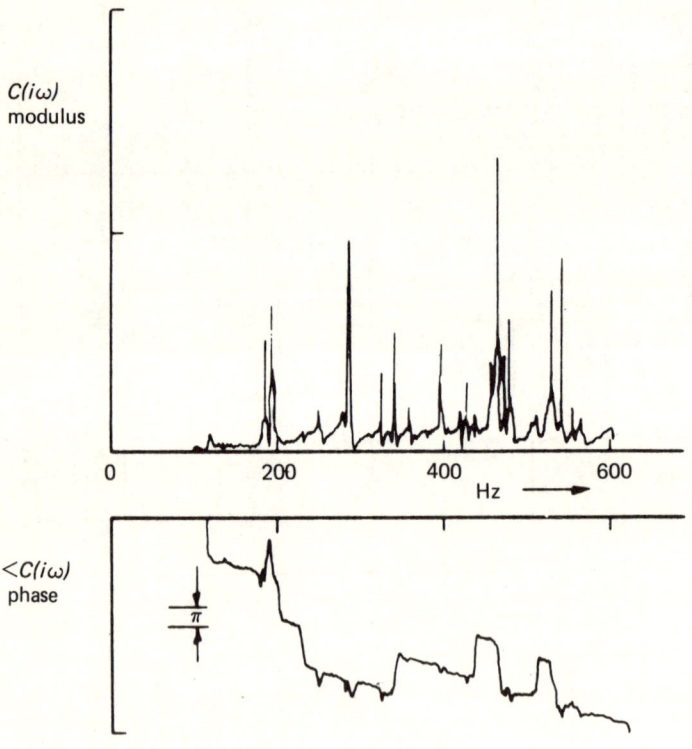

(b) Frequency response function

Fig. 27.16 – Single-path sound transmission to resonant plate, path length 2.0 m.

The above approach using lumped element representations gives very simple models in which systems are represented by combinations of resonant and delay elements, and non-dispersive transmission was assumed to occur. In practice, flexural wave propagation in structures is dispersive, that is the wave propagation velocity is frequency dependent; pulses containing many frequencies, such as the rapid frequency sweep, are therefore dispersed with distance.

The use of pulses with broadband spectra is therefore precluded as measurement of delays from conventional cross-correlograms is impossible. Further work has therefore been carried out to develop techniques which are suitable for use with dispersive systems. A transient excitation function consisting of a sequence of bursts produced by intermitting a pure tone has been developed, and it has been shown that this essentially single frequency excitation function exhibits little dispersion with distance in the flexural vibration of structures [27.41]. The analysis technique involves cross-correlation of the intermitting sequence with the squared response signal [27.41, 27.42] and yields correlation functions with clear triangular peaks from which group velocities and attenuation factors can be measured.

The above studies concerned 'ideal' test situations, however, the effects of noise in path identification experiments have been examined [27.43]. The case of both periodic and random extraneous noise was considered, and the errors associated with the measurement of attentuation factors have been studied. An inverse Fourier transform technique has been used [27.44] to obtain a function with which to modify the cross-correlogram obtained in the conventional way in order to produce cross-correlograms from dispersive systems which could be understood.

27.5 POWER FLOW MEASUREMENT IN VIBRATING STRUCTURES

In vibration transmission problems, measurement of transfer functions does not give information concerning power flow in the structure. It is this quantity which is of primary importance if vibration control procedures are to be attempted. Very often, the vibration field in a structure is highly reverberent, and path importance determination would indicate good wave propagation in all directions. The power flow measurement technique should therefore be able to indicate the small net power flow in such a field. Repetition of measurements across a structure would determine the paths of major power transport and thus indicate where vibration control measures should be applied. Power flow measurement in both one-dimensional and two-dimensional wavefields has been studied and the necessary experimental techniques developed [27.45]. In the two-dimensional case it is necessary to introduce the concept of flexural wave intensity as power per unit width in a given direction.

By using the Bernoulli theory of bending for a plate an expression for the wave intensity in each coordinate direction can be formulated. In order to measure the required quantities, i.e. time and spatial derivatives of the deflection, a finite difference method can be used with signals from an array of suitably placed transducers. Ten transducers can be arranged such that simultaneous measurements of directional components of intensity can be made.

The use of a finite difference method for estimation of variables involves approximation, and inherent errors have been evaluated. For example, if it is required to measure the intensity in a 3 cm thick plate at frequencies up to 5 kHz with a maximum error of 5% the spacing between transducers must not exceed 1.7 cm.

The method outlined above has been used with inverse Fourier transformation to obtain velocity time histories, and the necessary calculations carried out by digital computer estimated the power flow in a plate. Intensities due to shear, bending, and twisting components were obtained for each measurement position in two directions. The power flow into an absorbing termination was measured and found to agree within 20% of the averaged input power. The method therefore shows great promise as a structural analysis procedure.

27.6 ENVIRONMENTAL TESTING

Environmental testing implies subjecting a system to a specified dynamic excitation. Specifications are drawn up for various environments such as are encountered in aircraft or ships and are most often applied to electronic equipment, it usually being required to test the item to determine whether malfunction or damage occurs in the simulated environment. The required tests fall into two categories, steady-state and transient. The former involves sinewave and random excitation; the latter involves shock excitation.

Steady-state vibration tests are carried out to determine whether or not systems will withstand sustained dynamic loading, that is, a type of fatigue test; resonance searches are also carried out with slow sweep excitation to determine resonance frequencies and amplitudes, the sweep rate being specified. Sinusoidal excitations are defined, for example, as displacement or acceleration amplitude versus frequency, and in random testing, acceleration spectral density is defined. In both cases the specified excitation is applied at the equipment mounting points, although in resonance searches it is usually specified that the excitation is applied along each of the three coordinate axes of the equipment. In the random excitation test, the power spectral density of the excitation is specified with tolerance limits in dB for various frequency ranges. The required spectrum shape is achieved by use of a spectrum shaper consisting of many narrowband filters between the random voltage generator and the power amplifier if an electrodynamic vibration table is used. The characterisitics of analysers for checking the excitation spectrum are often also laid down in the specification. Equipment testing may be carried out with the device mounted on isolators or unmounted according to the particular application. Mounting techniques are quite important, and if rigid fixing is specified, the mounting must be carefully designed. The effects of unbalanced loads on the vibrator table are also significant.

The objective of shock tests is to simulate transient excitation likely to be encountered in practice, the integrity of the test item after testing being of primary importance. A repeated shock or 'bump' test is also often called for. Shock testing is often carried out mechanically, using drop test machines or pendulum-type exciters. It is required that the testing machine should generate a shock whose waveform approximates to a simple pulse shape. Half-sine, trapezoidal, or sawtooth waveforms are specified [27.46] to be applied to the test equipment along the three coordinate axes. The shock generated must fall within the limits specified and, equally important, the measurement instrumentation must have sufficiently wide bandwidth of amplitude and phase response to adequately record the shock. The shock waveform limits in terms of amplitude and duration tolerances are laid down [27.46], and also in the same document the necessary frequency response characteristics of the measurement instrumentation are specified. In the drop test machine, the required acceleration pulse is generated by arresting the falling equipment with mechanical elements such as a spring, lead block, or crushable material, It is also possible to use electrodynamic vibration tables, and considerable effort has been expended to develop techniques for generating well-defined shock waveforms. This has been achieved by measuring the complex transfer characteristic of the excitation system and calculating the required input to generate the specified shock waveform [27.47]. Such methods have not been widely used. In both the shock and bump test the method of mounting the test item is again important, and directions are given in specification; for example, mounting instructions for electronic components are given in [27.46].

It is apparent in the discussion of shock testing that, unlike vibration testing, shock testing is predominantly concerned with time domain analysis. The response of a system is a function of pulse shape, pulse duration, and the system characteristics. The analysis of time data may be simplified by using the shock spectrum approach based on the single degree of freedom analogy. The maximum and residual response amplitudes are plotted against the ratio of the pulse duration to the natural period of the system. This approach is useful in the estimation of likely damage and may be applied to damped or undamped systems, the latter representation often being used for simplicity and generally giving a 'worst case' criterion.

Now, as already noted above, in the shock testing of equipment, excitation pulses of simple form are often used. Such excitations have modulus spectra which fluctuate with frequency, exhibiting zeros or regions of very low spectral content, with the result that significant resonances of a test item may not be strongly excited into free vibration. This quality is clearly observable in the residual shock spectra of such functions. In practice, of course, the attachment points of equipment to structures are rarely, if at all, subjected to excitation by force or acceleration pulses of such simple form. Generally, shock excitation occurs at some remote point, and transient wave motion propagates through the structure to arrive at the equipment mounting in the form of a complicated pulse with decaying oscillatory components due to the stimulation of local resonances which are usually very lightly damped. Now the oscillatory motion may excite resonances of the equipment, even if it is mounted on isolators. If there is coincidence or near-coincidence between strong oscillatory components in the local shock waveform and resonances of the equipment, then very high response levels can occur. It can thus be seen that in practical circumstances the dynamic load factor produced by the oscillatory shock may be much greater than the limiting factor of two imposed by the simple shock pulse shapes commonly used. This effect is illustrated by the use of 'shock design numbers' which lead to criteria such that structural and equipment attachments, etc. must be designed to withstand loads of up to the order of one hundred times the static load.

The need to use oscillatory shocks in environmental testing can be met by using a waveform which is composed of a combination of decaying sinusoids. This in part overcomes the limitations of testing using simple pulse shapes. The spectrum of such a function is, however, peaked in nature, and not all significant resonances of the test items will be excited in a frequency range of interest. This problem may be overcome if a transient excitation is used which will excite all resonances in a selected frequency range, and the desirability of achieving controlled response levels, i.e. dynamic load factors, is of paramount importance in shock testing. The criterion of excitation without zeros in the Fourier transform modulus has been met by development of the rapid frequency sweep forcing function, defined in equation (27.12), as discussed in section 27.2.4.2. The shock spectrum shape of the rapid frequency sweep may obviously be controlled by influencing the Fourier transform modulus as a function of frequency. This may be achieved in two ways:

(a) by variation of amplitude with time, the sweep rate remaining constant;

or (b) by variation of sweep rate with time, the amplitide remaining constant.

The third possibility would cause unnecessary complication. Spectrum control via method (b) has been demonstrated in [27.34] and has the considerable advantage that the dynamic range of the forcing function is limited in the time domain. Method (a) has been examined in [27.48]; it has the disadvantage that high dynamic range of the envelope of the function will be required if a wide frequency range is to be investigated, but that required in environmental shock tests is very often less than two decades in frequency. Also, a desirable shock spectrum which is independent of frequency is very easily achieved via use of

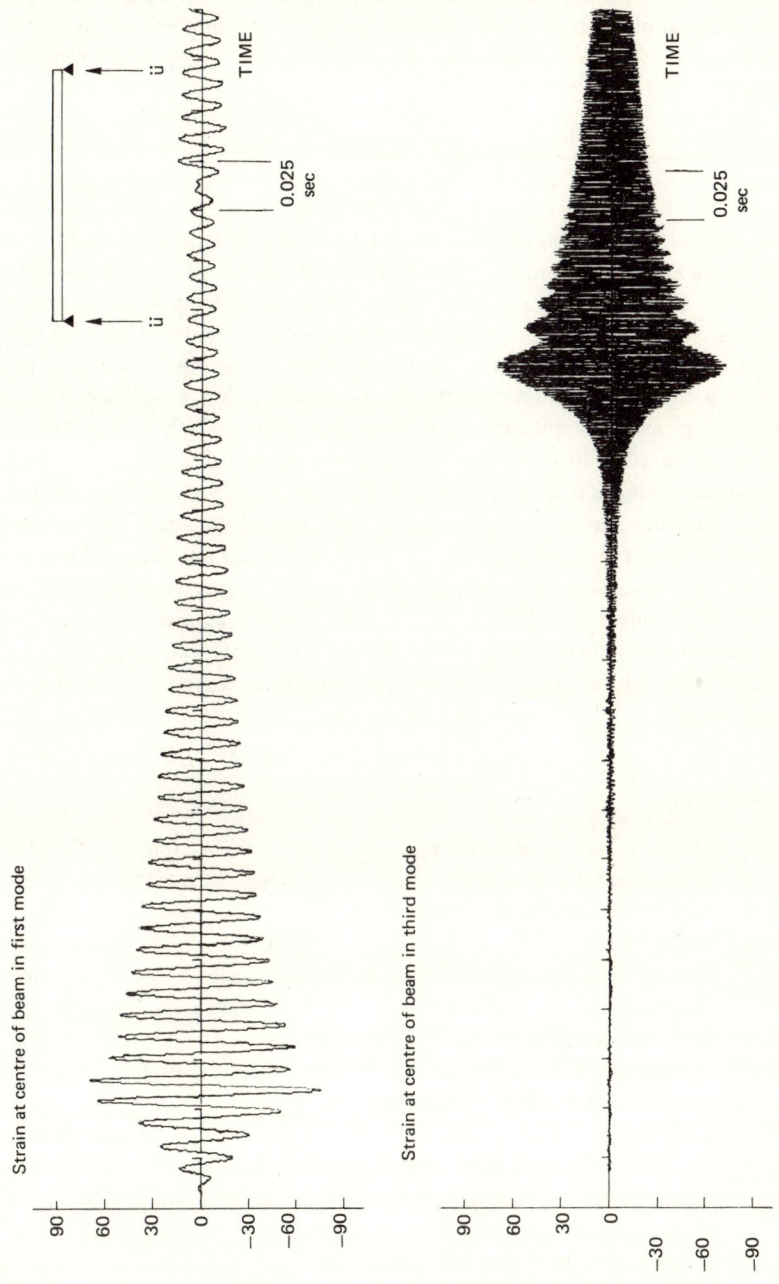

Fig. 27.17 – Uniform beam with linear acceleration sweep from 40 Hz to 1 kHz in 0.5 s, $N = -0.75$.

method (a). The approach developed in [27.48], and implemented in the shock testing of a simple structure via acceleration, $\ddot{u}$, of its supports on an electro-dynamic vibrator [27.49], was to use a rapid frequency sweep of the following form.

$$\ddot{u}(t) = \frac{b\,\ddot{u}_o}{(2at + b)^N} \sin(at^2 + bt) \; . \tag{27.21}$$

This is a signal of the type given in equation (27.12) but which is 'weighted' by the instantaneous frequency to the power $-N$, the initial frequency being included in the numerator. A very simple theoretical analysis together with consideration of the coupled vibrator/power amplifier characteristics, enabled experimental peak modal strains in the test item to be controlled to a high degree of accuracy using this type of rapid sweep excitation, as is desirable in an environmental test. For example, one objective set in [27.49] was to excite a beam via acceleration of its supports using a previously calculated value of N in the excitation function of equation (27.21); the objective was that the peak strains in the first and third modes of the beam should be equal. The resulting strain waveforms, reproduced in Fig. 27.17, show that the desired condition was very easily and accurately achieved. The results in [27.49] lead to the observation that the amplitude controlled, rapid frequency sweep acceleration waveform shows considerable advantages as an oscillatory environmental shock test signal compared with the simple pulse or decaying waveform currently used in speci-fication testing.

REFERENCES

[27.1]   Clarkson, B. L. & Mercer, C. A. (1965) *AIAA Journal*, **3**, 2287-2291. Use of cross correlation in studying the response of lightly damped structures to random forces.

[27.2]   White, R. G. (1971) *J. Sound Vib.*, **16(2)**, 255-267. Effects of non-linearity due to large deflections in the resonance testing of structures.

[27.3]   Bishop, R. E. D. & Gladwell, G. M. L. (1963) *Phil. Trans. A*, **255**, 241-280. An investigation into the theory of resonance testing.

[27.4]   Broadbent, E. G. & Hartley, E. V. (1961) *ARC R and M 3125*. Vec-torial analysis of flight flutter test results.

[27.5]   Kennedy, C. C. & Pancu, C. D. P. (1947), *Journal of Aero Science*. Use of vectors in vibration measurement and analysis.

[27.6]   Mead, D. J. (1958) *ARC 19870*. The internal damping due to struc-tural joints and techniques for general damping measurements.

[27.7]   Pendered, J. W. (1965) *J. Mech. Eng. Sci.*, **7**, 372-379. Theoretical investigation into the effects of close natural frequencies in resonance testing.

[27.8] Gaukroger, D. R., Skingle, C. W. & Heron, K. H. (1973) *J. Sound Vib.*, **29**, 341-353. Numerical analysis of vector response loci.

[27.9] Wright, G. C. (1972) *J. Sound Vib.*, **21(2)**. The dynamic properties of glass and carbon fibre reinforced plastic beams.

[27.10] White, R. G. (1975) *The Aeronautical Journal of the Royal Aero Soc.*, 318-325. Some measurements of the dynamic properties of mixed, carbon fibre reinforced plastic beams and plates.

[27.11] Mason, J. M. & Leventhall, H. G. (1969) *J. Sci Inst.*, **2**, 1104-1106. Rapid measurement of the decay rate of vibrations.

[27.12] Wright, G. C. (1972) *Symposium – Applications of experimental and theoretical structural dynamics.* Southampton University. Dynamic behaviour of fibre reinforced materials.

[27.13] Lewis, R. C. & Wrisely, D. L. (1950) *Journal of Aero Sci.*, 705-722. A system for the excitation of pure natural modes of a complex structure.

[27.14] Hawkins, F. J. (1969) *ARC R and M 3588*, GRAMPA – An automatic technique for exciting the principal modes of vibration of complex structures.

[27.15] Taylor, G. A., Gaukroger, D. R. & Skingle, C. W. (1969) *ARC R and M 3590*. MAMA – A semi automatic technique for exciting the principal modes of vibration of complex structures.

[27.16] Anderson, D. & Mills, B. (1971) *Environmental Eng.*, **51**, 12-16. Multipoint excitation techniques.

[27.17] British Standard 3015 (1976). *Glossary of terms relating to mechanical vibration and shock.*

[27.18] Ewins, D. J. (1975) *Journal of the Society of Environmental Engineers*, 3-12. Measurement and application of mechanical impedance data: Part I, Introduction and Ground Rules.

[27.19] Ewins, D. J. (1976) *Journal of the Society of Environmental Engineers*, 23-33. Measurement and application of mechanical impedance data: Part II. Measurement techniques.

[27.20] Ewins, D. J. (1976) *Journal of the Society of Environmental Engineers*, 7-17. Measurement and application of mechanical impedance data: Part III, Interpretation and application of measured data.

[27.21] Ewins, D. J. & Gleeson, P. T. (1975) *Shock and Vibration Bulletin*, **45(5)**, 158-173. Experimental determination of multi-directional mobility data for beams.

[27.22] Hok, G. (1948) *Journal of App. Phys.* **19**, 242-250. Response of linear resonant systems to excitation of a frequency varying slowly with time.

[27.23] Reed, W. H. (1958) *Journal Aero/Space Sci.*, 435-443. Effects of a time-varying test environment on the evaluation of dynamic stability with application to flutter testing.

[27.24]   Reed, W. H., Hall, A. W. & Barker, L. E. (1960) *NASA TN D508.* Analog techniques for measuring the frequency response of linear systems excited by frequency sweep inputs.

[27.25]   Lorenzo, C. F. (1970) *NASA TN D7022.* Variable sweep rate testing; a technique to improve the quality and acquisition of frequency response and vibration data.

[27.26]   White, R. G. (1969) *The Aeronautical Journal of the Royal Aeronautical Society,* **73**, 1047-1050. Use of transient excitation in the dynamic analysis of structures.

[27.27]   Holmes, P. J. & White, R. G. (1972) *J. Sound Vib.,* **25(2).** Data analysis criteria and instrumentation requirements for the transient measurement of mechanical impedance.

[27.28]   Skingle, C. W. (1966) *R.A.E. Tech. Rep. TR 66379.* A method of analysing the response of a resonant system to a rapid frequency sweep input.

[27.29]   Kandianis, F. (1971) *J. Sound Vib.,* **15**, 203. Frequency response of structures and the effects of noise on its estimates.

[27.30]   Kandianis, F. (1974) *J. Sound Vib.,* **36(2)**, 207. Notes on the properties of correlation functions in linear system analysis.

[27.31]   White, R. G. (1972) *J. Royal Inst. Nav. Arch.,* **114**. The use of a transient test method in the investigation of the local vibration characteristics of ship structures.

[27.32]   White, R. G. & Pinnington, R. J. (1982) *Aeronautical Journal of the Royal Aero Society,* May, 179-199. Practical application of the rapid frequency sweep technique for structural frequency response measurement.

[27.33]   White, R. G. (1973) *J. Sound Vib.,* **29(3)**, 295-307. Effects of non-linearity due to large deflections in the derivation of frequency response data from the impulse response of structures.

[27.34]   White, R. G. (1972) *J. Sound Vib.,* **23(3)**, 307-318. Spectrally shaped transient forcing functions for frequency response testing.

[27.35]   White, R. G. (1971) *J. Sound Vib.,* **15(2)**, 147-161. Evaluation of the dynamic characteristics of structures by transient testing.

[27.36]   Timoshenko, S. P. & Goodier, J. N. (1951). McGraw-Hill. *Theory of elasticity.*

[27.37]   Brownjohn, J. M. W., Steel, G. H., Cawley, P. & Adams, R. D. (1980). *J. Sound Vib.,* **73(3)**, 461-468. Errors in mechanical impedance data obtained with impedance heads.

[27.38]   White, M. F. & White, R. G. (1976) *J. Sound Vib.,* **48(4).** 543-557. Frequency response testing in a noisy environment or with a limited power supply.

[27.39]   Goyder, H. G. D. (1980) *J. Sound Vib.,* **68(2)**, 209-230. Methods and application of structural modelling from measured structural frequency response data.

[27.40] Holmes, P. J. (1974) *J. Sound Vib.*, **35(2)**, 253. The experimental characterisation of wave propagation systems: 1, non-dispersive waves in lumped systems.

[27.41] Holmes, P. J. (1974) *J. Sound Vib.*, **35(2)**, 277. The experimental characterisation of wave propagation systems: 2, continuous systems and the effects of dispersion.

[27.42] Aoshima, N. & Igarashi, J. (1969) *ISAS Report 436*. University of Tokyo. The measurement of flexural wave propagation by correlation techniques.

[27.43] Pavic, G. (1974) *ISVR Tech. Rep. 52*. Path identification in non-dispersive media by cross correaltion: effects of noise and error analysis.

[27.44] Pavic, G. & White, R. G. (1977) *Acustica,* **38(1)**, 76–80. On the determination of transmission path importance in dispersive systems.

[27.45] Pavic, G. (1976) *J. Sound Vib.,* **49(2)**, 221–230. Measurement of structure borne wave intensity, Part 1: Formulation of the methods.

[27.46] BS 2011 (1973). *Methods for the environmental testing of electronic components and electronic equipment.*

[27.47] Favour, J. D., Le Brun, J. M. & Young, J. P. (1969) *Shock and Vib. Bull.,* **40/2.** Transient waveform control of electromagnetic test equipment.

[27.48] White, R. G. (1981) *ISVR Memornadum 612.* Shock spectrum control of an oscillatory waveform for environmental testing.

[27.49] White, R. G. (1981) *ISVR Memorandum 619.* The response of a non-uniform beam, representative of an electronic component, to oscillatory shock motion of its supports.

CHAPTER 28

Subjective acoustics

C. G. Rice and J. G. Walker

Institute of Sound and Vibration Research, University of Southampton

28.1 INTRODUCTION

An observer's auditory impression of sound may be categorised in many ways, and considerable research and other effort has been spent in trying to define and quantitatively relate such human responses to the physical characteristics of the noise exposure. The intensity, spectral, and durational characteristics of noise play differing roles depending upon the type of interference under consideration, and the combination of these physical parameters into a scale of noise measurement is therefore of considerable interest.

The predictive ability of such noise measures can be estimated from the results of social survey data and from laboratory studies. It is obvious, however, that human responses to noise are complex and can take differing forms depending upon attitudinal and environmental factors. These latter factors are hard to quantify, but their relative importance is particularly interesting to note in the overall context of the formulation of criteria for the prediction of annoyance due to noise.

28.2 TERMINOLOGY

28.2.1 Loudness

Loudness is defined as the observer's auditory impression of the strength of a sound. In other words, it is a subjective impression related to the intensity component of the sound, and is different in concept from other abstractions such as noisiness and annoyance. Some psychophysical experiments have demonstrated differences between loudness and noisiness in that loudness is not as responsive to instructional cues (e.g., paying particular attention to duration or spectral complexities) as is noisiness.

The *sone* scale is used in determining whether or not a sound is one-third, one-half, twice, or three times as loud as another sound. This provides a numerical designation of the loudness of sounds that is proportional to the subjective magnitude as estimated by normal observers (see ISO/R226 [28.1]). One sone is the loudness of a sound for which the loudness level is 40 *phon*.

The loudness level of a sound is expressed as n phon when it is judged by normal observers to be equally loud to a pure tone of frequency 1000 Hz consisting of a plane progressive sound wave, coming from directly in front of the observer, the sound pressure level of which is n dB. The relationship between the loudness in sones N and loudness level in phons L_N is given by the following formulae [28.2]

$$N = 2^{0.1(L_N - 40)} \text{ or } \log_{10}N = 0.03(L_N - 40).$$

The term loudness should be used only when accompanied by a quantitative statement in sones (not phons) or when such a statement would be relevant in context but no numerical value is actually stated. Loudness should not be confused with loudness level.

There are several other phon scales in use, including phon(OD) [28.3], phon(GD) and (GF) [28.4]. All phon scales are based on empirical relationships which have related the physical properties of sounds to their perceived attributes. Loudness level values in phons do not immediately convey the magnitude of the loudness sensation. They have to be interpreted by the user on the basis of his experience of previously heard sounds to which phon values have been attached. It is generally understood, however, that a change in loudness level of 10 phon is approximately equal to a doubling or halving of loudness sensation.

28.2.2 Perceived Noisiness
The subjective impression of the unwantedness of a not unexpected, non-pain or fear-producing sound as part of one's environment is defined (Kryter [28.5]) as the attribute of perceived noisiness. Kryter further states that the measurement or estimation of this subjective attribute or quality is of central importance to the evaluation of environmental sounds or noises with regard to their physical content. Descriptor terms such as disturbing, unwantedness, unacceptableness, objectionableness, or noisiness fit the total attribute of 'perceived noisiness' and are fairly consistently used by subjects in psychological judgement tests.

The concept of perceived noisiness and its historical development into rating scale units such as PNdB for Perceived Noise Level and EPNdB for Effective Perceived Noise Level has depended upon a clear distinction between loudness and noisiness. In parallel to the use of the sone for loudness, the *noy* became the subjective unit of noisiness. A sound that is judged to be subjectively equal in noisiness to an octave band of random noise centred at 1000 Hz and of sound pressure level 40 dB re 2.10^{-5} Pa is designated a value of one noy; sounds judged to be twice or three times is noisy are 2 noys and 3 noys, respectively, etc.

28.2.3 Annoyance
Annoyance has been defined (Borsky, [28.6]) as being a feeling of displeasure associated with any agent or condition realised or believed by an individual or a group to be adversely affecting them. Borsky furthermore continues "While it is

often useful or necessary from an analytical point of view to focus attention on a single environmental agent (such as noise, for example) it should be recognised that the single agent appears in real life as one of a complex of environmental stresses. The degree of the physical exposure as well as intervening psychosocial variables determine the occurrence of the annoyance response. All these variables must be measured or controlled in experimental studies in order to arrive at an appropriate judgement concerning annoyance effects due to noise".

Therefore annoyance may be taken to include contributions that are dependent not only on the physical characteristics of the noise, but also on the psychosocial variables manifested by the attitudinal and environmental influences of the observer.

28.2.4 Nuisance
This term tends to have a legal connotation and means 'injurious' or 'obnoxious' to the community. In the case of noise it implies 'annoying' or 'disturbing'.

28.3 QUANTITIES
A noise 'scale' refers to a combination of the physical quantities (sound pressure, time, etc.) which may be used to express noise impact. Noise 'index' is reserved for the numerical description of noise in which other factors are superimposed on the scale numbers describing the physical quantity defined above. An index may be considered as an adjusted scale to be used as a basis for 'rating' or assessment in planning and in regulations [28.7]. A useful selection of scales and indices will be described.

28.3.1 Weighting networks
The A, B, and C-weighted sound pressure level networks (see Fig. 28.1) are derived from the smoothed 40, 70, and 90 loudness levels [28.8]. They were originally used for the measurement of sounds in the <55 dB, 55–80 dB, and >80 dB regions respectively, but the success of the L_A in the subjective acoustics field has made it a very popular and easy means of measuring noise.

28.3.2 The phon
As discussed earlier, the loudness level in phons of a sound is the sound pressure level of a pure tone of 1kHz which is judged to be equal in loudness to the sound in question. The equal loudness contours for pure tones as obtained by Robinson & Dadson [28.9] have been standardised [28.1] and can be seen in Fig. 28.2. Although of major academic interest, these contours in fact play little part in the rating of community noises.

28.3.3 Phon (OD)–Stevens, Mk V1
The work of Stevens [28.3] provided a method for calculating a loudness level

of a sound from its octave band sound pressure levels. The relationship is based on three factors: (a) the expression of a loudness function in sones in terms of a loudness level in phons; (b) an empirical set of equal loudness contours for bands of noise in a diffuse sound field; and (c) a rule relating total loudness of a sound to the loudness indices of the frequency bands of which it is comprised.

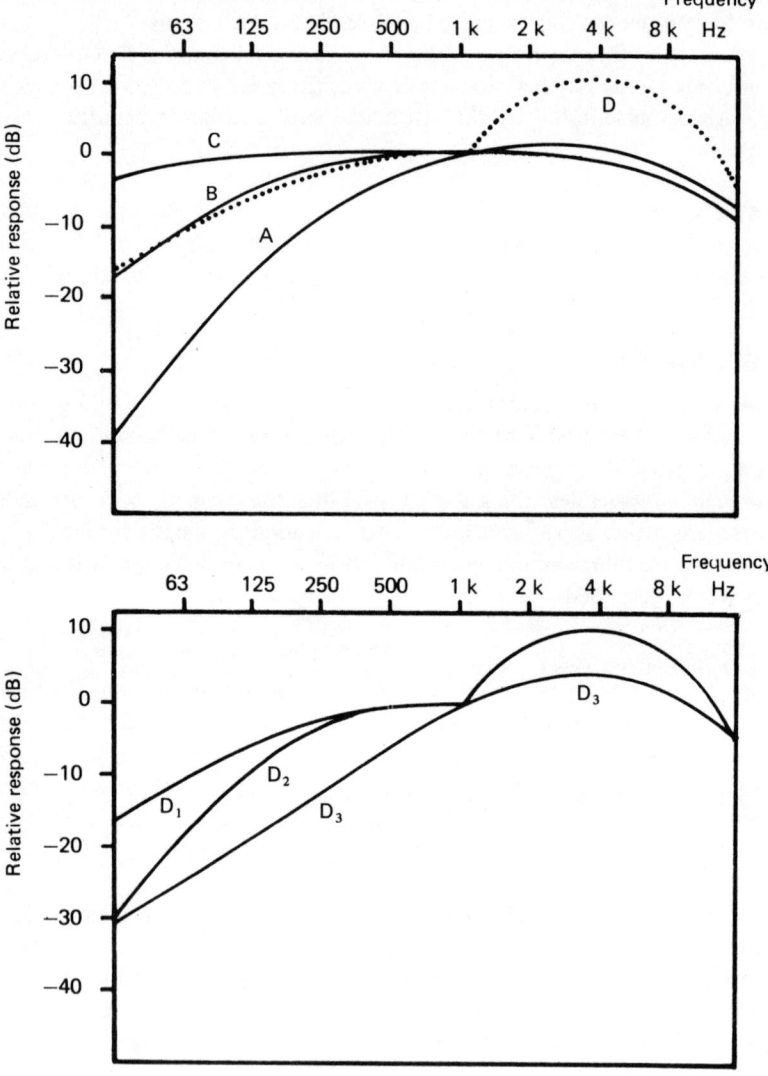

Fig. 28.1 – Frequency weighting networks for sound level meters (after Kryter [28.5]).

The procedural steps involve the determination of a loudness index in each spectral band and the computation of the total sone value, S_t, where $S_t = S_m + 0.3(S - S_m)$. S_m is the maximum sone value in the spectrum, and S is the sum of the loudness indices of all the octave bands. The phon value (P) is computed from the relationship $P = 40 + 10 \log_2 S_t$. The procedure for calculation has been standardised [28.10].

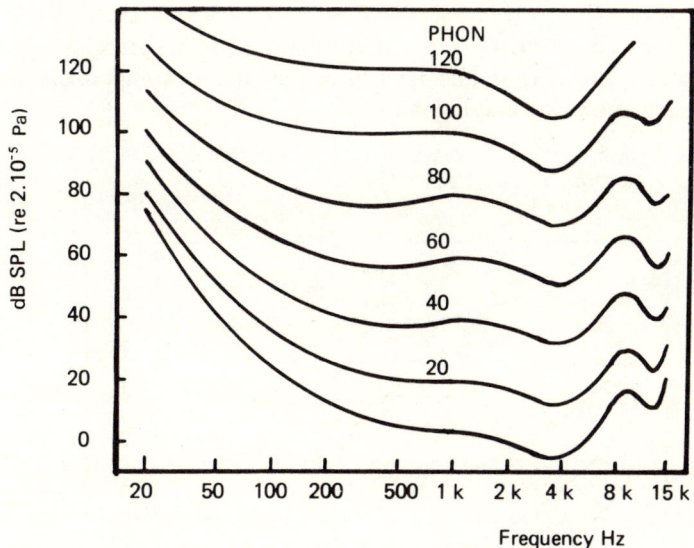

Fig. 28.2 – Equal loudness contours for pure-tones using free-field listening (after ISO R226, 1961 [28.1]).

28.3.4 Phon (GD) and (GF)–Zwicker

The Zwicker [28.4] procedure for calculating loudness level is based on: (a) the use of the critical band concept; (b) a rule relating the total loudness of a sound to the contributions from the critical bands of which it is composed; (c) a relation between the part of the loudness appropriate to each band and the midband frequency; (d) the difference between equal loudness contours for frontal sound (GF) and of a diffuse sound (GD) field; (e) a loudness relation relating the loudness in sones to loudness level in phons similar to that used in the Stevens Mk VI method.

The phon value is obtained from the area under graphical plots of the one-third octave sound pressure level band values, suitably combined to approximate critical bands. The procedure for calculation has been standardised [28.10]. A discussion of the critical band concept is given by Kryter [28.5].

28.3.5 Perceived noise level

The scale of perceived noise level L_{PN} (in PNdB) is based on the noy and the fact that a 10 PNdB change is proportional to a doubling in the noisiness of the sound. The procedure for calculating L_{PN} values from one-third octave band spectral data is due to Kryter & Pearsons [28.11, 28.12]. Noy values for each band (n_i) are obtained (see Fig. 28.3) and combined to form a noy value N for the total where $N = n_{max} + 0.15(\Sigma n_i - n_{max})$. This value is converted into perceived noise level by the relationship $L_{PN} = 40 + 33.3 \log_{10} N$.

Of all the scales discussed so far, apart from L_A, L_{PN} has probably been most widely used, particularly in the field of aircraft noise. An approximation to weighting network values is as follows:

$$L_{PN} = L_A + 13 \, (\pm 3)$$

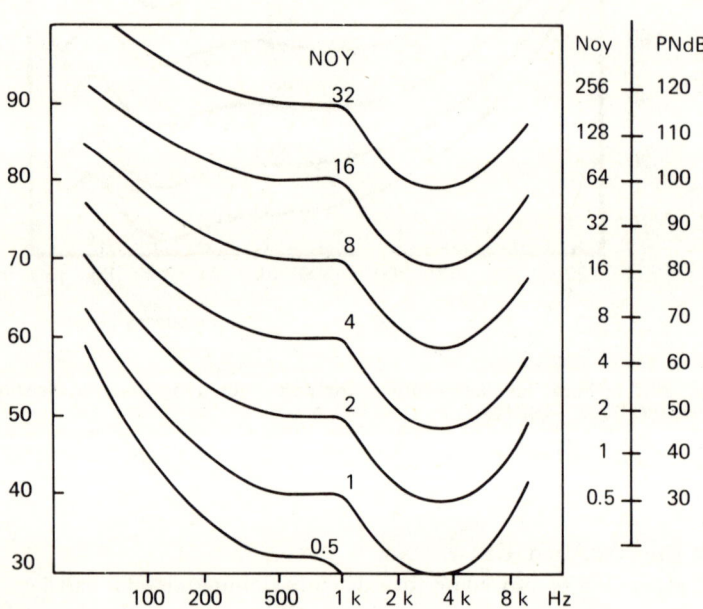

Fig. 28.3 – Equal noisiness contours and chart for the calculation of perceived noisiness (after Kryter [28.5]).

28.3.6 D-weighting response

It was recommended by Kryter [28.13] that the inverse of the 40 noy contour could be used as a means of rating noises and designated D. The frequency weighting is shown in Fig. 28.1 as D_1. The D_2 network is a modification due to Kryter [28.14], and the D_3 network is a later modification due to Young & Peterson [28.15]. The Kryter modification takes account of the grouping of the one-third octave bands below 355 Hz in order to conform to the critical band concept.

L_D usually refers to the D_1 network and it is now incorporated in some sound level meters.

The noise level measured in L_D can be used to approximate to the perceived noise level, where

$$L_{PN} = L_D + 7$$

28.3.7 Effective perceived noise level

This scale is used for the noise certification of aircraft [28.16]. It is calculated from the one-third octave band values obtained for each one-half second of the complete aircraft flyover noise cycle.

For each half-second the third octave band values are combined to calculate the $L_{PN(k)}$ value, to which is added an appropriate tone correction factor $C(k)$. This tone correction factor is obtained from a somewhat simple if not straight-forward relationship of the SPL values in adjacent one-third octave bands. The maximum value of $L_{PN(k)} + C(k)$ across the whole flyover is obtained and denoted L_{PNTM}.

A duration correction factor (D) is computed by integration under the curve of tone corrected perceived noise level versus time. Hence a value for the effective perceived noise level becomes

$$L_{EPN} = L_{PNTM} + D.$$

The use of real-time analysis and computer facilities are ideally required to compute these values.

28.3.8 Perceived level – Stevens Mk VIII

The calculation procedure Mark VII (Stevens, [28.17]) gives the perceived level of loudness or noisiness L_{PL}. It was derived from a critical review of all the existing work that had been performed on the subjective magnitude judgements of complex noises.

The essential features of L_{PL} as compared to phon (OD) are: (a) the introduction of a new reference signal of a one-third octave band of noise centred at 3150 Hz; (b) the unit of subjective or perceived magnitude (loudness/noisiness) of the reference signal at 32 dB is the sone which has a loudness equal to a 1 kHz tone of 40 dB SPL; (c) a power law relationship for the perceived magnitude $S = k(E - E_o)^{1/3}$ where $E_o = -3$ dB SPL. Above 20 dB perceived magnitude doubles every 9 dB; (d) new frequency weighting functions of equal loudness contours for bands of noise (see Fig. 28.4); (e) a new summation rule $S_t = S_m + F(S - S_m)$ where F has tabulated values.

This procedure has yet to be experimentally verified, although it was designed with the object of superseding other existing scales.

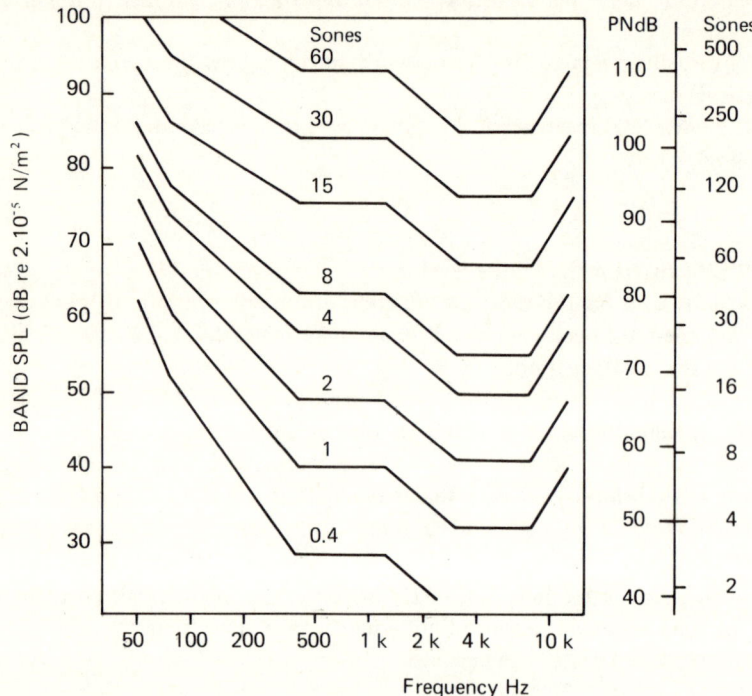

Fig. 28.4 — Countours of equal perceived magnitude in sones and chart for the calculation of perceived level (after Stevens [28.17]).

28.3.9 E-weighting response
The E-weighting (ear-weighting) network is the 20 sone contour from the Stevens Mk VII PLdB method (Fig. 28.4). It is equalised at 1 kHz in the same way as the other weighting networks shown in Fig. 28.1.

28.3.10 The equivalent continuous sound level (L_{eq} or L_{Aeq})
The equivalent continuous sound level has been adopted in a number of countries as a means of measuring and assessing noise. It is sometimes referred to by various other terms than equivalent continuous sound level, such as mean energy level and equivalent sound level. The equivalent continuous sound level is given by the level of steady noise that has the same energy as the actual time-varying noise in question.

The true L_{eq} of a sound measured over a time period T is given by

$$L_{eq} = 10 \log_{10} \frac{1}{T} \int_0^T \frac{p^2}{p_0^2} \, dt$$

where p = instantaneous sound pressure
p_o = 20 μPa

The equivalent continuous sound level may also be calculated from:

$$L_{eq} = 10 \log_{10} \left[\frac{1}{100} \, \Sigma f_i \cdot 10^{L_i/10} \right]$$

where, L_{eq} = the A-weighted equivalent continuous sound level.
L_i = the sound level corresponding to the class midpoint of the class i.
f_i = the time interval expressed as a percentage of the relevant time period for which the sound level is within the limits of class i.

Although the sound level can be expressed in any units, it is normally measured in dB(A) for environmental purposes.

Modification of L_{eq} has taken place in order to account for day-night differences in response. The day-night average level (L_{dn}) is based on a modification of the 24 hr L_{eq} formula,

$$L_{dn} = 10 \log \frac{1}{24} \left[15.10^{L_d/10} + 9.10^{(L_n + 10)/10} \right]$$

where L_d = 15 hr L_{eq} (0700-2200 hrs)
and L_n = 9 hr L_{eq} (2200-0700 hrs).

28.3.11 Noise pollution level (L_{NP})

Robinson [28.18] specifically devised the noise pollution level (L_{NP}) to take account of more complex time-varying noises. The scale takes account of the equivalent continuous sound level over a particular period of time together with the variability of the noise environment. The scale is expressed as

$$L_{NP} = L_{eq} + 2.56\sigma$$

where L_{eq} is the equivalent continuous sound level over the time period under consideration and σ is the standard deviation of the instantaneous sound level fluctuations over the same period. The factor 2.56 was chosen as giving a good fit with available social survey data.

For Gaussian distributions of noise $L_{NP} = L_{eq} + (L_{10} - L_{90})$

or $$L_{NP} = \frac{(L_{10} - L_{90})^2}{56} - (L_{10} - L_{90})$$

where L_{10}, L_{50}, L_{90} are the noise levels exceeded for 10%, 50%, or 90% of the time period respectively.

28.4 FORMULATION OF ANNOYANCE CRITERIA

Although the major source contributor to the community noise problem is transportation noise and in particular road traffic noise, there is nevertheless a need to measure and evaluate intrusion from other noise sources in the community. These include industrial noise, noise from construction sites, domestic appliance noise, and recreational disturbances. The annoyance responses created by such noises often include interference with speech, television and radio listening, rest, relaxation, and sleep.

In the past each identifiable noise source was considered in isolation, but in recent years it has become apparent that responses to individual noises are influenced by noise from other sources. It is therefore necessary to consider the total environmental noise problem, and now research effort is aimed at providing methods of measuring and evaluating the total noise environment.

28.4.1 Industrial noise

The disturbing qualities of noise emitted by industrial premises are generally its loudness, its distinguishing features such as tonal or impulsive components, and its intermittency and duration. A British Standard [28.19] has been drawn up against a background of case histories of complaints which describes a method of rating such noise. The method is based on a measure in L_A of the noise to which corrections to take account of the characteristics of the noise are added.

The use of the L_{Aeq} is recommended by the International Organisation for Standardisation [28.20]. Corrections are again applied for the characteristics of the noise.

28.4.2 Traffic noise

The penetration of vehicle noise into residential areas is known to occur over considerable distances from the major road routes. Furthermore, it is known to be a function of the type of vehicle (diesel truck, bus, passenger car, etc.), the type of tyres used and the road surface over which the vehicle is driven, the way in which the vehicle is driven, and the condition of the silencer or muffler. The varying traffic conditions (motorway, urban, city), the concentration of vehicles (heavy commercial, private, etc.), and the speeds at which the traffic is allowed to flow all influence the noise patterns.

Measurement and analysis of such noise sources therefore inevitably present considerable problems, and numerous attempts have been made to quantify/model the noise exposure patterns in various environmental situations. Social surveys showed that dissatisfaction towards traffic noise expressed by people in their houses sometimes depends on the variability of the noise. Griffiths & Langdon

[28.21] derived the traffic noise index (TNI) to correlate with dissatisfaction. TNI is given by

$$TNI = 4(L_{10} - L_{90}) + L_{90} - 30$$

where L_{10}, L_{90} are measured in dB(A). A TNI value of 74 dB(A) was found to be likely to give a 50% community dissatisfaction with the traffic noise.

However, the measurement of TNI is difficult because of the uncertainty arising from background noise coming from sources other than traffic on the road being considered. Prediction is also difficult because of problems in predicting the background noise at large distances from the road. Instead the L_{10} (18-hour) index was proposed by Scholes & Sargent [28.22] and adopted [28.23] in the UK for noise legislation purposes. L_{10} (18-hour) is the average of the values of L_{10} in dB(A) for each hour between 0.600 and 24.00 hours on a normal working day. Methods for its prediction have been proposed [28.24], its measurement is relatively easy, and recent surveys have shown that it gives as good a correlation with dissatisfaction as does TNI, although it does not take account of the background noise level.

28.4.3 Aircraft noise
Many attempts have been made to formulate criteria for the prediction of aircraft noise annoyance in the community. Of these, two seem to have some merit [28.25] [28.26], and although their limitations are evident [28.27], have been used with some success in land planning considerations.

The Composite Noise Rating (CNR) was originally proposed by Bolt, Beranek and Newman Inc in the early 1950s when attempts were made to relate concepts of the noise stimulus to expected community response. Over the years, various modifications have been made, and the currently used version is the Noise Exposure Forecast (NEF) [28.25]. The procedure is comprehensive and takes account of the different kinds of aircraft movements, percentage usage of different runways, background noise levels, times of day, adaptation of community by previous exposure, etc. Annoyance is predicted from a scale related to the NEF value.

The Noise and Number Index (NNI) [28.26] is derived from noise measurements correlated with social survey questionnaire responses of people exposed to aircraft noise around London (Heathrow) Airport. Specific disturbance resulting from such exposure are that it wakes you up, makes TV picture flicker, interferes with conversation, and makes the house shake or vibrate. The NNI is given by

$$NNI = L_{PN} + 15 \log_{10} N - 80$$

where L_{PN} is the logarithmic average of the maximum perceived levels attained during the passage of successive aircraft, and N is the number of aircraft heard in the defined daytime period. For an aircraft to be classified as 'heard' the maximum L_{PN} at the position concerned must exceed 80 PNdB. NNI has been adopted as indicating the extent of the disturbance from aircraft noise at busy urban airports. However, it is also widely used for other airports, even though the index may not give a good correlation between noise exposure and response for these airports.

In discussing the community response to aircraft noise as viewed from the evidence of sample surveys, McKennell [28.28] raises the point that it is unlikely that improving the objective characterisation of the stimulus for annoyance will account for more than a few percentage points in the amount of annoyance variance explained. Even when all available acoustical knowledge is utilised to the full, by far the greatest proportion of variance in annoyance will remain attributable to the psychosocial factors independent of exposure.

However, in spite of this comment, it appears evident that the physical data obtained during such social surveys in no way match the quality of the annoyance data. Therefore it is imperative that comprehensive noise measurement data, however complex and costly, be obtained during future researches in this area.

28.4.4 Railway noise

Noise from railway operations is by no means as serious a nuisance as noise from other transportation systems. However, concern has been expressed over the past few years about the effect of high-speed trains running through residential areas, particularly in view of the adverse reaction against the high-speed trains in Japan. This type of reaction also occurred when the route for the Channel Tunnel Rail Link was proposed in South-Eastern England, when noise was one of the major environmental issues.

A major national survey of railway noise in Great Britain has been completed by Fields & Walker [28.29, 28.30, 28.31]. The more important results are summarised below. Railway noise bothers 2% of the UK population. Approximately 170 00 people live where railway noise levels are above an $L_{Aeq,24\ hr}$ of 65 dB(A). Annoyance increases steadily with noise level; thus there is no particular 'acceptable' noise level. Railway noise is less annoying than aircraft or traffic noise of equivalent noise level, at least above an $L_{Aeq,\ 24\ hr}$ of 50 to 65 dB(A) (see Fig. 28.5). Noise is rated as the most serious environmental nuisance caused by railways, and maintenance noise is rated as a bigger problem than passing train noise. Vibration is the most important non-noise problem. Reactions to vibrations are related to distance from route, train speed and number of trains.

The railway survey's highly stratified, probability sample design with many study areas made it possible to evaluate the effects of area characteristics on reactions. The $L_{Aeq,\ 24\ hr}$ noise index is more closely related to annoyance than are other accepted noise indices examined.There is no support for ambient noise level or night-time corrections. Of thirteen railway operation characteristics examined, one, the type of traction, has a strong effect on reactions after controlling for L_{eq} (overhead electrified routes are the equivalent of about 10 dB less annoying at high noise levels).

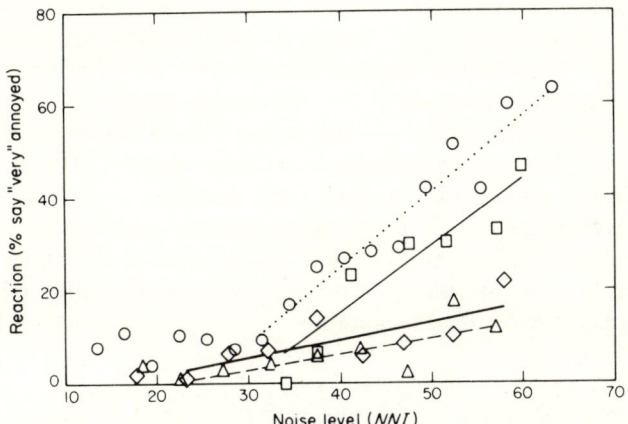

Fig. 28.5a – Railway and 1967 and 1976 Heathrow Reactions ("very" annoyed). · · · ○ · · ·, Aircraft–1967 Heathrow; ‒ □ ‒, aircraft–1976 Heathrow; ‒◇‒, railway (middle correction); ‒ ‒ △ ‒ ‒, railway (low correction).

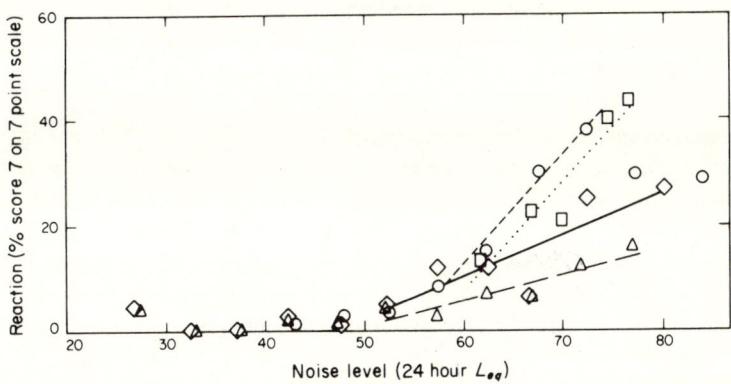

Fig. 28.5b – Railway and BRS and England road traffic reactions (dichotomy 7 point scale). ‒ ‒ ○ ‒ ‒, Road traffic England (middle correction); · · □ · ·, road traffic BRS; ‒◇‒, railway (high correction); ‒ ‒ △ ‒ ‒, railway (low correction).

Fig. 28.5 – Response to railway, road traffic, and aircraft noise (from Fields & Walker [28.46]).

Among the other available information which makes any attempt to relate community response to railway noise are results from surveys carried out in Japan [28.32] and in France [28.33]. The Japanese survey was restricted to noise measurements and social surveys carried out alongside a very high-speed track. Their data suggest that it is satisfactory to describe the noise in terms of the maximum L_A during a train pass-by. This seems reasonable as all residents alongside the high-speed railway line were exposed to the same number of trains of a single type. Obviously the railway system that exists in Britain is very different from that considered in the Japanese study. The survey carried out in the suburbs of Paris may bear more similarity to the British situation. The results from the French survey also suggest that the equivalent continuous sound level is a suitable scale to use to describe railway noise with respect to annoyance. The data showed that if the L_{Aeq} exceeded to 68 dB(A) then dissatisfaction was likely to increase significantly.

28.4.5 Construction and demolition sites

A recent British Standard [28.34] advocated the use of the L_{Aeq} scale to describe noise from construction and demolition sites. The Control of Pollution Act (1974) does not set any noise standards, and it is left to the Local Authority to establish realistic noise limits in the light of their local needs, the practical possibilities of the situation, and the existing background noise, etc. Due regard should be taken of the location of the site, the duration of the operation, and the times of the operation. The standards set by the Local Authority should be agreed between themselves and the Contractor and should take due notice of special requirements of the operation and location of the site.

28.4.6 Measuring the total noise exposure

It has been mentioned previously that environmental noise is often a complex of noises from several sources. Difficulties therefore arise when a mix of, for example, traffic noise and aircraft noise is to be evaluated, because the L_{10} concept is of little value in rating aircraft noise, whilst NNI is meaningless when applied to traffic noise.

The ISO R 1996 [28.20] applies to most community noises except aircraft noise, and in several counties aircraft noise has been rated with L_{eq}. The use of L_{Aeq}, normalised for background, for daytime and night-time exposures, etc., is also used in the USA to assess environmental noise and is referred to as the L_{dn} (Eldred, [28.35]; US EPA, [28.36] and US NRC, [28.37]), and other L_{eq} based units have also been proposed.

The noise pollution level was intended by Robinson [28.18] to overcome the need for separate criteria for each type of noise. Again it contains a L_{Aeq} based term with an added term to take account of the variability of the noise, which Robinson suggested should affect the subjective response. Robinson has shown that L_{NP} can account for both the NNI and TNI concepts, and a survey into annoyance caused by the interaction of aircraft and traffic noise [28.38] supports the idea that L_{NP} may have value in the prediction of community response. However, many reports now indicate that L_{NP} gives a worse correlation between response and noise exposure than does L_{eq} alone. It will require much further research to assess the true value of the L_{NP} index and to determine whether more accurate response predictions result from the added sophistication of the index.

28.5 ANNOYANCE DOSE-RESPONSE RELATIONSHIPS

The concept of a unique dose-response relationship for the prediction of all noise annoyance in residential areas is appealing to scientists and administrators alike.

Considerable effort has been expended in formulating methods for quantifying single noise sources in residental areas, and relating these to the human responses evoked. This has resulted in a variety of noise scales and indices being introduced for use in different situations when noise is subject to planning regulation and control. For example, in the UK the Noise and Number Index (NNI) is used for aircraft, L_{10} (18-hour) for road traffic, Corrected Noise Level (CNL) for noise from industrial premises and other fixed installations, and L_{Aeq} for trains, construction sites, and noise abatement zones. This approach is not too dissimilar from those followed in other countries, although recently efforts have been made to unify all noise exposures in terms of equivalent continuous sound level, i.e., an L_{Aeq} based scale (US EPA, [28.36]; NAC, [28.7]; US NRC, [28.37]; NAC, [28.39]). The move towards the incorporation of L_{Aeq} into all indices raises a number of very profound scientific issues which are related to the concept of the unique dose-response relationship.

A noise index is a composite measure of noise over a period of time, derived from a noise scale to allow for additional psychosciological factors which are relevant to rating or assessment for purposes of planning and regulation. The major requirements of a unified noise scale are that it should: (a) concur in numerical terms with people's responses to noise; (b) permit the influence of background noise to be determined; (c) be applicable not only to individual noises of various kinds but also to combinations of these noises; (d) allow predictions of scale values for use in planning; (e) take account of the time distribution of the noise; and (f) ideally, be acceptable internationally. The major requirements of the psychosociological factors relevant to the index are that account should be taken of: (a) health effects; (b) interference with valued ongoing human activities; and (c) improving the perceived environmental quality of life. That a single factor index could account for all of these variables is almost inconceivable.

The US National Research Council [28.37] believes, however, that sufficient of these requirements are met in their guidelines for preparing environmental impact statements on noise. Using USA Environmental Protection Agency [28.36] assertions, NRC infers that a day-night sound level (L_{dn}) of 55 dB(A) in residential areas should result in negligible impact on public health and welfare, and that the degree of impact will increase as L_{dn} increases. This allows human response (expressed as a percentage of the population highly annoyed) to be characterised by a single functional relationship of the noise environment. Without discussing the detail of this relationship (which involves the use of population density statistics), two important issues are raised. These are that exposure to equal L_{dn} scale values of different noise sources and their combination evoke equal annoyance responses, and that these annoyance responses can be adequately represented by the single heuristic concept of the 'percentage of people highly annoyed'.

Whilst it may be scientifically and administratively desirable to formulate such a unique noise index, it is believed that the principles involved in the NRC approach have not been substantiated. It is at present therefore considered inadvisable to place too much store on achieving a viable single dose-response relationship in the near future. The scientific data currently available are insufficient to completely specify the form such an index should take, and whilst the experts continue to debate these issues the premature promulgation of a unique noise index would only lead to confusion and misuse amongst environmental planners.

28.5.1 Acoustical factors

Whilst it may be argued by the Noise Advisory Council [28.39] that the different methods used to evaluate environmental noise in the UK lead to confusion and make comparisons difficult, they nevertheless have considerable merit, quite apart from being incorporated into current legislation. They do, for example, in a unique way, attempt to account for the various qualities of noise found to be disturbing to the community exposed to that particular noise. These qualities include the patterns of activity, day-night relationships, and the contribution made by other noise sources. Furthermore, although it is known that these methods have limitations, adjustment for them can be made when planning decisions are being taken.

The comparison argument put forward by the NAC is twofold. Firstly there is no doubt that the use of the same noise scale for all situations would enable better comparisons of the noise environments to be made. Scientifically it does not matter too much what that choice is, although there is considerable evidence to favour L_{Aeq}. Secondly there seems to be just as much difficulty in comparing annoyance responses because there is no set of core questions asked in all surveys. It might be justifiably concluded at the present time, therefore, that because one cannot sensibly compare data from field studies, there is little foundation for the formulation of the unique index.

This is certainly true if the results of laboratory studies are considered alongside those of field studies. Although in the laboratory practical difficulties preclude the use of long noise exposures equivalent to those in real-life situations, the judicious use of test periods lasting up to one hour allow comparison of the merits or otherwise of the physical parameters used in noise scales and indices. The value of such studies should not be underestimated, because if the 'lab-field gap' could be satisfatorily bridged the laboratory provides a means of obtaining much more detailed information about the physical parameters than can be achieved from field studies, and in addition constitutes a much cheaper method of obtaining scientific data. It is therefore interesting to see where such studies have so far led us.

The USA dose-response relationship. This approach relies heavily on the work of Schultz [28.40] who compared the reports from eighteen social surveys dealing with the noise of aircraft, street traffic, expressway traffic, and railway traffic, spanning a period of fourteen years and a range of nine countries. By making a number of assumptions about the relationships between the various noise scales and L_{dn} and the various annoyance scales and the percentage of people highly annoyed, he produced graphs of L_{dn} against percentage of highly annoyed. On this basis he put forward the hypothesis that a single relationship between noise exposure and annoyance exists for all kinds of noise. This relationship is shown in Fig. 28.6, although as Schultz himself states, "the largest uncertainty in the results of this study are associated with the judgement as to whom is 'highly annoyed'."

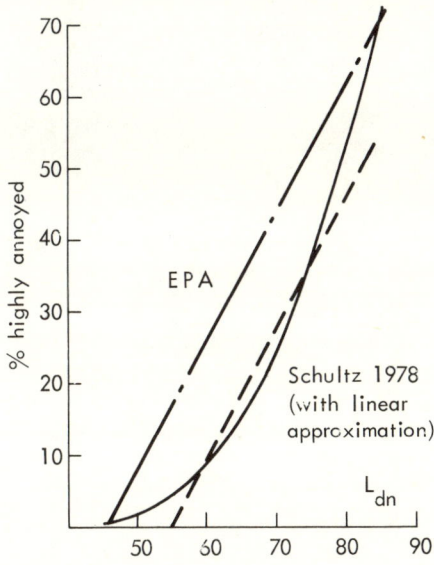

Fig. 28.6 – Examples of USA dose-response relationships.

Importance of prior experience. The 1963 Report of the Committee on the problem of Noise [28.26] stated that people's judgement of the noisiness of a noise from a recognisable source seems to be related to their previous experience of it. Simple corrections for this effect have been built into some general noise criteria [28.41] [28.19] [28.35] [28.20] . Usually in these cases prior experience of a noise will involve a penalty against the person; that is, allow a higher noise level. However, such penalties are equivocal, and in reality within a noise source the reaction will depend upon the exposure patterns, as well as the attitudes of the people concerned. For example, if the Schultz data are re-analysed taking account of the 'non-clustering surveys' which he omitted, it can be argued that the dose-response relationships follow those shown in Fig. 28.7, where total numbers of movements became important. Furthermore, laboratory studies [28.42] [28.43] [28.44] have shown that annoyance responses to aircraft noise can correlate best with either average peak noise level L_A, L_{Aeq}, NNI, or L_{NP} (the generalised form of which will approximate to $L_A \pm k \log N$ where k varies from 0 to 30 and N is the number of aircraft heard in a one-hour period) depending upon prior experience. It is therefore not surprising that field data show a weak dependence upon the value of k, which has traditionally been either 10 (as in L_{Aeq}) or 15 (as in NNI). Such results argue against the concept of a unique index.

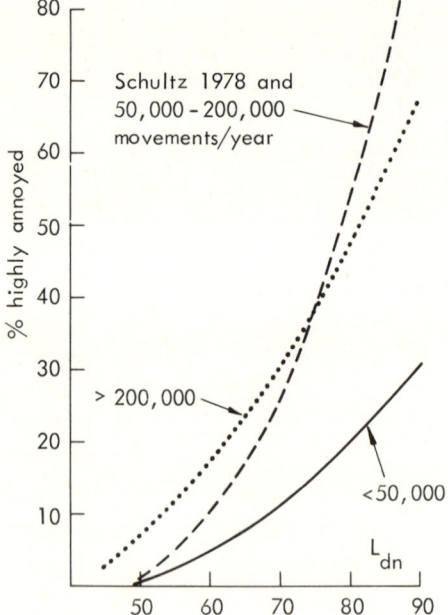

Fig. 28.7 – Variations in dose-response with airport size.

Comparison between noise scales. Many problems arise when making comparisons of noise data obtained in different surveys; Schultz [28.40], House [28.45], and Fields & Walker [28.46] have noted this problem. House has calculated that the conversion from NNI to L_{Aeq} is dependent not only on the number of aircraft operations, but also on aircraft type and whether one is considering landings or take-offs. Fig. 28.8 shows the influence of these parameters, which highlight the fact that simple conversion factors advocated by such bodies as the NAC [28.39] are not universally applicable. It would seem, therefore, that if a particular noise exposure has not been measured directly on the scale under consideration, extrapolation between scales is a procedure that should be entered into only with extreme caution. The problem has been discussed in detail also by Fields & Walker in comparing the results of five noise surveys [28.46].

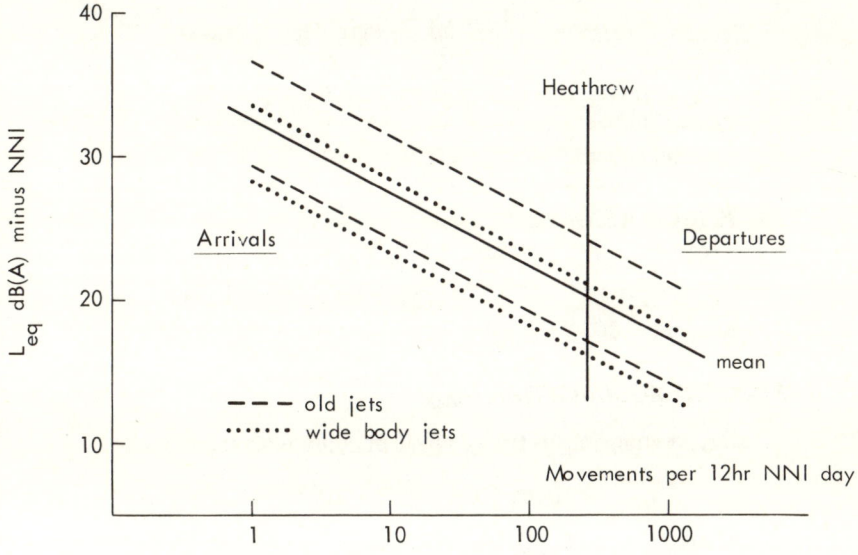

Fig. 28.8 – Difference between L_{eq} and NNI (after House [28.45]).

Day-evening-night corrections. Several indices weight noise levels according to the time of day in which they occur. These penalties, as Galloway [28.47] reports, seem to have been arbitrarily assigned on the basis of results obtained from complaint studies and social survey data, firm evidence to support the actual choice of numbers being hard to come by. Robinson [28.48], commenting on the arbitrary nature of such refinements, particularly when used in conjunction with "the unsubstantiated assumption of the veracity of the energy principle", concludes that such measures (e.g., L_{dn}, etc) should be regarded as convenient expressions and not taken as faithful representations of human reaction.

$$L_{dn} = 10 \log \frac{1}{24} \left[15.10^{L_{d_1}/10} + 9.10^{(L_n + 10)/10} \right]$$

$$\text{CNEL} = 10 \log \frac{1}{24} \left[12.10^{L_{d_2}/10} + 3.10^{(L_e + 5)/10} + 9.10^{(L_n + 10)/10} \right]$$

where $L_{d_1} = L_{eq}(0700\text{-}2200 \text{ hrs})$ $L_{d_2} = L_{eq}(0700\text{-}1900 \text{ hrs})$

$L_e = L_{eq}(1900\text{-}2200 \text{ hrs})$ $L_n = L_{eq}(2200\text{-}0700 \text{ hrs})$

NOISE LEVEL PENALTY (dB)

	day	*evening*	*night*
L_{dn}	0	0	10
CNEL	0	5	10
Recent studies (mean)	0	5	5
Recommended	−5	0	0

TIME PERIOD WEIGHTING

L_{dn}	15/24		9/24
CNEL	12/24	3/24	9/24
Recommended	1	1	1

PROPOSED MODIFICATION (L_{den})

Day, evening, night-time equivalent continuous sound level

$$L_{den} = 10 \log \left[10^{(L_d - 5)/10} + 10^{L_e/10} + 10^{L_n/10} \right]$$

where $L_d = L_{eq}(0700\text{-}1800 \text{ hrs})$
$L_e = L_{eq}(1800\text{-}2200 \text{ hrs})$
$L_n = L_{eq}(2200\text{-}0700 \text{ hrs})$

Table 28.1
Influence of day, evening, and night-time periods on L_{eq} based noise scales.

Table 28.1 shows the formulation of L_{dn} and Community Noise Equivalent Level (CNEL) and illustrates the arbitrariness of the time periods, the weights attributed to each, and the penalties assigned to the noise levels. Taking account of these, Galloway [28.47] argues for the L_{dn} approach, principally on the basis that the numerical differences between L_{dn} and CNEL are no more than about 1 dB.

The reality of these weights means that $L_{dn} \propto L_d + 6 L_n$ and that CNEL $\propto L_d + 0.75L_e + 7.5 L_n$, which highlights the fact that there does not appear to be any justifiable reason for weighting the separate parts of the day according to their duration. In fact recent laboratory and field studies (Borsky [28.49]; Ollerhead, [28.50]; Powell, [28.51]; Rice, [28.44]) suggest that evening and night-time noise levels should be penalised by 5 dB with respect to the daytime, although Fields & Walker [28.30] found no evidence for a night-time weighting for railway noise in Great Britain. This would lead to an L_{den} scale similar to that proposed in Table 28.1 which summarises these findings and infers that the evening period may be the most important as far as annoyance responses are concerned.

Such a view was given particular prominence during the 1980 NASA/FAA Workshop on time-of-day metrics [28.52]. In this workshop the continued use of energy summation models (i.e., L_{dn}, CNEL) was queried by Fields, who proposed that independent effects models might be more appropriate. In a synthesis of sleep disturbance due to noise, Rice [28.53] preferred this latter approach and suggested that daytime annoyance should be predicted from $L_{de} = 10 \log(10^{L_d/10} + 10^{(L_e + 5)/10})$ and night-time annoyance (which is a function of the night-time L_{Aeq}, the maximum sound pressure level of individual events and the number of occurrences during the period in question) from

$$L_n = f_n(L_{Aeq}, L_{Amax} \text{ and } N).$$

The threshold of onset for source-specific sleep disturbance due to noise may be quantified in terms of the finding that about 25% of the population suffer sleep disturbance from causes other than noise, which means that noise is often wrongly attributed as the causal factor. For the onset of specific noise-induced sleep disturbance the following outdoor values are recommended:

$L_{n, \, 2200-0600 \, h} > 60$ dB where for steady noises $L_{Amax} < 75$ dB and for aircraft and trains $L_{Amax} < 85$ dB providing the number of events $N < 20$.

Basic uncertainties such as those discussed here, however, make it very difficult to properly account for the wide variety of noise response patterns and environmental interferences which will inevitably occur over long periods of time.

Equivalence of noise sources. Assuming that a convenient expression such as L_{Aeq} could form the basis of a normalising scale for all noise sources, then it would be gratifying if equal L_{Aeq} values of different noises evoked equal annoyance reactions. Unfortunately this is not the case as is evidenced in Fig. 28.5. Other field and laboratory data [28.26] [28.42] [28.43] [28.44] [28.51] [28.54] support this finding and so appear to destroy one of the cornerstones of the concept of the unique index.

Combinations of noise sources. Several models which account for the subjective response of people exposed to noise sources in combination have recently been reviewed by Taylor [28.55]. Ollerhead [28.54] takes account of the subjective differences between noise sources and then sums their noise levels according to the energy principle. Powell [28.56], however, uses classical psycho-acoustical principles to account for the mutual summation and inhibition of annoyance present in the separate noise sources. The difference between these approaches is significant, and depending upon which form of expression is used for the relationship between noise exposure and the percentage of people highly annoyed, differences of up to 20% from the energy model can be obtained. This magnitude of uncertainty is clearly unacceptable to the environmental planner, and the whole issue requires considerable further research.

The problem is further confounded by the types of question which are asked when soliciting annoyance reactions to combinations of noise. For example, people think that aircraft noise should be more annoying in quiet than in noise areas, despite the fact that the total noise exposure (and hence the overall annoyance) will be greater for the combined situation. Therefore different responses will be elicited if source-specific rather than total noise annoyance questions are asked. This means that aircraft noise may be shown to be both responsive and passive to the influence of background noise depending on the way in which the annoyance responses are elicited.

28.5.2 Discussion

It is quite clear that situational factors influence annoyance relationships in a very significant manner. It is also quite clear that whilst we are not yet in possession of sufficient scientific data to formulate a unique prediction noise index, we are aware of the problems. These problems will be systematically studied in future researches, and until such time as these results are available, retention of the status quo is recommended. Change for change's sake at this time could be counterproductive. However, considerable help could be given to the research worker if all future noise measurements included information on day, evening, and night-time L_{Aeq} levels, and if social surveys could include a standardised set of core questions.

28.6 CONCLUSIONS

Several predictive measures of annoyance have been discussed, each of which contains a physical exposure term such as L_A or L_{PN}. It has been seen that in many instances noise sources are classified and rated separately, the major groupings being aircraft noise, traffic noise, and industrial noise. The particular predictive measure is usually governed by the noise under consideration, and this leads to considerable confusion when comparing different noise nuisance situations.

The major difficulties currently presented in criteria formulation are: (a) the inadequate physical representation of complex and time-varying noises; (b) the long time periods over which criteria are meant to be applicable (e.g. 12 hours for NNI, 18 hours for L_{10}, etc.); (c) the weakness of rating scale units and equivalent energy concepts in accounting for the way in which a noise varies in a given time period; (d) the central tendency annoyance response measures derived from social surveys are not specific enough to allow accurate assessment of individual noise nuisance situations.

Even so, well-documented procedures are available within each noise source classification to allow standardised instrumentation, measurement, and analysis techniques to be used to formulate central tendency conclusions regarding subjective reactions. Moreover, theoretical methods are available which allow noise exposure predictions to be made when changes or extensions to existing operations are required. However, each noise index must be judged as a compromise, and as such it has error and limitation associated with it.

The review has indicated that human responses such as annoyance reactions are likely to provide a most suitable basis for meeting environmental quality noise control goals. Research suggests that for a given noise source in index incorporating an L_{Aeq} based scale explains human reactions to noise as satisfactorily as any other widely used index. This might seem to imply that noise problems could be solved by the adoption of a single L_{Aeq} based index for all regulatory purposes. This approach would ignore the strong research evidence that annoyance reactions are strongly affected by other acoustic and non-acoustic characteristics of the environment. This being the case, the simple application of a unique dose-response relationship to all sources and conditions would certainly lead to a highly inefficient and unsatisfactory noise control strategy.

It is clear, therefore, that current research findings cannot firmly establish the adjustment which should be incorporated into noise criteria in order to account for situational factors. Given this existing state of knowledge it is perhaps best, from a regulatory point of view, to continue with the present UK system of different noise descriptors for different noise sources. This approach gives flexibility by allowing for occasional changes in noise criteria for one source without having to revise the whole system of noise control. At a future date with additional research findings it should be possible to design a firmly based noise criterion (perhaps based on L_{Aeq}) which takes into account differences in noise sources and situational factors. Researchers could hasten that day by: (1) using comparable noise measures (based on L_{Aeq}); (2) using a core set of comparable annoyance measures; (3) studying reactions to noise under varying acoustical and environmental conditions; and (4) developing methods of combining the unique contributions of laboratory and field studies.

REFERENCES

[28.1] International Organisation for Standardisation (1961) R 226. *Normal equal loudness contours for pure tones and threshold of hearing under free-field listening conditions.*

[28.2] International Organisation for Standardisation (1979) R 131. *Acoustics – Expression of physical and subjective magnitudes of sound or noise in air.*

[28.3] Stevens, S. S. (1961) *J. Acoust. Soc. Am.* **33**, 1577. Procedure for calculating loudness; Mk. VI.

[28.4] Zwicker, E. (1961) *J. Acoust. Soc. Am.* **33**, 248. Subdivision of the audible frequency range into critical bands (Frequensgruppen).

[28.5] Kryter, K. D. (1970) *The effects of noise on man.* Academic Press, New York.

[28.6] Borsky, P. N. (1972) *J. Sound Vib.* **20**, 527. Sonic boom exposure effects.

[28.7] HMSO (1975) *Noise Units* Report by Working Party for the Research Sub-Committee of the Noise Advisory Council.

[28.8] Fletcher, H. & Munson, W. A. (1933) *J. Acoust. Soc. Am.* **5**, 82. Loudness, its definition, measurement and calculation.

[28.9] Robinson, D. W. & Dadson, R. S. (1956) *Brit. J. Applied Physics* 7, 166. A re-determination of the equal loudness relation for pure tones.

[28.10] International Organisation for Standardisation (1966) R 532. *Method for calculating loudness level.*

[28.11] Kryter, K. D. & Pearsons, K. S. (1963) *J. Acoust. Soc. Am.* **35**, 886. Some effects of spectral content and duration on perceived noise level.

[28.12] Kryter, K. D. & Pearsons, K. S. (1964) *J. Acoust. Soc. Am.* **36**, 394. Modification of noy tables.

[28.13] Kryter, K. D. (1968) *J. Acoust. Soc. Am.* **43**, 344. Concepts of perceived noisiness, their implementation and application.

[28.14] Kryter, K. D. (1969) Possible modifications to procedures for the calculation of perceived noisiness. *NASA Report CR-1635.*

[28.15] Young, R. W. & Peterson, A. (1969) *J. Acoust. Soc. Am.* **45**, 834. On estimating noisiness of aircraft sounds.

[28.16] Federal Aviation Authority (1970) Regulations, Volume III, Part 36 in Federal Register 34 FR 19025. *Noise Standards: Aircraft type certification.*

[28.17] Stevens, S. S. (1972) *J. Acoust. Soc. Am.* **51**(2), 575. Perceived level of noise by Mark VII and Decibels (E).

[28.18] Robinson, D. W. (1971) *J. Sound Vib.* **14**, 279. Towards a unified system of noise assessment. (See also *NPL Aero Report Ac 38,* March 1969).

[28.19] British Standards Institution (1967) BS.4142. *Method of rating industrial noise affecting mixed residential and industrial areas* (Amended 1975

AMD 1661).

[28.20] International Organisation for Standardisation (1971) R 1996. *Noise assessment with respect to community response.*

[28.21] Griffiths, I. S. & Langdon, F. J. (1968) *J. Sound Vib.* **8**, 16. Subjective response to traffic noise.

[28.22] Scholes, W. F. & Sargent, J. W. (1971) *Applied Acoustics* **4**, 203. Designing against noise from road traffic.

[28.23] Department of the Environment (1973) *Noise insulation regulations.* HMSO, London.

[28.24] Department of the Environment (1975) *Calculation of road traffic noise.* HMSO, London.

[28.25] Galloway, W. J. & Bishop, D. E. (1970) Bolt, Beranek and Newman, Inc. *Final report No. FAA NO-70-9.* Noise exposure forcasts: Evolution, evaluation, extensions and land use interpretations.

[28.26] HMSO (1963) *Noise, final report.* Committee on the problem of noise. London.

[28.27] Department of Trade and Industry (1971) *Second survey of aircraft noise annoyance around London (Heathrow) Airport.* HMSO. London.

[28.28] McKennell, A. C. (1971) *WHO Conference on Noise Standards, Geneva, Dec. 1971.* Community response to noise; the evidence from sample surveys.

[28.29] Fields, J. M. & Walker, J. G. (1978) Reactions to railway noise in Great Britain. *Proceedings of Internoise 78,* San Francisco.

[28.30] Fields, J. M. & Walker, J. G. (1978) Reactions to railway noise; a survey near railway lines in Great Britain. *ISVR Technical Report No. 102.* University of Southampton.

[28.31] Fields, J. M. & Walker, J. G. (1982). *J. Sound Vib.* **85**(1). The response to railway noise in residential areas in Great Britain (in press).

[28.32] Nimura, T., Sone, T. & Kono, S. (1973) Some considerations on noise problems of high speed railways in Japan. *Internoise 73, Copenhagen, Paper D22Y 17.*

[28.33] Aubree, D. (1973) *Acoustic and sociological survey to define a scale of annoyance felt by people in their homes due to the noise of railroad trains.* CSTB. Nantes, France.

[28.34] British Standards Institution (1975) BS.5228. *Code of Practice for noise control on construction and demolition sites.*

[28.35] Eldred, K. M. (1975) *J. Sound Vib.* **43**, 137-147. Assessment of community noise.

[28.36] US Environmental Protection Agency (1974) *Report 550/9-74-004.* Information on levels of environmental noise requisite to public health and welfare with an adequate margin of safety.

[28.37] US National Research Council (1977) *Report of Working Group 69.* Guidlines for preparing environmental impact statements on noise.

[28.38] Bottom, C. G. (1971) *J. Sound Vib.* **19**, 473. A social survey into annoyance caused by the interaction of aircraft noise and traffic noise.

[28.39] Noise Advisory Council (1978) *A guide to measurement and prediction of the equivalent continuous sound level L_{eq}*. HMSO, London.

[28.40] Schultz, T. J. (1978) *J. Acoust. Soc. Am.* **64**(2), 377–405. Synthesis of social surveys on noise annoyance.

[28.41] Bastenier, H., Klosterkoetter, W. & Large, J. B. (1975) *Damage and annoyance caused by noise*. Commission of the European Communities. EUR.5398e.

[28.42] Rice, C. G. (1977) *J. Sound Vib.* **52**, 325-344. Investigation of the trade-off effects of aircraft noise and number.

[28.43] Rice, C. G. (1977) *J. Sound Vib.* **52**, 345-364. Development of cumulative noise measure for the prediction of general annoyance in an average population.

[28.44] Rice, C. G. (1978) Trade-off effects of aircraft noise and number. *Proc. Intern. Commission of Biological Effects of Noise* – Frieburg, Germany. September 1978.

[28.45] House, M. E. (1978) University of Southampton. Personal communication.

[28.46] Fields, J. M. & Walker, J. G. (1982) *J. Sound Vib.* **81**(1), 51-80. Comparing the relationships between noise level and annoyance in different surveys: a railway noise *vs* aircraft and road traffic comparison.

[28.47] Galloway, W. J. (1974) AMRC–TR–73–106. *Community noise exposure resulting from aircraft operations: technical review.*

[28.48] Robinson, D. W. (1970) *Aeronautical Journal* April, 147-160. The assessment of noise, with particular reference to aircraft.

[28.49] Borsky, P. N. (1976) *Sound & Vibration,* December, 18-21. Sleep interference and annoyance by aircraft noise.

[28.50] Ollerhead, J. B. (1977) *Internoise 77, B692-697.* Variation of community noise sensitivity with time of day.

[28.51] Powell, C. A. (1979) *NASA Technical Paper 1478.* Laboratory study of annoyance to combined airplane and road-traffic noise.

[28.52] NASA/FAA (1980) *Proceedings of Workshop.* Time of day day corrections to aircraft noise metrics.

[28.53] Rice, C. G. (1982) *ISVR Memorandum 623, University of Southampton.* A synthesis of studies on noise induced sleep disturbance.

[28.54] Ollerhead, J. B. (1978) *Internoise 78, 579-584.* Predicting public reaction to noise from mixed sources.

[28.55] Taylor, S. M. (1982) *J. Sound Vib.* **81**(1), 123-138. A comparison of models to predict annoyance reactions to noise from mixed sources.

[28.56] Powell, C. A. (1979) *NASA Technical Paper 1479.* A summation and inhibition model of annoyance response to multiple community noise sources.

CHAPTER 29

Occupational hearing loss and hearing conservation

A. M. Martin

Institute of Sound and Vibration Research, University of Southampton

29.1 INTRODUCTION

Modern technology has created many environmental pollutants of which noise is an immediate and identifiable example. Many industrial processes generate noise of sufficient sound level to cause deafness, and many of these processes have existed since the industrial revolution. Occupational hearing loss is not a new phenomenon; for example 'boilermaker's deafness' was well known among foundry workers in the nineteenth century, and gunpowder explosions are reported to have caused deafness amongst gun-crews in the Battle of Trafalgar.

This chapter is concerned with the general subject of hearing conservation, and gives a brief description of the physiology of the ear, noise-induced hearing loss, social handicap, assessment of hearing hazard, and the principles of hearing protectors and audiometry as integral parts of a hearing conservation programme.

29.2 THE EAR

The transmission of sound through the ear and the mechanism of hearing are described briefly in this section. As illustrated in Fig. 29.1, the ear can be considered in three parts which are generally called the outer, middle, and inner ears. The ear canal (auditory meatus) extends from the pinna to the ear-drum (tympanic membrane). The canal itself is approximately oval in cross-section and extends for about 25 mm into the head.

The middle ear is an air-filled cavity which contains a chain of three small movable bones called the auditory ossicles (the malleus, which is partly within the ear-drum, the incus, and the stapes), as well as two minute muscles which contract in an involuntary reflex response to intense sounds. The original function of this reflex contraction, which reduces the transmission of sound across the middle ear, is not clear but is thought to be related to the improvement of speech discrimination rather than the protection of the inner ear from loud noises. The Eustachian tube connects the middle ear cavity to the pharynx and allows the equalisation of air pressure on either side of the ear-drum.

The footplate of the stapes is located in the oval window leading to the inner ear. The inner ear is a complex system of fluid-filled cavities positioned deep in the temporal bone of the skull so that it is physically well protected. The system of cavities includes the cochlea and three semicircular canals oriented at right angles to each other. These canals form part of the body's balancing mechanism and need not concern us further.

The cochlea transduces the mechanical vibrations of the ossicles into neural impulses. It is extremely small (about 8 mm in diameter), filled with fluid, and similar in shape to a snail's shell, having about $2\frac{3}{4}$ turns coiled round a central bony pillar which also carries the auditory nerve fibres from the cochlea to the brain.

The cochlea itself is split longitudinally into two fluid-filled sections by the cochlear partition which is itself a tube filled with another fluid. This is illustrated in Fig. 29.2 which is a simplified representation of the cross-section of the cochlea. The sections of the cochlea are clearly shown. The lower part of the partition is formed by a flexible membrane called the basilar membrane which supports the hair cells as illustrated in Fig. 29.3. There are three rows of outer hair cells and one row of inner hair cells. They are the sensors which initiate neural impulses to the brain to produce the sensation of hearing. Immediately above the hair cells is the tectorial membrane which is relatively rigid, and the hairs of the hair cells are probably attached to its undersides. The upper part of the cochlear partition is called Reissner's membrane.

29.3 THE MECHANISM OF HEARING

The chain of events that leads to the sensation of hearing starts with the passage of sound waves down the ear canal. The ear-drum vibrates when these pressure fluctuations impinge upon it. The movement of the ear-drum is transmitted by the ossicles to the fluid in the cochlea via the oval window. The ossicular chain may be considered as an impedance-matching device (or mechanical transformer) which greatly enhances the flow of sound energy across the air/fluid interface. This comes about partly owing to a mechanical advantage and partly to the difference in areas between the ear-drum and the footplate of the stapes. Movement of the stapes sets up a series of complex travelling waves within the cochlear fluids. There is a small hole called the helicotrema in the cochlear partition at the apex of the cochlea, and another small flexible membrane called the round window, in the lower part of the cochlea. These enable the fluid motion to take place. This is best illustrated if the cochlea is considered as a straight tube rather than the spiral it is. The round window and the helicotrema are shown in Fig. 29.4. The basilar membrane is displaced as a result of the complex fluid motion, and the relative motion of the flexible basilar membrane and rigid tectorial membrane bends the hairs of the hair cells. Movement of the hairs causes electrochemical changes to take place in the hair cells which are transmitted as bursts of neural impulses along the auditory nerve to the brain. The

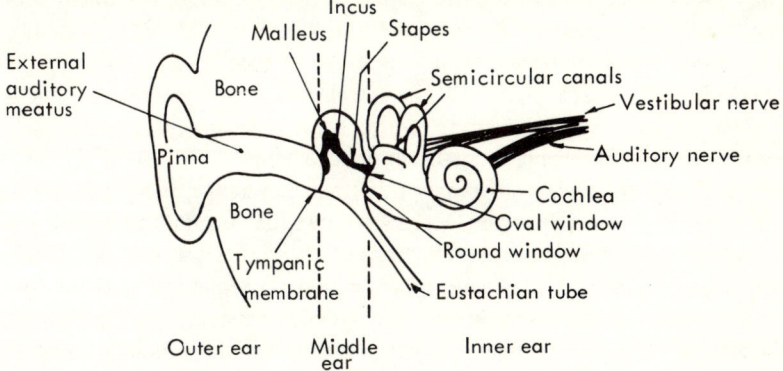

Fig. 29.1 – The main components of the ear.

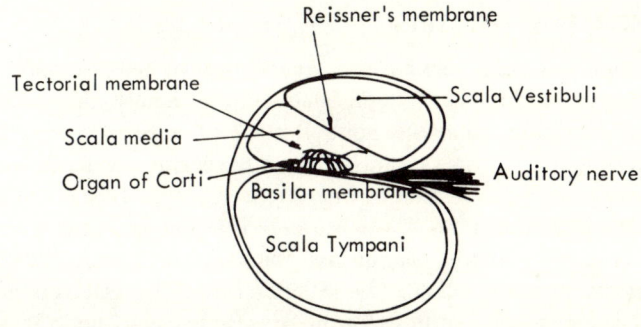

Fig. 29.2 – Section through one turn of the cochlea showing its three sections. The Basilar membrane, tectorial membrane, and hair cells are not shown in any detail.

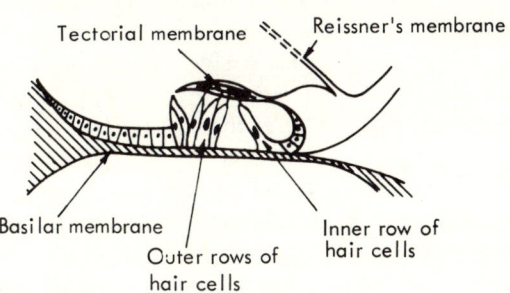

Fig. 29.3 – Diagram of the hair cells on the Basilar membrane together with the tectorial membrane (Organ of Corti).

ability of the ear to discriminate between different sounds it understood to result partly from analysis of the sound in the cochlea and partly from analysis in the brain. High frequencies are sensed at the basal end (round window) of the cochlea, and low frequencies at the apical (helicotrema) end.

The ear of a normal healthy young adult is generally regarded as being sensitive to sounds having frequencies in the range from 20 Hz to 20 kHz. The ear is not equally sensitive over this frequency range, the sensitivity being greatest in the 1 to 4 kHz region. Fig. 29.5 shows the binaural free-field threshold of hearing at different frequencies in terms of sound pressure level, and illustrates the particular sensitivity of hearing in the speech frequency range of sounds. The cochlea is an extremely sensitive and delicate mechanism. The amplitude of vibration of the basilar membrane at normal threshold is thought to be of the order of 10^{-10} mm, which is of the order of the diameter of a hydrogen molecule. The dynamic range of a normal young ear is considered to be from about -15 to 150 dB SPL, although damage to the cochlea may start to occur above about 85 dB.

29.4 NOISE–INDUCED HEARING LOSS

Any reduction in hearing sensitivity is generally called 'hearing loss'. There are many causes of hearing loss, but the main concern here is that which results from habitual exposure to intense noise. It is important to appreciate that noise-induced hearing loss damages the hair cells in the cochlea, while the ear-drum is rarely if ever damaged as a result of exposure to industrial-type noise. The exact nature of the cellular changes in the hair cells is not clear, but it is certain that the majority of noise-induced hearing loss results from their gradual deterioration, due to habitual excessive stimulation. Deterioration of hair cells is thought to be caused by some or all of the following processes: metabolic exhaustion and depletion of important nutrients, ionic changes in the surrounding fluids, vascular insufficiency and oxygen depletion, and at high intensities mechanical damage to the Organ of Corti. Whatever the process the damage is irreversible and cannot be improved by any known surgical or drug treatment. Affected hair cells die off and are not replaced.

A further cause of irreversible deafness in the inner ear is that resulting from the ageing process which is termed presbyacusis. As age increases, hearing sensitivity is reduced, particularly at the high frequencies. Fig. 29.6 shows the average effect of ageing on hearing sensitivity and includes the range of the most important speech frequencies. It can be seen that even in old age, the hearing loss in this region is not severe. Whilst presbyacusis may not be a serious disability in itself, when added to noise-induced hearing loss, which may have been caused at an early age and was then no great disability, the combined effect can cause a serious handicap in later life.

The earliest manifestation of hearing damage resulting from exposure to intense noise is a change in the threshold sensitivity of hearing of high-frequency

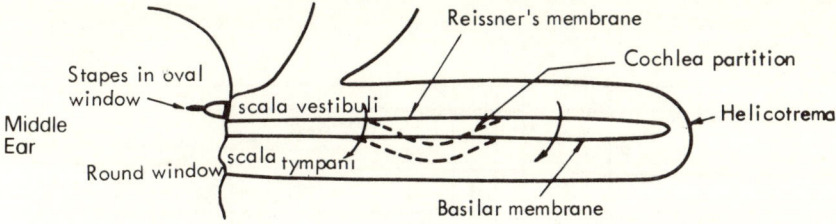

Fig. 29.4 – Simplified illustration of the cochlea considered as a straight tube. A positive pressure wave moves the foot of the stapes into the oval window causing the fluid to move and to bend the cochlear partition. The fluid moves through the helicotrema and displaces the membrane in the oval window; the dashed lines show in an exaggerated manner the deflection caused by a low-frequency sound wave.

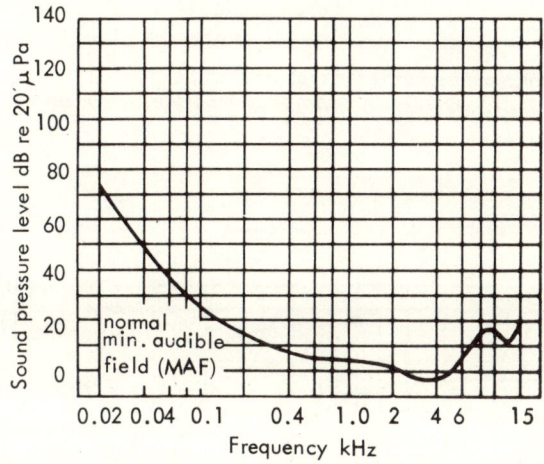

Fig. 29.5 – Normal binaural minimum audible field (MAF) at different frequencies.

sounds in the region of 4 kHz. However, it is unlikely that in the early stages of noise-induced hearing loss the person will notice any real changes in his hearing ability.

Before noise-induced hearing loss becomes permanent, exposure to intense noise will result in temporary hearing loss. This is noticeable as a dullness in hearing, and it is sometimes accompanied by a ringing or whistling sensation in the ears called tinnitus, although the latter may only be noticeable in quiet surroundings. These effects may disappear a few hours after the noise exposure

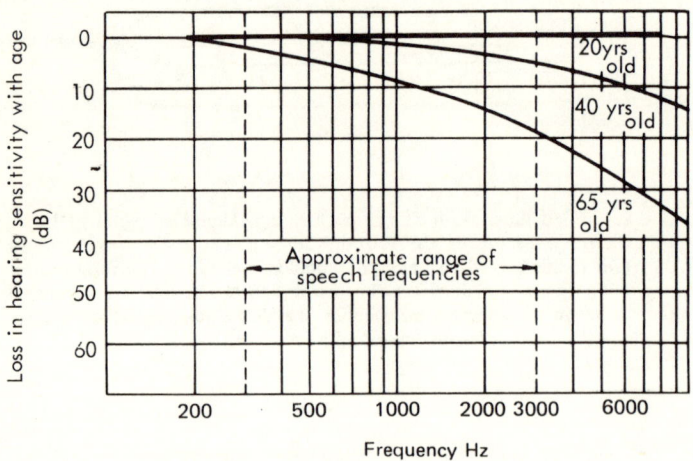

Fig. 29.6 – Audiograms showing the typical reduction of hearing sensitivity with increasing age. The approximate range of speech frequencies is also shown.

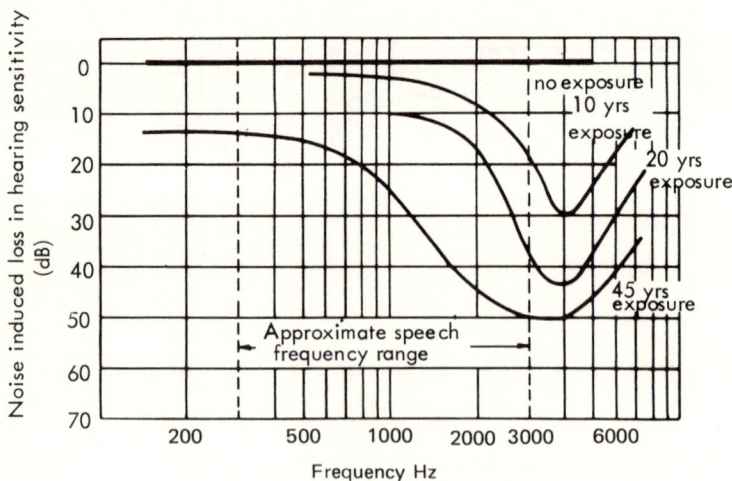

Fig. 29.7 – Audiograms showing the typical reduction of hearing sensitivity as a result of long-term exposure to industrial noise, excluding effects of age. The approximate range of speech frequencies is shown.

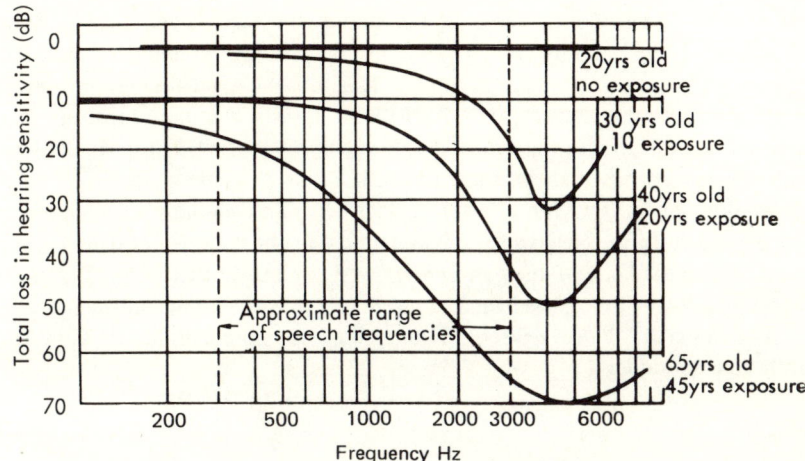

Fig. 29.8 — Audiograms showing typical hearing losses resulting from both
presbyacusis and noise-induced hearing loss.

has ceased, but repeated exposure to noise can result in the hearing loss becoming
permanent.

An audiogram taken from a person suffering from noise-induced hearing
loss will usually show the greatest loss of hearing sensitivity in the 4 kHz region,
which is typically the region most sensitive to damage resulting from many
types of industrial noise. The hearing loss starts in this region and with continued
exposure, the *4 kHz dip,* as it is widely known, deepens and spreads to the lower
and higher frequencies. An example of the growth of hearing loss with continued
exposure to noise (but excluding presbyacusis) is shown in Fig. 29.7.

The noise-induced hearing loss of the type illustrated in Fig. 29.7 results in
difficulties in hearing high-pitched sounds which can include, for instance, some
industrial warning signals. As the 4 kHz dip widens and the lower frequencies
are affected, speech reception is increasingly affected. Since most information
is borne by the consonant components of speech and many of these contain
high-frequency sound energy, speech intelligibility is reduced. Eventually, whilst
the affected person will still be able to hear the speech he will not be able to
discriminate sufficiently well to enable him to understand the information it
contains. This problem becomes particularly severe when noise-induced deafness
is added to presbyacusis, typical resultant hearing losses being shown in
Fig. 29.8. It can be seen that an elderly person who also suffers from noise-
induced hearing loss will have a more serious hearing handicap in the speech
frequency range.

29.5 HAZARDOUS NOISE EXPOSURE

Much of the impetus for the establishment of an accepted method for assessing hearing hazard in Britain is derived from a study on steady-state noise by Burns & Robinson [29.1] and by Martin [29.2, 29.3] on industrial impulse noise. This work showed that a good measure of noise hazard is the A-weighted sound energy of the noise. This work has since formed the basis for a method of assessing potential hazard to hearing from noise, and a number of European national bodies have also based their standard methods of assessment on this type of information (e.g. ISO R1996 [29.4]). A single and relatively simple principle may now be said to form the basis for the assessment of the majority of occupational noises. A-weighted sound energy may be considered to be the unifying factor in what, up to recent times, has been a somewhat confused area of knowledge.

The concept of A-weighted sound energy is embodied in an equivalent continuous sound level L_{eq}, which may be defined as

$$L_{eq} = 10 \log_{10} \frac{1}{8} \int_0^T \frac{(P_a(t))^2 \, dt}{P_0}$$

where P_a is the instantaneous A-weighted sound pressure in Pascals and T is the total daily working period in hours. Alternatively L_{eq} can also be described as the level of continuous noise, in dB(A), which in the course of a working day would cause the same sound energy to be received as that due to actual noise over a typical day. The parameter L_{eq} may also be considered as a measure of *noise dose,* a concept familiar to many other aspects of occupational medicine but relatively new to the field of acoustics. The symbol $L_{Aeq, T}$ is increasingly and more precisely used instead of L_{eq} in International Standards; 'A' denotes that A-weighted sound pressure level is being used; 'T' refers to the time period over which the L_{eq} is determined.

A-weighted sound energy received during a noise exposure may be deduced from the product of the noise level in dB(A) and the duration of exposure. A doubling of energy represents an increase in noise level of 3 dB. Thus, for example, exposure to a steady sound of 90 dB(A) for a given period is equivalent in terms of sound energy, and therefore hazard, to an exposure to 93 dB(A) for half that period. Table 29.1 gives examples of values of sound level and exposure duration which, when considered together, represent an L_{eq} of 90 dB(A) for 8 hours. These figures are considered at the present time in Britain to be maximum permissible limits for exposure to steady noise.

The Health and Safety Executive *Code of Practice for reducing the exposure of employed persons to noise* [29.5] is the most suitable document for *practical*

Table 29.1
Permissible equivalent continuous sound level
(L_{eq}) for an 8-hour working day

Sound level dB(A)	Permissible duration of exposure per day
90	8 hours
93	4
96	2
99	1
102	30 minutes
105	15
108	$7\frac{1}{2}$
111	225 seconds
114	112
117	56
120	28
123	14
126	7
129	$3\frac{1}{2}$
132	2
135	less than 1 second

application to the industrial situation in the United Kingdom at present. This specifies a limit for exposure to noise of an L_{eq} of 90 dB(A) and describes methods of measurement which can be used to determine whether the limit is exceeded. The *Code* makes the assumption that steady-state noise, fluctuating, intermittent, and impact noise may all be described in terms of L_{eq}. Furthermore, it recommends overriding limits for the unprotected ear, these being a sound pressure level of 135 dB for steady-state noise and 150 dB for impulse noise. The *Code* is important in Britain because it specifies safe conditions of work, and HM Factory Inspectorate relies upon its recommendations to define the auditory safety of a working environment. It is about to become 'approved' under the Health and Safety at Work Etc. Act (1974) and thereby will have the force of law behind it. The *Code* states that its primary aim is the reduction of hazardous noise by engineering means and only by personal hearing protection as a last resort. It is intended to supplement the new draft noise regulations of the Health and Saftey at Work Act (Health and Safety Commision, 1982) [29.6], which are also about to be published in completed and final form.

29.6 PERSONAL HEARING PROTECTION

The most obvious and efficient method of reducing exposure to noise is to prevent the generation of the noise in the first place. When it is not possible to reduce noise levels to within safe limits by treatment of the source using noise control techniques or by replacement with an inherently quieter machine, the problem can sometimes be solved by covering the source with an acoustic hood or by the use of barriers. Another obvious means of noise reduction is to remove either the offending machine or the persons exposed to another location; and similar results can be achieved by limiting the time that persons are exposed to the noise.

There are many situations, however, where such noise control techniques are either impracticable, uneconomic, or not sufficient. In these cases, personal hearing protective devices must be provided and must be worn. Nevertheless, hearing protection should be considered as a temporary measure only; engineering methods of noise control are the only satisfactory solution in the long term [29.7].

Where a hazard to hearing exists, the onus is on the employer to provide hearing protectors that attenuate noise sufficiently to remove the hazard. In the eyes of the law, as it stands at the moment in the UK, it is not good enough merely to supply hearing protection. The devices provided must be capable of removing the hazard to hearing from the noise environment in which they are being worn. Consequently, knowledge is required of the physical characteristics of the offending noise as well as the acoustic attenuation characteristics of the hearing protectors provided. Furthermore, according to the Health and Safety at Work Act, there is a certain degree of responsibility placed on the employer to educate the employee about the need to wear hearing protection and to persuade him to do so. Hence other non-acoustical properties such as comfort and wearer-acceptability should also be taken into account. Employers and other persons responsible for providing hearing protection should therefore have a good working knowledge of the many aspects and requirements of a hearing conservation programme.

29.6.1 Types of hearing protector

There are many brands and types of hearing protector available on the market, and there are several factors to be considered, in addition to the protection to hearing they provide, in selecting the most suitable type for any given situation. Some of these are comfort, cost, durability, availability, wearer acceptance, and hygiene. A more detailed account can be found in Harris & Martin [29.8], and Alberti [29.0].

Prefabricated earplugs
The best prefabricated earplugs for general industrial use are available in three to five different sizes. They are made from soft flexible material that will conform

to the many different ear canal shapes, thus providing a snug, airtight, and comfortable fit.

Disposable and malleable earplugs

These types of earplug are usually fashioned from low-cost materials such as cotton, wax, glass wool, expanded plastic foam, and mixtures of these and other substances. Typically, a small cone of the material is hand-formed and the apex of the cone is inserted into the ear canal with sufficient force so that it conforms to the shape of the canal and expands in position. In general, malleable and disposable earplugs made correctly of easily formed materials are comfortable and capable of providing attenuation values similar to or greater than those afforded by prefabricated plugs.

Individually moulded earplugs

These are usually made from some form of silicone rubber and are actually moulded in a permanent form within the ear canal. Usually the earplug material is supplied with a curing agent, and the two are mixed to a putty-like consistency and then this is inserted into the ear canal of the person to be fitted. Having cured, the earplugs are in a permanent form and may be removed and re-inserted any number of times without affecting their performance. In situations where difficulty is encountered in persuading men to wear hearing protection, the provision of a personal and individually moulded device that will fit only the person for whom they are intented (like spectacles and false teeth!) is a psychological advantage.

Earmuffs

Most types of earmuff are of similar design and are made from rigid cups specially designed to cover the external ear completely. They are held against the sides of the head by a spring-loaded adjustable band and are sealed to the head with soft circumaural cushion seals.

Earmuff seals may be liquid filled or plastic foam filled. Liquid filled seals usually provide marginally better protection with less head band tension, all things being equal, but suffer from the problem of leakage if treated roughly. Modern foam filled seals are almost as good as the liquid seals and have the additional advantage of robustness. In any case earmuffs should be provided with seals that are easily and separately replaceable in the factory environment. It is usually necessary to replace earmuff seals every 3 to 6 months to maintain good hygiene and attenuation.

29.6.2 Noise reduction

The prime function of hearing protectors is to reduce the noise level at the wearer's ears to within safe limits. Information on the ability and consistency of hearing protectors to attenuate sound should be closely examined when consider-

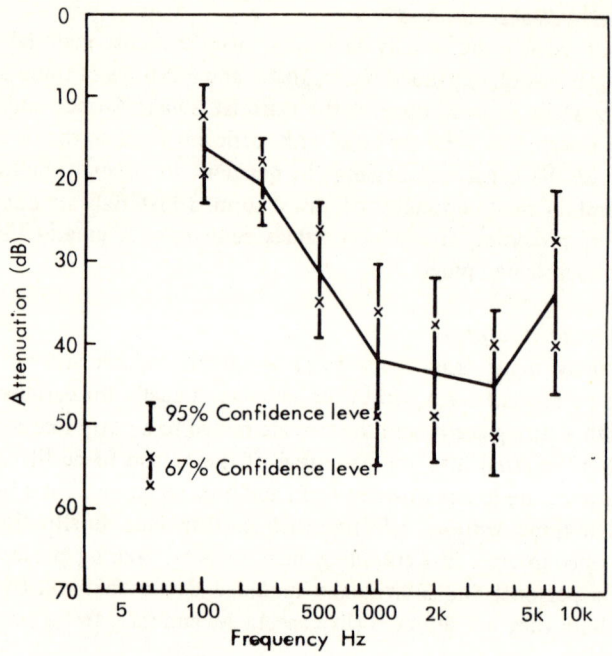

Fig. 29.9 – Attenuation characteristics of a good fluid-seal earmuff.

ing which type is most suitable for a particular noise environment.

The acoustic attenuation of hearing protectors is usually expressed in decibels attenuation at various test frequencies and is usually described in graphical form, as for example in Figs. 29.9 and 29.10, or in tabular form as given in Table 29.2. To estimate the protection supplied by a particular hearing protector, its attenuation-frequency characteristics must be compared with (and subtracted from) the sound level–frequency characteristics of the noise concerned. This, according to the *Code of Practice*, will then provide the sound level–frequency characteristics of the noise present at the ear canal.

A second, but equally important, measure associated with attenuation is the degree of scatter of the attenuation data as measured on different subjects. This is usually expressed as the standard deviation about the grand mean, or as the interquartile range about the median. This figure should accompany each attenuation datum point when expressing the attenuation (see for example

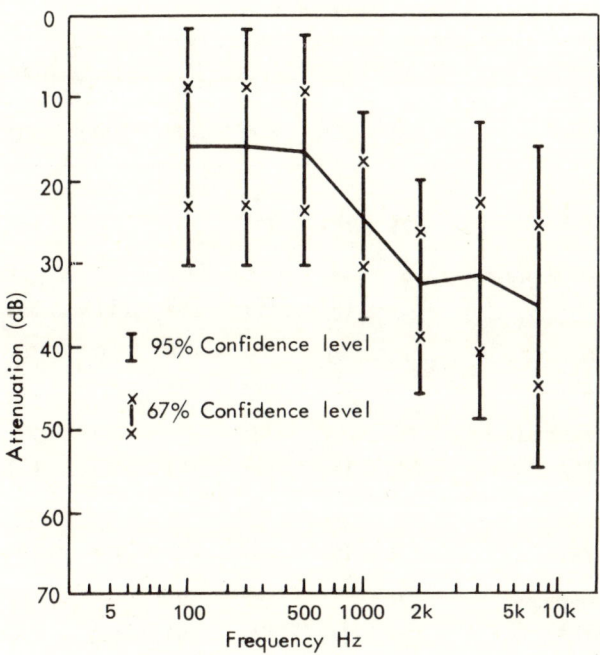

Fig. 29.10 – Attenuation characteristics of a well fitting earplug.

Table 29.2). Depending upon the measurement technique used, this figure tends to give, among other things, a measure of the hearing protector's ability to fit different individuals. The smaller the spread, the better the hearing protectors adapt to different head or ear canal shapes and sizes. It also provides a measure of the accuracy with which the attenuation measurements were carried out.

Fig. 29.9 shows an example of the mean attenuation characteristics of an earmuff, plotted with one and two standard deviations representing a spread of 67% and 95% of the attenuation data respectively. Thus, at a frequency of 4.0 kHz, 67% of the people wearing the muff will be supplied with 40 to 50 dB protection and 95% with 35 to 55 dB protection. In other words, there will be 5% of persons who will receive less than 35 dB protection and 33% who get less than 40 dB protection at this frequency for this particular earmuff. Fig. 29.10 shows similar attenuation characteristics for an earplug. As can be seen, the spread of the attenuation data in this case is greater than that for earmuffs.

Table 29.2
Typical mean attenuation and standard deviation (S.D.)
characteristics, in dB, of different types of hearing protection

Test frequency Hz	125	250	500	1000	2000	4000	8000
Dry cotton wool plugs:	2	3	4	8	12	12	9
(S.D.)	(2)	(2)	(3)	(3)	(6)	(4)	(5)
Waxed cotton wool plugs:	6	10	12	16	27	32	26
(S.D.)	(7)	(9)	(9)	(8)	(11)	(9)	(9)
Glass down plugs:	7	11	13	17	29	35	31
(S.D.)	(4)	(5)	(4)	(7)	(6)	(7)	(8)
Personalised earmould plugs:	15	15	16	17	30	41	28
(S.D.)	(7)	(8)	(5)	(5)	(5)	(5)	(7)
V–51R type plugs:	21	21	22	27	32	32	33
(S.D.)	(7)	(9)	(9)	(7)	(5)	(8)	(9)
Expanding foam plugs:	26	27	29	30	33	44	44
(S.D.)	(8)	(7)	(7)	(6)	(5)	(5)	(6)
Foam-seal muffs:	8	14	24	35	36	43	31
(S.D.)	(6)	(5)	(6)	(8)	(7)	(8)	(8)
Fluid-seal muffs:	13	20	33	35	38	47	41
(S.D.)	(6)	(6)	(6)	(6)	(7)	(8)	(8)
Flying helmet:	14	17	29	32	48	59	54
(S.D.)	(4)	(5)	(4)	(5)	(7)	(9)	(9)

29.6.3 Evaluation of attenuation of hearing protectors

There are a number of national standard methods of measurements of attenuation available, published by different countries, as for example that specified in the American Standard ASA 24.22-1957 [29.10]. This requires that threshold measurements are carried out three times on 10 subjects under free-field conditions, with and without the protectors being worn, using a single loudspeaker to generate a pure-tone audiometric signal. The mean difference between these two values is taken as a measure of the attenuation. However, it has been found that measurements carried out by different laboratories on the same protector, following this same procedure, can produce widely differing results. Consequently, the American National Standards Institute has rewritten this standard, and the new standard ASA-STD1-1975 [29.11] is replacing the previous one. The British Standards Institution has also published a standard method BS 5108:1974 [29.12] applicable in the UK. This, although based on the threshold shift technique, attempts to provide a more consistent evaluation of attenuation. This is attained by improvements in experimental technique, such as a more diffuse test sound field, $\frac{1}{3}$-octave bands of random noise as the test signal, rigorously specified fitting procedures, and the revised estimate of the number of measurements required.

When using attenuation data as one of the guidelines for deciding upon the type and make of hearing protector required, the method by which this information is obtained and the laboratory carrying out the measurements should be carefully noted. It should be ensured that the attenuation data published by manufacturers are measured following a recognised standard procedure such as one of those mentioned above [29.13].

29.6.4 Amount of protection provided in practice

The Health and Safety Executive's *Code of Practice on reducing the exposure of employed persons to noise* [29.5] recommends that the actual sound level in dB(A) at the wearer's ear inside the protector should be calculated for the particular noise environment in which it is to be worn. The *Code* provides a scheme for performing this calculation, which requires knowledge of the means and standard deviations of the attenuation and also an octave-band frequency analysis of the noise. The measure of the attenuation used by the Code is called "the assumed protection", which is the "mean minus one standard deviation" which covers 84% of wearers. These data are also published by the HSE in the form of *Guidance notes*.

Such calculations make the assumption that the hearing protectors in question are worn continuously throughout the time the users are exposed to the noise. If, however, the protectors are not worn all the time, their effective protection is severely reduced. Even if they are not worn for only a few minutes in a day, their effective attenuation may be halved.

Fig. 29.11 from Else [29.14] illustrates the maximum protection to be expected from hearing protectors providing differing amounts of attenuation as a function of the percentage of the time they are worn. As can be seen, even as 'infinite' protector, which allows no sound to reach the ears at all, is effectively reduced in efficacy to about 20 dB(A) protection if worn for 99% of the time. In the case of protectors encountered in practice this effect is even worse. For example, if hearing protectors with an effective attenuation of 30 dB(A) are worn in a noise environment of 115 dB(A), the sound level at the wearer's ears will be 85 dB(A) which may be considered as a non-hazardous exposure. If,

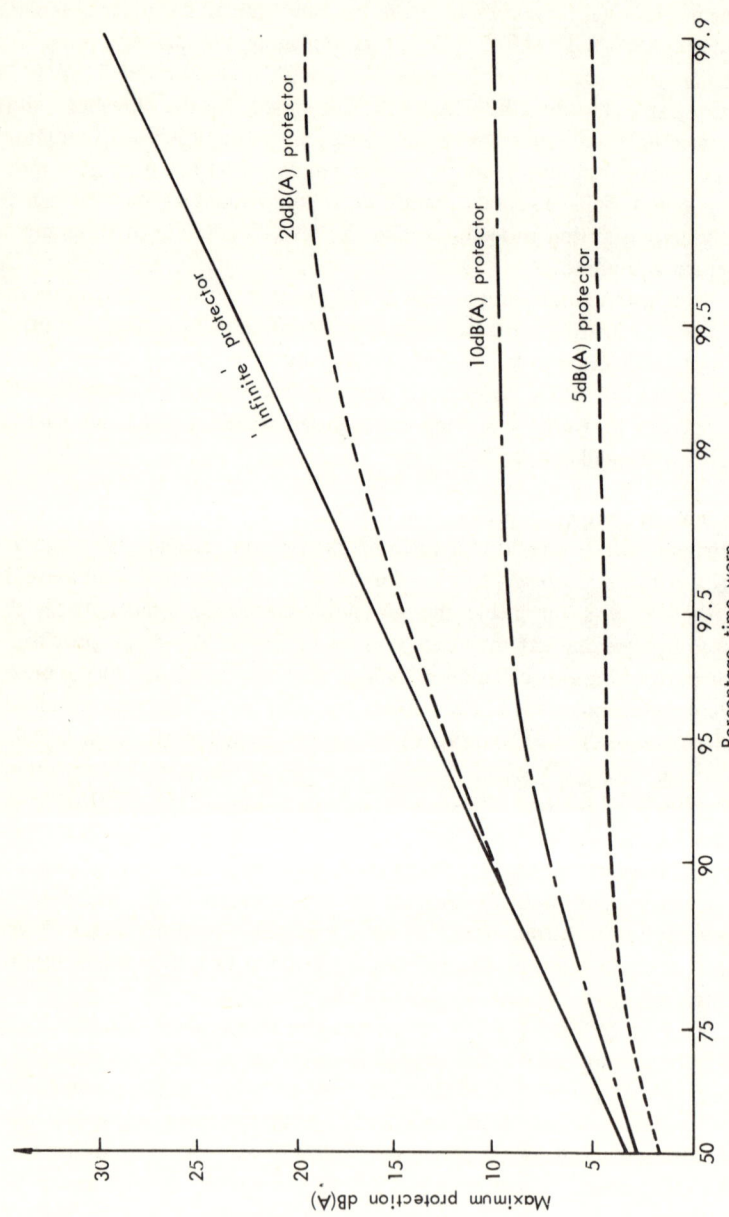

Fig. 29.11 — Maximum protection as a function of percentage time worn for different hearing protectors.

however, the user fails to wear them for a total of only 2% of the working day, i.e. about 10 minutes in 8 hours, he will be exposed to 115 dB(A) for 10 minutes, which is equal to an equivalent continuous noise level of about 98 dB(A) for 8 hours. Consequently, the effective protection has been reduced from 30 dB(A) to 17 dB(A) merely because the user failed to wear the hearing protection for a total time of 10 minutes in one working day. Furthermore, he is now exposed to an effectively more hazardous noise.

These calculations make the most important point that, to be effective, all hearing protectors must be worn 100% of the time the user is exposed to noise. Even a few minutes a day of unprotected exposure to noise very severely reduces the protection provided for the rest of the time. Remember the adage: "The only effective hearing protector is the one that is worn".

29.6.5 Education of the workforce

This is most important if protection is to be effective, and is a requirement of UK Government noise regulations. The presence of 'noise hazard' warning signs, and the personal instruction which can be carried out during monitoring audiometry, are the major methods of persuasion. The latter is considered to be particularly effective, as the individual worker may be instructed in the method of fitting the earplugs and the importance of wearing them all the time in noise, at the same time as his audiogram is measured when his awareness of his hearing acuity and the problems of noise are enhanced.

Another important method of education is the insistence that all managerial staff and visitors wear protection when in, or passing through, a noise hazardous area. Although such exposures may not be hazardous, owing to the short duration of exposure, the worker will not appreciate the fact and will take the view that if the manager does not wear protection, why should he? The regular use of films and talks on noise and the hazards to hearing should also form a basic part of the education programme. These may also be backed up by the display of warning posters in prominent positions to remind the worker about hearing protection and its importance.

29.6.6 The selection of hearing protectors

Both insert-type and muff-type hearing protectors are suitable for particular situations and noise environments. They both have distinct advantages and disadvantages, and a summary of some of these is given below:

(i) Insert-type protectors
(a) Advantages
1. Earplugs are small and easily carried.
2. They can be worn conveniently and effectively with other personal safety equipment, such as glasses and helmets, and do not interfere with hair styles.

3. They are relatively comfortable to wear in hot environments.
4. They are convenient to wear when the head may be manoeuvred in confined spaces.
5. The initial cost of prefrabricated plugs is usually less than muffs, although malleable, disposable, or earmould protectors may cost as much or even more than muffs over a period.

(b) Disadvantages

1. Earplugs usually require more time and effort in fitting than muffs, especially initially.
2. The degree of protection provided is generally less and more variable between wearers than that provided by earmuffs.
3. Dirt and other contaminants may be inserted more easily into the ear canal during fitting and re-inserting.
4. Earplugs are not easily seen in the ear canal from any distance and thus make it more difficult to monitor groups of persons wearing them.
5. They can only be worn in healthy ear canals, and, even in some healthy canals, a period of time is necessary for acceptance.

(ii) Muff-type protectors

(a) Advantages

1. The protection of a good earmuff is generally greater and less variable than a good earplug.
2. A single size of earmuffs fits a large percentage of heads.
3. Earmuffs can be easily seen at a distance, thus making it relatively easy to check that they are being worn.
4. They are generally accepted more readily at the beginning of a hearing conservation programme than plugs.
5. Muffs can be worn even with many minor ear infections.
6. They are not misplaced or lost as easily as earplugs.

(b) Disadvantages

1. Muffs may become uncomfortable in hot environments.
2. They are not so easily carried or stored as earplugs.
3. They are less compatible with other personally worn items such as glasses and headgear.
4. The suspension forces may be reduced by usage, or by deliberate bending of the headband, so that the protection provided may be less than expected.
5. The relatively large muff size may not be acceptable where the head is manoeuvred in confined spaces.
6. Earmuffs are usually more expensive than most types of earplugs.

It is doubtful whether earplugs or earmuffs alone can satisfy all the needs of a hearing protection programme in any one organisation. The obvious advantages of each should be utilised wherever possible, so that a hearing protection programme may be made as acceptable as possible to the potential wearers.

the practical details of industrial audiometry are also described in the Health and Safety discussion documents on *Audiometry in industry* [29.15] and *Protection of hearing at work* [29.6].

29.7 MONITORING AUDIOMETRY

Damage risk criteria specify relatively safe levels for persons of average noise-sensitivity, i.e. hearing losses will not be serious in the majority of persons exposed to noise of this level. People with more-than-average noise sensitivity (for which there is no useful prognostic test) may suffer more serious hearing losses. For this reason, and because ear protection is seldom used to its best advantage by all concerned and its protective effect varies even if worn, it is desirable to monitor changes in hearing as they occur. These requirements and A good monitoring audiometric programme might consist of:

(a) Pre-employment audiogram, preferably repeated, over a wide range of audiometric frequencies, e.g. 250, 500, 1000, 2000, 3000, 4000, 6000, 8000 Hz.

(b) Retests at end of 6 months, 1 year, 2 years, and then every second year; more frequently in individual cases needing closer watch.

If hearing losses are found to have developed or advanced, then stricter enforcement of ear protection or change of nature of employment is indicated.

The initial hearing tests must be done at the beginning of a working day, preferably after a week-end's rest away from the noise. Otherwise the hearing levels recorded are liable to include temporary changes in threshold due to auditory fatigue. A compromise for retests is to ask the worker to wear well-fitting good-quality earmuffs during *all* noise exposure in the 24 hours preceding the retest; if no significant deterioration is detected, the test result is accepted; if an apparent deterioration is found, he is retested again at the beginning of another working day.

A monitoring audiometry programme should be considered as an integral part of any hearing protection and hearing conservation programme. If hearing protection is supplied to personnel as the only means of reducing their noise exposure, it is essential that their hearing acuity is monitored regularly to ensure that the hearing protection is being used effectively. It also provides an extremely good propaganda and educational medium, whereby employees may be instructed in the need to wear protection and to be made aware of the hazard from noise. Moreover, it has several medico-legal advantages, not least being the detection of hearing loss on entry to a particular employment, so it cannot subsequently be attributed to noise exposure in that employment.

It is expected that industrial monitoring audiometry will be a legal requirement in the UK in noise levels above 105 dB(A) in the near future.

Audiometric information obtained from regular tests in industry should be examined by a fully qualified person who is expert in the interpretation of audiograms (e.g. a medical officer). Furthermore, it is probable that in any group of individuals, a certain percentage (about 5%) will be found to have disorders of the ear other than those produced by noise, and in these cases individuals should be referred to their own general practitioner for further medical advice and help.

Consideration should also be given to the ethical requirements of such medical information. Depending upon the persons carrying out the audiometry and the agreed policies of the medical department and management, there should be certain limitations placed upon the general availability of data of this type. It is obviously essential that the confidence of the work-people is maintained in the confidential nature of such medical information and those who have access to it.

It is apparent that many larger organisations may have the facilities and finances to set up a monitoring audiometry programme, and that it may be difficult for smaller firms to do this economically. However, there is available at least one commercial audiometric service which will carry out such a programme on site. This type of service makes it feasible for small organisations, and indeed the larger ones as well, to provide a thorough audiometric programme for the assessment of the hearing acuity of the work-people exposed to noise [29.8].

29.8 THE MANAGEMENT OF AN INDUSTRIAL HEARING CONSERVATION PROGRAMME

The basic procedures involved in the instigation of the Hearing Conservation Programme may be summarised as follows:

(i) *Measure the noise.* A detailed noise survey should be carried out in those areas which are thought to present a possible hazard to hearing.

(ii) *Evaluate the hazard.* Measured A-weighted equivalent continuous noise levels should be obtained and compared with the current limit of 90 dB(A), and all machines, workshops, and areas where this level is exceeded should be designated as 'Noise Hazardous Areas'.

(iii) *Noise reduction.* All hazardous noise sources should be reduced to a level below the limit following the engineering noise and vibration control techniques described elsewhere in this book. If reduction of the noise at source or by the use of acoustic and vibration barriers is neither economic, practical, nor sufficient, then as a last and temporary resort a *Hearing protection programme* should be instigated. This should be backed up by a monitoring audiometry programme.

An effective hearing conservation programme is not assured merely by making good properly-fitting hearing protectors available to those persons exposed to high-level noise. Management, medical, industrial hygiene, and safety personnel should all be made aware of the problems as well as those being exposed to the noise. They must all support the programme to the full if it is to be effective. In addition, such a programme will be much more efficient if a responsible member of the organisation is designated as a 'coordinator' to initiate the programme and follow it through, thereby sustaining management and workers' support.

The coordinator must determine the noise exposure patterns, both in terms of sound level and duration of exposure, for all persons under his responsibility, and then pinpoint those areas where noise control and hearing protection are necessary. He should evaluate other existing environmental factors such as temperature, humidity, and possible communication and warning-signal requirements. Also he should ascertain the need for other personal safety equipment such as goggles, helmets, and gloves in that these may affect the final choice of hearing protector. With such information and by personally experimenting with different types of hearing protection in the various environments, he will be able to decide upon the most suitable types of protector required for the different situations existing in his organisation.

Having started a hearing conservation programme, it is important that all efforts are made to maintain it. Education of the worker about noise and the need to wear hearing protectors is paramount. This should take the form of films and lectures on the hazards of noise, personal issues of leaflets, and posters and prominent warning signs displayed in the vicinity of all noise hazardous areas and on all noisy machinery, including hand-held tools.

The introduction of monitoring audiometry will soon be a legal requirement in the UK (Health and Safety Commission) [29.6] and must also be considered as a most powerful medium for education and propaganda, as well as a check upon the workers' hearing and thus the effectiveness of the hearing protection supplied.

A close relationship must be maintained by the coordinator with management, medical staff, and employees of all grades in order to be aware of reactions to various stages of the hearing conservation programme. In this way any problem will become evident early on, and corrective measures can be taken before any serious damage to the programme results.

29.9 CONCLUSIONS

A hearing conservation and audiometric programme, if carried out with enthusiasm and perseverance and given full backing by management, trade unions, and the medical staff, should drastically reduce the incidence of occupational noise-induced hearing loss within a particular organisation. Perseverance and propaganda are often required to overcome the reluctance to wear hearing protection *properly* that is shown by some employees, and this must be achieved if the programme is to be effective.

It should be apparent that the cost and effort required to ensure successful hearing conservation by hearing protection and audiometry may be equal to or even greater than that involved in other engineering noise control measures. Furthermore, a hearing protection programme should not be considered as an easy or cheap alternative to the more desirable elimination of the noise from the working environment.

29.10 REFERENCES

[29.1] Burns, W. & Robinson, D. W. (1970) HMSO London. Hearing and noise in industry.

[29.2] Martin, A. M. (1971) *Annals of Occupational Hygiene* **14** 11-23. Equivalent continuous noise level as a measure of injury from impact and impulse noise.

[29.3] Martin, A. M. (1976) Raven Press, New York. *Effects of noise on hearing,* Ed. by Henderson, D., Hamernik, R. P., Dosanjh, D. S. & Mills, J. M. The equal energy concept applied to impulse noise.

[29.4] International Standards Organisation (1982) ISO R1999. *Determination of occupational noise exposure and estimation of noise-induced hearing impairment.*

[29.5] Health and Safety Executive (1972) HMSO London. *Code of practice for reducing the exposure of employed persons to noise.*

[29.6] Health and Safety Commision (1982) HMSO London. *Protection of hearing at work.* Proposed noise regulations, draft approved code of practice and guidance note.

[29.7] Martin, A. M. (1976) Heineman, London. *Scientific foundations of otolaryngology,* Ed. by Hinchcliffe, R. & Harrison, E. F. Physical aspects of hearing conservation.

[29.8] Harris, G. G. & Martin, A. M. (1980) Trade and Technical Press, *Handbook of industrial safety.* Hearing conservation.

[29.9] Alberti, P. W. (1982) (Ed.) Raven Press. New York. Personal Hearing protection in Industry.

[29.10] American Standards Association (1957) ASAZ24-22. *Measurement of real-ear attenuation of ear protectors at threshold.*

[29.11] Acoustical Society of America (1975) ASA-STD-1. *Method for the measurement of real-ear protection of hearing protectors and physical attenuation of earmuffs.*

[29.12] British Standards Institution (1974) BS5108. *Method of measurement of the attenuation of hearing protectors at threshold.*

[29.13] Martin, A. M. (1977) *Annals of Occupational Hygiene* **20** 229–246. The acoustic attenuation of 26 hearing protectors evaluated following the British Standard procedure.

[29.14] Else, D. (1973) *Annals of Occupational Hygiene* **10** 415–418. A note on the protection afforded by hearing protectors.

[29.15] Health and Safety Executive (1978) HMSO London. *Audiometry in industry.*

CHAPTER 30

Effects of noise on people

P. A. Wilkins

Institute of Sound and Vibration Research, University of Southampton

30.1 INTRODUCTION

There have been numerous claims of a number of adverse non-auditory effects of noise on people. However, the majority of these claims are difficult to evaluate because of inadequate or conflicting information.

The phrase 'non-auditory effects' has been coined to encompass those effects on the body other than direct damage to the auditory system which is covered in Chapter 29. The distinction is clarified by considering the two distinct systems in the auditory pathways of the nervous system (see Fig. 30.1). One is a specific projection system which transmits coded neural impulses from the ear receptors to the higher brain centres for perception and interpretation. The

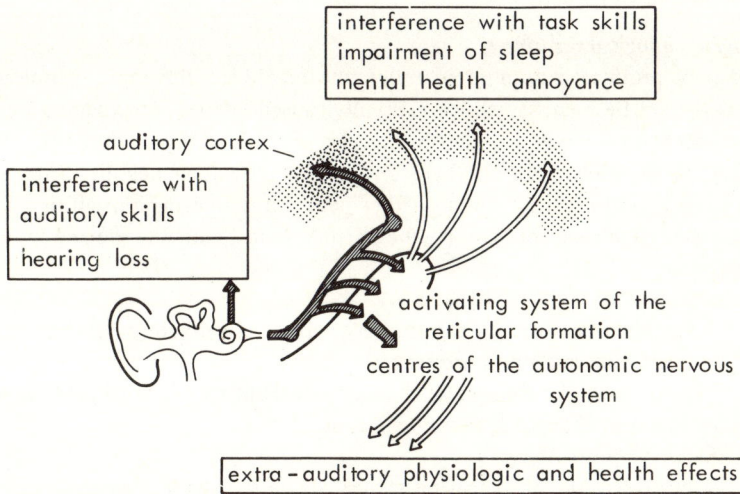

Fig. 30.1 – The physiologic organisation of the auditory pathways in the brain, and their relation to the effects of noise on man (reproduced with permission and modified from Grandjean [30.1]).

other is a non-specific projection system which branches off the main auditory pathway into the reticular formation, and then spreads diffusely into different areas of the brain as well as affecting the autonomic nervous system. This non-specific projection system is concerned with the state of arousal and various sensory, motor, and autonomic activities.

Very high levels of sound (above approximately 130 dB) can have direct effects on the body; however, in this chapter consideration will only be given to those effects which occur via the auditory system. These will be discussed in three sections:

(i)   HEALTH EFFECTS – physiological responses, sleep disturbance, and mental health.
(ii) PERFORMANCE EFFECTS – the results of field and laboratory studies, the mechanisms underlying these effects.
(iii) EFFECTS ON AUDITORY SKILLS – the masking of one sound by another, reduced loudness, and interference with speech communications.

This chapter aims to review what is known of these various effects with particular reference to occupational noise conditions and exposures. It is but an introduction, and the interested reader should consult more comprehensive reviews [30.1-30.6]. A particular aim is to highlight the limitations and deficiencies of current knowledge as a basis for assessing the results of future research studies.

30.2 HEALTH EFFECTS

30.2.1 Physiological responses

Sound may produce a number of reflex-like reactions. The most common of these is the startle response which is usually caused by loud, unexpected sounds and is evident as a flexing of various muscles and a blinking of the eyes. These reactions in themselves do not have any known effect on health, although in many work situations they could startle a person into injuring himself or others. Physiological changes accompanying the startle response and other noise exposures include:

(i)   A vascular response characterised by peripheral vaso-constriction, changes in heart rate and blood pressure.
(ii) Various glandular changes (such as increased output of adrenalin) evidenced as chemical changes in blood and urine.
(iii) Slow, deep breathing.
(iv) A change in the electrical resistance of the skin with changes in activity of the sweat glands.
(v)   Brief changes in skeletal muscle tension.

There is, however, no direct evidence that these changes are associated with any harmful effects. Instead, a variety of indirect evidence is often cited: for instance exposure of rats and rabbits to a mixture of sounds with average level of 83 dB has been found to cause enlarged ventricles and increased heart weight relative to groups of non-exposed control animals over a period of three weeks [30.7]. Extrapolation of animal studies to humans is, however, fraught with uncertainities.

By contrast, laboratory studies with people are constrained by the relatively low noise levels and short durations which can be used. The majority of tests reported are of short duration, so it remains open to question whether the body adapts to noise with prolonged exposure.

In an evolutionary context these physiological responses appear to have a survival value – an increase in blood supply to the brain and improved grip due to moistening of the palms being of value during an emergency. Whilst noise is clearly a stressor in evoking these changes, it need not necessarily have harmful effects. An increased output of adrenalin during a period performing a simulated industrial task in 'realistic industrial noise' may cause no greater harm than similar increases when viewing films depicting violence [30.8]. Alternatively, it is possible that repeated evoking of these changes with regular prolonged exposure to noise may contribute to health disorders. Evidence for such long-term effects could be provided by field studies which are considered next.

30.2.2 Epidemiological studies

Several researchers have conducted field studies testing industrial workers and/or collating their health records in an attempt to overcome the limitations of duration and realism in laboratory studies. A classic study by Jansen [30.9] examined the occurrence of physiological problems amongst two groups of workers he classified as from "very noisy industries" and "less noisy industries" respectively. The differences between the two groups, shown in Fig. 30.2, indicate a higher incidence of problems amongst the high-noise group than in the low-noise group. Noise level, however, is not isolated in such a study as the *causative* factor. The two groups were in fact based on two separate industries, the high-noise group being steelworkers whereas the low-noise group came from lighter industries. Between these industries there would be numerous other differences which could have effects on health, such as heat, physical work load, anxiety, and the type of people first attracted to the respective industries. At most, therefore, one can conclude that noise may have been a *contributory* factor in the differences found.

A study by Cohen [30.10] partly overcame this problem by locating different areas within the same boiler manufacturing complex. Comparing information from medical records of two noise groups as shown in Fig. 30.3, indicated that the high-noise group had a substantially higher proportion of diagnosed disorders.

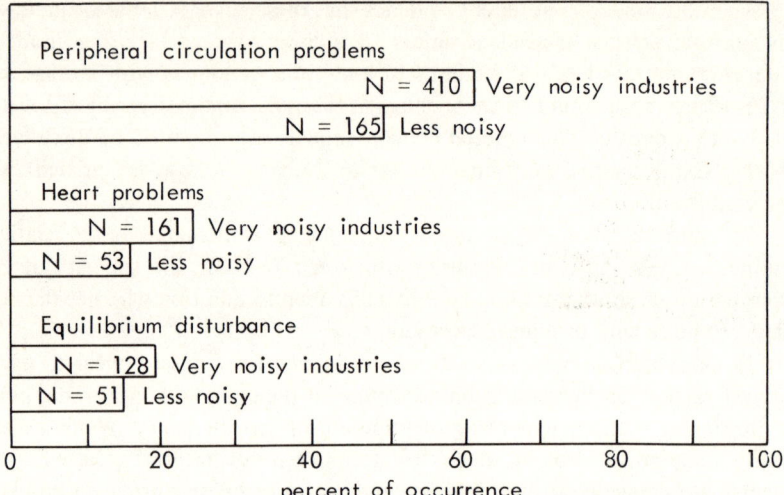

Fig. 30.2 – Occurrence of physiological problems in 1005 industrial workers (reproduced with permission from Kryter [30.2], after Jansen [30.9]).

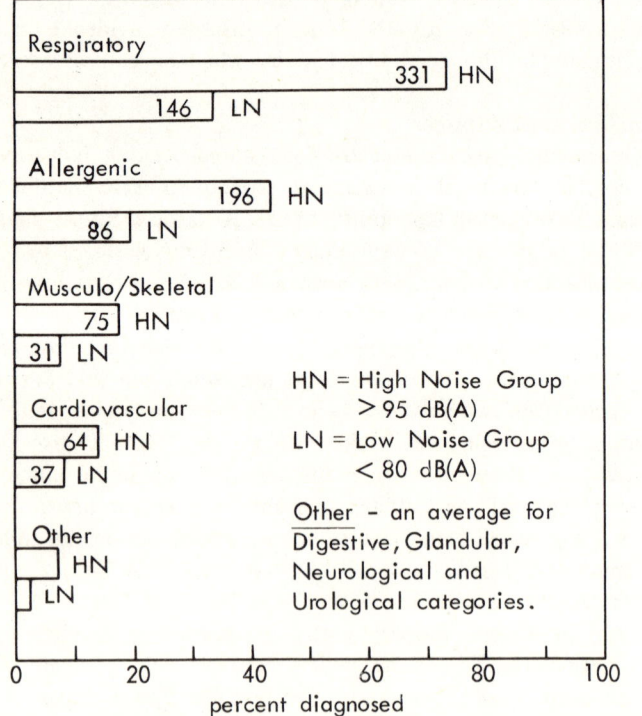

Fig. 30.3 – Occurrence of diagnosed disorder in a five-year period for 903 workers as a function of medical category. (Data from Cohen, [30.10]).

In a continuation of this study Cohen [30.11] reported an apparent reduction in the medical problems of the high-noise group after the introduction of hearing protectors, thus providing some further support for the hypothesis that the noise levels were at least a contributory factor to the differences observed.

Using measures of blood pressure, Parvizpoor [30.12] reported three-to-four times as many hypertensive amongst a group of weavers than in a control group working in 'light' less noisy industries. By contrast a similar study by Malchaire & Mullier [30.13] did not reveal any significant differences between two groups of noise-exposed workers (categorised by degree of hearing loss) and a control group who were not exposed to noise. Conflicting results are also found in studies which employed measures of hearing loss as direct estimates of the noise exposure [30.14, 30.15].

The limited number of studies in this area prevents the drawing of any firm conclusions as to whether noise is, or is not, associated with adverse health effects. It would appear from these and other studies that noise may be evident (if at all) as an additional stressor. For instance in the study by Jansen noise appeared to cause an additional 10% of peripheral circulation problems on an already high (50%) occurrence rate amongst the low-noise group. Noise may therefore have an effect by increasing the probability of the appearance of symptoms amongst susceptible individuals.

30.2.3 Mental health

A number of researchers have sought to establish whether there is any link between noise and mental health. Abey-Wickrama *et al.* [30.16] found that there was a higher psychiatric hospital admission rate from people living in the "maximum noise area" (levels in excess of 100 PN dB, which is approximately 87 dB(A), or NNI of 55), around Heathrow Airport, compared to that from outside this area. The authors carefully avoided the suggestion that aircraft noise itself causes mental illness, but concluded that "high intermittent noise levels from aircraft using Heathrow Airport may be a factor in increased rates of admission". Nevertheless, the findings of this study have been questioned, and other data have not indicated a similar relationship. As with the epidemiological studies discussed in section 30.2.2 there are numerious possible confounding factors in such investigations including differences other than aircraft noise between the two groups and the fear and anxiety produced by aircraft flyovers. A pilot study using interviews suggested that particular groups in the population may be vulnerable, and that factors such as the annoyance associated with the noise and the individual's sensitivity may also have to be considered [30.17].

The relationship between noise and annoyance is discussed in Chapter 28. In a related area extensive research has also investigated the disturbance of sleep by noise, most commonly in relation to environmental noise sources (a useful review of this area is provided by Lukas [30.18]). There is, however, no direct evidence of any deleterious effects on health due to such sleep disturbance, although it can impair performance at a variety of work tasks the next day.

It is worth noting that of course sound often has beneficial effects on our mental wellbeing. The appreciation of natural sounds can be very relaxing, and listening to music can be very pleasurable.

30.2.4 Summary

As yet, the case for specific harmful health effects of noise remains unproven. On the present evidence it seems unlikely that any major effects of noise on health could have escaped detection. Contributory effects of noise as an additional stressor still, however, remain as a strong possibility.

In a broader context, the World Health Organization definition that "health is a state of complete physical, mental, and social wellbeing and not merely an absence of disease and infirmity" led Burns [30.19] to comment in relation to noise that "this enviable condition is obviously vulnerable in a number of ways; health is prejudiced by interference with peace of mind, privacy, the pursuit of work or pleasure, or with the basic requirements of a natural and undisturbed nightly sleep".

30.3 PERFORMANCE EFFECTS

Another major area of concern is the effect that noise may have on task performance. It is likely that any new sound or change in an existing sound may result in at least momentary distraction, and that this may impair a person's ability to perform some tasks. More generally, changes via the non-specific connections in the cortex (Fig. 30.1) associated with prolonged exposure to even relatively steady noise may have some effect on such task performance. Extensive research both through field and laboratory studies has failed, to date, to produce definite criteria; however, recent research has brought to light the mechanisms whereby noise may interfere with task performance.

30.3.1 Field studies

A classic early study by Weston & Adams [30.20] found that loom operators working in a 96 dB noise environment had a greater output when wearing earplugs, as shown in Fig. 30.4. Those individuals who were most aware of a distracting influence of noise showed the greatest improvement with earplugs, but an improvement was evident even amongst those who believed themselves to be unaffected by noise. In spite of this, Broadbent argues that the differences observed were probably due to suggestion and a morale boost to the weavers [30.21].

Many other such studies have been plagued by this 'Hawthorne Effect', named after a series of studies at the GEC works in Hawthorne, Chicago, where

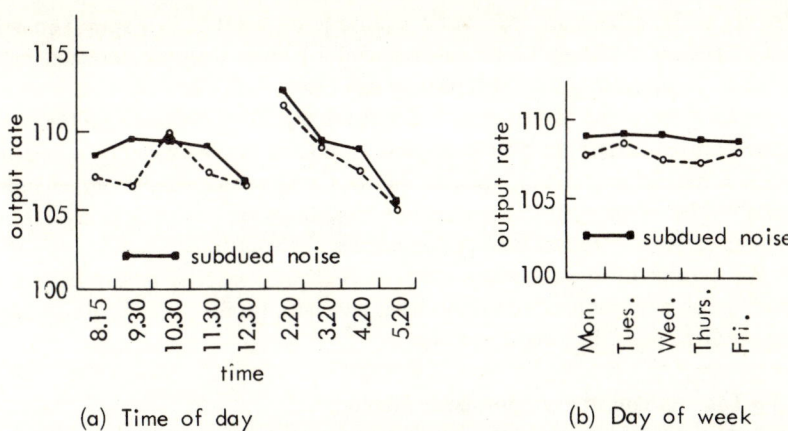

Fig. 30.4 – Average output rate (hundreds of picks per hour) for 10 weavers
with and without wearing earplugs [30.20].

it was found that any change of work conditions, be it an improvement or a
degradation, resulted in increased production [30.22]. The experiment was
confounded by the improved motivation of the workers due to the interest
shown in them by the experimenters.

Similarly a study by Broadbent & Little [30.23] found that there was an
increased rate of working by operators of film-perforating machines when
working in rooms where the noise level had been reduced from approximately
98 to 90 dB. However, the same workers were also studied in rooms which had
not been acoustically treated, and similar improvements were observed. This
was attributed to improved morale of the workforce, an effect which appeared
whichever room they were working in at the time. In the same study it was
observed that the number of broken rolls of film attributable to faults of the
operators was significantly reduced by the noise treatment compared with the
change in the untreated rooms. The results provide some evidence that noise
reduction reduced human error due to momentary lapses of attention, whilst
it did not improve work rate, except perhaps by a general morale boost. These
findings indicate that our understanding of noise effects will require consideration
of details of the type of task and the measure of task performance used.

30.3.2 Laboratory studies

A large number of laboratory-based experiments have investigated under controlled conditions the effects of noise on task performance. A review of these
experiments is provided by Broadbent [30.24].

Recent research has indicated that noise at levels as low as 80 dB(C) may
significantly impair task performance relative to quiet conditions [30.25].

More generally, effects are reported for noise levels greater than approximately 90 dB(C). Larger effects have been observed for noise with predominant high-frequency components, and for time-varying noise.

Tasks most prone to disturbance are those that have little margin for error, require constant attention over long periods, involve complex mental manipulations, or require interacting with more than one source or sensory channel. Typically, the impairment in performance takes the form of signals being missed, increased errors in response, and prolonged response times.

The results of these studies tend to be highly specific to the particular experimental conditions used. They have, however, brought to light the underlying mechanism of the effects of noise.

30.3.3 The mechanism of performance effects

Any explanation of why noise affects performance must account for the fact that whilst noise may, under certain conditions, impair performance, under other conditions it may actually improve performance. An example of this is the finding of Davies & Hockey shown in Fig. 30.5 that performance at a visual vigilance task was better in 95 dB of noise than that in 70 dB of noise [30.26]. Similarly it is reported that judicious use of background music may improve efficiency in work situations involving long duration and repetitive tasks [30.27].

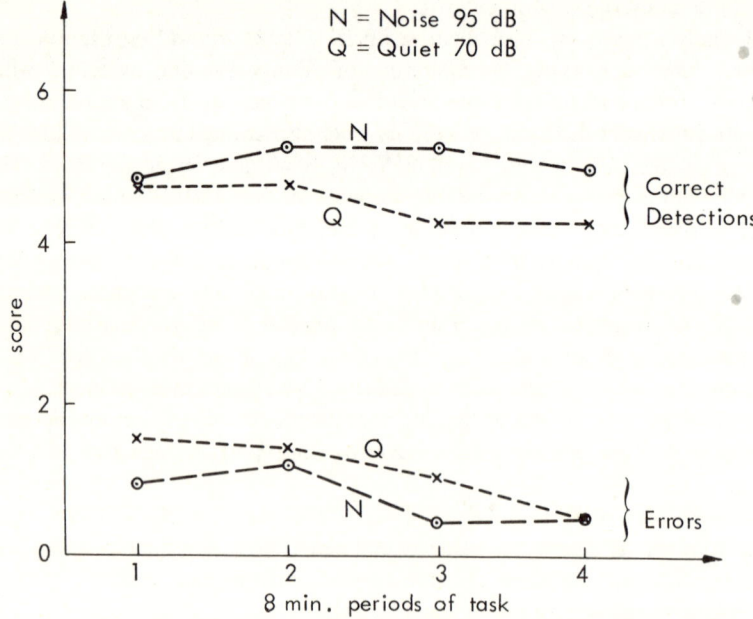

Fig. 30.5 – Mean numbers of correct detections and errors in a visual vigilance task with 6 signals per 8-min period. (Data from Davies & Hockey [30.26]).

The concept of arousal has been invoked to explain this and related phenomena. Arousal in this context refers to the level of behavioural activity on a hypothetical continuum from sleep to the highly alert state (and even mania). An 'inverted U' relationship is postulated between performance and arousal as shown in Fig. 30.6. It is generally agreed that noise increases arousal in the same manner as incentives or anxiety. It can be seen that an increase of arousal from A_1 to A_2 increases performance; however, over-arousal from A_2 to A_3 may in fact decrease performance at the task by going 'over the top' of the inverted U function.

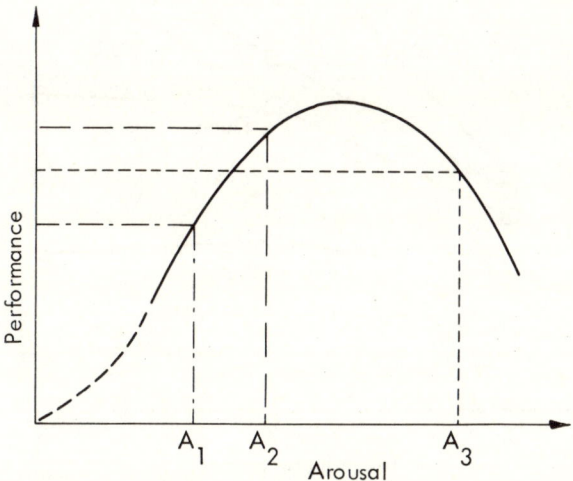

Fig. 30.6 – Hypothetical performance-arousal function.

This function explains the results shown in Fig. 30.7 that whereas white noise at 100 dB increased errors relative to quiet conditions in a serial reaction task by a process of over-arousal, after sleep deprivation (which results in low arousal) noise acts to improve performance in terms of a reduced number of errors [30.28]. Similar functions can explain the results of McGrath & Hatcher [30.29] shown in Fig. 30.8 that performance at an easy vigilance task can be improved by (more arousing) varied auditory stimulation than by white noise, whereas the reverse is true on a more difficlut version of the same task. Postulating two inverted U functions as shown in Fig. 30.9 shows that a change in arousal from A_1 to A_2 would cause improved performance on the easy task but degrade performance on the difficult task.

Under-arousal is believed to impair performance by moderating sensory inputs, lack of attention, and a slowing of the response process. Over-arousal would be intuitively expected to cause tense and jittery responses. It has also

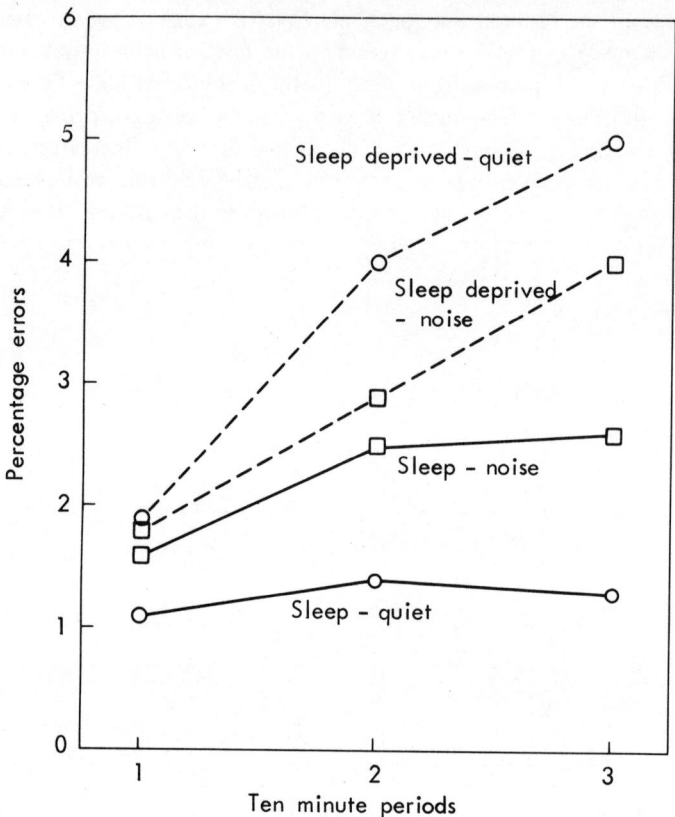

Fig. 30.7 — Errors in a serial reaction task with sleep and under sleep deprivation, both with noise and in quiet. (After Wilkinson [30.28]).

been suggested that it induces a less flexible mode of attending, Hockey [30.30] having found that peripheral, less likely visual signals were missed more often in the presence of noise. Similarly Cohen & Lezak found that whilst noise (95 dB(A) compared with unspecified no-noise condition) did not affect performance at a memory task, it was associated with significantly poorer memory of irrelevant cues [30.31].

Evidence of a 'psychic cost' associated with the performance of tasks in moderate levels of noise has been indicated by several laboratory studies. For instance, Wohlwill *et al.* [30.32] reported reduced resistance to frustration after performing a dial-monitoring task in noise. This was indicated by less persistence in attempts to solve insoluble puzzles amongst the noise-exposed group. It is of note that this after-effect of the noise was not attributable to the disruption of performance at the primary task by the noise, as a noise-exposed group who had

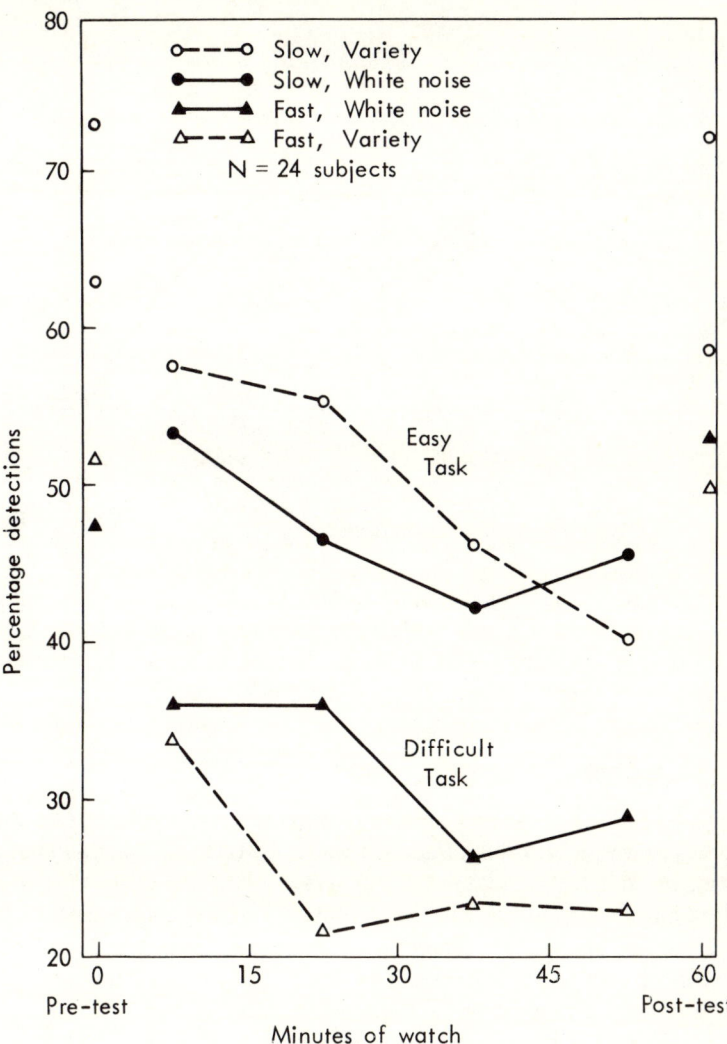

Fig. 30.8 – Percentage of signals detected as a function of time on a visual watch-standing task under two rates of stimulus presentation and two conditions of auditory stimulation [30.29].

not performed the task showed a similar difference relative to a corresponding no-noise group. More generally, the cost associated with noise in the workplace may be seen not in terms of reduced performance but as a reduced ability to react to additional demands, and increased fatigue after completion of the task.

A recent debate has considered the possibility that noise may affect task performance by masking auditory cues used by subjects to aid their performance

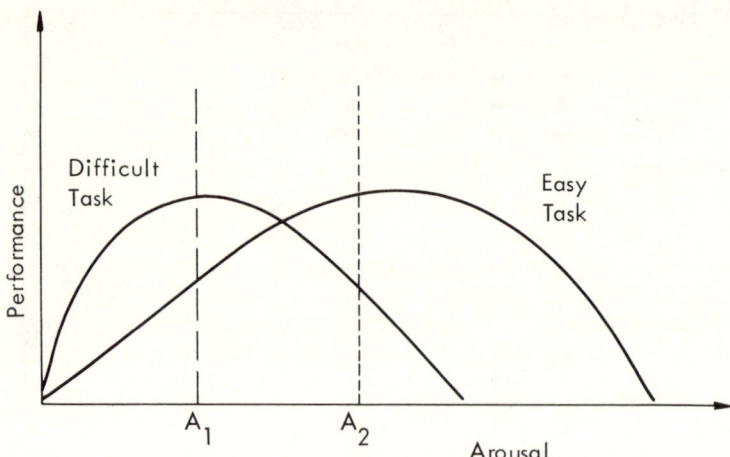

Fig. 30.9 — The effect of a change of arousal on performance at tasks of different difficulty.

of the tasks and by masking the 'inner speech' people use in memory rehearsal [30.33–30.35]. Whilst this may explain the decrement in performance in some specific cases, it does not appear to be an adequate explanation of the effects which have been reported in such a wide variety of laboratory tasks.

30.4 EFFECTS ON AUDITORY SKILLS

Hearing plays an important role not only in speech communication but in the perception of a variety of important sounds. It provides information about our environment, and as the 'sentinel of our senses' often acts as the first warning of impending danger. It can allow identification of the source of the sound, give information about the location of the source, and in a more subtle manner the degree of reverberation gives a 'feel' for the acoustic environment. Noise can impair or eliminate perception of this information.

A recent review has indicated that whilst it has often been said that noise can be the cause of accidents, there is relatively little data available to quantify such as an effect [30.36]. More generally, a recent World Health Organization report noted that "communication requirements in industrial situations frequently do not receive adequate attention, particularly with reference to the accident risk" [30.6]. Studies such as those reported by Cohen (see section 30.2.2) have indicated higher accident rates in high-noise areas of factories [30.10, 30.11]. Whilst results such as these do not establish a causal link, there are several ways in which noise could give rise to such an effect. Firstly, the 'psychic costs' discussed in the preceding section associated with working in noise could make a person more likely to be involved in an accident by increasing fatigue and impairing the perception of unexpected hazards. Other mechanisms could be associated with the masking effect of the noise, which could impair the perception of warning sounds, auditory cues, and speech messages. These effects are considered next.

30.4.1 Masking

The ear has a remarkable ability to 'hear out' some sound from a background of other sounds. There are, however, definite limits to this ability. Through the process of masking, noise can make a sound inaudible. A considerable proportion of research into the theory of hearing has focused on masking effects, and useful reviews are provided by Kryter [30.2] and Scharf [30.37].

The ear can be thought of as analysing sounds through a set of filters which can be tuned to any centre frequency. The bandwidth of these 'critical bands' is shown in Fig. 30.10, where it can be seen that below 500 Hz the bandwidth is approximately constant and above 500 Hz it is approximately proportional to frequency. Above 500 Hz the critical bandwidth is reasonably approximated by one-third octave bands, and so these can be used in approximate masking calculations.

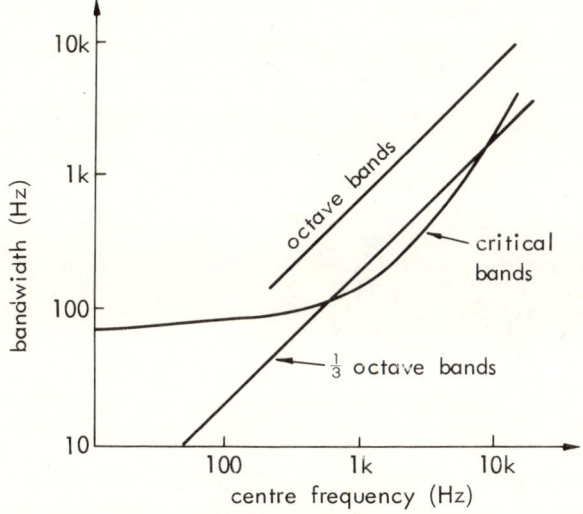

Fig. 30.10 – Comparison of critical bandwidths of the ear with octave and one-third octave bandwidths.

The ear summates the sound energy within a critical band, and therefore the detectability of a signal in noise is determined by the total signal energy within a critical band relative to the total noise energy within the same band. The signal-to-noise ratio (S/N) within a critical band at which the signal can just be heard is dependent on frequency; however, to a first approximation a signal will be audible if the one-third octave band S/N is −5 dB.

An example calculation of the masked threshold of a siren warning sound is illustrated in Fig. 30.11. The signal-to-noise ratio is calculated for each one-third octave band from spectra of the siren and the background noise (Fig. 30.11a). For the combination shown the most detectable band is centred at 2 kHz, and this optimal one-third octave S/N of + 5dB indicates that with the siren at an

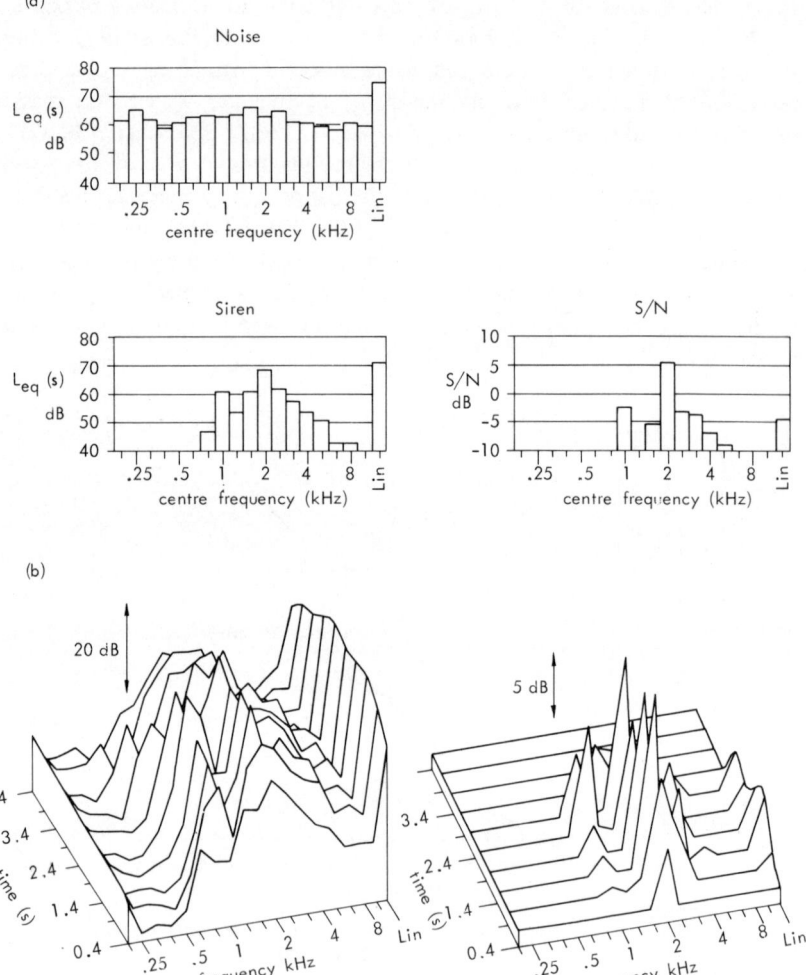

Fig. 30.11 – One-third octave band spectra of the signal, noise, and signal-to-noise ratio; the S/N indicates approximately those components of the siren audible, (a) sample equivalent continuous sound level, $L_{eq(s)}$, using a 5 s sample duration. (b) Three-dimensional displays indicating the temporal variation of the one-third octave band spectrum; integration periods 0.1s with 0.5s between spectra. (Reproduced with permission of the Almquist and Wiksell Periodical Company).

overall intensity of 70 dB SPL it is approximately 10 dB above its masked threshold in the background noise of overall intensity 75 dB SPL [30.38].

The three-dimensional displays in Fig. 30.11b indicate the complex variation with time of the one-third octave band and S/N levels. The use of measures based on the sample equivalent continuous sound level ($L_{eq(s)}$) obtained with suitable sampling duration adequately takes account of such temporal variations.

At higher intensities the critical bandwidths are believed to be increased, and the threshold S/N of tonal signals may rise by 5–10 dB [30.39]. In addition, low-frequency noise can have a greater masking effect on higher frequency tones than that indicated by the size of the critical bands, an effect known as the 'upward spread of masking.' For noise spectra with pronounced low-frequency components (differences between successively higher frequency one-third octave bands greater than approximately − 10 dB) it would also be necessary to take account of this factor.

30.4.2 Loudness
The loudness of a sound close to its masked threshold will be considerably reduced by the presence of a noise. This process of 'partial masking' results in a growth of loudness with signal level similar to that observed with many hearing-impaired people. It can be seen in Fig. 30.12 that for a sound in a masking noise which raises the threshold of the sound by 60 dB, the loudness of the sound rapidly increases at levels above this threshold relative to the loudness of the sound in quiet conditions [30.40]. At 70 dB (10 dB above its masked threshold) the sound is already as loud as the same sound at 55 dB in quiet conditions. Within the partial masking zone there may be difficulties in hearing sounds directly, in distinguishing between similar sounds, or in being alerted by the sounds as warnings. This provides some support for the generally accepted requirement that to be effective a warning sound should be at a level approximately 15 dB above its masked threshold [30.41, 30.42].

30.4.3 Speech interference
Speech reception is the most important and also the most complex use of the auditory system. Noise can either mask speech to make it inaudible, or by masking only some frequencies leave it audible but of reduced intelligibility. The interfering effect of noise on face-to-face speech communication can be assessed in terms of the maximum possible distance between the speaker and the listener in a given steady background noise level for a particular voice level (Fig. 30.13). For personal conversation where a separation of 2 m is typical it can be seen that normal communications are possible in noise levels up to approximately 60 dB(A). At higher levels the talker will unconsciously raise his voice level to compensate for the high-noise environment, and with this greater effort communications may still be just reliable in noise levels up to approximately 70–75 dB(A).

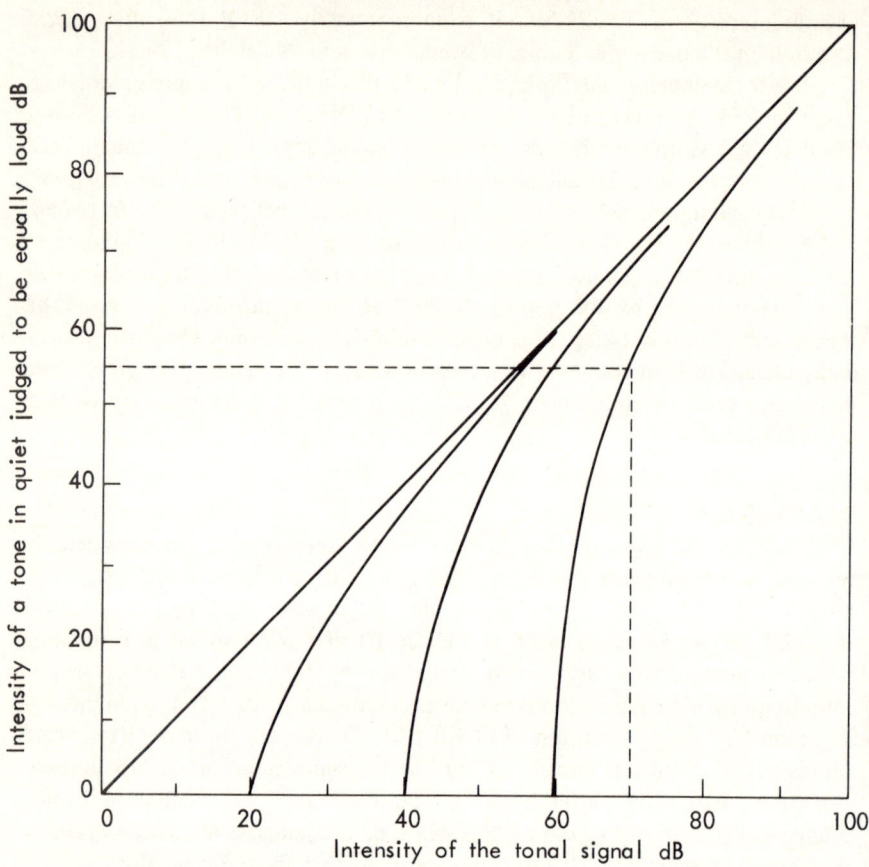

Fig. 30.12 – Curves showing the estimated equivalent loudness of a tone heard in three different intensities of background noise. (Based on the equations of Macrae [30.40]).

In still higher noise levels it would be necessary to move closer together if possible, or to provide some aid to speech communication. For discussions involving groups, such as during training and instruction, separations of 4 m may be more typical, and in these cases noise levels should not exceed approximately 55 dB(A).

The data used in preparing Fig. 30.13 were for young adults with normal speech and hearing and in a diffuse noise field. Variations from these conditions, such as use of dialects, use of a limited vocabulary, obscuring of visual cues, or if the listeners have a marked hearing loss, could considerably affect these criteria. The exact spectrum of the noise is also an important factor, and for more accurate assessment of a noise environment, reference should be made to either the Articulation Index (AI) or the Preferred Speech Interference Level (PSIL).

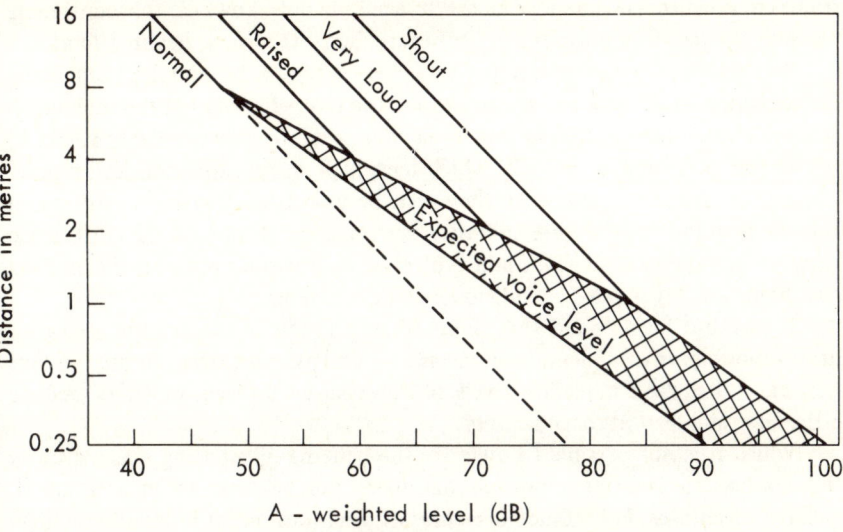

Fig. 30.13 – The approximate relationship between talker-to-listener distances for just reliable communication, the A-weighted noise level, and the relative voice level. The region below each curve shows the combination of conditions for which just reliable face-to-face communication is possible. (This figure is excerpted with permission and modified from ANSI S3.14 – 1977 [30.43]).

More general assessments of indoor noise levels are provided by the Preferred Noise Criteria (PNC) curves or by Noise Rating (NR) curves. For details of these indices see Kryter [30.2].

There are important implications of the reduced opportunity for relaxed, reliable speech communication. In a noisy industrial environment, for instance, it will be considerably more difficult to instruct new employees on the job. Whilst conditions may be adequate for the predictable information associated with some jobs, important unexpected messages may not be comprehended which could be a hazard to safety and the efficient operation of plant equipment. It has been suggested that as workers get used to the noise environment they adopt a non-communicating life style, relying more on non-verbal communication through gestures and facial expression, which impairs the development of social relations and degrades their work environment [30.2].

30.5 CONCLUSIONS

The evidence reviewed has failed to establish any direct effects of noise on health, excluding, of course, noise-induced hearing loss which is discussed separately in Chapter 29. Research has been hampered by the inability of laboratory studies to simulate typical noise exposures realistically, and the inability of epidemiological studies to isolate noise as a single factor. It seems possible, however, that continuing research may confirm noise as a contributory factor to adverse health effects.

The effects of noise on task performance are seen to be complex, depending on the nature of the task and the noise, and the level of arousal of the individual. The evidence, however, is clear that levels of noise above approximately 80 dB(A) can impair performance at tasks which have little margin for error and require continuous attention, and that there can be associated 'psychic costs' which become manifest both during and after performance of the task. The difference between laboratory conditions and typical work situations prevents the existing data being reliably applied to individual work situations.

It has also been found that noise often prevents or impairs the ability to hear important environmental sounds and to understand speech. In this context it is clear that even moderate levels of noise can be a threat to safety, reduce efficiency, and disrupt social activity.

Whilst it is not possible to quantify the benefits of reducing noise levels in this context, it is often reported that there can be financial in addition to welfare advantages. H.M. Chief Inspector of Factories recently described a firm where engineering noise controls were introduced: there was a subsequent reduction in absenteeism and an increase in production from the machines involved of approximately 20% [30.44].

In meeting requirements for the protection of hearing at work [30.45] it is likely that many of the other possible adverse effects of noise will also be reduced. However, the evidence currently available supports a requirement that noise should be reduced to the lowest level reasonably practicable. Future research should aim to establish the basis and extent of these effects. In particular, clarification is required in areas such as possible long-term adaptation, the extent of individual differences, the effects of noise on those with pre-existing disorders, and the indirect costs of living or working in noise.

REFERENCES

[30.1]   Grandjean, E. undated *Biological effects of noise,* Zurich: Swiss Federal Institute of Technology.

[30.2]   Kryter, K. D. (1970) *The effects of noise on man.* New York: Academic Press.

[30.3]   Miller, J. D. (1974) *Journal of the Acoustical Society of America* **56** 729-764. Effects of noise on people.

[30.4]   Burns, W. (1979) *In: Handbook of noise control (C. M. Harris editor)* New York: McGraw-Hill. Physiological effects of noise.

[30.5]   Taylor, S. M., Young, P. J., Birnie, S. E. & Hall, F. L. (1980) *McMaster University Report to Motor Vehicle Manufacturers Association of the United States, Inc.* Health effects of noise — a review of existing evidence. Hamilton, Ontario.

[30.6] Environmental Health Criteria 12 (1980) *Noise.* Geneva: World Health Organisation.

[30.7] Geber, W. F. & Anderson, T. A. (1967) *Comparative Biochemistry and Physiology* **21** 273-277. Cardiac hypertrophy due to chronic audiogenic stress in the rat, Rattus Norvegicus Albinus, and rabbit, Lapus Cuniculus.

[30.8] Levi, L. (1967) *In: Introduction to Clinical Neuroendocrinology (E. Bajusz editor)* Basel: Karger. Sympatho-adrenomedullary responses to emotional stimuli.

[30.9] Jansen, G. (1961) *Stahl und Eisen* **81** 217-220. Adverse effects of noise in iron and steel workers (German).

[30.10] Cohen, A. (1973) *In: Proceedings of the International Congress on Noise as a Public Health Hazard,* Report EPA 550/9-73-008. Industrial noise and medical absence, and accident record data on exposed workers.

[30.11] Cohen, A. (1976) *Journal of Safety Research* **8** 146-162. The influence of a company hearing conservation program on extra-auditory problems in workers.

[30.12] Parvizpoor, D. (1976) *Journal of Occupational Medicine* **18** 730-731. Noise exposure and prevalence of high blood pressure among weavers in Iran.

[30.13] Malchaire, J. B. & Mullier, M. (1979) *Annals of Occupational Hygiene* **22** 63-66. Occupational exposure to noise and hypertension: A retrospective study.

[30.14] Jonsson, A. & Hansson, L. (1977) *The Lancet* 8 Jan 1977, 86-87. Prolonged exposure to a stressful stimulus (noise) as a cause of raised blood pressure in man.

[30.15] Hedstrand, H., Drettner, B., Klockhoff, & Svedberg, A. (1977) *The Lancet,* 17 Dec 1977, 1291. Noise and blood pressure.

[30.16] Abey-Wickrama, I., O'Brook, M. F., Gattoni, F. E. G. & Herridge, C. F. (1969) *The Lancet* 13 Dec 1969, 1275-1277. Mental hospital admissions and aircraft noise.

[30.17] Tarnapolsky, A., Barker, S. M., Wiggins, R. D. & McLean, E. K. (1978) *Psychological Medicine* **8** 219-233. The effect of aircraft noise on the mental health of a community sample: A pilot study.

[30.18] Lukas, J. S. (1975) *Journal of the Acoustical Society of America* **58** 1232-1242. Noise and sleep: A literature review and a proposed criterion for assessing effect.

[30.19] Burns, W. (1973) *Noise and man.* London: John Murray, second edition.

[30.20] Weston, H. C. & Adams, S. (1932) *In: Industrial Health and Research Board Report No. 65,* London: HMSO. The effects of noise on the performance of weavers.

[30.21] Broadbent, D. E. (1971) *Decision and stress.* London: Academic Press.

[30.22] Roethlisberger, F. J. & Dickson, W. J. (1939) *Management and the worker: An account of a research program conducted by Western Electric Company.* Boston: Havard University Press.

[30.23] Broadbent, D. E. & Little, E. A. J. (1960) *Occupational Psychology* **34** 113-140. Effects of noise reduction in a work situation.

[30.24] Broadbent, D. E. (1979) *In: Handbook of noise control (C. M. Harris editor).* New York: McGraw-Hill. Human performance and noise.

[30.25] Jones, D. M., Smith, A. P. & Broadbent, D. E. (1979) *Journal of Applied Psychology* **64** 627-634. Effects of moderate intensity noise on the Bakan vigilance task.

[30.26] Davies, D. R. & Hockey, G. R. J. (1966) *British Journal of Psychology* **57** 381-389. The effect of noise and doubling the signal frequency on individual differences in visual vigilance performance.

[30.27] Anon (1972) *Report of the Ergonomics Information Analysis Centre, University of Birmingham.* Background music as an aid to productivity – Improving fault detection in high ambient noise conditions.

[30.28] Wilkinson, R.T. (1963) *Journal of Experimental Psychology* **66** 332-337. Interaction of noise with knowledge of results and sleep deprivation.

[30.29] McGrath, J. J. & Hatcher, J. F. (1961) *Human Factors Research Inc., Technical Report No. 7.* Irrelevant stimulation and vigilance under fast and slow stimulus rates.

[30.30] Hockey, G. R. J. (1970) *Quarterly Journal of Experimental Psychology* **22** 28-36. Effect of loud noise on attentional selectivity.

[30.31] Cohen, S. & Lezak, A. (1977) *Environmental and Behaviour* **9** 559-572. Noise and inattentiveness to social cues.

[30.32] Wohlwill, J. F., Nasar, J. L., Dejoy, D. M. & Foruzani, H. H. (1976) **61** 67-74. Behavioural effects of a noisy environment: Task involvement versus passive exposure.

[30.33] Poulton, E. C. (1979) *Psychological Review* **86** 361-375. Composite model for human performance in continuous noise.

[30.34] Hartley, L. R. (1981) *Psychological Review* **88** 86-89. Noise does not impair by masking: A reply to Poulton's "Composite model for human performance in continuous noise".

[30.35] Poulton, E. C. (1981) *Psychological Review* **88** 90-92. Not so! Rejoinder to Hartley on masking by continuous noise.

[30.36] Wilkins, P. A. & Acton, W. I. (1982) Annals of Occupational Hygiene (in press). Noise and accidents – A review.

[30.37] Scharf, B. (1970) *In: Foundations of Modern Auditory Theory Volume 1* (J. V. Tobias editor) New York: Academic Press. Critical bands.

[30.38] Wilkins, P. A. & Martin, A. M. (1980) *Scandinavian Audiology* **10** 37-43. The effects of hearing protectors on the attention demand of warning sounds.

[30.39] Wilkins, P. A. & Martin, A. M. (1977) *Paper presented at 9th International Congress of Acoustics, Madrid.* The effects of hearing protectors on the masked thresholds of acoustic warning signals.

[30.40] Macrae, J. H. (1979) *Australian Journal of Audiology* 1 15-18. An improved form of the loudness function.

[30.41] Wilkins, P. A. & Martin, A. M. (1978) *Institute of Sound and Vibration Research Technical Report 98.* The effect of hearing protectors on the perception of warning and indicator sounds — A general review.

[30.42] Wilkins, P.A. (1982) *British Journal of Audiology* 15 263-274. Assessing the effectiveness of auditory warnings.

[30.43] American National Standards Institute (1977) *ANSI S3.14 – 1977.* Rating noise with respect to speech interference.

[30.44] Health and Safety Executive (1981) *Health and Safety: Manufacturing and service industries 1979.* London: HMSO.

[30.45] Health and Safety Commission (1981) *Consultative Document: Protection of hearing at work.* London: HMSO.

CHAPTER 31

Human response to vibration

M. J. Griffin

Institute of Sound and Vibration Research, University of Southampton

31.1 INTRODUCTION

Exposure of the human body to vibration may cause a change of comfort, a change in performance, or a reduction in health and safety. The three criteria corresponding to these changes are composed of many individual effects. Comfort reactions may include perception and annoyance as well as a general subjective sensation of the magnitude of a vibration stimulus. The effects of vibration on performance may include central or cognitive interference or direct mechanical effects due to movement of the limbs or eyes. Health effects may arise due to direct damage caused by the vibration intensity, or indirect damage due to instability or falls. Health effects include chronic injury due to cumulative long-term exposures over a working life in addition to the more immediate effects. Very low frequency oscillation of the body can produce all of the above effects and, most notably, the symptoms often described as motion sickness.

The response of an individual due to a particular vibration condition is dependent on a complex combination of very many variables. These variables may be conveniently divided into intrinsic and extrinsic factors. The intrinsic variables are those concerning the subject and are in turn divided into inter- and intra-subject variabilities. Intra-subject variability includes all those factors which change the response of an individual with time – the largest changes are often due to alterations of body posture, motivation, and arousal. The inter-subject variabilities are associated with changes between individuals and include both those variables which affect an individual's response and factors such as age, sex, height, weight, etc. Although studies have been conducted to determine the influence of some of these individual changes it is not usually possible to allow for their effect in system design or evaluation (see Griffin & Whitham [31.1], Griffin et al. [31.2]). The principal exception concerns body posture where this has a very large effect and may be greatly influenced by, for example, seating conditions.

The extrinsic variables include vibration level, vibration frequency, vibration axis and direction, vibration duration, and the criteria by which the vibration is to be evaluated. There are often many additional extrinsic variables including physical factors such as the seating and task dynamics, and psychological variables including subject instructions. Combined effects of vibration and other environmental stresses are also sometimes considered important, although little substantial evidence of large effects has been reported.

Any precise guidance concerning the effects of vibration on man should express a relation between the above intrinsic and extrinsic variables and the effects on man. In practice, the variables are too many and too complex, and only a small number may reasonably be measured. The general guidance provided in existing vibration standards and proposed standards is usually restricted to the effects of the extrinsic variables and, in consequence, is limited to the effect of the magnitude, frequency, axis, and duration of vibration. It will be seen below that the specification of the effects of these variables is only simple if gross approximations are made. Since other unspecified variables have an effect it is reasonable that existing standards should claim only to provide general approximations. The scientific literature must be consulted, or research conducted, if more detailed guidance is required.

This Chapter is intended to introduce the reader to some of the principal effects of vibration on the body. The next section summarises the principal existing and proposed vibration standards and limits. The later section indicates some of the additional research data that are available and also provides some examples of the application of both the information in standards and that obtained from more recent research.

31.2 STANDARDS

31.2.1 Whole-body vibration standards

Whole-body vibration is generally considered to be that due to vibration from a platform or surface which is the principal supporting surface for the body. This definition is primarily intended to distinguish whole-body vibration from local vibration due to, for example, the vibration of hand-held tools. Since 1974 the principal human vibration standard has been International Standard ISO 2631 [31.3]. A similar document (BSI Draft for Development DD32 [31.4]) was published at about the same time. More recently, many other documents have appeared to amend or extend the guidance provided in ISO 2631. These standards, draft standards, and proposed standards generally express a consensus view of the relevant cause-and-effect relationship. In some cases a standard has been drafted in the recognized absence of complete data on the effect to which it refers. Several of the standards have the declared intention of stimulating the provision of further information in addition to standardizing the evaluation of vibration and the collection of vibration dose–effect data. In Britain none of the standards have any statutory significance — although they contain information that a prudent employer might be expected to act upon.

31.2.1.1 International Standard ISO 2631(1978)

The *Guide for the evaluation of human exposure to whole-body vibration* (ISO 2631 [31.3]) was first published in 1974 and re-issued with minor editorial changes in 1978. Amendments and Addenda to this standard have been issued and are discussed in separate sections below. The document issued by the British Standards Institution (BSI DD32) is essentially the same as ISO 2631. A major revision of the International Standard is currently being considered, but this is not expected to be completed for several years.

The International Standard gives numerical limits for exposure to vibration transmitted from solid surfaces to the human body over the frequency range 1–80 Hz. It applies to the three axes of translational vibration which are centred in the body as shown in Fig. 31.1. These axes move with the body as it changes its orientation with respect to gravity – z-axis vibration, for example, is only vertical when man is sitting or standing. The vibration levels refer to those at the point of entry of vibration into the body and not, for example, at the substructure of a resilient seat.

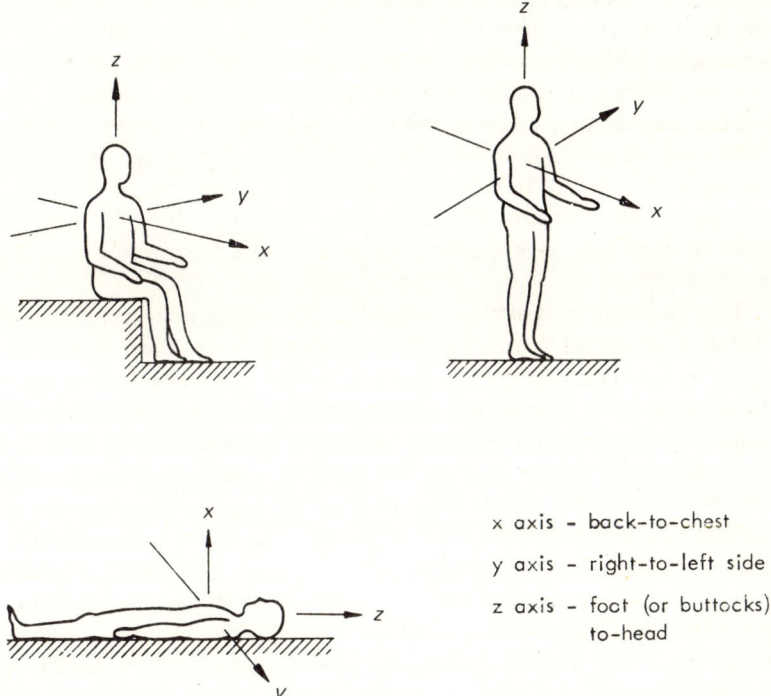

x axis – back-to-chest
y axis – right-to-left side
z axis – foot (or buttocks) to-head

Fig. 31.1 – Directions of co-ordinate system for mechanical vibrations influencing humans as defined in ISO 2631 [31.3].

The limits are given according to the three principal criteria mentioned in section 31.1 above:

"exposure limits"
— concerned with the preservation of health or safety;
"fatigue — decreased proficiency boundary"
— concerned with the preservation of working efficiency;
"reduced comfort boundary"
— concerned with the preservation of comfort.

The limits corresponding to these three criteria are given in a simple hierarchical relationship such that for any particular vibration frequency, axis, and duration:

Exposure limits
= 2 times fatigue — decreased proficiency (FDP) limits;
Fatigue — decreased proficiency limits
= 3.15 times reduced comfort boundary.

Vibration levels should be measured in ms^{-2} r.m.s. The standard indicates that when vibration contains occasional high peaks the crest factor (ratio of maximum peak to r.m.s. value) should be determined, and states that the guidance is very tentative in the case of vibrations having crest factors greater than about 3. (The Amendments outlined in section 31.2.1.2 below modify this recommendation.)

For all three general effects of vibration, the body is considered to exhibit the same sensitivity to different vibration frequencies. In the z-axis (often the vertical axis) maximum sensitivity to acceleration occurs in the range 4 to 8 Hz as shown in Fig. 31.2. This increase in sensitivity is largely due to a tendency for the body to exhibit a vertical resonance in this region. In the x- and y-axes (often the horizontal axes) maximum sensitivity is in the range of 1–2 Hz (see Fig. 31.2).

The frequency sensitivities reflected in Fig. 31.2 may be simply applied if a vibration condition contains a single frequency. In practice, however, most realistic vibrations in transport or industry contain many components, and some random vibration is often present. The 'preferred' method for evaluating complex vibration spectra given in the Standard amounts to assessing the worst frequency component in a vibration spectra. For example, a random vibration condition would be evaluated by determining the r.m.s. acceleration in each third-octave band and then assessing the severity of vibration from the one band which was highest in relation to the limits given in Fig. 31.2. An 'alternative' procedure by which an overall weighted vibration acceleration magnitude is obtained is also specified in the Standard. As this produces a value determined by the complete spectra rather than a single third-octave band it necessarily gives higher values. (The Amendments discussed in section 31.2.1.2 below modify the method of evaluating complex spectra.) It has been common practice in many areas to interpret the frequency sensitivity contours as frequency weightings. In Fig. 31.3 a mathematical definition of the two frequency weightings is provided.

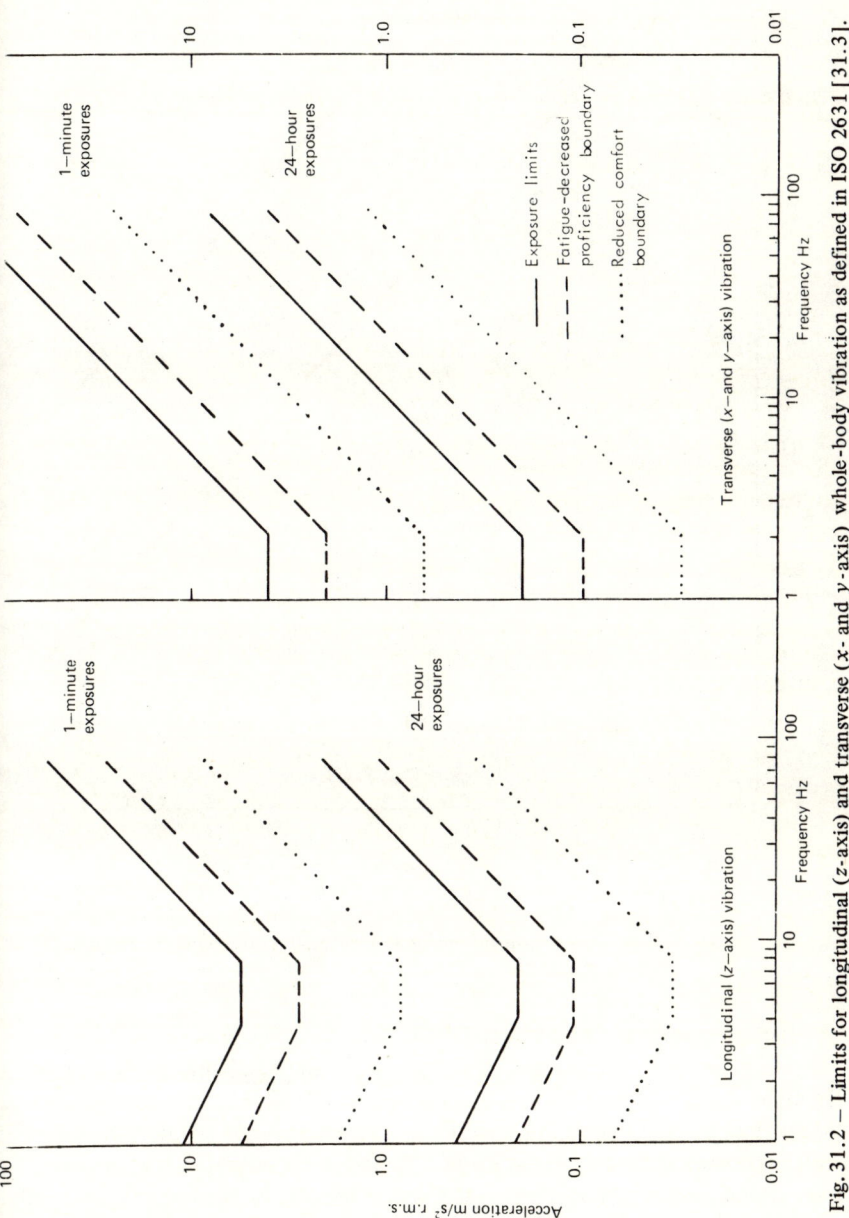

Fig. 31.2 — Limits for longitudinal (z-axis) and transverse (x- and y-axis) whole-body vibration as defined in ISO 2631 [31.3].

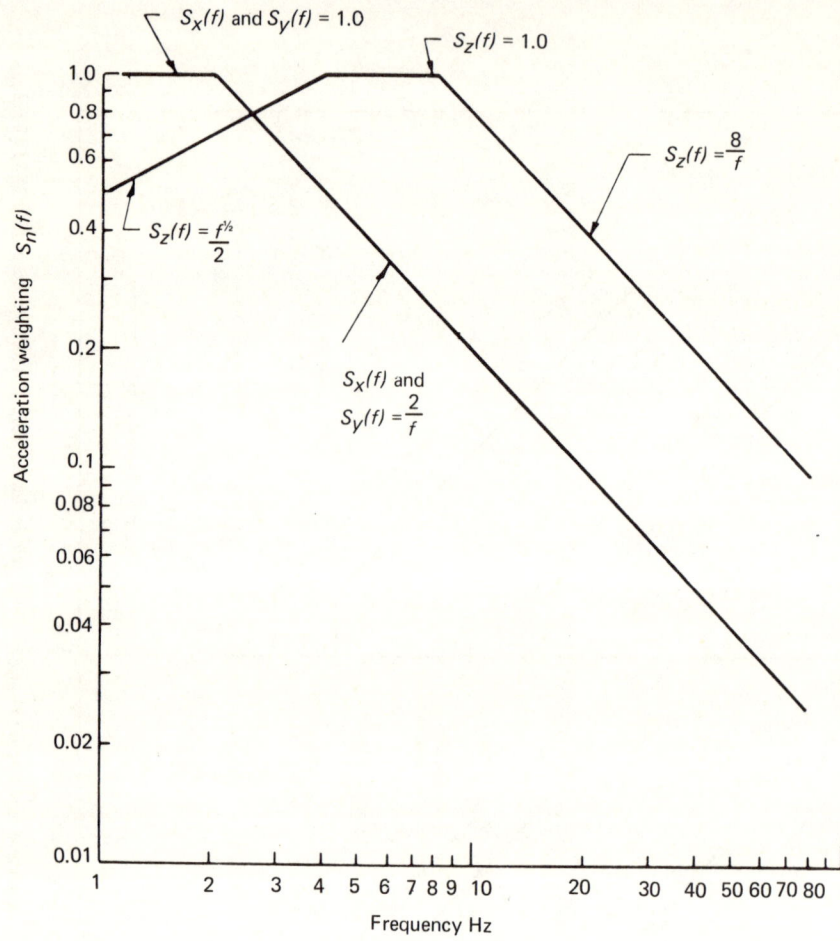

Fig. 31.3 – Values of the human response to vibration frequency weightings
defined in ISO 2631 for x-, y- and z-axis acceleration.

It may reasonably be expected that the effect of vibration increases with
increasing duration of exposure. This is certainly the case with discomfort produced by short periods of whole-body vibration [31.5]. For vibration durations
longer than about 1 minute there is little conclusive evidence of increasing effect
with increasing duration, and some results have even suggested that adaptation
gives rise to greater tolerance with increasing exposure. Notwithstanding the lack
of experimental data, the International Standard presents a time dependency
from 1 minute to 24 hours which reflects a considerable increase in sensitivity
for durations greater than 4 minutes. The form of the time dependency is

complex as shown in Fig. 31.4. In this figure the time dependency corresponding to r.m.s. integration (a^2t = constant), and that corresponding to r.m.q. integration (a^4t = constant) are also shown. It may be seen that over a period from about 10 minutes to 8 hours the ISO time dependency approximates to the r.m.s. integration slope, but for short periods there is a strong contradiction. The Amendments (see section 31.2.1.2 below) provide a more simple approximation to the ISO time dependency shown in Fig. 31.4. The approximation may help some practical measurements, but it is important to recognize that the ISO time dependency is a gross approximation. In many cases the vibration conditions may be adequately summarized by the use of the frequency and axis weightings and expressed as weighted r.m.s. vibration magnitude without reference to the time dependency.

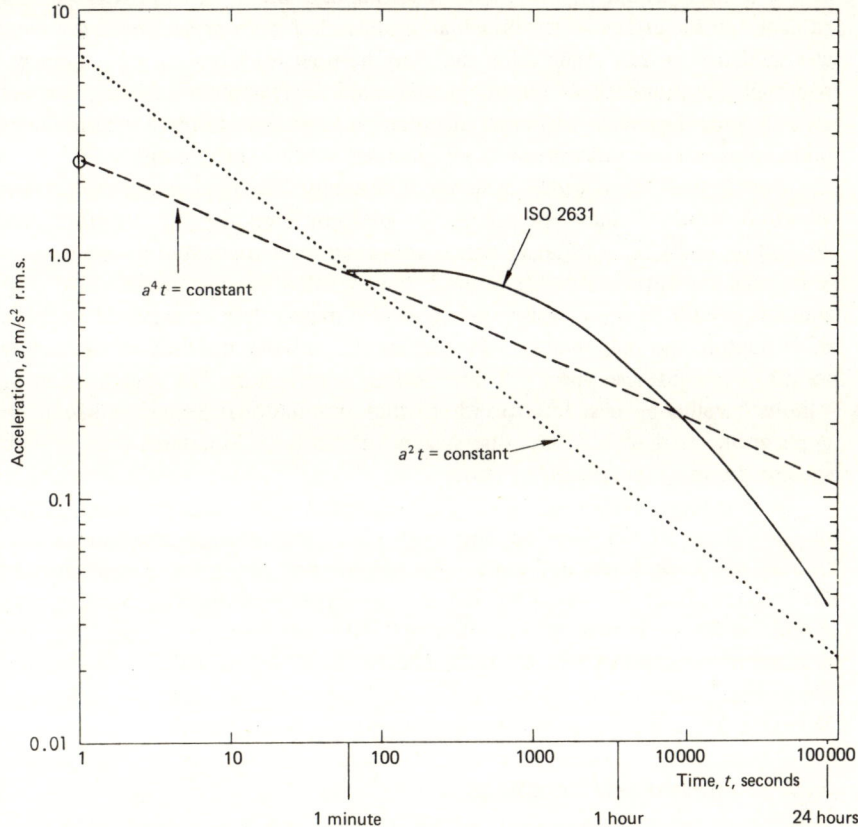

Fig. 31.4 – Comparison of the time dependency given in ISO 2631 with a^2t = constant and a^4t = constant, time dependencies. (ISO curve is for 4 Hz reduced comfort boundary; curves drawn to coincide at one minute.)

The principal problems with the current International Standard are partly those which arise through lack of evidence of the effects of certain variables. There are also problems due to insufficient definition and consequent ambiguity in certain areas. As mentioned above, time dependency is a major problem. The definition of limits for only the translation axes at the seat, or other principal supporting surface, is another major limitation. In many systems other inputs, either rotational vibration at the principal supporting surface or vibration at the feet, seat back, or other body interfaces can contribute to discomfort and loss of performance. A further restriction concerns the treatment of impulsive vibrations and other conditions which give rise to high crest factors. The definition of the crest factor limit is unsatisfactory, since crest factors are only a crude measure largely influenced by a single peak in the motion which is difficult to determine reliably. A measure which is based on a more continuous assessment of the motion would be preferable. The Standard attempts to provide general guidance on the effects of the principal variables, but in its present form does not give guidance on instrumentation that may be used for vibration measurement. Although the Standard should not be expected to give precise guidance in these areas it should provide sufficient information to enable standard measurement procedures and instrumentation to be specified in other documents.

Contrary to a now widely accepted philosophy, ISO 2631 appears to present vibration "limits". Since the acceptable vibration level for any condition will depend on many circumstances which are varied from application to application it is usually inappropriate for limits to be specified in a general standard. The standard would be more easily applicable if it merely defined standard methods of evaluating the combination of variables and uniform methods of expressing results. It might also specify a dose–effect relationship. The specification of "limits" would be best left to either other International Standards having an application restricted to, say, one system, or National Standards dealing with specific groups of people or systems.

A long-term revision of ISO 2631 is expected to separate the specification of the effects of the variables and dose–effect relationships from the definition of allowable levels and limits. The revision will also more clearly separate the various effects of vibration: effects on health, performance, comfort, and motion sickness. It may be expected that different frequency weightings and assessment procedures will be applicable to the different criteria. The simple hierarchical relationship shown in Fig. 31.2 may not be retained in the new Standard, and a change in the time dependency can be expected.

31.2.1.2 Amendment 1 to ISO 2631

A series of brief Amendments to the principal standard have recently been published [31.6]. The purpose of these Amendments is to clarify and assist the application of the standard. The Amendments provide additional information in six areas. It is emphasized that the standard is concerned with the provision of only general guidance, and states that factors not specified in the standard can have large effects. The crest factor limit of 3 given in the standard is raised to 6, and crest factors are defined less ambiguously. The frequency weighting method is advocated as the recommended procedure when assessing the effect of vibration on comfort and performance. It is proposed that when the vibration occurs in three axes the effect on comfort and performance should be determined by taking the square root of the sums of the squares of the weighted values in each axis. A convenient approximation to the ISO time dependency is offered to assist its use. The final addition is a series of references which, primarily, provide the basis for the Amendment.

31.2.1.3   Draft Addendum 1 to ISO 2631 (ISO DAD 1 (1980) [31.7])

Draft Addendum 1 is a guide to the evaluation of human exposure to vibration and shock in buildings. It presents limits of vibration acceptability for various building types and for vibration frequency in the range 1–80 Hz. The frequency weightings are based on those in ISO 2631, but the limits are adjusted such that the lowest values for the most critical buildings are at about the same level as the threshold of human perception of vibration. Since vibration measurements will be made on a part of the building (normally on the floor at the point of greatest vibration) the axis of vibration with respect to the body will depend on the orientation of the body. For example, vertical building vibration will be z-axis for standing and seated persons but x-axis for persons lying on their backs. A combined standard has therefore been proposed which consists of a combination of the lowest levels of the limits for z-axis and x- and y-axis vibration. This therefore consists of the limits for x- and y-axis vibration from 1–2 Hz and the limits for z-axis vibration from 8–80 Hz. Between 2 Hz and 8 Hz there is an interpolation between the two curves. Levels for the separate axes may be determined by reference to Figs 31.2 and 31.3 using a baseline level of 3.6×10^{-3} ms^{-2} r.m.s. for x- and y-axis vibration from 1 to 2 Hz and a baseline level of 5.0×10^{-3} ms^{-2} r.m.s. for z-axis vibration from 4 to 8 Hz. Various curves corresponding to multiplying factors from 1 to 128 times the threshold are defined. These correspond to the acceptable building vibration levels as tabulated in Table 31.1. The levels corresponding to these multiplying factors are said to represent "good environmental standards". Vibration levels up to a factor 2 greater are said to give rise to "moderate complaint". An increase above the basic levels by a factor of 4 will give rise to "major complaints" unless prior warning is given.

Although the Draft Addendum may offer the best possible general guidance, several problems exist. It is sometimes suggested that greater emphasis should be placed on the value of warnings before vibration occurs, and that far greater levels may be allowed, particularly for impulsive shock due to blasting, than are given in the Table. There are certainly problems in the measurement of impulsive motions, and the Draft Addendum presents a method of measurement which is unsatisfactory. The vibration levels are based on a combination of the ISO frequency weighting and some data on vibration thresholds. However, existing evidence suggests that the frequency sensitivity of vibration thresholds is somewhat different from the ISO frequency weighting, and in consequence the guidance in DAD1 may tend to over-emphasize sensitivity at low frequencies and relatively under-estimate sensitivity to high frequencies. Notwithstanding this evidence a review of vibration standards in other countries [31.8] shows that the range of levels specified in ISO 2631 encompasses those used elsewhere. This Draft Addendum is now being modified.

Table 31.1 – Multiplying factors for acceptable building vibration with respect to human response as defined in ISO 2631 DAD 1 [31.7]

Place	Time	Continuous or intermittent vibration and repeated impulsive shock	Impulsive shock excitation with approximately three occurences per day
Hospital operating theatre	Day	1	1
and critical working areas	Night	1	1
Residential	Day	2	16
(minimum complaint level)	Night	1.4	1.4
Office	Day	4	128
	Night	4	128
Workshop	Day	8	128
	Night	8	128

31.2.1.4 Addendum 2 to ISO 2631 (ISO 2631 Add 2 (1982) [31.9])

This Addendum presents a guide to the evaluation of human exposure to vertical z-axis low-frequency vibration. It presents guidance on the levels of motion at frequencies between 0.1 and 0.63 Hz that may cause motion sickness (termed severe discomfort). The motion sickness data are provided for periods of half an hour, two hours, and a tentative eight-hour limit. The relationship between the levels for different times is determined by an r.m.s. integration procedure (a^2t = constant). The guidance for eight hours is tentative since some adaptation may occur by this time – in which case sensitivity would decrease and the levels rise (see Fig. 31.5).

The guidance applies strictly to vertical z-axis vibration (i.e. standing or seated persons only), and the levels are thought to correspond to approximately 10% motion sickness in the fit general public.

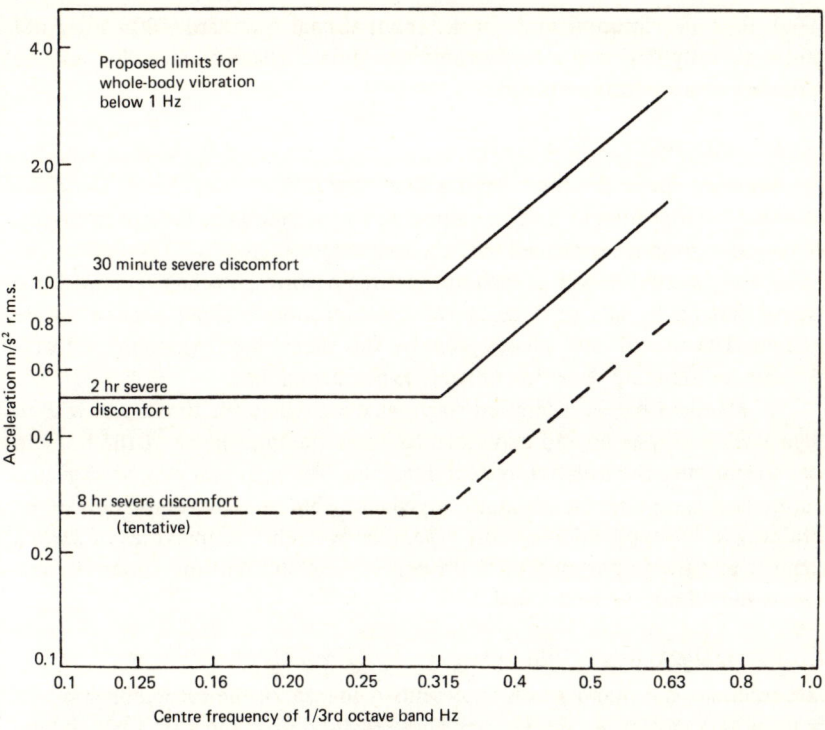

4.0

Proposed limits for
whole-body vibration
below 1 Hz

2.0

30 minute severe discomfort

1.0

0.8

0.6

2 hr severe
discomfort

0.4

8 hr severe discomfort
(tentative)

0.2

0.1

0.1   0.125   0.16   0.20   0.25   0.315   0.4   0.5   0.63   0.8   1.0

Acceleration m/s² r.m.s.

Centre frequency of 1/3rd octave band Hz

Fig. 31.5 – Severe discomfort boundaries for 0.1 to 0.63 Hz z-axis vertical
vibration as defined in Addendum 2 to ISO 2631 [31.9].

The Addendum provides guidance where none previously existed, but is necessarily based on less than complete data. Very few studies of the oscillatory motions that produce motion sickness have been conducted, and data are not available in the quantity required for the formulation of a complete standard. In particular, there are doubts as to the importance of other axes of motion. Some data suggest that sensitivity is decreased if the body lies down such that vertical motion is in the x-axis. Roll and pitch vibration of a ship is certainly important since it produces translational motion in the z-axis which varies according to the distance from the centre of rotation of the ship. It is not clear whether roll and pitch in isolation cause motion sickness. From experimental data it is not clear how motions which vary in level with time should best be integrated, but the standard presents a strong implication that r.m.s. integration should be used.

31.2.1.5 ISO 5805 (1981) [31.10]

This International Standard defines some of those terms which relate to human exposure to mechanical vibration and shock. It is intended to supplement an existing standard (ISO 2041-1975) which presents a vocabulary for use in the general field of vibration and shock. International Standard 5805 does not attempt to fully define the biodynamic coordinate systems since this may be the subject of a separate standard.

31.2.1.6   ISO 5982 (1981) [31.11]

This Standard presents the approximate mechanical impedance (modulus and phase) of the human body in three postures: standing, sitting, and lying. Mechanical impedance is defined over the frequency range 1 to 30 Hz and for the vertical axis (z-axis) only. For each of the three postures, mechanical impedance is approximated by a 2 or 3 degree of freedom model whose parameters are specified. The moduli and phases given by this model are then compared with the mean and standard deviation of some experimental data.

The Standard has been drafted to provide assistance for those who wish to design isolation systems and may need to know the impedance of the body in order to optimize the isolation system dynamics. The Standard may be expected to stimulate the gathering of more impedance data and the evaluation of its usefulness in an application of this type. However, the complexities of seating dynamics and the importance of multi-axis motion and multiple input motions on seats should not be overlooked.

31.2.1.7   ISO DIS 6897 (1982) [31.12]

Draft International Standard 6897 presents guidance on the acceptable levels of vibration and motion at frequencies between 0.063 Hz and 1 Hz for buildings and off-shore fixed structures. The guidance is presented as a set of curves.

High levels of low-frequency building vibration are most usually caused by storms and high wind, and they therefore occur intermittently. Some of the guidance presented in this draft standard applies to the worst ten minutes of any storm during one-year and five-year periods. Separate curves indicate what may be expected to be perceived, and give rise to problems when the motion occurs more frequently. Separate curves in this standard relate to vibration on off-shore structures and are mainly concerned with performance.

31.2.1.8   BSI DD 23 (1973) [31.13]

This Draft for Development is based upon the guidance given in ISO 2631 [31.3] and BSI DD 32 [31.4]. It indicates the safety precautions that should be taken when exposing subjects to vibration on laboratory apparatus and indicates the various categories of precautions that should be taken for different vibration levels. Particular guidance is given on the safety systems appropriate when using electrohydraulic vibrators.

31.2.2   Hand/arm vibration

Several types of injury or disease may be caused by prolonged exposure to high levels of local vibration. The most easily observed condition is *Vibration White*

Finger or *Raynaud's Phenomenon of Occupational Origin.* This condition is most commonly found amongst users of pneumatic hammers and drills in mining, the users of percussive metal-working tools, users of grinders and some other rotary tools, and users of some chain saws. Typically it affects those fingers which are in contact with high levels of vibration over prolonged periods.

In the very early stages of injury, vibration may cause only slight tingling and numbness. Later, the tips of one or more fingers most exposed to vibration suffer attacks of blanching. The attacks are usually precipitated by cold and often occur early in the morning. Continued exposure to vibration may result in an increase in the area affected, with the blanching extending to the base of those fingers exposed to vibration. Attacks of White Finger may last about one hour and are terminated with a reactive hyperaemia (seen as a red flush) and often considerable pain. During an attack, touch, pain, and temperature sensitivity are often greatly reduced. After continued and prolonged further exposure to vibration, the condition sometimes advances and the fingers may take on a permanent blue-black cyanotic appearance. In some cases there may be signs of skin necrosis and, very exceptionally, gangrene. In some persons exposed to very high vibration levels attacks of blanching may begin to occur within a few months, but more often the progression of symptoms is far slower, taking five, ten, or twenty years for a significant number of persons to be affected.

Bone and joint deformities are also often detected when vibration-exposed groups are subjected to x-ray diagnosis. It is also often reported that neural activity of the whole arm can be affected and that nerve conduction velocity is reduced in persons exposed to high levels of local vibration. In some countries a condition called *Vibration Disease* is described. This term is often applied to a wide range of effects which are considered to occur throughout the body owing to local exposures to vibration. Although effects other than Vibration White Finger have been reported in many countries it is Vibration White Finger that is most commonly and easily observed and has usually been the subject of vibration standards and limits,

Vibration standards for local vibration exist in many countries (see Griffin [31.14]), and in some countries the standards are said to have statutory force. In Britain the British Standards Institution published a Draft for Development DD 43 in 1975 [31.15]. Earlier, the Forestry Commission specified their own limits for the purchase of new chain saws. However, there is no statutory vibration limit in the United Kingdom at present.

31.2.2.1 *BSI DD 43-1975* [31.15]

The two limits proposed in BSI DD 43 are shown as the upper and lower curves in Fig. 31.6. The limits are defined as constant acceleration levels from 4-16 Hz and constant velocity from 16 Hz-1000 Hz. Analysis is to be conducted in octave-bands, and it is implied that the limits should be evaluated separately for each octave and separately in each of the three orthogonal axes of the hand.

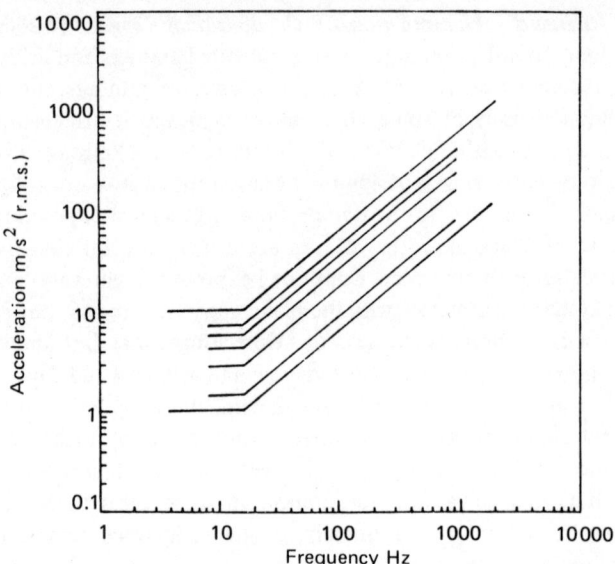

Fig. 31.6 – Comparison of octave band vibration limits defined in BSI DD 43
[31.15] and ISO DIS 5349 [31.18].

The two limits defined in the standard are said to apply to 150- and 400-minute exposures. However, this labelling is confusing. The Draft for Development states: "it is recommended that cumulative exposure time to vibration should never exceed 400 minutes; for intermediate periods between 150 minutes and 400 minutes some interpolation should be made; the values given for 150 minutes apply for all shorter periods. These latter values should not be exceeded by regular users of hand-held equipment". The interpretation of these durations has been discussed elsewhere [31.14]. The Draft for Development gives insufficient guidance on the interpolation method by which to calculate limits for periods between 2½ hours and 6 hours 40 minutes. There is also a considerable problem in deciding how to evaluate exposures which change in level from moment to moment, and how to sum exposures of various levels and durations [31.16] [31.17].

Although a strict interpretation of the working in BSI DD 43 is not easy it may be sufficient to consider the two levels as defining a band above which continued regular exposure to vibration is likely to cause injury, and below which long-term exposures are not expected to be hazardous. Within the band there is an area of uncertainty. It has been shown [31.14] that the standards in most other countries fall somewhere within the 10 to 1 range caused by the band. The principal limitation on the Draft Development then becomes the absence of clear methods of specifying how complex, multi-axis, time-varying, impulsive motions should be measured and compared with the band.

Table 31.2 – Tentative correction factors for hand-arm vibration exposures during daily shifts (8 h) as defined in ISO DIS 5349 [31.18]

VIBRATION EXPOSURE

Exposure time during daily shift (8 h)	Uninterrupted or not regularly interrupted	Regularly interrupted				
		Duration of recurrent time interval without vibration (in minutes per working hour)				
		Up to 10	More than 10 up to 20	More than 20 up to 30	More than 30 up to 40	More than 40
Up to 30 min	5	5	–	–	–	–
More than 30 min up to 1 h	4	4	–	–	–	–
More than 1 h up to 2 h	3	3	3	4	5	5
More than 2 h up to 4 h	2	2	2	3	4	5
More than 4 h up to 8 h	1	1	1	2	3	4

31.2.2.2 ISO DIS 5349 [31.18]

Draft International Standard DIS 5349 presents guidance over the frequency range 8 to 1000 Hz. Vibration spectra are to be analysed in either octave or one-third octave bands (the latter for discrete frequency and narrow-band vibration spectra). It is stated that the limits (shown as the centre five curves in Fig. 31.6) should not be exceeded in any frequency band or in any of the three orthogonal axes of the hand. A table (see Table 31.2) is provided to show multiplying factors from 1 to 5 (corresponding to the five curves in the figure) according to the exposure pattern of vibration.

Comparing the octave-band limits from ISO DIS 5349 with those from BSI DD 43 it is seen that the five ISO curves fall between the two BSI curves. If the upper limit in BSI DD 43 is considered to apply to continuous 150-minute exposures it follows that for all short durations it allows much higher levels than permitted in ISO DIS 5349. The ISO Draft Standard, however, allows higher levels than the Draft for Development when the daily exposure time is very long.

The selection of multiplying factors as given in Table 31.2 is exceedingly difficult for some patterns of exposure. In considering the evaluation of complex exposure patterns the Draft International Standard suggests a possible dose (summation) procedure. This requires the integration of the square of the acceleration in each octave (or third-octave) band during the daily exposure. The equivalent constant levels which would give the same energies in each band in an eight-hour period are then to be calculated. The equivalent levels are to be compared with the four- to eight-hour exposure boundaries.

The performance of an electronic network which could be used for the frequency weighting of spectra, and thus avoid spectrum analysis, is also outlined in DIS 5349. However, the incorporation of such a weighting network into a device to determine the dose from a complex exposure pattern is not mentioned. The Draft Standard does not define measurement procedures, although this is widely recognized as a major area of difficulty when measuring vibration exposures of the hand. This Draft International Standard is now being revised.

31.2.2.3 Compensation for vibration injuries of the hand and arm

Possible prescription for vibration injuries under the National Insurance (Industrial Injuries) Act has been considered by the Industrial Injuries Advisory Council at several times during the last 25 years. Until 1981 there was a majority decision against compensation for three principal reasons: the apparent "triviality" of the condition in most cases, the difficulties associated with diagnosis, and difficulties in establishing that the symptoms are due to vibration exposure and not some other cause of White Finger. The Minister of Social Security referred the question to the Industrial Injuries Advisory Council again in 1980 and their report in 1981 [31.19] proposed prescription when there is "episodic blanching of at least the two distal phalanges of three or more fingers of one hand occurring throughout the year". Government action on the report is pending.

There have been several attempts to obtain compensation for vibration injuries at common law. An unknown number of claims have been settled out of court, some have been successful in the courts and some have been unsuccessful. The increased knowledge, documentation, and publicity given to Vibration White Finger during recent years must increase the probability of court cases of this type.

31.3 RESEARCH DATA

Although a lot of useful general guidance is provided in the available standards it is often more limited, less precise, and less powerful than that which can be obtained by a careful study of research publications. However, the literature on the effects of vibration on man is highly diffuse and, for example, relevant publications have appeared in several hundred different scientific journals and a wide variety of technical reports and memoranda.

To aid the potential user of the literature the Institute of Sound and Vibration Research operates a Human Response to Vibration Literature Collection which consists of more than 4000 papers which are available for use by visitors to the University. Advice and copies of papers can also be provided by post.

In the following sections a brief guide to a few of the systematic studies of the effects of some variables on human vibration discomfort, performance, and health is presented.

31.3.1 Vibration discomfort

Although most studies have assumed that vibration is primarily vertical at the seat or vertical and horizontal at the seat this is by no means the general case. Fig. 31.7 shows a more complex model in which twelve input axes are defined: three translational and three rotational axes on the main supporting seat surface; three translational axes of motion of the backrest; and three translational axes of the footrest. Studies of the discomfort produced by different vibration frequencies in each of these axes have led to the formulation of twelve standard frequency weightings. The weightings are arranged such that when applied to motion in any axis they give the level (in ms^{-2} r.m.s.) of 10 Hz vertical (z-axis) sinusoidal vibration which will produce an equivalent degree of discomfort. The derivation of these frequency weightings and their interpretation is presented elsewhere [31.20]. The application of frequency weightings of this type neccessarily requires information on how to sum across frequencies, across axes, and over time.

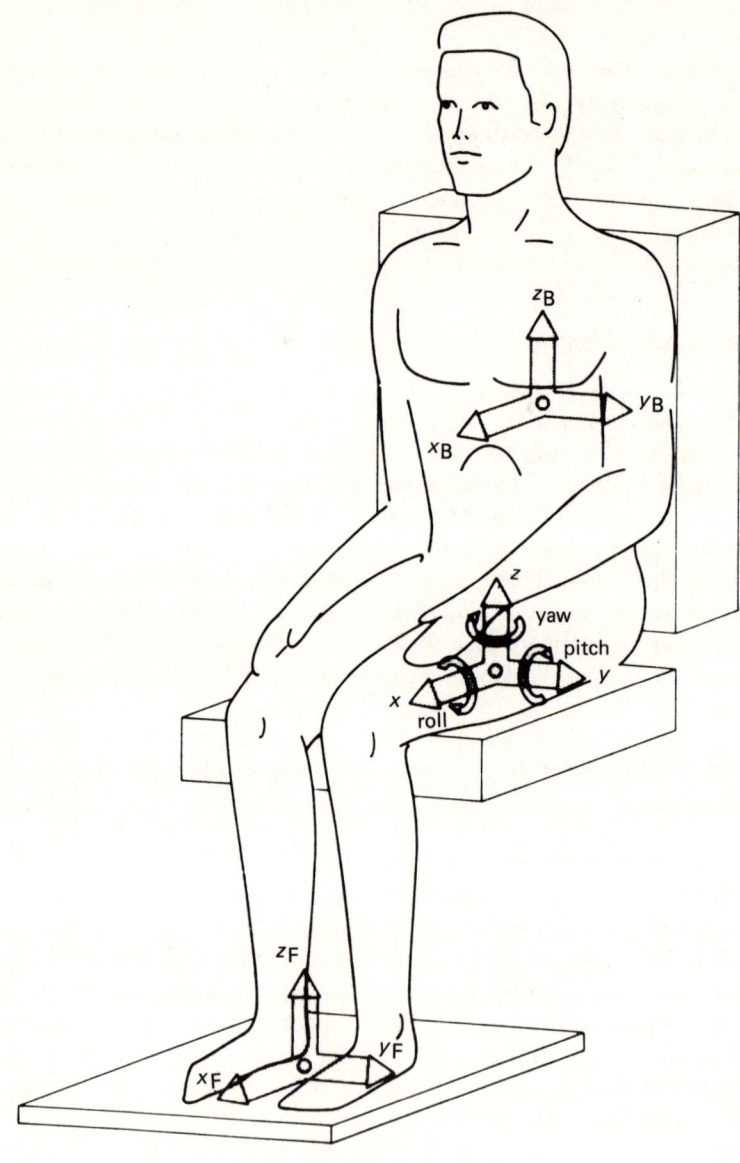

Fig. 31.7 – Coordinate system for vibration inputs to the feet, ischial tuberosities, and back of a seated person used in the ISVR comfort studies.

Studies of response to impulsive vibration have indicated that motions having short periods of high-level vibration are more uncomfortable than predicted by their r.m.s. value. Conversely, longer periods are less uncomfortable than predicted by their r.m.s. level. Studies have suggested that the change in discomfort with vibration duration for short exposures (up to half a minute) are better predicted from the root-mean-quad (using the fourth root of the fourth power of the acceleration) in preference to the root mean square value [31.5]. If this method of summation is applied to very long periods it produces a time dependency which gives far less effect of time than either the r.m.s. summation or ISO time dependency with increasing duration (see Fig. 31.3). Although this application of the r.m.q. is unproven it may be far more acceptable than the use of the other alternative procedures currently available [31.21].

A variety of studies of human response to complex vibration conditions have shown that a motion containing more than one component is more uncomfortable than a motion containing a single component of motion [31.22] [31.23] [31.24]. It has been found that the r.m.s. sum of the frequency components, whether they be discrete sinusoids or broadband random motions, generally provides a good estimate of the total discomfort. There have been fewer studies of multi-axis or multiple input motion, but some of these suggest that a similar procedure which involves the square roots of the sums of the squares of the levels in each axis provides a reasonable estimate of total vibration discomfort [31.25] [31.26].

31.3.2 Vehicle ride comfort

The procedures mentioned above may be used to predict ride comfort in vehicles. Fig. 31.8 shows, for a passenger car, short segments of acceleration time history recorded in the twelve axes defined by Fig. 31.7. Also shown are the frequency weightings mentioned in section 31.3.1 above and the form of the twelve time histories after they have been weighted by these networks. The weighted time histories provide a simple visual indication of both the axes and frequencies which are contributing to discomfort in this vehicle – the unweighted time histories provide little indication. It may be seen, for the 1s period, that the principal components in this vehicle are z-axis and pitch axis vibration on the seat and fore-and-aft vibrations on the seatback. In other vehicles or with different road conditions vibration of the feet may dominate or other axes may assume greater importance. The effect of variations in the ride comfort procedure (different methods of summation across frequencies, across axes, and over time) have been reported by Griffin & Parsons [31.27]. These procedures allow the calculation of a single value indicating the ride in a vehicle. The effect on ride comfort of changing vehicle dynamics, road conditions, etc. may be simply investigated by monitoring this value.

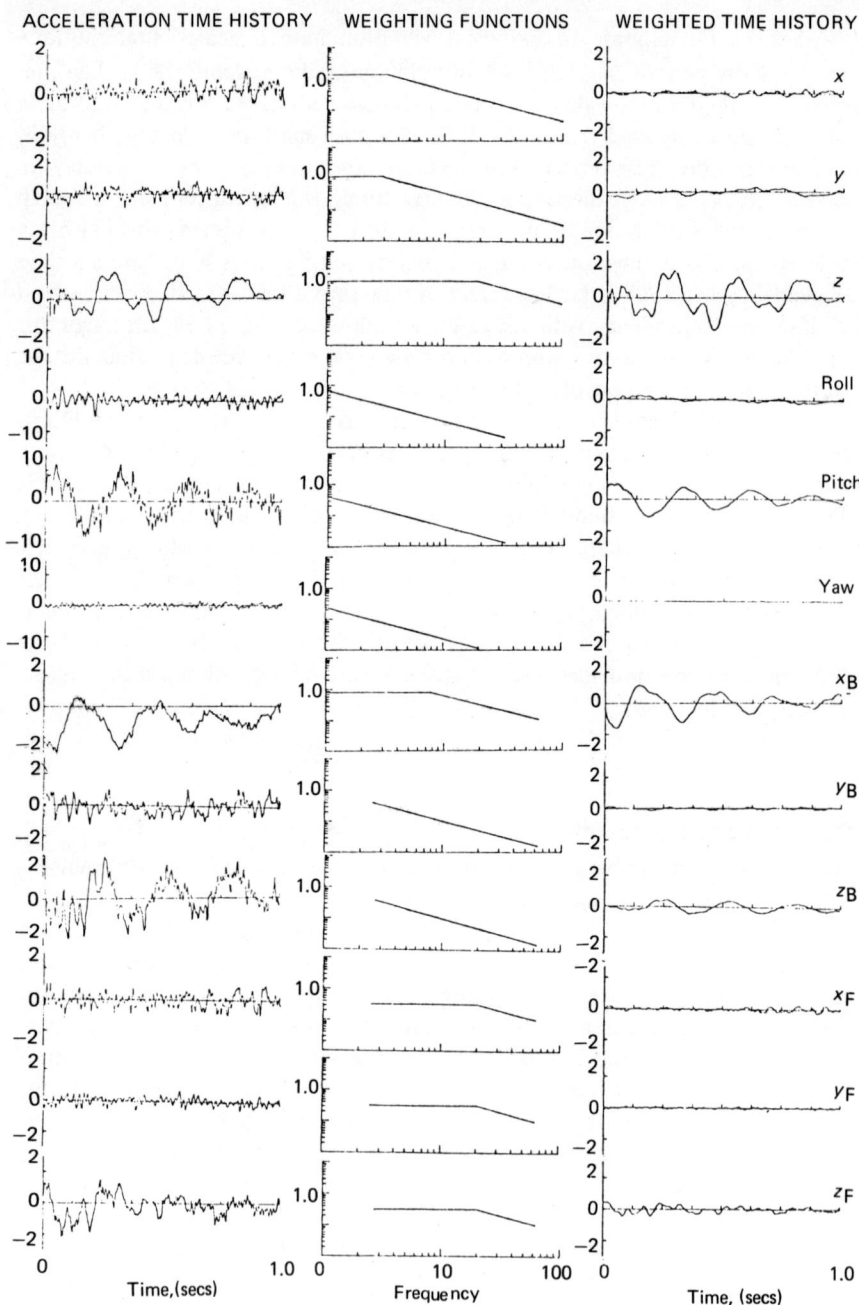

ACCELERATION TIME HISTORY   WEIGHTING FUNCTIONS   WEIGHTED TIME HISTORY

Fig. 31.8 – Example 1 second acceleration time histories from a small car, ISVR comfort weightings and weighted time histories for the twelve axes of vibration shown in Fig. 31.7.

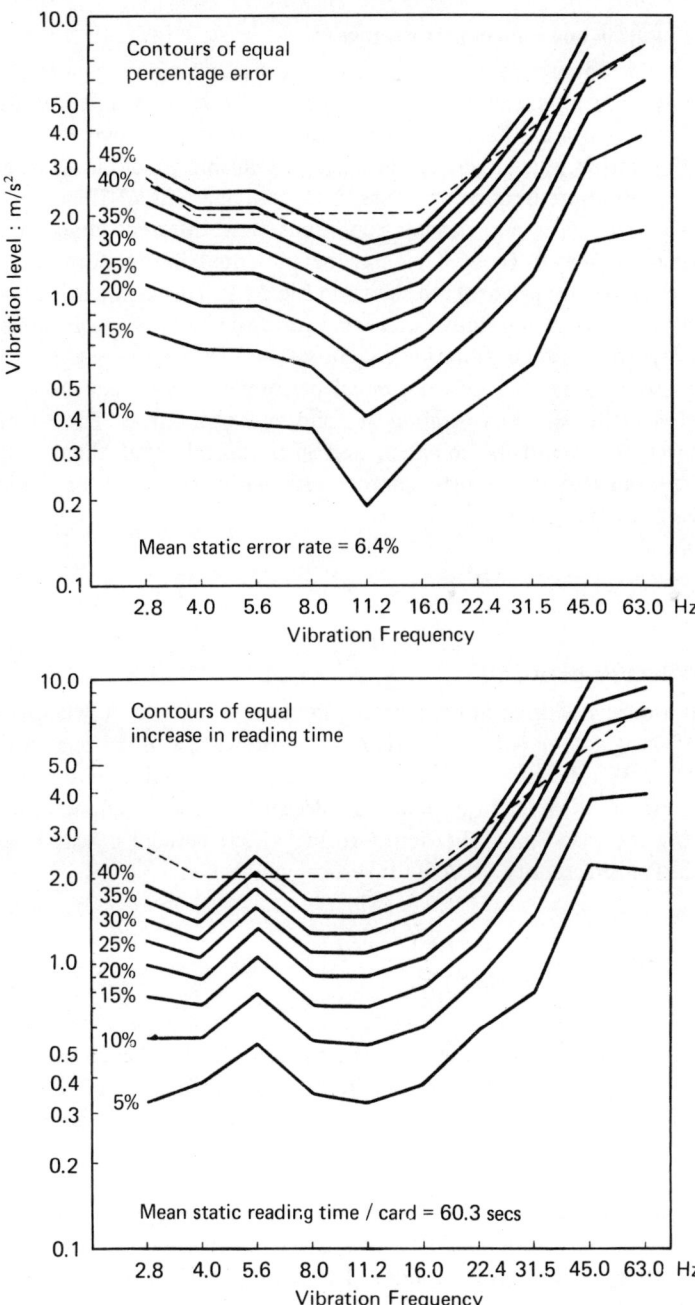

Fig. 31.9 – Equal performance contours obtained from subjects exposed to vertical sinusoidal vibration in a simulated helicopter seat. (Lewis and Griffin [31.29].)

31.3.3 Vibration and human performance

Conclusive experimental data concerning an effect of vibration on performance are probably only available for the effects of vibration on vision and manual control. These two areas have been reviewed by Griffin & Lewis [31.28] and Lewis & Griffin [31.29]. Experimental data are available which show the effect of different vibration frequencies, axes, and combinations of frequencies on reading errors for different viewing conditions (e.g. different character sizes and viewing distances). These data may be presented in the form of equal performance contours such as those shown in Fig. 31.9. This information may be used to construct vibration limits, or it may form the basis of design guides for visual displays to be used in vibration conditions.

It is also possible to construct equal performance contours showing the effects of vibration on the operation of hand or foot controls. The effects on performance are potentially complex, and it is possible that changes in the control or display dynamics may improve task performance during vibration. (See Lewis & Griffin [31.30].)

31.3.4 Vibration isolation

To design a means of isolating man from vibration requires a knowledge of the motions from which he is to be isolated *and* a knowledge of the relative effect of the different motions on man. It is, for example, essential to have equivalent comfort contours in order to optimize the design of a seat to improve comfort by isolating the man from vibration. Griffin [31.31] defined a Seat Effective Amplitude Transmissibility (SEAT) for vertical vibration:

$$\text{SEAT\%} = \frac{(\int G_s(f) S^2(f) \, df)^{\frac{1}{2}}}{(\int G_f(f) S^2(f) \, df)^{\frac{1}{2}}} \times 100$$

where $G_s(f)$, $G_f(f)$ are the power spectra for vertical vibration on the seat and the floor respectively. $S(f)$ is the frequency function of human response to vibration in the vertical axis (defined by the appropriate equivalent comfort contour).

When the SEAT value is 100% the motions on the floor and on the seat produce equivalent discomfort — even though the frequency content of the two motions may be very different. If SEAT is greater than 100% the motion of the seat is worse than that of the floor. The degree to which SEAT is less than 100% indicates the amount of useful isolation provided by the seat.

Values of SEAT determined in 15 different road vehicles and an electric train are shown in Fig. 31.10. Seat transmissibilities for the vehicles in the vertical axis are shown in Fig. 31.11. It may be seen that very few seats provided much useful attenuation and that several seats produced discomfort greater than that which would be experienced when sitting on the floors of the vehicles.

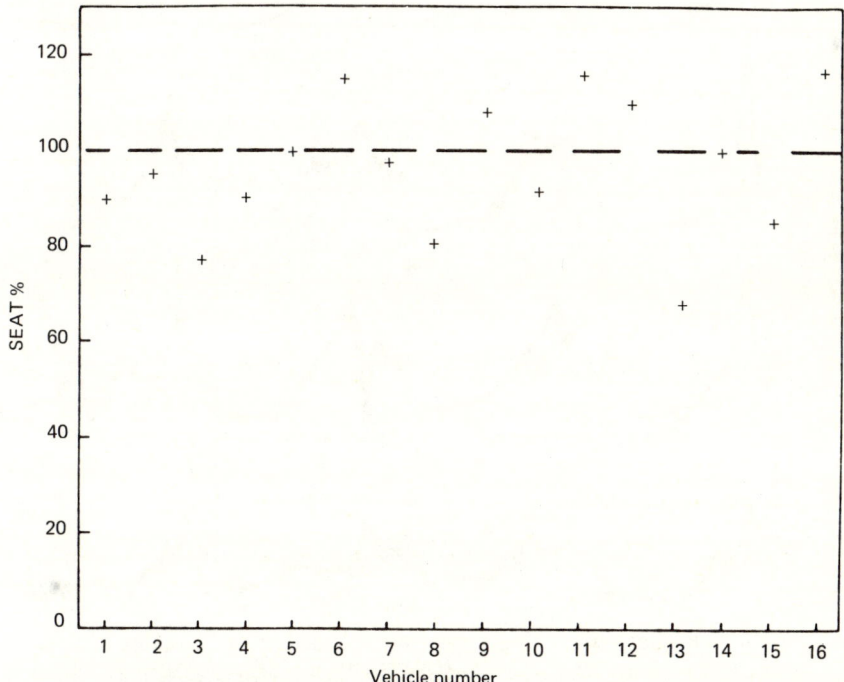

Fig. 31.10 – Seat effective transmissibility (SEAT) for 16 vehicles – values greater than 100% suggest that vertical vibration on the seat is worse than vertical floor vibration. (Based on frequency weightings given in ISO 2631.)

The SEAT values shown in Fig. 31.10 were obtained by applying the ISO weighting to vertical vibration. The values change if other weighting procedures are used. Lower values will tend to be found when using ISVR frequency weightings shown in Fig. 31.8, and lower values are also found when the r.m.q. unit of integration is used in place of r.m.s. integration. The correct 'tuning' of seats is therefore clearly dependent on a knowledge of the effects of vibration on man, and cannot be achieved without such information – the transmissibilities shown in Fig. 31.11 are not sufficient indicators of the quality of the seat. The design of efficient seating may involve a dynamic model of the seat and the man but, as mentioned in section 31.2.1.6 above, a model of the system may be complex.

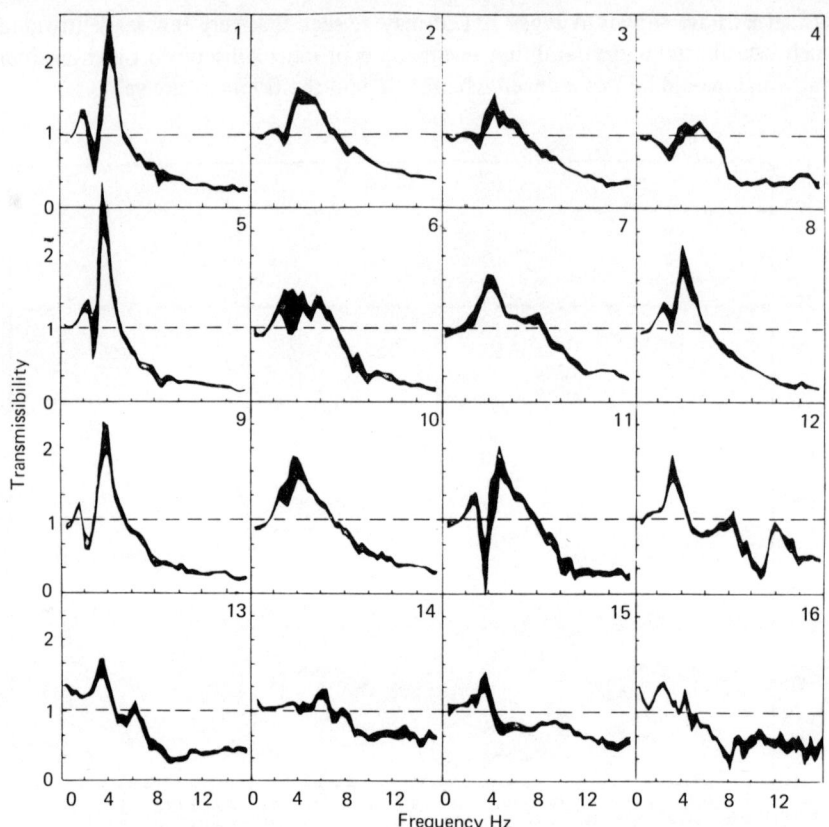

Fig. 31.11 – Vertical (z-axis) seat transmissibility in 16 vehicles – black bands indicate the 10th to 90th percent confidence intervals. (Transmissibility = (cross spectrum of floor and seat vibration)/(power spectrum of floor vibration)).

31.4 SUMMARY

This chapter provides a brief guide to some of the current knowledge concerning human response to vibration. The subject is inherently multidisciplinary and potentially complex. Human response is variable both between and within individuals. One vibration situation may differ very greatly from another, and the guidance required may also differ. To some extent a different vibration limit and a different technical solution is required for every situation. However, some general guidance is possible, and the existing standards often provide sufficient information to indicate whether any potential problem requires more detailed study. Future standards may be expected to provide somewhat more precise guidance and less ambiguity. Optimum solutions may always require more detailed consideration than will be available in general vibration standards.

REFERENCES

[31.1] Griffin, M. J. & Whitham, E. M. (1978) *J. Sound Vib.* **58** (2), 239–250. Individual variability and its effect on subjective and biodynamic response to whole-body vibration.

[31.2] Griffin, M. J., Lewis, C. H., Parsons, K. C. & Whitham, E. M. (1978) The biodynamic response of the human body and its application to standards. *AGARD – CP-253, Paper A28.*

[31.3] International Organization for Standardization (1974) ISO 2631. *Guide for the evaluation of human exposure to whole-body vibration.*

[31.4] British Standards Institution (1974) BSI DD 32. Draft for Development: *Guide to the evaluation of human exposure to whole-body vibration.*

[31.5] Griffin, M. J. & Whitham, E. M. (1980) *J. Acoust. Soc. Am.* **68** (5), 1277–1284. The discomfort produced by impulsive whole-body vibration.

[31.6] International Organization for Standardization (1980) ISO 2631. *Guide for the evaluation of human exposure to whole-body vibration. Amendment 1.*

[31.7] International Organization for Standardization (1980) ISO 2631 DAD 1. *Guide to the evaluation of human exposure to vibration and shock in buildings. Addendum 1: Acceptable magnitudes of vibration.*

[31.8] Griffin, M. J. (1978) *Proc. Symp. on Control of Odours and Smells in the Process Industries.* May 23, London. "It smells rotten and it sounds awful", pp. 27–38. Methods of predicting community response to building vibration.

[31.9] International Organization for Standardization (1982) ISO 2631 DAD 2. *Guide for the evaluation of human exposure to whole-body vibration. Addendum 2: evaluation of exposure to whole-body z-axis vertical vibration in the frequency range 0.1 to 1.0 Hz.*

[31.10] International Organization for Standardization (1980) ISO 5805. *Mechanical vibration and shock affecting man – vocabulary.*

[31.11] International Organization for Standardization (1981) ISO 5982. *Vibration and shock – mechanical impedance of the human body.*

[31.12] International Organization for Standardization (1982) ISO DIS 6897. *Guide to the evaluation of the response of occupants of fixed structures, expecially buildings and off-shore structures, to low-frequency horizontal motion (0.063 Hz to 1 Hz).*

[31.13] British Standards Institution (1973) BSI DD 23. Draft for Development: *Guide to the safety aspects of human vibration experiments.*

[31.14] Griffin, M. J. (1980) *Health & Safety Executive Research Paper 9* ISBN 011 883271 9. Vibration injuries of the hand and arm: their occurrence and the evolution of standards and limits.

[31.15] British Standards Institution (1975) BSI DD 43. Draft for Development. *Guide to the evaluation of exposure of the human hand-arm system to vibration.*

[31.16] Griffin, M. J. (1981) *Third International Symposium on Hand-Arm Vibration.* 18–20 May, Ottawa. Hand-arm vibration standards and dose-effect relationships.

[31.17] Griffin, M. J. (1981) *U. K. Informal Group Meeting on Human Response to Vibration.* 9–11 September, Heriot–Watt University, Edinburgh. Dose-effect relationships for vibration-induced white finger.

[31.18] International Organization for Standardization (1979) ISO DIS 5349. *Principles for the measurement and evaluation of human exposure to vibration transmitted to the hand.*

[31.19] Industrial Injuries Advisory Council (1981) *Vibration White Finger.* Cmnd. 8350, HMSO.

[31.20] Griffin, M. J., Parsons, K. C. & Whitham, E. M. (1982) *Ergonomics* **25 (8)**. Vibration and comfort: IV Application of experimental results.

[31.21] Griffin, M. J. & Whitham, E. M. (1980) *J. Acoust. Soc. Am.* **68 (5)**, 1522–1523. Time-dependency of whole-body vibration discomfort.

[31.22] Fothergill, L. C. & Griffin, M. J. (1977) *Ergonomics* **20 (3)**, 263–276. The evaluation of discomfort produced by multiple frequency whole-body vibration.

[31.23] Griffin, M. J. (1976) *J. Acoust. Soc. Am.* **60 (5)**, 1140–1145. Subjective equivalence of sinusoidal and random whole-body vibration.

[31.24] Shoenberger, R. W. (1978) *Aviat. Space Environ. Med.* **49 (11)**, 1327–1330. Intensity judgements of non-sinusoidal vibrations: Support for the ISO weighting method.

[31.25] Griffin, M. J. & Whitham, E. M. (1977) *J. Sound Vib.* **54 (1)**, 107–116. Assessing the discomfort of dual-axis whole-body vibration.

[31.26] Parsons, K. C. & Griffin, M. J. (1978) *J. Sound Vib.* **58 (1)**, 127–141. The effect of the position of the axis of rotation on the discomfort caused by whole-body roll and pitch vibrations of seated persons.

[31.27] Griffin, M. J. & Parsons, K. C. (1980) *Time series models of human response to vibration.* In Chapter 9, Applications of time series analysis. 14–18 April. Short Course at ISVR, Southampton.

[31.28] Griffin, M. J. & Lewis, C. H. (1978) *J. Sound Vib.* **56 (3)**, 383–413. A review of the effects of vibration on visual acuity and continuous manual control, Part I: Visual acuity.

[31.29] Lewis, C. H. & Griffin, M. J. (1978) *J. Sound Vib.* **56 (3)**, 415–457. A review of the effects of vibration on visual acuity and continuous manual control, Part II: Continuous manual control.

[31.30] Lewis, C. H. & Griffin, M. J. (1979) *Ergonomics* **22 (7)**, 855–890. Mechanisms of the effects of vibration frequency, level and duration on continuous manual control.

[31.31] Griffin, M. J. (1978) *Applied Ergonomics* **9 (1)**, 15–21. The evaluation
of vehicle vibration and seats.

Index